**Public Health in the 21st Century**

**Life Sciences Research and Development**

nova
Medicine & Health
New York

# Public Health in the 21st Century

**Interdisciplinary Perspectives on Occupational Safety and Health**
Mehmed Zahid Çögenli, PhD (Editor)
2023. ISBN: 979-8-89113-085-2 (Hardcover)
2023. ISBN: 979-8-89113-120-0 (eBook)

**Critical Care and COVID-19**
Salim Surani (Editor)
Syed Anjum Khan, MD (Editor)
Reena Shah (Editor)
2023. ISBN: 979-8-88697-940-4 (Hardcover)
2023. ISBN: 979-8-88697-969-5 (eBook)

**Barakoa (Masks) During COVID-19: Malevolent Pathogens and Pandemic Responses in Kenya After 600 Days**
Stephen M. Magu, PhD (Editor)
2022. ISBN: 978-1-68507-798-3 (Hardcover)
2022. ISBN: 979-8-88697-014-2 (eBook)

More information about this series can be found at https://novapublishers.com/product-category/series/public-health-in-the-21st-century/

# Life Sciences Research and Development

**J-Shaped Distributions and Their Applications**
Mohammad Ahsanullah, PhD (Editor)
Mohammad Shakil (Editor)
2023. ISBN: 979-8-89113-013-5 (Softcover)
2023. ISBN: 979-8-89113-182-8 (eBook)

**Transcriptome Analysis and Why it Matters**
Sagarika Mishra, PhD (Editor)
Michael Campbell, PhD (Editor)
2023. ISBN: 979-8-89113-098-2 (eBook)

**Snakes: Morphology, Function, and Ecology**
David Penning, PhD (Editor)
2023. ISBN: 979-8-88697-855-1 (Hardcover)
2023. ISBN: 979-8-88697-921-3 (eBook)

More information about this series can be found at https://novapublishers.com/product-category/series/life-sciences-research-and-development/

**Kuladip Jana**
Editor

# The Role of Reactive Oxygen Species in Health and Disease

Medicine & Health
New York

## Library of Congress Cataloging-in-Publication Data

Names: Jana, Kuladip, editor.
Title: The role of reactive oxygen species in health and disease / Kuladip Jana.
Description: New York : Nova Science Publishers, [2024] | Series: Life sciences research and development | Includes bibliographical references and index. |
Identifiers: LCCN 2023049576 (print) | LCCN 2023049577 (ebook) | ISBN 9798891132832 (hardcover) | ISBN 9798891133303 (adobe pdf)
Subjects: LCSH: Active oxygen--Pathophysiology. | Oxidative stress--Pathophysiology.
Classification: LCC RB170 .R66 2024 (print) | LCC RB170 (ebook) | DDC 615.8/36--dc23/eng/20231208
LC record available at https://lccn.loc.gov/2023049576
LC ebook record available at https://lccn.loc.gov/2023049577

Published by Nova Science Publishers, Inc. † New York

*This Book dedicated to my respected Parents:*
*Sukanta Kumar Jana & Rina Rani Jana;*

*My beloved wife:*
*Moumita Khatua (Jana)*
*&*
*My lovely Son and Daughter:*
*Sharanyo and Shreenika*

***Dr. Kuladip Jana***
***Editor***

# Contents

# Preface

A double-edged sword of ROS in health and disease is well accepted, although the underlying mechanisms are still incompletely understood. ROS is tightly regulated in the physiological processes and plays an essential role in molecular signaling, gene transcription, immune response, and cell and tissue homeostasis. The overproduction of reactive oxygen species (ROS) can result in oxidative damage to biomolecules such as lipids, proteins, and DNA, which has been implicated in the development of aging and numerous chronic and degenerative ailments such as cancer, diabetes, digestive and neurodegenerative diseases. On the other hand, excess levels of ROS generation and/or impaired antioxidant activity may induce oxidative stress associated with cellular damage/dysfunctions, leading to several diseases or death. It has been established that dietary antioxidant supplementation has been shown to attenuate endogenous antioxidant depletion thus alleviating associated oxidative damage. In this scenario, further advancement is necessary in personalized redox medicine or antioxidant therapy to alleviate oxidative stress.

The implication of oxidative stress in the etiology of several chronic and inflammatory diseases specifies that antioxidant-based therapy could be promising for these disorders. A therapeutic strategy that increases a person's antioxidant capacity may be beneficial for long-term treatment. However, many difficulties remain intangible regarding antioxidant supplements in disease prevention. It remains to be elucidated about the precise roles of ROS in the pathogenesis of various diseases progression. Recent recommendations are based on the intake of a balanced diet (rich in fruits, vegetables, grains, and low amount of red meat and avoiding intake of alcohol/smoking) and maintaining a healthy lifestyle, which has been demonstrated the ability to reduce the risk of death and diseases. Moreover, further research is warranted before using the antioxidants as an adjuvant therapy. In the meantime, cigarette smoke and alcohol intake which are considered oxidant sources must be avoided while consuming dietary antioxidants.

The clinical awareness of nutritional imbalance in disease occurrence, progression, and outcomes is limited. This book discusses the roles of ROS in common oxidative stress-associated health issues, different diseases, and aging, and the effects of nutritional antioxidants as treatments or adjuvants have been discussed.

# Acknowledgment

I would like to acknowledge the extraordinary debt I owe to my teachers who have told me to act as an Editor of a book entitled "Reactive Oxygen Species in Health and Disease" which is now a very relevant and emerging research area in drug discovery.

I would not be able to get my work done without the continuous help and support of my Ph.D. student, Satyajit Halder. He helped me a lot in writing, formatting the chapters, and designing the front cover of this book.

I would also like to thank all my friends who are the authors of this book for their acceptance of my invitation and timely submission of their chapters. They all kept me going; this book would have been impossible without them.

Finally, I would like to thank the support and love of my family members, parents, and mother Rina Rani Jana for her untiring support and inspiration.

I would also like to acknowledge significantly my two sisters, Madhumita Jana and Monalisa Jana for their immense love and support.

I am grateful to my wife, Moumita Khatua (Jana), who is a constant source of inspiration and motivation.

Last, I would like to express my sincere love and affection to our lovely son and daughter, Sharanyo and Shreenika, for their support and sacrifice.

**Dr. Kuladip Jana**
**Editor**

## Chapter 1

# Reactive Oxygen Species Act as Double-Edged Swords in Cancer Progression and Therapy

**Satyajit Halder**
**and Kuladip Jana***
Division of Molecular Medicine, Bose Institute, Bidhan Nagar, Kolkata, West Bengal, India

## Abstract

Reactive oxygen species (ROS), a class of highly reactive molecules, have been extensively studied in numerous cancers. Classically, cancer cells exhibit higher levels of ROS in comparison with normal cells and as a result of an imbalance among oxidants and antioxidants. ROS at low to moderate levels acts as a signal transducer to activate cell proliferation, differentiation, invasion, migration, and angiogenesis. In contrast, high levels of ROS may cause damage to lipids, proteins, nucleic acids, cell membranes, and organelles leading to cell death and apoptosis. Various studies have revealed that anticancer therapies like immunotherapies, which manipulate ROS levels, show very promising results in *in vitro* as well as *in vivo* experiments. Very recently, the researchers are interested to know the molecular mechanisms of the oncogenic functions of ROS and their useful roles for the development of cancer therapeutics. This Chapter provides some insights into the future development of some new effective agents that regulate the redox system and ROS for cancer therapy.

**Keywords:** mitochondria, mitochondrial ROS, endoplasmic reticulum, endoplasmic reticulum ROS, apoptosis, autophagy, NRF-2, cancer immunity

## 1. Introduction

Despite significant scientific and social developments, cancer is still a prevalent and alarming disease, a primary source of pain and death. The occurrence of new cancer forms is rising quickly, creating a genuine public health emergency on a global scale [1]. Numerous human disorders, including cancer, have been thoroughly researched about reactive oxygen species

---

* Corresponding Author's Email: kuladip@jcbose.ac.in; kuladip.jana@gmail.com.

In: The Role of Reactive Oxygen Species in Health and Disease
Editor: Kuladip Jana
ISBN: 979-8-89113-283-2

(ROS). ROS are common consequences of biological functions, including oxygen consumption and are unstable, reactive, and partially reduced oxygen derivatives. These species include singlet oxygen ($1O_2$), hypochlorous acid (HOCl), superoxide anion ($O_2$), hydroxyl radical (•OH), and hydrogen peroxide ($H_2O_2$) [2-4]. In addition to acting as second messengers in cell signalling, ROS are crucial for a number of biological activities in both healthy and malignant cells. A growing body of research indicates that ROS may operate as a "double-edged sword" in cancer [5]. When ROS levels are low to moderate, they function as signalling transducers that promote various cellular processes in cancer cells, including proliferation, migration, invasion, angiogenesis, and drug resistance [2, 6, 7]. In simpler terms, adequate ROS levels are critical for maintaining cancer cell homeostasis and facilitating cellular processes such as proliferation, differentiation, migration, and cell death. On the other hand, high levels of ROS are harmful to cancer cells and ultimately lead to cell death [8].

Strategies that either reduce or increase ROS levels in cancer cells appear to be promising treatments due to the bidirectional nature of ROS. Antioxidants are commonly considered beneficial for cancer prevention and treatment as they can reduce oxidative stress by quenching ROS levels. In fact, previous research has shown that deferasirox (DFX), an oral iron chelator, can induce apoptosis in multiple myeloma (MM) cells by suppressing intracellular ROS levels. Mechanistically, DFX inhibits proline-rich tyrosine kinase 2 phosphorylation, which leads to a reduction in ROS production and anti-MM activity [9]. However, in reality, antioxidants such as N-acetyl l-cysteine (NAC) and vitamin E have been found to significantly increase tumor progression and cancer metastasis in previous studies [10, 11].

In cancer cells, the production of ROS is often increased due to various factors, including elevated metabolic rate, gene mutations, and relative hypoxia [12]. To counteract this, cancer cells activate antioxidant pathways to increase ROS clearance [13]. However, the higher endogenous ROS levels in cancer cells make them more vulnerable to ROS-inducing therapy. Many chemotherapeutic agents are known to increase ROS production in cancer cells [14, 15, 16]. Previous studies have shown that increased intracellular ROS levels in multiple myeloma and colorectal cancer cells can lead to suppressed cell growth and apoptosis [17, 18]. Therefore, inducing ROS production appears to be a promising strategy for combating cancer.

In this chapter, we initially focus on the molecular mechanisms and oncogenic functions of ROS, and subsequently discuss several potential therapeutic strategies that modulate ROS levels in cancer cells.

## 2. Molecular Mechanism of ROS Production in Cancer Cells

### 2.1. Generation of ROS

ROS is a group of unstable and reactive oxygen derivatives that can be generated by various endogenous and exogenous factors [19]. One significant source of ROS is mitochondria, which produce endogenous ROS and contribute to intracellular oxidative stress [20]. Mitochondria have a four-layer structure, including an outer mitochondrial membrane, inter-membrane space, inner mitochondrial membrane, and matrix. The process of oxidative phosphorylation occurs on the inner mitochondrial membrane via five protein complexes: NADH ubiquinone reductase (complex I), succinate ubiquinone reductase (complex II), ubiquinol cytochrome c reductase (complex III), cytochrome c oxidase (complex IV), and F1F0-ATP synthase (complex V), which create ATP from oxygen and simple sugars [21, 22]. Complexes I and III are the primary sites for ROS generation in the electron transport chain, where significant changes occur in the potential energy of electrons related to oxygen reduction [23]. The leakage of electrons at these complexes leads to the production of superoxide, which is quickly dismutated to hydrogen peroxide by SOD1 in the inter-membrane space or by SOD2 in the matrix. Superoxide and hydrogen peroxide are recognized as ROS that are generated in mitochondria [21].

Transition metals such as iron can generate ROS non-enzymatically through the Fenton reaction, which involves the reaction between Fe2+ and hydrogen peroxide to form a highly reactive hydroxy radical that can cause damage to DNA and other biomolecules [24, 25]. This mechanism contributes to the production of ROS and oxidative stress in cancer cells.

Furthermore, several environmental factors can induce the generation of exogenous ROS. Air pollution, such as fine particulate matter and ozone, has been shown to promote oxidative stress in the lungs and other tissues. Dietary habits, such as high-fat and high-sugar diets, can increase ROS production and lead to oxidative stress. Additionally, certain drugs, such as anthracyclines used for chemotherapy, can generate ROS as a side effect [26-28]. Overall, both endogenous and exogenous factors can stimulate ROS formation in cancer cells, contributing to their development and progression (Figure 1).

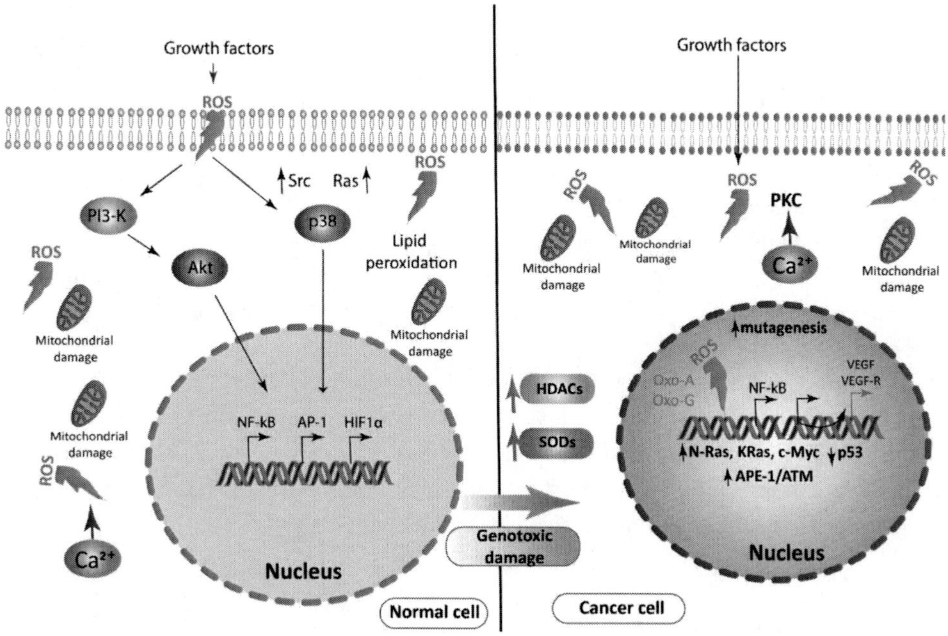


**Figure 1.** ROS signalling in Normal and cancer cells. Elevated ROS induce oncogenes and inhibit tumor suppressor genes. ROS also causes mitochondrial damage and induces apoptosis in cancer cells.

## 2.2. Redox Homeostasis

This activation of NRF2 and subsequent induction of antioxidant enzymes is a common adaptive mechanism used by cancer cells to mitigate the negative effects of excess ROS. However, this also presents a challenge for cancer treatment, as the up-regulation of these enzymes can confer resistance to chemotherapy and radiotherapy [29-31]. As a result, targeting the NRF2 pathway has emerged as a potential strategy for cancer therapy. Several small-molecule inhibitors of NRF2 have been developed and shown to sensitize cancer cells to chemotherapy and radiotherapy. By blocking NRF2-mediated antioxidant responses, these inhibitors increase intracellular ROS levels and promote cancer cell death, thereby enhancing the efficacy of traditional cancer therapies. Normally, the expression and activity of NRF2 are tightly controlled by kelch-like ECH-associated protein 1 (KEAP1). However, under oxidative conditions, NRF2 dissociates from KEAP1 and translocates to the nucleus where it binds and activates the antioxidant response element (ARE) in various target genes [31]. This results in the regulation of downstream antioxidant enzymes such as NAD(P)H quinonedehydrogenase 1, hemeoxygenase 1, thioredoxinreductase 1, superoxide dismutase, glutathione peroxidase, and catalase [32]. The cellular antioxidant system is largely regulated by NRF2 and its associated genes, which are activated and overexpressed in cancer to promote cancer cell survival [30]. In previous studies, it has been reported that STEAP1, a six-transmembrane epithelial antigen of the prostate1, which functions as an intracellular transporter and is overexpressed in colorectal cancer cells compared to normal cells, is involved in regulating ROS levels by manipulating NRF2 and its downstream genes (Figure 2) [18]. These findings suggest that similar to previous reports, NRF2 plays a critical role in protecting cancer cells from excessive ROS.

## 3. Oncogenic Functions of ROS

ROS can cause oxidative DNA damage, leading to DNA double-stranded breaks and the formation of mutagenic 8-oxo-7-hydro-2′-deoxyguanosine (8-oxodG). This molecule is a major cause of spontaneous mutagenesis, as it can induce guanine to thymine transversion by pairing with both cytosine and adenine. Therefore, the accumulation of 8-oxodG in cellular genomes can lead to carcinogenesis [33-35]. Prior research has established a strong correlation between hepatic iron overload-induced oxidative DNA damage and the development of hepatocarcinogenesis in individuals with chronic hepatitis C (CHC) [36-37]. Studies have shown that reducing therapeutic iron levels leads to a significant decrease in elevated hepatic ROS levels, which can be evaluated through immunostaining 8-oxodG. This reduction in ROS levels is often accompanied by an improvement in hepatitis symptoms and a decreased risk of developing hepatocellular carcinoma (HCC) in CHC patients [38, 39].

## 4. Cell Death Induced by ROS

### 4.1. ROS and Apoptosis

Excess ROS levels can damage various cellular components and induce apoptosis. Mitochondria are particularly susceptible to ROS-induced damage and play a crucial role in initiating intrinsic apoptosis [40]. High levels of ROS in mitochondria can trigger the release of cytochrome c into the cytosol, which then engages APAF1 and forms an apoptosome that activates caspase-9, subsequently activating effector caspases and leading to apoptosis.

Additionally, ROS can activate transmembrane death receptors like Fas, TRAIL-R1/2, and TNF receptor 1, which recruit FADD and procaspase-8/-10 to form DISCs, leading to caspase-8/-10 activation and apoptosis. Caspase-8/-10 also cleaves Bid to produce tBid, which translocates to mitochondria, blocking anti-apoptotic Bcl-2 and Bcl-XL, activating Bax and Bak, and leading to mitochondrial pathway-mediated apoptosis. Overall, ROS can activate both intrinsic and extrinsic pathways of apoptosis in cancer cells [41, 42].

## 4.2. ROS and Autophagy

Autophagy is a cellular process that helps in recycling damaged organelles and cellular components under stress conditions. The role of autophagy in cancer is complex and can either promote or suppress tumors. The relationship between ROS and autophagy is reflected in two ways: induction of autophagy by oxidative stress and reduction of ROS by autophagy. During nutrient starvation, hydrogen peroxide is produced and oxidizes ATG4, a protease that regulates autophagy by processing and deconjugating ATG8, leading to increased formation of light chain 3-associated autophagosomes [43-48]. Moreover, in addition to the above, which is regarded as a direct mechanism, an indirect induction of autophagy by ROS can also occur ROS can activate AMPK, which suppresses the activity of mTOR and leads to the induction of autophagy. Therefore, the interplay between ROS and autophagy in cancer cells is intricate [49] (Figure 2).

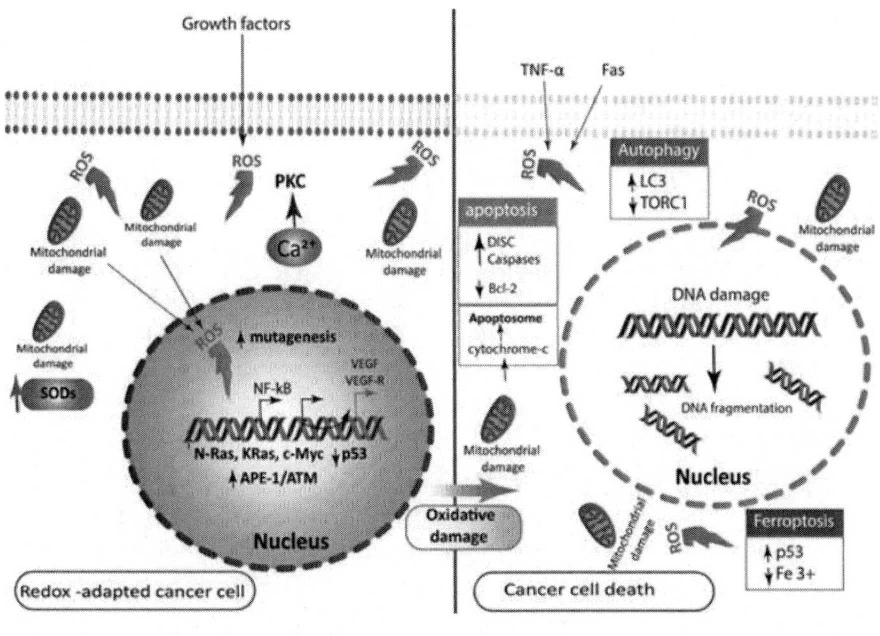


**Figure 2.** ROS and programmed cell death in cancer cells. ROS enhances the assembly of DISCs with TNF-α/Fas and the activation of effector caspases and reduces Bcl-2 activity or, as a consequence of increased permeability of mitochondrial permeability transition pore, stimulates the intra-cytoplasmic release of cytochrome c, which interacts with Apaf-1 and induce cell death (apoptosis). ROS can also inhibit the negative regulators of autophagy (TORC1) and increase the formation of LC3-dependent autophagosomes (autophagy). Elevated levels of ROS, induced by several receptor-interacting protein kinases (RIPs), increase p53 expression, which increases ROS levels *via* a mechanism that depends on intracellular iron (ferroptosis).

# 5. ROS in Cancer Treatment

## 5.1. ROS and Cancer Chemotherapy

Chemotherapy is a commonly used treatment for cancer patients. Many chemotherapeutic agents produce reactive oxygen species (ROS) and can disrupt redox homeostasis in cancer cells [50]. Examples of such drugs include anthracyclines (such as doxorubicin and epirubicin), alkylating agents, platinum coordination complexes (including cisplatin, carboplatin, and oxaliplatin), and camptothecins (such as topotecan and irinotecan). Chemotherapeutic agents generate ROS through two main mechanisms: mitochondrial ROS production and inhibition of cellular antioxidant systems. For example, cisplatin is a widely used chemotherapeutic agent that induces mitochondria-dependent ROS generation, which contributes to cell death by causing nuclear DNA damage. Cisplatin also impairs the synthesis of electron transport chain proteins by directly affecting mitochondrial DNA [51].

Multidrug resistance is a major cause of chemotherapy failure in cancer treatment. P-glycoprotein (P-gp) and other efflux pumps located in the plasma membrane are strongly associated with multidrug resistance in cancer cells [52-54]. P-gp, encoded by the MDR1/ABCB1 gene, is a member of the ATP-binding cassette family of proteins and acts as a barrier to the uptake of xenobiotics, including chemotherapeutic agents [55]. NRF2 is activated and overexpressed in cancer cells as a protective mechanism against excessive ROS. MDR1/ABCB1 is an NRF2 target gene that contributes to the increased expression of P-gp and multidrug resistance in cancer cells. Overall, ROS levels induced by chemotherapy play a crucial role in damaging cancer cells and also contribute to multidrug resistance [56-58].

## 5.2. ROS in Cancer Immunity

In this section, we focus on the impacts of $CD4^{+}Foxp3^{+}$ regulatory T (Treg) cells and dendritic cells (DCs) in anti-tumor immune responses as both cell types play essential roles in the tumor microenvironment (TME). Studies have shown that Treg cells act as potent immune suppressors within the TME [59-61]. The TME, composed of cancer cells, cancer-associated fibroblasts, and immune cells, produces high levels of ROS [62-63], which is one of the main mechanisms responsible for resistance to immunotherapies, including immune checkpoint blockade [64, 65]. Treg cells within the TME are susceptible to apoptosis due to the high concentration of ROS. These apoptotic Treg cells have a weak NRF2-mediated antioxidant system and can convert ATP to adenosine, which undermines anti-tumor T cell immunity induced by programmed death-ligand 1-blockade. Strikingly, apoptotic Treg cells exhibit a more powerful immunosuppressive effect compared to live Treg cells. Decreasing ROS levels within the TME can lead to improved immunotherapies for cancer. For instance, a novel nano-scavenger anchored on the extracellular matrix has been shown to alleviate suppressive immunogenic cell death by scavenging ROS [66]. ROS are important regulators of anti-tumor immunity not only in Treg cells but also in DC-mediated and cytotoxic T cell-mediated responses. In DCs, ROS are critical for initiating a STING-mediated anti-tumor response in a SENP3-dependent manner. Specifically, ROS induce SENP3 accumulation in DCs, leading to SENP3-IFI204 interaction and IFI204 deSUMOylation, which triggers STING-mediated anti-tumor activities. In contrast, ROS can also suppress anti-tumor immunity through ER stress-XBP1 signaling in DCs [67-70] Overall, the relationship between ROS and cancer immunity is still elusive, underlining the continuing need to explore this to improve the efficacies of immunotherapies.

## 6. ROS Promote Pro-tumorigenic Signaling

Studies conducted in the mid-1990s demonstrated that reactive oxygen species (ROS) play a role in promoting tumorigenesis [71]. Research has shown that oncogenic Ras and growth factor signaling can activate $H_2O_2$, leading to the activation of the PI3K/Akt/mTOR and MAPK/ERK signalling pathways [72, 73]. Mutations in Ras can cause an increase in ROS production from NOX4, leading to enhanced proliferation [74]. The survival pathway of PI3K/Akt/mTOR can be hyperactivated by ROS, which oxidizes and inactivates the phosphatases PTEN and PTP1B, both of which are negative regulators of PI3K/Akt signalling [75]. PTEN is frequently inactivated in a wide range of human cancers, such as glioblastomas, melanoma, prostate, breast, and endometrial cancers. Consequently, the ROS-induced PI3K/Akt survival pathway has been shown to be activated in most cancers. Furthermore, the oncogenic activation of Akt can increase ROS production, further promoting cancer cell survival and proliferation. ROS can also oxidize and inactivate MAPK phosphatases, resulting in growth factor receptor activation and MAPK/ERK pro-proliferative signalling [76].

The MAPK/ERK pathway is regulated by mitochondrial reactive oxygen species (ROS), which play a crucial role in the Kras-induced anchorage-independent growth of lung cancer cells. Moreover, studies have shown that reducing ROS levels by completely disrupting the mitochondrial respiratory chain can decrease tumorigenesis. An investigation using mitochondrial transcription factor A heterozygous mice, which have reduced mitochondrial DNA and increased mitochondrial ROS (mROS) levels, has demonstrated that these mice exhibit enhanced tumorigenesis in an APCMin/+ mouse model of intestinal cancer, providing further evidence that mROS contributes to tumorigenesis [77-79].

Reactive oxygen species (ROS) can also contribute to tumor cell survival by activating transcription factors such as NF-κB and NRF2, which increase the expression of antioxidants to evade ROS-mediated cell death. Recent studies have shown that oncogenic Kras-derived mitochondrial ROS activate NF-κB through PKD1 to up-regulate pro-proliferative EGFR signaling, resulting in the formation of pancreatic precancerous lesions. Additionally, research has demonstrated that the loss of the tumor suppressor APC leads to abnormal Wnt signalling and intestinal proliferation through the up-regulation of RAC1 and TIGAR. Interestingly, RAC1 paradoxically increases ROS generation by inducing NOX activation, while TIGAR reduces ROS by increasing cytosolic NADPH levels. It is believed that TIGAR-induced NADPH restores antioxidant function to scavenge the pool of damaging ROS, allowing RAC1 to increase the pool of pro-proliferative signaling ROS and promote optimal intestinal proliferation [80-81].

Furthermore, reactive oxygen species (ROS) play a role in promoting tumor angiogenesis and metastasis, both of which are often associated with poor prognoses. As highly proliferative tumors outgrow their blood supply, hypoxia, and glucose deprivation can occur within solid tumor regions, which leads to ROS production. Cancer cells can adapt to this harsh tumor microenvironment through metabolic changes such as hypoxia-inducible factor (HIF) stabilization and activation of the AMPK and one-carbon metabolism pathways. These adaptations can enhance NADPH production and maintain redox balance, ultimately allowing cancer cells to survive and proliferate in this environment [82].

During hypoxia, mitochondria-derived reactive oxygen species (mROS) can stabilize the oxygen-sensitive HIF-α subunit by inhibiting PHD2 activity. Under normoxic conditions with ample oxygen, HIF-α is hydroxylated by PHD2 and degraded via the E3 ubiquitin ligase von Hippel-Lindau protein [83]. However, in hypoxia, the HIF-α protein can dimerize with the HIF-β subunit and translocate to the nucleus to induce the expression of proangiogenic genes, such as vascular endothelial growth factor, as well as genes involved in cell survival, glycolysis, and metastasis. ROS-dependent stabilization of HIF-α has been shown to promote the tumorigenesis of certain cancer cells. Due to low glucose supply, PPP-dependent NADPH production is reduced, leading to increased ROS levels in the solid tumor. To prevent ROS-mediated tumor cell death and raise the antioxidant capacity, AMPK is activated to promote NADPH production and prevent anabolic processes that require NADPH consumption. The loss of AMPK has been shown to prevent oncogenic transformation. Additionally, during hypoxia, HIF-dependent up-regulation of the one-carbon metabolism enzyme SHMT2 promotes mitochondrial serine catabolism and NADPH production [84-86].

Metastasis, the process by which cancer cells break away from the primary tumor and invade other parts of the body, requires the loss of cell-to-cell adhesion. This loss results in cancer cell dissociation from the primary tumor and a breakthrough into the basement membrane. Research has shown that reactive oxygen species (ROS) can regulate several signaling pathways, such as the MAPK and PI3K/Akt pathways, and transcriptional activities, such as HIF and Snail, to enhance cancer cell migration and invasion. ROS generated by NADPH oxidases (NOX) can promote the formation of invadopodia, plasma membrane protrusions that tumor cells use to invade. Additionally, ROS-dependent v-Src oxidation and activation can enhance the invasion potential and anchorage- and growth factor-independent growth of Src-transformed cells [87, 88].

Anoikis, the apoptosis or programmed cell death that occurs when cells detach from the extracellular matrix, plays an important role in preventing metastasis. However, ROS can confer resistance to anoikis in cancer cells by oxidizing and activating Src, which leads to constitutive, ligand-independent EGFR activation and promotes cell survival. Interestingly, elevated ROS levels resulting from mutations in mitochondrial DNA that impair complex I activity have been shown to promote the metastasis of mouse and human tumor cell lines. These findings suggest that an increase in ROS levels, as long as it stays below the threshold for cell death, is necessary for modulating structural changes and adaptive responses in tumor cells to promote local invasion and metastasis [89, 90].

## 7. ROS Promote Anti-tumorigenic Signaling

High levels of ROS can lead to cell cycle arrest, senescence, and cancer cell death. To trigger apoptosis, ROS activates the ASK1/JNK and ASK1/p38 signaling pathways. In its inactive form, ASK1 binds to reduced TRX. However, when TRX undergoes H2O2-mediated oxidation, it dissociates from ASK1 and activates it, leading to sustained activation of the downstream MAP kinase (MKK)4/MKK7/JNK and MKK3/MKK6/p38 pathways [91]. Mutations that inactivate the p38 and JNK pathways are observed in various human tumors, suggesting that these signaling pathways promote cancer cell death. P38 MAPK also functions as a tumor suppressor by inhibiting malignant transformation by oncogenic Hras. Furthermore, ROS-induced activation of the JNK and p38 signaling pathways can cause cell cycle arrest, halting cancer cell growth and division [92].

As cancer cells break away from the extracellular matrix and enter the basement membrane, the level of ROS increases and the amount of GSH decreases. However, the oxidizing nature of the blood makes it difficult for many circulating cancer cells to survive and proliferate, leading to a decrease in distant metastasis. To support anchorage-independent growth and prevent cell death during metastasis, cancer cells undergo metabolic changes to increase their antioxidant capacity [93]. One such change is the cytosolic IDH1-dependent reductive carboxylation, which increases the amount of mitochondrial NADPH and lowers mROS levels, promoting anchorage-independent growth. Additionally, the folate-mediated one-carbon metabolism pathway's NADPH-generating enzymes are essential for human melanoma cells to promote distant metastasis. Therefore, targeting these ROS-mitigating pathways could be a potential therapeutic strategy to suppress cancer cell proliferation and metastasis [94].

## 8. Targeting ROS for Cancer Therapy

There are two faces of ROS in cancer, pro- and anti-tumorigenic. Although the increase in ROS allows for localized signaling to promote cancer cell proliferation and survival, the increase in ROS scavenging reduces distant ROS-mediated oxidative damage and maintains ROS homeostasis.

### 8.1. Decreasing ROS Levels to Prevent Cancer Cell Proliferation

Reactive oxygen species (ROS) are known to contribute to the development of cancer by promoting cancer cell growth. To prevent cancer cell proliferation, it is important to develop methods that can decrease intracellular ROS levels. Antioxidants are one potential therapy for cancer treatment as they can reduce ROS levels. For instance, in a mouse model of Myc-dependent lymphoma, the antioxidants NAC and vitamin C have been found to inhibit tumorigenesis by reducing the expression of the HIF-1$\alpha$ protein [95]. In addition, overexpression of SOD3 has been shown to reduce breast cancer metastasis in vivo. Despite these promising results, large-scale clinical studies have demonstrated that dietary antioxidant supplementation with $\beta$-carotene, vitamin A, or vitamin E increased the incidence of cancer. This is consistent with mouse models of Kras- and Braf-induced lung cancer, where NAC and vitamin E supplementation markedly accelerated tumor development and mortality. In another study, NAC administration increased melanoma metastasis in oncogenic Braf/Pten-null [96, 97]. Dietary antioxidants may not effectively scavenge the pool of ROS that promotes tumor growth. Instead, they may decrease the toxicity of ROS far from their origin, leading to tumor progression. To address this issue, a targeted antioxidant, such as a mitochondria-targeted antioxidant, may be useful in detoxifying the locally generated ROS involved in pro-tumorigenic signaling. For example, the mitochondria-targeted scavenger MitoTEMPO has been found to inhibit tumor cell migration and reduce spontaneous tumor metastasis in both human and murine models. Additionally, the mitochondria-targeted antioxidant MitoQ has been shown to decrease the formation of Kras-driven pancreatic precancerous lesions, indicating that mitochondrial ROS propagates the pro-tumorigenic signal [98]. Furthermore, a recent high-throughput screening of 635,000 small molecules against mitochondrial H2O2 production identified three compounds that selectively suppress complex III O2− production, preventing HIF-1α protein stabilization without impairing the bioenergetic function of the mitochondria, which is promising [99].

## 8.2. Increasing ROS Levels to Promote Oxidative Stress–Induced Cancer Cell Death

Cancer cells typically have higher levels of reactive oxygen species (ROS) compared to normal cells. This makes them more susceptible to oxidative stress-induced cell death caused by ROS-generating treatments. By increasing ROS production or reducing ROS scavenging, cancer cells may be pushed towards cell death without harming normal cells. Recent studies have shown that high doses of vitamin C, traditionally thought of as an antioxidant, can actually function as a pro-oxidant and selectively induce cell death in Kras- and Braf-mutant colon cancer cells. In addition, a synthetic lethal screen with small molecules identified lanperisone as a selective killer of Kras-mutant cancer cells that increased steady-state ROS levels [100].

Compounds that quickly bind to glutathione (GSH) and deplete its pool can induce oxidative stress and promote cancer cell death [101]. One such compound is buthioninesulfoximine (BSO), which inhibits glutamate cysteine ligase (GCL), the enzyme required for GSH synthesis, thereby reducing the levels of GSH and exhibiting anticancer activity [102]. In vivo studies have shown that genetic loss of the modifier subunit of GCL (GCLM) or administration of BSO before tumor onset impairs mammary tumor initiation and progression [103]. Furthermore, a recent study found that the combination of BSO and the glucose analog 2-deoxy-d-glucose (2DG) selectively induced cell death in metabolically reprogrammed mutant Kras copy gain–specific lung cancer cells both in vitro and in vivo [104]. This combination also had a similar effect on human breast cancer cells. Another small molecule, piperlongumine, which was discovered through an unbiased screen, increased ROS levels and selectively induced cancer cell death by reducing the abundance of reduced GSH. Targeting cancer stem cells (CSCs), a highly tumorigenic subpopulation of cancer cells that are often resistant to conventional therapies such as chemotherapy and radiation, presents a promising therapeutic approach [105]. CSCs have lower levels of ROS than regular cancer cells and non-tumorigenic cells, partly due to increased PPP-dependent NADPH production and overexpression of antioxidants such as GSH. For instance, CD44v-expressing gastrointestinal CSCs have been shown to stabilize the cysteine-glutamate exchange transporter and increase GSH synthesis, contributing to the heightened antioxidant capacity that supports their self-renewal and differentiation and allows them to evade ROS-mediated oxidative damage and cell death induced by radiation and chemotherapy [106, 107]. Inhibiting GSH synthesis with BSO has been demonstrated to reduce the colony-forming ability and radioresistance of breast CSCs. Additionally, CSCs maintain their survival and stem-like properties by up-regulating several ROS-dependent signaling pathways, such as the PI3K/Akt/mTOR and Notch pathways. As low ROS levels appear to support the survival of CSCs, treatments that increase ROS levels could effectively induce CSC death. [108].

Various antioxidants have been identified as potential targets for selectively killing cancer cells. In addition to GSH, other targets include TRX, TR, and SOD1 inhibitors. For example, studies have shown that the SOD1 inhibitor ATN-224 induces cell death in non-small-cell lung cancer cells with an oncogenic Kras mutation by activating the p38/MAPK signaling pathway and downregulating the anti-apoptotic factor myeloid cell leukemia 1. Another small SOD1 inhibitor, LCS-1, impairs mitochondrial integrity in breast cancer cells, suggesting its potential as a therapeutic approach for breast cancer. Additionally, targeting the antioxidant CAT in combination with ROS-generating CDK4 and CDK6 inhibitors elevates ROS levels above the threshold and reduces pancreatic ductal adenocarcinoma cell growth [109, 110].

The inhibition of the antioxidant master regulator NRF2 has the potential to be an effective anticancer therapy. However, the development of NRF2 inhibitors is challenging due to a lack of specificity. Therefore, targeting specific NRF2-regulated metabolic pathways that produce NADPH may be a more effective approach for anticancer therapeutics. RNA profiling of various tumor types has shown that methylenetetra-hydrofolate dehydrogenase 2, an enzyme involved in mitochondrial one-carbon metabolism, is highly expressed in cancer cells compared to normal proliferating cells. Thus, targeting the mitochondrial one-carbon metabolism pathway could provide a selective and effective anticancer therapy, particularly in hypoxic tumors where treatment options are limited [111, 112].

One promising therapeutic strategy for cancer treatment is to manipulate nutrient availability since cancer cells consume nutrients, such as serine, to promote their proliferation. Serine depletion has been shown to impair proliferation in HCT116 p53-null cancer cells and increase intracellular ROS levels while decreasing levels of GSH. However, wild-type p53 during serine starvation allows cell cycle arrest via p21-dependent mechanisms and redirects de novo serine synthesis to GSH synthesis to maintain cellular antioxidant capacity, preventing oxidative stress. Thus, serine depletion may be an effective treatment for p53-deficient tumors, which is a common genetic alteration in many cancer types. Combining ROS-producing agents with ROS-scavenging inhibitors could also reduce cancer cell adaptation and resistance to chemotherapeutics [113].

## Conclusion

The role of reactive oxygen species (ROS) in cancer has shifted from being viewed as harmful molecules that promote cancer development to critical signaling molecules that can participate in both pro- and anti-tumorigenic signaling events. Traditional ROS-manipulation strategies have focused on antioxidant therapy; however, the failure of dietary antioxidants in clinical trials has led to alternative approaches. Antioxidants or inhibitors targeting the source of ROS generation near the site of pro-tumorigenic signaling may prevent tumorigenesis. In addition, pro-oxidant cancer therapy, where ROS production is increased, or ROS scavenging is decreased, is an attractive approach, given the higher levels of ROS production in cancer cells compared with normal cells. ROS accumulation due to pro-oxidant therapy may be enough to tip the balance toward oxidative stress-induced cell death. However, normal cells, particularly stem cells, are sensitive to ROS levels, so it will be important to identify and target specific antioxidant pathways used by cancer cells to maintain redox balance. A better understanding of the molecular mechanisms of ROS signaling in cancer and the identification of specific ROS targets may provide novel therapeutic avenues for treating cancer.

## References


[1] Sung H, Ferlay J, Siegel RL, Laversanne M, Soerjomataram I, Jemal A, Bray F. Global Cancer Statistics 2020: GLOBOCAN Estimates of Incidence and Mortality Worldwide for 36 Cancers in 185 Countries. *CA Cancer J Clin.* 2021 May; 71(3):209-249.

[2] Sena LA, Chandel NS. Physiological roles of mitochondrial reactive oxygen species. *Mol Cell.* 2012; 48(2):158-67.

[3] Cairns RA, Harris IS, Mak TW. Regulation of cancer cell metabolism. *Nat Rev Cancer.* 2011; 11(2):85-95.

[4] Yang H, Villani RM, Wang H, Simpson MJ, Roberts MS, Tang M, Liang X. The role of cellular reactive oxygen species in cancer chemotherapy. *J Exp Clin Cancer Res.* 2018; 37(1):266.

[5] Chio IIC, Tuveson DA. ROS in Cancer: the burning question. *Trends Mol Med.* 2017; 23(5):411-29.

[6] Wiseman H, Halliwell B. Damage to DNA by reactive oxygen and nitrogen species: role in inflammatory disease and progression to cancer. *Biochem J.* 1996; 313(1):17-29.

[7] Okon IS, Zou MH. Mitochondrial ROS and cancer drug resistance: implications for therapy. *Pharmacol Res.* 2015; 100:170-4.

[8] Gorrini C, Harris IS, Mak TW. Modulation of oxidative stress as an anticancer strategy. *Nat Rev Drug Discov.* 2013; 12(12):931-47.

[9] Kamihara Y, Takada K, Sato T, Kawano Y, Murase K, Arihara Y, Kikuchi S, Hayasaka N, Usami M, Iyama S, Miyanishi K, Sato Y, Kobune M, Kato J. The iron chelator deferasirox induces apoptosis by targeting oncogenic Pyk2/β-catenin signaling in human multiple myeloma. *Oncotarget.* 2016; 7(39):64330-64341.

[10] Sayin VI, Ibrahim MX, Larsson E, Nilsson JA, Lindahl P, Bergo MO. Antioxidants accelerate lung cancer progression in mice. *Sci Transl Med.* 2014; 6(221):221ra15.

[11] Piskounova E, Agathocleous M, Murphy MM, Hu Z, Huddlestun SE, Zhao Z, Leitch AM, Johnson TM, DeBerardinis RJ, Morrison SJ. Oxidative stress inhibits distant metastasis by human melanoma cells. *Nature.* 2015 Nov 12;527(7577):186-91.

[12] Peiris-Pagès M, Martinez-Outschoorn UE, Sotgia F, Lisanti MP. Metastasis and Oxidative Stress: Are Antioxidants a Metabolic Driver of Progression? *Cell Metab.* 2015; 22(6):956-8.

[13] Schafer ZT, Grassian AR, Song L, Jiang Z, Gerhart-Hines Z, Irie HY, Gao S, Puigserver P, Brugge JS. Antioxidant and oncogene rescue of metabolic defects caused by loss of matrix attachment. *Nature.* 2009;461(7260):109-13.

[14] DeAtley SM, Aksenov MY, Aksenova MV, Harris B, Hadley R, Cole Harper P, Carney JM, Butterfield DA. Antioxidants protect against reactive oxygen species associated with adriamycin-treated cardiomyocytes. *Cancer Lett.* 1999;136(1):41-6.

[15] Siomek A, Tujakowski J, Gackowski D, Rozalski R, Foksinski M, Dziaman T, Roszkowski K, Olinski R. Severe oxidatively damaged DNA after cisplatin treatment of cancer patients. *Int J Cancer.* 2006; 119(9):2228-30.

[16] Dougan SJ, Habtemariam A, McHale SE, Parsons S, Sadler PJ. Catalytic organometallic anticancer complexes. *Proc Natl Acad Sci U S A.* 2008; 105(33):11628-33.

[17] Arihara Y, Takada K, Kamihara Y, Hayasaka N, Nakamura H, Murase K, Ikeda H, Iyama S, Sato T, Miyanishi K, Kobune M, Kato J. Small molecule CP-31398 induces reactive oxygen species-dependent apoptosis in human multiple myeloma. *Oncotarget.* 2017; 8(39):65889-65899.

[18] Nakamura H, Takada K, Arihara Y, Hayasaka N, Murase K, Iyama S, Kobune M, Miyanishi K, Kato J. Six-transmembrane epithelial antigen of the prostate 1 protects against increased oxidative stress via a nuclear erythroid 2-related factor pathway in colorectal cancer. *Cancer Gene Ther.* 2019;26(9-10):313-322.

[19] Galadari S, Rahman A, Pallichankandy S, Thayyullathil F. Reactive oxygen species and cancer paradox: to promote or to suppress? *Free Radic Biol Med.* 2017; 104:144-64.

[20] Starkov AA. The role of mitochondria in reactive oxygen species metabolism and signaling. *Ann N Y Acad Sci.* 2008; 1147:37-52.

[21] Li X, Fang P, Mai J, Choi ET, Wang H, Yang XF. Targeting mitochondrial reactive oxygen species as novel therapy for inflammatory diseases and cancers. *J Hematol Oncol.* 2013; 6:19.

[22] Schägger H, Pfeiffer K. The ratio of oxidative phosphorylation complexes I-V in bovine heart mitochondria and the composition of respiratory chain supercomplexes. *J Biol Chem.* 2001; 276(41):37861-7.

[23] Balaban RS, Nemoto S, Finkel T. Mitochondria, oxidants, and aging. *Cell.* 2005; 120(4):483-95.

[24] Beyersmann D, Hartwig A. Carcinogenic metal compounds: recent insight into molecular and cellular mechanisms. *Arch Toxicol.* 2008; 82(8):493-512.

[25] Bystrom LM, Guzman ML, Rivella S. Iron and reactive oxygen species: friends or foes of cancer cells? *Antioxid Redox Signal.* 2014;20(12):1917-24.

[26] Bhattacharyya A, Chattopadhyay R, Mitra S, Crowe SE. Oxidative stress: an essential factor in the pathogenesis of gastrointestinal mucosal diseases. *Physiol Rev.* 2014; 94(2):329-54.

[27] van der Toorn M, Rezayat D, Kauffman HF, Bakker SJ, Gans RO, Koëter GH, Augustine MK Choi, Antoon JM van Oosterhout, Dirk-Jan Slebos. Lipid-soluble components in cigarette smoke induce mitochondrial production of reactive oxygen species in lung epithelial cells. *Am J Physiol Lung Cell Mol Physiol.* 2009; 297(1):L109-14.

[28] Riley PA. Free radicals in biology: oxidative stress and the effects of ionizing radiation. *Int J Radiat Biol.* 1994;65(1):27-33.

[29] Motohashi H, Yamamoto M. Nrf2-Keap1 defines a physiologically important stress response mechanism. *Trends Mol Med.* 2004; 10(11):549-57.

[30] Sporn MB, Liby KT. NRF2 and cancer: the good, the bad and the importance of context. *Nat Rev Cancer.* 2012; 12(8):564-71.

[31] Kansanen E, Kuosmanen SM, Leinonen H, Levonen AL. The Keap1-Nrf2 pathway: mechanisms of activation and dysregulation in cancer. *Redox Biol.* 2013; 1(1):45-9.

[32] Ma Q. Role of Nrf2 in oxidative stress and toxicity. *Annu Rev Pharmacol Toxicol.* 2013; 53: 401- 426.

[33] Sallmyr A, Fan J, Rassool FV. Genomic instability in myeloid malignancies: increased reactive oxygen species (ROS), DNA double strand breaks (DSBs) and error-prone repair. *Cancer Lett.* 2008; 270: 1-9.

[34] Shibutani S, Takeshita M, Grollman AP. Insertion of specific bases during DNA synthesis past the oxidation-damaged base 8-oxodG. *Nature.* 1991; 349: 431- 434.

[35] Oka S, Nakabeppu Y. DNA glycosylase encoded by MUTYH functions as a molecular switch for programmed cell death under oxidative stress to suppress tumorigenesis. *Cancer Sci.* 2011; 102: 677-682.

[36] Dixon SJ, Stockwell BR. The role of iron and reactive oxygen species in cell death. *Nat Chem Biol.* 2014; 10: 9- 17.

[37] Toyokuni S. Oxidative stress as an iceberg in carcinogenesis and cancer biology. *Arch Biochem Biophys.* 2016; 595: 46- 49.

[38] Kato J, Kobune M, Nakamura T, Kuroiwa G, Takada K, Takimoto R, Sato Y, Fujikawa K, Takahashi M, Takayama T, Ikeda T, Niitsu Y. Normalization of elevated hepatic 8-hydroxy-2'-deoxyguanosine levels in chronic hepatitis C patients by phlebotomy and low iron diet. *Cancer Res.* 2001;61:8697-702.

[39] Kato J, Miyanishi K, Kobune M, Nakamura T, Takada K, Takimoto R, Kawano Y, Takahashi S, Takahashi M, Sato Y, Takayama T, Niitsu Y. Long-term phlebotomy with low-iron diet therapy lowers risk of development of hepatocellular carcinoma from chronic hepatitis C. *J Gastroenterol.* 2007; 42(10):830-6.

[40] Redza-Dutordoir M, Averill-Bates DA. Activation of apoptosis signalling pathways by reactive oxygen species. *Biochim Biophys Acta.* 2016; 1863: 2977- 2992.

[41] Simon HU, Haj-Yehia A, Levi-Schaffer F. Role of reactive oxygen species (ROS) in apoptosis induction. *Apoptosis.* 2000; 5: 415- 418.

[42] Ow YP, Green DR, Hao Z, Mak TW. Cytochrome c: functions beyond respiration. *Nat Rev Mol Cell Biol.* 2008; 9: 532- 542.

[43] Filomeni G, De Zio D, Cecconi F. Oxidative stress and autophagy: the clash between damage and metabolic needs. *Cell Death Differ.* 2015; 22: 377- 388.

[44] Yun CW, Lee SH. The roles of autophagy in cancer. *Int J Mol Sci.* 2018; 19: 3466.

[45] Li L, Tan J, Miao Y, Lei P, Zhang Q. ROS and autophagy: interactions and molecular regulatory mechanisms. *Cell Mol Neurobiol.* 2015; 35: 615- 621.

[46] Maruyama T, Noda NN. Autophagy-regulating protease Atg4: structure, function, regulation and inhibition. *J Antibiot* (Tokyo). 2017; 71: 72- 78.

[47] Scherz-Shouval R, Shvets E, Fass E, Shorer H, Gil L, Elazar Z. Reactive oxygen species are essential for autophagy and specifically regulate the activity of Atg4. *EMBOJ.* 2007; 26: 1749- 1760.

[48] Poillet-Perez L, Despouy G, Delage-Mourroux R, Boyer-Guittaut M. Interplay between ROS and autophagy in cancer cells, from tumor initiation to cancer therapy. *Redox Biol.* 2015; 4: 184- 192.

[49] Emerling BM, Weinberg F, Snyder C, Burgess Z, Mutlu GM, Viollet B, Budinger GR, Chandel NS. Hypoxic activation of AMPK is dependent on mitochondrial ROS but independent of an increase in AMP/ATP ratio. *Free Radic Biol Med*. 2009;46(10):1386-91.

[50] Conklin KA. Chemotherapy-associated oxidative stress: impact on chemotherapeutic effectiveness. *Integr Cancer Ther.* 2004; 3: 294- 300.

[51] Marullo R, Werner E, Degtyareva N, Moore B, Altavilla G, Ramalingam SS, Doetsch PW. Cisplatin induces a mitochondrial-ROS response that contributes to cytotoxicity depending on mitochondrial redox status and bioenergetic functions. *PLoS One.* 2013 Nov 19; 8(11):e81162.

[52] Szakács G, Paterson JK, Ludwig JA, Booth-Genthe C, Gottesman MM. Targeting multidrug resistance in cancer. *Nat Rev Drug Discov.* 2006; 5: 219- 234.

[53] Wang H, Gao Z, Liu X, Agarwal P, Zhao S, Conroy DW, Ji G, Yu J, Jaroniec CP, Liu Z, Lu X, Li X, He X. Targeted production of reactive oxygen species in mitochondria to overcome cancer drug resistance. *Nat Commun.* 2018; 9(1):562.

[54] Ambudkar SV, Kimchi-Sarfaty C, Sauna ZE, Gottesman MM. P-glycoprotein: from genomics to mechanism. *Oncogene.* 2003; 22: 7468- 7485.

[55] Jeddi F, Soozangar N, Sadeghi MR, Somi MH, Shirmohamadi M, Eftekhar-Sadat AT, Samadi N. Nrf2 overexpression is associated with P-glycoprotein up-regulation in gastric cancer. *Biomed Pharmacother.* 2018; 97:286-292.

[56] Sadeghi MR, Jeddi F, Soozangar N, Somi MH, Shirmohamadi M, Khaze V, Samadi N. Nrf2/P-glycoprotein axis is associated with clinicopathological characteristics in colorectal cancer. *Biomed Pharmacother.* 2018 Aug; 104:458-464.

[57] Zheng H, Nong Z, Lu G. Correlation Between Nuclear Factor E2-Related Factor 2 Expression and Gastric Cancer Progression. *Med Sci Monit.* 2015; 21:2893-9.

[58] Panieri E, Saso L. Inhibition of the NRF2/KEAP1 Axis: A Promising Therapeutic Strategy to Alter Redox Balance of Cancer Cells. *Antioxid Redox Signal.* 2021; 34(18):1428-1483.

[59] Liyanage UK, Moore TT, Joo HG, Tanaka Y, Herrmann V, Doherty G, Drebin JA, Strasberg SM, Eberlein TJ, Goedegebuure PS, Linehan DC. Prevalence of regulatory T cells is increased in peripheral blood and tumor microenvironment of patients with pancreas or breast adenocarcinoma. *J Immunol.* 2002; 169(5):2756-61.

[60] Curiel TJ, Coukos G, Zou L, Alvarez X, Cheng P, Mottram P, Evdemon-Hogan M, Conejo-Garcia JR, Zhang L, Burow M, Zhu Y, Wei S, Kryczek I, Daniel B, Gordon A, Myers L, Lackner A, Disis ML, Knutson KL, Chen L, Zou W. Specific recruitment of regulatory T cells in ovarian carcinoma fosters immune privilege and predicts reduced survival. *Nat Med.* 2004; 10(9):942-9.

[61] Sakaguchi S. Naturally arising Foxp3-expressing CD25+CD4+ regulatory T cells in immunological tolerance to self and non-self. *Nat Immunol.* 2005; 6(4):345-52.

[62] Weinberg F, Ramnath N, Nagrath D, Nagrath D. Reactive oxygen species in the tumor microenvironment: an overview. *Cancers* (Basel). 2019; 11(8):1191.

[63] Malla R, Surepalli N, Farran B, Malhotra SV, Nagaraju GP. Reactive oxygen species (ROS): Critical roles in breast tumor microenvironment. *Crit Rev Oncol Hematol.* 2021; 160:103285.

[64] Cubillos-Ruiz JR, Silberman PC, Rutkowski MR, Chopra S, Perales-Puchalt A, Song M, Zhang S, Bettigole SE, Gupta D, Holcomb K, Ellenson LH, Caputo T, Lee AH, Conejo-Garcia JR, Glimcher LH. ER Stress Sensor XBP1 Controls Anti-tumor Immunity by Disrupting Dendritic Cell Homeostasis. *Cell.* 2015; 161(7):1527-38.

[65] Maj T, Wang W, Crespo J, Zhang H, Wang W, Wei S, Zhao L, Vatan L, Shao I, Szeliga W, Lyssiotis C, Liu JR, Kryczek I, Zou W. Oxidative stress controls regulatory T cell apoptosis and suppressor activity and PD-L1-blockade resistance in tumor. *Nat Immunol.* 2017; 18(12):1332-1341.

[66] Deng H, Yang W, Zhou Z, Tian R, Lin L, Ma Y, Song J, Chen X. Targeted scavenging of extracellular ROS relieves suppressive immunogenic cell death. *Nat Commun.* 2020; 11(1):4951.

[67] Gehrke N, Mertens C, Zillinger T, Wenzel J, Bald T, Zahn S, Tüting T, Hartmann G, Barchet W. Oxidative damage of DNA confers resistance to cytosolic nuclease TREX1 degradation and potentiates STING-dependent immune sensing. Immunity. 2013; 39(3):482-95.

[68] Chamoto K, Chowdhury PS, Kumar A, Sonomura K, Matsuda F, Fagarasan S, Honjo T. Mitochondrial activation chemicals synergize with surface receptor PD-1 blockade for T cell-dependent antitumor activity. *Proc Natl Acad Sci U S A.* 2017; 114(5):E761-E770.

[69] Hu Z, Teng XL, Zhang T, Yu X, Ding R, Yi J, Deng L, Wang Z, Zou Q. SENP3 senses oxidative stress to facilitate STING-dependent dendritic cell antitumor function. *Mol Cell.* 2021; 81(5):940-952.e5.

[70] Wang L, Azad N, Kongkaneramit L, Chen F, Lu Y, Jiang BH, Rojanasakul Y. The Fas death signaling pathway connecting reactive oxygen species generation and FLICE inhibitory protein down-regulation. *J Immunol.* 2008; 180(5):3072-80.

[71] Safford SE, Oberley TD, Urano M, St Clair DK. Suppression of fibrosarcoma metastasis by elevated expression of manganese superoxide dismutase. *Cancer Res.* 1994;54(16):4261-65.

[72] Sundaresan M, Yu ZX, Ferrans VJ, Irani K, Finkel T. Requirement for generation of $H_2O_2$ for platelet-derived growth factor signal transduction. *Science.* 1995; 270(5234):296-99.

[73] Irani K, Xia Y, Zweier JL, Sollott SJ, Der CJ, Fearon ER, Sundaresan M, Finkel T, Goldschmidt-Clermont PJ. Mitogenic signaling mediated by oxidants in Ras-transformed fibroblasts. *Science.* 1997; 275(5306):1649-52.

[74] Irani K, Goldschmidt-Clermont PJ. Ras, superoxide and signal transduction. *Biochem Pharmacol.* 1998; 55(9):1339-46.

[75] Salmeen A, Andersen JN, Myers MP, Meng TC, Hinks JA, Tonks NK, Barford D. Redox regulation of protein tyrosine phosphatase 1B involves a sulphenyl-amide intermediate. *Nature.* 2003; 423(6941):769-73.

[76] Son Y, Cheong YK, Kim NH, Chung HT, Kang DG, Pae HO. Mitogen-activated protein kinases and reactive oxygen species: how can ROS activate MAPK pathways? *J Signal Transduct.* 2011; 2011:792639.

[77] Weinberg F, Hamanaka R, Wheaton WW, Weinberg S, Joseph J, Lopez M, Kalyanaraman B, Mutlu GM, Budinger GR, Chandel NS. Mitochondrial metabolism and ROS generation are essential for Kras-mediated tumorigenicity. *Proc Natl Acad Sci U S A.* 2010; 107(19):8788-93.

[78] Weinberg SE, Chandel NS. Targeting mitochondria metabolism for cancer therapy. *Nat Chem Biol.* 2015; 11(1):9-15.

[79] Woo HA, Chae HZ, Hwang SC, Yang KS, Kang SW, Kim K, Rhee SG. Reversing the inactivation of peroxiredoxins caused by cysteine sulfinic acid formation. *Science.* 2003; 300(5619):653-6.

[80] Kobayashi M, Yamamoto M. Nrf2–Keap1 regulation of cellular defense mechanisms against electrophiles and reactive oxygen species. *Adv Enzym Regul.* 2006; 46:113-40.

[81] Cheung EC, Ludwig RL, Vousden KH. Mitochondrial localization of TIGAR under hypoxia stimulates HK2 and lowers ROS and cell death. *Proc Natl Acad Sci U S A.* 2012; 109(50):20491-96.

[82] Jeon SM, Chandel NS, Hay N. AMPK regulates NADPH homeostasis to promote tumour cell survival during energy stress. *Nature.* 2012;485(7400):661-65.

[83] Bell EL, Emerling BM, Ricoult SJ, Guarente L. SirT3 suppresses hypoxia inducible factor 1α and growth by inhibiting mitochondrial ROS production. *Oncogene.* 2011; 30(26):2986-96.

[84] Horak P, Crawford AR, Vadysirisack DD, Nash ZM, DeYoung MP, Sgroi D, Ellisen LW. Negative feedback control of HIF-1 through REDD1-regulated ROS suppresses tumorigenesis. *Proc Natl Acad Sci U S A.* 2010; 107(10):4675-80.

[85] Ren Y, Shen HM. Critical role of AMPK in redox regulation under glucose starvation. *Redox Biol.* 2019; 25:101154.

[86] Ye J, Fan J, Venneti S, Wan YW, Pawel BR, Zhang J, Finley LW, Lu C, Lindsten T, Cross JR, Qing G, Liu Z, Simon MC, Rabinowitz JD, Thompson CB. Serine catabolism regulates mitochondrial redox control during hypoxia. *Cancer Discov.* 2014;4(12):1406-17.

[87] Tochhawng L, Deng S, Pervaiz S, Yap CT. Redox regulation of cancer cell migration and invasion. *Mitochondrion.* 2013;13(3):246-53.

[88] Diaz B, Shani G, Pass I, Anderson D, Quintavalle M, Courtneidge SA. Tks5-dependent, Nox-mediated generation of reactive oxygen species is necessary for invadopodia formation. *Sci Signal.* 2009; 2(88):ra53.

[89] Giannoni E, Buricchi F, Raugei G, Ramponi G, Chiarugi P. Intracellular reactive oxygen species activate Src tyrosine kinase during cell adhesion and anchorage-dependent cell growth. *Mol Cell Biol.* 2005; 25(15):6391-403.

[90] Ishikawa K, Takenaga K, Akimoto M, Koshikawa N, Yamaguchi A, Imanishi H, Nakada K, Honma Y, Hayashi J. ROS-generating mitochondrial DNA mutations can regulate tumor cell metastasis. *Science.* 2008; 320(5876):661-4.

[91] Ichijo H, Nishida E, Irie K, ten Dijke P, Saitoh M, Moriguchi T, Takagi M, Matsumoto K, Miyazono K, Gotoh Y. Induction of apoptosis by ASK1, a mammalian MAPKKK that activates SAPK/JNK and p38 signaling pathways. *Science.* 1997; 275(5296):90-4.

[92] Thornton TM, Rincon M. Non-classical p38 map kinase functions: cell cycle checkpoints and survival. *Int J Biol Sci.* 2009; 5(1):44-51.

[93] Piskounova E, Agathocleous M, Murphy MM, Hu Z, Huddlestun SE, Zhao Z, Leitch AM, Johnson TM, DeBerardinis RJ, Morrison SJ. Oxidative stress inhibits distant metastasis by human melanoma cells. *Nature.* 2015; 527(7577):186-91.

[94] Wu X, Zhu Y, Yan H, Liu B, Li Y, Zhou Q, Xu K. Isothiocyanates induce oxidative stress and suppress the metastasis potential of human non-small cell lung cancer cells. *BMC Cancer.* 2010; 10:269.

[95] Gao P, Zhang H, Dinavahi R, Li F, Xiang Y, Raman V, Bhujwalla ZM, Felsher DW, Cheng L, Pevsner J, Lee LA, Semenza GL, Dang CV. HIF-dependent antitumorigenic effect of antioxidants in vivo. *Cancer Cell.* 2007; 12(3):230-8.

[96] Sayin VI, Ibrahim MX, Larsson E, Nilsson JA, Lindahl P, Bergo MO. Antioxidants accelerate lung cancer progression in mice. *Sci Transl Med.* 2014; 6(221):221ra15.

[97] Le Gal K, Ibrahim MX, Wiel C, Sayin VI, Akula MK, Karlsson C, Dalin MG, Akyürek LM, Lindahl P, Nilsson J, Bergo MO. Antioxidants can increase melanoma metastasis in mice. *Sci Transl Med.* 2015 Oct 7; 7(308):308re8.

[98] Liou GY, Döppler H, DelGiorno KE, Zhang L, Leitges M, Crawford HC, Murphy MP, Storz P. Mutant KRas-Induced Mitochondrial Oxidative Stress in Acinar Cells Upregulates EGFR Signaling to Drive Formation of Pancreatic Precancerous Lesions. *Cell Rep.* 2016; 14(10):2325-36.

[99] Orr AL, Vargas L, Turk CN, Baaten JE, Matzen JT, Dardov VJ, Attle SJ, Li J, Quackenbush DC, Goncalves RL, Perevoshchikova IV, Petrassi HM, Meeusen SL, Ainscow EK, Brand MD. Suppressors of superoxide production from mitochondrial complex III. *Nat Chem Biol.* 2015;11(11):834-6.

[100] Shaw AT, Winslow MM, Magendantz M, Ouyang C, Dowdle J, Subramanian A, Lewis TA, Maglathin RL, Tolliday N, Jacks T. Selective killing of K-ras mutant cancer cells by small molecule inducers of oxidative stress. *Proc Natl Acad Sci U S A.* 2011; 108(21):8773-8.

[101] Dvorakova K, Payne CM, Tome ME, Briehl MM, McClure T, Dorr RT. Induction of oxidative stress and apoptosis in myeloma cells by the aziridine-containing agent imexon. *Biochem Pharmacol.* 2000; 60(6):749-58.

[102] Griffith OW, Meister A. Potent and specific inhibition of glutathione synthesis by buthionine sulfoximine (S-n-butyl homocysteine sulfoximine). *J Biol Chem.* 1979; 254(16):7558-60.

[103] Harris IS, Treloar AE, Inoue S, Sasaki M, Gorrini C, Lee KC, Yung KY, Brenner D, Knobbe-Thomsen CB, Cox MA, Elia A, Berger T, Cescon DW, Adeoye A, Brüstle A, Molyneux SD, Mason JM, Li WY, Yamamoto K, Wakeham A, Berman HK, Khokha R, Done SJ, Kavanagh TJ, Lam CW, Mak TW. Glutathione and thioredoxin antioxidant pathways synergize to drive cancer initiation and progression. *Cancer Cell.* 2015; 27(2):211-22.

[104] Raj L, Ide T, Gurkar AU, Foley M, Schenone M, Li X, Tolliday NJ, Golub TR, Carr SA, Shamji AF, Stern AM, Mandinova A, Schreiber SL, Lee SW. Selective killing of cancer cells by a small molecule targeting the stress response to ROS. *Nature.* 2011; 475(7355):231-4.

[105] Meacham CE, Morrison SJ. Tumour heterogeneity and cancer cell plasticity. *Nature.* 2013; 501(7467):328-37.

[106] Diehn M, Cho RW, Lobo NA, Kalisky T, Dorie MJ, Kulp AN, Qian D, Lam JS, Ailles LE, Wong M, Joshua B, Kaplan MJ, Wapnir I, Dirbas FM, Somlo G, Garberoglio C, Paz B, Shen J, Lau SK, Quake

SR, Brown JM, Weissman IL, Clarke MF. Association of reactive oxygen species levels and radioresistance in cancer stem cells. *Nature.* 2009; 458(7239):780-3.

[107] Ishimoto T, Nagano O, Yae T, Tamada M, Motohara T, Oshima H, Oshima M, Ikeda T, Asaba R, Yagi H, Masuko T, Shimizu T, Ishikawa T, Kai K, Takahashi E, Imamura Y, Baba Y, Ohmura M, Suematsu M, Baba H, Saya H. CD44 variant regulates redox status in cancer cells by stabilizing the xCT subunit of system xc (-) and thereby promotes tumor growth. *Cancer Cell.* 2011; 19(3):387-400.

[108] McAuliffe SM, Morgan SL, Wyant GA, Tran LT, Muto KW, Chen YS, Chin KT, Partridge JC, Poole BB, Cheng KH, Daggett J Jr, Cullen K, Kantoff E, Hasselbatt K, Berkowitz J, Muto MG, Berkowitz RS, Aster JC, Matulonis UA, Dinulescu DM. Targeting Notch, a key pathway for ovarian cancer stem cells, sensitizes tumors to platinum therapy. *Proc Natl Acad Sci U S A.* 2012;109(43):E2939-48.

[109] Glasauer A, Sena LA, Diebold LP, Mazar AP, Chandel NS. Targeting SOD1 reduces experimental non-small-cell lung cancer. *J Clin Invest.* 2014;124(1):117-28.

[110] Franco J, Balaji U, Freinkman E, Witkiewicz AK, Knudsen ES. Metabolic reprogramming of pancreatic cancer mediated by CDK4/6 inhibition elicits unique vulnerabilities. *Cell Rep.* 2016;14(5):979-90.

[111] Jaramillo MC, Zhang DD. The emerging role of the Nrf2–Keap1 signaling pathway in cancer. *Genes Dev.* 2013;27(20):2179-91.

[112] Nilsson R, Jain M, Madhusudhan N, Sheppard NG, Strittmatter L, Kampf C, Huang J, Asplund A, Mootha VK. Metabolic enzyme expression highlights a key role for MTHFD2 and the mitochondrial folate pathway in cancer. *Nat Commun.* 2014;5:3128.

[113] Maddocks OD, Berkers CR, Mason SM, Zheng L, Blyth K, Gottlieb E, Vousden KH. Serine starvation induces stress and p53-dependent metabolic remodeling in cancer cells. *Nature.* 2013;493(7433):542-6.

## Chapter 2

# The Interplay Between Endoplasmic Reticulum Stress Mediated ROS Generation and Apoptosis in Human Diseases

**Satyajit Halder**[1]
**Nikhilesh Joardar**[2,3]
**Rajkumar Maiti**[4]
**Shanti Prasad Sinha Babu**[3]
**Debidas Ghosh**[5]
**and Kuladip Jana**[1,*]

[1]Division of Molecular Medicine, Bose Institute, Calcutta Improvement Trust Scheme VII M, Salt Lake City, Bidhannagar, Kolkata, India
[2]Infectious Disease Division, Washington University School of Medicine in Saint Louis, Saint Louis, Missouri, USA
[3]Parasitology Laboratory, Department of Zoology, Siksha Bhavana, Visva Bharati, Santiniketan, West Bengal, India
[4]Department of Physiology, Bankura Christian College, Bankura, India
[5]Department of Bio-Medical Laboratory Science and Management, Vidyasagar University, Midnapur, India

## Abstract

Reactive oxygen species (ROS) play an essential role in different biological functions, including physiological regulatory mechanisms to damaging alterations, which is involved in the pathogenesis of an increasing number of different human diseases. ROS are metabolic products arising from numerous cells but two vital cellular organelles are intimately involved in their production and metabolism, i.e., the endoplasmic reticulum and the mitochondria. ER chaperones aid in the correct folding of proteins created in the endoplasmic reticulum (ER). Miss-folded proteins are disposed of by ER-associated protein degradation (ERAD). Human cells activate a defense mechanism known as the ER stress response when the amount of unfolded protein exceeds the ER's capacity to fold it. This response causes the production of ER chaperones and ERAD components and temporarily slows protein synthesis to lessen the load on the ER. A range of cellular

* Corresponding Author's Email: kuladip@jcbose.ac.in; kuladip.jana@gmail.com.

In: The Role of Reactive Oxygen Species in Health and Disease
Editor: Kuladip Jana
ISBN: 979-8-89113-283-2

signalling systems, including reduction-oxidation (redox) homeostasis, are triggered by changes in the protein-folding environment. Overproduction of ROS within the cell may damage proteins, nucleic acids, lipids, membranes, and organelles, which may activate cell death processes like apoptosis. Apoptosis is a highly regulated process that is essential for the development and survival of multicellular organisms. These organisms often need to discard cells that are superfluous or potentially harmful, having accumulated mutations or become infected by pathogens. Cells undergo a cascade of self-destruction during apoptosis, which is characterized by a specific set of morphological and biochemical properties. To maintain appropriate cellular homeostasis, apoptosis must be appropriately regulated. In addition to mitochondrial ROS (mtROS), endoplasmic ROS (erROS) is crucial for cell signalling and for controlling the primary routes of apoptosis, which are controlled by mitochondria, death receptors, and the endoplasmic reticulum. The focus of this book chapter is on the current understanding of the role of erROS in each of these three primary pathways of apoptosis connected to various human diseases, with a particular focus on metabolic, inflammatory, and neoplastic diseases. The role of erROS in the complex interplay and crosstalk between these different signalling pathways apoptosis-related with different human diseases remains to be unraveled further in future research.

**Keywords**: endoplasmic reticulum, endoplasmic reticulum ROS, apoptosis, diabetes mellitus, atherosclerosis, renal disease, liver disease, inflammation, heart diseases

## 1. Introduction

Proteins are the macromolecules that are essential to life and require certain three-dimensional shapes to operate. Protein folding is the stage of gene expression that is most prone to mistakes. Secretory or membrane proteins are folded within the endoplasmic reticulum (ER). Only correctly folded proteins are delivered to the Golgi apparatus. Nascent proteins are folded with the help of molecular chaperones and folding enzymes found in the ER (together referred to as ER chaperones). Proteins that are misfolded or are not folded correctly are maintained in the ER, retro-translocated to the cytoplasm by the ER-associated degradation (ERAD) machinery, and then destroyed by the proteasome. To deal with developing proteins, the ER expresses ER chaperones and ERAD components. Unfolded proteins build up in the ER when cells produce secretory proteins in quantities greater than the ability of the folding apparatus and ERAD machinery to handle. Improperly folded proteins expose hydrophobic amino acid residues that are usually concealed inside the protein and are prone to creating protein clumps. Protein aggregates eventually leads to apoptotic cell death and "conformational diseases" like cancer and diabetes mellitus as well as neurodegenerative disorders. Eukaryotic cells employ a set of defense mechanisms collectively known as the ER stress response or unfolded protein response to ameliorate such a stressful situation (ER stress) [1-4]. The four processes that make up the mammalian ER stress response. The first is a reduction in protein synthesis, which stops any additional build-up of improperly folded proteins. The transcriptional induction of ER chaperone genes to boost folding capacity is the second, and the transcriptional induction of ERAD component genes to enhance ERAD capability is the third. The fourth is the safe disposal of cells damaged by ER stress by the activation of apoptosis in order to ensure the survival of the organism. The homeostasis of ER protein folding is tightly correlated with the ER redox status. However, pro-apoptotic signalling pathways are activated in the cell when ER stress is too great or persistent or the UPR is chemically or genetically damaged and unable to reduce protein-folding errors. Since both reducing and oxidizing chemicals interfere with protein folding and result in ER stress, disulfide bond formation in the ER lumen is very sensitive to changing redox balance [5]. The thiol groups on the cysteines of the substrate peptides are oxidized during oxidative protein folding in the ER and form disulfide bonds with hydrogen peroxide (H2O2) produced as a by-product. Reactive oxygen species (ROS) build-up and oxidative stress may be brought on by dysregulated disulfide bond production and breakdown in a stressed ER [6-8]. In this chapter, we will describe the basics of unfolded protein response mediated ER stress and oxidative stress in the cell and its role in activating apoptotic cell death.

## 2. Chemicals Inducing ER Stress

For experimental reasons, substances like tunicamycin, thapsigargin, and dithiothreitol are frequently employed to induce ER stress in cultured cells or animals. Here we will briefly summarize the ER stress-inducing chemicals below.

The first class of ER stressors includes inhibitors of glycosylation. N-glycosylation, which occurs on the majority of proteins made in the ER, is frequently necessary for proper protein folding. Therefore, substances that interfere with N-glycosylation may cause ER stress. Tunicamycin is an antibiotic made by Streptomyces lysosuperificus that blocks UDP-GlcNAc-dolichol phosphate GlcNAc-phosphate transferase activity, hence inhibiting N-glycosylation [9].

Disruptors of Ca2+ metabolism are another category of ER stressors. Chemicals that disrupt Ca2+ metabolism in the ER cause ER stress because the concentration of Ca2+ ions in the ER is maintained at a high level, and ER chaperones like BiP need Ca2+ ions to function. ER stress is frequently induced with Ca2+ ionophores like A23187 and the Ca2+ pump inhibitor thapsigargin [10]. Reducing drugs falls under the third group of ER stressors. Proteins produced in the ER can generate intramolecular or intermolecular disulfide bonds between their cysteine residues because the ER lumen is highly oxidative. Disulfide bond formation is crucial for the folding of secretory proteins; hence reducing drugs that prevent disulfide bond formation causes ER stress. For this purpose, 2-mercaptoethanol was frequently employed [11].

ER stress has also been linked to hypoxia, albeit the exact mechanism is yet unclear. Hypoxia is thought to reduce glucose concentration because it activates glycolytic enzymes to maintain ATP synthesis, and when cells consume glucose, N-glycosylation is inhibited, which causes ER stress [12].

## 3. ER Chaperones

The folding of newly synthesized proteins is carried out by ER chaperones, which are molecular chaperones and folding enzymes based within the ER. Additionally, they play a role in ERAD's unfolding of misfolded proteins.

The heat shock protein (HSP) 70 family includes a well-known ER chaperone called binding protein (BiP)–glucose-regulated protein (GRP) 78. BiP uses a substrate binding domain to bind to the hydrophobic area of unfolded proteins and promotes folding by inducing conformational changes through ATP hydrolysis by its ATPase domain. Another ER chaperone, oxygen-regulated protein (ORP) 150 or GRP170, belongs to the HSP110 family, a subgroup of HSP70, and utilizes a similar mechanism as BiP to promote protein folding. It was initially identified as a protein that responds to hypoxia. Several ER chaperones belonging to the HSP40 family, including ER-associated dnaJ (HEDJ), ERdj1, ERdj3, ERdj4, ERdj5, and p58IPK, function as co-chaperones that control BiP's ATPase activity. BiP-associated protein (BAP), a member of the GrpE family, promotes nucleotide exchange and thus also impacts BiP's performance. GRP94, a member of the HSP90 family, is another ER chaperone that promotes protein folding through ATP hydrolysis. Peptidyl-prolyl isomerase FKBP13, which belongs to the FKBP family, also plays a vital role in the folding of secretory proteins. Overall, these ER chaperones are critical in facilitating the folding of secretory proteins [12, 13].

ER chaperones called calnexin and calreticulin are explicitly engaged in the folding of glycoproteins. Most proteins created in the ER have a high mannose-type oligosaccharide linked to them, which is subsequently successively trimmed. Calnexin and calreticulin bind to and fold the nascent protein when two glucose residues are removed by glucosidase I or II and the protein contains only one glucose residue. The nascent protein is freed from calnexin and calreticulin when glucosidase II trims the last glucose residue, at which point it binds to UDP-glucose-glycoprotein glucosyltransferase. If the protein is folded, it is taken to the Golgi apparatus after being released from the enzyme. UDP-glucose-glycoprotein glucosyltransferase attaches one glucose residue and returns it to calnexin and calreticulin if it is not folded. The calnexin is responsible for this folding process. This folding process is called the calnexin cycle [14].

The endoplasmic reticulum (ER) is responsible for the production of disulfide bonds, facilitated by several folding enzymes such as protein disulfide isomerase (PDI), ERp72, ERp61, GRP58-ERp57, ERp44, ERp29, and PDI-P5. These folding enzymes help proteins generate the appropriate disulfide linkages by oxidizing cysteine residues during protein development. ER oxidoreductin (ERO1) is capable of using molecular oxygen as a terminal electron acceptor to reoxidize reduced folding enzymes [15].

## 4. ERAD

The ERAD mechanism captures misfolded or unfolded proteins and transports them to the cytoplasm [16]. The proteasome in the cytosol ubiquitinates and breaks down retro-translocated proteins. As a result, the four phases of the ERAD process are recognition, retro-translocation, ubiquitination, and destruction.

### 4.1. Recognition

During the calnexin cycle, developing polypeptides contain nine mannose residues in their oligosaccharide. Once the nascent polypeptide has eight mannose residues, it dissociates from calnexin or calreticulin and binds to ER degradation-enhancing α-mannosidase-like protein (EDEM), which can differentiate between unfolded and folded proteins by trimming one mannose residue with α-mannosidase 1 [17-18]. EDEM is comprised of three genes, and ERAD requires both EDEM1 and EDEM2, while EDEM2 and EDEM3 are luminal proteins [19]. EDEM1 is an ER membrane protein, and all EDEM proteins contain the mannosidase-like domain, which recognizes mannose residues. In addition to EDEM, unfolded protein recognition is performed by osteosarcoma 9 (OS9) and XTP3-transactivated gene B (XTP3B) [20]. OS9 recognizes unfolded glycoproteins with eight (or five) mannose residues. Furthermore, OS9 binds to unfolded proteins without glycosylation, indicating that it is responsible for recognizing both types of proteins. OS9 and XTP3B contain the mannose-6-phosphate receptor-like domain [21], which could play a role in recognizing mannose residues.

### 4.2. Retro-Translocation

When newly formed glycoproteins are identified by EDEM and OS9 as misfolded, they are intended for the retro-translocation machinery [22]. Prior to retro-translocation, nascent proteins bind with PDI and BiP to cleave disulfide bonds and, respectively, unfold the partially folded structure [23]. While it is believed that unfolded ER proteins are retro-translocated via the translocon, the molecular makeup of the retro-translocation apparatus remains unknown. Derlin-1, a human homolog of yeast Der1, is thought to be a critical component of the mechanism. Derlin-1 links to p97 through an adapter protein called valocin-containing protein (VIMP1) and may form a retro-translocation channel in the ER membrane [24]. Other Der1 homologs, Derlin-2 and Derlin-3, are also involved in ERAD [25, 26], although the precise underlying mechanism is still unknown. Unfolded ER proteins are attracted to the cytosol by a cytosolic AAA-ATPase called p97, cdc48, and VCP [27]. Nuclear protein localization 4 (Npl4) and ubiquitin fusion degradation protein 1 (Ufd1) bind to p97 as cofactors and assist p97 in extracting unfolded proteins.

### 4.3. Ubiquitination

The ubiquitination process plays a critical role in the degradation of retro-translocated proteins. The E1-E2-E3 system of enzymes is responsible for adding ubiquitin molecules to these proteins. Initially, E1 activates ubiquitin and transfers it to E2, which subsequently transfers it to the substrate with the aid of E3 ligases. E3 ligases involved in ERAD include gp78, TEB4, Doa10, and HRD1 [28]. Membrane-anchored E3 ligases play a crucial role in the ubiquitination of substrates in the ER. UBC7 is an E2 conjugase that is specific to ERAD. UBX2 and Herp are additional molecules that assist in ubiquitination [29]. UBX2 binds to both p97 and E3 ligases such as HRD1 and Doa10, while Herp interacts with a complex of proteins that includes HRD1, p97, Derlin-1, and VCP-interacting membrane protein [30, 31].

### 4.4. Degradation

Prior to being degraded by the proteasome, retro-translocated and ubiquitinated proteins are deglycosylated by peptide-N-glycanase because bulky glycan chains may prevent substrates from entering the proteasome pore. Given that Derlin-1 and peptide-N-glycanase are related, correlative retro-translocation of deglycosylation may take place [32]. Substrates that have been deglycosylated are subsequently brought to the proteasome. This transport of ERAD substrates is facilitated by Dsk2 and Rad23 [33].

## 5. ER Stress Response Mechanisms

Four mechanisms comprise the mammalian ER stress response: translational attenuation, increased expression of ER chaperones, ERAD component expression, and apoptosis activation. The regulatory mechanisms that are discussed here control these four reactions (Figure 1).

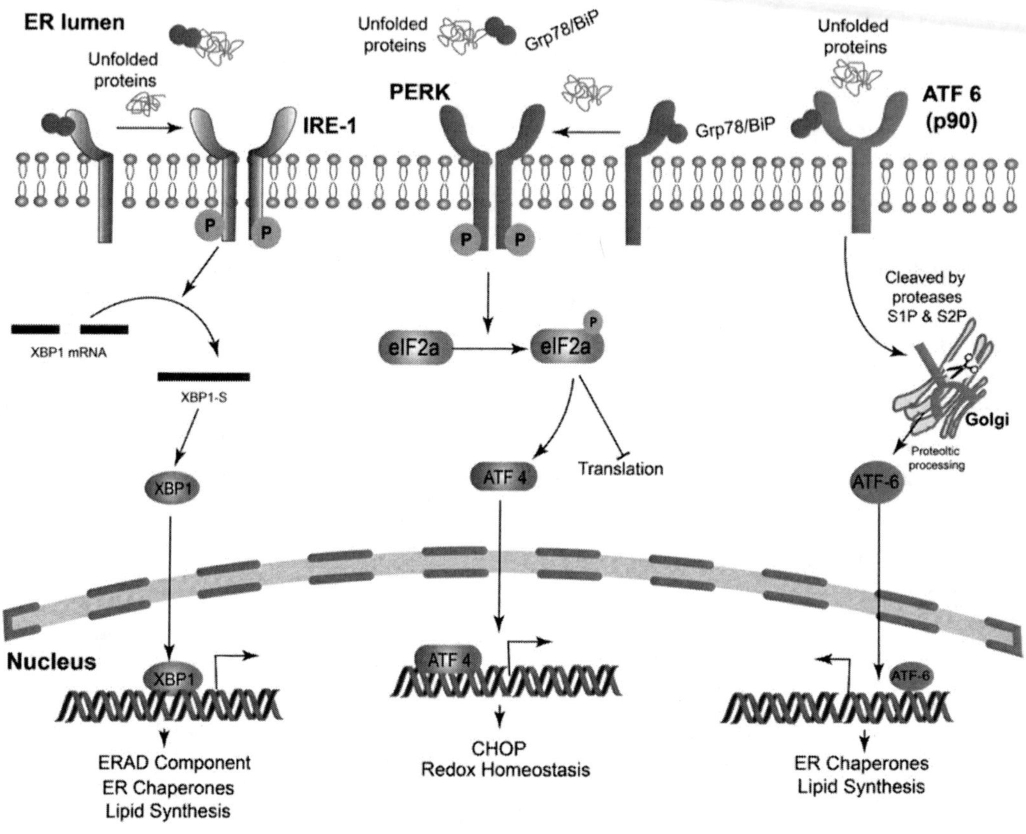


**Figure 1.** Mammalian response pathways for ER stress. Three response pathways (PERK, ATF6, and IRE1 pathways) regulate the mammalian ER stress response. PERK is a transmembrane kinase that is responsible for phosphorylating eIF2a. This helps attenuate translation and also leads to the upregulation of ATF4 expression. This, in turn, enhances the transcription of target genes such as CHOP. Another response pathway is the ATF6 pathway. ATF6 is a transmembrane transcription factor that is translocated to the Golgi apparatus. It is then cleaved by proteases such as S1P and S2P, which enhances the transcription of ER chaperone genes. The IRE1 pathway is also involved in the mammalian ER stress response. IRE1 is a transmembrane RNase that splices XBP1 pre-mRNA. The pXBP1(S) protein, translated from mature XBP1 mRNA, then activates the transcription of ERAD component genes.

### 5.1. PERK Pathway

PERK is one of the three transmembrane sensors of the unfolded protein response (UPR) in the ER. The other two sensors are IRE1 and ATF6. These sensors activate downstream signaling pathways that aim to restore ER homeostasis by reducing the burden of unfolded proteins or by increasing the capacity of the ER to cope with unfolded proteins. The luminal segment of the PERK protein plays a crucial role in detecting unfolded proteins within the endoplasmic reticulum (ER). On the other hand, the cytoplasmic part contains a kinase domain [34]. Under normal ER conditions without stress, BiP acts as a binding partner to the luminal domain of PERK, preventing its activation. However, when the ER experiences stress, BiP dissociates from PERK, activating PERK through oligomerization and trans-phosphorylation [35].

Upon activation, PERK phosphorylates eIF2α at serine 51, which inhibits the initiation of general translation, thereby decreasing the influx of newly synthesized proteins into the ER [36]. However, the translation of a few specific mRNAs, such as ATF4, is increased due to the presence of upstream open reading frames (uORFs) in their 5' untranslated regions (UTRs). ATF4 upregulates the expression of genes involved in amino acid metabolism, redox control, and autophagy, among others. Chronic or severe ER stress leads to the activation of CHOP, which promotes apoptosis by downregulating anti-apoptotic genes and upregulating pro-apoptotic genes. The balance between pro- and anti-apoptotic genes determines the fate of the cell: either recovery from ER stress or cell death.

### 5.2. ATF6 Pathway

ATF6 is a type II transmembrane protein located on the ER membrane that detects unfolded proteins through its luminal domain. The cytoplasmic region of ATF6 contains a DNA-binding domain with a transcriptional activation domain and a bZIP motif. BiP binds to the luminal domain of ATF6 and inhibits its translocation to the Golgi apparatus in the absence of ER stress. However, in response to unfolded proteins, BiP dissociates from ATF6, which is transported to the Golgi apparatus and subsequently cleaved by S1P and S2P processing proteases. The resulting cytoplasmic fragment of ATF6, pATF6 (N), translocates into the nucleus and binds to the ERSE, promoting the transcription of ER chaperone genes like BiP, GRP94, and calreticulin. ATF6 cleavage is unique in that S2P causes a second cleavage in the transmembrane region, a process known as regulated intramembrane proteolysis (RIP). SREBP, a protein involved in regulating cholesterol metabolism, is also a substrate of RIP [37-39].

### 5.3. IRE1 Pathway

IRE1 (inositol requirement 1) serves as the third sensor molecule located within the endoplasmic reticulum (ER) membrane [40, 41]. IRE1 contains a cytoplasmic domain with both a kinase domain and an RNase domain, while its luminal domain shares similarities with PERK and plays a role in detecting unfolded proteins. There are two gene variants of IRE1: IRE1α and IRE1β. When the ER undergoes stress, BiP's inhibitory effect on IRE1 activation is released, leading to IRE1's activation through dimerization and transphosphorylation. Once activated, IRE1α utilizes an unconventional splicing mechanism to convert XBP1 (x-box binding protein 1) pre-mRNA into mature mRNA. XBP1 plays a critical role as a transcriptional activator for numerous UPR (unfolded protein response) genes that oversee various cellular processes, including ER protein folding, intracellular trafficking, ERAD (ER-associated protein degradation), phospholipid biosynthesis, and expansion of the ER membrane. Besides its involvement in the cleavage of Xbp1 mRNA, the RNase domain of IRE1 is also responsible for degrading a specific group of mRNAs localized within the ER during ER stress. This process is referred to as regulated IRE1-dependent decay of mRNA (RIDD) [42].

## 6. Mechanism of ER Stress-Induced ROS during the ER Folding Process

Protein disulfide isomerase (PDI) is a crucial and well-studied enzyme for the synthesis of disulfide bonds in the ER. It is one of the most prevalent proteins in cells and exists in both plants and fungi in addition to animals. Twenty oxidoreductases with varied redox potentials and substrate specificities are known to catalyse disulfide bonding [43-45]. They are the proper sequence of pairs of cysteine residues. Two electrons are given to the cysteine residue in the PDI active site during the creation of disulfide bonds between polypeptide chain substrates with the assistance of chaperones [44]. The PDI active site is reduced as a result of this electron transfer, and the substrate is oxidized. Figure 2 shows the principal ER-originating ROS generation process and the related ROS release mechanisms. The preservation of the ER redox environment has previously been investigated in relation to cysteine and glutathione [46, 47]. Endoplasmic reticulum oxidoreductin-1 (ERO1), a protein that is associated with the ER membrane and oxidizes PDI, was discovered through genetic research on yeast [48, 49]. An FAD-dependent process is used in the pathway of oxidative proteins to transport electrons from PDI to molecular oxygen and ERO1. This electron transfer shows that ERO1 has a significant relationship with protein load in the ER and can cause ROS production and ER stress. PDI has a chaperone function and, along with thiol oxidoreductase, suppresses protein aggregation and preserves hydrophobic pockets inside the primary protein core. The movable arm of the PDI opens in an oxidized state and shuts in a reduced state [50]. To prepare nascent proteins for membrane insertion or secretion, PDI creates or rearranges disulfide bonds in the ER lumen. The ER redox homeostasis and signalling mechanisms congregate at the PDI and other isoforms. A transition in PDI's reduced to oxidized state, which eventually affects its substrate, occurs in the overexpressed ERO1 model. During disulfide bond modifications, electrons go through a number of thiol-disulfide exchange events just before molecular oxygen produces ROS. These superoxide anion radicals contribute to increased ROS production. According to an estimation analysis, disulfide linkages formed during oxidative protein folding in the ER p reduce around 25% of all ROS [51-53]. As a result, proteins with more disulfide bonds than those with less of them play a more significant role in the production of ROS.

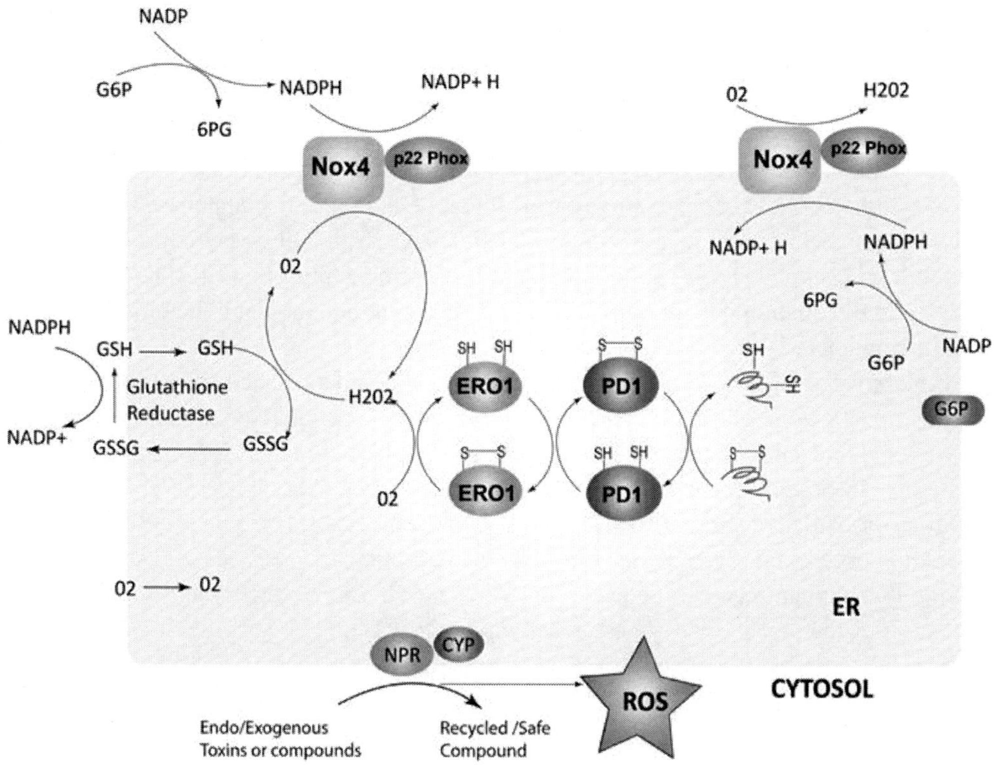


**Figure 2**. The communication of ER redox by mediators is illustrated in the following depiction. During the process of protein folding, reactive oxygen species (ROS) can be produced by various sources in the vicinity of the ER, such as NADPH oxidase 4 (Nox4), NADPH-P450 reductase (NPR), and GSH topologies. These sources perform specific functions in the outer ER environment. NADPH stands for Nicotinamide adenine dinucleotide phosphate, while G6P refers to glucose-6-phosphate and 6PG to 6-phospho gluconate. Nox4 represents NADPH oxidase 4, and GSH denotes glutathione, whereas GSSG is glutathione disulphide. The ER oxidoreductin 1 (ERO1) and protein disulfide isomerase (PDI) are also present, and CYP/p450 refers to Cytochrome p450.

### 6.1. Mechanism of ER Stress-Induced ROS: NADPH Oxidase 4 (Nox4)

The protein family known as nicotinamide adenine dinucleotide phosphate (NADPH) oxidase has seven members, including Nox 1 to 5 and Duox 1 and 2 [54]. NADPH oxidase 4 is the isoform that is most frequently linked to the ER (Nox4). Nox4 has been found in a number of different places, including the mitochondria, nucleus, focal adhesions, and the cytoskeleton. It has been demonstrated that Nox4 physically interacts with PDI; without PDI, Akt phosphorylation decreased, leading to cell death [55-57]. Macrophages were also found to interact with PDI and p22phox. However, the downstream function of ROS produced by PDI-Nox4 may be related to ER-mediated phagocytosis, much like PDI-ERO1. The basic elements of the acute and chronic states of UPR signalling are ER stress and ROS production. However, it has been demonstrated that in a few instances, ROS generation alone can start the UPR [58-61]. Similarly, peripheral vascular cells undergo an increase in Nox4 levels following UPR activation, which increases ROS production (Figure 2).

### 6.2. Glutathione Coupled in the ER

As previously mentioned, molecular oxygen oxidizes ERO1 during the formation of disulfide bonds, producing H2O2 and oxidized glutathione from glutathione (GSH) as a byproduct. Following oxidative stress, both produce an alteration in the redox state of the ER lumen [62, 63]. Indicators of the redox environment include the ratio of glutathione's oxidized and reduced forms, which is crucial for molecular processes such cell division, proliferation, and death following apoptosis [64]. Although it has been demonstrated that ER flavoprotein ERO1 catalyzes PDI oxidation both in vitro and in vivo, it is still thought that glutathione plays a role in PDI oxidation [65].

The ability of GSH to keep ER oxidoreductases in a reduced state and so enable them to catalyse reduction or isomerization processes has been demonstrated [66, 67]. It is being looked into why and how an oxidizing balance of GSHGSH is needed in the ER lumen. In addition to its function in producing disulfide bonds, GSH also serves as a redox buffer source during conditions of oxidative stress. ERO1 oxidises PDI during disulfide exchange, which results in the creation of reduced ERO1 [68]. ERO1 may be oxidised in the presence of FAD and oxygen, showing that oxygen serves as an electron acceptor [65]. The production of ROS in the ER is brought on by the molecular oxygen's reactivation of ERO1.

Research on the mechanism of cellular defense against damage brought on by ER-associated ROS is ongoing, however, recent studies indicate that GSH plays a critical part in this process. Reducing equivalents are continuously introduced into the protein translocation process as cysteine residues, which are oxidized and diminish PDI during the disulfide bond formation process. Later, ROS are produced as a result of ERO1 oxidizing PDI. The cytosolic distribution of these already-produced ROS molecules, their interactions with other organelles, and their reactions with GSH in the ER to raise glutathione disulphide (GSSG) levels are all possible. The [GSH: GSSG] ratio is subsequently disturbed as a result of this. As was previously mentioned, in both yeast and mammalian systems, the creation of disulfide bonds results in the production of ROS. The transmembrane kinases ATF6, IRE1, and PERK are involved in ER stress in mammalian cells [66]. PERK phosphorylates eIF2 in ER stress situations, which inhibits mRNA translation and lessens stress. ATF4 and NF-E2-related factor 2 (Nrf2) are two other transcription factors whose production is stimulated by PERK [67, 68]. These two transcription factors boost GSH production and amino acid metabolism. As a result, less protein is synthesized, but more GSH is produced. Additionally, cells lacking PERK are more vulnerable to ER stress and hydrogen peroxide, making them more susceptible to these conditions. The most crucial requirement for GSH is to raise ROS levels in order to preserve homeostasis and redox balance. In order to lower ROS levels and maintain redox balance, mammalian, and yeast cells boost GSH synthesis in response to stressful situations. The production of ROS is also associated with the misfolding of the protein carboxypeptidase Y (CPY*); however, research has shown that overexpression of CPY* (which has a mutant cysteine) does not result in comparable increases in ROS levels. However, when glutathione is added to the media, cells exhibit considerable proliferation and a decline in ROS levels. As a result, when mammalian and yeast cells are stressed, a rise in glutathione levels supports the reduction of ROS and the facilitation of the creation of disulfide bonds.

### 6.3. ER and Mitochondria Connection and Relationship to ROS

Oxidative stress drives calcium out of the ER and triggers mitochondrial reuptake in the early stages of ER stress disorders. Therefore, increasing the calcium concentration also increases the production of ROS and metabolic activity in mitochondria. Calcium ions further increase the sensitivity of calcium channels through feedback processes [69, 70]. Calcium release and migration are closely related to ER stress situations. Under ER stress, GSH and xanthine/xanthine oxidase have been proposed to promote calcium release via inositol triphosphate receptors (IP3R) [71]. According to reports, 1-methyl-4-phenyl-1,2,3,6-tetrahydropyridine (MPTP) opening controlled by thiols and IP3R's redox sensitivity increases mitochondrial ROS formation and causes the release of ER calcium [72, 73]. Cyclic-ADP ribose (cADPR) causes an increase in ER calcium release in the presence of ROS [74]. This might increase the production of ROS, which would further open the MPTP [75]. At more significant quantities, ROS can block the ryanodine receptor, which inhibits calmodulin action (RyR). Calmodulin-induced RyR is involved in the stimulation of cADPR synthesis, which results in calcium release at low ROS concentrations [76]. It has been demonstrated that calcium spikes caused by ROS-based sensitization that releases calcium occur repeatedly because MPTP opening is constantly induced by ER stress. Signaling systems are significantly impacted by ER stress; if calcium in the mitochondrial matrix results in MPTP opening and fast ATP dissipation or stoppage, mitochondria enlarge. ER, calcium moves toward the mitochondria and speeds up metabolism, increasing the amount of ATP available for use. It also increases the production of ROS signalling in the mitochondria, which then cycles back to the ER and causes more calcium to be released [75].

Since variations in calcium levels are closely related to ER stress-linked ROS creation, ROS generation is also a result of calcium leakage in both ER and oxidative stress [77]. The PDI and ERO1 system, intra ER-GSSG/GSH, or NPR system are thought to be the sources of ROS in the ER-electron coupling system. Calcium reserves are decreased during ER stress because of ER release. Increased mitochondrial calcium levels change metabolism and ultimately lead to the formation of ROS [78]. The mitochondria operate more quickly and use more oxygen as a result of the rise in oxidative phosphorylation and ROS production. Nitric oxide synthase is increased by calcium under conditions of high mitochondrial ROS creation, which limits the action of complex IV, further increasing ROS production as oxygen consumption rises [79-81]. Typically, this signalling pathway functions at normal nitric oxide concentrations (NO). When NO and high-calcium mitochondria work together, they can inhibit complex I, open MPTP to release cytochrome c and obstruct complex III's respiratory chain, which starts the production of ROS [82]. By preventing GSH inhibition, calcium alters the redox environment as well.

The ER releases calcium in response to the rise in ROS levels in the mitochondria, which also makes a calcium-releasing channel in the ER membrane more sensitive and sends a feedback signal [83]. By either causing cytoplasmic calcium to increase or by raising ROS levels in the mitochondria, which subsequently affect the ER calcium release channels, ER and oxidative stress have strong relationships in the ROS-generating process.

## 7. Apoptosis-Inducing Pathways

Unfolded protein buildup in the ER is harmful to cells. As a result, an apoptotic pathway is activated as a last resort to ensure the organism's survival if the PERK, ATF6, and IRE1 pathways cannot control ER stress. Several pathways have been implicated in ER stress-induced apoptosis, and the full induction of apoptosis appears to need the concurrent activation of several death pathways, even though there are still many disagreements regarding this phenomenon [84- 86]. The known death processes will be briefly reviewed in this section, emphasizing recent advancements. The CHOP pathway has the most distinct characteristics. ATF6 and PERK pathways following ER stress stimulate the production of CHOP-GADD153 (growth arrest and DNA damage 153) which is a transcription factor [87]. When exposed to ER stress, CHOP-deficient cells have a lower rate of programmed cell death [88], indicating that the CHOP pathway plays a significant role in controlling ER stress-induced apoptosis. Regarding CHOP's target genes, these appear to be in charge of apoptosis: GADD34, ERO1, DR5 (death receptor 5), and carbonic anhydrase VI. In contrast, ERO1, which encodes an ER oxidase, causes the ER to become a more hyper-oxidizing environment [89]. GADD34, which is connected with protein phosphatase 2C, improves eIF2a's dephosphorylation and stimulates the manufacture of ER client proteins [90]. Caspase cascades may be activated by DR5, which encodes a cell surface death receptor [91]. Numerous cellular processes may be impacted by a shift in cellular pH caused by carbonic anhydrase VI [92, 93]. The precise signalling mechanism from CHOP to apoptosis is still unknown, though, the IRE1-TRAF2-ASK1 route is the second apoptotic pathway. To connect the plasma membrane death receptor to the adaptor proteins Jun kinase (JNK) and stress-activated protein kinase (SAPK), the cytoplasmic portion of IRE1 binds to TRAF2 (tumour necrosis factor receptor-associated factor 2) [94]. The IRE1-TRAF2-ASK1 complex, which includes TRAF2 and the mitogen-activated protein kinase kinase ASK1 (apoptosis signal-regulating kinase 1), is what phosphorylates and activates JNK. [95] Additionally, TRAF2 interacts with caspase-12 and controls its activation [96]. The transcriptional repressor ATF3 is also activated by IRE1- TRAF2, which triggers apoptosis [97]. These indicate that a significant regulator of ER stress-induced apoptosis is the IRE1-TRAF2-ASK1 pathway.

Caspases are well-known pro-apoptotic components, and caspases 2, 3, 4, 7, 9 and 12 reportedly involve ER stress-induced cell death [98, 99]. Caspase-12 is associated with the ER membrane, and activated by ER stress, possibly by calpain [100]. Then caspase-12 activates caspase-9, which in turn activates caspase-3 leading to cell death. Caspase-12−/− mice are resistant to ER stress-induced apoptosis but sensitive to other death stimuli, suggesting that caspase-12 is a regulator specific to ER stress-induced apoptosis [101-102]. The activation of caspase-12 by ER stress is observed in models of various diseases such as Alzheimer's disease, polyglutamine disease, ischemia, and viral infection [103], suggesting that ER stress-induced apoptosis is closely involved in these diseases (see below). Since the human caspase-12 gene has several essential alterations for its function, the role of caspase-12 in the death of human cells is still up for debate [104]. In human cells, it's possible that a caspase other than caspase-12 that hasn't been found is to blame.

Some essential Bcl2 family proteins are involved in ER stress-induced apoptosis. They are well-known parts of the machinery that triggers programmed cell death. The Bcl2 family appears to recruit pro-apoptotic members to the ER surface and activate caspase-12, whereas the anti-apoptotic members block this recruitment. Upon ER stress, the transcription factor

CHOP inhibits the anti-apoptotic protein Bcl2, which increases oxidative damage and apoptosis [105]. Caspase-12 activation and apoptosis are inhibited by Bcl2 overexpression when there is ER stress. BAD (Bcl2 antagonist of cell death), a pro-apoptotic factor, is dephosphorylated and activated in response to ER stress [106], while Bax (Bcl2-associated X protein) and Bak (Bcl-2 homologous antagonist kil- ler), other pro-apoptotic factors, are present in both the ER membrane and the mitochondrial membrane [107, 108]. Bax and Bak oligomerize and trigger caspase-12 under ER stress. It's interesting to note that under ER stress, Bax and Bak connect with IRE1a and modify its function [109]. For the majority of apoptosis types, Bax and Bak are necessary [110]. Pro-apoptotic members of the BH3 (homology domain-3) domain-only family PUMA (p53 up-regulated modulator of apoptosis) and NOXA (neutrophil NADPH oxidase factor) and their transcription are up-regulated by p53 during ER stress, and PUMA- and NOXA-deficient cells are resistant to ER stress-induced apoptosis [111-112].

Apoptosis brought on by ER stress has also been linked to members of the inhibitor of apoptosis (IAP) family. An IAP that inhibits caspase-3 and caspase-7 is called human inhibitor of apoptosis 2 (HIAP2). The eukaryotic initiation factor, p97 DAP5 NAT1, is cleaved explicitly by caspases activated by ER stress, and the cleavage product activates the HIAP2 internal ribosome entry site, enhancing HIAP2 translation. This results in the induction of HIAP2 expression at the level of translation [113]. Two other IAPs, IAP-2 and XIAP (inhibitor of apoptosis, X-linked), have increased transcription under ER stress, and cells lacking these IAPs are more susceptible to apoptosis caused by ER stress [114]. XIAP or HIAP1 overexpression makes cells resistant to ER stress [115-116]. These findings imply that IAP proteins are involved in ER stress-induced apoptosis. The transcription factors nuclear respiratory factor (NRF)1 and NRF2 control the oxidative stress response. Through its interaction with the microtubule-associated protein Keap1, NRF2 is transported throughout the cytoplasm (Kelch-like Ech-associated protein 1). NRF2 is recruited to the nucleus as a result of PERK phosphorylating NRF2 and separating it from Keap1 in response to ER stress [117]. NRF1 is found in the ER membrane and moves to the nucleus in response to ER stress, whereas NRF2-- cells are sensitive to ER stress-induced apoptosis [118], suggesting that these proteins are implicated in ER stress-specific apoptosis.

During myogenesis, ATF6 is also engaged in the apoptotic process. The ATF6 pathway is activated in developing myoblasts, and the expression of BiP and CHOP is increased, which may activate caspase-12. Furthermore, the ATF6 pathway may contribute to apoptosis during myogenesis because AEBSF [4-(2-aminoethyl) benzene sulfonyl fluoride hydrochloride], an inhibitor of ATF6 activation, prevents apoptosis [119].

## 8. ER Stress Related Human Diseases

The ER controls protein synthesis and appropriate folding. ERAD, on the other hand, eliminates improperly folded proteins. The ER stress response for cytoprotection is activated when the workload is excessive compared to the ER's capability. This reaction induces the expression of chaperone and ERAD components to lessen the burden on the ER. Although both the ER and the cytoplasm can experience unfolded protein aggregation, ROS-associated tiny protein aggregates are more dangerous than bigger ones in ER stress situations because they bind transcription factors like CREB and TATA and disrupt the ubiquitin-proteasome cascade [120]. However, by further suppressing the formation of protein aggregates, bigger protein aggregates and molecular chaperones, such as the TCP1-ring complex (TRiC) and Heat Shock Protein 70 (HSP70), serve to promote cytoprotection [121]. Additionally, it has been noted that oestrogen receptors and Breast Cancer 1 (BRCA1, a tumour suppressor) interact to synergistically influence and support the regulation of Nrf2, which regulates ROS levels, cell metabolism, and antioxidant activities [122]. Different diseases can arise from a failure of the ER stress response brought on by various circumstances, including old age, mutations, or idiosyncratic or environmental variables. The disorders resulting from cellular protein misfolding are collectively called "conformational diseases" or "folding diseases." Given that protein misfolding and protein aggregation can cause ER stress, it has been hypothesized that ER stress is responsible for most conformational disorders. We will briefly review recent research on the role of ER stress in conformational disorders in this section.

## 8.1. Diabetes Mellitus

Insulin is primarily produced by pancreatic β cells, producing nearly half of the proteins in these specialized secretory cells. High protein production makes β cells more vulnerable to the difficulties of ER stress. β cells, on the other hand, are vulnerable to oxidative stress because of their high energy requirements and low amounts of antioxidant enzymes. β -cell malfunction and/or apoptosis, which are connected to ER stress and oxidative stress, are involved in both type I and type II diabetes. Insulin-secreting β -cells are specifically destroyed in type I diabetes mellitus (IDDM). Still, insulin resistance is a clinical feature of type II diabetes mellitus (NIDDM), which is not insulin-dependent. The relationship between ER stress and IDDM, particularly the PERK pathway, has been carefully studied since β -cells appear to experience ER stress brought on by the synthesis of a lot of insulin. Wolcott-Rallison syndrome, an early-onset IDDM, is caused by PERK, and PERK -/- deficient animals exhibit IDDM-like symptoms [123]. Furthermore, CHOP-/- mice have delayed development of diabetes compared to animals with a homozygous mutation at the eIF2a phosphorylation site (Ser51Ala), but both mice exhibit identical clinical characteristics [124]. It's interesting to note that IDDM involves ER stress pathways other than the PERK pathway. First, in the Akita diabetes mouse model, BiP, GRP94, ORP150, and HRD1 expression are all up-regulated, and ATF6 and XBP1 are both activated [125]. Second, p58IPK -/- deficient mice exhibit symptoms comparable to IDDM [126]. Lastly, ER stress induces the expression of the WFS1 gene, which causes Wolfman syndrome (juvenile diabetes), and WFS1 knockdown or deletion produces ER stress in pancreatic β -cells [127]. These results strongly imply that ER stress plays a significant role in the development of IDDM.

Several data indicate that ER stress may potentially play a role in NIDDM. First, although the underlying mechanism is uncertain, obesity, one of the causes of NIDDM, induces ER stress and leads to the development of insulin resistance in XBP1 -/- deficient mice. Second, β -cells that express ORP150 ectopically have improved insulin tolerance [128]. This might be explained by the fact that ORP150 expression enhances ER folding capacity and guard's cells against ER stress-induced apoptosis. A polymorphism in ATF6 is related to NIDDM in Pima Indians, according to a small nuclear polymorphism analysis [129].

### 8.2. Liver Disease

Through elevated levels of homocysteine, ER stress-mediated ROS plays a role in the etiology of numerous illnesses, including chronic viral hepatitis, alcoholic hepatic steatosis, and non-alcoholic hepatic steatosis. Homocysteine builds up due to alcohol's inhibition of the sulphur amino acid metabolism's two key enzymes, methionine synthetase and betaine-homocysteine methyltransferase. Research is still being done to determine how alcohol promotes ER stress [130]. By activating ATF6 and IRE1 in addition to the main ER stress chaperone, BiP/Grp78, ER stress is also involved in hepatic cancer and human liver carcinoma. Cells react to the ER stress response in this way [131, 132]. The GSH level is markedly reduced in acute hepatitis, and the buildup of ROS speeds up protein aggregation and ER stress [133]. This may increase the development of disease and reduce hepatic functional efficiency.

### 8.3. Renal Disease

The enhanced expression of the chaperone Bip/Grp78 serves as evidence for the excessive ROS production brought on by PAT-activated ER stress. Additionally, it has recently been discovered that crocin and quercetin suppress the ROS-mediated ER stress pathway in a human cell model, preventing PAT-induced apoptosis [134]. Acetaminophen, for example, produces renal tubular damage [135] by activating the PERK pathway, inducing CHOP, and cleaving caspase-12. This is only one of many factors that generate ROS-mediated ER stress and result in cell death. Age-related renal fibrosis is also impacted by ER stress.

Kidney impairment caused by ER stress is caused by the age-related aggregation of oxidative carbonylated Bip/Grp78 or PDI [136]. The ER marker Bip/Grp78 was consistently shown to be substantially elevated in the samples from seven patients with uromodulin-linked renal illness [137]. Aggregation of factor VIII, a protein that is lacking in the ER in haemophilia A, can also cause oxidative stress. These results imply that the ER stress response is one of the safeguards and protective mechanisms against kidney injury and that ER stress is a significant contributor to chemically induced renal illness.

### 8.4. Atherosclerosis

Atherosclerosis is a condition that causes arteries to harden and narrow due to the buildup of fatty substances, cholesterol, cellular waste, Ca2+, and other compounds in the inner lining of the arteries, which can result in heart attack or stroke. Homocysteine buildup, an intermediate created during the metabolism of sulphur amino acids, is one of the risk factors for atherosclerosis [138]. It's interesting to note that homocysteine causes ER stress through an unidentified mechanism before upregulating the expression of BiP, GRP94, CHOP, Herp, and TDAG51 [139]. By activating the transcription factor SREBP, homocysteine-induced ER stress also boosts cholesterol production. Inducing CHOP and TDAG51 (a member of the pleckstrin homology-related domain family) as well as the formation of cholesterol seem to promote the apoptosis of macrophages. This results in the deposition of macrophage debris in blood vessels, which aids in developing the atherosclerosis seen in hyperhomocysteinemia [140]. Actually, CHOP -/- macrophages are less susceptible to apoptosis spurred on by cholesterol. These results provide compelling evidence that one of the primary causes of atherosclerosis is ER stress-induced apoptosis.

### 8.5. Inflammation

The initial immunological reaction to an infection is inflammation. However, the mechanisms behind inflammation are diverse, and ER stress plays a role in various inflammations. Interferon-gamma can be produced by CNS inflammation, and oligodendrocytes may undergo apoptosis as a result of ER stress. Oligodendrocyte loss and increased hypomyelination were observed in PERK+/+ animals, confirming the PERK pathway's protective effect against interferon-gamma-induced cell death. Lipopolysaccharide-induced ER stress causes CHOP expression in the lungs, which leads to a buildup of ROS and cell death [141]. Additionally, ER stress-related inflammation plays a significant role in the emergence of insulin resistance [142]. Diclofenac and other anti-inflammatory medications have demonstrated the capacity to decrease ER stress [143]. In inflammatory bowel disease (IBD) and several autoimmune disorders, misfolded protein buildup has been noticed. On the other hand, in response to NLR family pyrin domain-containing 3 activators, calcium and calcium detecting receptors limit inflammatory activation. Calcium-detecting receptors trigger inflammasomes via an inositol triphosphate (IP3)-mediated process, which releases calcium that has been held in the ER and causes an accumulation of ROS. It is still unclear how inflammasome activation and ER calcium release are related.

### 8.6. Heart Diseases

ER stress has reportedly been linked to heart-related disorders. Heart hypertrophy results from ER stress-induced apoptosis of cardiac myocytes and pressure overload caused by transverse aortic constriction [144]. Interestingly, dilated cardiomyopathy was caused by a breakdown of the ER quality control system and elevated ER stress in transgenic mice expressing a mutant KDEL receptor, a retrieval receptor for ER chaperones in the early secretory pathway [145]. Additionally, cardiomyocytes were shielded from ER stress-related apoptosis by up-regulating ER chaperones [146]. These results strongly imply that the ER stress response is necessary for maintaining cardiomyocyte homeostasis.

## Conclusion

Recent discoveries suggest that oxidative stress and ER stress are closely connected biological processes that control various signalling pathways in the cell. This is shown by the coexistence and induction of the two stresses as well as by the multifunctional stress responses that target both ER protein misfolding and redox imbalance. The two cellular stressors have a significant impact on human pathologies such as metabolic, neurodegenerative, inflammatory, and neoplastic illnesses as well as normal physiology. Recent research suggests that the two stressors and the signalling pathways that they activate are attractive targets for new treatments.

For a comprehensive understanding of the route and to clarify the processes behind their interactions in both physiological and pathological settings, more research is required to explain the relationship between ER and ROS. A real-time in vitro and in vivo imaging system, innovative conditional genetic animal models, and additional mechanistic and physiological research using these newly developed high-throughput technologies are required. Such knowledge is anticipated to aid in the more precise and efficient treatment or prevention of illnesses linked to ER stress.

## References


[1] Harding HP, Calfon M, Urano F, Novoa I, Ron D. Transcriptional and translational control in the Mammalian unfolded protein response. *Annu Rev Cell Dev Biol*. 2002; 18:575-99.

[2] Mori K. Tripartite management of unfolded proteins in the endoplasmic reticulum. Cell. 2000 May 26; 101(5):451-4.

[3] Patil C, Walter P. Intracellular signaling from the endoplasmic reticulum to the nucleus: the unfolded protein response in yeast and mammals. *Curr Opin Cell Biol*. 2001; 13(3):349-55.

[4] Schröder M, Kaufman RJ. ER stress and the unfolded protein response. *Mutat Res*. 2005; 569(1-2):29-63.

[5] Malhotra JD, Kaufman RJ. Endoplasmic reticulum stress and oxidative stress: a vicious cycle or a double-edged sword? *Antioxid Redox Signal*. 2007;9(12):2277-93.

[6] Cao SS, Kaufman RJ. Unfolded protein response. *Curr Biol*. 2012 Aug 21; 22(16):R6226.

[7] Ushioda R, Hoseki J, Araki K, Jansen G, Thomas DY, Nagata K. ERdj5 is required as a disulfide reductase for degradation of misfolded proteins in the ER. *Science*. 2008; 321(5888):569-72.

[8] Dorner AJ, Wasley LC, Raney P, Haugejorden S, Green M, Kaufman RJ. The stress response in Chinese hamster ovary cells. Regulation of ERp72 and protein disulfide isomerase expression and secretion. *J Biol Chem*. 1990; 265(35):22029-34.

[9] Takatsuki A, Arima K, Tamura G. Tunicamycin, a new antibiotic. I. Isolation and characterization of tunicamycin. *J Antibiot* (Tokyo). 1971; 24(4):215-23.

[10] Price BD, Mannheim-Rodman LA, Calderwood SK. Brefeldin A, thapsigargin, and AIF4- stimulate the accumulation of GRP78 mRNA in a cycloheximide dependent manner, whilst induction by hypoxia is independent of protein synthesis. *J Cell Physiol*. 1992; 152(3):545-52.

[11] Fernandez F, Jannatipour M, Hellman U, Rokeach LA, Parodi AJ. A new stress protein: synthesis of Schizosaccharomycespombe UDP--Glc:glycoprotein glucosyltransferase mRNA is induced by stress conditions but the enzyme is not essential for cell viability. *EMBO J*. 1996; 15(4):705-713.

[12] Ikeda J, Kaneda S, Kuwabara K, Ogawa S, Kobayashi T, Matsumoto M, Yura T, Yanagi H. Cloning and expression of cDNA encoding the human 150-kDa oxygen-regulated protein, ORP150. *Biochem Biophys Res Commun*. 1997; 230(1):94-9.

[13] Gething MJ. *Guidebook to molecular chaperones and protein-folding catalysts*. Vol. T9. Oxford: Oxford University Press; 1997. [p. T10].

[14] Deprez P, Gautschi M, Helenius A. More than one glycan is needed for ER glucosidase II to allow entry of glycoproteins into the calnexin/calreticulin cycle. *Mol Cell*. 2005; 19(2):183-195.

[15] Gross E, Kastner DB, Kaiser, Fass, Fass D. Structure of Ero1p, source of disulfide bonds for oxidative protein folding in the cell. *Cell*. 2004; 117(5):601-10.

[16] Jarosch E, Lenk U, Sommer T. Endoplasmic reticulum-associated protein degradation. *Int Rev Cytol*. 2003; 223:39-81.

[17] Eriksson KK, Vago R, Calanca V, Galli C, Paganetti P, Molinari M. EDEM contributes to maintenance of protein folding efficiency and secretory capacity. *J Biol Chem*. 2004; 279(43):44600-5.

[18] Oda Y, Hosokawa N, Wada I, Nagata K. EDEM as an acceptor of terminally misfolded glycoproteins released from calnexin. *Science*. 2003; 299(5611):1394-7.

[19] Olivari S, Galli C, Alanen H, Ruddock L, Molinari M. A novel stress-induced EDEM variant regulating endoplasmic reticulum-associated glycoprotein degradation. *J Biol Chem.* 2005;280(4): 2424-8.

[20] Szathmary R, Bielmann R, Nita-Lazar B, Jakob, Jakob CA. Yos9 protein is essential for degradation of misfolded glycoproteins and may function as lectin in ERAD. *Mol Cell.* 2005; 19(6):765-75.

[21] Cruciat CM, Hassler C, Niehrs C. The MRH protein Erlectin is a member of the endoplasmic reticulum synexpression group and functions in N-glycan recognition. *J Biol Chem.* 2006; 281(18):12986-93.

[22] Lee RJ, Liu CW, Harty C, McCracken AA, Latterich M, Römisch K, DeMartino GN, Thomas PJ, Brodsky JL. Uncoupling retro-translocation and degradation in the ER-associated degradation of a soluble protein. *EMBO J.* 2004; 23(11):2206-15.

[23] Kabani M, Kelley SS, Morrow MW, Montgomery DL, Sivendran R, Rose MD, Gierasch LM, Brodsky JL. Dependence of endoplasmic reticulum-associated degradation on the peptide binding domain and concentration of BiP. *Mol Biol Cell.* 2003; 14(8):3437-48.

[24] Ye Y, Shibata Y, Yun C, Ron D, Rapoport TA. A membrane protein complex mediates retro-translocation from the ER lumen into the cytosol. *Nature.* 2004; 429(6994):841-7.

[25] Bordallo J, Plemper RK, Finger A, Wolf DH. Der3p/Hrd1p is required for endoplasmic reticulum-associated degradation of misfolded lumenal and integral membrane proteins. *Mol Biol Cell.* 1998; 9(1):209-22.

[26] Oda Y, Okada T, Yoshida H, Kaufman RJ, Nagata K, Mori K. Derlin-2 and Derlin-3 are regulated by the mammalian unfolded protein response and are required for ER-associated degradation. *J Cell Biol.* 2006; 172(3):383-93.

[27] Ye Y, Meyer HH, Rapoport TA. The AAA ATPase Cdc48/p97 and its partners transport proteins from the ER into the cytosol. *Nature.* 2001; 414(6864):652-6.

[28] Bays NW, Gardner RG, Seelig LP, Joazeiro CA, Hampton RY. Hrd1p/Der3p is a membrane anchored ubiquitin ligase required for ER-associated degradation. *Nat Cell Biol.* 2001; 3(1):24-9.

[29] Schuberth C, Buchberger A. Membrane-bound Ubx2 recruits Cdc48 to ubiquitin ligases and their substrates to ensure efficient ER-associated protein degradation. *Nat Cell Biol.* 2005; 7(10):999-1006.

[30] Zhong X, Shen Y, Ballar P, Apostolou A, Agami R, Fang S. AAA ATPase p97/valosin-containing protein interacts with gp78, a ubiquitin ligase for endoplasmic reticulum-associated degradation. *J Biol Chem.* 2004; 279(44):45676-84.

[31] Schulze A, Standera S, Buerger E, Kikkert M, van Voorden S, Wiertz E, Koning F, Kloetzel PM, Seeger M. The ubiquitin-domain protein HERP forms a complex with components of the endoplasmic reticulum associated degradation pathway. *J Mol Biol.* 2005; 354(5):1021-7.

[32] Katiyar S, Joshi S, Lennarz WJ. The retrotranslocation protein Derlin-1 binds peptide: N-glycanase to the endoplasmic reticulum. *Mol Biol Cell.* 2005; 16(10):4584-94.

[33] Medicherla B, Kostova Z, Schaefer A, Wolf DH. A genomic screen identifies Dsk2p and Rad23p as essential components of ER-associated degradation. *EMBO Rep.* 2004; 5(7):692-7.

[34] Harding HP, Zhang Y, Ron D. Protein translation and folding are coupled by an endoplasmic-reticulum-resident kinase. *Nature.* 1999; 397(6716):271-4.

[35] Bertolotti A, Zhang Y, Hendershot LM, Harding HP, Ron D. Dynamic interaction of BiP and ER stress transducers in the unfolded-protein response. *Nat Cell Biol.* 2000; 2(6):326-32.

[36] Boyce M, Bryant KF, Jousse C, Long K, Harding HP, Scheuner D, Kaufman RJ, Ma D, Coen DM, Ron D, Yuan J. A selective inhibitor of eIF2alpha dephosphorylation protects cells from ER stress. *Science.* 2005; 307(5711):935-9.

[37] Yoshida H, Matsui T, Yamamoto A, Okada T, Mori K. XBP1 mRNA is induced by ATF6 and spliced by IRE1 in response to ER stress to produce a highly active transcription factor. *Cell.* 2001; 107(7):881-91.

[38] Shen J, Chen X, Hendershot L, Prywes R. ER stress regulation of ATF6 localization by dissociation of BiP/GRP78 binding and unmasking of Golgi localization signals. *Dev Cell.* 2002; 3(1):99-111.

[39] Ye J, Rawson RB, Komuro R, Chen X, Davé UP, Prywes R, Brown MS, Goldstein JL. ER stress induces cleavage of membrane-bound ATF6 by the same proteases that process SREBPs. *Mol Cell.* 2000; 6(6):1355-64.

[40] Yoshida H, Haze K, Yanagi H, Yura T, Mori K. Identification of the cis-acting endoplasmic reticulum stress response element responsible for transcriptional induction of mammalian glucose-regulated proteins. Involvement of basic leucine zipper transcription factors. *J Biol Chem.* 1998; 273(50):33741-9.

[41] Kaneko M, Nomura Y. ER signaling in unfolded protein response. *Life Sci.* 2003; 74(2-3):199-205.

[42] Sriburi R, Jackowski S, Mori K, Brewer JW. XBP1: a link between the unfolded protein response, lipid biosynthesis, and biogenesis of the endoplasmic reticulum. *J Cell Biol.* 2004; 167(1):35-41.

[43] Ellgaard L, Ruddock LW. The human protein disulphide isomerase family: substrate interactions and functional properties. *EMBO Rep.* 2005; 6(1):28-32.

[44] Braakman I, Bulleid NJ. Protein folding and modification in the mammalian endoplasmic reticulum. *Annu Rev Biochem.* 2011; 80:71-99.

[45] Sriburi R, Jackowski S, Mori K, Brewer JW. XBP1: a link between the unfolded protein response, lipid biosynthesis, and biogenesis of the endoplasmic reticulum. *J Cell Biol.* 2004; 167(1):35-41.

[46] Alanen HI, Williamson RA, Howard MJ, Lappi AK, Jäntti HP, Rautio SM, Kellokumpu S, Ruddock LW. Functional characterization of ERp18, a new endoplasmic reticulum-located thioredoxin superfamily member. *J Biol Chem.* 2003; 278(31):28912-20.

[47] Braakman I, Bulleid NJ. Protein folding and modification in the mammalian endoplasmic reticulum. *Annu Rev Biochem.* 2011; 80:71-99.

[48] Kramer B, Ferrari DM, Klappa P, Pöhlmann N, Söling HD. Functional roles and efficiencies of the thioredoxin boxes of calcium-binding proteins 1 and 2 in protein folding. *Biochem J.* 2001; 357(1):83-95.

[49] Frand AR, Kaiser CA. The ERO1 gene of yeast is required for oxidation of protein dithiols in the endoplasmic reticulum. *Mol Cell.* 1998; 1(2):161-70.

[50] Carelli S, Ceriotti A, Cabibbo A, Fassina G, Ruvo M, Sitia R. Cysteine and glutathione secretion in response to protein disulfide bond formation in the ER. *Science.* 1997; 277(5332):1681-4.

[51] Pollard MG, Travers KJ, Weissman JS. ERO1p: A novel and ubiquitous protein with an essential role in oxidative protein folding in the endoplasmic reticulum. *Mol Cell.* 1998; 1(2):171-82.

[52] Wang C, Li W, Ren J, Fang J, Ke H, Gong W, Feng W, Wang CC. Structural insights into the redox-regulated dynamic conformations of human protein disulfide isomerase. *Antioxid Redox Signal.* 2013; 19(1):36-45.

[53] Wang C, Yu J, Huo L, Wang L, Feng W, Wang CC. Human protein-disulfide isomerase is a redox-regulated chaperone activated by oxidation of domain a. *J Biol Chem.* 2012; 287(2):1139-49.

[54] Xiao R, Wilkinson B, Solovyov A, Winther JR, Holmgren A, Lundström-Ljung J, Gilbert HF. The contributions of protein disulfide isomerase and its homologues to oxidative protein folding in the yeast endoplasmic reticulum. *J Biol Chem.* 2004; 279(48):49780-6.

[55] Tu BP, Weissman JS. Oxidative protein folding in eukaryotes: mechanisms and consequences. *J Cell Biol.* 2004; 164(3):341-6.

[56] Kleniewska P, Piechota A, Skibska B, Gorąca A. The NADPH oxidase family and its inhibitors. *Arch Immunol Ther Exp* (Warsz). 2012; 60(4):277-94.

[57] Anilkumar N San Jose, G, Sawyer, I, Santos. CX, Sand, C, Brewer, AC, Warren, D, Shah, AM. A 28-kDa splice variant of NADPH oxidase-4 is nuclear-localized and involved in redox signaling in vascular cells. *Arterioscler. Thromb. Vasc. Biol.* 2013, 33, e104–e112.

[58] Bedard K, Lardy B, Krause KH. Nox family NADPH oxidases: not just in mammals. *Biochimie.* 2007; 89(9):1107-12.

[59] Nisimoto Y, Jackson HM, Ogawa H, Kawahara T, Lambeth JD. Constitutive NADPH-dependent electron transferase activity of the Nox4 dehydrogenase domain. *Biochemistry.* 2010; 49(11):2433-42.

[60] Janiszewski M, Lopes LR, Carmo AO, Pedro MA, Brandes RP, Santos CX, Laurindo FR. Regulation of NAD(P)H oxidase by associated protein disulfide isomerase in vascular smooth muscle cells. *J Biol Chem.* 2005 Dec 9; 280(49):40813-9.

[61] Santos CX, Stolf BS, Takemoto PV, Amanso AM, Lopes LR, Souza EB, Goto H, Laurindo FR. Protein disulfide isomerase (PDI) associates with NADPH oxidase and is required for phagocytosis of Leishmania chagasi promastigotes by macrophages. *J Leukoc Biol.* 2009; 86(4):989-98.

[62] Santos CX, Tanaka LY, Wosniak J, Laurindo FR. Mechanisms and implications of reactive oxygen species generation during the unfolded protein response: roles of endoplasmic reticulum oxidoreductases, mitochondrial electron transport, and NADPH oxidase. *Antioxid Redox Signal.* 2009; 11(10):2409-27.
[63] Haynes CM, Titus EA, Cooper AA. Degradation of misfolded proteins prevents ER-derived oxidative stress and cell death. *Mol Cell.* 2004; 15(5):767-76.
[64] Malhotra JD, Miao H, Zhang K, Wolfson A, Pennathur S, Pipe SW, Kaufman RJ. Antioxidants reduce endoplasmic reticulum stress and improve protein secretion. *Proc Natl Acad Sci USA.* 2008; 105(47):18525-30.
[65] Dickinson DA, Forman HJ. Glutathione in defense and signaling: lessons from a small thiol. *Ann N Y Acad Sci.* 2002; 973:488-504.
[66] Tu BP, Weissman JS. The FAD- and O2-dependent reaction cycle of ERO1-mediated oxidative protein folding in the endoplasmic reticulum. *Mol Cell.* 2002; 10(5):983-94.
[67] Chakravarthi S, Bulleid NJ. Glutathione is required to regulate the formation of native disulfide bonds within proteins entering the secretory pathway. *J Biol Chem.* 2004; 279(38):39872-9.
[68] Sitia R, Molteni SN. Stress, protein (mis) folding, and signaling: the redox connection. *Sci STKE.* 2004; 2004(239):pe27.
[69] Reimann D, Dachs D, Meye C, Gross P. Amino acid-based peritoneal dialysis solution stimulates mesothelial nitric oxide production. *Perit Dial Int.* 2004; 24(4):378-84.
[70] Gerasimenko JV, Gerasimenko OV, Palejwala A, Tepikin AV, Petersen OH, Watson AJ. Menadione-induced apoptosis: roles of cytosolic Ca2+ elevations and the mitochondrial permeability transition pore. *J Cell Sci.* 2002; 115(3):485-97.
[71] Li G, Mongillo M, Chin KT, Harding H, Ron D, Marks AR, Tabas I. Role of ERO1-alpha-mediated stimulation of inositol 1,4,5-triphosphate receptor activity in endoplasmic reticulum stress-induced apoptosis. *J Cell Biol.* 2009; 186(6):783-92.
[72] Cioffi DL. Redox regulation of endothelial canonical transient receptor potential channels. *Antioxid Redox Signal.* 2011; 15(6):1567-82.
[73] Luciani DS, Gwiazda KS, Yang TL, Kalynyak TB, Bychkivska Y, Frey MH, Jeffrey KD, Sampaio AV, Underhill TM, Johnson JD. Roles of IP3R and RyR Ca2+ channels in endoplasmic reticulum stress and beta-cell death. *Diabetes.* 2009;58(2):422-32.
[74] Lock JT, Sinkins WG, Schilling WP. Protein S-glutathionylation enhances Ca2+-induced Ca2+ release via the IP3 receptor in cultured aortic endothelial cells. *J Physiol.* 2012; 590(15):3431-47.
[75] Yamasaki-Mann M, Demuro A, Parker I. Modulation of endoplasmic reticulum Ca2+ store filling by cyclic ADP-ribose promotes inositol trisphosphate (IP3)-evoked Ca2+ signals. *J Biol Chem.* 2010; 285(32):25053-61.
[76] Jacobson J, Duchen MR. Mitochondrial oxidative stress and cell death in astrocytes—requirement for stored Ca2+ and sustained opening of the permeability transition pore. *J Cell Sci.* 2002; 115(6):1175-88.
[77] Okabe E, Tsujimoto Y, Kobayashi Y. Calmodulin and cyclic ADP-ribose interaction in Ca2+ signaling related to cardiac sarcoplasmic reticulum: superoxide anion radical-triggered Ca2+ release. *Antioxid Redox Signal.* 2000; 2(1):47-54.
[78] Malhotra JD, Kaufman RJ. Endoplasmic reticulum stress and oxidative stress: A vicious cycle or a double-edged sword? *Antioxid Redox Signal.* 2007; 9(12):2277-93.
[79] Cao SS, Kaufman RJ. Endoplasmic reticulum stress and oxidative stress in cell fate decision and human disease. *Antioxid Redox Signal.* 2014; 21(3):396-413.
[80] Görlach A, Klappa P, Kietzmann T. The endoplasmic reticulum: folding, calcium homeostasis, signaling, and redox control. *Antioxid Redox Signal.* 2006; 8(9-10):1391-418.
[81] Meares GP, Hughes KJ, Naatz A, Papa FR, Urano F, Hansen PA, Benveniste EN, Corbett JA. IRE1-dependent activation of AMPK in response to nitric oxide. *Mol Cell Biol.* 2011; 31(21):4286-97.
[82] Xu W, Liu L, Charles IG, Moncada S. Nitric oxide induces coupling of mitochondrial signalling with the endoplasmic reticulum stress response. *Nat Cell Biol.* 2004; 6(11):1129-34.
[83] Murphy MP. How mitochondria produce reactive oxygen species. *Biochem J.* 2009; 417(1):1-13.

[84] Jekabsone A, Ivanoviene L, Brown GC, Borutaite V. Nitric oxide and calcium together inactivate mitochondrial complex I and induce cytochrome c release. *J Mol Cell Cardiol.* 2003; 35(7):803-9.
[85] Kadowaki H, Nishitoh H, Ichijo H. Survival and apoptosis signals in ER stress: the role of protein kinases. *J Chem Neuroanat.* 2004; 28(1-2):93-100.
[86] Kim R, Emi M, Tanabe K, Murakami S. Role of the unfolded protein response in cell death. *Apoptosis.* 2006; 11(1):5-13.
[87] Ma Y, Brewer JW, Diehl JA, Hendershot LM. Two distinct stress signaling pathways converge upon the CHOP promoter during the mammalian unfolded protein response. *J Mol Biol.* 2002; 318(5):1351-65.
[88] Zinszner H, Kuroda M, Wang X, Batchvarova N, Lightfoot RT, Remotti H, Stevens JL, Ron D. CHOP is implicated in programmed cell death in response to impaired function of the endoplasmic reticulum. *Genes Dev.* 1998; 12(7):982-95.
[89] Marciniak SJ, Yun CY, Oyadomari S, Novoa I, Zhang Y, Jungreis R, Nagata K, Harding HP, Ron D. CHOP induces death by promoting protein synthesis and oxidation in the stressed endoplasmic reticulum. *Genes Dev.* 2004; 18(24):3066-77.
[90] Novoa I, Zeng H, Harding HP, Ron D. Feedback inhibition of the unfolded protein response byGADD34-mediated dephosphorylation of eIF2alpha. *J Cell Biol.* 2001; 153(5):1011-22.
[91] Yamaguchi H, Wang HG. CHOP is involved in endoplasmic reticulum stress-induced apoptosis by enhancing DR5 expression in human carcinoma cells. *J Biol Chem.* 2004; 279(44):45495-502.
[92] Sok J, Wang XZ, Batchvarova N, Kuroda M, Harding H, Ron D. CHOP-Dependent stress-inducible expression of a novel form of carbonic anhydrase VI. *Mol Cell Biol.* 1999; 19(1):495-504.
[93] Wang XZ, Kuroda M, Sok J, Batchvarova N, Kimmel R, Chung P, Zinszner H, Ron D. Identification of novel stress-induced genes downstream of chop. *EMBO J.* 1998; 17(13):3619-30.
[94] Urano F, Wang X, Bertolotti A, Zhang Y, Chung P, Harding HP, Ron D. Coupling of stress in the ER to activation of JNK protein kinases by transmembrane protein kinase IRE1. *Science.* 2000; 287(5453):664-6.
[95] Nishitoh H, Matsuzawa A, Tobiume K, Saegusa K, Takeda K, Inoue K, Hori S, Kakizuka A, Ichijo H. ASK1 is essential for endoplasmic reticulum stress-induced neuronal cell death triggered by expanded polyglutamine repeats. *Genes Dev.* 2002; 16(11):1345-55.
[96] Yoneda T, Imaizumi K, Oono K, Yui D, Gomi F, Katayama T, Tohyama M. Activation of caspase-12, an endoplastic reticulum (ER) resident caspase, through tumor necrosis factor receptor-associated factor 2-dependent mechanism in response to the ER stress. *J Biol Chem.* 2001; 276(17):13935-40.
[97] Zhang C, Kawauchi J, Adachi MT, Hashimoto Y, Oshiro S, Aso T, Kitajima S. Activation of JNK and transcriptional repressor ATF3/LRF1 through the IRE1/TRAF2 pathway is implicated in human vascular endothelial cell death by homocysteine. *BiochemBiophys Res Commun.* 2001; 289(3):718-24.
[98] Cheung HH, Lynn Kelly N, Liston P, Korneluk RG. Involvement of caspase-2 and caspase-9 in endoplasmic reticulum stress-induced apoptosis: A role for the IAPs. *Exp Cell Res.* 2006; 312(12):2347-57.
[99] Dahmer MK. Caspases-2, 3, and 7 are involved in thapsigargin-induced apoptosis of SH-SY5Y neuroblastoma cells. *J Neurosci Res.* 2005; 80(4):576-83.
[100] Tan Y, Dourdin N, Wu C, De Veyra T, Elce JS, Greer PA. Ubiquitous calpains promote caspase-12 and Jnk activation during ER stress-induced apoptosis. *J Biol Chem.* 2006; 281(23):16016-24.
[101] Morishima N, Nakanishi K, Takenouchi H, Shibata T, Yasuhiko Y. An endoplasmic reticulum stressspecific caspase cascade in apoptosis. Cytochrome cindependent activation of caspase-9 by caspase-12. *J Biol Chem.* 2002; 277(37):34287-94.
[102] Nakagawa T, Yuan J. Cross-talk between two cysteine protease families. Activation of caspase-12 by calpain in apoptosis. *J Cell Biol.* 2000; 150(4):887-94.
[103] Nakagawa T, Zhu H, Morishima N, Li E, Xu J, Yankner BA, Yuan J. Caspase-12 mediates endoplasmic-reticulum-specific apoptosis and cytotoxicity by amyloid-beta. *Nature.* 2000; 403(6765):98-103.
[104] Fischer H, Koenig U, Eckhart L, Tschachler E. Human caspase 12 has acquired deleterious mutations. *Biochem Biophys Res Commun.* 2002; 293(2):722-6.

[105] McCullough KD, Martindale JL, Klotz LO, Aw TY, Holbrook NJ. Gadd153 sensitizes cells to endoplasmic reticulum stress by down-regulating Bcl2 and the cellular redox state. *Mol Cell Biol.* 2001; 21(4):1249-59.

[106] Elyaman W, Terro F, Suen KC, Yardin C, Chang RC, Hugon J. BAD and Bcl-2 regulation are early events linking neuronal endoplasmic reticulum stress to mitochondria-mediated apoptosis. *Brain Res Mol Brain Res.* 2002; 109(1-2):233-8.

[107] Scorrano L, Oakes SA, Opferman JT, Cheng EH, Sorcinelli MD, Pozzan T, Korsmeyer SJ. BAX and BAK regulation of endoplasmic reticulum Ca2+: a control point for apoptosis. *Science.* 2003; 300(5616):135-9.

[108] Zong WX, Li C, Hatzivassiliou G, Lindsten T, Yu QC, Yuan J, Thompson CB. Bax and Bak can localize to the endoplasmic reticulum to initiate apoptosis. *J Cell Biol.* 2003; 162(1):59-69.

[109] Hetz C, Bernasconi P, Fisher J, Lee AH, Bassik MC, Antonsson B, Brandt GS, Iwakoshi NN, Schinzel A, Glimcher LH, Korsmeyer SJ. Proapoptotic BAX and BAK modulate the unfolded protein response by a direct interaction with IRE1alpha. *Science.* 2006; 312(5773):572-6.

[110] Zong WX, Lindsten T, Ross AJ, MacGregor GR, Thompson CB. BH3-only proteins that bind pro-survival Bcl-2 family members fail to induce apoptosis in the absence of Bax and Bak. *Genes Dev.* 2001; 15(12):1481-6.

[111] Li J, Lee B, Lee AS. Endoplasmic reticulum stress-induced apoptosis: multiple pathways and activation of p53-up-regulated modulator of apoptosis (PUMA) and NOXA by p53. *J Biol Chem.* 2006; 281(11):7260-70.

[112] Reimertz C, Kögel D, Rami A, Chittenden T, Prehn JH. Gene expression during ER stress-induced apoptosis in neurons: Induction of the BH3-only protein Bbc/ PUMA and activation of the mitochondrial apoptosis pathway. *J Cell Biol.* 2003; 162(4):587-97.

[113] Warnakulasuriyarachchi D, Cerquozzi S, Cheung HH, Holcík M. Translational induction of the inhibitor of apoptosis protein HIAP2 during endoplasmic reticulum stress attenuates cell death and is mediated via an inducible internal ribosome entry site element. *J Biol Chem.* 2004; 279(17):17148-57.

[114] Hu P, Han Z, Couvillon AD, Exton JH. Critical role of endogenous Akt/ IAPs and MEK1/ERK pathways in counteracting endoplasmic reticulum stress induced cell death. *J Biol Chem.* 2004; 279(47):49420-9.

[115] Zhang K, Wong HN, Song B, Miller CN, Scheuner D, Kaufman RJ. The unfolded protein response sensor IRE1alpha is required at 2 distinct steps in B cell lymphopoiesis. *J Clin Invest.* 2000; 115(2):268-81.

[116] Wootz H, Hansson I, Korhonen L, Lindholm D. XIAP decreases caspase-12 cleavage and calpain activity in spinal cord of ALS transgenic mice. *Exp Cell Res.* 2006; 312(10):1890-8.

[117] Cullinan SB, Zhang D, Hannink M, Arvisais E, Kaufman RJ, Diehl JA. Nrf2 is a direct PERK substrate and effector of PERK-dependent cell survival. *Mol Cell Biol.* 2003; 23(20):7198-209.

[118] Wang W, Chan JY. Nrf1 is targeted to the ER membrane by an N-terminal trans- membrane domain: inhibition of nuclear translocation and transacting function. *J Biol Chem.* 2006; 281(28):19676-87.

[119] Nakanishi K, Sudo T, Morishima N. Endoplasmic reticulum stress signaling transmitted by ATF6 mediates apoptosis during muscle development. *J Cell Biol.* 2005; 169(4):555-60.

[120] Schaffar G, Breuer P, Boteva R, Behrends C, Tzvetkov N, Strippel N, Sakahira H, Siegers K, Hayer-Hartl M, Hartl FU. Cellular toxicity of polyglutamine expansion proteins: mechanism of transcription factor deactivation. *Mol Cell.* 2004; 15(1):95-105.

[121] Behrends C, Langer CA, Boteva R, Böttcher UM, Stemp MJ, Schaffar G, Rao BV, Giese A, Kretzschmar H, Siegers K, Hartl FU. Chaperonin TRiC promotes the assembly of polyQ expansion proteins into nontoxic oligomers. *Mol Cell.* 2006; 23(6):887-97.

[122] Wang L, Di LJ. BRCA1 and estrogen/estrogen receptor in breast cancer: where they interact? *Int J Biol Sci.* 2014; 10(5):566-75.

[123] Zhang P, McGrath B, Li S, Frank A, Zambito F, Reinert J, Gannon M, Ma K, McNaughton K, Cavener DR. The PERK eukaryotic initiation factor 2 alpha kinase is required for the development of the skeletal system, postnatal growth, and the function and viability of the pancreas. *Mol Cell Biol.* 2002; 22(11):3864-74.

[124] Scheuner D, Song B, McEwen E, Liu C, Laybutt R, Gillespie P, Saunders T, Bonner-Weir S, Kaufman RJ. Translational control is required for the unfolded protein response and in vivo glucose homeostasis. *Mol Cell*. 2001; 7(6):1165-76.
[125] Nozaki Ji, Kubota H, Yoshida H, Naitoh M, Goji J, Yoshinaga T, Mori K, Koizumi A, Nagata K. The endoplasmic reticulum stress response is stimulated through the continuous activation of transcription factors ATF6 and XBP1 in Ins2+/Akita pancreatic beta cells. *Genes Cells*. 2004; 9(3):261-70.
[126] Ladiges WC, Knoblaugh SE, Morton JF, Korth MJ, Sopher BL, Baskin CR, MacAuley A, Goodman AG, LeBoeuf RC, Katze MG. Pancreatic beta-cell failure and diabetes in mice with a deletion mutation of the endoplasmic reticulum molecular chaperone gene P58IPK. *Diabetes*. 2005; 54(4):1074-81.
[127] Fonseca SG, Fukuma M, Lipson KL, Nguyen LX, Allen JR, Oka Y, Urano F. WFS1 is a novel component of the unfolded protein response and maintains homeostasis of the endoplasmic reticulum in pancreatic beta-cells. *J Biol Chem*. 2005; 280(47):39609-15.
[128] Ozcan U, Cao Q, Yilmaz E, Lee AH, Iwakoshi NN, Ozdelen E, Tuncman G, Görgün C, Glimcher LH, Hotamisligil GS. Endoplasmic reticulum stress links obesity, insulin action, and type 2 diabetes. *Science*. 2004; 306(5695):457-61.
[129] Thameem F, Farook VS, Bogardus C, Prochazka M. Association of amino acid variants in the activating transcription factor 6 gene (ATF6) on 1q21-q23 with type 2 diabetes in Pima Indians. *Diabetes*. 2006; 55(3):839-42.
[130] Kaplowitz N, Ji C. Unfolding new mechanisms of alcoholic liver disease in the endoplasmic reticulum. *J Gastroenterol Hepatol*. 2006; 21; Suppl 3 (Suppl. S3):S7-9.
[131] Arai M, Kondoh N, Imazeki N, Hada A, Hatsuse K, Kimura F, Matsubara O, Mori K, Wakatsuki T, Yamamoto M. Transformation-associated gene regulation by ATF6 alpha during hepatocarcinogenesis. *FEBS Lett*. 2006; 580(1):184-90.
[132] Shuda M, Kondoh N, Imazeki N, Tanaka K, Okada T, Mori K, Hada A, Arai M, Wakatsuki T, Matsubara O, Yamamoto N, Yamamoto M. Activation of the ATF6, XBP1 and grp78 genes in human hepatocellular carcinoma: a possible involvement of the ER stress pathway in hepatocarcinogenesis. *J Hepatol*. 2003; 38(5):605-14.
[133] Galligan JJ, Smathers RL, Shearn CT, Fritz KS, Backos DS, Jiang H, Franklin CC, Orlicky DJ, Maclean KN, Petersen DR. Oxidative Stress and the ER Stress Response in a Murine Model for Early-Stage Alcoholic Liver Disease. *J Toxicol*. 2012; 2012:207594.
[134] Boussabbeh M, Prola A, Ben Salem I, Guilbert A, Bacha H, Lemaire C, Abis-Essefi S. Crocin and quercetin prevent PAT-induced apoptosis in mammalian cells: Involvement of ROS-mediated ER stress pathway. *Environ Toxicol*. 2016; 31(12):1851-1858.
[135] Lorz C, Justo P, Sanz A, Subirá D, Egido J, Ortiz A. Paracetamol-induced renal tubular injury: A role for ER stress. *J Am Soc Nephrol*. 2004; 15(2):380-9.
[136] Naidoo N. The endoplasmic reticulum stress response and aging. *Rev Neurosci*. 2009; 20(1):23-37.
[137] Adam J, Bollée G, Fougeray S, Noël LH, Antignac C, Knebelman B & Pallet, N. Endoplasmic reticulum stress in UMOD-related kidney disease: A human pathologic study. *Am J Kidney Dis*. 2012; 59(1):117-21.
[138] Lawrence de Koning AB, Werstuck GH, Zhou J, Austin RC. Hyperhomocysteinemia and its role in the development of atherosclerosis. *Clin Biochem*. 2003; 36(6):431-41.
[139] Werstuck GH, Lentz SR, Dayal S, Hossain GS, Sood SK, Shi YY, Zhou J, Maeda N, Krisans SK, Malinow MR, Austin RC. Homocysteine-induced endoplasmic reticulum stress causes dysregulation of the cholesterol and triglyceride biosynthetic pathways. *J Clin Invest*. 2001; 107(10):1263-73.
[140] Feng B, Yao PM, Li Y, Devlin CM, Zhang D, Harding HP, Sweeney M, Rong JX, Kuriakose G, Fisher EA, Marks AR, Ron D, Tabas I. The endoplasmic reticulum is the site of cholesterol-induced cytotoxicity in macrophages. *Nat Cell Biol*. 2003; 5(9):781-92.
[141] Endo M, Mori M, Akira S, Gotoh T. C/EBP homologous protein (CHOP) is crucial for the induction of caspase-11 and the pathogenesis of lipopolysaccharide-induced inflammation. *J Immunol*. 2006; 176(10):6245-53.
[142] Li S, Jiang W, Hu S, Song W, Ji L, Wang Y, Cai L. Fucosylated chondroitin sulphate from Cusumariafrondosa mitigates hepatic endoplasmic reticulum stress and inflammation in insulin resistant mice. *Food Funct*. 2015; 6(5):1547-56.

[143] Yamazaki T, Muramoto M, Oe T, Morikawa N, Okitsu O, Nagashima T, Nishimura S, Katayama Y, Kita Y. Diclofenac, a non-steroidal anti-inflammatory drug, suppresses apoptosis induced by endoplasmic reticulum stresses by inhibiting caspase signaling. *Neuropharmacology*. 2006; 50(5):558-67.

[144] Okada K, Minamino T, Tsukamoto Y, Liao Y, Tsukamoto O, Takashima S, Hirata A, Fujita M, Nagamachi Y, Nakatani T, Yutani C, Ozawa K, Ogawa S, Tomoike H, Hori M, Kitakaze M. Prolonged endoplasmic reticulum stress in hypertrophic and failing heart after aortic constriction: possible contribution of endoplasmic reticulum stress to cardiac myocyte apoptosis. *Circulation*. 2004; 110(6):705-12.

[145] Hamada H, Suzuki M, Yuasa S, Mimura N, Shinozuka N, Takada Y, Suzuki M, Nishino T, Nakaya H, Koseki H, Aoe T. Dilated cardiomyopathy caused by aberrant endoplasmic reticulum quality control in mutant KDEL receptor transgenic mice. *Mol Cell Biol*. 2004; 24(18):8007-17.

[146] Martindale JJ, Fernandez R, Thuerauf D, Whittaker R, Gude N, Sussman MA, Glembotski CC. Endoplasmic reticulum stress gene induction and protection from ischemia/reperfusion injury in the hearts of transgenic mice with a tamoxifen-regulated form of ATF6. *Circ Res*. 2006; 98(9):1186-93.

## Chapter 3

# The Role of Misfolded Proteins on Oxidative Stress Mediated Neurodegeneration

**Nikhilesh Joardar[1,*]**
**Satyajit Halder[2]**
**Santi Prasad Sinha Babu[3]**
**and Kuladip Jana[2]**

[1]Infectious Disease Division, Washington University School of Medicine in Saint Louis, Saint Louis, Missouri, USA
[2]Division of Molecular Medicine, Bose Institute, Bidhan Nagar, Kolkata–West Bengal, India
[3]Parasitology Laboratory, Department of Zoology, Siksha Bhavana, Visva Bharati, Santiniketan, West Bengal, India

## Abstract

Proteins with improper folding are thermodynamically unstable and mostly toxic in nature. These misfolded toxic proteins have the ability to interact and catalyze the same native protein for their conversion into the toxic state. Accumulation of these misfolded proteins is very much responsible for the neurodegenerative/amyloid diseases. Specifically, accrual of misfolded proteins such as β-amyloid, tau, huntingtin, and α-synuclein is found to be the underlying reason behind the vast range of neurodegenerative disorders (Alzheimer's and Parkinson's). Reactive oxygen species (ROS) play a significant role in human neurodegenerative disorders (NDDs). Our present article seeks to reveal up-to-date information regarding the association among ROS, misfolded protein and NDDs.

**Keywords:** neurodegenerative disorder, protein misfolding, oxidative stress, reactive oxygen species

## 1. Introduction

Neurodegenerative diseases (NDDs) are an assorted group of neurological maladies that are predominantly regarded as off rapid loss of neurons in different functional anatomic systems

* Corresponding Author's Email: nikhileshjoardar@gmail.com.

In: The Role of Reactive Oxygen Species in Health and Disease
Editor: Kuladip Jana
ISBN: 979-8-89113-283-2

of the brain or the spinal cord [1]. Although those diseases are infrequent, they do have genetic factors playing behind disease pathogenesis. With enhanced in life expectancy, the commonness of age-related neurodegenerative diseases has dramatically increased across the globe. Moreover, lack of effective therapies for these diseases has made the situation worse [2]. Studies have revealed that deposition and spreading of aggregated proteins, oxidative stress, metal ions accumulation, protein misfolding, cell-cell diffusion of disease-related proteins and apoptotic neuronal death, chronic neuroinflammation, and mitochondrial dysfunction, are some of the major reasons behind the occurrence of neuronal loss leading to NDDs [3, 4]. Several studies have fortified the fact that misfolding of the proteins followed by aggregation is the dominant pathogenic event in various protein misfolding disorders (PMDs), including NDDs [5, 6]. Upon misfolding the proteins lose their native physical properties and tend to aggregate, become insoluble, resistant to proteolysis, and resilient to cellular clearance mechanisms [7, 8]. Those protein misfolding related anomalies appear late in life, progressive and chronic by nature, and the misfolded proteins are commonly dumped in tissues and form aggregates [9].

Here in this chapter, we will focus on the effect of redox imbalance-mediated protein misfolding in neuronal damage. Redox disequilibrium or oxidative stress has been largely accepted as a crucial issue for NDDs [9, 10]. Fundamentally, oxidative stress occurs due to oxidative damage of important biomolecules that hamper the normal functioning of the same gradually leading to instigation of cell death events [11]. There should always be equilibrium between the generated free radicals (reactive oxygen species) and the antioxidant system [11, 12]. An imbalance between the two could push the entire system towards physiological distress. Among various biomolecules, proteins are one of them. It is well known that all life activities are the cumulative manifestation of some among the millions of proteins. The correct functioning of them entirely depends on their proper three-dimensional conformation i.e., correct folding [13]. However, at time misfolding of proteins may occur due to incorrect posttranslational modifications, hostile environmental situations, accidental mutations, oxidative stress, and trafficking error etc. [14]. Amplified oxidative stress fuels augmentation of various neurodegenerative diseases. Subsequent enhancement of the redox imbalance inside the body may severely affect the living organisms by introducing cellular damage, diminishing DNA repair system, and mitochondrial dysfunction [15, 16]. These factors are fairly responsible for quickening senescence and occurrence of NDDs. Therefore, the primary objective of this manuscript will be to reveal up-to-date information regarding the occurrence of NDDs due to oxidative stress-mediated protein misfolding.

## 2. Overview of the Neurodegenerative Disorder

The chronic and progressive neurodegenerative disorders are annotated by their characteristic selective and symmetric loss of neurons in the body especially in cognition [17, 18]. Additionally, NDDs can influence an individual's drive, dialog, remembrance, intellect, etc. Therefore, these diseases are equally frustrating as well as fatal in certain cases. Each of the major NDD has been characterized depending upon the specific pattern they exhibit. Martin (1999) has excellently described that Delineation of the patterns of cell loss and the identification of disease-specific cellular markers have aided in nosologic classification: senile plaques, neurofibrillary tangles, neuronal loss, and acetylcholine deficiency define Alzheimer's disease; Lewy bodies and depletion of dopamine characterize Parkinson's disease; cellular inclusions and swollen motor axons are found in amyotrophic lateral sclerosis, and γ-aminobutyric acid–containing neurons of the neostriatum are lost in Huntington's disease." In majority, proteins are involved in neurodegenerative diseases [19] (Table 1):

**Table 1.** Different proteins which are involved in neurodegenerative diseases

| Disease | Name of the protein | Function | Reference |
|---|---|---|---|
| Alzheimer's disease, Dementia with Lewy bodies | Beta-amyloid | It is a large membrane protein that normally plays an essential role in neural growth and repair. Amyloid beta peptide (Aβ) is produced through the proteolytic processing of a transmembrane protein, amyloid precursor protein (APP), by β- and γ-secretases. Aβ accumulation in the brain is proposed to be an early toxic event in the pathogenesis of Alzheimer's disease and dementia | [20] |
| Alzheimer's disease, Progressive supranuclear palsy, Croticobasal syndrome, types of Frontotemporal Lobar Degeneration and Chronic Traumatic Encephalopathy | Tau | Tau is a protein that helps stabilize the internal skeleton of nerve cells (neurons) in the brain. In Alzheimer's disease (AD), there are intraneuronal neurofibrillary tangles composed of the microtubule-associated protein tau (MAPT) | [21] |
| Parkinson's disease | Synuclein | Missense mutation in the α-synuclein gene is a rare genetic cause of Parkinson's disease. Second, the identification of the α-synuclein protein as the main component of Lewy bodies and Lewy neurites, the defining neuropathological characteristics of all cases of Parkinson's and several other diseases. The filamentous inclusions of multiple system atrophy are also made of α-synuclein. These findings have placed α-synuclein dysfunction at the centre of several common neurodegenerative diseases | [22] |

**Table 1.** (Continued)

| Disease | Name of the protein | Function | Reference |
|---|---|---|---|
| Amyotrophic Lateral Sclerosis | Transactive response DNA-binding protein 43 (TDP-43) | nuclear TDP-43 when depleted, mislocalized, and modified that generate insoluble cytoplasmic inclusions marked by hyperphosphorylation, ubiquitination, and fragmentation | [23] |
| Huntington's disease | Huntington protein | This protein is found to be involved in chemical signaling, transporting materials, attaching (binding) to proteins and other structures, and protecting the cell from self-destruction | [24] |
| Creutzfeldt-Jakob disease | PrPSc | oxidative stress, disruption of proteostasis and induction of the unfolded protein response | [25] |

## 3. Misfolded Protein and Oxidative Stress

Proteins are huge, intricately folded biomolecules that play indispensable and varied roles in all living beings. To perform well, proteins must attain and maintain a precise 3D conformation. Disruption in the folding event will lead to generation of sticky surfaces on the protein that will eventually aggregate through several stages. These assemblages are nonfunctional and toxic in nature [26]. Studies have confirmed that misfolding of proteins usually arise due to impairments in endoplasmic reticulum or the mistrafficking of that particular protein [27]. Cells have efficiently developed surveillance system (ubiquitin (Ub)-proteasome system (UPS), chaperone mediated autophagy (CMA) and macroautophagy) to monitor production and aggregation of misfolded proteins inside the cytoplasm, nucleus and endoplasmic reticulum (ER) [28, 29]. Of note, the primary mechanism for the dilapidation is the UPS. *Figure 1* depicts the general mechanism of protein misfolding along with its degradation by the host surveillance sentinels. To be precise, most of the misfolded and aggregated proteins can be degraded by cellular protein quality control, however, interestingly a few of them have likely to attain β-sheet-enriched oligomers and become resilient to all degradation mechanisms [30, 31]. Other intriguing theories regarding protein misfolding are polymerization hypothesis, conformational hypothesis, and conformation/oligomerization hypothesis [5].

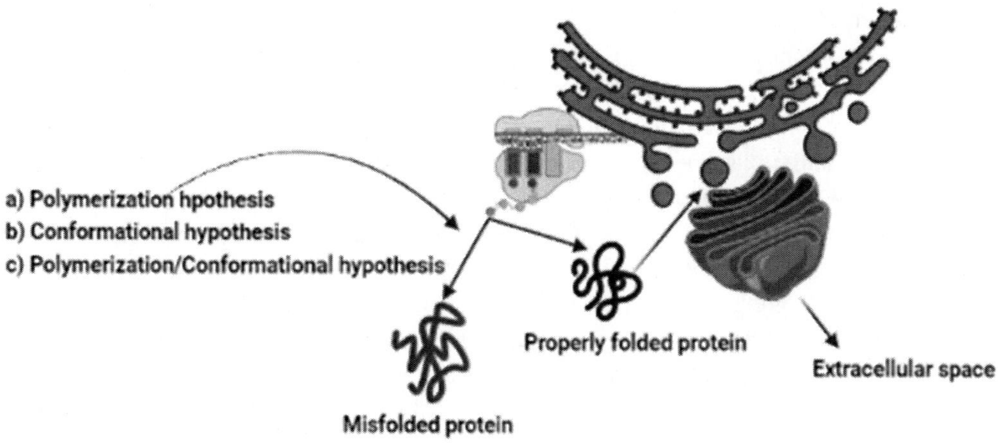


**Figure 1.** Pictorial representation of protein misfolding hypothesis.

Surprisingly, subsequent building up of protease-resistant misfolded and aggregated proteins is one of the various mechanisms behind the onset of NDDs [3, 32]. Here in this chapter, we are particularly focused on the development of misfolded protein-mediated NDDS (*Figure 2*).

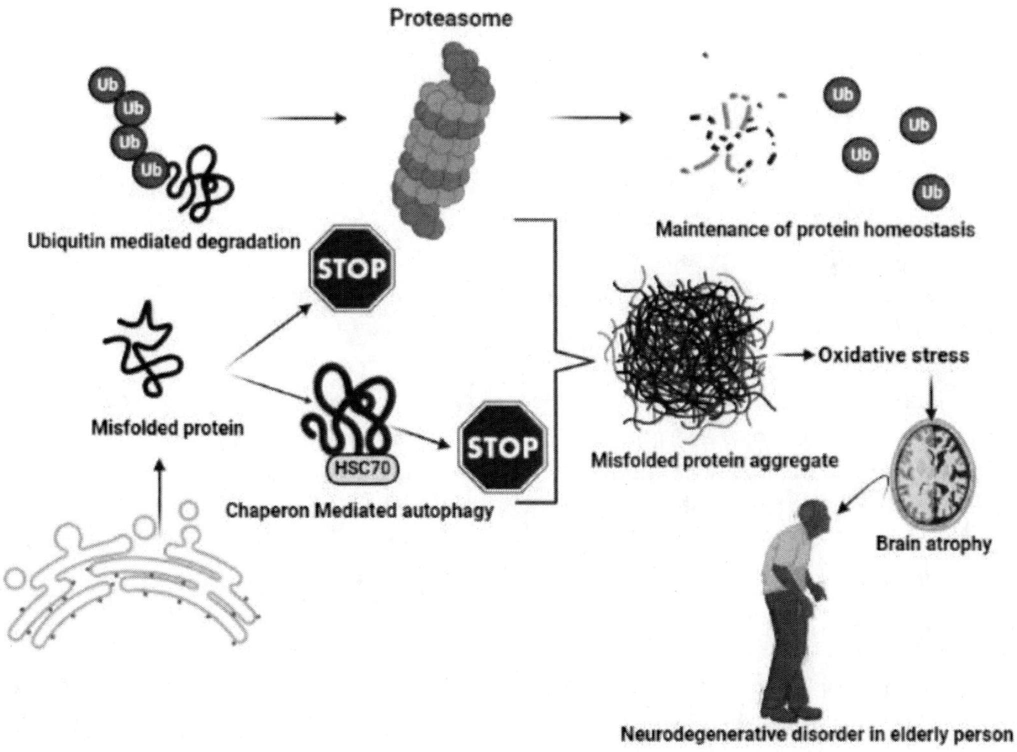


**Figure 2.** Misfolded protein generation and development of neurodegenerative disorder.

Misfolded protein generation and subsequent development of NDDs can be inhibited either through ubiquitin mediated degradation or chaperon mediated autophagy. During NDD those misfolded protein can accumulate, and forms aggregate which results in oxidative stress mediated brain atrophy. Brain atrophy in turn manifests as NDDs.

Information regarding misfolded protein mediated oxidative stress generation is very inadequate. Presently available knowledge states that the fibrilar misfolded proteins permeabilise through plasma membranes and upsurge intracellular calcium ions that activate mitochondrial metabolic processes and eventual assembly of ROS [33, 34]. In addition to that, the fibril-forming misfolded proteins as well as the accumulated misfolded proteins inhibit the proteasome [35, 36]. Organelles such as endoplasmic reticulum (ER) are profoundly involved in protein misfolding events. The proteins that have entered into the ER for post translational modifications (glycosylation and or disulfide bond formation) but are unable to fold rightly cause ER stress that activate the unfolded protein response (UPR) [35, 37]. UPR plays a plethora of roles (upregulating genes required for protein folding, ER expansion, ER-Golgi trafficking, and endoplasmic reticulum associated protein degradation (ERAD) in relieving stress within the ER [38, 39]. There are a number of surveillance systems that track and ensure correct post processing of the proteins within the ER. ERAD system plays the chief tasks in recognizing and degrading abnormally folded proteins to avert the clumping and transfer of potentially dysfunctional and toxic proteins [40]. Functionally, there are two distinct types of ERAD viz. HRD- and Hip-dependent ERAD. HIP-dependent ERAD is essential for the degradation of elevated levels of this misfolded protein. Studies have revealed that the HIP pathway degrades the misfolded proteins as the increased accumulation consequently results into ER stress allied with the NDDs [41].

## 4. Role of Misfolded Protein in Neurodegenerative Diseases

It has been observed long ago that the appearance of the misfolded protein plaques in affected tissues but their starring role in the disease etiology was completely under shaded area. Convincing experimental results from genetic, neuropathological, cellular, and biochemical with transgenic mouse models have confirmed that protein misfolding, oligomerization, and accretion in the brain are the major procedures that assist in pathological abnormalities responsible for the NDDs. Misfolded proteins are the sinister contributors to human diseases. Particularly, various human NDDs are allied with the accretion of misfolded proteins (α-synuclein, tau, β-amyloid, and prion) [42]. Although most of the misfolded protein related diseases appear infrequently with seemingly no hereditary phenotype, but gene-based studies have attested that a part of the disease cases can be inherited [43, 44]. Intriguingly, these proteins display no structural as well as functional resemblances to their functional counterparts [5]. Regardless of their native structural and functional conformations they get misfolded to form small oligomers to large fibrillar aggregates, in the diseased brain [45, 46]. Prior concept regarding those protein deposits supports their neurotoxic nature however, recent reports entail those smaller, soluble misfolded oligomers, precursors of the fibrillar aggregates contribute mostly for the neurodegeneration [5, 47, 48]. It has been perceived that mutations in the genes coding for the protein constituents of the misfolded masses can be inherited through generations in certain forms of these diseases (AD, TSEs, HD, PD, T2D, ALS and various rarer amyloid-related diseases such as familial amyloid polyneuropathy, cardiac amyloidosis, visceral amyloidosis, cerebral hemorrhage with amyloidosis of the Dutch and Icelandic type, and cerebral amyloidosis of the British and Danish type) [49]. In case of Alzhimers disease models proteins associated with familial AD such as the amyloid precursor protein (APP) and/or the presenilin 1 (PS1) were found to be high [50]. Sigurdson et al., (2011) have shown that insertion of the misfolded cellular prion protein, $PrP^{C}$, into a β-sheet-rich, aggregated isoform, $PrP^{Sc}$ in transgenic mice leads to emergence of spontaneous neurodegeneration accompanied by brain vacuolization [51]. Basically, during course of time these proteins cause synaptic dysfunction and eventual loss that in turn results into death of the neuronal cells [52]. Interestingly, toxic gain-of-function and/or dominant-negative effects, loss-of-function effects are the various reasons behind the augmentation of the misfolded proteins [53, 54]. Experimental evidence has been revealed for the acute effects of misfolded proteins on neuronal function after administering purified oligomers or transfecting the neurons with vectors that produce toxic proteins [55, 56, 57]. Misfolded proteins exhibit their toxic nature by targeting other cytosolic proteins as well [58, 59]. Clearly, the relationship between the accumulation of misfolded disease proteins and augmentation of neurodegenerative diseases is bidirectional, and in many cases mutually aggravating. Evidence has been gathered for the misfolded proteins that cause acute oxidative stress in nerve cells (astrocytes) and eventually destroy the anti-oxidant responses within them [60]. Damage in the antioxidant system eventually results in oxidative stress which in turn endorses the accumulation of NDD related proteins [61]. In addition to this, neuroinflammation also contributes to age-related diseases [62- 65].

## 5. Challenges and Therapeutic Strategies

Although a bunch of research works on the NDDs have been performed, unfortunately no cures or efficient treatments are yet available. It has been hypothesized that efficient and/or selective targeting of the misfolded protein aggregates might be the best approaches among all for therapeutic intrusion [66]. Soto and Pritzkow (2018) have specifically pointed out five distinct strategies, viz. (a) Elimination of the source of exogenous seeds to which an individual may be exposed; (b) Prevent the formation of or to remove misfolded aggregates; (c) Prevent the spread of seeds; (d) Arresting the elongation and multiplication of seeds, and most interestingly (e) development of antiprion therapeutic molecules [5] Regrettably, synthesizing therapeutic molecules or chemical compounds that particularly target misfolded protein aggregates are certainly very difficult and somewhat challenging. Another hurdle in developing successful therapy is the inability to identify the toxic species of misfolded proteins [67]. However, recent instrumental advancements have supported in identifying structural anomalies in the proteins which in turn are helpful for developing inhibitors [68] (Table 2). Some non-specific targeting of the upstream molecules as well as the pathway intermediates can be advantageous. Chaperons can be the other alternatives for therapeutic intervention. Chaperons play important roles in developing protein related diseases (proteostasis). Small molecule inhibitors or activators of specific chaperones can be synthesized to deal with protein related diseases. Several research works were conducted but unfortunately the molecules which seemed to be effective did not enter the clinical trial phase. Protein engineering has been an important tool for devising therapeutic approaches [69, 70].

**Table 2.** Drug targets and their effective molecules in neurodegeneration

| Molecules | Target | Mechanism of action | References |
|---|---|---|---|
| Rolipram | Ubiquitin-proteasome system | stimulates the phosphorylation and activity of the 26S proteasome, decreased tau aggregation, and improved cognition | [71] |
| Rapamycin and other mTOR inhibitors (curcumin, resveratrol) | Mammalian target of rapamycin (mTOR) pathway | inhibits autophagy by inhibiting the ULK1 complex, which is required for the biogenesis of autophagosomes, increase the clearance of abnormal protein aggregates and slow neurodegeneration | [72, 73] |
| Monoclonal antibodies | targeting clearance of misfolded proteins | 1) by penetrating the blood brain barrier (BBB) to bind Aβ in the extracellular space, and 2) through "a" peripheral sink" effect | [74] |
| LXR beta receptor agonists | Harnessing the endogenous mechanisms of protein clearance into the extracellular space | promote clearance of Aβ into the extracellular space by the ABCA1 transporter | [75] |
| Small molecule activators of HSF1 | Heat shock factor 1 (HSF1) | Over-expression of human HSF1 is directly related to neuroprotective functions, reduce polyglutamine aggregate formation and prolong lifespan. Reduction in pathogenic androgen receptor accumulation and neurotoxicity | [76] |
| Hsp90 inhibitors | HSP90 | induces the heat shock response in HD model mice | [77] |

Moreover, Sweeney et al., (2017) have beautifully described the drug-target pairs for NDDs associated with misfolded proteins [67].

## Conclusion

Regardless of the outstanding research outcomes and knowledge it should be admitted that NDDs are not curable. With growing days, the number of infected individuals is increasing. It has been widely acknowledged that this protein misfolding event and their accumulation in the brain is central to all NDDs. Several therapeutic options have been reconnoitered but unfortunately, due to the dearth of knowledge regarding specific molecular players and reliable biomarkers for monitoring therapeutic responses in relevant animal models, in vivo authentication of novel therapeutic compounds is obstructed. Soto and Pritzkow, (2018) have portrayed about a recent invention, where they mentioned that, "misfolded protein aggregates can self-propagate their pathological properties using the prion principle of transmission of biological information by seeding of protein misfolding." This innovation will not only help in understanding the process behind augmentation and advancements of NDDs but also will be beneficial for scheming innovative approaches for disease identification and management [5].

## References

[1] Przedborski S, Vila M, Jackson-Lewis V. Neurodegeneration: what is it and where are we? *J. Clin. Invest.* 2003; 111(1):3-10.

[2] Kovacs GG. Molecular pathological classification of neurodegenerative diseases: turning towards precision medicine. *Int. J. Mol. Sci.* 2016; 17(2):189. .

[3] Ashraf GM, Greig NH, Khan TA, Hassan I, Tabrez S, Shakil S, Sheikh IA, Zaidi SK, Akram M, Jabir NR, Firoz CK, Naeem A, Alhazza IM, Damanhouri GA, Kamal MA. Protein misfolding and aggregation in Alzheimer's disease and type 2 diabetes mellitus. *CNS Neurol. Disord. Drug Targets.* 2014; 13(7):1280-93.

[4] Abramov A Y, Potapova E V, Dremin V V, Dunaev A V. Interaction of oxidative stress and misfolded proteins in the mechanism of neurodegeneration. *Life (Basel, Switzerland).* 2020; 10(7):101.

[5] Soto C, Pritzkow S. Protein misfolding, aggregation, and conformational strains in neurodegenerative diseases. *Nat. Neurosci.* 2018; 21(10):1332-40.

[6] Chiti F, Dobson C M. Protein misfolding, functional amyloid, and human disease. *Annu. Rev. Biochem.* 2006; 75:333-66.

[7] Sgarbossa A. Natural biomolecules and protein aggregation: emerging strategies against amyloidogenesis. *Int. J. Mol. Sci.* 2012; 13(12):17121-37.

[8] Gsponer J, Babu MM. Cellular strategies for regulating functional and nonfunctional protein aggregation. *Cell Rep.* 2012; 2(5):1425-37.

[9] Singh A, Kukreti R, Saso L, Kukreti S. Oxidative stress: A key modulator in neurodegenerative diseases. *Molecules.* 2019; 24(8):1583.

[10] Liguori I, Russo G, Curcio F, Bulli G, Aran L, Della-Morte D, Gargiulo G, Testa G, Cacciatore F, Bonaduce D, Abete P. Oxidative stress, aging, and diseases. *Clin. Interv. Aging.* 2018; 13:757-772.

[11] Pizzino G, Irrera N, Cucinotta M, Pallio G, Mannino F, Arcoraci V, Squadrito F, Altavilla D, Bitto A. Oxidative Stress: Harms and Benefits for Human Health. *Oxid. Med. Cell Longev.* 2017; 2017:8416763.

[12] Poljsak B, Šuput D, Milisav I. Achieving the balance between ROS and antioxidants: when to use the synthetic antioxidants. *Oxid Med. Cell Longev.* 2013; 2013:956792.

[13] Uversky V N, Dunker A K. Understanding protein non-folding. *Biochim. Biophys. Acta.* 2010; 1804(6):1231-64.

[14] Uversky V N. Introduction to intrinsically disordered proteins (IDPs). *Chem. Rev.* 2014; 114(13):6557-60.

[15] Zorov D B, Juhaszova M, Sollott S J. Mitochondrial reactive oxygen species (ROS) and ROS-induced ROS release. *Physiol. Rev.* 2014; 94(3):909-50.

[16] Guo C, Sun L, Chen X, Zhang D. Oxidative stress, mitochondrial damage and neurodegenerative diseases. *Neural Regen Res.* 2013; 8(21):2003-14.

[17] Dugger B N, Dickson D W. Pathology of neurodegenerative diseases. *Cold Spring Harb. Perspect. Biol.* 2017; 9(7):a028035.

[18] Erkkinen M G, Kim M O, Geschwind M D. Clinical neurology and epidemiology of the major neurodegenerative diseases. *Cold Spring Harb. Perspect Biol.* 2018; 10(4):a033118.

[19] Martin JB. Molecular basis of the neurodegenerative disorders. N Engl J Med. 1999; 340(25):1970-80.

[20] Chen G F, Xu T H, Yan Y, Zhou YR, Jiang Y, Melcher K, Xu H E. Amyloid beta: structure, biology and structure-based therapeutic development. *Acta Pharmacol. Sin.* 2017; 38(9):1205-1235.

[21] Iqbal K, Liu F, Gong C X, Grundke-Iqbal I. Tau in Alzheimer disease and related tauopathies. *Curr. Alzheimer Res.* 2010; 7(8):656-64.

[22] Goedert M, Spillantini M G, Serpell L C, Berriman J, Smith M J, Jakes R, Crowther R A. From genetics to pathology: tau and alpha-synuclein assemblies in neurodegenerative diseases. *Philos. Trans. R. Soc. Lond B Biol. Sci.* 2001; 356(1406):213-27.

[23] de Mena L, Lopez-Scarim J, Rincon-Limas DE. TDP-43 and ER stress in neurodegeneration: friends or foes? *Front. Mol. Neurosci.* 2021; 14:772226.

[24] Schulte J, Littleton J T. The biological function of the huntingtin protein and its relevance to Huntington's disease pathology. *Curr. Trends Neurol.* 2011; 5:65-78.

[25] Hughes D, Halliday M. What is our current understanding of PrPSc-associated neurotoxicity and its molecular underpinnings? *Pathogens.* 2017; 6(4):63.

[26] Valastyan J S, Lindquist S. Mechanisms of protein-folding diseases at a glance. *Dis. Model Mech.* 2014; 7(1):9-14.

[27] Rao R V, Bredesen D E. Misfolded proteins, endoplasmic reticulum stress and neurodegeneration. Curr Opin Cell Biol. 2004; 16(6):653-62.

[28] Li Y, Li S, Wu H. Ubiquitination-Proteasome System (UPS) and autophagy two main protein degradation machineries in response to cell stress. *Cells.* 2022; 11(5):851.

[29] Kocaturk N M, Gozuacik D. Crosstalk between mammalian autophagy and the ubiquitin-proteasome system. *Front Cell Dev. Biol.* 2018; 6:128.

[30] Chen B, Retzlaff M, Roos T, Frydman J. Cellular strategies of protein quality control. *Cold Spring Harb. Perspect Biol.* 2011; 3(8):a004374.

[31] Ciechanover A, Kwon Y T. Degradation of misfolded proteins in neurodegenerative diseases: therapeutic targets and strategies. *Exp. Mol. Med.* 2015; 47(3):e147.

[32] Jellinger K A. Basic mechanisms of neurodegeneration: a critical update. *J. Cell Mol. Med.* 2010; 14(3):457-87.

[33] Görlach A, Bertram K, Hudecova S, Krizanova O. Calcium and ROS: A mutual interplay. *Redox Biol.* 2015; 6:260-71.

[34] Baev A Y, Vinokurov A Y, Novikova I N, Dremin V V, Potapova EV, Abramov AY. Interaction of mitochondrial calcium and ROS in neurodegeneration. *Cells.* 2022; 11(4):706.

[35] Rao R V, Bredesen D E. Misfolded proteins, endoplasmic reticulum stress and neurodegeneration. *Curr. Opin. Cell Biol.* 2004; 16(6):653-62.

[36] Ciechanover A, Kwon Y T. Degradation of misfolded proteins in neurodegenerative diseases: therapeutic targets and strategies. *Exp. Mol. Med.* 2015; 47(3):e147.

[37] Chong W C, Shastri M D, Eri R. Endoplasmic reticulum stress and oxidative stress: A vicious nexus implicated in bowel disease pathophysiology. *Int. J. Mol. Sci.* 2017; 18(4):771.

[38] Corazzari M, Gagliardi M, Fimia G M, Piacentini M. Endoplasmic reticulum stress, unfolded protein response, and cancer cell fate. *Front. Oncol.* 2017; 7:78.

[39] Costa C A D, Manaa W E, Duplan E, Checler F. The endoplasmic reticulum stress/unfolded protein response and their contributions to Parkinson's disease physiopathology. *Cells.* 2020; 9(11):2495.

[40] Verghese J, Abrams J, Wang Y, Morano K A. Biology of the heat shock response and protein chaperones: budding yeast (Saccharomyces cerevisiae) as a model system. *Microbiol. Mol. Biol. Rev.* 2012; 76(2):115-58.

[41] Haynes CM, Titus EA, Cooper AA. Degradation of misfolded proteins prevents ER-derived oxidative stress and cell death. Mol Cell. 2004; 15(5):767-76.

[42] Crestini A, Santilli F, Martellucci S, Carbone E, Sorice M, Piscopo P, Mattei V. Prions and Neurodegenerative Diseases: A Focus on Alzheimer's Disease. *J. Alzheimers Dis.* 2022; 85(2):503-518.

[43] Silverman G A, Pak S C, Perlmutter D H. Disorders of protein misfolding: alpha-1-antitrypsin deficiency as prototype. *J. Pediatr.* 2013; 163(2):320-6.

[44] Tzekov R, Stein L, Kaushal S. Protein misfolding and retinal degeneration. *Cold Spring Harb Perspect Biol.* 2011; 3(11):a007492.

[45] Ross CA, Poirier MA. Protein aggregation and neurodegenerative disease. *Nat. Med.* 2004; 10; Suppl: S10-7.

[46] Soto C. Unfolding the role of protein misfolding in neurodegenerative diseases. *Nat. Rev. Neurosci.* 2003; 4(1):49-60.

[47] Caughey B, Lansbury PT. Protofibrils, pores, fibrils, and neurodegeneration: separating the responsible protein aggregates from the innocent bystanders. *Annu. Rev. Neurosci.* 2003; 26:267-98.

[48] Gadad BS, Britton G B, Rao K S. Targeting oligomers in neurodegenerative disorders: lessons from α-synuclein, tau, and amyloid-β peptide. *J. Alzheimers Dis.* 2011; 24; Suppl 2:223-32.

[49] Moreno-Gonzalez I, Soto C. Misfolded protein aggregates: mechanisms, structures and potential for disease transmission. *Semin. Cell Dev. Biol.* 2011; 22(5, July):482-7.

[50] O'Brien RJ, Wong PC. Amyloid precursor protein processing and Alzheimer's disease. *Annu. Rev. Neurosci.* 2011; 34:185-204.

[51] Sigurdson C J, Joshi-Barr S, Bett C, Winson O, Manco G, Schwarz P, Rülicke T, Nilsson KP, Margalith I, Raeber A, Peretz D, Hornemann S, Wüthrich K, Aguzzi A. Spongiform encephalopathy in transgenic mice expressing a point mutation in the β2-α2 loop of the prion protein. *J. Neurosci.* 2011; 31(39): 13840-7.

[52] Poggiolini I, Saverioni D, Parchi P. Prion protein misfolding, strains, and neurotoxicity: an update from studies on Mammalian prions. *Int. J. Cell Biol.* 2013; 2013:910314.

[53] Winklhofer K F, Tatzelt J, Haass C. The two faces of protein misfolding: gain- and loss-of-function in neurodegenerative diseases. *EMBO J.* 2008; 27(2):336-49.

[54] Valastyan JS, Lindquist S. Mechanisms of protein-folding diseases at a glance. *Dis. Model Mech.* 2014; 7(1):9-14.

[55] Kopeikina K J, Hyman B T, Spires-Jones T L. Soluble forms of tau are toxic in Alzheimer's disease. *Transl. Neurosci.* 2012; 3(3):223-33.

[56] Ingelsson M. Alpha-synuclein oligomers-neurotoxic molecules in Parkinson's disease and other Lewy body disorders. *Front. Neurosci.* 2016; 10:408.

[57] Thies E, Mandelkow EM. Missorting of tau in neurons causes degeneration of synapses that can be rescued by the kinase Mark2/Par-1. *J. Neurosci.* 2007; 27(11):2896-907.

[58] Olzscha H, Schermann S M, Woerner A C, Pinkert S, Hecht M H, Tartaglia G G, Vendruscolo M, Hayer-Hartl M, Hartl F U, Vabulas R M. Amyloid-like aggregates sequester numerous metastable proteins with essential cellular functions. *Cell.* 2011; 144(1):67-78.

[59] Vidal R L, Matus S, Bargsted L, Hetz C. Targeting autophagy in neurodegenerative diseases. *Trends Pharmacol. Sci.* 2014; 35(11):583-91.

[60] Angelova P R, Abramov A Y. Functional role of mitochondrial reactive oxygen species in physiology. *Free Radic. Biol. Med.* 2016; 100:81-5.

[61] Höhn A, Weber D, Jung T, Ott C, Hugo M, Kochlik B, Kehm R, König J, Grune T, Castro JP. Happily (n) ever after: Aging in the context of oxidative stress, proteostasis loss and cellular senescence. *Redox Biol.* 2017; 11:482-501.

[62] Heppner F L, Ransohoff R M, Becher B. Immune attack: the role of inflammation in Alzheimer disease. *Nat. Rev. Neurosci.* 2015; 16(6):358-72.

[63] Guillot-Sestier MV, Doty KR, Gate D, Rodriguez J Jr, Leung BP, Rezai-Zadeh K, Town T. Il10 deficiency rebalances innate immunity to mitigate Alzheimer-like pathology. *Neuron.* 2015; 85(3):534-48.

[64] Deleidi M, Jäggle M, Rubino G. Immune aging, dysmetabolism, and inflammation in neurological diseases. *Front. Neurosci.* 2015; 9:172.

[65] Currais A, Huang L, Goldberg J, Petrascheck M, Ates G, Pinto-Duarte A, Shokhirev MN, Schubert D, Maher P. Elevating acetyl-CoA levels reduces aspects of brain aging. *Elife.* 2019; 8:e47866.

[66] Hussain R, Zubair H, Pursell S, Shahab M. Neurodegenerative diseases: regenerative mechanisms and novel therapeutic approaches. *Brain Sci.* 2018; 8(9):177.

[67] Sweeney P, Park H, Baumann M, Dunlop J, Frydman J, Kopito R, McCampbell A, Leblanc G, Venkateswaran A, Nurmi A, Hodgson R. Protein misfolding in neurodegenerative diseases: implications and strategies. *Transl. Neurodegener.* 2017; 6:6.

[68] Lee Yoonji, Basith S, Choi S. Recent advances in structure-based drug design targeting class A G protein-coupled receptors utilizing crystal structures and computational simulations. *J. Med. Chem.* 2018;61(1):1-46.

[69] Fontaine S N, Zheng D, Sabbagh J J, Martin M D, Chaput D, Darling A, Trotter JH, Stothert AR, Nordhues B A, Lussier A, Baker J, Shelton L, Kahn M, Blair L J, Stevens SM Jr, Dickey C A.

DnaJ/Hsc70 chaperone complexes control the extracellular release of neurodegenerative-associated proteins. *EMBO J.* 2016; 35(14):1537-49.

[70] DeSantis M E, Leung E H, Sweeny E A, Jackrel M E, Cushman-Nick M, Neuhaus-Follini A, Vashist S, Sochor M A, Knight MN, Shorter J. Operational plasticity enables hsp104 to disaggregate diverse amyloid and nonamyloid clients. *Cell.* 2012; 151(4):778-793.

[71] Myeku N, Clelland C L, Emrani S, Kukushkin N V, Yu W H, Goldberg A L, Duff K E. Tau-driven 26S proteasome impairment and cognitive dysfunction can be prevented early in disease by activating cAMP-PKA signaling. *Nat. Med.* 2016; 22(1):46-53.

[72] Harris H, Rubinsztein DC. Control of autophagy as a therapy for neurodegenerative disease. *Nat. Rev. Neurol.* 2011; 8(2):108-17.

[73] Martinez-Vicente M. Autophagy in neurodegenerative diseases: from pathogenic dysfunction to therapeutic modulation. *Semin. Cell Dev. Biol.* 2015; 40:115-26.

[74] Wisniewski T, Goñi F. Immunotherapeutic approaches for Alzheimer's disease. *Neuron.* 2015; 85(6):1162-76.

[75] Boehm-Cagan A, Michaelson DM. Reversal of apoE4-driven brain pathology and behavioral deficits by bexarotene. *J. Neurosci.* 2014; 34(21):7293-301.

[76] Neef D W, Jaeger AM, Thiele DJ. Heat shock transcription factor 1 as a therapeutic target in neurodegenerative diseases. *Nat. Rev. Drug Discov.* 2011; 10(12):930-44.

[77] Labbadia J, Cunliffe H, Weiss A, Katsyuba E, Sathasivam K, Seredenina T, Woodman B, Moussaoui S, Frentzel S, Luthi-Carter R, Paganetti P, Bates GP. Altered chromatin architecture underlies progressive impairment of the heat shock response in mouse models of Huntington disease. *J. Clin. Invest.* 2011; 121(8):3306-19.

Chapter 4

# The Mechanism of Mitochondrial Reactive Oxygen Species (mtROS) Generation and Its Role in the Apoptosis, Pathogenesis of Inflammatory and Neurodegenerative Disorders

**Chirantan Majumder**
**Anirban Manna**
**Satyajit Halder**
**and Kuladip Jana***
Division of Molecular Medicine, Kolkata, India
Unified Academic Campus, Division of Molecular Medicine, Salt Lake City, Bidhannagar, Kolkata, India

## Abstract

Mitochondrial electron transport chain (mtETC) is the major source of cellular reactive oxygen species (ROS). Dysfunctions associated with ETC generates mitochondrial ROS implicates cell cycle progression in cancer cells, cellular apoptosis, inflammatory and degenerative disorders. Mitochondrial ROS (mtROS) is now believed as integral component of cellular signalling. Complex I and III of mtETC are the major sources of ROS, released initially into mitochondrial matrix and intermembrane space respectively. In this review we summarized the molecular mechanism of mtROS production and its role in cellular apoptosis, inflammatory and neurodegenerative disorders.

**Keywords:** mitochondria, mitochondrial ROS, mitochondrial ETC, apoptosis, rheumatoid arthritis, amyotrophic lateral sclerosis (ALS), Parkinson's disease

## 1. Introduction

Reactive oxygen species (ROS) are highly reactive molecules that can be produced from molecular oxygen by reduction like superoxide anion ($O_2^{\cdot-}$), hydroxyl radical ($OH^{\cdot}$) and/or non-

* Corresponding Author's Email: kuladip@jcbose.ac.in; kuladip.jana@gmail.com.

In: The Role of Reactive Oxygen Species in Health and Disease
Editor: Kuladip Jana
ISBN: 979-8-89113-283-2

radical oxides, hydrogen peroxide molecules ($H_2O_2$) and singlet oxygen ($^1O_2$) [1–3]. Cellular ROS production always remains in an equilibrium to maintain cellular homeostasis which is regulated by both enzymatic (e.g., superoxide dismutase, catalase, glutathione peroxidase) and non-enzymatic ways (glutathione, vitamin C, vitamin E and iron chelators) [2]. According to historical prejudice ROS was considered as cell destructive molecules, but studies carried out later on indicated that low or moderate ROS has important cellular functions like cell cycle progression, cell proliferation, differentiation, migration and death [4, 5]. A balanced level of cellular ROS regulates different types of cellular signalling pathways like phosphoinositide 3-kinase (PI3K)/PKB, nuclear factor (erythroid-derived 2)-like 2 (Nrf2)/Kelch like-ECH-associated protein 1 (Keap1) which can lead to cell survival or death. Failing to maintain the balance of cellular ROS production leads to physiological to pathological shift like cancer, diabetes, Parkinson's and Alzheimer's diseases [6]. The primary source of intracellular ROS is mitochondrial electron transport chain (ETC) and its activity. About 0.2-2% of total electrons in mitochondrial ETC is not associated to ATP production but for ROS generation. In mitochondria complex I and III of ETC are the major sites of ROS production [7–10]. Except ETC, xanthene oxidase, NADPH oxidase and even cytochrome p450 oxidases can produce ROS. In physiological conditions, Superoxides that are produced in mitochondria are scavenged to molecular oxygen and water by dismutation through superoxide dismutases (SODs) or to only water by glutathione peroxidase 1(GPx1), otherwise $O_2^{\cdot-}$ can interact with nitric oxide molecules to produce highly reactive peroxynitrite molecules ($ONO_2^-$) which are considered as reactive nitrogen species (RNS). Unregulated overproduction of mitochondrial ROS, known as oxidative stress, precedes mitochondrial membrane depolarization and it causes the opening of Bax/Bak channels at the outer mitochondrial membrane (OMM), allowing apoptogenic factors like cytochrome C, Smac/Diablo and AIPs to release out from inner mitochondrial space (IMS) to cytosol which further binds to caspase-9 and Apaf-1 to produce apoptosome (wheel of death) leads to apoptosis by activation of caspase-3. Excessive mitochondrial ROS production can activate p53 and/or c-Jun N-terminal kinase (JNK) pathway leading to the inactivation of ani-apoptotic proteins like Bcl-2 and can activate apoptotic proteins like Bax, Bak, Puma, and Noxa that initiates the processes of apoptosis. mtROS can oxidize cardiolipin leading to the opening of mitochondrial permeability transition pore(mptp), which leads to the release of cytochrome C, which can induce apoptosis [6]. Interestingly mitochondrial ROS generation can be regenerative and can induce the release of such toxic ROS outside of mitochondria through the opening of MPTP is called ROS-induced ROS release (RIRR) [11].

## 2. Molecular Mechanism of Mitochondrial ROS Generation

Jensen was one of the people who demonstrated mitochondria can produce ROS in the early sixties. Later on, Britton Chance and co-workers did a more general study on mitochondrial ROS generation in the early seventies [11]. There are eleven distinct sites found in mitochondria where ROS can be produced and the mitochondrial electron transport chain and its components (complex I and III) are the major sources [12].

## 2.1. Complex I

NADH-ubiquinone oxido-reductase or complex I is the largest complex among mitochondrial ETC complexes. It's an L- shaped multi-subunit enzyme complex and in eukaryotes, it has two chemically distinct domains: one is a hydrophobic transmembrane domain, which remains embedded in the inner-mitochondrial membrane known as the membrane arm and the hydrophilic matrix arm protrudes into matrix [12, 13]. These two domains of complex I are composed of 14 distinct subunits which are conserved in bacterial CI and are the core of enzymatic reactions. 45 distinct proteins are involved in the formation of these core subunits. Seven hydrophobic subunits (ND1-6 and ND4L) form the membrane arm, all encoded by the mitochondrial genome. On the other hand the matrix arm is composed of seven core subunits (NDUFS1, NDUFS2, NDUFS3, NDUFS7, NDUFS8, NDUFV1 and NDUFV 2), having cofactors: A flavin mononucleotide (FMN) molecule; 7-9 iron-sulfur (FeS) clusters [(2Fe-2S)N1b, (4Fe-4S)N3, (4Fe-4S)N4, (4Fe-4S)N5, (4Fe-4S)N6a/b and (4Fe-4S)N2 clusters] and the final electron accepting iron-sulfur cluster (N2 cluster) which is now believed to deliver electrons to ubiquinone binding domain [14, 15]. The cofactor flavin mononucleotide (FMN) located at the pinnacle of matrix arm could reduce itself by accepting a pair of electrons from NADH, produced from the tri-carboxylic acid cycle (TCA cycle) occurring in the mitochondrial matrix. Once that occur electrons pass to ubiquinone via a series of iron-sulphur clusters arranged in a gradient from higher to lower i.e., FMN→N3→N1b→N4→N5→N6a→N6b→N2 and finally to ubiquinone (CoQ) located at the juncture of membrane and matrix arm and reduced to ubiquinol ($QH_2$) [16,17]. A strong hypothesis by Ohnishi, says that the energy released in such electron transfer through CI causes a conformational change in CI that indirectly transfers a pair of protons from matrix to IMS and another pair of protons are directly transferred, which is coupled to redox of CoQ [18].

### *2.1.1. ROS Production by CI Under Physiological Conditions*

A pre-existing dogma indicates that complex I (CI) is the primary source of ROS production in mitochondria but in reality, it depends upon the situations and circumstances. Practically CI is the primary source of mitochondrial ROS in pathological conditions rather than healthy and physiological conditions. Under physiological conditions, ROS can be produced in the mitochondrial matrix at FMN and ubiquinone (CoQ) binding sites when electrons are transferred through them. Small molecules like rotenone and piericidin are ubiquinone (CoQ) inhibitors, interrupt electron transfer to ubiquinone (CoQ) and this then leaks out from FMN that reacts with molecular oxygen ($O_2$) to produce superoxide($O_2^{\cdot}$-), a ROS [11, 12].

### *2.1.2. ROS Production by CI Under Pathological Conditions*

A wide variety of neurodegenerative diseases are associated with mutations in CI which is associated with abnormal overproduction of mitochondrial ROS, including Parkinson's disease, characterized by loss of dopaminergic neurons and imbalanced glutathione activity. Loss of Cell bodies of dopaminergic neurons found in substantianigra pars compacta and nerve terminals in the striatum [19]. In physiological conditions, oxidative deamination of dopamine by monoamine oxidase (MAO) occurs for the production of catecholamines which produce $H_2O_2$, and can produce harmful toxic oxidants through an alternative pathway. In this pathway, superoxide-producing dopamine quinone oxidizes dopamine which can be reduced by mitochondrial CI that produces semiquinone that transfers electrons to molecular oxygen and forms superoxides [20, 21]. Hypoxia and or reoxygenation has the most powerful and deleterious ROS-generating capabilities in mitochondria. Three major factors of mitochondria are believed to be responsible in hypoxia-induced mitochondrial ROS generation: complex I, xanthin oxidase and NADPH oxidase but the latter two are probably not responsible in this because in vivo inhibition of these two did not provide any protection [22–26]. De Jong and colleagues by using electron paramagnetic resonance spectroscopy studied those conformational changes that occur in NADH CoQ oxidoreductase in an energy-dependent manner leading to the formation of ubisemiquinone. So pathological conditions by structural rearrangements can change the ROS-producing efficiency of CI [27].

## 2.2. Complex II

Complex II (CII) or succinate-ubiquinone oxidoreductase, also commonly known as succinate dehydrogenase is a component of tricarboxylic acid cycle (TCA cycle) or Krebs cycle and ETC. CII is a tetrameric iron sulphur flavoprotein located in the inner mitochondrial membrane that links metabolism and OXPHOS [28, 29]. As a part of Krebs cycle, it oxidizes succinate and produces fumarate and yields reduced equivalent in the form of reduced flavin adenine nucleotide ($FADH_2$) [11]. CII is also another entry site of electrons, which is delivered from succinate to CoQ via Fe-S [29]. FAD is reduced to $FADH_2$ by receiving electrons from succinate and then transferring the electrons to FE-S. Finally, CoQ is reduced to $QH_2$ by receiving electrons from FE-S [30]. Electron transfer through CII is not coupled to proton translocation from matrix to IMS, unlike CI [12]. CII comprises 4 subunits: two of them are hydrophobic membrane anchor proteins CybL and CybS, which anchor the entire complex in the membrane and also contain CoQ binding domain. The other two subunits are located at the matrix side of the IMM and contain a binding domain for succinate, three FeSclusters [(2Fe-2S), (4Fe-4S) and (3Fe-4S)], and a flavoprotein covalently attached to FAD [31].

### *2.2.1. ROS Production by CII Under Physiological Conditions*

Under physiological conditions, ROS production by CII is negligible and it may be due to low tissue levels of succinate. Studies have shown that the FAD of succinate dehydrogenase instead of electron carriers like iron-sulphur clusters and quinones generates ROS [11].

### *2.2.2. ROS Production by CII Under Pathological Conditions*

The tissue level of succinate increases from 100/500 micromolar (physiological concentration of succinate) to 500/5000 micromolar in hypoxia/ischemia [23, 32, 33]. Significant increase of succinate levels observed in macrophages (in millimolar ranges) exposed to lipopolysaccharides. Succinate is believed to be the metabolite of the innate immune system which stabilizes hypoxia-inducible factor1α (HIF1α) and IL-1β in inflammation [34]. Activation of macrophages in inflammation induces ROS generation but the role of succinate in it is still unknown [11].

## 2.3. Complex III

Co-Q cytochrome oxidoreductase or CIII transfers electrons carried by ubiquinone ($QH_2$) to cytochrome c. CIII is a symmetrical dimer in which each monomer is composed of 11 subunits [35]. There are 3 catalytically active subunits in CIII: cytochrome b ($b_L$ and $b_H$), cytochrome $c_1$ and a high potential (2Fe+ 2S) cluster that exists as an iron-sulphur protein [36]. At both ends of Cytb, there are two CoQ binding domains, and one binding domain is for $QH_2$ oxidation ($Q_o$) located at the side of IMS and related lower potential containing cytochrome $b_L$. The other domain is for CoQreduction ($Q_i$) located at matrix side and related to high potential $b_H$ [37]. The reducing equivalents are received by CIII from CI and CII and processed through Q cycle. This cycle is initiated by ubiquinole ($QH_2$) that transfers one electron to Riske iron-sulfur protein (a, myxothiazol binding and target site) and converted to ubisemiquinone ($QH^-$) and at the same moment, transfers a pair of protons to IMS from the matrix. This Riske iron-sulphur protein then transferred the electron to cytochrome $c_1$ and finally to cytochrome c. At this moment, highly reactive $QH^-$ is produced at $Q_o$ and rapidly transfers the second electron to $b_L$and and then to $b_H$ at $Q_i$ and $b_H$ gets reduced, which finally transfers the electron to CoQ and $QH^-$ is formed. Completion of Q-cycle occurs by oxidation of second $QH_2$ at $Q_o$ and transfers two protons to IMS from matrix. Likewise, one electron is transferred to iron-sulphur cluster and the other one to $QH^-$via cytochrome $b_H$ and produces $QH_2$ at $Q_i$ [11, 12].

At physiological conditions, ROS generation from CIII is relatively low because of rapid oxidation and less stability of semiquinone ($Q^{-\cdot}$). Inhibitors like antimycin A which reduces both heme and cytochrome c and increases the production and stability of semiquinone ($Q^{-\cdot}$) can produce significant amounts of ROS from CIII [11].

## 2.4. Other Mitochondrial Sites of ROS Production

### *2.4.1. Aconitase*

Aconitase converts citrate to isocitrate in the Krebs cycle. Aconitase contains iron-sulphur center [4Fe-4S] with 3 iron atoms interacting with cysteine residues and sulphur atoms and the remaining one [$Fe^{2+}$-α] is exposed to a solvent that dehydrates citrate to form cis-aconitate (intermediate) followed by hydration of cis-aconitate to form isocitrate [34, 38]. The prosthetic group of aconitase is highly susceptible of superoxide attacks and get inactivated and forms an inactive form of [$Fe^{2+}$-α] and $H_2O_2$ which further interacts with each other (Fenton's reaction) and produces more harmful hydroxyl radicals [38–40]. Doxorubicin-induced cardiotoxicity is believed to be a consequence of reduced aconitase activity and ROS release. So, aconitase inactivation is an example of ROS-induced ROS production and release, where superoxide is responsible for developing and releasing more toxic hydroxyl radicals [41].

### *2.4.2. Monoamine Oxidase (MAO)*

Monoamine oxidase is a flavoenzyme located outside of the mitochondrial membrane. This enzyme has two isoforms, MAO-A and MAO-B which have different substrate specificity and are inhibited by different inhibitors [42]. MAO oxidizes biologically formed amines to aldehydes, $H_2O_2$, and ammonium base. MAO activity is more critical in brain tissues compared to other parts of the body because of the low activity of catalase and peroxidase to fully neutralize $H_2O_2$ [43]. The $H_2O_2$ forming capacity of MAO is the highest in among mitochondrial components, and it is about 48 times greater than mitochondrial CIII [11]. People have shown through experiments that MAO-dependent ROS production is enhanced in some pathophysiological conditions and aging. A 24-month-old rat heart shows about eight times more MOA activity than a 1-month-one, indicating more MAO-induced ROS generation in aging heart [44, 45]. ROS, produced as a by-product of MAO activity, might be detrimental to cells, e.g., the deamination of dopamine in neuronal cells by MAO produces reactive quinones that could damage cellular components through direct interactions [46].

#### *2.4.3. P66^shc^*

P66$^{shc}$ is a cytosol-resident adapter protein. Under oxidative stress, it could be translocated from cytosol to mitochondria in a PKC-β dependent manner [47]. Once in mitochondria, it could oxidize cytochrome c and produce $H_2O_2$ in the N-terminal domain of p66$^{shc}$. It is believed that in mitochondria, p66$^{shc}$ exists in a large complex composed of mtHSP70 and TIM-TOM complex and remains inactive. Once the apoptotic signal is propagated, this complex is dissociated and makes p66$^{shc}$ free and active, which could then be involved in electron transport and producing $H_2O_2$ [48]. NADPH-oxidase, α- Glycerophosphate dehydrogenase, electron transfer flavoprotein (ETF) and ETF quinine oxidoreductase are also responsible for mitochondrial ROS generation [11].

## 3. Molecular Mechanism of Mitochondrial ROS Induced Apoptosis

Mitochondrial ROS induced apoptosis occurs through three major pathways.

1. ROS-induced mitochondrial pathway (intrinsic pathway)
2. ROS-induced death-receptor pathway (extrinsic pathway)

### 3.1. ROS-Induced Mitochondrial Pathway (Intrinsic Pathway)

Mitochondria are the 'power-house' of cell that converts nutrients to ATP, are used for all biological activities, and are essential for maintaining cellular entity in every aspect [49, 50]. Beside the energy source of a cell mitochondria have other important cellular functions like absorption and storage of calcium ions, non-shivering thermogenesis and apoptosis [13, 51, 52]. Most of this function depends on the permeability of IMM and OMM. The Permeability of OMM is up to 5 KDa which allows ETC components (substrates and products). The inner mitochondrial membrane is highly impermeable [6]. Permeability of IMM is one of the critical measures of mitochondria dependent-apoptosis. Mitochondria dependent-apoptosis requires mitochondrial permeability transition pore (MPTP) complex formation, which is composed of cyclophilin D, a chaperon located in the mitochondrial matrix associated with adenine–nucleotide–translocator (ANT) in the IMM, interacts with voltage dependent-anion-channel (VDAC) in the OMM and makes the connection between IMM and OMM [53–55]. Under specific cues (e.g., stress), the permeability of IMM increases and allows passage of molecules having a molecular weight less than 1.5KDa, including protons into the matrix and altering

OXPOHS [54], leading to osmotic swelling of the mitochondrial matrix, causing inward folds in mitochondrial cristae that compress mitochondrial vesicles. This combined effect of compressed mitochondrial vesicles and outer mitochondrial membrane permeabilization (MOMP) leads to the release of pro-apoptotic proteins like, apoptosis-inducing factor (AIF), cytochrome c endonuclease G (endo G), and Smac/Diablo into cytosol. After releasing of cytochrome c into cytosol it makes a complex with apoptosis activating factor-1 (Apaf-1) and pro-caspase9, known as apoptosome or 'wheel of death' in a ATP-dependent manner that leads to auto-conversion of pro-caspase9 to active caspase9 that leads to activation of central executioner caspase, caspase3 leads to apoptosis. Smac/Diablo, after entering into cytosol binds and inactivates ani-apoptotic proteins, inhibitors of apoptotic proteins (IAP), which inhibits caspase activation and promotes apoptosis. On the other hand, apoptosis-inducing factor (AIF) and endoG after releasing into the cytosol, are translocated into the nucleus and trigger apoptosis in a caspase-independent way. The exact mechanism of endoG-mediated apoptosis is still unknown. Apoptosis-inducing factor (AIF), after entering into the nucleus, binds to DNA and leads to heterochromatinization and DNA fragmentation [56].

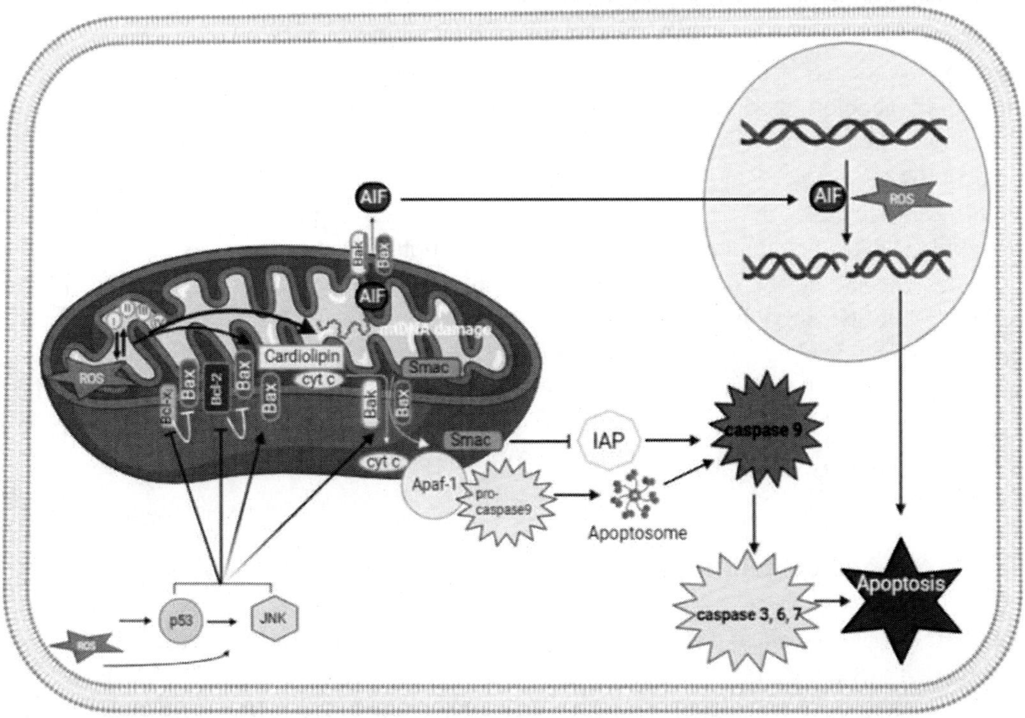


**Figure 1**. ROS-induced activation of the mitochondrial (intrinsic) pathway of apoptosis. Exogenous or endogenous ROS can activate p53 and/ or c-Jun N-terminal kinase (JNK), and can activate pro-apoptotic proteins like bax/bak and inhibits anti-apoptotic proteins like Bcl-2 and/or Bcl-xL. ROS can lead to peroxidation of cardiolipin, resulting in the release of cytochrome c (Cyt c) into cytosol. ROS can cause mitochondrial membrane depolarization, which leads to the opening of Bax/Bak channels on OMM, and causes the release of several apoptotic factors like AIF, Endo G, cytochrome c, and Smac/Diablo into the cytosol. Cyt c then makes a complex with Apaf-1 and procaspase-9, known as apoptosome complex in the cytosol, leading to caspase-9 activation. Caspase-9 then activates executioner caspases like caspase-3,6 and 7, resulting in apoptosis. AIF and Endo G, after entering into the cytosol, translocate to the nucleus and cause DNA damage. ROS can directly cause damage to nuclear and mitochondrial DNA leads to apoptosis.

Mitochondrial outer membrane permeabilization (MOMP) is tightly regulated by the B-cell-lymphoma protein 2(Bcl-2) family of proteins. Bcl-2 family has both anti-apoptotic and pro-apoptotic proteins. These proteins may obtain up to four conserved Bcl-2 homology domains (BH1-4) and among them BH3 is the death domain. There are six anti-apoptotic Bcl-2 family proteins: Bcl-2, Bcl-XL, Bcl-W, Bcl-B, A1 and Mcl-1 (all of which may have 3 to 4 BH domains) and three pro-apoptotic proteins Bax, Bak and Bok (all of which may have 2 to 3 BH domains). Other pro-apoptotic proteins contain BH3-only domains like Bid, Bim, Puma and Noxa, Bad, Bik, Bmf, Hrk and SOUL. The balanced activity of these two kinds of proteins decides cell fate. At physiological conditions, pro-apoptotic proteins of Bcl-2 family (Bax, Bak) are localized in the cytosol and anti-apoptotic protein are localized in OMM, and produces heterodimer with pro-apoptotic proteins, Bax and Bak and inhibits their activity. Under stress conditions, the homeostatic balance between these two types of proteins is disrupted, leading to apoptosis [57–60] (Figure 1).

Stressful cellular conditions that lead to mitochondrial ROS generation can lead to peroxidation of cardiolipin and release cyt c into the cytosol [61, 62]. Mitochondrial ROS can interact with MPT pore proteins and induce oxidative modifications that lead to anion influx, causing the release of apoptogenic factors like cyt c, into the cytosol [63]. Another mechanism of mtROS-induced apoptosis was observed in neutrophils, in which pyocyanin-mediated mtROS generation dissipates mitochondrial membrane potential, leads to activation of sphingomyelinases that produces ceramide from sphingomyelin, induces release cyt c into the cytosol [64].

Once in cytosol cyt c interacts with apoptosis-promoting factor-1(Apaf-1) and pro-caspase9 and forms apoptosome, leading to formation of active caspase 9, which further activates executioner caspases i.e., caspase 3, 6, 7, leads to apoptosis. Smac/Diablo after releasing into cytosol, inhibits inhibitors of apoptotic proteins (IAP) which is an inhibitor of caspase activation, leading to apoptosis. On the other hand, released AIP or endoG induces non-caspase-dependent apoptosis by DNA condensation and degradation. mtDNA highly susceptible to mtROS because of less histone content, which causes a decreased production of mitochondrial ETC elements leads to more mtROS production, and causes apoptosis [56].

### 3.2. ROS-Induced Death-Receptor Pathway (Extrinsic Pathway)

Extrinsic apoptosis pathway is achieved by death receptors of tumor necrosis factor receptor (TNF-R) superfamily [65]. This family include Death receptor 1 (DR1) (also known as TNF-R1, CD120a, p55), DR2 (Fas, CD95 or Apo-1), DR3 (Apo-3, TRAMP, LARD or TNFRSF25 (TNF receptor superfamily 25), DR4 (TNF related apoptosis-inducing ligand (TRAIL)-R1 or Apo-2), DR5 (TRAILR2 or TRICK2) and DR6 (TNF receptor superfamily member 21(TNFRSF21)). All of these members have death domains (DD) [66]. DR1 or TNFR1 gets activated by trimerization when its ligand TNF-$\alpha$ binds to it. Trimerization of TNFR1, recruits TNF-R-associated adaptor protein with death domain (TRADD) to its cytoplasmic domain [66]. RIP and TNF-associated factor-2 (Traf-2) are then recruited to TRADD, causing activation of nuclear factor kappa B(NF-$\kappa$B) [67]. NF-$\kappa$B activation either causes inflammation or activation of anti-apoptotic factors like Bcl-$X_L$, X-linked inhibitor of apoptosis (XIAP) and cellular inhibitors of apoptosis (cIAP) [67, 68]. But the association of FADD with TRADD domain sequentially recruits and activates caspase-8, which can directly activate executioner caspases 3,-6,-7 or indirectly by activation of t-Bid mediated mitochondrial pathway [69]. Cell survival or death due to TNFR1 activation depends on a regulatory protein called FLICE inhibitory protein (FLIP) [67, 68, 70, 71].

Activation of Fas receptor (DR2) through trimerization by FasL recruits FADD which activates caspase-8, which further activates central executioner caspases, caspase3,-6 and 7, leads to apoptosis. Recruitment of Daxx to Fas leads to activation of ASK-1, which further activates JNK, causes the inactivation of Bcl-2 and Bcl-$X_L$ and promotes apoptosis [6].

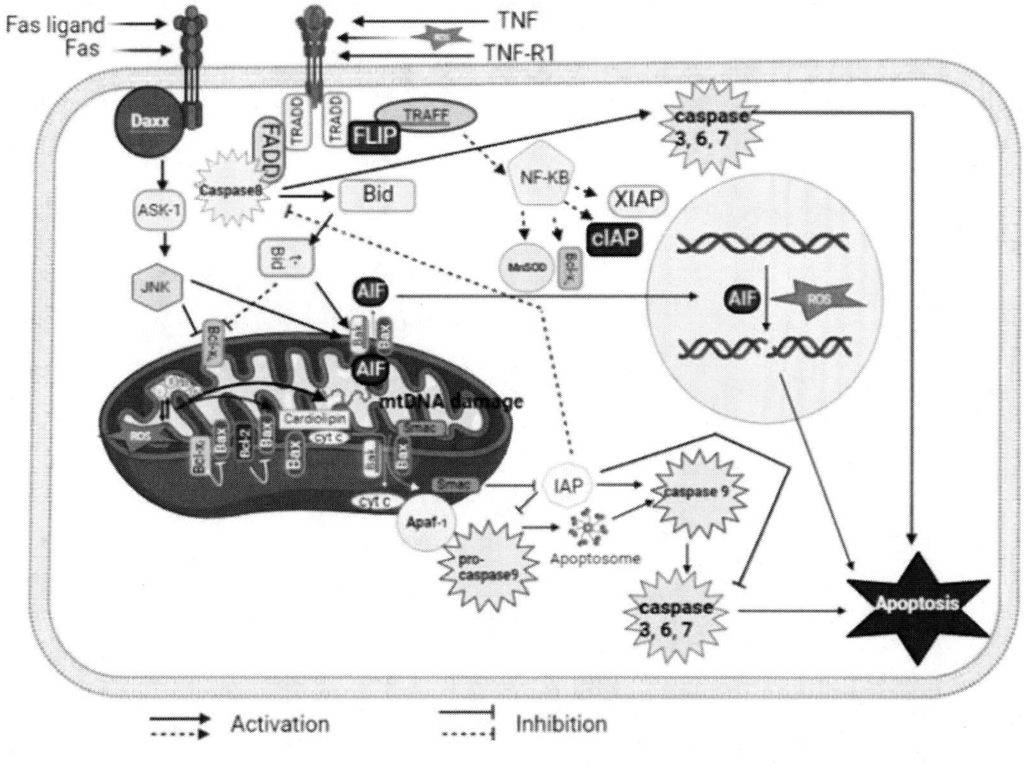


**Figure 2.** ROS-induced activation of the death receptor (extrinsic) pathway of apoptosis and its crosstalk with intrinsic pathway. ROS can activate transmembrane death receptors such as Fas, and TNF-R1. Activation of TNF-R1 by TNF causes recruitment of adaptor protein FADD and procaspase-8 to the receptor to form the DISC, causes activation of caspases-8, which can directly activate executioner caspases-3, 6 and7 and triggers apoptosis. Activated caspases can also cleave Bid to form t-Bid, which activates the mitochondrial pathway of apoptosis via a cross-talk pathway. t-Bid can block the anti-apoptotic activity of Bcl-2 and Bcl-XL, and activates Bax and Bak, leads to the release of cytochrome c and Smac/Diablo, which causes apoptosis. Activation of Fas by Fas ligand (Fas L) can recruit Daxx and causes activation of ASK-1 and JNK, leading to apoptosis. TNF-R1 can activate survival pathways mediated by adaptor protein TRADD, RIP and Traf-2 causes the formation of survival genes through the production of NF-κB.

There is a significant debate, on whether mtROS activates or inhibits NF-κB through TNFR1. But people have found that ROS can activate NF-κB through TNFR1 which causes the formation of catalase and MnSOD [68], that neutralizes mtROS-mediated TNFR1 activation and apoptosis by inhibiting caspase-8 activation [72]. The mechanism by which TNF increases mtROS production is still unknown but when activated by TNFR1, caspase-8 binds to ROS modulator-1 (ROMO-1), localized at OMM [68] and sequesters Bcl-$X_L$, causes loss of mitochondrial membrane potential resulting ROS formation that further activates ASK-1 [67], which further activates downstream targets JNK and p38 MAPK, leads to TNF-mediated apoptosis [68] (Figure 2).

## 4. Mitochondrial ROS Is a Master Player of Rheumatoid Arthritis (RA), an Inflammatory Disorder

Rheumatoid arthritis is a systemic autoimmune inflammatory disorder that is characterized by the formation of new blood vessels, synovial hyperplasia associated with local infiltration of different inflammatory and immune cells resulting in damage of cartilage and bone leading to poor quality of life in both physiological and psychological regards [73, 74]. Elevated mtROS levels is directly associated with increased TNF-α synthesis and release in plasma resulting in inflammation, increased synovial proliferation and vasculogenesis in NFκβ dependent manner, leading to the cause of RA joints [75] (Figure 3).

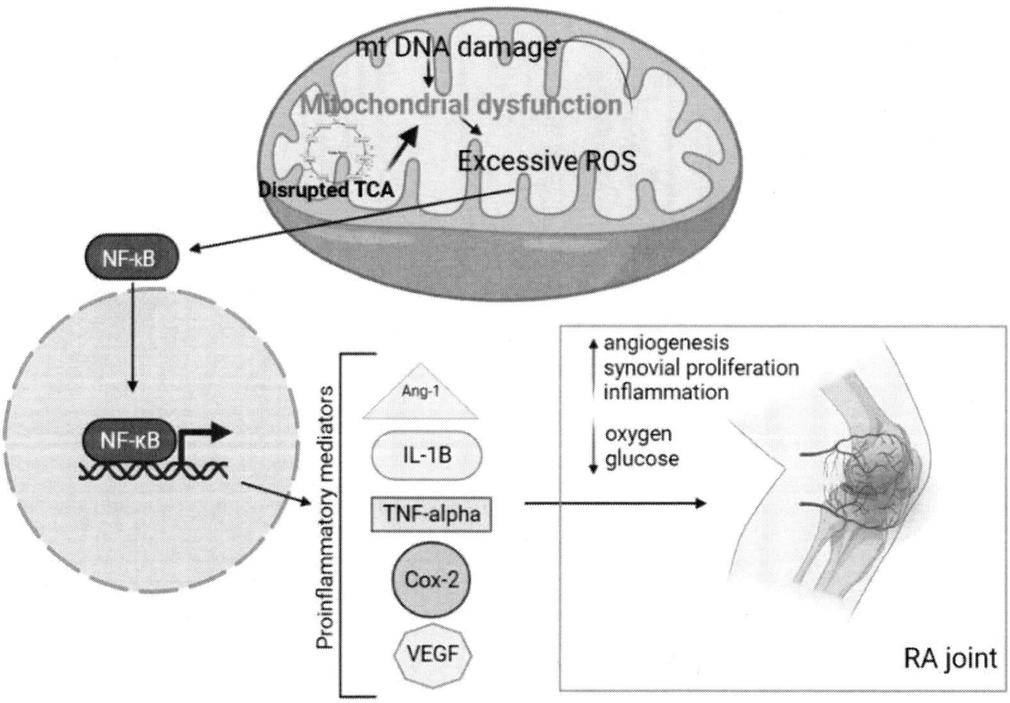


**Figure 3.** Mitochondrial ROS (mtROS) is a critical factor of rheumatoid arthritis (RA). Genetic predisposition, altered TCA cycle, obesity and many other factors can induce excessive mtROS generation, leading to the formation and release of different proinflammatory mediators in plasma in a NF-κβ mediated way, which in turn causes inflammation, synovial proliferation and angiogenesis in joints leading to RA.

## 5. Role of Mitochondrial ROS in Neurodegenerative Disorders

### 5.1. Amyotrophic Lateral Sclerosis (ALS)

Amyotrophic lateral sclerosis (ALS) is a highly progressive, neurodegenerative diseases, characterized by the severe loss of upper and lower motor neurons. Physiological characterization of ALS includes severe denegation of limbs, followed by muscle atrophy and paralysis (respiratory paralysis ultimately leads to death) [76]. There are two types of ALS found, familial ALS (fALS) (occurrence rate-10%) and sporadic ALS (sALS). More than 20 genes like superoxide dismutase 1(SOD1), transactive response DNA binding protein 43 kDa (TDP-43) [77], fused in sarcoma/transfer in liposarcoma (FUS),tubulin, alpha 4A (TUBA4A), cyclin F (CCNF) and many others [78].

Cues that trigger oxidative stress (like heavy metals or toxic agents, genetic abnormalities) cause PINK1-parkin mediated mitophagy and mitochondrial damage, which further leads to, thiol oxidation, calcium imbalance and mitochondrial membrane rupture, resulting in motor neuronal damage, causing ALS [76]. Oxidative damage can promote aggregation of TDP-43 through cysteine oxidation as well as disulphide bond formation. On the other side mutated TDP-43 can induce mtROS generation as well [79, 80]. People have shown increased expression of wild-type as well as mutant variants of TDP-43 like, mutant TDP-43 M337V and Q331K or its C-terminal fragments in NSC-34 can induce excessive ROS generation and leads to oxidative damage [81].

### 5.2. Parkinson's Disease (PD)

Parkinson's disease is a progressive, age-induced neurodegenerative disorder characterized by loss of dopaminergic neurons in a specialized domain of mid-brain, called substantianigra pars compacta (SNPC). PD is the second most occurring, incurable neurodegenerative disorder and its pathogenesis is still unknown. ROS-induced peroxidation of membrane lipids and proteins is a major cause of Parkinson's disease [82]. Lipids, more importantly polyunsaturated fatty acids (PUFA) are susceptible to ROS-induced damage [83]. The hydrogen moiety of unsaturated carbon in fatty acids are attacked by ROS leads to production of water and leaving an unpaired electron in PUFA, ultimately converted into peroxyl radicals [84] which then turned into malondialdehyde (MDA), 4-hydroxynonenal (4-HNE), and other toxic compounds [85]. 4-hydroxynonenal (4-HNE) is potent activator of caspases in dopaminergic neurons, leads to apoptosis, causes PD pathogenesis and progression [86, 87].

ROS-induced DNA damage is also believed as a major cause of PD pathogenesis. When ROS, 8-hydroxy-deoxyguanosine (8-OHdG), oxidizes DNA is produced as essential product, causes transcriptional mutagenesis, leads to the formation of mutated proteins which has contributions in PD pathogenesis [88, 89]. Damage of mtDNA by ROS could trigger aberrant mt-protein production, could produce more ROS can lead to apoptosis of dopaminergic neurons in SNPC, causing PD pathogenesis [82] (Figure 4).

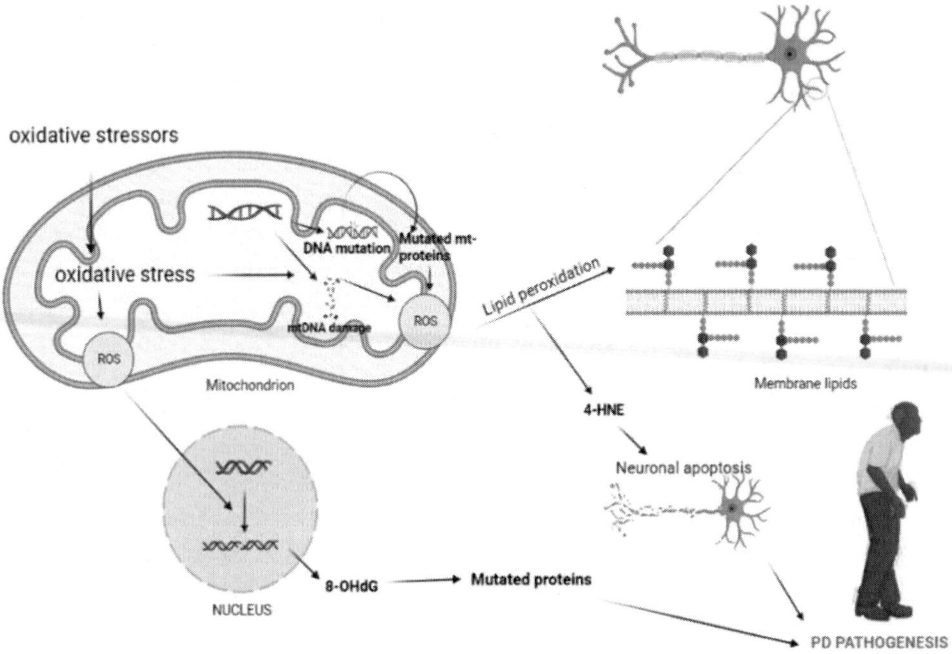


**Figure 4**. Role of mtROS in PD pathogenesis. Oxidative stressors (e.g., heavy metals) induce oxidative stress that could induce mtDNA damage and mutation, which in turn induces excessive mtROS production .mtROS can lead to membrane lipid peroxidation, which is further converted into 4-HNE can induce neuronal apoptosis leading to PD pathogenesis. ROS induce DNA damage can produce 8-OHdG, leads to mutated PD inducive protein production, and leads to PD pathogenesis.

## Conclusion

Mitochondria, the 'power house' of cell along with ATP production, play many other critical biological functions like calcium homeostasis, apoptosis, ROS production and others. Cells produce basal level of ROS which is essential for maintaining cellular homeostasis. Still, the overproduction of cellular ROS in several conditions can induce mitochondrial oxidative stress, may be a leading cause of different pathophysiological conditions like inflammatory disorders, cancer or neurogenerative disorders. But the entire scenario and molecular mechanisms of mtROS generation and its role in pathophysiology are still unknown. So, considering the importance of mtROS in different pathophysiological conditions, it is believed by different independent researchers that mtROS could be an essential drug target but needs more research.

## References

[1] Hayyan M, Hashim MA, AlNashef IM. *Superoxide Ion: Generation and Chemical Implications* [Internet]. ACS Publications. American Chemical Society; 2016 [cited 2022 Sep 14]. Available from: https://pubs.acs.org/doi/full/10.1021/acs.chemrev.5b00407.

[2] Halliwell B. Free radicals and antioxidants - quo vadis? *Trends Pharmacol Sci*. 2011 Mar;32(3): 125–30.

[3] Winterbourn CC. Are free radicals involved in thiol-based redox signaling? *Free Radic Biol Med.* 2015 Mar 1;80:164–70.

[4] Covarrubias L, Hernández-García D, Schnabel D, Salas-Vidal E, Castro-Obregón S. Function of reactive oxygen species during animal development: passive or active? *Dev Biol.* 2008 Aug 1;320(1):1–11.

[5] Sena LA, Chandel NS. Physiological roles of mitochondrial reactive oxygen species. *Mol Cell.* 2012 Oct 26;48(2):158–67.

[6] Redza-Dutordoir M, Averill-Bates DA. Activation of apoptosis signalling pathways by reactive oxygen species. *Biochim Biophys Acta BBA - Mol Cell Res.* 2016 Dec;1863(12):2977–92.

[7] Turrens JF. Mitochondrial formation of reactive oxygen species. *J Physiol.* 2003 Oct 15;552(Pt 2):335–44.

[8] Cadenas E, Davies KJ. Mitochondrial free radical generation, oxidative stress, and aging. *Free Radic Biol Med.* 2000 Aug;29(3–4):222–30.

[9] Kowaltowski AJ, de Souza-Pinto NC, Castilho RF, Vercesi AE. Mitochondria and reactive oxygen species. *Free Radic Biol Med.* 2009 Aug 15;47(4):333–43.

[10] Brand MD. The sites and topology of mitochondrial superoxide production. *Exp Gerontol.* 2010 Aug;45(7–8):466–72.

[11] Zorov DB, Juhaszova M, Sollott SJ. Mitochondrial Reactive Oxygen Species (ROS) and ROS-Induced ROS Release. *Physiol Rev.* 2014 Jul;94(3):909–50.

[12] Zhao R, Jiang S, Zhang L, Yu Z. Mitochondrial electron transport chain, ROS generation and uncoupling (Review). *Int J Mol Med* [Internet]. 2019 May 8 [cited 2022 Sep 14]; Available from: http://www.spandidos-publications.com/10.3892/ijmm.2019.4188.

[13] Osellame LD, Blacker TS, Duchen MR. Cellular and molecular mechanisms of mitochondrial function. *Best Pract Res Clin Endocrinol Metab.* 2012 Dec;26(6):711–23.

[14] Hunte C, Zickermann V, Brandt U. Functional Modules and Structural Basis of Conformational Coupling in Mitochondrial Complex I. *Science.* 2010 Jul 23;329(5990):448–51.

[15] Formosa LE, Dibley MG, Stroud DA, Ryan MT. Building a complex complex: Assembly of mitochondrial respiratory chain complex I. *Semin Cell Dev Biol.* 2018 Apr;76:154–62.

[16] Sazanov LA, Hinchliffe P. Structure of the Hydrophilic Domain of Respiratory Complex I from *Thermus thermophilus*. *Science.* 2006 Mar 10;311(5766):1430–6.

[17] Berrisford JM, Sazanov LA. Structural Basis for the Mechanism of Respiratory Complex I. *J Biol Chem.* 2009 Oct 23;284(43):29773–83.

[18] Ohnishi T. Piston drives a proton pump. *Nature.* 2010 May;465(7297):428–9.

[19] Perry TL, Godin DV, Hansen S. Parkinson's disease: a disorder due to nigral glutathione deficiency? *Neurosci Lett.* 1982 Dec 13;33(3):305–10.

[20] Maker HS, Weiss C, Silides DJ, Cohen G. Coupling of dopamine oxidation (monoamine oxidase activity) to glutathione oxidation via the generation of hydrogen peroxide in rat brain homogenates. *J Neurochem.* 1981 Feb;36(2):589–93.

[21] Zoccarato F, Toscano P, Alexandre A. Dopamine-derived dopaminochrome promotes H(2)O(2) release at mitochondrial complex I: stimulation by rotenone, control by Ca(2+), and relevance to Parkinson disease. *J Biol Chem.* 2005 Apr 22;280(16):15587–94.

[22] Abramov AY, Scorziello A, Duchen MR. Three distinct mechanisms generate oxygen free radicals in neurons and contribute to cell death during anoxia and reoxygenation. *J Neurosci Off J Soc Neurosci.* 2007 Jan 31;27(5):1129–38.

[23] Wiesner RJ, Rösen P, Grieshaber MK. Pathways of succinate formation and their contribution to improvement of cardiac function in the hypoxic rat heart. *Biochem Med Metab Biol.* 1988 Aug;40(1):19–34.

[24] Zorov DB, Filburn CR, Klotz LO, Zweier JL, Sollott SJ. Reactive Oxygen Species (Ros-Induced) Ros Release: A New Phenomenon Accompanying Induction of the Mitochondrial Permeability Transition in Cardiac Myocytes. *J Exp Med.* 2000 Oct 2;192(7):1001–14.

[25] Chaudhari T, McGuire W. Allopurinol for preventing mortality and morbidity in newborn infants with suspected hypoxic-ischaemic encephalopathy. *Cochrane Database Syst Rev* [Internet]. 2008 [cited 2022

Oct 8];(2). Available from: https://www.cochranelibrary.com/cdsr/doi/10.1002/14651858.CD006817.pub2/full.

[26] Oxypurinol administration fails to prevent hypoxic–ischemic brain injury in neonatal rats- *ScienceDirect* [Internet]. [cited 2022 Oct 8]. Available from: https://www.sciencedirect.com/science/article/abs/pii/S0361923002009632?via%3Dihub.

[27] de Jong AMPh, Kotlyar AB, Albracht SPJ. Energy-induced structural changes in NADH: Q oxidoreductase of the mitochondrial respiratory chain. *Biochim Biophys Acta BBA - Bioenerg*. 1994 Jul 29;1186(3):163–71.

[28] Cecchini G. Function and Structure of Complex II of the Respiratory Chain. *Annu Rev Biochem*. 2003;72(1):77–109.

[29] Sun F, Huo X, Zhai Y, Wang A, Xu J, Su D, Bartlam M, Rao Z. Crystal structure of mitochondrial respiratory membrane protein complex II. *Cell.* 2005 Jul 1;121(7):1043-57. doi: 10.1016/j.cell.2005.05.025.

[30] Iverson TM. Catalytic mechanisms of complex II enzymes: a structural perspective. *Biochim Biophys Acta*. 2013 May;1827(5):648–57.

[31] Bezawork-Geleta A, Rohlena J, Dong L, Pacak K, Neuzil J. Mitochondrial Complex II: At the Crossroads. *Trends Biochem Sci*. 2017 Apr 1;42(4):312–25.

[32] Benzi G, Pastoris O, Dossena M. Relationships between gamma-aminobutyrate and succinate cycles during and after cerebral ischemia. *J Neurosci Res*. 1982;7(2):193–201.

[33] Sato K, Kashiwaya Y, Keon CA, Tsuchiya N, King MT, Radda GK, Chance B, Clarke K, Veech RL. Insulin, ketone bodies, and mitochondrial energy transduction. *FASEB J.* 1995 May;9(8):651-8. doi: 10.1096/fasebj.9.8.7768357.

[34] Tannahill GM, Curtis AM, Adamik J, Palsson-McDermott EM, McGettrick AF, Goel G, Frezza C, Bernard NJ, Kelly B, Foley NH, Zheng L, Gardet A, Tong Z, Jany SS, Corr SC, Haneklaus M, Caffrey BE, Pierce K, Walmsley S, Beasley FC, Cummins E, Nizet V, Whyte M, Taylor CT, Lin H, Masters SL, Gottlieb E, Kelly VP, Clish C, Auron PE, Xavier RJ, O'Neill LA. Succinate is an inflammatory signal that induces IL-1β through HIF-1α. *Nature.* 2013 Apr 11;496(7444):238-42. doi: 10.1038/nature11986.

[35] Schägger H, Link TA, Engel WD, von Jagow G. Isolation of the eleven protein subunits of the bc1 complex from beef heart. *Methods Enzymol*. 1986;126:224–37.

[36] Yang XH, Trumpower BL. Purification of a three-subunit ubiquinol-cytochrome c oxidoreductase complex from Paracoccus denitrificans. *J Biol Chem*. 1986 Sep;261(26):12282–9.

[37] Gao X, Wen X, Esser L, Quinn B, Yu L, Yu CA, Xia D. Structural basis for the quinone reduction in the bc1 complex: a comparative analysis of crystal structures of mitochondrial cytochrome bc1 with bound substrate and inhibitors at the Qi site. *Biochemistry*. 2003 Aug 5;42(30):9067-80.

[38] Flint DH, Tuminello JF, Emptage MH. The inactivation of Fe-S cluster containing hydro-lyases by superoxide. *J Biol Chem*. 1993 Oct 25;268(30):22369–76.

[39] Liang LP, Patel M. Iron-sulfur enzyme mediated mitochondrial superoxide toxicity in experimental Parkinson's disease. *J Neurochem*. 2004 Sep;90(5):1076–84.

[40] Vasquez-Vivar J, Kalyanaraman B, Kennedy MC. Mitochondrial aconitase is a source of hydroxyl radical. An electron spin resonance investigation. *J Biol Chem*. 2000 May 12;275(19):14064–9.

[41] Minotti G, Recalcati S, Menna P, Salvatorelli E, Corna G, Cairo G. Doxorubicin cardiotoxicity and the control of iron metabolism: quinone-dependent and independent mechanisms. *Methods Enzymol*. 2004;378:340–61.

[42] Ekstedt B. Substrate specificity of the different forms of monoamine oxidase in rat liver mitochondria. *Biochem Pharmacol*. 1976 May 15;25(10):1133–8.

[43] Sandri G, Panfili E, Ernster L. Hydrogen peroxide production by monoamine oxidase in isolated rat-brain mitochondria: its effect on glutathione levels and $Ca^{2+}$ efflux. *Biochim Biophys Acta*. 1990 Sep 14;1035(3):300–5.

[44] Di Lisa F, Kaludercic N, Carpi A, Menabò R, Giorgio M. Mitochondrial pathways for ROS formation and myocardial injury: the relevance of p66(Shc) and monoamine oxidase. *Basic Res Cardiol*. 2009 Mar;104(2):131–9.

[45] Maurel A, Hernandez C, Kunduzova O, Bompart G, Cambon C, Parini A, Francés B. Age-dependent increase in hydrogen peroxide production by cardiac monoamine oxidase A in rats. *Am J Physiol Heart Circ Physiol.* 2003 Apr;284(4):H1460-7. doi: 10.1152/ajpheart.00700.2002.

[46] Stokes AH, Hastings TG, Vrana KE. Cytotoxic and genotoxic potential of dopamine. *J Neurosci Res.* 1999 Mar 15;55(6):659–65.

[47] Pinton P, Rimessi A, Marchi S, Orsini F, Migliaccio E, Giorgio M, Contursi C, Minucci S, Mantovani F, Wieckowski MR, Del Sal G, Pelicci PG, Rizzuto R. Protein kinase C beta and prolyl isomerase 1 regulate mitochondrial effects of the life-span determinant p66Shc. *Science.* 2007 Feb 2;315 (5812):659-63. doi: 10.1126/science.1135380.

[48] Giorgio M, Migliaccio E, Orsini F, Paolucci D, Moroni M, Contursi C, Pelliccia G, Luzi L, Minucci S, Marcaccio M, Pinton P, Rizzuto R, Bernardi P, Paolucci F, Pelicci PG. Electron transfer between cytochrome c and p66Shc generates reactive oxygen species that trigger mitochondrial apoptosis. *Cell.* 2005 Jul 29;122(2):221-33. doi: 10.1016/j.cell.2005.05.011.

[49] Murphy E, Ardehali H, Balaban RS, DiLisa F, Dorn GW 2nd, Kitsis RN, Otsu K, Ping P, Rizzuto R, Sack MN, Wallace D, Youle RJ; American Heart Association Council on Basic Cardiovascular Sciences, Council on Clinical Cardiology, and Council on Functional Genomics and Translational Biology. Mitochondrial Function, Biology, and Role in Disease: A Scientific Statement From the American Heart Association. *Circ Res.* 2016 Jun 10;118(12):1960-91. doi: 10.1161/RES.00000000 00000104.

[50] Youle RJ, van der Bliek AM. Mitochondrial fission, fusion, and stress. *Science.* 2012 Aug 31;337(6098):1062–5.

[51] Santulli G, Xie W, Reiken SR, Marks AR. Mitochondrial calcium overload is a key determinant in heart failure. *Proc Natl Acad Sci.* 2015 Sep 8;112(36):11389–94.

[52] Mozo J, Emre Y, Bouillaud F, Ricquier D, Criscuolo F. Thermoregulation: What Role for UCPs in Mammals and Birds? *Biosci Rep.* 2005 Jun 8;25(3–4):227–49.

[53] Bernardi P. The mitochondrial permeability transition pore: a mystery solved? *Front Physiol.* 2013;4:95.

[54] Halestrap AP. What is the mitochondrial permeability transition pore? *J Mol Cell Cardiol.* 2009 Jun;46(6):821–31.

[55] Grimm S. The ER-mitochondria interface: the social network of cell death. *Biochim Biophys Acta.* 2012 Feb;1823(2):327–34.

[56] Orrenius S, Gogvadze V, Zhivotovsky B. Calcium and mitochondria in the regulation of cell death. *Biochem Biophys Res Commun.* 2015 Apr 24;460(1):72–81.

[57] Luna-Vargas MPA, Chipuk JE. Physiological and Pharmacological Control of BAK, BAX, and Beyond. *Trends Cell Biol.* 2016 Dec;26(12):906–17.

[58] Leibowitz B, Yu J. Mitochondrial signaling in cell death via the Bcl-2 family. *Cancer Biol Ther.* 2010 Mar;9(6):417–22.

[59] Kaufmann T, Strasser A, Jost PJ. Fas death receptor signalling: roles of Bid and XIAP. *Cell Death Differ.* 2012 Jan;19(1):42–50.

[60] Glory A, Bettaieb A, Averill-Bates DA. Mild thermotolerance induced at 40 °C protects cells against hyperthermia-induced pro-apoptotic changes in Bcl-2 family proteins. Int J Hyperth Off J Eur Soc Hyperthermic *Oncol North Am Hyperth Group.* 2014 Nov;30(7):502–12.

[61] Iverson SL, Orrenius S. The cardiolipin-cytochrome c interaction and the mitochondrial regulation of apoptosis. *Arch Biochem Biophys.* 2004 Mar 1;423(1):37–46.

[62] Wiswedel I, Gardemann A, Storch A, Peter D, Schild L. Degradation of phospholipids by oxidative stress--exceptional significance of cardiolipin. *Free Radic Res.* 2010 Feb;44(2):135–45.

[63] Cook SA, Sugden PH, Clerk A. Regulation of Bcl-2 Family Proteins During Development and in Response to Oxidative Stress in Cardiac Myocytes. *Circ Res.* 1999 Nov 12;85(10):940–9.

[64] Managò A, Becker KA, Carpinteiro A, Wilker B, Soddemann M, Seitz AP, Edwards MJ, Grassmé H, Szabò I, Gulbins E. Pseudomonas aeruginosa pyocyanin induces neutrophil death via mitochondrial reactive oxygen species and mitochondrial acid sphingomyelinase. *Antioxid Redox Signal.* 2015 May 1;22(13):1097-110. doi: 10.1089/ars.2014.5979.

[65] Kumar R, Herbert PE, Warrens AN. An introduction to death receptors in apoptosis. *Int J Surg.* 2005 Jan 1;3(4):268–77.

[66] Mahmood Z, Shukla Y. Death receptors: targets for cancer therapy. *Exp Cell Res*. 2010 Apr 1;316(6):887–99.
[67] Circu ML, Aw TY. Glutathione and modulation of cell apoptosis. *Biochim Biophys Acta.* 2012 Oct;1823(10):1767–77.
[68] Blaser H, Dostert C, Mak TW, Brenner D. TNF and ROS Crosstalk in Inflammation. *Trends Cell Biol.* 2016 Apr;26(4):249–61.
[69] Russo M, Mupo A, Spagnuolo C, Russo GL. Exploring death receptor pathways as selective targets in cancer therapy. *Biochem Pharmacol.* 2010 Sep 1;80(5):674–82.
[70] Micheau O, Lens S, Gaide O, Alevizopoulos K, Tschopp J. NF-κB Signals Induce the Expression of c-FLIP. *Mol Cell Biol*. 2001 Aug 15;21(16):5299–305.
[71] Tschopp J, Irmler M, Thome M. Inhibition of fas death signals by FLIPs. *Curr Opin Immunol*. 1998 Oct;10(5):552–8.
[72] West JD, Marnett LJ. Endogenous reactive intermediates as modulators of cell signaling and cell death. *Chem Res Toxicol*. 2006 Feb;19(2):173–94.
[73] McInnes IB, Schett G. Pathogenetic insights from the treatment of rheumatoid arthritis. *Lancet Lond Engl.* 2017 Jun 10;389(10086):2328–37.
[74] Goma SH, Razek MRA, Abdelbary NM. Impact of rheumatoid arthritis on the quality of life and its relation to disease activity. *Egypt Rheumatol Rehabil*. 2019 Oct;46(4):304–12.
[75] López-Armada MJ, Fernández-Rodríguez JA, Blanco FJ. Mitochondrial Dysfunction and Oxidative Stress in Rheumatoid Arthritis. *Antioxidants* [Internet]. 2022 Jun [cited 2023 Feb 18];11(6). Available from: https://www.ncbi.nlm.nih.gov/pmc/articles/PMC9220001/.
[76] Zhao J, Wang X, Huo Z, Chen Y, Liu J, Zhao Z, Meng F, Su Q, Bao W, Zhang L, Wen S, Wang X, Liu H, Zhou S. The Impact of Mitochondrial Dysfunction in Amyotrophic Lateral Sclerosis. *Cells.* 2022 Jun 28;11(13):2049. doi: 10.3390/cells11132049.
[77] Mejzini R, Flynn LL, Pitout IL, Fletcher S, Wilton SD, Akkari PA. ALS Genetics, Mechanisms, and Therapeutics: Where Are We Now? *Front Neurosci. 2019*;13:1310.
[78] Chia R, Chiò A, Traynor BJ. Novel genes associated with amyotrophic lateral sclerosis: diagnostic and clinical implications. *Lancet Neurol.* 2018 Jan;17(1):94–102.
[79] Cohen TJ, Hwang AW, Unger T, Trojanowski JQ, Lee VMY. Redox signalling directly regulates TDP-43 via cysteine oxidation and disulphide cross-linking. *EMBO J*. 2012 Mar 7;31(5):1241–52.
[80] Cohen TJ, Hwang AW, Restrepo CR, Yuan CX, Trojanowski JQ, Lee VMY. An acetylation switch controls TDP-43 function and aggregation propensity. *Nat Commun*. 2015 Jan 5;6(1):5845.
[81] Hong K, Li Y, Duan W, Guo Y, Jiang H, Li W, Li C. Full-length TDP-43 and its C-terminal fragments activate mitophagy in NSC34 cell line. *Neurosci Lett.* 2012 Nov 21;530(2):144-9.
[82] Weng M, Xie X, Liu C, Lim KL, Zhang C wu, Li L. The Sources of Reactive Oxygen Species and Its Possible Role in the Pathogenesis of Parkinson's Disease. *Park Dis*. 2018 Sep 2;2018:1–9.
[83] Su LJ, Zhang JH, Gomez H, Murugan R, Hong X, Xu D, Jiang F, Peng ZY. Reactive Oxygen Species-Induced Lipid Peroxidation in Apoptosis, Autophagy, and Ferroptosis. *Oxid Med Cell Longev*. 2019 Oct 13;2019:5080843. doi: 10.1155/2019/5080843.
[84] Patten DA, Germain M, Kelly MA, Slack RS. Reactive oxygen species: stuck in the middle of neurodegeneration. *J Alzheimers Dis JAD*. 2010;20 Suppl 2:S357-367.
[85] Barrera G, Pizzimenti S, Daga M, Dianzani C, Arcaro A, Cetrangolo GP, Giordano G, Cucci MA, Graf M, Gentile F. Lipid Peroxidation-Derived Aldehydes, 4-Hydroxynonenal and Malondialdehyde in Aging-Related Disorders. *Antioxidants* (Basel). 2018 Jul 30;7(8):102. doi: 10.3390/antiox7080102.
[86] Shoeb M, Ansari NH, Srivastava SK, Ramana KV. 4-hydroxynonenal in the pathogenesis and progression of human diseases. *Curr Med Chem*. 2014;21(2):230–7.
[87] Li Y, Zhao T, Li J, Xia M, Li Y, Wang X, Liu C, Zheng T, Chen R, Kan D, Xie Y, Song J, Feng Y, Yu T, Sun P. Oxidative Stress and 4-hydroxy-2-nonenal (4-HNE): Implications in the Pathogenesis and Treatment of Aging-related Diseases. *J Immunol Res*. 2022 Mar 23;2022:2233906. doi: 10.1155/2022/2233906.
[88] Brégeon D, Peignon PA, Sarasin A. Transcriptional Mutagenesis Induced by 8-Oxoguanine in Mammalian Cells. *PLOS Genet*. 2009 Jul 24;5(7):e1000577.

[89] Gmitterová K, Gawinecka J, Heinemann U, Valkovič P, Zerr I. DNA versus RNA oxidation in Parkinson's disease: Which is more important? *Neurosci Lett.* 2018 Jan 1;662:22–8.

## Chapter 5

# Reactive Oxygen Species: Its Role in the Aging of the Brain and Neurodegenerative Diseases

**Anirban Manna**
**Chirantan Majumder**
**Satyajit Halder**
**and Kuladip Jana***

Division of Molecular Medicine, Centenary Campus, Bose Institute P-1/12 C.I.T. Scheme VII-M, Kolkata and Unified Academic Campus, Division of Molecular Medicine, Salt Lake City, Bidhannagar, Kolkata, India

## Abstract

Reactive Oxygen Species (ROS) are byproducts of cellular metabolism required for maintaining normal physiological functions of cells. However, ROS produced in excess can cause deleterious changes to cells, which impair their proper functioning and lead to various pathological conditions. Oxidative stress caused by excess ROS increases as we age, compromising mitochondrial functions and adversely affecting all parts of the body in general and the central nervous system in particular. Oxidative stress is thought to be one of the main factors behind cognitive aging and neurodegenerative diseases like Alzheimer's disease. In the present chapter, we have discussed the sites of cellular ROS production with emphasis on mitochondrial ROS. ROS production by different cell types and brain parts has also been outlined. Finally, the impact of neuronal oxidative stress on mitochondrial functioning, cellular DNA damage, and epigenetic modifications and their roles in the progression of aging and Alzheimer's disease have been discussed.

**Keywords:** oxidative stress, aging, mitochondria, neurodegenerative diseases

---

* Corresponding Author's Email: kuladip@jcbose.ac.in; kuladip.jana@gmail.com.

In: The Role of Reactive Oxygen Species in Health and Disease
Editor: Kuladip Jana
ISBN: 979-8-89113-283-2

## 1. Introduction

Aging is a natural phenomenon that is marked by the loss of physiological functions of a living organism with time. Modern science has increased the life expectancy of human beings; as a result, the occurrence of age-associated disorders is on the rise, placing a huge burden on our society. A detailed understanding of the aging process is therefore crucial to improving the quality of life for elderly people.

Most of our body's functions are controlled by the brain, a highly complex organ that lies at the center of our nervous system. It is broadly divided into three main parts: cerebrum, cerebellum, and the brain stem, and consists of two major types of cells: nerve cells (neurons) and glial cells. This organ has high metabolic activity owing to the functions it needs to perform, taking up one-fourth of our body's glucose and oxygen consumption. Such high oxygen consumption sets the stage for increased generation of Reactive Oxygen Species (ROS), which results in oxidative damage to the brain as we age. This is the proposition of the Free Radical Theory of Aging [1].

Mitochondria are the major source of ROS, known as 'mitochondrial ROS' in cells. ROS plays a crucial role in cellular signalling [2]. However, an alteration in mitochondrial functions which leads to an increased production of ROS, is detrimental to cells and is thought to be a major driving force behind aging. During the aging process, mitochondrial function is impaired, and there occurs a huge accumulation of damaged mitochondria and ROS in the brain. An increase in mitochondrial DNA mutations and by-products of cellular oxidative damage is observed as a consequence. Increasing oxidative damage to biomolecules causes functional impairment of cellular organelles. Accumulation of damaged mitochondria is thought to be the most important hallmark of aging [3, 4].

ROS was previously considered a single entity; its cellular level alone was considered for determining its physiological effects. Recent findings suggest ROS be an important player in cellular signalling, which helps in maintaining cellular homeostasis and aids in cellular differentiation [2]. Superoxide and hydrogen peroxide, the so-called loROS or less reactive oxygen species, are the leading players in cellular signalling. However, hiROS or highly reactive oxygen species like hydroxyl radical and peroxynitrite cause oxidative damage of biomolecules [5]. Interestingly, cells have been shown to possess enzymatic systems for scavenging loROS only: Superoxide Dismutases (SOD) scavenge superoxide; peroxiredoxins and glutathione peroxidases deal with hydrogen peroxide. It has been suggested that hiROS is most probably generated from loROS by Fenton and Haber-Weiss reactions, and cells, instead of evolving pathways for scavenging hiROS can efficiently prevent their formation by using enzyme systems to take care of low ROS [6].

## 2. ROS Production in the Mitochondria

Three sites in the mitochondrial Electron Transport Chain (ETC) are reportedly involved in ROS generation in vivo. These are respiratory Complex I (NADH: ubiquinone oxidoreductase), Complex II (succinate dehydrogenase), and Complex III (ubiquinol: cytochrome c reductase). Complex IV (cytochrome c oxidase), which carries out the final reduction step of oxygen to water, has been shown not to produce ROS.

ROS produced by CII has been implicated in certain types of cancer [7]. On the other hand, regulating various physiological and pathological processes, e.g., cellular differentiation, ischemia/reperfusion damage, etc., are thought to be regulated by the ROS generated from CI and CII [2, 8]. CI-generated ROS is directed at the mitochondrial matrix, and CIII-generated ROS is received partly by the matrix and partly by the inter-membrane space [9]. This suggests the reason for the different downstream effects of the ROS produced by these two sites.

CI-generated ROS are known to be involved in the following: i) differentiation of myocytes to myotubes, ii) oxygen sensing by cells of the carotid body, iii) organization of ETC, iv) ischemia/reperfusion damage, v) metabolic reprogramming of macrophages upon bacterial infection, vi) regulation of lifespan [10]. CI has been reported to generate ROS both in forward and reverse directions and all the above-mentioned functions can be performed by ROS produced through Reverse Electron Transport (RET) [10]. Over-reduction of Coenzyme Q (CoQ) with electrons and the consequent back flow of electrons to CI from ubiquinol (reduced form of CoQ) is referred to as RET [11]. A considerable amount of ROS is produced through RET, which is involved in communicating information between mitochondria and the rest of the cell. Possibly, RET conveys information about ATP synthesis and electron flow in mitochondria to the rest of the cell.

CIII-generated ROS has been reported to be involved in the hypoxia response and human adipocyte differentiation [12].

## 3. Extra-mitochondrial ROS Generation

Certain extra mitochondrial enzymes, namely NADPH oxidases (NOXs), xanthine oxidase, monoamine oxidases, and peroxisomal enzymes, are able to generate ROS. NOXs are the only family of enzymes specifically dedicated to ROS (superoxide/hydrogen peroxide) generation [13]. They are involved in various signalling pathways pertaining to growth, inflammation, aging, and related disorders [14]. Changes in cellular ROS levels can occur by continuous crosstalk between mitochondrial ROS and NOXs [15].

Xanthine oxidase is found in the cytosol, and monoamine oxidase is present on the outer mitochondrial membrane facing the cytosol. These have been reported to increase ROS production during aging [16, 17, 18]. A correlation between enhanced ROS production by MAO and neuronal death in the *substantia nigra* has been reported [19].

Certain cellular organelles apart from mitochondria, such as peroxisomes and lysosomes, can also give rise to significant levels of ROS in cells. Peroxisomes, involved in generating and detoxifying hydrogen peroxide in cells, may cause induction of cellular senescence when defective redox homeostasis in this organelle leads to an enhanced level of hydrogen peroxide [20, 21]. Lysosomes, involved in the degradation of non-functional proteins and organelles, may sometimes release their contents into the cell, such as free metals (like iron), resulting in oxidative damage to cellular components. Undigested lysosomal stuff accumulates in the cell with increasing age in the form of lipofuscin [22].

## 4. Mitochondrial ROS Generation in Neurons and Glial Cells

A variation in the shape, chemical composition, and level of ATP generation is observed among mitochondria depending on the organ's function in which it is found [23, 24]. Heart mitochondria, with its more significant need for ATP generation to power the pumping of the heart, have a biomolecular composition that is quite distinct from the liver mitochondria. As a result, they show different levels of ROS generation, too [23, 25]. Mitochondrial diversity is also observed inside a cell, depending on the location and function of the mitochondria within that cell. In the case of highly polarized cells like neurons, synaptic and nonsynaptic mitochondria differ significantly in their biomolecular composition, rate of respiration, and level of ROS production [26].

Neurons and glia use different pathways for energy generation, the former being dependent on oxidative phosphorylation and the latter on glycolysis. Glia utilizes glucose to produce lactate which is then transferred to the neurons and other metabolites. Neuronal mitochondria oxidize these to produce energy [27]. Hence, mitochondria in these two kinds of cells produce different amounts of ROS. In rats and mice, complex I have been shown to generate more ROS in glia compared to their neuronal counterparts. Due to a limitation of complex III in glia, complex I is less incorporated in respiratory super-complexes. Glia probably generates more ROS due to an increase in free complex I. Modulation of the levels of the complex I subunit named NADH-ubiquinone oxidoreductase 75 kDa subunit, mitochondrial (NDUFS1) by knockdown/overexpression caused changes in mitochondrial ROS levels in neurons and glia. When this subunit was overexpressed in glia, there was less generation of ROS due to the incorporation of complex I in the respiratory super-complexes. On the other hand, the knockdown of this subunit in neurons caused an increase in the generation of mitochondrial ROS [28].

A change in the levels of various mitochondrial substrates like pyruvate, fatty acids, amino acids, etc. has been shown to alter the incorporation of respiratory complexes into super-complexes. The respirasome (a super-complex formed by the association of complex I with complex III and -IV) has been shown to have complex I as an important part when mitochondria use pyruvate as the substrate. This complex I is, however, released from the respirasome when the mitochondria switch to fatty acid as the substrate [29], or glucose oxidation occurs via anaerobic glycolysis [28]. Studies suggest that incorporating Complex I in the respiratory complexes causes a decrease in the ROS levels [28, 30, 31]. Disorganization of respiratory super-complexes causes an increase in electron leak and a consequent enhancement of ROS generation, which accelerates the aging process [32]. Indeed, aged rat brains have been found to possess disorganized super-complexes, clearly explaining their excess production of mitochondrial ROS [33, 34].

## 5. ROS Generation in Various Parts of the Brain

The brain, with its greater oxygen consumption, presence of large amounts of polyunsaturated fatty acids (that can be easily oxidized), and concentration of transition metals like iron and copper, have a greater susceptibility to damage by oxidative stress. Certain areas of the brain are more prone to oxidative damage than others due to the presence of various types of cells and their subtypes [35]. Unlike glia, neurons produce a lesser amount of mitochondrial ROS, yet they are more susceptible to oxidative damage. This can be attributed to the presence of fewer antioxidants in neurons than in glia [36]; and to the fact that as neurons are post-mitotic cells, once damaged, they are rarely replenished. Differences in topology, myelination, and the number of synaptic connections among neurons are key factors that determine their sensitivity to ROS levels, a phenomenon called selective neuronal vulnerability [37]. Neurons having longer axons have greater susceptibility to oxidative stress [37]. Also, neurons which generate more ROS are more prone to oxidative damage [38]. Areas such as hippocampus, substantianigra, amygdala, and frontal cortex are more sensitive to ROS-mediated damage. Within these regions, certain neurons, such as the hippocampal CA1 neurons and those in the substantianigra pars compacta, have greater sensitivity to oxidative stress [37]. ROS levels in the hippocampus and striatum are significantly increased during severe hypoglycaemia [39], and the highest levels of protein carbonyls are found in these two regions during aging [40]. The hippocampus is the brain region most susceptible to ischemia-reperfusion damage and exhibits the least oxygen consumption during aging [41, 42].

The above discussion indicates that some areas of the brain can be more affected by ROS-mediated neuronal and glial cell death than others. Changes in redox signalling can damage some areas and cells in the brain; on the other hand, there are areas that suffer damage due to increased oxidative stress. Surprisingly, age-related enhancement of ROS production can play a protective role by activating quality control mechanisms via 'mitohormesis'. Studies in certain lower organisms contradict the general notion that an increase in ROS generation brings about aging. In worms and flies, several studies show that enhancement of ROS levels can actually extend lifespan, a phenomenon known as mitohormesis [10]. Low to moderate levels of ROS induction can, in fact, activate protective and repair mechanisms which can manage oxidative stress in the long run [43]. Indeed, overexpression and feeding of antioxidants have been shown to block the beneficial effects of lifespan extension via mitohormesis [44, 45, 46].

Taken together, it can be said that ROS playing a detrimental role in some parts of the brain and a protective role in other parts could be the possible reason for the ineffectiveness of antioxidant therapies in extending lifespan and offering protection against age-related disorders [47].

## 6. Origin of Oxidative Stress

In neuronal cells, the mitochondrial electron transport chain gives rise to most free radicals as byproducts. For example, decreased electron transport caused by an excess of ATP may convert oxygen to superoxide radicals [48]. Superoxide is converted to hydrogen peroxide by the antioxidant enzyme Superoxide Dismutase (SOD). Hydrogen peroxide is subsequently broken down into hydroxyl radicals and anions by the Fenton reaction [49]. Redox imbalance, neurotoxicity, genomic instability, the transcription of proinflammatory genes, and the release of cytokines like IL-1, IL-6, and TNF-α are the hallmarks of mitochondrial ROS generation [50]. The generated ROS can damage parts of the mitochondrial electron transport chain which then leads to increased ROS production by reduction of oxygen to superoxide radical, thereby creating a positive feedback loop [50, 48, 51]. In particular, mitochondrial DNA being not bound to histones, is more susceptible to oxidative damage and mutates at a higher rate. These mitochondria with damaged DNA are removed by mitophagy to ensure the proper functioning of the neuronal cells [50, 52]. Oxidative stress has been shown to play a dual role in mitophagy: it decreases mitochondrial membrane potential and causes induction of mitophagy on one hand; it also modifies and inhibits key mitophagy regulators on the other hand (e.g., S-nitrosylation and consequent inactivation of mitophagy regulator Parkin, which is associated with Parkinson's disease) [52].

A theory of aging has been proposed encompassing mitochondrial ROS and DNA damage which runs as follows. DNA damage can cause activation of kinases and Poly (ADP-ribose) Polymerase [PARP], which leads to depletion of NAD+ levels. This leads to an increased need for oxygen consumption and ATP generation. Mitochondrial coupling occurs to meet such high energy demands, which increases mitochondrial membrane potential and impairs mitophagy-mediated removal of damaged mitochondria. Further ROS is generated by these mitochondria which lead to even more DNA damage [53, 54]. Thus, mitochondria are thought to play a key role in aging and associated neurodegenerative diseases (Figure 1).

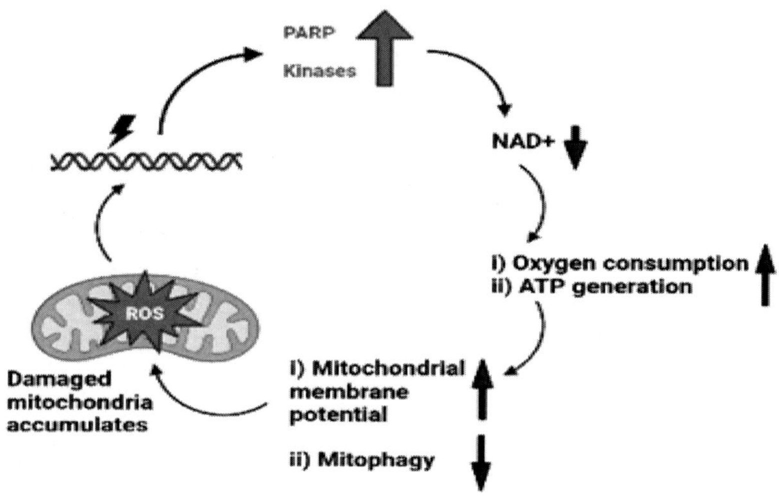


**Figure 1.** A theory of aging: mitochondrial ROS and DNA damage.

## 7. Mitochondrial ROS and the Aging Brain

One of the hallmarks of the normal aging process is mitochondrial dysfunction. There is a marked reduction in the α subunit of the mitochondrial F1 ATP synthase that couples oxidative phosphorylation and ATP synthesis, which results in decreased ATP synthesis and increased oxidative stress, causing damage to cellular macromolecules [55, 56, 57]. Impaired mitochondrial metabolism leads to damage of nuclear DNA as well, in addition to mtDNA. The promoters of certain genes which are downregulated with age are particularly affected. These genes encode proteins that are involved in various cellular processes, such as synaptic plasticity and vesicular and mitochondrial functions [56]. Overexpression of antioxidants could be beneficial in reversing the deleterious effects of ROS in aging. Indeed, the lifespan of mice has been shown to be considerably extended upon overexpression of mitochondrial catalase (mCAT). An overall decrease in oxidative stress, proper insulin signalling, and delayed onset of cardiac and cataract complications was observed in mCAT mice [58, 59, 60]. Mitochondrial catalase prevented the damage of both mitochondrial and nuclear DNA, as evident from a significant reduction of DNA damage in aged skeletal muscles and mitochondrial DNA from asbestos-exposed lungs in mCAT mice [61, 62, 60].

## 8. Role of Mitochondrial ROS in Alzheimer's Disease

Oxidation of mitochondrial DNA is one of the first biomarkers of Alzheimer's disease [55, 57]. Deterioration of mitochondrial functioning with age may lead to sporadic, late-onset AD. Impaired mitochondrial functioning adversely affects the expression and processing of Amyloid Precursor Protein (APP), leading to the formation of amyloid beta oligomers that accumulate into senile plaques [63, 64].

Amyloid beta itself acts as a source of oxidative stress. Its hydrophobic 25-35 region is responsible for generating ROS and causes neuronal toxicity [65, 66]. Neurons incubated with amyloid beta 1-42 oligomers showed signs of lipid peroxidation as exhibited by 4-hydroxy-2 transnonenal (HNE) bound to proteins [67]. HNE hampers the functioning of glucose and glutamate transporters, $Na^+$-$K^+$ ATPase, and decreases mitochondrial transmembrane potential [68]. Amyloid beta has been shown to form complexes with redox-active metals like copper, iron, and zinc, which promotes plaque formation. The copper amyloid beta complex is the most stable and it generates superoxide and hydrogen peroxide [64, 69].

A high level of amyloid beta peptide leads to S-nitrosylation and hyperactivation of DRP1 (Dynamin-related protein 1), a mitochondrial fission protein. This causes excessive mitochondrial breakdown and synaptic damage in AD [70].

Removal of damaged mitochondria might be a good strategy to combat oxidative stress in AD. Khandelwal et al. (2011) have reported a decrease in amyloid beta accumulation and damaged mitochondria and restoration of neurotransmitter biosynthesis upon overexpression of Parkin (an E3 ubiquitin ligase) in 3x-Tg mice (a mouse model of AD) [71]. Overexpression of PARK2 in cultured fibroblasts from spontaneous AD patients caused an increase in the number of autophagic vacuoles, mitochondrial targeting, and mitochondrial recycling [72].

## 9. Oxidative Stress and DNA Damage

Oxidative stress has been shown to cause DNA damage, which is characterized by apurinic and apyrimidinic sites on DNA, oxidation of purine and pyrimidine bases, and single-stranded and double-stranded DNA breaks. ROS causes an oxidation-mediated breakdown of the sugar-phosphate backbone of DNA, giving rise to Single-Stranded DNA Breaks (SSBs). Double-stranded DNA Breaks (DSBs) are reported to be more toxic because they last longer and cause deleterious changes in the transcription of promoters near the DSB sites [73, 74].

There is a relation between NAD+ levels and the capability of a cell's DNA repair. Higher NAD+ levels in a cell is associated with good DNA repair capacity, while lower levels of NAD+ implies excess ROS. NAD+ has been reported to activate SIRT1, which in turn reduces oxidative stress and its consequent DNA damage by increasing mitophagy, autophagy and generation of protective methylation of histones [75, 76]. NMN (nicotinamide mononucleotide), an intermediate of NAD+, has also been shown to activate SIRT1 and thereby increase NAD+ and Mn-SOD (manganese Superoxide Dismutase) levels [77]. It is worth mentioning here that compared to other cell types, neuronal cells are more likely to sustain DNA damage. Also, mitochondrial DNA is more prone to damage than nuclear DNA, as it is not bound with protective histones and lies very close to the Electron Transport Chain [57].

The very first step in the repair of DNA double-strand breaks is the phosphorylation of H2AX; it is a well-known marker of DSB. Quite surprisingly, H3K9me3, an epigenetic modification associated with age-related cognitive decline, has also been shown to recruit the DSB repair complex and cause a reduction in DNA damage [78, 79]. In spite of their deleterious effects, DSBs are reported to play an essential role in the expression of critical Immediate Early Genes (IEGs). Stimulation of neurons can cause DSBs [80], especially in the promoter regions of IEGs, thereby inducing their expression even in the absence of external stimuli [81]. Gadd45γ (Growth arrest and DNA damage 45γ), a protein associated with DNA repair, has been shown to regulate the expression of certain IEGs like Arc, cFOS, Npas4 and Cyr61 in the prelimbic prefrontal cortex of adult male C57BL/6 mice [82].

## 10. DNA Damage in the Aging Brain

An increase in DNA damage is observed both in healthy aged brains and the brains of AD patients. Possibly, diminished repair of damaged DNA could be a major factor that brings about aging [53, 74]. Downregulation of the promoters of certain cortical genes involved in synaptic plasticity, vesicular transport, and mitochondrial functions has been observed as age increases. This starts after 40 years of age and reaches maximum levels at 70. It shows the direct impact of DNA damage on neuronal functions. Brain aging has been shown to upregulate certain genes, which can deal with age associated deleterious changes. Among them are genes like (i) heat shock protein 70 and α-crystallin (which help in protein folding); (ii) non-selenium glutathione peroxidase, paraoxonase, and selenoprotein P (involved in antioxidant defense), and (iii) genes involved in metal ion homeostasis. Two base excision repair enzymes: 8-oxoguanine DNA glycosylase and uracil DNA glycosylase, are upregulated, which indicates both increased ssDNA damage and neuroprotection [83].

In order to find out whether promoters of age-downregulated genes have defective DNA repair, promoter-reporter constructs were exposed to hydrogen peroxide, followed by transfection in SH-SY5Y cell line. Upon in vitro oxidative damage by hydrogen peroxide, decreased activation rates of reporters from promoters of age-downregulated genes were observed compared to that from age-stable genes. This shows an increased incidence of ssDNA breaks and reduced base excision repair in age-downregulated genes [83].

In hippocampal pyramidal and granule cells and cerebellar granule cells of aged brains, an increased incidence of single-strand DNA breaks has been reported; however, such observation has not been made in cerebellar Purkinje cells [84]. An increase in DNA damage with age has been reported to be 10-fold higher in mitochondrial DNA compared to nuclear DNA in subjects of 42 years and above. This is 15-fold higher in subjects above 70 years of age [57]. An increased incidence of age-related single-strand DNA breaks adversely affects mitochondrial DNA, causing impairment of metabolism, energetics, and neuronal survival; functions such as synaptic plasticity, learning, and memory are also diminished.

## 11. Role of DNA Damage in Alzheimer's Disease

Impairment in the base excision repair pathway has been reported in AD brains, similar to that found in normal-aged brains [85, 81]. Recruitment of 53BP1 to double-strand breaks is decreased. There is also downregulation of DSB repair proteins like DNA protein kinase catalytic subunits and MRN complex proteins [81]. There is upregulation of γH2AX levels in the neurons and astrocytes of AD and Mild Cognitive Impairment (MCI) patients, compared to control individuals of the same age [86,73]. Interestingly, DSB can occur due to the physiological activity of the brain in mice; however, this becomes aggravated in transgenic mouse models of AD with higher amyloid beta levels. Mice transgenic for human amyloid precursor protein (hAPP) exhibited enhanced levels of DNA double-strand breaks compared to age-matched controls prior to exploration of a novel environment. After exploration, the DSBs became more long-lasting and severe [87]. DNA damage probably occurs early during AD progression because increased levels of nitrotyrosine and DNA damage are followed by the formation of neurofibrillary tangles in the visual cortex of AD brains [88]. Diminished repair of DNA is a hallmark of AD, as increased levels of 8-hydroxy-2'deoxy guanosine (8-OHdG) in intact DNA (a biomarker of oxidative DNA damage) has been found in the CSF of AD patients. Along with this, the levels of free 8-OHdG, a marker of DNA repair, are low [85]. In a study, Sykora et al. 2014 crossed mice that were haploin sufficient for the base excision repair polymerase Pol B with 3XTg mice. The progeny mice exhibited increased cognitive decline, loss of synapse, and neuronal death compared to the parent 3XTg animals [89].

Thus, enhanced oxidative stress hampers DNA repair mechanisms in AD. The severity of DNA damage has been found to be greater in AD brains compared to healthy aged brains, and this is attributed to both DNA damage and defective DNA repair mechanisms.

## 12. ROS Induced Epigenetic Modifications

Oxidative stress has been shown to have a direct impact on epigenetic changes. Epigenetic modifications can alter gene expression patterns. This can have an influence on aging and the progression of various diseases, including neurodegenerative diseases [90]. ROS causes oxidation of DNA and hence reduces DNA methylation. ROS can also decrease DNA methylation by increasing TET (Ten Eleven Translocation family of enzymes)-mediated hydroxymethylation of DNA and by compromising the binding of DNA methyltransferases [91, 74]. ROS-mediated hydroxylation of pyrimidines and 5-methyl cytosine can form DNA lesions; this interferes with 5-hydroxymethylcytosine epigenetic signals [54, 92]. ROS can decrease the levels of metabolites like acetyl CoA, $NAD^+$, and ketoglutarate, which are essential for the functioning of histone modifying enzymes, and thus changes post-translational modifications of histones which impacts chromatin structure, gene expression patterns, gene stability, and DNA replication [54]. In bronchial epithelial cells, hydrogen peroxide treatment caused an increase in methylation of lysine residues 4, 9, and 27 of histone H3. It also decreased acetylation of lysine 9 of H3 and lysine 8 of H4 histones. Pretreatment of the cells with the antioxidant L-ascorbate prevented these changes. The ROS-induced methylation changes disappeared after the washout of $H_2O_2$ [93]. Gu et al. (2013) treated the human neuroblastoma cell line SH-SY5Y with hydrogen peroxide, which appreciably increased the intracellular level of amyloid β. Reduction of methyltransferases and demethylation of the gene promoters caused upregulation of APP and BACE1 (β-site Amyloid β Precursor Protein Cleaving Enzyme 1) gene expression, which was responsible for amyloid β peptide overproduction. Hydrogen peroxide was also shown to cause the upregulation of histone acetyltransferase (p300/cAMP Response Element Binding Protein) and downregulation of histone deacetylases [94].

Snigdha et al. (2016) have reported that there is an upregulation in the levels of methylation of histone H3 at Lys9 position (H3K9me3) in the hippocampi of aged brains. When the H3K9me3 levels were pharmacologically decreased by inhibition of its catalyzing enzyme SUV39H1 by the inhibitor ETP69, an improvement in object location memory, and an increase in BDNF levels and synaptic density were observed in aged mice [79]. Further, oxidative stress upregulates SIRT1 and this indirectly causes an upregulation of global H3K9me3 levels. This offers protection against oxidative stress by an increase in mitophagy and autophagy [76]. Upregulation of SIRT1 leads to upregulation and stabilization of SUV39H1 enzyme, which in turn increases the levels of H3K9me3 [95, 96]. Another research conducted by Grinan-Ferre et al. (2019) has revealed the benefits of inhibition of G9a, a histone methyltransferase, by the small molecule UNC0642 in 5X-FAD mice (a mouse model of AD). G9a inhibition resulted in improvement in the memory of objects and locations, a decrease in 5methyl cytosine and amyloid β plaque levels, along with an increased antioxidant level in the said mouse model [97].

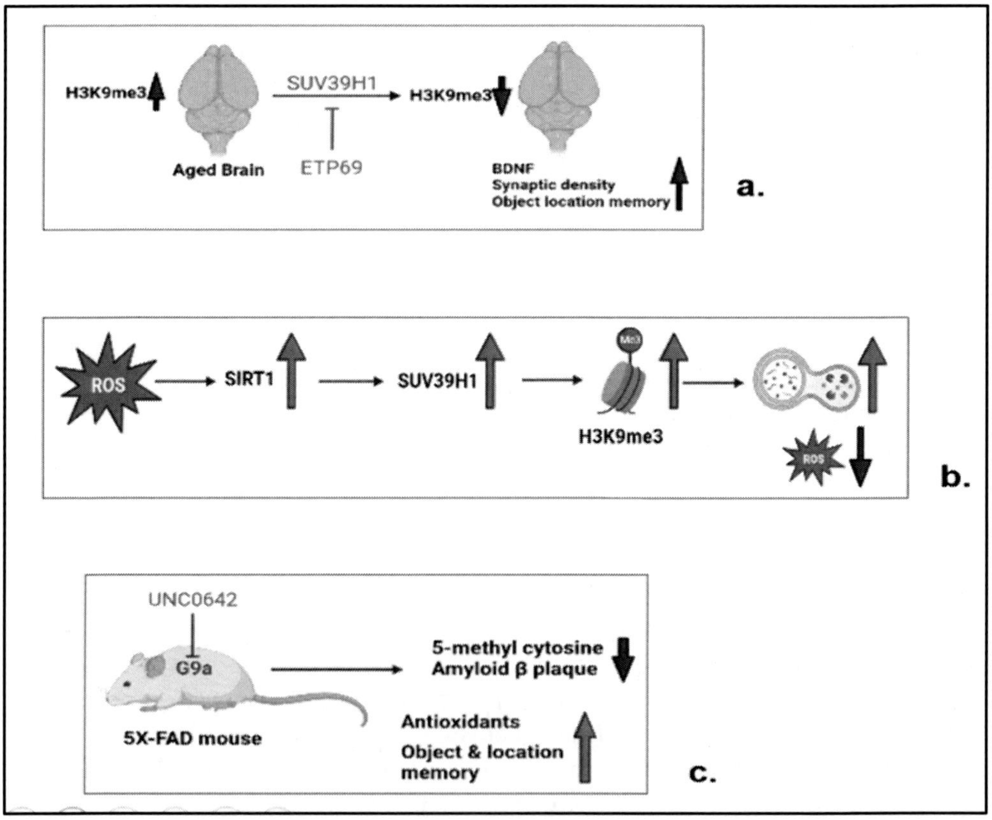


**Figure 2.** Oxidative Stress (OS) and epigenetic modifications.

Taken together, it can be said that oxidative stress increases H3K9 methylation, and pharmacological inhibition of this helps in maintaining cognitive abilities and reduces AD pathology in mice models (Figure 2).

a) Reduction in H3K9me3 levels in the hippocampi of aged mice brain by small molecule mediated inhibition of the enzyme SUV39H1 increases BDNF levels and synaptic density and also improves object location memory.
b) OS-mediated upregulation of SIRT1 upregulates and stabilizes SUV39H1, which in turn increases H3K9me3 levels. As a result, autophagy increases, thereby decreasing OS.
c) In 5X-FAD mice, small molecule mediated inhibition of the histone methyltransferase G9a decreases 5mC and amyloid β plaque levels and concomitantly increases antioxidant levels. Object and location memory is also improved.

## Conclusion

The primary role of oxidative stress in the progression of brain aging and Alzheimer's disease has been discussed in the present chapter. Mitochondria produce ROS as a byproduct of cellular respiration in healthy cells. This ROS plays an important role in cellular signalling. Mitochondrial functions are known to become compromised in aged individuals and AD patients. Accumulation of damaged mitochondria raises the production of amyloid β, which itself causes oxidative stress to the cells. Oxidative stress and DNA damage can be minimized in aged mice by overexpression of mitochondrial antioxidants. Also, in AD models, reinstatement of mitophagy shows a positive effect. ROS can cause damage to cellular DNA, particularly mtDNA, as it is histone-free and lies close to the ETC. A marked increase in double-stranded DNA breaks found in AD brains compared to normal-aged brains, and improper DNA repair machinery exacerbates the disease condition. Oxidative stress-induced epigenetic modifications down-regulate the expression of cognitive genes and activate the APP promoter, thereby worsening the disease pathology. Oxidative stress is a major contributing factor to the progression of brain aging and neurodegenerative diseases, although it may not be the only factor.

## References

[1] Barja G. Updating the mitochondrial free radical theory of aging: an integrated view, key aspects, and confounding concepts. *Antioxidants & Redox Signalling*. 2013 Oct 20; 19(12):1420-45.

[2] Reczek CR, Chandel NS. ROS-dependent signal transduction. *Current Opinion in Cell Biology*. 2015 Apr 1; 33:8-13.

[3] López-Otín C, Blasco MA, Partridge L, Serrano M, Kroemer G. The hallmarks of aging. *Cell*. 2013 Jun 6;153(6):1194-217.

[4] Chakrabarti S, Munshi S, Banerjee K, Thakurta IG, Sinha M, Bagh MB. Mitochondrial dysfunction during brain aging: role of oxidative stress and modulation by antioxidant supplementation. *Aging and Disease*. 2011 Jun; 2(3):242.

[5] Scialo F, Mallikarjun V, Stefanatos R, Sanz A. Regulation of lifespan by the mitochondrial electron transport chain: reactive oxygen species-dependent and reactive oxygen species-independent mechanisms. *Antioxidants & Redox Signalling.* 2013 Dec 1;19(16):1953-69.

[6] Stefanatos R, Sanz A. The role of mitochondrial ROS in the aging brain. *FEBS Letters.* 2018 Mar;592(5):743-58.

[7] Ishii T, Yasuda K, Akatsuka A, Hino O, Hartman PS, Ishii N. A mutation in the SDHC gene of complex II increases oxidative stress, resulting in apoptosis and tumorigenesis. *Cancer Research.* 2005 Jan 1; 65(1):203-9.

[8] Scialò F, Fernández-Ayala DJ, Sanz A. Role of mitochondrial reverse electron transport in ROS signalling: potential roles in health and disease. *Frontiers in Physiology.* 2017 Jun 27; 8:428.

[9] Muller FL, Liu Y, Van Remmen H. Complex III releases superoxide to both sides of the inner mitochondrial membrane. *Journal of Biological Chemistry.* 2004 Nov 19;279(47):49064-73.

[10] Sanz A. Mitochondrial reactive oxygen species: do they extend or shorten animal lifespan?. *Biochimica et Biophysica Acta (BBA)-Bioenergetics.* 2016 Aug 1; 1857(8):1116-26.

[11] Chouchani ET, Pell VR, James AM, Work LM, Saeb-Parsy K, Frezza C, Krieg T, Murphy MP. A unifying mechanism for mitochondrial superoxide production during ischemia-reperfusion injury. *Cell Metabolism.* 2016 Feb 9; 23(2):254-63.

[12] Sena LA, Chandel NS. Physiological roles of mitochondrial reactive oxygen species. *Molecular Cell.* 2012 Oct 26; 48(2):158-67.

[13] Drummond GR, Selemidis S, Griendling KK, Sobey CG. Combating oxidative stress in vascular disease: NADPH oxidases as therapeutic targets. *Nature Reviews Drug Discovery.* 2011 Jun; 10(6):453-71.

[14] Krause KH. Aging: a revisited theory based on free radicals generated by NOX family NADPH oxidases. *Experimental Gerontology.* 2007 Apr 1; 42(4):256-62.

[15] Dikalov S. Cross talk between mitochondria and NADPH oxidases. *Free Radical Biology and Medicine.* 2011 Oct 1; 51(7):1289-301.

[16] Vida C, Rodríguez-Terés S, Heras V, Corpas I, De la Fuente M, González E. The aged-related increase in xanthine oxidase expression and activity in several tissues from mice is not shown in long-lived animals. *Biogerontology. 2011* Dec; 12:551-64.

[17] Vida C, Corpas I, De la Fuente M, González EM. Age-related changes in xanthine oxidase activity and lipid peroxidation, as well as in the correlation between both parameters, in plasma and several organs from female mice. *Journal of Physiology and Biochemistry.* 2011 Dec; 67:551-8.

[18] Alper G, Girgin FK, Ozgönül M, Menteş G, Ersöz B. MAO inhibitors and oxidant stress in aging brain tissue. *European neuropsychopharmacology.* 1999 Mar 1; 9(3):247-52.

[19] Mallajosyula JK, Kaur D, Chinta SJ, Rajagopalan S, Rane A, Nicholls DG, Di Monte DA, Macarthur H, Andersen JK. MAO-B elevation in mouse brain astrocytes results in Parkinson's pathology. *PloS one.* 2008 Feb 20; 3(2):e1616.

[20] Titorenko VI, Terlecky SR. Peroxisome metabolism and cellular aging. *Traffic.* 2011 Mar; 12(3):252-9.

[21] Legakis JE, Koepke JI, Jedeszko C, Barlaskar F, Terlecky LJ, Edwards HJ, Walton PA, Terlecky SR. Peroxisome senescence in human fibroblasts. Molecular Biology of the *Cell.* 2002 Dec 1; 13(12):4243-55.

[22] Terman A, Kurz T, Navratil M, Arriaga EA, Brunk UT. Mitochondrial turnover and aging of long-lived postmitotic cells: the mitochondrial–lysosomal axis theory of aging. *Antioxidants & Redox Signalling.* 2010 Feb 15; 12(4):503-35.

[23] Forner F, Foster LJ, Campanaro S, Valle G, Mann M. Quantitative proteomic comparison of rat mitochondria from muscle, heart, and liver. *Molecular & Cellular Proteomics.* 2006 Apr 1; 5(4):608-19.

[24] Kuznetsov AV, Margreiter R. Heterogeneity of mitochondria and mitochondrial function within cells as another level of mitochondrial complexity. *International Journal of Molecular Sciences.* 2009 Apr 24; 10(4):1911-29.

[25] Sanz A, Caro P, Ayala V, Portero-Otin M, Pamplona R, Barja G. Methionine restriction decreases mitochondrial oxygen radical generation and leak as well as oxidative damage to mitochondrial DNA and proteins. *The FASEB Journal.* 2006 Jun; 20(8):1064-73.

[26] Stauch KL, Purnell PR, Fox HS. Quantitative proteomics of synaptic and nonsynaptic mitochondria: insights for synaptic mitochondrial vulnerability. *Journal of Proteome Research.* 2014 May 2; 13(5):2620-36.

[27] Bolaños JP, Almeida A, Moncada S. Glycolysis: a bioenergetic or a survival pathway?. *Trends in Biochemical Sciences.* 2010 Mar 1; 35(3):145-9.

[28] Lopez-Fabuel I, Le Douce J, Logan A, James AM, Bonvento G, Murphy MP, Almeida A, Bolaños JP. Complex I assembly into supercomplexes determines differential mitochondrial ROS production in neurons and astrocytes. *Proceedings of the National Academy of Sciences.* 2016 Nov 15; 113(46):13063-8.

[29] Lapuente-Brun E, Moreno-Loshuertos R, Acín-Pérez R, Latorre-Pellicer A, Colás C, Balsa E, Perales-Clemente E, Quirós PM, Calvo E, Rodríguez-Hernández MA, Navas P. Supercomplex assembly determines electron flux in the mitochondrial electron transport chain. *Science.* 2013 Jun 28; 340(6140):1567-70.

[30] Maranzana E, Barbero G, Falasca AI, Lenaz G, Genova ML. Mitochondrial respiratory supercomplex association limits production of reactive oxygen species from complex I. *Antioxidants &Redox Signalling.* 2013 Nov 1; 19(13):1469-80.

[31] Lenaz G, Baracca A, Barbero G, Bergamini C, Dalmonte ME, Del Sole M, Faccioli M, Falasca A, Fato R, Genova ML, Sgarbi G. Mitochondrial respiratory chain super-complex I–III in physiology and pathology. *Biochimicaet Biophysica Acta (BBA)-Bioenergetics.* 2010 Jun 1; 1797(6-7):633-40.

[32] Genova ML, Lenaz G. The interplay between respiratory supercomplexes and ROS in aging. *Antioxidants & Redox Signalling.* 2015 Jul 20; 23(3):208-38.

[33] Frenzel M, Rommelspacher H, Sugawa MD, Dencher NA. Ageing alters the supramolecular architecture of OxPhos complexes in rat brain cortex. *Experimental Gerontology.* 2010 Aug 1; 45(7-8):563-72.

[34] Sanz A, Caro P, Ibañez J, Gómez J, Gredilla R, Barja G. Dietary restriction at old age lowers mitochondrial oxygen radical production and leak at complex I and oxidative DNA damage in rat brain. *Journal of Bioenergetics and Biomembranes.* 2005 Apr; 37:83-90.

[35] Naudí A, Cabré R, Dominguez-Gonzalez M, Ayala V, Jové M, Mota-Martorell N, Piñol-Ripoll G, Gil-Villar MP, Rué M, Portero-Otín M, Ferrer I. Region-specific vulnerability to lipid peroxidation and evidence of neuronal mechanisms for polyunsaturated fatty acid biosynthesis in the healthy adult human central nervous system. *Biochimica et Biophysica Acta (BBA)-Molecular and Cell Biology of Lipids.* 2017 May 1; 1862(5):485-95.

[36] Herrero-Mendez A, Almeida A, Fernández E, Maestre C, Moncada S, Bolaños JP. The bioenergetic and antioxidant status of neurons is controlled by continuous degradation of a key glycolytic enzyme by APC/C–Cdh1. *Nature Cell Biology.* 2009 Jun; 11(6):747-52.

[37] Wang X, Michaelis EK. Selective neuronal vulnerability to oxidative stress in the brain. *Frontiers in Aging Neuroscience.* 2010; 2:12.

[38] Wang X, Pal R, Chen XW, Limpeanchob N, Kumar KN, Michaelis EK. High intrinsic oxidative stress may underlie selective vulnerability of the hippocampal CA1 region. *Molecular Brain Research.* 2005 Oct 31; 140(1-2):120-6.

[39] Amador-Alvarado L, Montiel T, Massieu L. Differential production of reactive oxygen species in distinct brain regions of hypoglycemic mice. *Metabolic Brain Disease.* 2014 Sep; 29:711-9.

[40] Forster MJ, Dubey A, Dawson KM, Stutts WA, Lal H, Sohal RS. Age-related losses of cognitive function and motor skills in mice are associated with oxidative protein damage in the brain. *Proceedings of the National Academy of Sciences.* 1996 May 14; 93(10):4765-9.

[41] Candelario-Jalil E, Mhadu NH, Al-Dalain SM, Martínez G, León OS. Time course of oxidative damage in different brain regions following transient cerebral ischemia in gerbils. *Neuroscience Research.* 2001 Nov 1; 41(3):233-41.

[42] Navarro A, López-Cepero JM, Bández MJ, Sánchez-Pino MJ, Gómez C, Cadenas E, Boveris A. Hippocampal mitochondrial dysfunction in rat aging. *American Journal of Physiology-Regulatory, Integrative and Comparative Physiology.* 2008 Feb; 294(2):R501-9.

[43] Ristow M, Zarse K. How increased oxidative stress promotes longevity and metabolic health: The concept of mitochondrial hormesis (mitohormesis). *Experimental Gerontology.* 2010 Jun 1; 45(6):410-8.

[44] Scialo F, Sriram A, Fernandez-Ayala D, Gubina N, Lohmus M, Nelson G, Logan A, Cooper HM, Navas P, Enríquez JA, Murphy MP. Mitochondrial ROS produced via reverse electron transport extend animal lifespan. *Cell Metabolism.* 2016 Apr 12; 23(4):725-34.

[45] Schulz TJ, Zarse K, Voigt A, Urban N, Birringer M, Ristow M. Glucose restriction extends Caenorhabditiselegans life span by inducing mitochondrial respiration and increasing oxidative stress. *Cell Metabolism.* 2007 Oct 3; 6(4):280-93.

[46] Yang W, Hekimi S. A mitochondrial superoxide signal triggers increased longevity in Caenorhabditiselegans. *PLoS Biology.* 2010 Dec 7; 8(12):e1000556.

[47] Sanz A, Pamplona R, Barja G. Is the mitochondrial free radical theory of aging intact? *Antioxidants &Redox Signalling.* 2006 Mar 1; 8(3-4):582-99.

[48] Sinha K, Das J, Pal PB, Sil PC. Oxidative stress: the mitochondria-dependent and mitochondria-independent pathways of apoptosis. *Archives of Toxicology.* 2013 Jul; 87:1157-80.

[49] Singh A, Kukreti R, Saso L, Kukreti S. Oxidative stress: a key modulator in neurodegenerative diseases. *Molecules.* 2019 Apr 22; 24(8):1583.

[50] Islam MT. Oxidative stress and mitochondrial dysfunction-linked neurodegenerative disorders. *Neurological Research.* 2017 Jan 2; 39(1):73-82.

[51] Sies H, Berndt C, Jones DP. Oxidative stress. *Annual Review of Biochemistry.* 2017 Jun 20; 86:715-48.

[52] Lee J, Giordano S, Zhang J. Autophagy, mitochondria and oxidative stress: cross-talk and redox signalling. *Biochemical Journal.* 2012 Jan 15; 441(2):523-40.

[53] Maynard S, Fang EF, Scheibye-Knudsen M, Croteau DL, Bohr VA. DNA damage, DNA repair, aging, and neurodegeneration. *Cold Spring Harbor Perspectives in Medicine.* 2015 Oct 1; 5(10):a025130.

[54] Guillaumet-Adkins A, Yañez Y, Peris-Diaz MD, Calabria I, Palanca-Ballester C, Sandoval J. Epigenetics and Oxidative Stress in Aging. *Oxid Med Cell Longev.* 2017 July; 2017:9175806.

[55] Grimm A, Eckert A. Brain aging and neurodegeneration: from a mitochondrial point of view. *Journal of neurochemistry.* 2017 Nov; 143(4):418-31.

[56] Lu T, Pan Y, Kao SY, Li C, Kohane I, Chan J, Yankner BA. Gene regulation and DNA damage in the ageing human brain. *Nature.* 2004 Jun 24; 429(6994):883-91.

[57] Mecocci P, Boccardi V, Cecchetti R, Bastiani P, Scamosci M, Ruggiero C, Baroni M. A long journey into aging, brain aging, and Alzheimer's disease following the oxidative stress tracks. *Journal of Alzheimer's Disease.* 2018 Jan 1; 62(3):1319-35.

[58] Dai DF, Santana LF, Vermulst M, Tomazela DM, Emond MJ, MacCoss MJ, Gollahon K, Martin GM, Loeb LA, Ladiges WC, Rabinovitch PS. Overexpression of catalase targeted to mitochondria attenuates murine cardiac aging. *Circulation.* 2009 Jun 2; 119(21):2789-97.

[59] Paglialunga S, Ludzki A, Root-McCaig J, Holloway GP. In adipose tissue, increased mitochondrial emission of reactive oxygen species is important for short-term high-fat diet-induced insulin resistance in mice. *Diabetologia.* 2015 May; 58:1071-80.

[60] Schriner SE, Linford NJ, Martin GM, Treuting P, Ogburn CE, Emond M, Coskun PE, Ladiges W, Wolf N, Van Remmen H, Wallace DC. Extension of murine life span by overexpression of catalase targeted to mitochondria. *Science.* 2005 Jun 24;308(5730):1909-11.

[61] Campisi J, Kapahi P, Lithgow GJ, Melov S, Newman JC, Verdin E. From discoveries in ageing research to therapeutics for healthy ageing. *Nature.* 2019 Jul 11;571(7764):183-92.

[62] Kim SJ, Cheresh P, Jablonski RP, Morales-Nebreda L, Cheng Y, Hogan E, Yeldandi A, Chi M, Piseaux R, Ridge K, Hart CM. Mitochondrial catalase overexpressed transgenic mice are protected against lung fibrosis in part via preventing alveolar epithelial cell mitochondrial DNA damage. *Free Radical Biology and Medicine.* 2016 Dec 1;101:482-90.

[63] Swerdlow RH, Burns JM, Khan SM. The Alzheimer's disease mitochondrial cascade hypothesis: progress and perspectives. *Biochimica et Biophysica Acta (BBA)-Molecular Basis of Disease.* 2014 Aug 1; 1842(8):1219-31.

[64] Cheignon C, Tomas M, Bonnefont-Rousselot D, Faller P, Hureau C, Collin F. Oxidative stress and the amyloid beta peptide in Alzheimer's disease. *Redox Biology.* 2018 Apr 1; 14:450-64.

[65] Hensley K, Carney JM, Mattson MP, Aksenova M, Harris M, Wu JF, Floyd RA, Butterfield DA. A model for beta-amyloid aggregation and neurotoxicity based on free radical generation by the peptide: relevance to Alzheimer disease. *Proceedings of the National Academy of Sciences.* 1994 Apr 12; 91(8):3270-4.

[66] Pike CJ, Cummings BJ, Cotman CW. beta-Amyloid induces neuritic dystrophy in vitro: similarities with Alzheimer pathology. *Neuroreport.* 1992 Sep 1; 3(9):769-72.

[67] Mark RJ, Lovell MA, Markesbery WR, Uchida K, Mattson MP. A role for 4-hydroxynonenal, an aldehydic product of lipid peroxidation, in disruption of ion homeostasis and neuronal death induced by amyloid β-peptide. *Journal of Neurochemistry.* 1997 Jan; 68(1):255-64.

[68] Keller JN, Mark RJ, Bruce AJ, Blanc E, Rothstein JD, Uchida K, Waeg G, Mattson MP. 4-Hydroxynonenal, an aldehydic product of membrane lipid peroxidation, impairs glutamate transport and mitochondrial function in synaptosomes. *Neuroscience.* 1997 Jul 28; 80(3):685-96.

[69] Reybier K, Ayala S, Alies B, Rodrigues JV, Bustos Rodriguez S, La Penna G, Collin F, Gomes CM, Hureau C, Faller P. Free superoxide is an intermediate in the production of H2O2 by copper (I)-Aβ peptide and O2. *Angewandte Chemie.* 2016 Jan 18; 128(3):1097-101.

[70] Nakamura T, Cieplak P, Cho DH, Godzik A, Lipton SA. S-nitrosylation of Drp1 links excessive mitochondrial fission to neuronal injury in neurodegeneration. *Mitochondrion.* 2010 Aug 1; 10(5):573-8.

[71] Khandelwal PJ, Herman AM, Hoe HS, Rebeck GW, Moussa CE. Parkin mediates beclin-dependent autophagic clearance of defective mitochondria and ubiquitinated Aβ in AD models. *Human Molecular Genetics.* 2011 Jun 1; 20(11):2091-102.

[72] Martín-Maestro P, Gargini R, Perry G, Avila J, García-Escudero V. PARK2 enhancement is able to compensate mitophagy alterations found in sporadic Alzheimer's disease. *Human Molecular Genetics.* 2016 Feb 15; 25(4):792-806.

[73] Shanbhag NM, Evans MD, Mao W, Nana AL, Seeley WW, Adame A, Rissman RA, Masliah E, Mucke L. Early neuronal accumulation of DNA double strand breaks in Alzheimer's disease. *Acta Neuropathologica Communications.* 2019 Dec; 7:1-8.

[74] Thanan R, Oikawa S, Hiraku Y, Ohnishi S, Ma N, Pinlaor S, Yongvanit P, Kawanishi S, Murata M. Oxidative stress and its significant roles in neurodegenerative diseases and cancer. *International journal of Molecular Sciences.* 2014 Dec 24;16(1):193-217.

[75] Bosch-Presegué L, Raurell-Vila H, Marazuela-Duque A, Kane-Goldsmith N, Valle A, Oliver J, Serrano L, Vaquero A. Stabilization of Suv39H1 by SirT1 is part of oxidative stress response and ensures genome protection. *Molecular Cell.* 2011 Apr 22;42(2):210-23.

[76] Fang EF, Lautrup S, Hou Y, Demarest TG, Croteau DL, Mattson MP, Bohr VA. NAD+ in aging: molecular mechanisms and translational implications. *Trends in Molecular Medicine.* 2017 Oct 1;23(10):899-916.

[77] de Picciotto NE, Gano LB, Johnson LC, Martens CR, Sindler AL, Mills KF, Imai SI, Seals DR. Nicotinamide mononucleotide supplementation reverses vascular dysfunction and oxidative stress with aging in mice. *Aging Cell.* 2016 Jun;15(3):522-30.

[78] Ji H, Zhou Y, Zhuang X, Zhu Y, Wu Z, Lu Y, Li S, Zeng Y, Lu QR, Huo Y, Shi Y. HDAC3 Deficiency Promotes Liver Cancer through a Defect in H3K9ac/H3K9me3 TransitionHDAC3 Targets the H3K9ac/H3K9me3 Transition. *Cancer Research.* 2019 Jul 15; 79(14):3676-88.

[79] Snigdha S, Prieto GA, Petrosyan A, Loertscher BM, Dieskau AP, Overman LE, Cotman CW. H3K9me3 inhibition improves memory, promotes spine formation, and increases BDNF levels in the aged hippocampus. *Journal of Neuroscience.* 2016 Mar 23; 36(12):3611-22.

[80] Suberbielle E, Sanchez PE, Kravitz AV, Wang X, Ho K, Eilertson K, Devidze N, Kreitzer AC, Mucke L. Physiologic brain activity causes DNA double-strand breaks in neurons, with exacerbation by amyloid-β. *Nature Neuroscience.* 2013 May; 16(5):613-21.

[81] Madabhushi R, Gao F, Pfenning AR, Pan L, Yamakawa S, Seo J, Rueda R, Phan TX, Yamakawa H, Pao PC, Stott RT. Activity-induced DNA breaks govern the expression of neuronal early-response genes. *Cell.* 2015 Jun 18; 161(7):1592-605.

[82] Li X, Marshall PR, Leighton LJ, Zajaczkowski EL, Wang Z, Madugalle SU, Yin J, Bredy TW, Wei W. The DNA repair-associated protein Gadd45γ regulates the temporal coding of immediate early gene

expression within the prelimbic prefrontal cortex and is required for the consolidation of associative fear memory. *Journal of Neuroscience.* 2019 Feb 6; 39(6):970-83.
[83] Lu T, Pan Y, Kao SY, Li C, Kohane I, Chan J, Yankner BA. Gene regulation and DNA damage in the ageing human brain. *Nature.* 2004 Jun 24; 429(6994):883-91.
[84] Rutten BP, Schmitz C, Gerlach OH, Oyen HM, de Mesquita EB, Steinbusch HW, Korr H. The aging brain: accumulation of DNA damage or neuron loss?. *Neurobiology of aging.* 2007 Jan 1;28(1):91-8.
[85] Lovell MA, Gabbita SP, Markesbery WR. Increased DNA oxidation and decreased levels of repair products in Alzheimer's disease ventricular CSF. *Journal of Neurochemistry.* 1999 Feb;72(2):771-6.
[86] Myung NH, Zhu X, Kruman II, Castellani RJ, Petersen RB, Siedlak SL, Perry G, Smith MA, Lee HG. Evidence of DNA damage in Alzheimer disease: phosphorylation of histone H2AX in astrocytes. *Age.* 2008 Dec; 30:209-15.
[87] Suberbielle E, Sanchez PE, Kravitz AV, Wang X, Ho K, Eilertson K, Devidze N, Kreitzer AC, Mucke L. Physiologic brain activity causes DNA double-strand breaks in neurons, with exacerbation by amyloid-β. *Nature Neuroscience.* 2013 May; 16(5):613-21.
[88] Su JH, Deng G, Cotman CW. Neuronal DNA damage precedes tangle formation and is associated with up-regulation of nitrotyrosine in Alzheimer's disease brain. *Brain research.* 1997 Nov 7; 774(1-2):193-9.
[89] Sykora P, Misiak M, Wang Y, Ghosh S, Leandro GS, Liu D, Tian J, Baptiste BA, Cong WN, Brenerman BM, Fang E. DNA polymerase β deficiency leads to neurodegeneration and exacerbates Alzheimer disease phenotypes. *Nucleic Acids Research.* 2015 Jan 30; 43(2):943-59.
[90] Kreuz S, Fischle W. Oxidative stress signalling to chromatin in health and disease. *Epigenomics.* 2016 Jun;8(6):843-62.
[91] Chia N, Wang L, Lu X, Senut MC, Brenner CA, Ruden DM. Hypothesis: environmental regulation of 5-hydroxymethylcytosine by oxidative stress. *Epigenetics.* 2011 Jul 1; 6(7):853-6.
[92] Lewandowska J, Bartoszek A. DNA methylation in cancer development, diagnosis and therapy—multiple opportunities for genotoxic agents to act as methylome disruptors or remediators. *Mutagenesis.* 2011 Jul 1; 26(4):475-87.
[93] Niu Y, DesMarais TL, Tong Z, Yao Y, Costa M. Oxidative stress alters global histone modification and DNA methylation. *Free Radical Biology and Medicine.* 2015 May 1; 82:22-8.
[94] Gu X, Sun J, Li S, Wu X, Li L. Oxidative stress induces DNA demethylation and histone acetylation in SH-SY5Y cells: potential epigenetic mechanisms in gene transcription in Aβ production. *Neurobiology of Aging.* 2013 Apr 1;34(4):1069-79.
[95] Bosch-Presegué L, Raurell-Vila H, Marazuela-Duque A, Kane-Goldsmith N, Valle A, Oliver J, Serrano L, Vaquero A. Stabilization of Suv39H1 by SirT1 is part of oxidative stress response and ensures genome protection. *Molecular Cell.* 2011 Apr 22; 42(2):210-23.
[96] Vaquero A, Scher M, Erdjument-Bromage H, Tempst P, Serrano L, Reinberg D. SIRT1 regulates the histone methyl-transferase SUV39H1 during heterochromatin formation. *Nature.* 2007 Nov 15; 450(7168):440-4.
[97] Griñán-Ferré C, Marsal-García L, Bellver-Sanchis A, Kondengaden SM, Turga RC, Vázquez S, Pallàs M. Pharmacological inhibition of G9a/GLP restores cognition and reduces oxidative stress, neuroinflammation and β-Amyloid plaques in an early-onset Alzheimer's disease mouse model. *Aging (Albany NY).* 2019 Dec 12; 11(23):11591.

Chapter 6

# Targeting Reactive Oxygen Species (ROS) in the Prevention of Non-Communicable Diseases: The Promise and Limitations of Antioxidant Therapy

**Satyajit Halder[1]**
**Chirantan Majumder[1]**
**Anirban Manna[1]**
**Sarita Sarkar[2]**
**Ankita Bhattachrya[3]**
**and Kuladip Jana[1,*]**
[1]Division of Molecular Medicine, Bose Institute, Bidhan Nagar, Kolkata, West Bengal, India
[2]Department of Biotechnology, The Neotia University, Sarisha, West Bengal, India
[3]Department of Biochemistry, K. P. C. Medical College and Hospital, Kolkata, India

## Abstract

Reactive oxygen species (ROS) overproduction has been concerned with developing various chronic and degenerative diseases such as cancer, diabetes, cardiovascular diseases, and inflammatory and neurodegenerative diseases. In normal physiological conditions, the concentrations of ROS are delicately controlled by antioxidants, which may be available endogenously or externally. A combination of antioxidant deficiency, food habits, alcohol intake, and smoking may create individuals more susceptible to oxidative stress thus increasing the risk of cancer and death. Moreover, antioxidant deficiency may be associated with chronic inflammation such as in COPD (Chronic obstructive pulmonary disease), IBD (inflammatory bowel disease), neurodegenerative disorders, cardiovascular diseases, cancer, and aging. Long-term exposure to increased levels of pro-oxidant factors like ROS can cause structural defects at a mitochondrial DNA level, as well as functional alteration of several enzymes and cellular structures, leading to aberrations in gene expression and that might be concomitant with several inflammatory diseases. However, the use of medicinal plants with antioxidant properties has been exploited for their ability to treat or prevent several human pathologies, of which oxidative stress seems to be one of the causes. Certain antioxidant vitamins, such as vitamins C, D, and E are essential in regulating biochemical pathways that lead to the proper functioning of the organs. This

* Corresponding Author's Email: kuladip@jcbose.ac.in; kuladip.jana@gmail.com.

In: The Role of Reactive Oxygen Species in Health and Disease
Editor: Kuladip Jana
ISBN: 979-8-89113-283-2

chapter highlights the potential roles of ROS and nutritional antioxidants in the pathogenesis of several redox imbalance-related diseases and the attenuation of oxidative stress-induced damages. Finally, both the beneficial and detrimental effects of antioxidant molecules that are used to reduce oxidative stress in several human disease conditions are discussed.

**Keywords**: ROS, oxidative stress, redox imbalance, inflammation, non-communicable diseases, antioxidants

## 1. Introduction

Reactive oxygen species (ROS) are produced in the cell, which is involved in numerous cellular activities. Overproduction of ROS may result in oxidative damage to biomolecules such as lipids, proteins, and DNA, which has been implicated in the progress of aging along with various diseases including cancer, asthma, heart attack, Alzheimer's and diabetes etc. It has been reported that the detrimental effects of excess ROS may cause oxidative stress (OS), eventually leading to cell death [1]. The cells are also well equipped with several machinery to counteract the detrimental effects of OS. Antioxidants, either endogenously produced or externally supplied, are capable of scavenging ROS and reducing the oxidation of biomolecules, accordingly alleviating OS [2-5]. The endogenous antioxidants obtained from the diet are essential for the neutralization of OS. Currently, the clinical awareness of nutritional imbalance in disease incidence and progression is limited. In this chapter, we will focus on the roles of ROS in common OS-associated diseases and aging, as well as discuss the effects of nutritional antioxidants as treatments or adjuvants and their possible promise and limitations.

## 2. ROS and Oxidative Stress (OS)

Oxidative stress (OS) is a result of an increase in the generation of ROS or a decrease in the level of antioxidants. OS is unstable and highly reactive. Cellular metabolic pathways produce ROS that can destroy the biological components of carbohydrates, lipids, proteins, and nucleic acid. Moreover, the accumulation of ROS damages the DNA, as it breaks down into single and double strands of DNA, degrades nitrogenous bases, and transforms and cross-links DNA with proteins [3-5]. These impaired DNA modifications can accelerate aging and cause neurodegenerative disorders, carcinogenesis, autoimmune, cardiovascular, respiratory and other metabolic diseases like diabetes [3-5]. ROS disrupts the cellular membrane, breaks down peptide bonds, oxidizes amino acids, as well as causes lipid peroxidation. Hence, the accumulation of ROS in the body could trigger an array of non-infectious diseases. As a result, antioxidants can help to keep them at the lowest concentration in the body.

## 3. Antioxidants: Different Types

Antioxidants are secondary cellular metabolites as well as highly present in fruits and vegetables. To avoid oxidation of the vulnerable substrate, the plants produce a significant number of antioxidants, such as alkaloids, polyphenols, and vitamins (C and E). The human body can't produce an adequate number of antioxidants to protect cells from the deleterious effects of ROS. Therefore, the plant is the best source of dietary antioxidants, which can play a crucial role in sustaining human health [5, 6]. Antioxidants may act as a scavenger of ROS and prevent disease progression by protecting cell and tissue damage [5, 6]. Nowadays antioxidants are generally classified into natural (Available in living organisms) and synthetic (Synthesized in the laboratory). However, they are further subdivided into groups based on their bioactivity (enzymatic and non-enzymatic antioxidants), solubility (water-soluble and fat-soluble antioxidants), and size (small molecule and large molecule antioxidants). There are other possible classifications of antioxidants, such as endogenous and exogenous antioxidants; and primary and secondary antioxidants [5, 7]. Enzymatic antioxidants neutralize the ROS by converting into hydrogen peroxide, which is further metabolized by catalase into water in a stepwise reaction in the presence of some cofactors. Non-enzymatic antioxidants inhibit ROS-mediated oxidative reactions [8]. Small molecule antioxidants scavenge ROS and on the other hand, large molecule antioxidants also scavenge ROS, but they protect oxidative damage to biomolecules such as lipids, proteins, and DNA [9]. Endogenous antioxidants are made by the human body, while exogenous antioxidants are obtained from dietary supplementation. The endogenous and exogenous antioxidants can work synergistically to maintain or establish a cellular redox balance. Besides this, primary antioxidants interact with the ROS and transform it into more stable, non-reactive products. Therefore, they are known as the chain-breaking antioxidants because they can directly interact with the radicals to transform them into non-radicals, consequently preventing them from further peroxidation. Secondary antioxidants are referred to as preventive antioxidants or hydro-peroxide decomposers and can convert hydro-peroxides into non-reactive products. They maintain the required antioxidant level in the body by reducing the level of peroxides and making nicotinamide adenine dinucleotide phosphate (NADPH) and glutathione available for the primary antioxidant enzymes [10]. Consuming antioxidants regularly will aid in eliminating excess ROS from the body, thus enhancing wellness by reducing the chance of non-communicable diseases like cancer, diabetes, coronary heart diseases, as well as liver, lung and kidney diseases. Nowadays antioxidants have gained huge attention in the clinical field and can be used in the management of different non-communicable diseases.

## 4. Antioxidants and Mechanisms of Action

The roles of antioxidants are to neutralize the excess of ROS, to protect the cells against their toxic effects, and to contribute to disease prevention. Endogenous cellular antioxidants can be classified as enzymatic antioxidants and non-enzymatic antioxidants. The major antioxidant enzymes directly involved in the neutralization of ROS are: superoxide dismutase (SOD), catalase (CAT), glutathione peroxidase (GPx), and glutathione reductase (GRx). SOD, the first line of defense against free radicals, catalyzes the dismutation of superoxide anion radical ($O_2^{\bullet -}$) into hydrogen peroxide ($H_2O_2$) by reduction reaction. The oxidant formed ($H_2O_2$) is transformed into water and oxygen ($O_2$) by catalase (CAT) or glutathione peroxidase (GPx). The selenoprotein GPx enzyme removes $H_2O_2$ by using it to oxidize reduced glutathione (GSH) into oxidized glutathione (GSSG). Glutathione reductase (GR), a flavoprotein enzyme, regenerates GSH from GSSG, with NADPH as a source of reducing agent. Besides hydrogen peroxide, GPx also reduces lipid or non-lipid hydroperoxides while oxidizing glutathione (GSH) [5, 11, 12]. The non-enzymatic antioxidants are also divided into metabolic antioxidants and nutrient antioxidants. Metabolic antioxidants belonging to endogenous antioxidants are produced by metabolism in the body, such as lipoid acid, glutathione, L-ariginine, coenzyme Q10, melatonin, uric acid, bilirubin, metal-chelating proteins, transferrin, etc. [13]. Nutrient antioxidants belonging to exogenous antioxidants are compounds that cannot be produced in the body and must be provided through foods or supplements, such as vitamin E, vitamin C, carotenoids, trace metals (selenium, manganese, zinc), flavonoids, omega-3 and omega-6 fatty acids, etc. When an antioxidant destroys a free radical, this antioxidant itself becomes oxidized. Therefore, the antioxidant resources must be constantly restored in the body. Also, in certain circumstances, an antioxidant may even act as a pro-oxidant e.g., it can generate toxic ROS [14]. The antioxidant process can function in one of two ways: chain-breaking or prevention. For the chain-breaking, when a radical releases or steals an electron, a second radical is formed. The last one exerts the same action on another molecule and continues until either the free radical formed is stabilized by a chain-breaking antioxidant (vitamin C, E, carotenoids, etc.) or it simply disintegrates into a harmless product. The classic example of such a chain reaction is lipid peroxidation. For the preventive way, an antioxidant enzyme like superoxide dismutase, catalase, and glutathione peroxidase can prevent oxidation by reducing the rate of chain initiation, e.g., either by scavenging ROS or by stabilizing transition metal radicals such as copper and iron [15]. Antioxidants from our diet play an important role in helping endogenous antioxidants for the neutralization of OS. Nutrient antioxidant deficiency is one of the causes of numerous chronic and degenerative pathologies. Each nutrient is unique in terms of its structure and antioxidant function.

## 5. Natural and Synthetic Antioxidants

Natural antioxidants are widely found in food and plants and exert numerous biological effects ranging from anti-aging, anticancer, anti-diabetic, and anti-inflammatory to anti-atherosclerosis [16]. Natural antioxidants coming from vegetables and fruits are of the utmost importance. Natural antioxidants from plants are primarily classified into three categories: polyphenols, carotenoids, and vitamins. Phenolic compounds are diverse in structure ranging from simple molecules, such as ferulic acid, vanillin, etc., to more complex polyphenols, such as flavonoids and tannins. Some phenolics, in addition to their antioxidant activities, also possess antimicrobial activities, thus exerting an effect on the food texture and flavors [17]. An array of vitamins has also been shown to possess antioxidant activity; the most important of these are vitamins E and C. Vitamin E is lipid soluble and comprises tocopherols and tocotrienols, which occur in four isomeric forms ($\alpha$, $\beta$, $\gamma$, and $\delta$), but only the $\alpha$-tocopherol is of nutritional importance. The $\alpha$-tocopherols can be found in legumes and cereals. Vitamin C is water-soluble and is present in fruits and vegetables. Vegetables and fruits are good sources of carotenoids as well, but the carotenoids that are known to possess antioxidant activities are $\beta$-carotene, $\alpha$-carotene, lycopene, and lutein. In addition to their antioxidant activity, carotenoids are used as coloring agents in food [18].

On the other hand, synthetic antioxidants are chemically produced compounds that do not occur in nature but can be added to food as preservatives to inhibit lipid oxidation. Because of the instability of natural antioxidants, synthetic antioxidants have been used in stabilizing fats and oil and have been used as antioxidants in human food for many decades. Synthetic antioxidants are commonly used in food production, and they include butylated hydroxyanisole (BHA), butylated hydroxytoluene (BHT), propyl gallate (PG), and *tert*-butyl hydroquinone (TBHQ). In addition, 2-naphthol (2NL), 4-phenyl phenol (OPP), and 2, 4-dichlorophenoxyacetic acid (2, 4-DA) are the ones mostly used in fruits and vegetables [19]. However, though synthetic antioxidants are still being used for the preservation of food, researchers have raised safety issues concerning their continued use. Such health issues that have been linked to protracted intake of synthetic antioxidants include triggering certain disease conditions [20]. It has been reported that an intake of a high dose of synthetic antioxidants in food could induce DNA damage and cause premature senescence [21]. Specifically, BHT and BHA have been found to cause a damaging effect on the liver and are carcinogenic as well [19]. Considering some of the non-desirable health effects of synthetic antioxidants, replacing synthetic antioxidants with larger natural antioxidants is important to meet dietary needs. However, synthetic antioxidants are cost-efficient, have a wide availability, and provide stability advantages over natural antioxidants, but have been associated with some deleterious side effects in human cells giving rise to controlled usage/consumption. As a consequence, this has fueled the search for safe natural antioxidants with characteristics of higher performance and sustainability, as consumers have a stronger preference for natural antioxidants, which can be broadly grouped into plants, animals, and microorganisms.

## 6. Antioxidants' Origin and Sources

### 6.1. Plant Sources of Antioxidants

Plants, particularly vegetables and fruits, are the major source of many natural antioxidants. According to preceding research, the reasons for the synthesis and accumulation of these antioxidants in plants are (1) To maintain normal physiologic functions as well as protection against pathogenic microbes and animal herbivores and (2) Capacity to build them in response to environmental stress conditions [22, 23]. Some of these antioxidants accumulate as colored pigments in the leaves, fruits, nuts, and roots of many plants. These colored pigments, such as beta-carotene, lutein, lycopene, and zeaxanthin, are all carotenoids either in their primary or secondary forms and abound in leafy vegetables and fruits. For instance, beta-carotene accumulates in carrots, leafy vegetables, spinach, and tomatoes. Lutein accumulates in green leafy vegetables, and zeaxanthin accumulates in spinach, while lycopene is abundant in tomatoes, watermelon, guava, pink grapefruit, and blood oranges. Likewise, studies have shown that some green tea leaves and herbs are better antioxidants than vegetables and fruits [24]. However, among the herbs tested, *C. quadrangularis* leaf extract exhibited superior antioxidant activity. Another type of antioxidant that accumulates in plants is vitamins like A, C, and E. Vitamin A is known to be abundant in sweet potatoes, vitamin C (ascorbic acid) in fruits and vegetables as well as in cereals, while vitamin E (*α-tocopherol*) accumulates in some plant oils, such as wheat germ oil, soybean oil, and corn oil. Essential oils from oregano and clove were found to have potential antioxidants, probably due to the high phenolic contents [25].

### 6.2. Animal Sources of Antioxidants

Animals or animal products are not significant sources of antioxidants in nature due to their low concentrations in comparison with plant-based antioxidants. Actually, the concentrations and types of antioxidants accumulated in animal/animal products depend on the constituents of the animal feed. Some animal products, such as fish lipid, egg, and milk, contain selenium and fat-soluble vitamins. Some antioxidants in milk, especially casein, occur as protein fractions whose antioxidant activities increase after digestion into peptides. The antioxidants from an animal are mainly from protein-rich compounds of animal products, which are digested by proteases either from the animal or by some microflora of the animal/animal products into peptides [26]. According to some reports in the literature, the antioxidants from animal and insect sources have not been commercialized because their antioxidant activities and safety profiles are still not validated.

### 6.3. Microbial Sources of Antioxidants

The microbial community probably contains the largest reservoir of antioxidants in nature. This may be due to the diversity of bacterial, fungal, and microalgae species. Both gram-positive and gram-negative bacteria have the potential to produce antioxidant compounds [27]. However, the *Streptomyces spp* and lactic acid bacteria are probably more prominent in antioxidant production. Probiotic bacteria (*Streptococcus thermophiles*) and many lactic acid bacteria (*Lactobacillus*spp) are excellent sources of many types of antioxidants, either as intact whole-cell or cell-free extracts [28]. Some species of *Streptomyces* also produce antioxidants like carotenoids. It was reported that *Streptomyces chrestomyceticus* produced lycopene, a type of carotenoid, which could be used as a coloring agent in the food industry [29]. Furthermore, *Xanthophyllomyces dendrorhous* could produce high amounts of antioxidants like carotenoids. The concentration and activities of antioxidants produced by microorganisms are dependent on some physicochemical parameters of the fermentation processes as well as the type of organic solvent used for antioxidant extraction [30].

The fungal species like *Aspergillus saitoi* and *Penicillium roquefortii* IFO 5956 contain a bioactive compound (2, 3-dihydroxybenzoic acid) with high antioxidant properties. *Aspergillus* spp and *Penicillium* spp can also produce antioxidants, such as citrinin, protocatechuic acid, curvilic acid, atroventin, and gallic acid [31]. The microalgal antioxidants are similar to plant antioxidants probably because of the similar characteristics shared by both in terms of physiology and pharmacology. For instance, both plants and microalgae obtain their food through the process of photosynthesis and receive the same or similar environmental stimulation that triggers OS. Microalgal species are very diverse and apparently are the richest source of antioxidants in nature. Microalgae accumulate different types of antioxidants, such as polysaccharides, carotenoids, sterols, vitamins (A, C, D, K, and E), flavonoids, amino acids, polyunsaturated fatty acids, minerals, sulfated polysaccharides, sulfolipids, peptides, coenzyme Q, phycocyanin, and scytonemin (a source of blue-green algae). Although all the antioxidants produced by microalgae are not commercialized, some of them that are widely produced with a high market value include lutein, astaxanthin, and $\beta$-carotene (31, 32). Under environmentally stressed conditions, every micro-algae species is a potential source of antioxidants, although the concentrations and types of antioxidants would depend on the type of stress to which the organism is subjected. Stressed conditions are known to induce enzymatic and non-enzymatic antioxidants in different microalgae species in response to ROS [32, 33]. Comparatively, microbial antioxidant productions have more prospects in terms of commercialization than plants or animal resources. This assertion is not just based on the many antioxidants produced by microbes but on the potential of optimizing productivity through genetic, metabolic, and environmental engineering [31, 32, 33].

## 7. Medicinal Uses of Antioxidants

Medicinal plants are possible sources of antioxidants and anti-inflammatory compounds that could be used in the management of different diseases (Figure 1). Preclinical and clinical studies of some antioxidants are summarized in Table 1 [34-45].

An imbalance of natural antioxidants causes ROS generation from numerous environmental and biological sources, which would result in a wide range of inflammatory illnesses. Additionally, oxidative stress and its accompanying components have now become a major public health concern. As such, therapeutic plant extracts and their isolated active ingredients have a wide range of therapeutic effects against a wide range of acute and chronic disorders. Anti-inflammatory studies have shown that extracts and their organic products exercise their bioactivity by inhibiting two main signaling pathways, mitogen-activated protein kinases (MAPKs) and nuclear factor kappa B (NF-κB), that are responsible for producing a variety of proinflammatory mediators [46]. Alkaloids, polyphenols, terpenoids, and flavonoids are well-studied phytonutrients for anti-inflammatory activity [47]. Plants that contain these phytoconstituents are used as anti-inflammatory agents. In ROS-mediated oxidative damage and the progression of carcinogenesis, two distinct processes are thought to be involved. The first mechanism is through gene expression regulation. Growth signals and proliferation can be stimulated by epigenetic changes in gene expression. ROS causes genetic changes, such as mutations and chromosomal conformational changes, which might contribute to the onset of carcinogenesis in the second mechanism [48]. Antioxidants can also act as anticancer agents because they can scavenge ROS that can cause DNA conformational changes, DNA/protein cross-links, and DNA damage, leading to cell mutation, transformation, and cancer [49]. As a consequence, antioxidants can inhibit the growth of oxidative cancer cells by neutralizing free radicals. A β-carotene, as an antioxidant, can protect against cancer development. Its photo-protective activities may guard against UV light-induced cell damage and cancer development, and its immune enhancement may help to protect against cancer progression [50]. Likewise, vitamin C may aid in the prevention of cancer because it has potent antioxidant activity to inhibit the development of nitrosamines, enhance the immunological response, and produce the detoxification of liver enzymes [51]. Additionally, vitamin E improves immune competence by boosting humoral immune responses, bacterial infection resistance, cellular immunity, tumor necrosis factor generation by inflammatory cells, and inhibiting mutagenesis, DNA repair, and microcell formation. As a result, vitamin E may be beneficial in preventing cancer and inhibiting carcinogenesis through regulating immune function [52]. Fruits and vegetables, which are rich in antioxidants, exert a protective effect against several different types of cancers. Plant foods that contain polyphenols have proven to be effective antioxidant agents for the body They have been shown to possess anti-cancer activity, which is effective against lung, breast, tongue, gastric, larynx, colon, and prostate cancers (Table 1) [53, 54]. Plant polyphenols induce apoptosis in cancer cells, inhibit the proliferation of mutated cells, reduce the production of cyclooxygenase-2 (COX-2), and down-regulate cancer gene expression [54] (Figure 1).

**Table 1.** Preclinical, and clinical studies of some antioxidants

| No. | Antioxidants | Pharmacokinetics | Medicinal Use |
|---|---|---|---|
| 1. | Ascorbic acid<br>Polyphenols<br>Tocopherols | Improved the endogenous ROS- ROS-scavenging system and prevented skin infections and aging. | Anti-aging |
| 2. | Lycopene | Decreases ROS and lipid peroxyl radicals. Lycopene has tremendous anti-oxidant properties. It also delays the activation of the mTOR/AMPK cascade pathway linked with aging. Lycopene quenches singlet oxygen, scavenges free radicals, and prevents the oxidative damage of DNA. Thus, it prevents the potential transformation of normal cells to cancer cells. | Anti-aging<br>Anti-Cancer |
| 3. | Cyanobacterial phycobiliproteins (PBPs) | Delayed aging and age-related diseases. | Anti-aging |
| 4. | Resveratrol | Inhibits pro-inflammatory cytokines like IL-1 and TNF-α. It acts as an antioxidant and stops liver steatosis. Resveratrol can promote cell cycle arrest, leading to apoptosis of tumor cells, prevent tumor-derived nitric oxide synthase expression to block tumor growth and migration, as well as act as an antioxidant to prevent DNA damage that can lead to tumor formation. | Anti-inflammatory<br>Anti-cancer<br>Hepato-protective |
| 5. | Curcumin | Curcumin exerts its immunomodulatory ability by interacting with several immune mediators such as cytokines, ROS, cyclooxygenase (COX-2), transcription factors including nuclear factor κB (NF-κB), protein kinases B (AKT), activator protein 1 (AP1), signal transducer and activator of transcription 3 (STAT3), causing the initiation and development of cancer, Curcumin can inhibit the formation of cancer and spread the cancerous cells by exerting anti-angiogenic effects, inducing apoptosis and interfering with the cell proliferation cycle. It helps in the maintenance of body function and mitochondrial redox balance. | Hepato-protective<br>Nephroprotective<br>Anti-inflammatory<br>Anti-Cancer |
| 6. | Epigallocatechin-3-gallate (EGCG) | EGCG functions as a powerful antioxidant, preventing oxidative damage in healthy cells, but also as an anti-angiogenic and antitumor agent. EGCG blocks carcinogenesis by affecting a wide array of signal transduction pathways, including JAK/STAT, MAPK, PI3K/AKT, Wnt, and Notch. | Anti-inflammatory<br>Anti-Cancer |
| 7. | Quercetin | Quercetin can significantly reduce the levels of inflammation moderators, including NO synthase, COX-2, as well as CRP in human hepatocytes. quercetin exerts antioxidant effects by free radical-scavenging activity, reducing the ROS level and inhibiting lipid peroxidation. Clinical studies have demonstrated the inhibitory effect of quercetin on inflammation and its ability to inhibit the risk of cancer and ovarian cancer. | Anti-inflammatory,<br>Hepato-protective<br>Anti-Cancer |
| 8. | Quercetin + Curcumin | They control hyperglycemia by promoting the stimulation and release of insulin. | Anti-diabetic |
| 9. | Trans-4-hydroxystilbene (THS), Piceatannol | Foraged DPPH and OH radicals and stop the accumulation of ROS in neurons | Neuro-protective |
| 10. | Quercetin 7-Rhamnoside (Q7R) | Q7R acts as an antioxidant and hepato-protective agent | Hepato-protective |

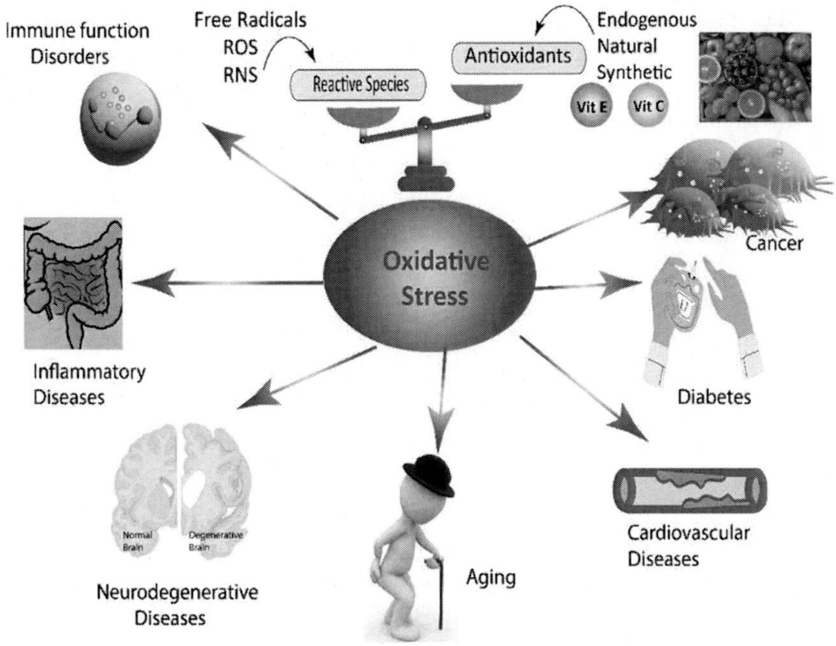


**Figure 1**. ROS and antioxidants in health and diseases.

Diabetes is a metabolic disorder marked by relative or absolute insulin secretion deficits leading to chronic hyperglycemia and carbohydrate, lipid, and protein metabolism abnormalities. Metabolic disorders cause a variety of complications. Additionally, diabetes mellitus has been linked to increased formation of ROS and a reduction in antioxidant activities, which also results in the imbalance between the generation of ROS and antioxidants, thus leading to oxidative damage to cell proteins, lipids, and nucleic acids. Glucose autoxidation is the most important factor in the generation of free radicals among the different variables that cause increased oxidative stress. Other factors also elevate the levels of pro-oxidants, imbalance in cellular oxidation, and a reduction in antioxidant defense. Low concentrations of ascorbate, glutathione, and superoxide dismutase are the most prevalent antioxidant deficits during the pathogenic process of diabetes. Therefore, plants, especially those with high quantities and potent antioxidant chemicals, can treat oxidative stress-related diseases like diabetes mellitus. Many studies have also examined the impact of their antioxidant components on diabetes complications to achieve promising results by demonstrating the benefits of plants with high antioxidant levels in the treatment of diabetes [55].

In addition, antioxidant consumption from fruits and vegetables aids in the management of cardiovascular illnesses. Because oxidative processes can alter cardiovascular disorders, they have the potential to deliver tremendous health and lifespan benefits. Polyunsaturated fatty acids make up a large portion of low-density lipoproteins (LDL) in the blood, and their oxidation plays an important function in atherosclerosis. With a high quantity of oxidized lipids in the blood, blood vessel damage occurs, which can result in the formation of foam cells and plaque, leading to atherosclerosis. Atherogenic oxidized LDL is thus thought to be essential in the production of atherosclerosis plaque. Moreover, oxidized LDL is cytotoxic and can directly harm endothelial cells. Likewise, antioxidants like β-carotene and vitamin E are important in preventing cardiovascular disorders [55, 56].

Due to the ever-increasing resistance to synthetic antibiotics, we must shift our focus to natural antioxidant-based antibacterial products, which have a range of scientific diversity and provide an effective therapeutic benefit while preventing microbes from replicating and developing resistance. Phenolics are also important antibacterial antioxidants because they inhibit the growth of bacteria and their pathogenic activity. Antioxidants act as antibacterial reagents by inhibiting nucleic acid production, the permeability of the outer membrane, and cytoplasm leakage. Antioxidants' antibacterial effect may be owing to their ability to chelate iron, which is essential for the existence of all bacteria. Polyphenols break the cell wall, enhance cytoplasm membrane permeability, as well as release lipopolysaccharides [56].

However, antioxidant efficacy against microbial infections is becoming more generally accepted. They also function in tandem with modern antimicrobial drugs to combat drug-resistant microbes. Hence, antioxidant structure determines its antimicrobial effect. Synthetic antibiotics have a speedy therapeutic impact when used to treat microbial infections, but they also represent a substantial risk of gastrointestinal and renal toxicity and microbial resistance. Nonetheless, natural antioxidants in their purest form have outstanding antibacterial activity against common microbes without evidence to develop resistance to these substances by reinforcing the desire to research natural products to replace synthetic antibiotics. Yet, even though antioxidants function slowly to inhibit microbial growth, their effect is consistent, and a thorough assessment in determining the antimicrobial profile of isolated antioxidants could aid in the use of phytochemicals against microbial infections with minimal toxicity as well as the risk of bacterial resistance [57, 58].

Liver diseases have also been linked to oxidative stress. Oxidized proteins and a reduction in antioxidant levels contribute to the development of liver diseases. However, antioxidants, both natural and synthetic, have been utilized to treat liver disease. When antioxidants were given orally or intraperitoneally, animal studies showed a significant beneficial effect on liver diseases. As a result, a large-scale model would be required to determine their safety and effective dose, optimum treatment absorption, duration, and bioavailability [58].

Resveratrol guards the liver against cholestasis, alcohol, and toxic damage by boosting the lipid profile and reducing liver fibrosis and cirrhosis. In diverse models, this would prevent hepatic steatosis by inducing insulin conflict and altering the lipid profile. In Wistar rats, preclinical research demonstrated that resveratrol had a therapeutic impact on liver disease. Yet, greater clinical trials would be needed to give scientific evidence for the management of liver disorders [38, 39]. Nevertheless, silymarin was discovered to have the greatest hepatoprotective impact, though additional clinical trials would be important to ascertain the significance of the antioxidants on liver function [59].

In addition, curcumin's nephroprotective effect was connected to bodily function and mitochondrial redox equilibrium [60]. Green tea extracts (phenolic chemicals) were also found to enhance kidney function and reduce the levels of blood urea nitrogen and serum creatinine. In patients with low antioxidant levels, administration of vitamin C protects the kidneys by lowering the ROS and oxidative damage. Additionally, endogenous vitamin C acts as an enzyme cofactor to have a renal protective impact [61].

Phenolic diet consumption provides protection against cardiovascular and neurological illnesses, such as Parkinson's, early onset dementia, and motor neuron disease. Antioxidant neuroprotective activities were investigated by suppressing the ROS in *in-vitro* and *in-vivo* studies. Resveratrol increased the enzyme activity like catalase, SOD, hemeoxygenase, and glutathione peroxidase while inhibiting xanthine oxidase. In addition, antioxidant supplementation could help people ameliorate stress-related mental and severe anxiety [38, 39].

Antioxidant micronutrients like retinol and ascorbic acid may also help prevent eye disorders. Natural antioxidants, including flavonoids, vitamins, phenolic acids, and carotenoids, could help prevent eye defects. It has also been found that nanotechnology-based formulations of antioxidant biomolecules could improve bioavailability, solubility, and stability, as well as activity. Treatment of eye defects, such as cataracts, could be conducted by using nanoparticle-based antioxidant biomolecules that would be effective, biodegradable, and non-toxic. Moreover, natural antioxidants have been used as therapeutic reagents to slow the progression of cataracts. Vitamins C and E and curcumin, having antioxidant and anti-cataract properties, can be included in the diet to help protect against free radicals [62, 63, 64].

In addition, free radicals could cause degenerative changes, which are linked with aging. DNA damage, or the buildup of physiological and structural damage, is the primary cause of aging. Reducing free radicals or slowing their formation with antioxidants would help delay the process of aging. Hence, increased oxidative stress would be prevalent in aged people, and antioxidants could have a major effect on oxidative damage. The antioxidant defense could minimize free radical damage, and adequate consumption of antioxidant nutrients could improve the quality of life and help people live longer [65].

## Summary and Prospective

The implication of OS in the etiology of several chronic and inflammatory diseases indicates that antioxidant-based therapy could be promising for these disorders. A therapeutic strategy that increases an individual's antioxidant capacity may be useful for long-term treatment. However, many problems remain elusive regarding antioxidant supplements in disease prevention. Various antioxidants can understand, prevent, and treat acute and chronic diseases like neurodegenerative diseases, eye defects, cancer, aging, diabetes, liver disease, and cardiovascular diseases. Despite discoveries in treatments and diagnosis of these diseases as well as the modifications of the old approaches, these diseases are still quite prevalent with high socio-economic impacts. Therefore, there is a need to continually discover this important therapeutic group of compounds to ignite the zeal for further research and development. The diversity and ubiquity of natural antioxidants in plants and animals are factors that elicit the continual discovery of their roles in these diseases. Therefore, the antioxidant effects of medicinal plants are required to allow their use in medical applications due to their efficacy and safety. Since a plethora of natural antioxidants have been discovered in plants and animals, there are still many more to be identified, considering the biodiversity of this source. Moreover, there are underutilized sources of antioxidants among the available ones. Therefore, more research on the potential of antioxidants may lead to a better understanding of how they work to prevent the oxidative process. Furthermore, antioxidant supplements can be recommended for patients who are suspected of having excessive levels of ROS. However, more research would be needed to confirm their efficacy. Because numerous habits and environmental factors enhance the formation of ROS and damage the body, it would sometimes be necessary to change certain lifestyle behaviours. Current recommendations are based on the intake of a healthy diet (rich in fruits, vegetables, and grains and low in red meat and alcohol) and a healthy lifestyle, which has demonstrated the ability to reduce the risk for diseases. Further research is warranted before using antioxidant supplements as an adjuvant therapy. In the meantime, avoiding oxidant sources such as cigarette smoke and alcohol must be considered when taking dietary antioxidants.

## Conclusion

Antioxidants can scavenge free radicals, hence preventing acute and chronic diseases like neurodegenerative diseases, eye defects, cancer, aging, diabetes, liver disease, and cardiovascular disease. Though oxidative stress affects biological systems in several ways, there are enough antioxidant defenses to slow the advancement of the damage. This review provided sufficient information on the classification (natural and synthetic), mechanisms (sequestration of free radicals and ROS), and sources (plants, animals, and microorganisms) of antioxidants, as well as their roles in preventing non-communicable diseases. There are still many unidentified, neglected, or underutilized sources of antioxidants, especially vegetables and fruits, which could be explored as a game-changing strategy in the prevention and treatment of non-communicable diseases. Diet is a key part of the antioxidant defense system since it provides important antioxidants like vitamin C, vitamin E, and carotenoids. As a result, foods high in these nutrients should be included in a regular diet.

## References

[1] Neha K, Haider MR, Pathak A, Yar MS. Medicinal prospects of antioxidants: A review. *Eur J Med Chem* 2019; 178: 687-704.

[2] Eboh AS. Biochemistry of Free Radicals and Antioxidants. *Sch Acad J Biosci* 2014;2(2):110-118.

[3] Alkadi H. A Review on Free Radicals and Antioxidants. *Infect Disord Drug Targets* 2020;20(1):16-26.

[4] Valko M, Rhodes CJ, Moncol J, Izakovic M, Mazur M. Free radicals, metals and antioxidants in oxidative stress-induced cancer. *Chem Biol Interact*. 2006; 160(1):1-40.

[5] Sharifi-Rad M, Anil Kumar NV, Zucca P, Varoni EM, Dini L, Panzarini E, Rajkovic J, TsouhFokou PV, Azzini E, Peluso I, Prakash Mishra A, Nigam M, El Rayess Y, Beyrouthy ME, Polito L, Iriti M, Martins N, Martorell M, Docea AO, Setzer WN, Calina D, Cho WC, Sharifi-Rad J. Lifestyle, Oxidative Stress, and Antioxidants: Back and Forth in the Pathophysiology of Chronic Diseases. *Front Physiol.* 2020; 11:694.

[6] Mehta SK, Gowder SJT. Basic Principles and Clinical Significance of Oxidative Stress. London: *Intech Open*; 2015.

[7] Atta EM, Mohamed NH, Abdelgawad AAM. Antioxidants: An Overview on the Natural and Synthetic Types. *Eur Chem Bull* 2017;6(8):374-384.

[8] Lourenço SC, Moldão-Martins M, Alves VD. Antioxidants of Natural Plant Origins: From Sources to Food Industry Applications. *Molecules* 2019;24(22):E4132.

[9] Shebis Y, Iluz D, Kinel-tahan Y, Dubinsky Z, Yehoshua Y. Natural Antioxidants: Function and Sources. *Food Nutr Sci* 2013;4(6):643-649.

[10] Ardabili G, Khaneghah M. An overview on some of important sources of natural antioxidants. *Int Food Res J* 2016; 23(3):928-933.

[11] Nimse SB, Pal D. Free radicals, natural antioxidants, and their reaction mechanisms. *RSC Adv* 2015; 5(35):27986-28006.

[12] Anwar H, Hussain G, Mustafa I. Antioxidants from Natural Sources. Antioxidants in Foods and Its Applications. Edited by Emad Shalaby and Ghada Mostafa Azzam, *Intech Open*; 2018.

[13] Xu DP, Li Y, Meng X, Zhou T, Zhou Y, Zheng J, Zhang JJ, Li HB. Natural Antioxidants in Foods and Medicinal Plants: Extraction, Assessment and Resources. *Int J Mol Sci.* 2017; 18(1):96.

[14] Hamid AA, Aiyelaagbe OO, Usman LA, Ameen OM, Lawal A. Antioxidants: Its medicinal and pharmacological applications. *Afr J Pure Appl Chem* 2010;4(8):142-151.

[15] Ali Raza Naqvi S, Nadeem S, Komal S, Ali Asad Naqvi S, SameeMubarik M, Yaqub Qureshi S, Ahmed S, Abbas A, Zahid M, Khan A-U-H, Raja SS, Aslam N. Antioxidants: Natural Antibiotics. Antioxidants. Edited by Emad Shalaby, *Intech Open*; 2019.

[16] Abbas M, Saeed F, Anjum FM, Afzaal M, Tufail T, Bashir MS, Ishtiaq A, Hussain S, Suleria HA. Natural polyphenols: An overview. *Int J Food Prop* 2017;20(8):1689-1699.

[17] Lobo V, Patil A, Phatak A, Chandra N. Free radicals, antioxidants and functional foods: Impact on human health. *Pharmacogn Rev* 2010;4(8):118-126.

[18] Soto-Vaca A, Gutierrez A, Losso JN, Xu Z, Finley JW. Evolution of phenolic compounds from color and flavor problems to health benefits. *J Agric Food Chem* 2012; 60(27):6658-6677.

[19] Botterweck AA, Verhagen H, Goldbohm RA, Kleinjans J, van den Brandt PA. Intake of butylated hydroxyanisole and butylated hydroxytoluene and stomach cancer risk: results from analyses in the Netherlands Cohort Study. *Food Chem Toxicol* 2000; 38(7):599-605.

[20] Sherwin ER. Antioxidants for vegetable oils. *J Am Oil Chem Soc* 1976;53(6):430-436.

[21] Kornienko JS, Smirnova IS, Pugovkina NA, Ivanova JS, Shilina MA, Grinchuk TM, Shatrova AN, Aksenov ND, Zenin VV, Nikolsky NN, Lyublinskaya OG. High doses of synthetic antioxidants induce premature senescence in cultivated mesenchymal stem cells. *Sci Rep.* 2019;9(1):1296.

[22] Saad B, Sing YY, Nawi MA, Hashim NH, Mohamed Ali AS, Saleh MI, Sulaiman S, Talib K, & Ahmad K. Determination of synthetic phenolic antioxidants in food items using reversed-phase HPLC. *Food Chem* 2007;105(1):389-394.

[23] Ballistreri G, Fabroni S, Romeo FV, Timpanaro N, Amenta M, Rapisarda P. Anthocyanins and Other Polyphenols in Citrus Genus: Biosynthesis, Chemical Profile, and Biological Activity. *Polyphenols in Plants* 2019;14(1):191-215.

[24] Kasote DM, Katyare SS, Hegde MV, Bae H. Significance of antioxidant potential of plants and its relevance to therapeutic applications. *Int J Biol Sci* 2015; 11(8):982-991.

[25] Carlsen MH, Halvorsen BL, Holte K, Bøhn SK, Dragland S, Sampson L, Willey C, Senoo H, Umezono Y, Sanada C, Barikmo I, Berhe N, Willett WC, Phillips KM, Jacobs DR Jr, Blomhoff R. The total antioxidant content of more than 3100 foods, beverages, spices, herbs and supplements used worldwide. *Nutr J.* 2010;9:3.

[26] Fardet A, Rock E. In vitro and in vivo antioxidant potential of milks, yoghurts, fermented milks and cheeses: a narrative review of evidence. *Nutr Res Rev* 2018; 31(1):52-70.

[27] Gupta C. Functional foods enhanced with Microbial antioxidants. *Acad J Nutr* 2013; 2(2):10-18.

[28] Dholakiya RN, Kumar R, Mishra A, Mody KH, Jha B. Antibacterial and Antioxidant Activities of Novel Actinobacteria Strain Isolated from Gulf of Khambhat, Gujarat. *Front Microbiol* 2017; 8:2420.

[29] Feng T, Wang J. Oxidative stress tolerance and antioxidant capacity of lactic acid bacteria as probiotic: a systematic review. *Gut Microbes* 2020; 12(1):1801944.

[30] Li M, Xia Q, Zhang H, Zhang R, Yang J. Metabolic Engineering of Different Microbial Hosts for Lycopene Production. *J Agric Food Chem* 2020; 68(48):14104-14122.

[31] Lin X, Xia Y, Wang G, Yang Y, Xiong Z, Lv F, Zhou W, Ai L. Lactic Acid Bacteria With Antioxidant Activities Alleviating Oxidized Oil Induced Hepatic Injury in Mice. *Front Microbiol.* 2018;9:2684.

[32] Gauthier MR, Senhorinho GNA, Scott JA. Microalgae under environmental stress as a source of antioxidants. *Algal Res* 2020; 52:102104.

[33] Sonani RR, Rastogi RP, Madamwar D. Antioxidant Potential of Phycobiliproteins: Role in Anti-Aging Research. *Biochem Anal Biochem* 2015;4(2):1-8.

[34] Masaki H. Role of antioxidants in the skin: anti-aging effects. *J Dermatol Sci* 2010; 58(2):85-90.

[35] Park K. The role of antioxidants in combating the aging process. *Inquiry* 2009; 13:1e6.

[36] Petyaev IM. Lycopene Deficiency in Ageing and Cardiovascular Disease. *Oxid Med Cell Longev* 2016; 2016:3218605.

[37] Kapała A, Szlendak M, Motacka E. The Anti-Cancer Activity of Lycopene: A Systematic Review of Human and Animal Studies. *Nutrients.* 2022; 14(23):5152.

[38] Faghihzadeh F, Hekmatdoost A, Adibi P. Resveratrol and liver: A systematic review. *J Res Med Sci* 2015; 20(8):797-810.
[39] Salehi B, Mishra AP, Nigam M, Sener B, Kilic M, Sharifi-Rad M, Fokou PVT, Martins N, Sharifi-Rad J. Resveratrol: A Double-Edged Sword in Health Benefits. *Biomedicines*. 2018; 6(3):91.
[40] Ko JH, Sethi G, Um JY, Shanmugam MK, Arfuso F, Kumar AP, Bishayee A, Ahn KS. The Role of Resveratrol in Cancer Therapy. *Int J Mol Sci*. 2017 Dec 1; 18(12):2589.
[41] Giordano A, Tommonaro G. Curcumin and Cancer. *Nutrients*. 2019; 11(10):2376.
[42] He Y, Yue Y, Zheng X, Zhang K, Chen S, Du Z. Curcumin, Inflammation, and Chronic Diseases: How Are They Linked? *Molecules*. 2015; 20(5):9183-9213.
[43] Yang CS, Lambert JD, Sang S. Antioxidative and anti-carcinogenic activities of tea polyphenols. *Arch Toxicol*. 2009; 83(1):11-21.
[44] Reyes-Farias M, Carrasco-Pozo C. The Anti-Cancer Effect of Quercetin: Molecular Implications in Cancer Metabolism. *Int J Mol Sci*. 2019; 20(13):3177.
[45] Huang ZQ, Chen P, Su WW, Wang YG, Wu H, Peng W, Li PB. Antioxidant Activity and Hepatoprotective Potential of Quercetin 7-Rhamnoside *In Vitro* and *In Vivo*. *Molecules*. 2018; 23(5):1188.
[46] Pepe G, Pagano F, Adesso S, Sommella E, Ostacolo C, Manfra M, Chieppa M, Sala M, Russo M, Marzocco S, Campiglia P. Bioavailable Citrus sinensis Extract: Polyphenolic Composition and Biological Activity. *Molecules*. 2017; 22(4):623.
[47] Tanase C, Coșarcă S, Muntean DL. A Critical Review of Phenolic Compounds Extracted from the Bark of Woody Vascular Plants and Their Potential Biological Activity. *Molecules* 2019; 24(6):E1182.
[48] Nworu CS, Akah PA, Okoye FB, Toukam DK, Udeh J, Esimone CO. The leaf extract of Spondiasmombin L. displays an anti-inflammatory effect and suppresses inducible formation of tumor necrosis factor-α and nitric oxide (NO). *J Immunotoxicol* 2011; 8(1):10-16.
[49] Venturella G, Ferraro V, Cirlincione F, Gargano ML. Medicinal Mushrooms: Bioactive Compounds, Use, and Clinical Trials. *Int J Mol Sci* 2021; 22(2):E634.
[50] Neira JL, Vega S, Martínez-Rodríguez S, Velázquez-Campoy A. The isolated GTPase-activating-protein-related domain of neurofibromin-1 has a low conformational stability in solution. *Arch Biochem Biophys* 2021; 700:108767.
[51] Isola G, Polizzi A, Muraglie S, Leonardi R, Lo Giudice A. Assessment of Vitamin C and Antioxidant Profiles in Saliva and Serum in Patients with Periodontitis and Ischemic Heart Disease. *Nutrients* 2019; 11(12):E2956.
[52] Singh N. Antioxidants in Oral Health and Diseases: Future Prospects. *IOSR J Dent Med Sci* 2013;10(3):36-40.
[53] Pham-Huy LA, He H, Pham-Huy C. Free radicals, antioxidants in disease and health. *Int J Biomed Sci.* 2008 Jun;4(2):89-96.
[54] Rudrapal M, Khairnar SJ, Khan J, Dukhyil AB, Ansari MA, Alomary MN, Alshabrmi FM, Palai S, Deb PK, Devi R. Dietary Polyphenols and Their Role in Oxidative Stress-Induced Human Diseases: Insights Into Protective Effects, Antioxidant Potentials and Mechanism(s) of Action. *Front. Pharmacol* 2022; 13:806470.
[55] Udeh NE, Nnadi CO, Anaga AO, Asuzu IU. Bioactivity-guided fractionation of a methanol leaf extract from Gnetumafricanum with potential anti-diabetic activity: (-)-epicatechin as the active principle. *J Res Pharm* 2021; 25(1):72-79.
[56] Forman HJ, Zhang H. Targeting oxidative stress in disease: promise and limitations of antioxidant therapy. *Nat Rev Drug Discov* 2021; 20(9):689-709.
[57] Parham S, Kharazi AZ, Bakhsheshi-Rad HR, Nur H, Ismail AF, Sharif S, RamaKrishna, S, & Berto, F. Antioxidant, Antimicrobial and Antiviral Properties of Herbal Materials. *Antioxidants* (Basel) 2020; 9(12):E1309.
[58] Hou HS, Bonku EM, Zhai R, Zeng R, Hou YL, Yang ZH, Quan C. Extraction of essential oil from Citrus reticulate Blanco peel and its antibacterial activity against Cutibacterium acnes (formerly Propionibacterium acnes). *Heliyon*. 2019;5(12):e02947.
[59] Nnadi CO, Nwodo NJ, Kaiser M, Brun R, Schmidt TJ. Steroid Alkaloids from Holarrhenaafricana with Strong Activity against Trypanosomabruceirhodesiense. *Molecules* 2017;22(7):E1129.

[60] Altemimi A, Lakhssassi N, Baharlouei A, Watson DG, Lightfoot DA. Phytochemicals: Extraction, Isolation, and Identification of Bioactive Compounds from Plant Extracts. Plants (Basel) 2017; 6(4):E42.

[61] Chen Z, Liu X, Gao W, Liu C. Studies on pharmacology, toxicology and pharmacokinetics of the chemical constituents in hemorrheologic agent. *Asian J Pharmacodyn Pharmacokinet* 2008; 8(1):15-30.

[62] Moyo B, Oyedemi S, Masika PJ, Muchenje V. Polyphenolic content and antioxidant properties of Moringaoleifera leaf extracts and enzymatic activity of liver from goats supplemented with Moringaoleifera leaves/sunflower seed cake. *Meat Sci* 2012;91(4):441-447.

[63] Al-Snafi AE. Medicinal plants with central nervous effects (Part 2): Plant based review. *Sch Acad J Pharm* 2016; 5(5):175-193.

[64] Del Rio D, Borges G, Crozier A. Berry flavonoids and phenolics: bioavailability and evidence of protective effects. *Br J Nutr* 2010; 104(Suppl 3):S67-S90.

[65] Halliwell B. Free radicals and antioxidants: updating a personal view. *Nutr Rev* 2012; 70(5):257-265.

# Chapter 7

# Parkinson's Disease and Oxidative Stress

**Nabanita Ghosh**[1]
**Krishnendu Sinha**[2]
**and Parames C. Sil**[3,*]
[1]Maulana Azad College, Kolkata, India
[2]Jhargram Raj College, Jhargram, India
[3]Division of Molecular Medicine, Bose Institute, Kolkata, India

## Abstract

Parkinson's disease is the second most prevalent age-related neurodegenerative disease worldwide with the three cardinal motor signs of this disease are tremor, rigidity, and bradykinesia which are accompanied by other non-motor symptoms including hyposmia, autonomic problems, cognitive disorders etc. Primarily, the demise of dopaminergic neurons in the substantia nigra pars compacta leads to this devastating disease. Parkinson's disease could be familial or idiopathic showing early or late onset respectively. Several etiological agents like exposure to neurotoxic substances, aggregation of pathological proteins, energy disturbances in the cell, decreased performance of the ubiquitin-proteasome system, apoptosis, autophagy, mutated genes etc. are shown to be responsible for the disease and all these etiological agents seem to be linked by the oxidative stress. Mutations in genes like PINK1, PRKN, SNCA, LRRK2, DJ1 etc., the oxidative metabolism of dopamine, presence of intracellular iron and ROS-producing enzymes like tyrosine hydroxylase and monoamine oxidase, mitochondrial disfunction etc. contribute to oxidative stress. Oxidative stress, in turn, initiates a cascade of events including the shutdown of the electron transport chain, blocking of complex I in the electron transport chain, induction of autophagy etc. which ultimately led to cell demise. Oxidative stress also severely affects available therapies like dopamine supplements, deep brain stimulation etc. ultimately the progression of the disease.

**Keywords:** oxidative stress, apoptosis, Parkinson's disease, xenobiotics

* Corresponding Author's Email: paremes@jcbose.ac.in.

In: The Role of Reactive Oxygen Species in Health and Disease
Editor: Kuladip Jana
ISBN: 979-8-89113-283-2

## 1. Introduction

Parkinson's disease (PD) is the second most prevalent chronic neurodegenerative disorder after Alzheimer's disease, responsible for severely affecting the quality of life and life expectancies of over 0.001% of individuals of any given population at a particular time which sharply increased up to 3% in elderly people above the age of 65 [1, 2]. After the onset of symptoms with the help of modern medicine the life expectancy of a patient hardly goes above 10 years. The three cardinal motor signs of this disease are tremor, rigidity, and bradykinesia which are accompanied by other non-motor symptoms including hyposmia, sleep disorders, mood disorders and autonomic problems especially constipation which may precede the development of motor symptoms for years or even decades severely affecting the quality of life [1–3]. In late-stage patients faced deadly cognitive impairment leading to social withdrawals and immense psychological impairment [2]. PD's motor deficits are usually associated with dopaminergic axonal neurodegeneration which initiates at the striatum and is then followed by dopaminergic neuronal demise in the substantia nigra pars compacta (SNpc) [1]. Dopaminergic neurons in PD have intracellular protein aggregates typically composed of unusually aggregated α-synuclein termed Lewy bodies and the striatum faces chronic neuroinflammation [1]. These cumulatively lead to severe dopamine deficiency in the striatum and chronic neuroinflammation.

PD could be idiopathic or familial. More and more current research indicates that idiopathic PD might develop due to environmental and professional hazards for which exposure to pesticides like rotenone, herbicides like paraquat etc. is likely to be responsible [4, 5]. Idiopathic PD is of late onset in nature. Whereas, familial PD is caused by specific mutations in different genes, which also involved in mitochondrial dynamics and neuronal transport. But whatever the etiological agents are, oxidative stress always plays a central role in the development of the disease [1].

Oxidative stress describes a disorder caused by an imbalance between the prooxidant-antioxidant balance in the cell and cellular biomolecules experiencing serious oxidative damage, eventually compromising cell viability [6]. Reactive oxygen species comprise free radicals, like superoxide anion ($O_2{\bullet}^-$), hydroxyl (•OH) radicals, etc., and nonradical species like $H_2O_2$ derived from oxygen ($O_2$) [6, 7]. In a living system, superoxide anion radicals are generated either through metabolic processes or subsequent activation of oxygen by physical irradiation. This superoxide anion is the primary ROS. It interacts with biomolecules directly or through enzyme- or metal-catalyzed processes and generates secondary ROS [8]. The adverse impact of ROS on the cellular system includes damage to biomolecules such as DNA mutation, amino acids oxidation in proteins, lipid peroxidation and inactivation of enzymes. ROS contributes significantly to the pathophysiology of numerous diseases like hypertension, diabetes, preeclampsia, atherosclerosis, acute renal failure, Alzheimer's disease, Parkinson's disease etc. In this chapter, we will emphasise the role of oxidative stress in the initiation and progression of PD.

## 2. Oxidative Stress

Professor Helmut Sies coined the term "oxidative stress" in 1985. Oxidative stress is an acute or chronic increase in the steady-state level of ROS, disturbing cellular metabolism and leading to damage to cellular constituents [6]. New outcomes from numerous research showing the pivotal role of oxidative stress in almost every pathophysiological condition one can think of. The fine balance between the oxidative damage induction and removal of its damaging effect determines the severity of oxidative stress. Free radicals are the oxidative stress-causing elements. Free radicals are molecular fragments or molecules holding one or more unpaired electrons in their molecular orbitals and due to these unpaired electrons, free radicals become notoriously reactive [6, 9, 10]. Oxygen-derived free radicals like superoxide anion ($O_2•^-$) and hydroxyl (•OH) radicals along with non-radical species like hydrogen peroxide ($H_2O_2$) are the most important class of radical species in living organisms and are known as reactive oxygen species (ROS) [6, 11–14]. For example, $H_2O_2$ is produced from superoxide anions by the dismutation reaction catalysed by III and IV of the mitochondrial respiratory chain (Figure 1). Likewise, the Fenton reaction or Haber–Weiss reaction produces extremely reactive and noxious hydroxyl radical (•OH) by reaction between $H_2O_2$ and $O_2•^-$or $H_2O_2$ cleavage by $Fe^{2+}$ (or $Cu^{2+}$), respectively [6, 15, 16]. Besides the respiratory chain, an outer mitochondrial membrane–resident flavoprotein monoamine oxidase (MAO), a mitochondrial intermembrane space enzyme p66Shc, matrix pH, and altered mitochondrial membrane potential are also proven important sources of ROS [6, 15–20]. Apart from hydroxyl (•OH), superoxide anion ($O_2•^-$), hydrogen peroxide ($H_2O_2$), and alkoxyl (RO•), other types of reactive radicals ensuing from oxygen in the living organisms are peroxyl radicals (ROO•). The protonated form of superoxide radicals ($O_2•^-$), the hydroperoxyl radical (HOO•), is the simplest peroxyl radical which presents as 0.3% of any superoxide in the cytosol in typical cells [6, 21]. Lipid peroxidation or fatty acid peroxidation is the main damaging effect of hydroperoxyl radical which is initiated by two parallel pathways called fatty acid hydroperoxide (LOOH)-independent or LOOH-dependent pathways [22].

Oxygen possesses a unique electronic configuration and potentially can act as a radical. The molecular oxygen forms a relatively stable intermediate, the superoxide anion radical ($O_2^-$ ), by accepting one electron which can arise in the living system by either metabolic procedures or by subsequent activation of oxygen by physical irradiation [6, 14, 15]. It is considered the primary ROS. Superoxide radical interacts with biomolecules directly or through enzyme- or metal-catalysed processes and produces secondary ROS [8, 9]. For an instance, SOD catalyses the dismutation reaction to produce $H_2O_2$ from superoxide anions. Similarly, Haber–Weiss reaction or Fenton reaction produces highly reactive and toxic hydroxyl radical (•OH), through interaction between $H_2O_2$ and $O_2•^-$ or $H_2O_2$ cleavage by $Fe^{2+}$ (or $Cu^{2+}$), respectively [6, 15, 16].

Mitochondria play a pivotal role in the creation of free radicals in living cells [23]. Upon entry of glucose into the cell, pyruvate is generated and is utilized in the TCA cycle which occurs in the mitochondrial matrix [6, 19]. Through NADH and $FADH_2$, electrons get transferred from TCA cycle intermediates to the electron transport chain. In the TCA cycle, NADH is formed by three dehydrogenases, namely isocitrate dehydrogenase, α-ketoglutarate dehydrogenase complex and malate dehydrogenase and $FADH_2$ is formed by the succinate dehydrogenase [24]. However, the respiratory chain is made up of five protein complexes which are encoded concurrently in the nuclear and mitochondrial genomes [6, 25]. When electron transfer through the respiratory chain occurs, the proton from the mitochondrial matrix gets exported across the inner membrane, which in turn establishes strong mitochondrial membrane potential, positive on the outside. In the inner mitochondrial membrane, complex I is the sole electron acceptor from NADH, whereas electrons are transferred from FADH2 to ubiquinone (Q) by complex II (succinate dehydrogenase) (Figure 1) [6, 24, 26]. On the other hand, water production and consumption of oxygen occur at complex IV [6, 24]. This is coupled to the ATP synthase (complex V) and promotes mitochondrial ATP formation from ADP and inorganic phosphate. The ATP, synthesized this way, is translocated to the cytosol. Adenine nucleotide translocator does this job in exchange for ADP [24]. Therefore, this electron transport chain is considered the utmost vital source of ATP in mammalian cells and hence essential for living organisms. Nevertheless, a small number of electrons escape to oxygen spontaneously during the process of energy transduction and consequently the oxygen-free radical superoxide is generated (Figure 1) [6, 24, 27]. ROS is a by-product of usual mitochondrial respiration [6]. 1–3% of all electrons in the electron transport chain of mitochondria outflow to produce superoxide radicals [6, 9, 15]. Various redox centres of the mitochondrial electron transport chain escape electrons to molecular oxygen and thus play the foremost role in superoxide production [20]. However, most of the superoxide radicals are generated by complex I in the intact mammalian mitochondria and released solely into the matrix by matrix-protruding iron-sulphur clusters of the hydrophilic arm of complex I [6, 15, 28]. Besides complex I, complex III is also considered a significant site of $O_2^-$ production [29, 30]. Ubiquinone and centre "o" within complex III is also considered a major contributor to $O_2^-$ formation by complex III [31]. Ubiquinone is a part of the respiratory chain and connects Complexes I and II with III [6, 15]. In the process of electron transfer between these complexes, ubiquinone is reduced to intermediate unstable semiquinone, which is accountable for $O_2^-$ formation [6] (Figure 1).

In addition to the respiratory chain complex, approximately 10 potential mitochondrial ROS-generating systems are identified as the source of mitochondrial $O_2^-$ and $H_2O_2$ [6, 20]. Among these, two systems are Krebs cycle enzyme complexes: a-ketoglutarate dehydrogenase (a-KGDH) and pyruvate dehydrogenase are considered major sources [23]. Noticeably, mitochondrial a-KGDH produces a high amount of $H_2O_2$ linked to the increased amount of nicotinamide adenine dinucleotide, and it is obvious that this elevated oxidant burden increases additional ROS production from mitochondrial complex I, which in turn imparts an elevated amount of oxidative stress to cells that promote cell death [6, 32].

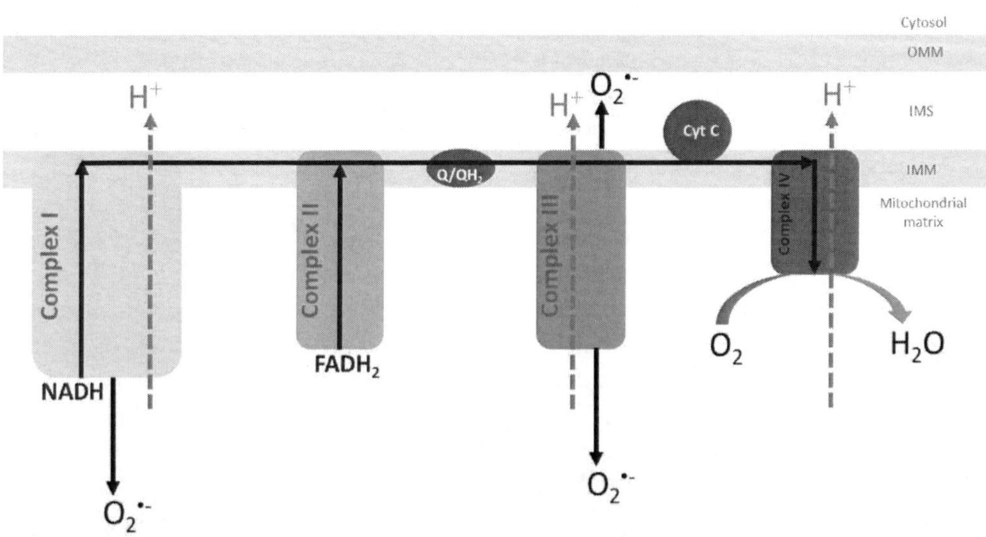


**Figure 1.** Mitochondrial electron transport chain and reactive species formation sites.

Being primary ROS generators mitochondria often suffer from severe ROS-mediated injuries like oxidative damage to the mitochondrial membrane and mitochondrial DNA (mtDNA) [6, 15, 33, 34]. Elevated levels of $O_2^-$ and •OH are major factors associated with mtDNA damage and play a major role in inducing apoptosis [6]. Also, oxidation and inactivation of iron-sulphur (Fe-S) proteins like aconitase, results in the release of iron which through the Fenton and Haber–Weiss reaction generate hydroxyl radical and significantly contributed to ROS [6, 15].

However, living organisms develop an antioxidant defence system to combat oxidative stress. This system protects from oxygen's deleterious effects. Phase II detoxification enzymes such as three superoxide dismutase isoforms namely SOD1, SOD2 and SOD3 along with glutathione peroxidases (GPxs), glutathione S-transferase (GST), catalase (CAT), glutathione reductase (GR), glutaredoxin, peroxiredoxins (Prxs), two thioredoxin isoforms (TRX1 and TRX2), non-enzyme glutathione (GSH) and glutathione disulfide (GSSG) are being used to combat oxidative stress [6, 35]. Superoxide dismutase neutralizes $O_2^-$ by converting it into $O_2$ and $H_2O_2$, and this $H_2O_2$ gets converted to $H_2O$ and $O_2$ by the catalase enzyme. On the other hand, the conversion of GSH from GSSG is catalysed by glutathione reductase while glutathione S-transferase aids in the radical-scavenging reaction of GSH [6]. GPx helps the consumption of $H_2O_2$ in combination with GSH [6].

## 3. Parkinson's Disease

Parkinson's disease (PD) is one of the most occurring age-related progressive neurodegenerative disorders [36]. Dopaminergic neurons' degeneration in substantia nigra pars compacta (SNpc), dopamine depletion in the striatum, cytoplasmic presence of aggregated proteins called Lewy bodies (LBs), and chronic neuroinflammation are the main physiological characteristics of PD [37–39]. In 2006, de Lau and Breteler reported that approximately 0.3% of the world population (i.e., approximately 10 million people worldwide) and 1% of those above 60 years are affected by PD [40]. It is highly prevalent in Western Europe, China, and America with an increasing rate in India. Population-based surveys in India have shown a crude prevalence rate (CPR) of PD ranging from 6-53/100000 [41]. This estimate excludes the Parsi community [41]. The epidemiological studies indicate a drastic increase in the onset of PD among the aged population from 1982 through 2007 [40]. Worldwide studies show that rural communities show greater susceptibility to PD in comparison to urban [40, 42]. This suggests that environmental factors may be of greater relevance than genetic factors in the pathogenesis of PD. With the growing occurrence in the elderly population, the management of PD is a progressively vital and challenging part of medical practice for neurologists and general physicians. Our understanding of the pathogenesis of the disease has been upgraded in the last decade with the identification of various gene mutations which may shed light on the mechanisms of pathogenesis in sporadic cases of PD. The diagnosis of PD remains fundamentally a clinical one, and it is imperative to identify the initial features together with symptoms and signs suggesting further causes of Parkinsonism. There has also been a quick development in the treatment choices both in the initial and in the delayed stages of the illness together with a better awareness of non-motor complications [36].

### 3.1. History of Understanding Parkinson's Disease

In his classic 1817 monograph "Essay on the Shaking Palsy," James Parkinson defined the basic clinical features of PD, the second most prevalent age-related neurodegenerative disease after Alzheimer's disease (AD) [43–46]. Even though more than a century passed before the essential pathological feature of PD was identified to be the loss of neurons in the substantia nigra pars compacta (SNpc), the pace of discovery augmented ensuing Arvid Carlsson's 1958 discovery of dopamine (DA) in the mammalian brain. SNpc neurons were then found to form the nigrostriatal dopaminergic pathway, and this line of research concluded with two crucial discoveries [47]. First, loss of SNpc neurons leads to striatal DA deficit, which is accountable for the key symptoms of PD. Second, the replacement of striatal DA through the oral administration of the DA precursor levodopa (L-3,4-dihydroxyphenylalanine) improves most of these symptoms [48]. Though the discovery of levodopa transformed the treatment of PD, it was soon learned that after some years of Levodopa treatment most patients develop involuntary movement, termed "dyskinesia," mostly denoted as 'Levodopa Induced Dyskinesia' (LID), which is tough to control and significantly harm the quality of life [49]. Current research is directed toward the prevention of dopaminergic neurodegeneration. Nevertheless, contempt advances toward this goal, all existing treatments are symptomatic; none halts or betters dopaminergic neurodegeneration.

The main obstacle in the development of neuroprotective drugs is the lack of knowledge about the exact reason that drives neurodegeneration in PD and the specific molecular events involved in it. To delve deeper into the mechanism of neurodegeneration, several studies have been conducted across the globe in the last decade by establishing animal models of PD using neurotoxic agents that cause PD-like symptoms by selectively destroying dopaminergic neurons. Such neurotoxic agents include herbicides and pesticides (Paraquat and Rotenone), 6-hydroxydopamine (6-OHDA) and 1-methyl-4-phenyl-1, 2, 3, 6-tetrahydropyridine (MPTP) [50–52]. Contact of humans to MPTP causes a syndrome that impersonates the neurological symptoms and dopaminergic neurodegeneration of PD, and MPTP toxicity in mice is the maximum studied animal model of PD [48]. These studies have focused primarily on three types of cellular dysfunction vital in the pathogenesis of PD: oxidative stress, mitochondrial respiration defect, and abnormal protein aggregation. Furthermore, the MPTP monkey model has produced valuable information regarding the functional changes in basal ganglia circuits that occur after striatal DA depletion, and this model remains the benchmark for the preclinical assessment of new therapies aimed at easing the symptoms of PD [48-50]. Many results from MPTP studies have been established in human PD brains to date [50].

Apart from sporadic PD (which is primarily idiopathic), several mutations have been identified. These have been shown to cause familial Parkinsonism. A few such mutations include PARK-3/PARK-4, LRRK2, PINK-1, Parkin etc. [50-53]. Most of these genes are found to be associated with α-synuclein synthesis (SNCA) or intracellular signalling of brain cells (LRRK2) or autophagy and mitochondrial health (Parkin and PINK-1) [54–57]. Familial Parkinsonism has been seen to have an early onset unlike sporadic PD [54]. Such mutations via inheritance make people genetically predisposed to PD [46].

### 3.2. Clinical Manifestations of Parkinson's Disease

PD is a progressive neurodegenerative disease with a mean age of onset at 55 (in the case of sporadic PD), which significantly increases with age, from 20/100,000 overall to 120/100,000 at age 70 [46, 58]. In most cases of sporadic PD (about 95%), a late onset with no apparent genetic reasons has been observed. But in familial PD an early onset at or before age of 40 is mostly observed and a strong inheritance pattern has been identified [46, 59, 60]. No blood biomarker of PD has yet been identified; diagnosis of this disease is largely symptomatic [46, 60]. It is very crucial to recognize early symptoms for a timely diagnosis and decelerating disease progression. The four cardinal signs of PD are generally referred to as the acronym TRAP, which stands for Tremor at rest, Rigidity, Akinesia (or bradykinesia) and Postural instability [61]. A recent rectification of the diagnostic criteria excludes postural instability as a hallmark. It is now considered a supportive criterion, and red flag [62]. Hence nowadays, PD is considered a movement disorder, judged by the three revised cardinal signs namely, tremors, rigidity, and bradykinesia [62]. Additionally, freezing (motor blocks) and flexed posture has been considered classic features of Parkinsonism, with PD as the most common form [61]. Because of the diverse profiles and lifestyles of those affected by PD, motor and nonmotor impairments should be evaluated in the context of each patient's needs and goals [61]. Non-motor symptoms like depression, anxiety, cognitive impairments etc. are drawing greater attention and along with motor symptoms, are now considered the supportive criteria [59–64]. Clinically, any illness associated with striatal DA deficiency, or any form of striatal injury may lead to "parkinsonism," which is typified by tremors at rest, rigidity, freezing, and slowness or absence of voluntary movement along with contextual postural instability [62]. Approximately 80% of cases of Parkinsonism occur due to PD [46]. In recent times many different treatment options have been developed in a stage-specific manner along with a better understanding of non-motor complications [36]. Modern medicines consist of dopamine agonists, MAO-B inhibitors, NMDA-receptor antagonist amantadine, and neuromodulation by deep brain stimulation. However, none of these has proven to have disease-modifying effects and the clinical outcome of the treatment may seem insignificant as the disease advances [36]. Even after some years of treatment, patients develop involuntary movements as side effects which are named "dyskinesia." Dyskinesia is difficult to manage and substantially worsens the quality of life. Despite some advances in modern research towards the goal of preventing dopaminergic neurodegeneration in PD brains, none have been found to cesses the neurodegeneration and mostly acts in a symptomatic manner [39].

### 3.3. Aetiology of Parkinson's Disease

Late-onset PD or sporadic PD is idiopathic [63]. The role of environmental toxins and genetic factors is uncertain in these cases [65]. According to some hypotheses, PD-related neurodegeneration is a result of exposure to any neurotoxin specific to dopaminergic neurodegeneration [60, 65–67]. Theoretically, exposure to chronic sub-acute neurotoxin could be a cause behind the neurodegeneration of PD which initiates a self-perpetuating cascade of damaging events [68]. Accidental exposure to MPTP develops Parkinsonism-like symptoms in humans [69]. Paraquat is also a reported dopaminergic neurotoxin which is widely used as a herbicide and structurally similar to the active metabolite of MPTP, 1-methyl-4-phenylpyridinium (MPP+) [70]. Like MPP+, rotenone, a mitochondrial poison which is used as an insecticide, causes Parkinsonism in men [71]. Worldwide studies show that rural communities show greater susceptibility to PD in comparison to urban [72]. This suggests that environmental factors may be of greater relevance than genetic factors in the pathogenesis of PD. However, high nicotine and caffeine intake lessen the risk of PD development, as they are reported to have beneficial effects on the health of dopaminergic neurons [73]. Along with the genetic or environmental aetiological factors related to PD neurodegeneration, the endogenous toxin may promote PD neurodegeneration. Normal metabolite of dopamine, which generates harmful reactive oxygen species (ROS) is considered as one of such endogenous toxins with detrimental effects on neuronal health. Certain environmental stress or inherited disorders in metabolic pathways could also give rise to neurotoxic metabolites. It has been reported that patients with certain polymorphisms in the gene encoding the xenobiotic detoxifying enzyme cytochrome P450, might be predisposed to young-onset PD [74]. Also, a derivative of isoquinoline has been found in PD brains, which is toxic to dopaminergic neurons [75].

## 3.4. Pathogenesis of Parkinson's Disease

Whatever the initial trigger of neurodegeneration in PD, studies of toxin-induced PD models and the genetically predisposed forms of PD suggest two major hypotheses about its pathogenesis. According to one hypothesis, the aggregation of misfolded proteins (especially α-synuclein) is responsible for the death of SNpc dopaminergic neurons, while the other postulates that mitochondrial dysfunction and the consequent oxidative stress along with oxidized toxic dopamine metabolites are the main culprits [74–76]. These pathogenic factors of PD are not mutually exclusive. Contemporary PD research is directed towards investigating the interaction points between the various pathways leading to the demise of SNpc dopaminergic neurons. One such interaction is supported by the report that oxidative damage to α-synuclein increases its capability to misfold and aggregate [77]. Whether the cell death pathways activated at PD's inception and progression involve common downstream machinery or remain divergent is debated to date.

### *3.4.1. Neuroinflammation and PD*

Neuroinflammation may not be the prime inductor of neurodegeneration. But epidemiological and pre-clinical data of age-related neurodegenerative diseases referred to chronic neuroinflammation as the "silent driver" of neuronal dysfunction and death during disease advancement [78–81]. Activated microglia and astrocytes are often used as an indicator of neuroinflammation in the central nervous system. In addition, elevated levels of proinflammatory factors like reactive oxygen and nitrogen species, cytokines, and chemokines occasionally result in disruption of the blood-brain barrier which leads to infiltration of the adaptive immune system in the brain resulting in adverse immune reactions contributing to disease pathogenesis [82]. Activated microglia are key in defence against inflammatory diseases and CNS infections and SN is comparatively rich in these cells compared to other brain regions [83]. Due to lower intracellular glutathione levels, nigral dopaminergic neurons are predisposed to microglial-mediated oxidative stress and injury [84, 85]. According to a recent report, after intra-striatal injection of 6-OHDA, nigrostriatal dopaminergic neurons die. This is complemented by activation of microglia detected by upregulation of the complement type 3 receptor [86]. Likewise, MPTP-caused neurodegeneration is also reported to be associated with activated microglia [87]. Recent research similarly reports that microglial activation and resultant neurotoxicity are a consequence of toxin-induced stimulation of NADPH oxidase [88]. Other factors like TGF-β, IFN-γ, brain-derived neurotrophic factor (BDNF), CD22, CX3CL1, neurotrophin-3, and neurotransmitters are also capable of microglial activation. Apart from neuroinflammatory molecules, activated microglia also secrete COX2, PGE2, etc. that contribute to the neurodegeneration in PD [89, 90]. Apart from that, the astrocytes, the most abundant glial cell in the brain, which play important role in the development and maintenance of CNS homeostasis, show altered function during progressive neuroinflammation [91]. Astrocytes have both protective and degenerative functions in the brain based on the microenvironment and factors released in reaction to it. Neurons are metabolically linked with astrocytes and depend on them to survive under any type of oxidative stress [92]. During any type of CNS damage (neurodegenerative disease or oxidative stress), the typical astrocytic function impairs temporarily or permanently leading to persistent neuroinflammation and pathogenesis in neurodegenerative diseases [91, 92]. The triggers behind astrogliosis may be manifold, such as infection, trauma, α-synuclein accumulation, and any type of neurotoxic insults [93, 94]. Reactive astrocytes contribute to neurodegeneration by producing ROS and releasing inflammatory cytokines (TNF-α, IL-1β, IL-6 etc.) [95].

Apart from brain resident immune cells, peripheral T cell infiltration has been reported to have a pivotal role PD brain [96]. Among T cell types, δT+ cells express CD25 mostly in Parkinsonian patients rather than in healthy populations [96, 97]. A similar result has been found in another study conducted with PD patients and normal individuals where the CD4+: CD8+ ratio and CD4+: CD25+ ratio decreased, whereas the number of IFN-γ producing T-cells increased in PD patients compared to unaffected controls [98]. In PD patients and MPTP-induced rodent models of Parkinsonism, CD4+ and CD8+ T-cells from systemic circulation have been found to invade the blood-brain barrier and enter the substantia nigra. This indicates increased blood-brain-barrier dysfunction with disease progression strengthening the possibility of infiltrated peripheral T-cells in the CNS to have a pivotal role in the inflammation with disease progression [99]. Interestingly, mice lacking T and B lymphocytes are found to be resistant to MPTP-induced Parkinsonism [100]. To date, the exact intensity and severity of chronic neuroinflammation and peripheral immune response during PD remains elusive.

### *3.4.2. Transcription Factors Involved in Neuroinflammation*

Transcription factors are proteins binding to specific DNA sequences, controlling the transcription of genetic information from DNA to mRNA transcript. NF-κB is such a contextually relevant transcription factor. NF-κB (nuclear factor kappa-light-chain-enhancer of activated B cells) proteins are a family of ubiquitous transcription factors which play a pivotal role in inflammation and immunity [101]. NF-κB transcription factors control growth, differentiation, and death of cell by controlling the expression level of several genes related to these pathways. The NF-κB proteins of the mammalian system consist of five different members that bind to 10-base pair κB sites as homodimers or heterodimers [102]. All these members have a Rel-homology (RHD) domain, essential for DNA binding and dimerization. The three Rel members of the family, RelA (p65), RelB, and cRel, have a C-terminal transcription activation domain (TAD) that positively regulates gene expression [103, 104]. All NF-κB proteins except for RelB (only form heterodimers) are capable of homodimerization or heteromerization with the other NF-κB proteins [104]. The ability to interact in various configurations results in their binding to κB sites in distinct DNA sequences with varying affinities, thus regulating unique and overlapping, gene sets [105, 106]. NF-κB activity is principally regulated by the IκB proteins, which, possess ankyrin repeats and inhibit their DNA binding ability. Three of these proteins namely IκBα, IκBβ, and IκBε are considered "typical" IκBs. These proteins upon binding to NF-κB proteins and mask their DNA binding domains. Their activity is controlled through phosphorylation by upstream IκB kinases (IKKs) [107].

Among many ways of activating NF-κB, two main signalling pathways lead to the activation of NF-κB target genes, the canonical (or classical) and non-canonical (or alternative) pathways [104, 106, 108]. These two pathways may usually be distinguished by the involvement of p50 (canonical) or p52 (non-canonical) [104]. The canonical NF-κB pathway activation signal mostly comes from the stimulation of proinflammatory receptors, such as the TNF Receptor superfamily, the Toll-Like receptor family (TLRs), and cytokine receptors for the Interleukins. Genotoxic agents also activate it [106]. These receptors typically activate an IKK complex consisting of IKKα and IKKβ (catalytic kinases) and IKKγ, the regulatory subunit (also called NEMO). IKK complex recruitment and activation are usually dependent on various TNF Receptor Associated Factor (TRAF) family proteins [106]. The p105 (*nfkb1* gene product) is constitutively processed by the proteasome into p50, whose heterodimer with RelA (or c-Rel) is held inactive by its interaction with the inhibitory IκB proteins. Subsequently, the ubiquitylated IκBα is degraded by the proteasome, thus unmasking the DNA binding activity of the p50/RelA heterodimer and allowing its translocation to the nucleus to bind κB sites and activate gene transcription [104, 106, 109]. On the other hand, noncanonical NF-κB activation is stimulated by specific TNF receptor family members which signal through the recruitment of TRAF2 and TRAF3. Such TNF receptor families include LTβR, CD40, CD27, CD30, BAFF-R, RANK, and a few others. The upstream kinase in this pathway is the NF-κB-inducing kinase (NIK). Constant degradation of NIK in resting cells prevents constitutive activation of the non-canonical NF-κB pathway [104, 109]. Apart from the known and widely studied NF-κB dimers of canonical and non-canonical pathways, another unusual heterodimer of p65 and p52 has been reported to be involved in GDNF-mediated neuroprotection of dopaminergic neurons of substantia nigra in the rat model of early PD [110].

### *3.4.3. Neurotransmitter Alterations in PD*

PD occurs when the dopaminergic neurons of the substantia nigra degenerate or die due to severe gliosis resulting in decreased dopaminergic input to the striatum and other basal ganglia nuclei [111]. The major reason behind loss of nigral dopaminergic neurons is probably their high sensitivity to excitotoxicity [112]. This resultant lack of dopamine supply causes major physiological disruptions in the basal ganglia-thalamocortical motor circuit, ultimately leading to altered motor activity, which mainly involves GABAergic and glutamatergic circuitry. This disruption is responsible for the clinical symptoms observed in PD [113]. Almost 70% of the afferent neurons to nigral dopaminergic neurons are GABAergic arising from the neostriatum, and the external globus pallidus (GPe) and efferent nigral GABAergic neurons in the substantia nigra [114, 115]. The substantia nigra pars reticulata (SNr) and the internal globus pallidus (GPi) are the output nuclei for the basal ganglia circuitry [116, 117]. These neurons release

GABA and have a significant role in the inhibition or disinhibition of neurons in the main target nuclei of thalamus and superior colliculus [118]. The level of GABA has been reported to increase in the SNr in hemi-parkinsonian rats [118]. The glutamatergic supply to SN comes specifically from the subthalamic nucleus (STN), and the ventral tegmental area (VTA) lying adjacent to it [119–121]. According to some previous studies in PD patients and animal models of PD, 6-OHDA-lesioned rats, and MPTP-treated monkeys, subthalamo-pallidal, subthalamo-nigral, and cortico-striatal pathways has shown hyper-activity due to loss of nigral dopaminergic neurons [122]. The motor disturbances during PD is probably caused due to changes in the firing of STN leading to abnormal activity of the basal ganglia output system the trans-subthalamic pathways [123]. Such changes ultimately alter basal glutamate levels in the SNr after the nigrostriatal denervation. Direct dopaminergic innervation of the STN might control STN neuron activity directly or via presynaptic modulation of neurotransmitter release [124]. Many subthalamic afferent neurons reach SNr and form asymmetrical synapses with tyrosine hydroxylase (TH)-negative (presumably GABAergic) dendrites, but over 10% contact TH-positive dendrites [125]. So, glutamate can directly excite DA-cells or indirectly inhibit them by exciting GABAergic neurons depending on the target cell and the receptor type (ionotropic or metabotropic) [126]. From the available literature, it is evident that the motor deficits that occur in PD patients are an outcome of the severe deregulation of the neurotransmitter system affecting various brain regions responsible for motor coordination [127].

### *3.4.4. Altered Neurotransmitter Receptor Signalling Mediated Cognitive Impairment in PD*

Although PD is primarily marked by motor impairment, accumulating evidence suggests that various non-motor symptoms, such as mild cognitive impairment (MCI), can significantly impact the quality of life and even disease prognosis [126, 128, 129]. According to longitudinal cohort studies, dementia is common in PD, affecting over 80% of those who survive for 20 years [130]. In addition to mechanisms like neuronal apoptosis, neuro-inflammation etc., recent reports on neurodegeneration and cognitive dysfunction and subsequent deterioration associated with a variety of mechanisms show significant involvement of pathological modifications in different receptors namely serotonin, adrenaline, dopamine receptors, N-methyl-D-aspartate, and acetylcholine receptors [131].

It is long reported that adrenergic α1 and β1 receptor density increases and α2 adrenergic receptors decrease in demented PD patients [132]. It was also reported that adrenergic α1 receptors increased in the synaptosomal fraction and adrenergic β receptors increased in both synaptosomal and microsomal fractions in the PD brains.

The relationship between acetylcholine receptors and PD is not well documented. Few available reports indicate an increased nicotinic acetylcholine receptor in the nigrostriatal area in PD brains [133]. On the contrary, a significant decrease in the density of α4β2-nAChRs in PD patients has been observed with progressive cognition [134]. The density of nAChRs has been reported to decrease significantly in the brainstem and frontal cortex of PD patients without dementia in comparison to healthy individuals, using (123) I-5IA SPECT [135].

Most studies focused on the correlation between dopamine receptors and dementia in PD. Several experimental outcomes suggest an association between dopaminergic receptor dysregulation in multiple brain regions and cognitive insufficiencies in older PD patients

[136–138]. Apart from acute nigrostriatal denervation, cognitive impairments in PD brains are associated with malformations in temporal processing and functioning of prefrontal cortex [139, 140]. 6-hydroxydopamine lesion in the medial forebrain bundle ofSprague–Dawley rats have shown radically reduced dopamine D1/D2 receptor [141]. The findings of Fang et al. and Sun et al. suggest a strong correlation between the downregulation of dopamine D2 receptors in the striatal PD rats the impaired learning and memory, mediated probably by unknown mechanisms involving altered signalling through dopaminergic receptors [142–144]. PD patients with dementia and cognitive decline exhibited an increase in D2 receptor density in the ventro-intermedius as measured by attentional deficits, and impaired cognition in 6-hydroxydopamine-lesioned PD rats, which showed significant recovery on sub-chronic treatment with the dopamine D2/3 receptor agonist, piribedil [145]. On the other hand, it is also reported that PD related dementia exhibited an increase in D2 receptor density in the ventro-intermedius in correlation with cognitive failure as quantified by MMSE scores [146]. The reasons behind these contradictory results are far from clear. Further investigations are required to dig deeper.

Apart from the neurotransmitter receptors mentioned in the previous sections, another important receptor needs special mention when it comes to cognitive impairment in correlation with a brain disorder, the NMDA receptor. There are very few reports indicating a correlation between NMDA receptors and cognitive decline in PD. A recent study observes that downregulation of NMDA receptors in the hippocampus, amygdala, CA1, and caudate putamen, in 6-OHDA-lesioned rats, caused anxiety, which was ameliorated by simvastatin, an NMDAR agonist [147]. Several studies investigated the changes of NMDA receptors in experimental animal models of PD and post-mortem tissue from PD patients.

However, contradictory results have been obtained regarding the changes of NMDA receptors experimental animal and human PD brains. 6-OHDA-lesioned rat's striatum showed increased an increase in NMDA binding sites [148]. A similar observation was reported in another study also where a bilateral increase in NMDA binding was seen which was more pronounced in the lesioned side of the striatum [149]. However, experiments with post-mortem PD brains present contrasting results. Significant decline in the expression and binding of striatum's NMDA receptor to glutamate in the post-mortem PD brain was observed. Also, a reduced expression of NMDA NR1 mRNA was detected in the prefrontal cortex of PD brains [150]. Experiments in animal models came up with different results. A 6-OHDA-lesioned rat demonstrated the richness of NR1 and a reduction in NR2B subunits in lesioned striatal synaptosomal membranes, with an unchanged expression of NR2A. The augmented NMDAR expression was restored upon chronic treatment with L-dopa [151]. Most of the work to date focused on NMDAR alteration in the striatum, frontal cortex, and substantia nigra.

Now a days, the plausible role of the hippocampus in nonmotor symptoms of PD has been of raising interest. The degree of dementia in PD patients is correlated to the degree of hippocampal LB pathology [152]. Diffusion tensor imaging in PD patients showed modified hippocampal connectivity which correlates with the occurrence of memory deficitsindicating that different plasticity may be one of the causes for structural differences in the hippocampus [153].

## 4. Xenobiotics and Parkinsonism

### 4.1. Paraquat

Paraquat (N, N′-dimethyl-4-4-4′-bypiridinium) (PQ) is an herbicide widely used in agriculture. It has a structural and behavioural resemblance to MPP+, a potent neurotoxin for inducing Parkinsonism in an animal model [154]. PQ is one of several chemicals used as herbicide is known to cause damage to the dopaminergic system [155]. According to epidemiological reports, herbicide use increases the risk of developing PD, but, in the case of PQ, only 95 cases of PD has been reported in human [154–157]. PQ exerts its deleterious effects through oxidative stress mediated by redox cycling, by ROS generation [154, 156]. In animal models, PQ application reduced motor activity and a dose-dependent loss of TH-positive striatal fibres and midbrain SNpc neurons [158]. Furthermore, it is suggested that the damage done by PQ is caused by PQ is mostly facilitated by glial cells. Its importance in PD research lies in its ability to induce an increase in α-synuclein accumulation in dopaminergic neurons in the SNpc and its ability to induce the formation of Lewy body-like structures in dopaminergic neurons of the SNpc [159]. The relation of dopaminergic neurodegeneration with α-synuclein increase and aggregation suggests the significance of this model in studying PD pathology.

### 4.2. Rotenone

Another herbicide that is reported to have detrimental effects on dopaminergic neurons besides PQ is rotenone [5]. Unlike PQ, rotenone is an herbicide and insecticide as well. It is a member of the rotenoid family of neurotoxins naturally occur in tropical plants [160]. It is highly lipophilic and easily crosses the blood-brain barrier. Rotenone inhibits the mitochondrial electron transport chain in the rat brain at chronic exposure to a low dose [161, 162]. The most used delivery route for rotenone in rodents is systemic administration by osmotic pumps. Intraperitoneal injections show high mortality and have been reported to induce behavioural and neurochemical deficits in an animal model [159]. Intravenous administration of rotenone is reported to damage nigral dopaminergic neurons along with ROS (reactive oxygen species) generation, α-synuclein aggregation and Lewy-like body formation [71]. Like paraquat, rotenone treatment also seems to replicate most of the hallmarks of Parkinsonism. Apart from dopaminergic cell death, it is associated with 26% depletion of noradrenergic, 35% serotonergic and 29% of cholinergic neurons. However, at chronic low-dose exposure, rotenone seems to produce a highly selective nigrostriatal neurodegeneration in about 50% of treated rats [159, 162].

### 4.3. 16-Hydroxydopamine (6-OHDA)

It is a classical and frequently used toxin used in an animal model of Parkinsonism [51, 159]. It is structurally closes to dopamine except for an extra hydroxyl group, which makes it toxic to dopaminergic neurons. Since it cannot cross the blood-brain barrier, the best route of administration is the direct injection into the SNpc, medial forebrain bundle (MFB), or the striatum. Injection of 6-OHDA into the SNpc of the rodent brain destroys approximately 60% of the tyrosine hydroxylase- (TH-) comprising neurons in this area and ensue damage to TH-positive terminals in the striatum [154]. This model has been used to study biochemical, behavioural, and physiological alterations in the CNS during Parkinsonism. 6-OHDA also obliterates catecholaminergic neurons by the combined effect of ROS and quinones induce inflammation in the brain which intensifies with time [51].

### 4.4. 1-methyl-4-phenyl-1, 2, 3, 6-tetrahydropyridine (MPTP)

Since its accidental discovery in 1982, 1-methyl-4-phenyl-1, 2, 3, 6-tetrahydropyridine (MPTP) has been the most frequently used toxin to successfully induce Parkinsonism in animal models [163–167]. Among various other toxins (discussed above) that damage the basal ganglia and the substantia nigra as well, MPTP appears to target dopaminergic neurons relatively specifically [166]. It has been applied to build animal models of PD to investigate the mechanism of neuronal damage and to test probable therapeutic strategies in patients. Investigations of the mechanisms of MPTP toxicity have also provided information regarding possible PD pathogenesis [164, 167].

MPTP is a lipid-soluble chemical, which enters brain cells by readily penetrating the blood-brain barrier. After entering the brain cells, it gets captured into acidic organelles (because it is amphiphilic), mostly lysosomes, of astrocytes [164]. Though MPTP itself is nontoxic, 1-methyl-4-phenylpyridinium ($MPP^+$), its oxidized product, is highly toxic. MPTP is converted to $MPP^+$ in astrocytes and serotonergic neurons in presence of MAO-B (monoamine oxidase-B). $MPP^+$ is transported into DA nerve terminals by the Dopamine transporter after it reaches the extracellular fluid [168, 169]. Inhibition of either MAO-B or DA transporter has been reported to be protective from $MPP^+$ toxicity (Figure 2) [170]. One mechanism of MPTP-induced neurotoxicity is the generation of oxidative stress [171]. The accumulation of MPP+ in dopaminergic neurons of SNpc leads to oxidative stress by mitochondria mediated generation of superoxide anion ($O_2\bullet^-$), hydrogen peroxide ($H_2O_2$), hydroxyl radicals (•OH), and nitric oxide (NO) [172]. Apart from this, $MPP^+$ leads to respiratory chain failure by inhibiting Complex I and thus energy failure, and inflammation is regularly reported as hallmarks of PD (Figure 2).

MPTP-induced neurotoxicity has been established in humans and other primate models like rabbits, cats, and rodent models [173]. In rodents, sensitivity towards MPTP is strain specific, which indicates genetic controls on the actions of MPTP [4, 174]. Strains of mice vary in their reaction to MPTP for the intensity of striatal DA exhaustion [4], the extent of dopaminergic neurodegeneration, and behavioural deficits. Regarding extent of SNpc neuron loss, several mouse strains can be considered as either "sensitive" (i.e., >50% SNpc neuron loss) or "resistant" (i.e., <25% SNpc loss) [175, 176].

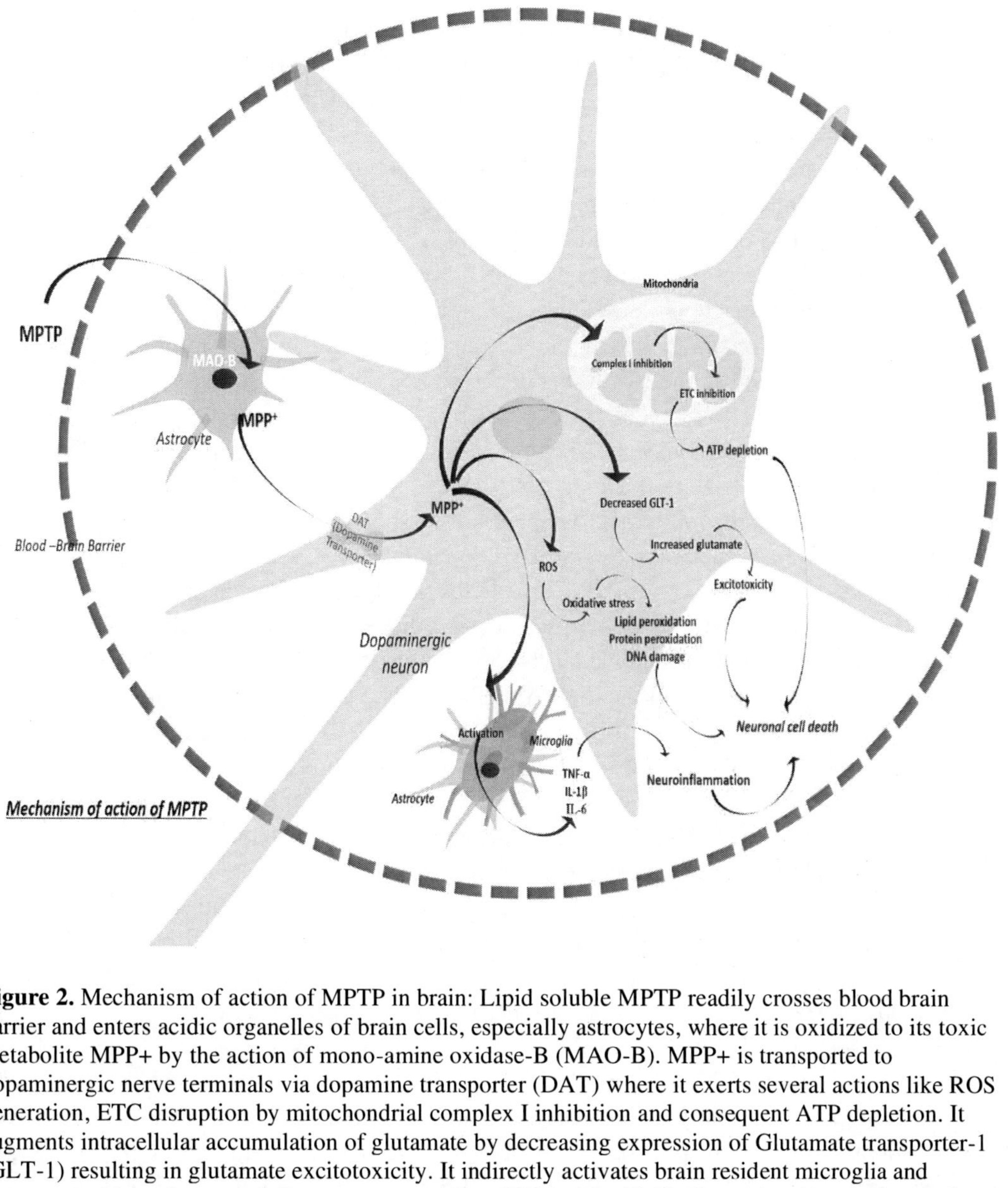


**Figure 2.** Mechanism of action of MPTP in brain: Lipid soluble MPTP readily crosses blood brain barrier and enters acidic organelles of brain cells, especially astrocytes, where it is oxidized to its toxic metabolite MPP+ by the action of mono-amine oxidase-B (MAO-B). MPP+ is transported to dopaminergic nerve terminals via dopamine transporter (DAT) where it exerts several actions like ROS generation, ETC disruption by mitochondrial complex I inhibition and consequent ATP depletion. It augments intracellular accumulation of glutamate by decreasing expression of Glutamate transporter-1 (GLT-1) resulting in glutamate excitotoxicity. It indirectly activates brain resident microglia and astrocytes which release proinflammatory cytokines (TNF-α, IL-1β, IL-6) causing neuroinflammation. All these factors collectively lead to severe injury to the health of dopaminergic neurons resulting in neurodegeneration.

## 5. Oxidative Stress and PD

Several etiological agents like exposure to neurotoxic substances, aggregation of pathological proteins, energy disturbances in the cell, decreased performance of the ubiquitin-proteasome system, apoptosis, autophagy, mutated genes etc., could be responsible for PD [177–179]. But oxidative stress is the common factor associated with all these causes. Studies have shown that dopamine, iron, calcium, mitochondria, and neuroinflammation are important contributors to the overwhelmed oxidative stress and related neurodegeneration in PD brains [180–185].

During neuroinflammatory responses activated microglia generates superoxide and nitric oxide which gets aggravated by the damaged dopaminergic neuron released molecules like α-synuclein, matrix metalloproteinase-3 and neuromelanin. The presence of ROS-producing enzymes tyrosine hydroxylase and monoamine oxidase make dopaminergic neurons prone to oxidative stress and higher concentrations of iron make the situation worse by fuelling the Fenton reaction. Thus, modest oxidative stress can lead to a chain of reactions terminating at neuronal death, which otherwise could be incapable of such harm [186].

Among many theories proposed to justify the causes of PD, the oxidant-antioxidant theory holds seminal importance. It says the formation of excessive ROS oxidizes various macromolecules like protein, lipids and nucleic acids which are evident from the accumulation of these macromolecules in the PD brains [177]. Hence oxidative stress is considered a major underlying cause of PD and related pathologies [177]. In PD brains, DNA oxidation may produce various genetic lesions and these mutations acquired because of DNA oxidation may further augment ROS production leading to enhanced neuronal deaths. This can be visualized as a combined effect of the mitochondrial electron transport chain fault, subsequent antioxidant exhaustion, and the presence of toxic oxidized dopamine [177]. As discussed earlier oxidative stress could be the outcome of faulty bioenergetics pathways of the cell and vice versa which intricately affects numerous cellular processes like highly energy-dependent processes like ubiquitination and ubiquitin-mediated protein degradation [177, 178, 187]. Failure of the respiratory chain complex I and specific PARKIN (PARK) gene variants might be a prominent contributor to oxidative stress in PD [177]. Mutations in PTEN-Induced Putative Kinase 1 (PINK1), α-Synuclein (SNCA), PRKN, DJ1, Leucine-Rich Repeat Kinase 2 (LRRK2), FBXO7 and ATP13A2 make individuals susceptible to PD [177, 180, 188]. Interestingly it has been found that PINK1, PRKN, SNCA, LRRK2 and DJ1 are most probably involved in oxidative stress generation [177]. These genes are correlated to disrupted oxidoreductive homeostasis in the PD brain. Oxidative stress can also act as an accelerator to α-synuclein aggregation and disease progression and acts trigger for apoptosis and autophagy which ultimately leads to the destruction of dopaminergic neurons [6, 189]. Oxidative stress severely impaired L-dopa pharmacotherapy and deep brain stimulation (DBS) therapy and thus impede the treatment for PD [177, 190].

As stated earlier, dopamine metabolism itself imposes a great threat to the cell by contributing oxidative stress which leads to functionally important intracellular macromolecule modifications. Though dopamine remains stable in the acidic environment of synaptic vesicles, it experiences rapid autoxidation at neutral pH of the extracellular space or cytosol and forms dopamine semiquinones like dopamine *o*-semiquinone radical. Interconversion between semiquinones and quinones constantly fuels the conversion of $\bullet O_2$ which becomes the driving force for the iron-dependent Harber-Weiss cyclewhere Fenton reactions create the very reactive •OH [191]. Dopamine can also get converted to the potent neurotoxin 6-OHDA in the presence of $Fe^{2+}$ and $H_2O_2$ and cyclize into a strong mitochondrial complex I inhibitor called aminochrome [191]. These in turn initiate deleterious cascading effects like scarce raised ROS levels, ATP production, mitochondrial fission/fragmentation, and mitochondrial dysfunction [191]. Simultaneously, nitrative stress, principally from the nitric oxide (•NO) may also subsidise neuronal dysfunction [191]. Oxidative and nitrative modifications initiate a positive feedback-driven destruction cascade by increasing the aggregation propensity of α-synuclein. Intracellular iron levels and noxious dopamine derivatives boost the destruction cascade and contribute to the dysfunction of the ubiquitin-proteasome system and the destruction of mitochondrial aerobic respiration [191]. A recent study also showed, that complex II or III in the electron transport chain (ETC) in addition to the respiratory chain complex I deficiency, can induce motor ailment and PD pathologies [76]. In neuronal Thy1-C/EBPβ transgenic mice, complex II inhibitor TTFA and complex III inhibitor atovaquone augment ROS generation and trigger C/EBPβ/AEP pathway, which in turn triggers dopaminergic neuronal cell death causing constipation and motor defects related to Lewy-body like inclusions [76] (Figure 3).

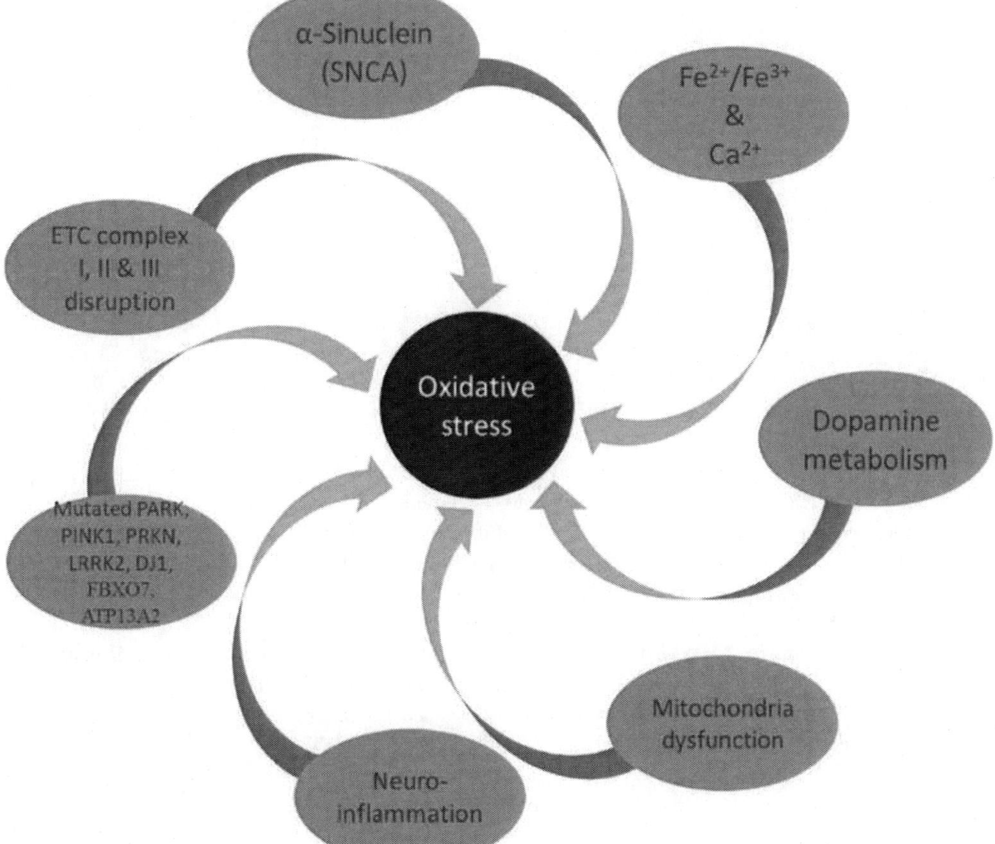


**Figure 3.** Contributing factors to oxidative stress related to PD.

## Conclusion

Though intensive studies present new insights related to PD pathogenesis every day, its underlying molecular mechanisms are still elusive to us. Nevertheless, the involvement of oxidative stress in the development and pathogenesis of PD is well accepted. Intrinsic factors like inherent mutations to crucial genes, mitochondrial dysfunctions, dopamine metabolism induced ROS generation, neuroinflammation etc. along with extrinsic factors like many xenobiotic neurotoxins are considered major sources of oxidative stress in PD [186]. Hence antioxidantssuch as vitamin E, creatine, pioglitazone, coenzyme Q10, melatonin and desferrioxamine could be thought of as promising agents with therapeutic potential in PD and have been tested in clinical trials, but unfortunately, till now, no significant breakthrough is achieved [180]. In addition to these issues, researchers are facing significant difficulties in formulating effective management of PD due to the absence of any premotor stage indicative biomarkers, prolonged progression of neurodegeneration and unsatisfactory drug delivery beyond the blood-brain barrier [180]. Hence more research worldwide is needed to demystify intricate pathogenic mechanisms of PD and develop effective therapies. Finding more genetic and environmental factors in addition to oxidative stress will have significance in the future for early diagnosis and better prognosis of PD. These also could help in the implementation of more efficient pharmacologic and nonpharmacologic treatments (Figure 3) [180, 190].

It seems rational that, the molecules involved in the process of oxidative stress, like coenzyme Q10, DJ-1, uric acid, homocysteine, 8-hydroxy-2'-deoxyguanosin, superoxide dismutase, glutathione peroxidase, retinoic acid/carotenes, vitamin E, xanthine oxidase and the products of lipid peroxidation, should serve as biomarkers for PD. But unfortunately, till now, none of these have proven to be clinically effective in diagnosis [180].

## References


[1] Dionísio PA, Amaral JD, Rodrigues CMP. Oxidative stress and regulated cell death in Parkinson's disease. *Ageing Res Rev.* 2021;67:101263.

[2] Poewe W, Seppi K, Tanner CM, Halliday GM, Brundin P, Volkmann J, Schrag AE, Lang AE. Parkinson disease. *Nat Rev Dis Primers.* 2017;3:17013.

[3] Chaudhuri KR, Schapira AH. Non-motor symptoms of Parkinson's disease: dopaminergic pathophysiology and treatment. *Lancet Neurol.* 2009;8(5):464-74.

[4] Sedelis M, Hofele K, Auburger GW, Morgan S, Huston JP, Schwarting RKW. MPTP susceptibility in the mouse: behavioral, neurochemical, and histological analysis of gender and strain differences. *Behav Genet.* 2000;30(3):171-82.

[5] Radad K, Al-Shraim M, Al-Emam A, Wang F, Kranner B, Rausch WD, Moldzio R. Rotenone: from modelling to implication in Parkinson's disease. *Folia Neuropathol.* 2019;57(4):317-326.

[6] Sinha K, Das J, Pal PB, Sil PC. Oxidative stress: the mitochondria-dependent and mitochondria-independent pathways of apoptosis. *Arch Toxicol.* 2013;87(7):1157-80.

[7] Farrugia G, Balzan R. Oxidative stress and programmed cell death in yeast. *Front Oncol.* 2012;2:64.

[8] Valko M, Morris H, Cronin MT. Metals, toxicity and oxidative stress. *Curr Med Chem.* 2005;12(10):1161-208.

[9] Valko M, Leibfritz D, Moncol J, Cronin MTD, Mazur M, Telser J. Free radicals and antioxidants in normal physiological functions and human disease. *Int J Biochem Cell Biol.* 2007;39(1):44-84.

[10] Halliwell B, Gutteridge JMC. Free radicals in biology and medicine. *Free Radic Biol Med.* 2015.

[11] Farrugia G, Balzan R. Oxidative stress and programmed cell death in yeast. *Front Oncol.* 2012;2:64.

[12] Matés JM, Segura JA, Alonso FJ, Márquez J. Oxidative stress in apoptosis and cancer: an update. *Arch Toxicol.* 2012;86(11):1649-65.

[13] Halliwell B, Cross CE. Oxygen-derived species: their relation to human disease and environmental stress. *Environ Health Perspect.* 1994;102;Suppl10:5-12.

[14] Miller DM, Buettner GR, Aust SD. Transition metals as catalysts of 'autoxidation' reactions. *Free Radic Biol Med.* 1990;8(1):95-108.

[15] Orrenius S, Gogvadze V, Zhivotovsky B. Mitochondrial oxidative stress: implications for cell death. *Annu Rev Pharmacol Toxicol.* 2007;47:143-83.

[16] Circu ML, Aw TY. Glutathione and modulation of cell apoptosis. *Biochim Biophys Acta.* 2012;1823(10):1767-77.

[17] Korshunov SS, Skulachev VP, Starkov AA. High protonic potential actuates a mechanism of production of reactive oxygen species in mitochondria. *FEBS Lett.* 1997;416(1):15-8.

[18] Lambert AJ, Brand MD. Superoxide production by NADH:ubiquinone oxidoreductase (Complex I) depends on the PH gradient across the mitochondrial inner membrane. *Biochem J.* 2004;382(2):511-7.

[19] Migliaccio E, Giorgio M, Pelicci PG. Apoptosis and aging: role of P66Shc redox protein. *Antioxid Redox Signal.* 2006;8(3-4):600-8.

[20] Andreyev AY, Kushnareva YE, Starkov AA. Mitochondrial metabolism of reactive oxygen species. *Biochemistry* (Mosc). 2005;70(2):200-14.

[21] De Grey AD. HO2*: the forgotten radical. *DNA Cell Biol.* 2002 Apr;21(4):251-7.

[22] Aikens J, Dix TA. Perhydroxyl radical (HOO.) initiated lipid peroxidation. The role of fatty acid hydroperoxides. *J Biol Chem.* 1991;266(23):15091-8.

[23] Circu ML, Aw TY. Reactive oxygen species, cellular redox systems, and apoptosis. *Free Radic Biol Med.* 2010;48(6):749-62.

[24] Supale S, Li N, Brun T, Maechler P. Mitochondrial dysfunction in pancreatic β cells. *Trends Endocrinol Metab.* 2012;23(9):477-87.

[25] Wallace DC. Mitochondrial diseases in Man and mouse. *Science.* 1999;283(5407):1482-8.

[26] Antinozzi PA, Ishihara H, Newgard CB, Wollheim CB. Mitochondrial metabolism sets the maximal limit of fuel-stimulated insulin secretion in a model pancreatic beta cell: A survey of four fuel secretagogues. *J Biol Chem.* 2002;277(14):11746-55.

[27] Kovacic P, Pozos RS, Somanathan R, Shangari N, O'Brien PJ. Mechanism of mitochondrial uncouplers, inhibitors, and toxins: focus on electron transfer, free radicals, and structure-activity relationships. *Curr Med Chem.* 2005;12(22):2601-23.

[28] Brand MD, Affourtit C, Esteves TC, Green K, Lambert AJ, Miwa S, Pakay JL, Parker N. Mitochondrial superoxide: production, biological effects, and activation of uncoupling proteins. *Free Radic Biol Med.* 2004;37(6):755-67.

[29] Turrens JF, Boveris A. Generation of superoxide anion by the NADH dehydrogenase of bovine heart mitochondria. *Biochem J.* 1980;191(2):421-7.

[30] Nishikawa, T, Edelstein, D, Du, XL, Yamagishi, S, Matsumura, T, Kaneda, Y, Yorek, MA, Beebe, D, Oates, PJ, Hammes, HP, Giardino, I, & Brownlee, M. Normalizing mitochondrial superoxide production blocks three pathways of hyperglycaemic damage. *Nature.* 2000;404(6779):787-90.

[31] Turrens JF, Alexandre A, Lehninger AL. Ubisemiquinone is the electron donor for superoxide formation by Complex III of heart mitochondria. *Arch Biochem Biophys.* 1985;237(2):408-14.

[32] Tretter L, Adam-Vizi V. Generation of reactive oxygen species in the reaction catalyzed by α-ketoglutarate dehydrogenase. *J Neurosci.* 2004;24(36):7771-8.

[33] Circu ML, Moyer MP, Harrison L, Aw TY. Contribution of glutathione status to oxidant-induced mitochondrial DNA damage in colonic epithelial cells. *Free Radic Biol Med.* 2009;47(8):1190-8.

[34] Rachek LI, Yuzefovych LV, Ledoux SP, Julie NL, Wilson GL. Troglitazone, but not rosiglitazone, damages mitochondrial DNA and induces mitochondrial dysfunction and cell death in human hepatocytes. *Toxicol Appl Pharmacol.* 2009;240(3):348-54.

[35] Masutani H, Yodoi J. Thioredoxin. Overview. *Methods Enzymol.* 2002; 347:279-86.

[36] Davie CA. A review of Parkinson's disease. *Br Med Bull.* 2008;86:109-27.

[37] Barcia, C, Bahillo, AS, Fernández-Villalba, E, Bautista, V, Poza Y Poza, M, Fernández-Barreiro, A, Hirsch, EC, & Herrero, MT. Evidence of active microglia in substantia nigra pars Compacta of parkinsonian monkeys 1 year after MPTP exposure. *Glia.* 2004; 46(4):402-9.

[38] McGeer PL, Schwab C, Parent A, Doudet D. Presence of reactive microglia in monkey substantia nigra years after 1-methyl-4-phenyl-1,2,3,6-tetrahydropyridine administration. *Ann Neurol.* 2003;54(5):599-604.

[39] McGeer PL, McGeer EG. Glial reactions in Parkinson's disease. *Mov Disord.* 2008;23(4):474-83.

[40] de Lau LM, Breteler MM. Epidemiology of Parkinson's disease. *Lancet Neurol.* 2006;5(6):525-35.

[41] Gourie-Devi M. Epidemiology of neurological disorders in India: review of background, prevalence and incidence of epilepsy, stroke, Parkinson's disease and tremors. *Neurol India.* 2014;62(6):588-98.

[42] Chen CC, Chen TF, Hwang YC, Wen YR, Chiu YH, Wu CY, Chen RC, Tai JJ, Chen TH, Liou HH. Different prevalence rates of Parkinson's disease in urban and rural areas: a population-based study in Taiwan. *Neuroepidemiology.* 2009;33(4):350-7.

[43] Kempster PA, Hurwitz B, Lees AJ. A new look at James Parkinson's essay on the shaking palsy. *Neurology.* 2007;69(5):482-5.

[44] Palacios-Sánchez L, Torres Nupan MT, Botero-Meneses JS. James Parkinson and his essay on 'shaking palsy', two hundred years later. *Arq Neuro Psiquiatr.* 2017;75(9):671-2.

[45] Parkinson J. An essay on the shaking palsy. 1817. *J Neuropsychiatry Clin Neurosci.* 2002;14(2):223-36; discussion 222.

[46] Dauer W, Przedborski S. Parkinson's disease: mechanisms and models. *Neuron.* 2003;39(6):889-909.

[47] Yeragani VK, Tancer M, Chokka P, Baker GB. Arvid Carlsson, and the story of dopamine. *Indian J Psychiatry.* 2010;52(1):87-8.

[48] Ghosh N, Mitra S, Sinha P, Chakrabarti N, Bhattacharyya A. TNFR2 mediated TNF-α signaling and NF-KB activation in hippocampus of 1-Methyl-4-Phenyl-1,2,3,6-Tetrahydropyridine-Treated mice. *Neurosci Res.* 2018;137:36-42.

[49] Pandey S, Srivanitchapoom P. Levodopa-induced dyskinesia: clinical features, pathophysiology, and medical management. *Ann Indian Acad Neurol.* 2017;20(3):190-8.

[50] Meredith GE, Rademacher DJ. MPTP mouse models of Parkinson's disease: an update. *J Parkinsons Dis.* 2011;1(1):19-33.

[51] Simola N, Morelli M, Carta AR. The 6-hydroxydopamine model of Parkinson's disease. *Neurotox Res.* 2007;11(3-4):151-67.

[52] Berry C, la Vecchia C, Nicotera P. Paraquat and Parkinson's disease. *Cell Death Differ.* 2010;17(7):1115-25.

[53] Klein C, Westenberger A. Genetics of Parkinson's disease. *Cold Spring Harb Perspect Med.* 2012;2(1):a008888.

[54] Valente EM, Salvi S, Ialongo T, Marongiu R, Elia AE, Caputo V, Romito L, Albanese A, Dallapiccola B, Bentivoglio AR. PINK1 mutations are associated with sporadic early-onset parkinsonism. *Ann Neurol.* 2004;56(3):336-41.

[55] Kitada T, Asakawa S, Hattori N, Matsumine H, Yamamura Y, Minoshima S, Yokochi M, Mizuno Y, Shimizu N. Mutations in the parkin gene cause autosomal recessive juvenile parkinsonism. *Nature.* 1998;392(6676):605-8.

[56] Zimprich A, Biskup S, Leitner P, Lichtner P, Farrer M, Lincoln S, Kachergus J, Hulihan M, Uitti RJ, Calne DB, Stoessl AJ, Pfeiffer RF, Patenge N, Carbajal IC, Vieregge P, Asmus F, Müller-Myhsok B, Dickson DW, Meitinger T, Strom TM, Wszolek ZK, Gasser T. Mutations in LRRK2 cause autosomal-dominant parkinsonism with pleomorphic pathology. *Neuron.* 2004;44(4):601-7.

[57] Polymeropoulos MH, Lavedan C, Leroy E, Ide SE, Dehejia A, Dutra A, Pike B, Root H, Rubenstein J, Boyer R, Stenroos ES, Chandrasekharappa S, Athanassiadou A, Papapetropoulos T, Johnson WG, Lazzarini AM, Duvoisin RC, Di Iorio G, Golbe LI, Nussbaum RL. Mutation in the alpha-synuclein gene identified in families with Parkinson's disease. *Science.* 1997;276(5321):2045-7.

[58] Reeve A, Simcox E, Turnbull D. Ageing and Parkinson's disease: why is advancing age the biggest risk factor? *Ageing Res Rev.* 2014;14(100):19-30.

[59] Tomiyama H, Lesage S, Tan EK, Jeon BS. Familial Parkinson's disease/Parkinsonism. *Bio Med Res Int.* 2015;2015:736915.

[60] Beitz JM. Parkinson's disease: a review. *Front Biosci* (Schol Ed). 2014;6(1):65-74.

[61] Jankovic J. Parkinson's disease: clinical features and diagnosis. *J Neurol Neurosurg Psychiatry.* 2008;79(4):368-76.

[62] Tysnes OB, Storstein A. Epidemiology of Parkinson's disease. *J Neural Transm* (Vienna). 2017;124(8):901-5.

[63] Bloem BR, Okun MS, Klein C. Parkinson's disease. *Lancet.* 2021;397(10291):2284-303.

[64] Zhong LL, Song YQ, Cao H, Ju KJ, Yu L. The non-motor symptoms of Parkinson's disease of different motor types in early stage. *Eur Rev Med Pharmacol Sci.* 2017;21(24):5745-50.

[65] Braak H, Rüb U, Gai WP, del Tredici K. Idiopathic Parkinson's disease: possible routes by which vulnerable neuronal types may be subject to neuroinvasion by an unknown pathogen. *J Neural Transm* (Vienna). 2003;110(5):517-36.

[66] Franke C, Ebersbach G. [Delirium in idiopathic Parkinson's disease]. *Nervenarzt.* 2020;91(2):107-13.

[67] Hayes MT. Parkinson's disease and Parkinsonism. *Am J Med.* 2019;132(7):802-7.

[68] Calne DB, Langston JW. Aetiology of Parkinson's disease. *Lancet.* 1983;2(8365-66):1457-9.
[69] Langston JW, Ballard PA. Parkinson's disease in a chemist W orking with 1-methyl-4-phenyl-1,2,5,6-tetrahydropyridine. *N Engl J Med.* 1983;309(5):310.
[70] Richardson JR, Quan Y, Sherer TB, Greenamyre JT, Miller GW. Paraquat neurotoxicity is distinct from that of MPTP and rotenone. *Toxicol Sci.* 2005;88(1):193-201.
[71] Cannon JR, Tapias V, Na HM, Honick AS, Drolet RE, Greenamyre JT. A highly reproducible rotenone model of Parkinson's disease. *Neurobiol Dis.* 2009;34(2):279-90.
[72] Chen CC, Chen TF, Hwang YC, Wen YR, Chiu YH, Wu CY, Chen RC, Tai JJ, Chen TH, Liou HH. Different prevalence rates of Parkinson's disease in urban and rural areas: a population-based study in Taiwan. *Neuroepidemiology.* 2009;33(4):350-7.
[73] Hernán MA, Takkouche B, Caamaño-Isorna F, Gestal-Otero JJ. A meta-analysis of coffee drinking, cigarette smoking, and the risk of Parkinson's disease. *Ann Neurol.* 2002;52(3):276-84.
[74] Sandy MS, Armstrong M, Tanner CM, Daly AK, Di Monte DA, Langston JW, Idle JR. CYP2D6 allelic frequencies in young-onset Parkinson's disease. *Neurology.* 1996;47(1):225-30.
[75] McNaught KS, Carrupt PA, Altomare C, Cellamare S, Carotti A, Testa B, Jenner P, Marsden CD. Isoquinoline derivatives as endogenous neurotoxins in the aetiology of Parkinson's disease. *Biochem Pharmacol.* 1998;56(8):921-33.
[76] Ahn EH, Lei K, Kang SS, Wang ZH, Liu X, Hong W, Wang YT, Edgington-Mitchell LE, Jin L, Ye K. Mitochondrial dysfunction triggers the pathogenesis of Parkinson's disease in neuronal C/EBPβ transgenic mice. *Mol Psychiatry.* 2021;26(12):7838-7850.
[77] Giasson BI, Uryu K, Trojanowski JQ, Lee VMY. Mutant and wild type human alpha-synucleins assemble into elongated filaments with distinct morphologies *in vitro*. *J Biol Chem.* 1999;274(12):7619-22.
[78] Pajares M, I Rojo AI, Manda G, Boscá L, Cuadrado A. Inflammation in Parkinson's disease: mechanisms and therapeutic implications. *Cells.* 2020;9(7).
[79] De Virgilio A, Greco A, Fabbrini G, Inghilleri M, Rizzo MI, Gallo A, Conte M, Rosato C, Ciniglio Appiani M, de Vincentiis M. Parkinson's disease: Autoimmunity and neuroinflammation. *Autoimmun Rev.* 2016;15(10):1005-11.
[80] Belarbi K, Cuvelier E, Bonte MA, Desplanque M, Gressier B, Devos D, Chartier-Harlin MC. Glycosphingolipids and neuroinflammation in Parkinson's disease. *Mol Neurodegener.* 2020;15(1):59.
[81] Rocha EM, de Miranda B, Sanders LH. Alpha-synuclein: pathology, mitochondrial dysfunction and neuroinflammation in Parkinson's disease. *Neurobiol Dis.* 2018;109(B):249-57.
[82] Ransohoff RM, Perry VH. Microglial physiology: unique stimuli, specialized responses. *Annu Rev Immunol.* 2009; 27:119-45.
[83] Rock RB, Gekker G, Hu S, Sheng WS, Cheeran M, Lokensgard JR, Peterson PK. Role of microglia in central nervous system infections. *Clin Microbiol Rev.* 2004;17(4):942-64, table of contents.
[84] Kim WG, Mohney RP, Wilson B, Jeohn GH, Liu B, Hong JS. Regional difference in susceptibility to lipopolysaccharide-induced neurotoxicity in the rat brain: role of microglia. *J Neurosci.* 2000;20(16):6309-16.
[85] Qian L, Flood PM, Hong JS. Neuroinflammation is a key player in Parkinson's disease and a prime target for therapy. *J Neural Transm* (Vienna). 2010;117(8):971-9.
[86] Walsh S, Finn DP, Dowd E. Time-course of nigrostriatal neurodegeneration and neuroinflammation in the 6-hydroxydopamine-induced axonal and terminal lesion models of Parkinson's disease in the rat. *Neuroscience.* 2011;175:251-61.
[87] Sugama S, Yang L, Cho BP, DeGiorgio LA, Lorenzl S, Albers DS, Beal MF, Volpe BT, Joh TH. Age-related microglial activation in 1-methyl-4-phenyl-1,2,3,6 tetrahydropyridine (MPTP)-induced dopaminergic neurodegeneration in C57BL/6 mice. *Brain Res.* 2003; 964(2):288-94.
[88] Wang X, Son YO, Chang Q, Sun L, Hitron JA, Budhraja A, Zhang Z, Ke Z, Chen F, Luo J, Shi X. NADPH oxidase activation is required in reactive oxygen species generation and cell transformation induced by hexavalent chromium. *Toxicol Sci.* 2011;123(2):399-410.
[89] Koppula S, Kumar H, Kim IS, Choi DK. Reactive oxygen species and inhibitors of inflammatory enzymes, NADPH oxidase, and INOS in experimental models of Parkinsons disease. *Mediators Inflamm.* 2012;2012:823902.

[90] Wang T, Pei Z, Zhang W, Liu B, Langenbach R, Lee C, Wilson B, Reece JM, Miller DS, Hong JS. MPP+-induced COX-2 activation and subsequent dopaminergic neurodegeneration. *FASEB J.* 2005 Jul;19(9):1134-6.

[91] Kimelberg HK, Nedergaard M. Functions of astrocytes and their potential as therapeutic targets. *Neurotherapeutics.* 2010;7(4):338-353.

[92] Hemmer K, Fransen L, Vanderstichele H, Vanmechelen E, Heuschling P. An in vitro model for the study of microglia-induced neurodegeneration: involvement of nitric oxide and tumor necrosis factor-alpha. *Neurochem Int.* 2001;38(7):557-565.

[93] Barreto G, Veiga S, Azcoitia I, Garcia-Segura LM, Garcia-Ovejero D. Testosterone decreases reactive astroglia and reactive microglia after brain injury in male rats: role of its metabolites, oestradiol and dihydrotestosterone. *Eur J Neurosci.* 2007;25(10):3039-3046.

[94] Gu XL, Long CX, Sun L, Xie C, Lin X, Cai H. Astrocytic expression of Parkinson's disease-related A53T -Synuclein causes neurodegeneration in mice. *Mol Brain.* 2010;3:12.

[95] Hamby ME, Sofroniew MV. Reactive astrocytes as therapeutic targets for CNS disorders. *Neurotherapeutics.* 2010;7(4):494-506.

[96] Seo, J, Park, J, Kim, K, Won, J, Yeo, HG, Jin, YB, Koo, BS, Lim, KS, Jeong, KJ, Kang, P, Lee, HY, Choi, WS, Baek, SH, Jeon, CY, Hong, JJ, Huh, JW, Kim, YH, Park, SJ, Kim, SU, ... Lee, Y. Chronic infiltration of T lymphocytes into the brain in a non-human primate model of Parkinson's disease. *Neuroscience.* 2020;431:73-85.

[97] Fiszer U, Mix E, Fredrikson S, Kostulas V, Link H. Parkinson's disease and immunological abnormalities: increase of HLA-DR expression on monocytes in cerebrospinal fluid and of CD45RO+ T cells in peripheral blood. *Acta Neurol Scand.* 1994;90(3):160-166.

[98] Baba Y, Kuroiwa A, Uitti RJ, Wszolek ZK, Yamada T. Alterations of T-lymphocyte populations in Parkinson disease. *Parkinsonism Relat Disord.* 2005;11(8):493-498.

[99] Brochard V, Combadière B, Prigent A, Laouar Y, Perrin A, Beray-Berthat V, Bonduelle O, Alvarez-Fischer D, Callebert J, Launay JM, Duyckaerts C, Flavell RA, Hirsch EC, Hunot S. Infiltration of CD4+ lymphocytes into the brain contributes to neurodegeneration in a mouse model of Parkinson disease. *J Clin Invest.* 2009;119(1):182-192.

[100] González H, Contreras F, Prado C, Elgueta D, Franz D, Bernales S, and Pacheco R. Dopamine receptor D3 expressed on CD4+ T cells favors neurodegeneration of dopaminergic neurons during Parkinson's disease. *J Immunol.* 2013;190(10):5048-5056.

[101] Liu T, Zhang L, Joo D, Sun SC. NF-KB signaling in inflammation. *Signal Transduct Target Ther.* 2017;2(1):17023.

[102] Lawrence T. The nuclear factor NF-kappaB pathway in inflammation. *Cold Spring Harb Perspect Biol.* 2009;1(6):a001651.

[103] Ahmed SMU, Luo L, Namani A, Wang XJ, Tang X. Nrf2 signaling pathway: pivotal roles in inflammation. *Biochim Biophys Acta Mol Basis Dis.* 2017;1863(2):585-97.

[104] Sun SC. The non-canonical NF-KB pathway in immunity and inflammation. *Nat Rev Immunol.* 2017;17(9):545-58.

[105] Nakano H, Nakajima A, Sakon-Komazawa S, Piao JH, Xue X, Okumura K. Reactive oxygen species mediate crosstalk between NF-kappaB and JNK. *Cell Death Differ.* 2006;13(5):730-7.

[106] Hayden MS, Ghosh S. NF-KB in immunobiology. *Cell Res.* 2011;21(2):223-44.

[107] D'Acquisto F, May MJ, Ghosh S. Inhibition of nuclear factor kappa B (NF-B):, *Mol Interv* 2. Vol. 22; 2002.

[108] Hayden MS, Ghosh S. Signaling to NF-KB. *Genes Dev.* 2004;18(18):2195-224.

[109] Ghosh S, Hayden MS. Celebrating 25 years of NF-KB research. *Immunol Rev.* 2012;246(1):5-13.

[110] Cao JP, Niu HY, Wang HJ, Huang XG, Gao DS. NF-KB P65/P52 plays a role in GDNF up-regulating Bcl-2 and Bcl-w expression in 6-OHDA-induced apoptosis of MN9D cell. *Int J Neurosci.* 2013;123(10):705-10.

[111] Fearnley JM, Lees AJ. Ageing and Parkinson's disease: substantia nigra regional selectivity. *Brain.* 1991;114(5):2283-301.

[112] Gao HM, Liu B, Zhang W, Hong JS. Synergistic dopaminergic neurotoxicity of MPTP and inflammogen lipopolysaccharide: relevance to the etiology of Parkinson's disease. *FASEB J.* 2003;17(13):1957-9.

[113] DeLong MR, Wichmann T. Circuits and circuit disorders of the basal ganglia. *Arch Neurol.* 2007;64(1):20-4.

[114] Celada P, Paladini CA, Tepper JM. GABAergic control of rat substantia nigra dopaminergic neurons: role of globus pallidus and substantia nigra pars reticulata. *Neuroscience.* 1999;89(3):813-25.

[115] Deniau JM, Kitai ST, Donoghue JP, Grofova I. Neuronal interactions in the substantia nigra pars reticulata through axon collaterals of the projection neurons. An electrophysiological and morphological study. *Exp Brain Res.* 1982;47(1):105-13.

[116] Alexander GE, Crutcher MD, Delong MR. Basal ganglia-thalamocortical circuits: parallel substrates for motor, oculomotor, "prefrontal" and "limbic" functions. *Prog Brain Res.* 1990, Chapter 6;85:119-46.

[117] Grillner S, Hellgren J, Ménard A, Saitoh K, Wikström MA. Mechanisms for selection of basic motor programs—roles for the striatum and pallidum. *Trends Neurosci.* 2005;28(7):364-370.

[118] Hikosaka O. GABAergic output of the basal ganglia. *Prog Brain Res.* 2007;160:209-226.

[119] Hnasko TS, Hjelmstad GO, Fields HL, Edwards RH. Ventral tegmental area glutamate neurons: electrophysiological properties and projections. *J Neurosci.* 2012;32(43):15076-15085.

[120] Yamaguchi T, Sheen W, Morales M. Glutamatergic neurons are present in the rat ventral tegmental area. *Eur J Neurosci.* 2007;25(1):106-118.

[121] Forster GL, Blaha CD. Pedunculopontine tegmental stimulation evokes striatal dopamine efflux by activation of acetylcholine and glutamate receptors in the midbrain and pons of the rat. *Eur J Neurosci.* 2003;17(4):751-762.

[122] Levy R, Ashby P, Hutchison WD, Lang AE, Lozano AM, Dostrovsky JO. Dependence of subthalamic nucleus oscillations on movement and dopamine in Parkinson's disease. *Brain.* 2002;125(6):1196-1209.

[123] Obeso JA, Lanciego JL. Past, present, and future of the pathophysiological model of the basal ganglia. *Front Neuroanat.* 2011;5:39.

[124] Shen KZ, Zhu ZT, Munhall A, Johnson SW. Synaptic plasticity in rat subthalamic nucleus induced by high-frequency stimulation. *Synapse.* 2003;50(4):314-319.

[125] Kita H, Kitai ST. Efferent projections of the subthalamic nucleus in the rat: light and electron microscopic analysis with the PHA-L method. *J Comp Neurol.* 1987;260(3):435-452.

[126] Lee CR, Tepper JM. Basal ganglia control of substantia nigra dopaminergic neurons. *J Neural Transm Suppl.* 2009;(73):71-90.

[127] di Michele F, Luchetti S, Bernardi G, Romeo E, Longone P. Neurosteroid and neurotransmitter alterations in Parkinson's disease. *Front Neuroendocrinol.* 2013;34(2):132-142.

[128] Ziemssen T, Reichmann H. Non-motor dysfunction in Parkinson's disease. *Parkinsonism Relat Disord.* 2007;13(6):323-332.

[129] Adler CH. Nonmotor complications in Parkinson's disease. *Mov Disord.* 2005;20 (suppl 11):S23-S29.

[130] Hely MA, Reid WGJ, Adena MA, Halliday GM, Morris JGL. The Sydney multicenter study of Parkinson's disease: the inevitability of dementia at 20 years. *Mov Disord.* 2008;23(6):837-844.

[131] Xu Y, Yan J, Zhou P, Li J, Gao H, Xia Y, Wang Q. Neurotransmitter receptors and cognitive dysfunction in Alzheimer's disease and Parkinson's disease. *Prog Neurobiol.* 2012 Apr;97(1):1-13.

[132] Cash R, Ruberg M, Raisman R, Agid Y. Adrenergic receptors in Parkinson's disease. *Brain Res.* 1984;322(2):269-275.

[133] Ward RJ, Lallemand F, de Witte P, Dexter DT. Neurochemical pathways involved in the protective effects of nicotine and ethanol in preventing the development of Parkinson's disease: potential targets for the development of new therapeutic agents. *Prog Neurobiol.* 2008;85(2):135-147.

[134] Meyer PM, Strecker K, Kendziorra K, Becker G, Hesse S, Woelpl D, Hensel A, Patt M, Sorger D, Wegner F, Lobsien D, Barthel H, Brust P, Gertz HJ, Sabri O, Schwarz J. Reduced alpha4beta2*-nicotinic acetylcholine receptor binding and its relationship to mild cognitive and depressive symptoms in Parkinson disease. *Arch Gen Psychiatry.* 2009;66(8):866-77.

[135] Oishi N, Hashikawa K, Yoshida H, Ishizu K, Ueda M, Kawashima H, Saji H, Fukuyama H. Quantification of nicotinic acetylcholine receptors in Parkinson's disease with (123)I-5IA SPECT. *J Neurol Sci.* 2007;256(1-2):52-60.

[136] Hallett PJ, Spoelgen R, Hyman BT, Standaert DG, Dunah AW. Dopamine D1 activation potentiates striatal NMDA receptors by tyrosine phosphorylation-dependent subunit trafficking. *J Neurosci.* 2006;26(17):4690-4700.

[137] Reeves SJ, Grasby PM, Howard RJ, Bantick RA, Asselin MC, Mehta MA. A positron emission tomography (PET) investigation of the role of striatal dopamine (D2) receptor availability in spatial cognition. *Neuroimage.* 2005;28(1):216-226.

[138] Fetsko LA, Xu R, Wang Y. Effects of Age and dopamine D2L receptor-deficiency on motor and learning functions. *Neurobiol Aging.* 2005;26(4):521-530.

[139] Nagano-Saito A, Kato T, Arahata Y, Washimi Y, Nakamura A, Abe Y, Yamada T, Iwai K, Hatano K, Kawasumi Y, Kachi T, Dagher A, Ito K. Cognitive- and motor-related regions in Parkinson's disease: FDOPA and FDG PET studies. *Neuroimage.* 2004 Jun;22(2):553-61.

[140] Harrington DL, Haaland KY, Hermanowicz N. Temporal processing in the basal ganglia. *Neuropsychology.* 1998;12(1):3-12.

[141] Wang Q, Wang PH, McLachlan C, Wong PTH. Simvastatin reverses the downregulation of dopamine D1 and D2 receptor expression in the prefrontal cortex of 6-hydroxydopamine-induced parkinsonian rats. *Brain Res.* 2005;1045(1-2):229-233.

[142] Fang X, Sugiyama K, Akamine S, Namba H. The stepping test and its learning process in different degrees of unilateral striatal lesions by 6-hydroxydopamine in rats. *Neurosci Res.* 2006;55(4):403-409.

[143] Sun W, Sugiyama K, Fang X, Yamaguchi H, Akamine S, Magata Y, Namba H. Different striatal D2-like receptor function in an early stage after unilateral striatal lesion and medial forebrain bundle lesion in rats. *Brain Res.* 2010;1317:227-35.

[144] Sun W, Sugiyama K, Asakawa T, Yamaguchi H, Akamine S, Ouchi Y, Magata Y, Namba H. Dynamic changes of striatal dopamine D2 receptor binding at later stages after unilateral lesions of the medial forebrain bundle in Parkinsonian rat models. *Neurosci Lett.* 2011;496(3):157-62.

[145] Turle-Lorenzo N, Maurin B, Puma C, Chezaubernard C, Morain P, Baunez C, Nieoullon A, Amalric M. The dopamine agonist piribedil with L-DOPA improves attentional dysfunction: relevance for Parkinson's disease. *J Pharmacol Exp Ther.* 2006;319(2):914-23.

[146] Piggott MA, Ballard CG, Dickinson HO, McKeith IG, Perry RH, Perry EK. Thalamic D2 receptors in dementia with Lewy bodies, Parkinson's disease, and Parkinson's disease dementia. *Int J Neuropsychopharmacol.* 2007;10(2):231-244.

[147] Yan J, Xu Y, Zhu C, Zhang L, Wu A, Yang Y, Xiong Z, Deng C, Huang XF, Yenari MA, Yang YG, Ying W, Wang Q. Simvastatin prevents dopaminergic neurodegeneration in experimental parkinsonian models: the association with anti-inflammatory responses. *PLoS One.* 2011;6(6):e20945.

[148] Wüllner U, Testa CM, Catania MV, Young AB, Penney JB. Glutamate receptors in striatum and substantia nigra: effects of medial forebrain bundle lesions. *Brain Res.* 1994;645(1-2):98-102.

[149] Hallett PJ, Spoelgen R, Hyman BT, Standaert DG, Dunah AW. Dopamine D1 activation potentiates striatal NMDA receptors by tyrosine phosphorylation-dependent subunit trafficking. *J Neurosci.* 2006;26(17):4690-4700.

[150] Meoni P, Bunnemann BH, Kingsbury AE, Trist DG, Bowery NG. NMDA NR1 subunit MRNA and glutamate NMDA-sensitive binding are differentially affected in the striatum and pre-frontal cortex of Parkinson's disease patients. *Neuropharmacology.* 1999;38(5):625-633.

[151] Dunah AW, Wang Y, Yasuda RP, Kameyama K, Huganir RL, Wolfe BB, Standaert DG. Alterations in subunit expression, composition, and phosphorylation of striatal N-methyl-D-aspartate glutamate receptors in a rat 6-hydroxydopamine model of Parkinson's disease. *Mol Pharmacol.* 2000 Feb;57(2):342-52.

[152] Churchyard A, Lees AJ. The relationship between dementia and direct involvement of the hippocampus and amygdala in Parkinson's disease. *Neurology.* 1997;49(6):1570-1576.

[153] Carlesimo GA, Piras F, Assogna F, Pontieri FE, Caltagirone C, Spalletta G. Hippocampal abnormalities and memory deficits in Parkinson disease: A multimodal imaging study. *Neurology.* 2012;78(24):1939-1945.

[154] Blandini F, Armentero MT. Animal models of Parkinson's disease. *FEBS Journal.* 2012;279(7):1156-1166.

[155] Konnova EA, Swanberg M. Animal models of Parkinson's disease. *Park Dis Pathog Clin Aspects.* 2018;83.

[156] Nandipati S, Litvan I. Environmental exposures and Parkinson's disease. *Int J Environ Res Public Health.* 2016;13(9).

[157] Chia SJ, Tan EK, Chao YX. Historical perspective: models of Parkinson's disease. *Int J Mol Sci.* 2020;21(7).

[158] Mitra S, Chakrabarti N, Bhattacharyya A. Differential regional expression patterns of α-synuclein, TNF-α, and IL-1β; and variable status of dopaminergic neurotoxicity in mouse brain after paraquat treatment. *J Neuroinflammation.* 2011; 8:163.

[159] Blesa J, Phani S, Jackson-Lewis V, Przedborski S. Classic and new animal models of Parkinson's disease. *J Biomed Biotechnol.* 2012; 2012:845618.

[160] Hamilton BT, Moore SE, Williams TB, Darby N, Vinson MR. Comparative effects of rotenone and antimycin on macroinvertebrate diversity in two streams in Great Basin national park, Nevada. *N Am J Fish Manag.* 2009; 29(6):1620-35.

[161] Dodiya HB, Forsyth CB, Voigt RM, Engen PA, Patel J, Shaikh M, Green SJ, Naqib A, Roy A, Kordower JH, Pahan K, Shannon KM, Keshavarzian A. Chronic stress-induced gut dysfunction exacerbates Parkinson's disease phenotype and pathology in a rotenone-induced mouse model of Parkinson's disease. *Neurobiol Dis.* 2020; 135:104352.

[162] Betarbet R, Canet-Aviles RM, Sherer TB, Mastroberardino PG, McLendon C, Kim JH, Lund S, Na HM, Taylor G, Bence NF, Kopito R, Seo BB, Yagi T, Yagi A, Klinefelter G, Cookson MR, Greenamyre JT. Intersecting pathways to neurodegeneration in Parkinson's disease: effects of the pesticide rotenone on DJ-1, alpha-synuclein, and the ubiquitin-proteasome system. *Neurobiol Dis.* 2006;22(2):404-20.

[163] Travagli RA, Browning KN, Camilleri M. Parkinson disease and the gut: new insights into pathogenesis and clinical relevance. *Nat Rev Gastroenterol Hepatol.* 2020;17(11):673-85.

[164] Liu WW, Wei SZ, Huang GD, Liu LB, Gu C, Shen Y, Wang XH, Xia ST, Xie AM, Hu LF, Wang F, Liu CF. BMAL1 regulation of microglia-mediated neuroinflammation in MPTP-induced Parkinson's disease mouse model. *FASEB J.* 2020;34(5):6570-6581.

[165] Lai F, Jiang R, Xie W, Liu X, Tang Y, Xiao H, Gao J, Jia Y, Bai Q. Intestinal Pathology and Gut Microbiota Alterations in a Methyl-4-phenyl-1,2,3,6-tetrahydropyridine (MPTP) Mouse Model of Parkinson's Disease. *Neurochem Res.* 2018;43(10):1986-1999.

[166] Mustapha M, Mat Taib CNM. MPTP-induced mouse model of Parkinson's disease: A promising direction of therapeutic strategies. *Bosn J Basic Med Sci.* 2021;21(4):422-33.

[167] Sian J, Youdim MBH, Riederer P, Gerlach M. MPTP-Induced Parkinsonian Syndrome. In: Siegel GJ, Agranoff BW, Albers RW, Fisher SK and Uhler MD, Editors. *Basic Neurochemistry: Molecular, Cellular and Medical Aspects.* 6th edition. Philadelphia: Lippincott-Raven; 1999.

[168] Pasqual L, Caldarazzo-Ienco E, Fornai F. MPTP neurotoxicity: actions, mechanisms, and animal modeling of Parkinson's disease, *Handbook of neurotoxicity.* 2014;1:237.

[169] Kopin IJ. Features of the dopaminergic neurotoxin MPTP. *Ann N Y Acad Sci.* 1992 May 11;648:96-104.

[170] Riederer P, Laux G. MAO-inhibitors in Parkinson's disease. *Exp Neurobiol.* 2011;20(1):1-17.

[171] Gerlach M, Riederer P. Animal models of Parkinson's disease: an empirical comparison with the phenomenology of the disease in Man. *J Neural Transm* (Vienna). 1996;103(8-9):987-1041.

[172] Muralikrishnan D, Mohanakumar KP. Neuroprotection by bromocriptine against 1-methyl-4-phenyl-1,2,3,6-tetrahydropyridine-induced neurotoxicity in mice. *FASEB J.* 1998;12(10):905-12.

[173] Donnan GA, Kaczmarczyk SJ, Solopotias T, Rowe P, Kalnins RM, Vajda FJ, Mendelsohn FA. The neurochemical and clinical effects of 1-methyl-4-phenyl-1,2,3,6-tetrahydropyridine in small animals. *Clin Exp Neurol.* 1986; 22:155-64. PMID: 3495376.

[174] Riachi NJ, Harik SI. Strain differences in systemic 1-methyl-4-phenyl-1,2,3,6-tetrahydropyridine neurotoxicity in mice correlate best with monoamine oxidase activity at the blood-brain barrier. *Life Sci.* 1988;42(23):2359-63.

[175] German DC, Nelson EL, Liang CL, Speciale SG, Sinton CM, Sonsalla PK. The neurotoxin MPTP causes degeneration of specific nucleus A8, A9 and A10 dopaminergic neurons in the mouse. *Neurodegeneration*. 1996;5(4):299-312.

[176] Hamre K, Tharp R, Poon K, Xiong X, Smeyne RJ. Differential strain susceptibility following 1-methyl-4-phenyl-1,2,3,6-tetrahydropyridine (MPTP) administration acts in an autosomal dominant fashion: quantitative analysis in seven strains of Mus musculus. *Brain Res.* 1999;828(1-2):91-103.

[177] Dorszewska J, Kowalska M, Prendecki M, Piekut T, Kozłowska J, Kozubski W. Oxidative stress factors in Parkinson's disease. *Neural Regen Res.* 2021;16(7):1383-91.

[178] Bragoszewski P, Turek M, Chacinska A. Control of mitochondrial biogenesis and function by the ubiquitin-proteasome system. *Open Biol.* 2017;7(4).

[179] Banerjee R, Beal MF, Thomas B. Autophagy in neurodegenerative disorders: pathogenic roles and therapeutic implications. *Trends Neurosci.* 2010;33(12):541-9.

[180] Chang KH, Chen CM. The role of oxidative stress in Parkinson's disease. *Antioxidants* (Basel). 2020;9(7):597.

[181] Monzani E, Nicolis S, Dell'Acqua S, Capucciati A, Bacchella C, Zucca FA, Mosharov EV, Sulzer D, Zecca L, Casella L. Dopamine, Oxidative Stress and Protein-Quinone Modifications in Parkinson's and Other Neurodegenerative Diseases. *Angew Chem Int Ed Engl.* 2019 May 13;58(20):6512-6527.

[182] Sziráki I, Mohanakumar KP, Rauhala P, Kim HG, Yeh KJ, Chiueh CC. Manganese: A transition metal protects nigrostriatal neurons from oxidative stress in the iron-induced animal model of Parkinsonism. *Neuroscience*. 1998;85(4):1101-11.

[183] Hurley MJ, Brandon B, Gentleman SM, Dexter DT. Parkinson's disease is associated with altered expression of CaV1 channels and calcium-binding proteins. *Brain*. 2013;136(7):2077-97.

[184] Langston JW, Ballard P, Tetrud JW, Irwin I. Chronic Parkinsonism in humans due to a product of meperidine-analog synthesis. *Science.* 1983;1979:219, 979-80.

[185] Hassanzadeh K, Rahimmi A. Oxidative stress and neuroinflammation in the story of Parkinson's disease: could targeting these pathways write a good ending? *J Cell Physiol.* 2018;234(1):23-32.

[186] Hwang O. Role of oxidative stress in Parkinson's disease. *Exp Neurobiol.* 2013;22(1):11-7.

[187] Puspita L, Chung SY, Shim JW. Oxidative stress and cellular pathologies in Parkinson's disease. *Mol Brain.* 2017;10(1):53.

[188] Kalinderi K, Bostantjopoulou S, Fidani L. The genetic background of Parkinson's disease: current progress and future prospects. *Acta Neurol Scand.* 2016;134(5):314-26.

[189] Filomeni G, de Zio D, Cecconi F. Oxidative stress and autophagy: the clash between damage and metabolic needs. *Cell Death Differ.* 2015;22(3):377-88.

[190] Dorszewska J, Prendecki M, Lianeri M, Kozubski W. Molecular effects of L-dopa therapy in Parkinson's disease. *Curr Genomics.* 2014;15(1):11-7.

[191] Dionísio PA, Amaral JD, Rodrigues CMP. Oxidative stress and regulated cell death in Parkinson's disease. *Ageing Res Rev*. 2021; 67:101263.

# Chapter 8

# The Role of ROS in Cancer Progression and Therapy

**Subhajit Ghosh***
**Shrestha Dutta***
**and Arghya Adhikary†**
Department of Life Science and Bio-Technology, Jadavpur University, Poddar Nagar, Jadavpur, Kolkata, West Bengal, India

## Abstract

Reactive oxygen species (ROS), which are widely known for their capacity to regulate cellular signaling, have two distinct roles in cancer. Mitochondria, endoplasmic reticulum (ER), and peroxisome take a crucial part in the process of ROS generation in cells. Enzymes such as xanthine oxidase, NADPH oxidase (NOX), and P450 cytochromes assist in this process. It is important for biological activities, and the steady generation of ROS in normal tissues is managed by the proper balance of effects mediated by low and high ROS concentrations. In comparison to normal cells, cancer cells regulate ROS production, which is necessary for the hyperactivation of the cell signaling pathways required for cellular transformation and carcinogenesis. Cancer cells simultaneously boost their antioxidant ability to preserve ROS equilibrium and avoid cell death. This change in the redox state of cancer cells may make them more susceptible to ROS-manipulated therapy than normal cells are.

**Keywords:** reactive oxygen species, ROS, endoplasmic reticulum, ER, xanthine oxidase, NADPH oxidase, NOX, P450 cytochromes

## 1. Introduction

Reactive oxygen species (ROS) are a class of extremely reactive chemicals that have developed into key signaling pathway regulators. It is also found that several biological processes, including gene expression, require moderate levels of ROS. Because of their higher metabolic

---

* Both the authors contributed equally.
† Corresponding Author's Email: arghyaa.lifesc@jadavpuruniversity.in.

In: The Role of Reactive Oxygen Species in Health and Disease
Editor: Kuladip Jana
ISBN: 979-8-89113-283-2

rate, gene mutations, and relative hypoxia, tumor cells produce more ROS than normal, which are then suppressed by their increased antioxidant enzymatic and nonenzymatic mechanisms. Due to their involvement in several signaling pathways and ability to cause DNA mutation, moderate elevations in ROS are a major factor in several pathologic diseases, including the promotion and development of tumors [1]. ROS play an important role in biological activities, and the steady generation of it in healthy cells is regulated by low and high ROS concentrations. It's interesting to note that ROS dynamically affects the tumor microenvironment and can promote cancer angiogenesis, metastasis, and survival at various concentrations [2].

The disruption of redox balance is one of the most significant factors behind the onset, progression, and metastasis of cancer in human cells over the past few decades [3]. It has been shown that an increase in free radicals, especially ROS, is what causes this disproportionality in redox homeostasis. Both intrinsic and extrinsic causes contribute to the production of these extremely energetic radicals. Extrinsic sources of ROS include pro-oxidant environmental pollutants, radiation, and a variety of chemical compounds, such as alcohol, tobacco use, and some medicines [4, 5]. Intrinsic sources of ROS are mostly mitochondria, inflammatory cells, and various enzymatic cellular complexes. These free radicals can harm numerous biomolecules, such as lipids, proteins, and nucleic acids, which can cause oxidative stress and harm to various human tissues.

Low levels of ROS act as natural regulators which are important for signaling pathways in cells that act like "Trojan horses" to kill cancer cells. Scientists in Italy suggested that producing natural ROS species can pave a new path in treating cancer. It has also been noticed that increasing ROS production in cancer cells can promote apoptosis. This will control cell turnover but can specifically stimulate to target diseased cells.

## 2. Generation of ROS

As we have mentioned, numerous exogenous (or extrinsic) and endogenous (or intrinsic) factors can be responsible for the generation of ROS in cells [6]. In organelles like mitochondria and peroxisomes (as endogenous sources of ROS), metabolic events take place during the respiration and photosynthesis processes, which generate ROS [7]. The structural components of mitochondria are divided into four levels i.e., the outer membrane, the inner membrane, intermembrane space (the region between the outer and inner membrane), and the matrix [8]. The electron transport chain (ETC) system is a network of proteins that transports electrons during oxidative phosphorylation to create ATP [9]. During oxidative phosphorylation, electrons are transported via a chain of proteins to produce ATP called the electron transport chain (ETC) system [9, 10]. This ETC includes five major protein complexes: (i) complex-I (NADH-ubiquinone reductase), complex-II (succinate-ubiquinone reductase), (iii) complex-III (ubiquinol-cytochrome c reductase), (iv) complex-IV (cytochrome c oxidase), and complex-v (F1-F0 ATP synthase) [8]. Both electrons and oxygen molecules pass through the protein complexes involved in the ETC system [9]. Normally, destined oxygen molecules along with electrons in ETC are completely reduced to produce water, but ROS (such as O2•) is generated only in the premature reduction of oxygen molecules (known as oxidative stress) [7, 11]. Similar to mitochondria, peroxisomes produce ROS as a by-product of their metabolic processes through several enzymes including xanthine oxidase, nitric oxide synthase, NADPH oxidase (NOX), and P450 cytochromes [12, 13] (Figure 1). Phagocytic cells also stimulate ROS production as they consume an increased amount of oxygen molecules, known as 'the respiratory burst' where NOX plays some significant roles [14, 15]. The intracellular environment is always oxidizing because, in a reducing state necessary molecules such as DNA, proteins, and enzymes lose their functionality and become useless as well as toxic for cells [16]. Some enzymes such as superoxide dismutase (SOD), catalase, and ascorbate peroxidase (APX) remove ROS and help cells to get relief from oxidative stress [17]. Additionally, various exogenous factors, for example, heavy metals (e.g., mercury, arsenic, lead, cadmium, etc.) tobacco compounds, xenobiotics, radiation, foods, excessive alcohol consumption, microplastics, etc. [18, 19] also responsible for the generation of ROS in human cells. These factors oxidize water both directly and indirectly to produce hydroxyl radicals and partial secondary ROS respectively. The mitochondrial genome lacks DNA-repairing machinery making it more vulnerable to oxidative stress-induced mutations [20]. Another report indicates that in human retinal pigment epithelial (RPE) cells, tumour necrosis factor (TNF), interleukin-1 (IL-1), and interferon (IFN) all promote intracellular and extracellular ROS generation in a time- and dose-dependent way [21].

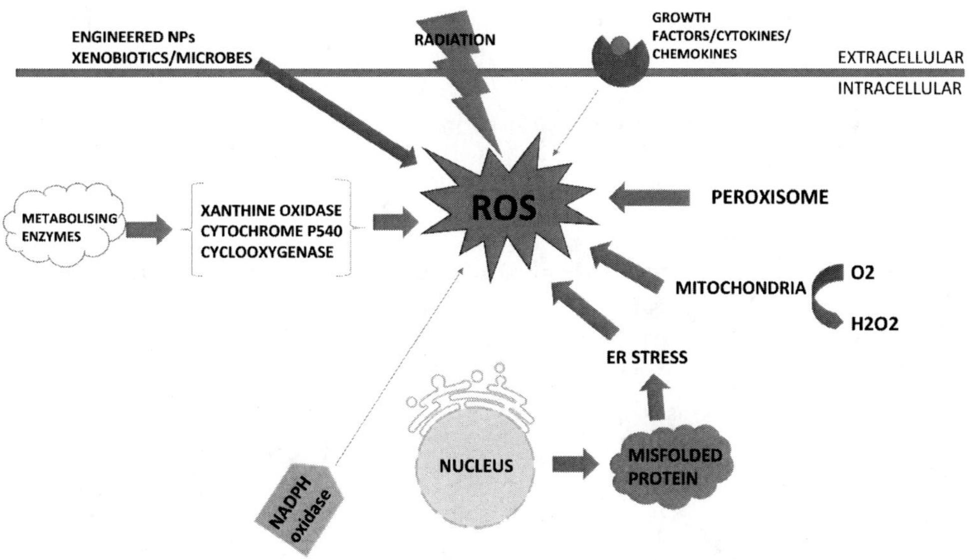


**Figure 1.** Diagram illustrating the extracellular and intracellular origins of the production of ROS. The extracellular sources of ROS are represented by environmental contaminants, radiation exposure, microbial infection, and exposure to engineered Nanoparticles (NPs). The endoplasmic reticulum (ER) stress, mitochondria, cellular-metabolizing enzymes (xanthine oxidase, cytochrome P450, cyclooxygenase), and NOX can all produce intracellular ROS.

## 3. ROS in Cancer Development

Numerous cellular activities, including immunological defence systems and obligatory secondary signaling, depend on ROS in normal somatic cells [22]. The final effect on cancer biology is influenced by the overall balance of ROS as well as the sum of its beneficial and harmful effects.

Both endogenous and external factors can generate ROS. The mitochondria and membrane-bound NADPH oxidases are the two primary endogenous producers of ROS in living things. Superoxide dismutase 1 and 2 (SOD1 and SOD2, respectively) in the mitochondria produce highly reactive superoxide anions (O2•), which are then quickly converted into hydrogen peroxide($H_2O_2$). The inflammatory cells contribute significantly to the generation of cellular ROS and are important for cellular defence systems. For instance, some exogenous and endogenous triggers an increase in ROS production in phagocytes through NOX. Pharmaceuticals, industrial chemicals, radiation, ionizing radiation, chlorinated compounds, metal ions, barbiturates, and phorbol esters are examples of exogenous sources of ROS [23, 24].

According to various comprehensive literature, reactive oxygen species (ROS) and their products are important for maintaining cell homeostasis, defending against cancer, and protecting tissue from further harm by regulating cell growth through inducing apoptosis. Although ROS have an impact on some of the most important cell survival mechanisms, such as proliferation, angiogenesis, and tumor invasion, at low and moderate levels, these agents can also expose cells to the negative effects of oxidative stress, such as DNA damage and apoptosis, which have therapeutic effects on cancer [25].

## 4. ROS-Associated Signaling Pathways in Cancer

### 4.1. ROS and MAPK Signaling Pathways

Mitogen-activated protein kinases (MAPKs) are a class of protein kinases that are negatively regulated by MAPK phosphatases (MKPs) and play an important role in conveying extracellular signals from the cell membrane to the nucleus through a cascade of phosphorylation events [26]. MAPK signaling controls numerous biological processes, including cell death and cell survival [27].

Extracellular signal-regulated kinases (ERKs, including ERK-1 and ERK-2 isoforms), c-Jun N-terminal kinases (JNKs, including JNK-1, JNK-2, and JNK-3 isoforms), and p38 MAPKs (containing p38-, p38-, p38-, and p38- isoforms) are three distinct subgroups of MAPKs A series of successive phosphorylation events are used to activate each subgroup of MAPKs, starting with the activation of MAPK kinase kinases (MAP3Ks). The dual phosphorylating threonine and tyrosine residues within a conserved tripeptide motif, the MAP3Ks phosphorylate and activate a downstream dual-specificity MAPK kinase (MAP2Ks), which in turn stimulates MAPK activity [26, 28]. It should be mentioned that the three MAPK subgroups (ERKs, JNKs, and p38 MAPKs) are important in regulating these pathways because they are implicated in both cell growth and cell death [29]. These kinases are required for intracellular transduction networks that are important regulators of a variety of cellular and pathological processes, including differentiation, immunological and stress responses, transformation, and apoptosis [30].

Different growth factors (GFs), cytokines, hormones, and oxidative stress all caused these MAPKs to become activated. When it comes to stress-related reactions, ROS is the key component that may promote the activation of growth factor receptors without the assistance of their appropriate ligands, which in turn triggers the activation of MAPK cascades [31]. For instance, ROS can activate the platelet-derived growth factor (PDGF) receptor alpha even in the absence of the corresponding ligand [32-34], hence promoting cell survival. Numerous studies have shown that MAPK cascades are crucial for the development of cancer, and elevated ROS levels can activate MAPK, which in turn promotes tumor metastasis by increasing the epithelial-to-mesenchymal transition (EMT), cell migration, and invasion, all of which can be prevented using MAPK inhibitors or ROS scavengers [35, 36].

### 4.2. ROS and PI3K/Akt Signaling Pathways

One of the most over-activated intracellular pathways in several human malignancies is the PI3K/AKT pathway. This pathway contributes to the carcinogenesis, proliferation, invasion, and metastasis of tumor cells by acting on many downstream target proteins [37].

Phosphatidylinositol-3-kinase (PI3K) and a serine/threonine protein kinase B (PKB) are the two parts of the PI3K/Aktsignaling pathway [38]. The phosphorylation of the 3- hydroxyl group of phosphatidylinositol on the plasma membrane initiates the activity of the phosphoinositide 3-kinases (PI3Ks), a significant family of lipid enzymes [39]. According to their affinity for lipid substrates and specific structures, PI3K is divided into four different classes which influence various elements of cell biology. One of the most activated intracellular pathways, the phosphatidylinositol-3-kinase PI3K/Akt/mammalian target of rapamycin (mTOR) pathway is regularly involved as a regulator for human cancer [40]. This pathway controls several physiological processes that occur within cells, most notably cell proliferation, cell cycle, and apoptosis [41]. Studies have suggested that the dysregulation of the PI3K/Akt pathway has a role in the occurrence and progression of several types of cancer, particularly solid tumors. As a result, great effort has been put into discovering new medications that target PI3K/AKT signaling, either downstream or upstream of the pathway [42-44].

In almost all types of cancer, the PI3K pathway is multistep, highly regulated, and broadly associated with survival cascade. It gets activated in response to different GFs and pro-inflammatory cytokines like vascular endothelial growth factor (VEGF), epidermal growth factor (EGF), and PDGF, as well as cytokines like L-1, IL-6, and IL-8 that alter a variety of cellular processes like cell growth, cell cycle progression, migration, invasion, and metastasis [45, 46]. The tightly coupled receptor tyrosine kinases (RTKs) or adaptor molecules along with p110 (catalytic) and p85 (regulatory), activates the downstream cascade components of the PI3K, which in turn regulates various cellular processes. Phosphatidylinositol 4,5-bisphosphate (PIP2) is converted by the catalytic subunit of PI3K to phosphatidylinositol 3,4,5-triphosphate (PIP3), which allows it to recruit and activate protein kinase B (PKB/AKT), which in turn mediates additional downstream signaling targets like mTOR, glycogen synthase kinase-3 (GSK3), p53, and nuclear factor kappa-B(NF-kB) [47, 48]. It's interesting to note that PTEN (phosphatase and tensin homolog), a tumor suppressor, blocks Akt activation by dephosphorylates PIP3 back to PIP2 [49, 50]. PTEN may commonly be deleted or mutated in most human cancers [51, 52]. In addition, ROS can encourage PTEN's proteolytic degradation by phosphorylating it with casein kinase II (CKII) [49, 53]. Additionally, it has been demonstrated that endogenously produced ROS from GFs (such as EGF or PDGF) trigger the oxidation of PTEN, which activates the PI3K signaling pathway [54]. Overall, ROS regulates the PI3K pathway and activates signaling processes via oxidizing protein phosphatases.

Notably, different ROS-producing substances may activate diverse antioxidant genes via a mechanism controlled by PI3K-Nrf2 pathways [55].

### 4.3. ROS and Src Signaling Pathways

Cell spreading into the extracellular matrix depends on Src tyrosine kinases, which are essential for adhesive reactions [56]. To govern the cytoskeletal organization and produce integrin-dependent signaling responses in fibroblasts, Src family kinases tyrosine phosphorylate a variety of signaling and cytoskeletal proteins [57]. The inhibitory phosphorylation of a carboxy-terminal tyrosine residue is regulated by the activity of the Src family of tyrosine kinases (Tyr527 in chicken c-Src). Through subsequent intramolecular interaction with the kinase's SH2 domain, this phosphorylation causes the kinase to adopt an inactive conformation. Tyr527 dephosphorylation transforms the closed/inactive state into the open/active state [58]. While CD45 tyrosine phosphatase [59] appears to be responsible for dephosphorylating the inhibitory tyrosine in lymphocytes, during integrin-mediated reactions, the receptor protein tyrosine phosphatase is in charge of activating the Src family kinase by dephosphorylating Tyr527 [60]. Following Tyr527's dephosphorylation, the conformational change encourages autophosphorylation at Tyr416. This, in turn, results in increased Src kinase activity and complete activation of the enzyme [61, 62]. A third method of controlling Src kinase activity that has been proposed is SH3 domain contact with the linker bridging the SH2 domain and the catalytic domain. This interaction results in negative inhibition of activity [63].

It has been reported that high ROS production is linked to cell transformation, which directly correlates with its tumorigenic/metastatic potentials, indicating the important role of ROS in cancer progression under the influence of the Src signal transduction pathway [64]. ROS are also known to play a role in the pathogenesis of other diseases, including cancer. The increased cell-to-ECM interaction augments ROS production and oxidizes the Src which results in the activation and induction of Src and associated signaling molecules that take part in signal transduction [65]. Integrins are important components of ECM which play a significant role in ROS-mediated oxidation-dependent Src activation, which not only affects how adhesion molecules and other ECM-associated components are expressed and function, but also promotes EMT and increases tumor cell proliferation and migration by activating several survival signaling cascades [56, 66, 67]. In addition to activating Akt, ROS-induced Src controls the activity of EGFR and receptor tyrosine kinases (RTKs) signaling to promote carcinogenesis and tumor metastasis [68, 69].

### 4.4. ROS and NF-kB Signaling Pathways

The transcription factor NF-kB was identified as a DNA binding protein in activated B cells, that regulates gene expression in various cellular processes like the development of the organ, innate immune response, embryogenesis, cell proliferation, cell death (apoptosis), and stress responses to several harmful stimuli [70, 71]. The REL subfamily proteins (p65/RELA, RELB, and c-REL) and the NF-kB subfamily proteins make up the homo- and heterodimers of five different proteins that make up the NF-kB (nuclear factor kappa-light-chain-enhancer of activated B cells) transcription factor (p50, and p52, and its precursors p105 and p100, respectively). It's interesting to note that it has a role in both oxidation and ROS generation [72]. c-Rel, RelA/p65, RelB, p50/p105/NF-B1, and p52/p100/NF-kB2 are all members of the NF-kB family. NF-kB is inhibited in the cytoplasm under physiologically normal conditions where it binds to NF-kB inhibitors (IBs), but upon activation with growth factors or cytokines, IKK complexes (IKK, IKK, and NEMO/IKK) are activated [73, 74].

According to reports, ROS can control the NF-kB cascade either positively (during the early phase) by potentiating NF-kB activation or negatively (during the late phase) by inhibiting it [75]. Studies show that ROS activates the NF-kB pathway in a variety of ways, most notably by increasing the phosphorylation and subsequent degradation of I-kBα [76]. Additionally, IKK and its kinases (MEKK1) are also the main targets for ROS, which deactivate MEKK1 and IKK by S- glutathionylation to affect the activity of NF-kB [77]. Finally, ROS causes the upstream kinase in the noncanonical NF-kB pathway, NF-kB-inducing kinase (NIK), to become active by inhibiting phosphatases and oxidation of cysteine residues. Additionally, ROS directly activates NF-kB, which is an important regulator in the development of cancer by altering the expression of genes linked to metastasis, such as matrix metalloproteinases (MMPs) and pro-inflammatory cytokines [78, 79]. Endogenous ROS induce several transcription factors, including NF-kB, in cancer cells through activating the Src, PI3K/AKT, and Ras/MAPK pathways. Thus, through promoting cell proliferation, survival, invasion, and angiogenesis, these activated redox-sensitive signaling cascades and associated transcription factors aid in the spread of cancer.

## 5. ROS Associated Apoptotic Pathways

Apoptosis is a well-conserved and highly regulated form of cell death in which a cell destroys itself [80]. It is a significant stage in multicellular organisms that neutralizes possibly lethal cells with DNA damage or eliminates unwanted or unnecessary cells during development, preventing carcinogenesis. It can initiate several extrinsic and intrinsic signaling. Among them are numerous stressors like ROS, RNS, DNA-damaging substances (like radiation), thermal shock, serum depletion, viral infection, and hypoxia [81, 82].

The upstream of the activation of caspases, three primary signaling pathways might cause apoptosis. These include – intrinsic signaling which involves mitochondria and the endoplasmic reticulum and extrinsic signaling which involves death receptors localized at the cell surface [83-85]. Here we are going to elaborate some of the important ROS-based apoptotic pathways in cells (Figure 2).

### 5.1. Mitochondrial Pathway

The fundamental role of mitochondria is to turn nutrients into ATP, which is then exported and utilized by the entire cell [86]. Different characteristics of the outer and inner mitochondrial membranes enhance this function [87]. The high impermeability of the inner mitochondrial membrane (IMM) is an essential property for producing the electrochemical potential required for oxidative phosphorylation and ATP synthesis. Apoptosis is triggered and controlled by mitochondria [88]. The regulation of IMM permeability is important in this process. The molecular nature of the mitochondrial permeability transition pore (MPTP) complex is still unknown and appears to be involved in mitochondrial-dependent apoptosis [89, 90]. Cyclophilin D which is hypothesized to make up the MPTP complex is a molecular chaperone in the mitochondrial matrix that binds to the adenine-nucleotide-translocator (ANT) in the IMM, which connects with the voltage dependent-anion-channel (VDAC) in the outer mitochondrial membrane (OMM), and cyclophilin D, a molecular chaperone in the mitochondrial matrix [91].

The points of connection between the IMM and OMM include VDAC and ANT. When under stress, the inner membrane becomes more permeable, allowing molecules smaller than 1.5 kDa, including protons, to freely enter the mitochondrial matrix and impair oxidative phosphorylation [80]. Further, this results in osmotic swelling of the mitochondrial matrix and compression of vesicles formed by infolding of intercristal space. Pro-apoptotic proteins like cytochrome c, apoptosis-inducing factor (AIF), endonuclease G (endoG), and second mitochondria-derived activator of caspases/direct inhibitor of apoptosis (IAP)-binding protein with low pI (Smac/Diablo) are released into the cytosol because of increased outer mitochondrial membrane permeabilization (MOMP). In the end, both caspase-dependent and caspase-independent pathways result in apoptotic cell death [78].

### 5.2. Death Receptor Pathway

The extrinsic apoptotic pathway is activated by binding of death ligands — TNF, TRAIL, and FasL with their corresponding cognate receptors (TNFR, TRAILR1/R2, and Fas), which results in the recruitment of adaptor proteins and pro-caspases, which triggers the formation of death-inducing signaling complex (DISC). This DISC triggers downstream caspases, which ultimately result in apoptotic cell death [92, 93]. Additionally, it has been shown that ROS can activate death receptors or suppress c-FLIP, which can enhance the extrinsic apoptotic pathway [94, 95].

### 5.3. ROS and Apoptosis Mediated by Other Pathways

Additionally, ROS can cause apoptosis by accelerating the proteasomal degradation (ubiquitination) of several anti-apoptotic proteins present in the intrinsic pathway, including Bcl-2, Bcl-w, Bcl-XL, IAPs, and survivin, and by reducing the intracellular levels of some pro-apoptotic proteins, including Bax, Bak, Bid, Bim, and Bad [96, 97]. The expression and operations of numerous effector molecules implicated in the intrinsic pathway, including Cyt c, may also be impacted by the increased ROS levels. In a variety of cancer types, the activation of the JNK pathway by excessive ROS generation can also result in cell death brought on by intrinsic or extrinsic apoptotic signaling pathways [98, 99] (Figure 2). Due to modifications in the mitochondrial function triggered by MAPKs, many cancer cell types exhibit enhanced ROS-mediated apoptosis in response to tyrosine kinase inhibitors [100]. Monoclonal antibodies/tyrosine kinase inhibitors are used to induce apoptosis by increasing the levels of ROS [100, 101]. Additionally, it has been stated that the cytotoxic effects of doxorubicin in several solid and non-solid tumor types are based on the ROS-mediated death of cancer cells [102].

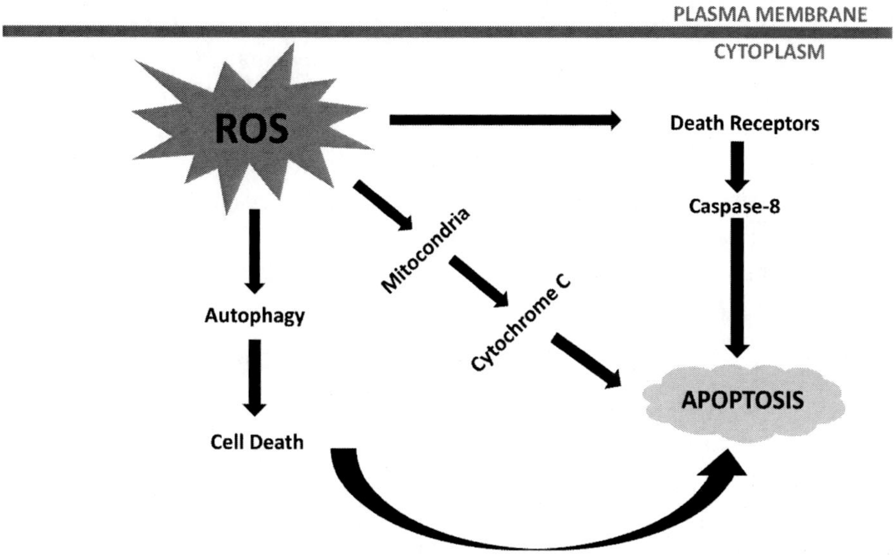


**Figur 2.** A schematic diagram of ROS-based apoptotic pathway in cancer cells. It summarizes how increased ROS production in cancer cells leads to death (or apoptosis) of those cells via mitochondrial pathway, death-receptor and other pathways.

## 6. ROS and Cancer Stem Cells (CSCs)

Carcinogenesis is a chain of events that is initiated by cells losing control over their growth due to accumulating mutations, which results in uncontrolled proliferation. Usually, this involves gene alterations like oncogenes, tumor suppressor genes, and genes involved in DNA repair pathways. Following subsequent mutation, clonal selection with more aggressive phenotypes is produced [103]. The presence of a subpopulation of stem-like cells with resistance qualities, which can re-initiate the tumor and cause tumor relapse, may also cause cancer to become more aggressive after a therapeutic intervention such as chemotherapy. Additional research on breast cancer [104], brain cancer [105], and subsequently other types of malignancies further confirmed the existence of stem cells in cancer. Their capability of self-renewal and differentiation initiates the growth of tumors. CSCs are derived from the normal stem cells and progenitor cells found in tumor tissues. Normal stem cells typically reside in a quiescent state under normal physiological conditions, which are maintained by a specialized habitat. The stem cells are dormant and do not divide or proliferate until they are stimulated. Stem cells will proliferate uncontrollably and may develop tumors if they inherit a genetic mutation that makes them immune to growth signals or resistant to antigrowth signals [106].

Tumor heterogeneity is a result of the variety of cancer cells that make up tumors. Among these cell types, cancer stem cells (CSCs) play an important role in the development and spread of cancer. CSCs can self-renew and differentiate. Mesenchymal stem cells (MSCs), which differentiate into cells with mesodermal features, make up a specific population of CSCs. The potential pro- or anti-tumorigenic impact of MSCs on tumor initiation and development has been documented in several investigations. MSCs are recruited to the area of a wound to help heal damaged tissues, a process that is also linked to the development of tumors [107].

Stem cells are famous for their high self-proliferative and differentiation capacity [108-110]. Cancer stem cells (CSCs) are only a set of a subpopulation of cancer cells, that have self-renewal ability [111]. Although CSCs were formerly thought to be a homogeneous population of cells, it has since been established that they are phenotypically and functionally heterogeneous [112]. CSCs display a variety of surface markers specific to different cell/tumor types, such as cluster of differentiation (CD); CD133, CD44, CD24, epithelial cell adhesion molecule (EpCAM), and aldehyde dehydrogenase-I (ALDH-I) [113, 114]. Aside from that, the expression of transcription factors like Octamer-binding transcription factor 4 (Oct4), Sex-determining region Y-box 2 (Sox2), Krüppel-like factor 4 (KLF4), and Homeobox transcription factor (Nanog), among others, is also considered as a marker for CSCs [115]. CSCs also show resistance against both chemotherapy and radiotherapy [116]. CSCs are carcinogenic by nature and are thought to be important contributors to tumor heterogeneity [117]. Additionally, new research suggests that CSCs are dynamic cell populations that are continuously altered by the factors within and outside of cells. In general, cells with low to moderate levels of ROS will proliferate, expand, and have a higher chance of survival [118]. However, excessive ROS can lead to cellular damage and initiate apoptosis [22]. Many cancer cells have heightened ROS levels in part as a result of their faster metabolism [75]. Cancer cells may undergo necrosis and apoptosis when ROS levels are out of whack. High antioxidant capability helps cancer cells fight and scavenge ROS [119]. Since high antioxidant capacity increases cell survival and reduces cellular responses to anticancer therapy. The idea of cancer stem cells (CSCs) has gained popularity recently as a result of the discovery and reporting of a subset of cancer cells having stem cell-like traits in several malignancies, including leukemia, breast cancer, and pancreatic cancer [120-122].

In cancer cells, high-rate metabolism at the mitochondria, endoplasmic reticulum, and cell membranes is primarily responsible for the production of ROS [123]. To produce ATP, cancer cells substitute glycolysis for oxidative phosphorylation to some extent [124]. To adapt to hypoxic circumstances with fewer mitochondrial faults and ROS generation, cancer cells require this metabolic transition [125]. It has been noticed that the stem-like population of human gastrointestinal cancer cells, as well as some breast tumors, have lower ROS levels [126]. This decreased ROS level in CSCs could be explained by increased ROS scavenging mechanisms [127]. In addition, many transcriptional activities and signaling pathways assist in scavenging ROS in CSCs and conventional stem cells [128]. Among them, in hematopoietic stem cells (HSCs), for instance, homeobox type O (FOXO) is crucial for maintaining low ROS levels [129], which are crucial for their stem-like property. To maintain low ROS levels and stemness, the ataxia telangiectasia mutant (ATM) can upregulate antioxidant enzymes while downregulating differentiation and proliferation genes [75]. The maintenance of stemness and treatment resistance in pancreatic cancer stem cells depends on JNK pathway activation, by reducing the production of ROS that those chemotherapeutic drugs cause [130, 131]. The increase in free radical scavenging systems, such as glutathione (GSH), was revealed to be the reason why the ROS was decreased [132]. GSH production was enhanced by the expression of CD44v [133], one of the CD44+ variant isoforms [134]. By interacting with and stabilizing the xCT, a subunit of the cysteine-glutamate transporter xc(-), the CD44v protein may facilitate cysteine uptake, further leading to increased GSH synthesis [135, 136]. Recent research suggests that the regulation of ROS in CSCs may also be significantly influenced by the epigenetic process. In basal-like breast cancer, fructose-1,6-biphosphatase (FBP1) was negatively regulated by epigenetic mechanisms, which enhanced glycolysis but decreased

Levels of ROS [137]. MicroRNA may also be crucial for controlling ROS generation and scavenging in CSCs [138, 139].

## 7. ROS-Dependent Transcription Factors in CSCs

In hypoxia, the transcriptional factors of the HIF family are elevated [140]. Both an oxygen-sensitive HIF-α subunit and a constitutively produced HIF-β subunit are found in HIFs [141]. According to certain research, ROS can control the expression of HIF [142]. HIF-α has three isoforms, i.e., HIF-1α, HIF-2α, and HIF-3α [143]. Additional research revealed that the HIF-1-mediated metabolic reprogramming boosted the survival of metastatic tumors and decreased ROS levels [144]. AML, breast cancer, and other malignancies have all been extensively researched for the potential role of NF-κB in chemotherapy resistance [145]. Mammary epithelial cells may exhibit fewer tumor stem cell markers and CSC populations when NF-kB is inhibited [146]. The Ras-related C3botulinum toxin 1 (RAC1) and ROS-induced activation of NF-κB plays a crucial role in the development of colorectal cancer [147]. Considering the low ROS levels, NF-kB upregulation in CSCs could aid in maintaining redox equilibrium [130]. Like HIF-family factors and NF-kB, p53 is another transcription factor that plays a crucial role in ROS regulatory mechanisms [148]. Depending on its levels, p53 can control the genes that produce or scavenge ROS and can have both pro- and antioxidant effects [149]. The release of mitochondrial ROS during stress-induced apoptosis is the basis of the prooxidant action of p53 [150]. However, the antioxidant activity of p53 is associated with the production of antioxidant gene products, which responded to decreased levels of p53 in cells that were neither stressed nor physiologically stressed [151]. By oxidizing p53 cysteine residues, ROS may also regulate the activity of p53 and make it inactive [152]. The control of the cell cycle and apoptosis depends heavily on the interaction between the p53 and ROS signaling pathways [83]. Several antioxidant and detoxifying genes are expressed by the nuclear factor erythroid 2-related factor (Nrf2), a crucial regulator of defense against endogenous and exogenous stresses [153, 154]. GSH synthase, GSH reductase, and peroxidase families are among the Nrf2-regulated antioxidant genes [155]. Although there is little knowledge on how ROS are regulated in CSCs, there is rapidly growing evidence that ROS may be crucial to the ability of CSCs to self-renew and differentiate [156]. ROS are notably a two-edged sword in the fight against cancer [157]. Targeting CSCs via antioxidant proteins and ROS control can significantly enhance cancer treatment. Targeting ROS manipulations may provide a novel method for treating cancer, although these different contradictions make it challenging.

## 8. ROS-Targeted Anticancer Therapeutic Strategy

The level of ROS in tumor cells is crucial for anticancer therapy. High ROS levels have been shown to cause cytotoxicity in tumorigenic cells and to reverse chemotherapy resistance. Numerous studies point to raised ROS levels as the fundamental mechanism of action and effectiveness of conventional cancer therapy, which is directly related to cancer cell death [139, 158, 159].

### 8.1. Targeting Antioxidant Enzymes for Cancer Cell Death

The restriction or suppression of cellular antioxidant enzymes increases the formation of ROS, which leads malignant cells to undergo apoptosis. ROS-enhancing anticancer medications primarily target the SOD, GSH system, and Thioredoxin 1 (Trx) enzymes [160-162]. Increased cellular antioxidant levels are thought to play a key role in the development of chemoresistance in cancer cells. Many chemotherapeutic agents and antitumor medicines have been created that particularly target intracellular GSH levels to combat chemotherapy resistance [163]. For instance, administering PEITC (phenethyl-isothiocyanate) together with GSH reduces the GSH pool, hence halting the oxidative stress that causes cytotoxicity in cancer cells. Glutathione peroxidase (GPX) has further been shown to be blocked in HRAS-transformed ovarian cancer cells, resulting in cell death [164, 165]. Apoptosis can be particularly induced in many tumors by targeting SOD1 in addition to GSH. Inhibition of SOD1 slowed the development of KRAS-mutant tumor cells, revealing in vitro research on lung adenocarcinoma [161, 166]. Nrf2, a crucial regulator of cellular antioxidants, has also been linked to the oncogenes KRAS and MYC, which are thought to mediate Nrf2 transcription. To encourage ROS-induced apoptosis in tumor cells, Nrf2 can thus be therapeutically addressed [167, 168].

### 8.2. Inducing Increased ROS Generation in Cancer Cells

As a higher concentration of ROS is responsible for cancer cell death, augmentation of ROS production in cancer cells can be an effective therapeutic strategy. Several research has proven it too. The release of cytochrome c and caspase-3-dependent apoptosis were directly induced by butyrate, which was also proven to be a ROS inducer via upregulating the expression of miR-22 and producing mitochondrial malfunction in hepatocellular carcinoma HCC [169]. Sulindac sulfide-triggered ROS reduced the expression of miR-27a, which inhibited colon cancer cell proliferation via the miR-27a/ZBTB10/specific protein (Sp) axis [170]. By boosting the production of ROS, phenyl-isothiocyanate (PEITC) reduced the expression of miR-27a, miR-20a, and miR-17 [171]. Quinacrine, on the other hand, activates p38 MAPK and inhibits ERK to operate as a ROS inducer to upregulate FOXP3 in human leukemia U937 cells [172]. Additionally, curcumin (by activation of miRNA-125/ERRα pathway) [173], betulinic acid (by activation of miRNA-21 dependent p53/sod2 pathway) [174], and Polygonatumodoratumlectin (POL) [175] can also exert their anticancer effects via miRNA mediated ROS buildup in cells. Nanoparticles such as mesoporous silica nanoparticles (MSNs), ZnO/CNT@FeO, and SASP/ZnO nanoparticles boost ROS levels and trigger cancer cell apoptosis [176-178].

### 8.3. ROS-Related Non-Coding RNA (ncRNA)

Many ROS-related ncRNAs are currently showing promise in cancer detection and treatment, with some of them being explored as therapeutic agents and targets. Contrasting to lncRNA or circRNA-based therapies, small ncRNAs like miRNAs and siRNAs have been used extensively in clinical trials for a range of diseases [179]. According to some evidence, miR-155 targets Nrf2-mediated ROS and is proven as oncogenic in many cancers [180]. Numerous oxidative stress (OS)-related oncogenes and signaling pathways, including SIRT1, IGF2BP3, FOXM1/eEF2K, and IL-6R/STAT3 signaling axis, were reported to be silenced by miR-34a, which also influenced the cancer drug resistance, proliferation, and migration [181-184]. Along with miR-21 (an oncogenic miRNA), hypoxia-inducible miR-210 has also been linked to the promotion of cancer growth in a hypoxic microenvironment [185, 186]. Cancer cell death and treatment resistance were reported to be reversed by polymeric CXCR4 antagonist (PCX) and anti-miR-210 nanoparticles [187]. The redox regulator miR-124 was shown to behave as a sponge for CircMMP9 (a circulatory RNA), speeding up the development and metastasis of glioblastoma multiforme (GBM) cells [188].

## Conclusion

The OS has two sides. Redox signaling is activated by oxidative stress, which also encourages growth and migration. On the contrary, oxidative distress promotes oxidative harm, growth inhibition, and cell death. The dual functions serve as a reminder of two categories of ROS-targeting treatment approaches. Therefore, ROS-based therapies have to be reviewed from a completely different angle. ROS-based ncRNA or ncRNA-targeted medication shows promise in the treatment of cancer, but further research is needed to ensure their effectiveness and safety.

## References

[1] Perillo B, Di Donato M, Pezone A, Di Zazzo E, Giovannelli P, Galasso G, Castoria G, Migliaccio A. ROS in cancer therapy: the bright side of the moon. *Exp Mol Med*. 2020 Feb;52(2):192-203.

[2] Aggarwal V, Tuli HS, Varol A, Thakral F, Yerer MB, Sak K, Varol M, Jain A, Khan MA, Sethi G. Role of Reactive Oxygen Species in Cancer Progression: Molecular Mechanisms and Recent Advancements. *Biomolecules*. 2019 Nov 13;9(11):735.

[3] Saikolappan S, Kumar B, Shishodia G, Koul S, Koul HK. Reactive oxygen species and cancer: A complex interaction. *Cancer letters*. 2019;452:132-43.

[4] Nourazarian AR, Kangari P, Salmaninejad A. Roles of oxidative stress in the development and progression of breast cancer. *Asian Pacific journal of cancer prevention: APJCP*. 2014;15(12):4745-51.

[5] Zhang J, Ahn KS, Kim C, Shanmugam MK, Siveen KS, Arfuso F, Samym RP, Deivasigamanim A, Lim LH, Wang L, Goh BC, Kumar AP, Hui KM, Sethi G. Nimbolide-Induced Oxidative Stress Abrogates STAT3 Signaling Cascade and Inhibits Tumor Growth in Transgenic Adenocarcinoma of Mouse Prostate Model. *Antioxid Redox Signal*. 2016 Apr 10;24(11):575-89.

[6] Zhou D, Shao L, Spitz DR. Reactive oxygen species in normal and tumor stem cells. *Advances in cancer research*. 2014;122:1-67.

[7] Zorov DB, Juhaszova M, Sollott SJ. Mitochondrial reactive oxygen species (ROS) and ROS-induced ROS release. *Physiological reviews*. 2014;94(3):909-50.

[8] Kühlbrandt W. Structure and function of mitochondrial membrane protein complexes. *BMC biology*. 2015;13:89.

[9] Zhao RZ, Jiang S, Zhang L, Yu ZB. Mitochondrial electron transport chain, ROS generation and uncoupling (Review). *International journal of molecular medicine*. 2019;44(1):3-15.

[10] Nolfi-Donegan D, Braganza A, Shiva S. Mitochondrial electron transport chain: Oxidative phosphorylation, oxidant production, and methods of measurement. *Redox biology*. 2020;37:101674.

[11] Turrens JF. Mitochondrial formation of reactive oxygen species. *The Journal of physiology*. 2003;552(Pt 2):335-44.

[12] Di Meo S, Reed TT, Venditti P, Victor VM. Role of ROS and RNS Sources in Physiological and Pathological Conditions. *Oxidative medicine and cellular longevity*. 2016;2016:1245049.

[13] Panday A, Sahoo MK, Osorio D, Batra S. NADPH oxidases: an overview from structure to innate immunity-associated pathologies. *Cellular & molecular immunology*. 2015;12(1):5-23.

[14] Robinson JM. Phagocytic leukocytes and reactive oxygen species. *Histochemistry and cell biology*. 2009;131(4):465-9.

[15] Moghadam ZM, Henneke P, Kolter J. From Flies to Men: ROS and the NADPH Oxidase in Phagocytes. *Frontiers in cell and developmental biology*. 2021;9:628991.

[16] Lobo V, Patil A, Phatak A, Chandra N. Free radicals, antioxidants and functional foods: Impact on human health. *Pharmacognosy reviews*. 2010;4(8):118-26.

[17] Das K, Roychoudhury A. *Reactive oxygen species (ROS) and response of antioxidants as ROS-scavengers during environmental stress in plants*. 2014;2.

[18] Tchounwou PB, Yedjou CG, Patlolla AK, Sutton DJ. Heavy metal toxicity and the environment. *Experientia supplementum* (2012). 2012;101:133-64.

[19] Balali-Mood M, Naseri K, Tahergorabi Z, Khazdair MR, Sadeghi M. Toxic Mechanisms of Five Heavy Metals: Mercury, Lead, Chromium, Cadmium, and Arsenic. *Frontiers in pharmacology*. 2021;12:643972.

[20] Druzhyna NM, Wilson GL, LeDoux SP. Mitochondrial DNA repair in aging and disease. *Mechanisms of ageing and development*. 2008;129(7-8):383-90.

[21] Yang D, Elner SG, Bian ZM, Till GO, Petty HR, Elner VM. Pro-inflammatory cytokines increase reactive oxygen species through mitochondria and NADPH oxidase in cultured RPE cells. *Experimental eye research*. 2007;85(4):462-72.

[22] Redza-Dutordoir M, Averill-Bates DA. Activation of apoptosis signalling pathways by reactive oxygen species. *Biochimica et biophysica acta*. 2016;1863(12):2977-92.

[23] Schieber M, Chandel NS. ROS function in redox signaling and oxidative stress. *Current biology: CB*. 2014;24(10):R453-62.

[24] Klaunig JE, Wang Z, Pu X, Zhou S. Oxidative stress and oxidative damage in chemical carcinogenesis. *Toxicology and applied pharmacology*. 2011;254(2):86-99.

[25] Kardeh S, Ashkani-Esfahani S, Alizadeh AM. Paradoxical action of reactive oxygen species in creation and therapy of cancer. *European journal of pharmacology*. 2014;735:150-68.

[26] Boutros T, Chevet E, Metrakos P. Mitogen-activated protein (MAP) kinase/MAP kinase phosphatase regulation: roles in cell growth, death, and cancer. *Pharmacological reviews*. 2008;60(3):261-310.

[27] Haagenson KK, Wu GS. The role of MAP kinases and MAP kinase phosphatase-1 in resistance to breast cancer treatment. *Cancer metastasis reviews*. 2010;29(1):143-9.

[28] Brown MD, Sacks DB. Protein scaffolds in MAP kinase signalling. *Cellular signalling*. 2009;21(4):462-9.

[29] Winter-Vann AM, Johnson GL. Integrated activation of MAP3Ks balances cell fate in response to stress. *Journal of cellular biochemistry*. 2007;102(4):848-58.

[30] Khavari TA, Rinn J. Ras/Erk MAPK signaling in epidermal homeostasis and neoplasia. *Cell cycle* (Georgetown, Tex). 2007;6(23):2928-31.

[31] Lei H, Kazlauskas A. A reactive oxygen species-mediated, self-perpetuating loop persistently activates platelet-derived growth factor receptor α. *Molecular and cellular biology*. 2014;34(1):110-22.

[32] Kyriakis JM, Avruch J. Mammalian MAPK signal transduction pathways activated by stress and inflammation: a 10-year update. *Physiological reviews*. 2012;92(2):689-737.

[33] Lei H, Kazlauskas A. Growth factors outside of the platelet-derived growth factor (PDGF) family employ reactive oxygen species/Src family kinases to activate PDGF receptor alpha and thereby promote proliferation and survival of cells. *The Journal of biological chemistry*. 2009;284(10):6329-36.

[34] Owens DM, Keyse SM. Differential regulation of MAP kinase signalling by dual-specificity protein phosphatases. *Oncogene*. 2007;26(22):3203-13.

[35] Wu WS, Wu JR, Hu CT. Signal cross talks for sustained MAPK activation and cell migration: the potential role of reactive oxygen species. *Cancer metastasis reviews*. 2008;27(2):303-14.

[36] McCubrey JA, Steelman LS, Chappell WH, Abrams SL, Wong EW, Chang F, Lehmann B, Terrian DM, Milella M, Tafuri A, Stivala F, Libra M, Basecke J, Evangelisti C, Martelli AM, Franklin RA. Roles of the Raf/MEK/ERK pathway in cell growth, malignant transformation and drug resistance. *Biochim Biophys Acta*. 2007 Aug;1773(8):1263-84.

[37] Rascio F, Spadaccino F, Rocchetti MT, Castellano G, Stallone G, Netti GS, Ranieri E. The Pathogenic Role of PI3K/AKT Pathway in Cancer Onset and Drug Resistance: An Updated Review. *Cancers* (Basel). 2021 Aug 5;13(16):3949.

[38] Xu F, Na L, Li Y, Chen L. Roles of the PI3K/AKT/mTOR signalling pathways in neurodegenerative diseases and tumours. *Cell & bioscience*. 2020;10(1):54.

[39] Shi X, Wang J, Lei Y, Cong C, Tan D, Zhou X. Research progress on the PI3K/AKT signaling pathway in gynecological cancer (Review). *Molecular medicine reports*. 2019;19(6):4529-35.

[40] Vanhaesebroeck B, Guillermet-Guibert J, Graupera M, Bilanges B. The emerging mechanisms of isoform-specific PI3K signalling. *Nature reviews Molecular cell biology*. 2010;11(5):329-41.

[41] Barra F, Evangelisti G, Ferro Desideri L, Di Domenico S, Ferraioli D, Vellone VG, De Cian F, Ferrero S. Investigational PI3K/AKT/mTOR inhibitors in development for endometrial cancer. *Expert Opin Investig Drugs*. 2019 Feb;28(2):131-142.

[42] Iida M, Harari PM, Wheeler DL, Toulany M. Targeting AKT/PKB to improve treatment outcomes for solid tumors. *Mutation research*. 2020;819-820:111690.

[43] Jiang N, Dai Q, Su X, Fu J, Feng X, Peng J. Role of PI3K/AKT pathway in cancer: the framework of malignant behavior. *Molecular biology reports*. 2020;47(6):4587-629.

[44] Shorning BY, Dass MS, Smalley MJ, Pearson HB. The PI3K-AKT-mTOR Pathway and Prostate Cancer: At the Crossroads of AR, MAPK, and WNT Signaling. *International journal of molecular* sciences. 2020;21(12).

[45] Qi S, Xin Y, Guo Y, Diao Y, Kou X, Luo L, Yin Z. Ampelopsin reduces endotoxic inflammation via repressing ROS-mediated activation of PI3K/Akt/NF-κB signaling pathways. *Int Immunopharmacol*. 2012 Jan;12(1):278-87.

[46] Park KR, Nam D, Yun HM, Lee SG, Jang HJ, Sethi G, Cho SK, Ahn KS. β-Caryophyllene oxide inhibits growth and induces apoptosis through the suppression of PI3K/AKT/mTOR/S6K1 pathways and ROS-mediated MAPKs activation. *Cancer Lett*. 2011 Dec 22;312(2):178-88.

[47] Zhang J, Wang X, Vikash V, Ye Q, Wu D, Liu Y, Dong W. ROS and ROS-Mediated Cellular Signaling. *Oxid Med Cell Longev*. 2016;2016:4350965.

[48] Papaiahgari S, Zhang Q, Kleeberger SR, Cho HY, Reddy SP. Hyperoxia stimulates an Nrf2-ARE transcriptional response via ROS-EGFR-PI3K-Akt/ERK MAP kinase signaling in pulmonary epithelial cells. *Antioxidants & redox signaling*. 2006;8(1-2):43-52.

[49] Leslie NR, Batty IH, Maccario H, Davidson L, Downes CP. Understanding PTEN regulation: PIP2, polarity and protein stability. *Oncogene*. 2008;27(41):5464-76.

[50] Leslie NR, Downes CP. PTEN: The down side of PI 3-kinase signalling. *Cellular signalling*. 2002;14(4):285-95.

[51] Wu X, Senechal K, Neshat MS, Whang YE, Sawyers CL. The PTEN/MMAC1 tumor suppressor phosphatase functions as a negative regulator of the phosphoinositide 3-kinase/Akt pathway. *Proceedings of the National Academy of Sciences of the United States of America*. 1998;95(26):15587-91.

[52] Li J, Yen C, Liaw D, Podsypanina K, Bose S, Wang SI, Puc J, Miliaresis C, Rodgers L, McCombie R, Bigner SH, Giovanella BC, Ittmann M, Tycko B, Hibshoosh H, Wigler MH, Parsons R. PTEN, a putative protein tyrosine phosphatase gene mutated in human brain, breast, and prostate cancer. *Science*. 1997 Mar 28;275(5308):1943-7.

[53] Zhang J, Grindley JC, Yin T, Jayasinghe S, He XC, Ross JT, Haug JS, Rupp D, Porter-Westpfahl KS, Wiedemann LM, Wu H, Li L. PTEN maintains haematopoietic stem cells and acts in lineage choice and leukaemia prevention. *Nature*. 2006 May 25;441(7092):518-22.

[54] Seo JH, Ahn Y, Lee SR, Yeol Yeo C, Chung Hur K. The major target of the endogenously generated reactive oxygen species in response to insulin stimulation is phosphatase and tensin homolog and not phosphoinositide-3 kinase (PI-3 kinase) in the PI-3 kinase/Akt pathway. *Molecular biology of the cell*. 2005;16(1):348-57.

[55] Sakamoto K, Iwasaki K, Sugiyama H, Tsuji Y. Role of the tumor suppressor PTEN in antioxidant responsive element-mediated transcription and associated histone modifications. *Molecular biology of the cell*. 2009;20(6):1606-17.

[56] Giannoni E, Buricchi F, Raugei G, Ramponi G, Chiarugi P. Intracellular reactive oxygen species activate Src tyrosine kinase during cell adhesion and anchorage-dependent cell growth. *Molecular and cellular biology*. 2005;25(15):6391-403.

[57] Tatosyan AG, Mizenina OA. Kinases of the Src family: structure and functions. *Biochemistry Biokhimiia*. 2000;65(1):49-58.

[58] Cooper JA, Howell B. The when and how of Src regulation. *Cell*. 1993;73(6):1051-4.

[59] Hurley TR, Hyman R, Sefton BM. Differential effects of expression of the CD45 tyrosine protein phosphatase on the tyrosine phosphorylation of the lck, fyn, and c-src tyrosine protein kinases. *Molecular and cellular biology*. 1993;13(3):1651-6.

[60] Su J, Muranjan M, Sap J. Receptor protein tyrosine phosphatase alpha activates Src-family kinases and controls integrin-mediated responses in fibroblasts. *Current biology: CB*. 1999;9(10):505-11.

[61] Brown MT, Cooper JA. Regulation, substrates and functions of src. *Biochimica et biophysica acta*. 1996;1287(2-3):121-49.

[62] Xu W, Harrison SC, Eck MJ. Three-dimensional structure of the tyrosine kinase c-Src. *Nature*. 1997;385(6617):595-602.

[63] Gonfloni S, Williams JC, Hattula K, Weijland A, Wierenga RK, Superti-Furga G. The role of the linker between the SH2 domain and catalytic domain in the regulation and function of Src. *The EMBO journal*. 1997;16(24):7261-71.

[64] Giannoni E, Chiarugi P. Redox circuitries driving Src regulation. *Antioxidants & redox signaling*. 2014;20(13):2011-25.

[65] Gianni D, Taulet N, DerMardirossian C, Bokoch GM. c-Src-mediated phosphorylation of NoxA1 and Tks4 induces the reactive oxygen species (ROS)-dependent formation of functional invadopodia in human colon cancer cells. *Molecular biology of the cell*. 2010;21(23):4287-98.

[66] Taddei ML, Parri M, Mello T, Catalano A, Levine AD, Raugei G, Ramponi G, Chiarugi P. Integrin-mediated cell adhesion and spreading engage different sources of reactive oxygen species. *Antioxid Redox Signal*. 2007 Apr;9(4):469-81.

[67] Palacios F, Tushir JS, Fujita Y, D'Souza-Schorey C. Lysosomal targeting of E-cadherin: a unique mechanism for the down-regulation of cell-cell adhesion during epithelial to mesenchymal transitions. *Molecular and cellular biology*. 2005;25(1):389-402.

[68] Haynes MP, Li L, Sinha D, Russell KS, Hisamoto K, Baron R, Collinge M, Sessa WC, Bender JR. Src kinase mediates phosphatidylinositol 3-kinase/Akt-dependent rapid endothelial nitric-oxide synthase activation by estrogen. *J Biol Chem*. 2003 Jan 24;278(4):2118-23.

[69] Mao W, Irby R, Coppola D, Fu L, Wloch M, Turner J, Yu H, Garcia R, Jove R, Yeatman TJ. Activation of c-Src by receptor tyrosine kinases in human colon cancer cells with high metastatic potential. *Oncogene*. 1997 Dec 18;15(25):3083-90.

[70] Sen R, Baltimore D. Multiple nuclear factors interact with the immunoglobulin enhancer sequences. *Cell*. 1986;46(5):705-16.

[71] Hayden MS, Ghosh S. Signaling to NF-kappaB. *Genes & development*. 2004;18(18):2195-224.

[72] Moniruzzaman M, Ghosal I, Das D, Chakraborty SB. Melatonin ameliorates H(2)O(2)-induced oxidative stress through modulation of Erk/Akt/NFkB pathway. *Biological research*. 2018;51(1):17.

[73] Nakajima S, Kitamura M. Bidirectional regulation of NF-κB by reactive oxygen species: a role of unfolded protein response. *Free radical biology & medicine*. 2013;65:162-74.

[74] Bu R, Hussain AR, Al-Obaisi KA, Ahmed M, Uddin S, Al-Kuraya KS. Bortezomib inhibits proteasomal degradation of IκBα and induces mitochondrial dependent apoptosis in activated B-cell diffuse large B-cell lymphoma. *Leukemia & lymphoma*. 2014;55(2):415-24.

[75] Liou GY, Storz P. Reactive oxygen species in cancer. *Free radical research*. 2010;44(5):479-96.

[76] Takada Y, Mukhopadhyay A, Kundu GC, Mahabeleshwar GH, Singh S, Aggarwal BB. Hydrogen peroxide activates NF-kappa B through tyrosine phosphorylation of I kappa B alpha and serine phosphorylation of p65: evidence for the involvement of I kappa B alpha kinase and Syk protein-tyrosine kinase. *The Journal of biological chemistry*. 2003;278(26):24233-41.

[77] Pineda-Molina E, Klatt P, Vázquez J, Marina A, García de Lacoba M, Pérez-Sala D, Lamas S. Glutathionylation of the p50 subunit of NF-kappaB: a mechanism for redox-induced inhibition of DNA binding. *Biochemistry*. 2001 Nov 27;40(47):14134-42.

[78] Svineng G, Ravuri C, Rikardsen O, Huseby NE, Winberg JO. The role of reactive oxygen species in integrin and matrix metalloproteinase expression and function. *Connective tissue research*. 2008;49(3):197-202.

[79] Radisky DC, Levy DD, Littlepage LE, Liu H, Nelson CM, Fata JE, Leake D, Godden EL, Albertson DG, Nieto MA, Werb Z, Bissell MJ. Rac1b and reactive oxygen species mediate MMP-3-induced EMT and genomic instability. *Nature*. 2005 Jul 7;436(7047):123-7.

[80] Kerr JF, Wyllie AH, Currie AR. Apoptosis: a basic biological phenomenon with wide-ranging implications in tissue kinetics. *British journal of cancer*. 1972;26(4):239-57.

[81] Redza-Dutordoir M, Averill-Bates DA. Activation of apoptosis signalling pathways by reactive oxygen species. *Biochimica et Biophysica Acta (BBA) - Molecular Cell Research*. 2016;1863(12): 2977-92.

[82] Fulda S, Gorman AM, Hori O, Samali A. Cellular stress responses: cell survival and cell death. *International journal of cell biology*. 2010;2010:214074.

[83] Green DR, Llambi F. Cell Death Signaling. *Cold Spring Harbor perspectives in biology*. 2015;7(12).

[84] Ferri KF, Kroemer G. Organelle-specific initiation of cell death pathways. *Nature cell biology*. 2001;3(11):E255-63.

[85] Jin Z, El-Deiry WS. Overview of cell death signaling pathways. *Cancer biology & therapy*. 2005;4(2):139-63.

[86] Youle RJ, van der Bliek AM. Mitochondrial fission, fusion, and stress. *Science* (New York, NY). 2012;337(6098):1062-5.

[87] Landes T, Martinou JC. Mitochondrial outer membrane permeabilization during apoptosis: the role of mitochondrial fission. *Biochimica et biophysica acta*. 2011;1813(4):540-5.

[88] Orrenius S, Gogvadze V, Zhivotovsky B. Calcium and mitochondria in the regulation of cell death. *Biochemical and biophysical research communications*. 2015;460(1):72-81.

[89] Bernardi P. The mitochondrial permeability transition pore: a mystery solved? *Frontiers in physiology*. 2013;4:95.

[90] Halestrap AP. What is the mitochondrial permeability transition pore? *Journal of molecular and cellular cardiology*. 2009;46(6):821-31.

[91] Grimm S. The ER-mitochondria interface: the social network of cell death. *Biochimica et biophysica acta*. 2012;1823(2):327-34.

[92] Wang L, Azad N, Kongkaneramit L, Chen F, Lu Y, Jiang BH, Rojanasakul Y. The Fas death signaling pathway connecting reactive oxygen species generation and FLICE inhibitory protein down-regulation. *J Immunol*. 2008 Mar 1;180(5):3072-80.

[93] Ashkenazi A, Dixit VM. Apoptosis control by death and decoy receptors. *Current opinion in cell biology*. 1999;11(2):255-60.

[94] Yodkeeree S, Sung B, Limtrakul P, Aggarwal BB. Zerumbone enhances TRAIL-induced apoptosis through the induction of death receptors in human colon cancer cells: Evidence for an essential role of reactive oxygen species. *Cancer research*. 2009;69(16):6581-9.

[95] Wilkie-Grantham RP, Matsuzawa S, Reed JC. Novel phosphorylation and ubiquitination sites regulate reactive oxygen species-dependent degradation of anti-apoptotic c-FLIP protein. *The Journal of biological chemistry*. 2013;288(18):12777-90.

[96] Hussain AR, Uddin S, Bu R, Khan OS, Ahmed SO, Ahmed M, Al-Kuraya KS. Resveratrol suppresses constitutive activation of AKT via generation of ROS and induces apoptosis in diffuse large B cell lymphoma cell lines. *PLoS One*. 2011;6(9):e24703.

[97] Luanpitpong S, Chanvorachote P, Stehlik C, Tse W, Callery PS, Wang L, Rojanasakul Y. Regulation of apoptosis by Bcl-2 cysteine oxidation in human lung epithelial cells. *Mol Biol Cell*. 2013 Mar;24(6):858-69.

[98] Kennedy NJ, Davis RJ. Role of JNK in tumor development. *Cell cycle* (Georgetown, Tex). 2003;2(3):199-201.

[99] Carmody RJ, Cotter TG. Signalling apoptosis: a radical approach. *Redox report: communications in free radical research*. 2001;6(2):77-90.
[100] Hartmann JT, Haap M, Kopp HG, Lipp HP. Tyrosine kinase inhibitors - a review on pharmacology, metabolism and side effects. *Current drug metabolism*. 2009;10(5):470-81.
[101] Dimitrov DS, Marks JD. Therapeutic antibodies: current state and future trends--is a paradigm change coming soon? *Methods in molecular biology* (Clifton, NJ). 2009;525:1-27, xiii.
[102] Kotamraju S, Chitambar CR, Kalivendi SV, Joseph J, Kalyanaraman B. Transferrin receptor-dependent iron uptake is responsible for doxorubicin-mediated apoptosis in endothelial cells: role of oxidant-induced iron signaling in apoptosis. *The Journal of biological chemistry*. 2002;277(19): 17179-87.
[103] Tysnes BB, Bjerkvig R. Cancer initiation and progression: involvement of stem cells and the microenvironment. *Biochimica et biophysica acta*. 2007;1775(2):283-97.
[104] Al-Hajj M, Wicha MS, Benito-Hernandez A, Morrison SJ, Clarke MF. Prospective identification of tumorigenic breast cancer cells. *Proceedings of the National Academy of Sciences of the United States of America*. 2003;100(7):3983-8.
[105] Singh SK, Hawkins C, Clarke ID, Squire JA, Bayani J, Hide T, Henkelman RM, Cusimano MD, Dirks PB. Identification of human brain tumour initiating cells. *Nature*. 2004 Nov 18;432(7015): 396-401.
[106] Sell S. On the stem cell origin of cancer. *The American journal of pathology*. 2010;176(6):2584-494.
[107] Papaccio F, Paino F, Regad T, Papaccio G, Desiderio V, Tirino V. Concise Review: Cancer Cells, Cancer Stem Cells, and Mesenchymal Stem Cells: Influence in Cancer Development. *Stem cells translational medicine*. 2017;6(12):2115-25.
[108] Zakrzewski W, Dobrzyński M, Szymonowicz M, Rybak Z. Stem cells: past, present, and future. *Stem cell research & therapy*. 2019;10(1):68.
[109] Kolios G, Moodley Y. Introduction to stem cells and regenerative medicine. *Respiration; international review of thoracic diseases*. 2013;85(1):3-10.
[110] Fuchs E, Chen T. A matter of life and death: self-renewal in stem cells. *EMBO reports*. 2013;14(1):39-48.
[111] Yu Z, Pestell TG, Lisanti MP, Pestell RG. Cancer stem cells. *The international journal of biochemistry & cell biology*. 2012;44(12):2144-51.
[112] Tang DG. Understanding cancer stem cell heterogeneity and plasticity. *Cell research*. 2012;22(3):457-72.
[113] Zhao W, Li Y, Zhang X. Stemness-Related Markers in Cancer. *Cancer translational medicine*. 2017;3(3):87-95.
[114] Walcher L, Kistenmacher AK, Suo H, Kitte R, Dluczek S, Strauß A, Blaudszun AR, Yevsa T, Fricke S, Kossatz-Boehlert U. Cancer Stem Cells-Origins and Biomarkers: Perspectives for Targeted Personalized Therapies. *Front Immunol*. 2020 Aug 7;11:1280.
[115] Gawlik-Rzemieniewska N, Bednarek I. The role of NANOG transcriptional factor in the development of malignant phenotype of cancer cells. *Cancer biology & therapy*. 2016;17(1):1-10.
[116] Phi LTH, Sari IN, Yang YG, Lee SH, Jun N, Kim KS, Lee YK, Kwon HY. Cancer Stem Cells (CSCs) in Drug Resistance and their Therapeutic Implications in Cancer Treatment. *Stem Cells Int*. 2018 Feb 28;2018:5416923.
[117] Rich JN. Cancer stem cells: understanding tumor hierarchy and heterogeneity. *Medicine*. 2016;95 (1 Suppl 1):S2-s7.
[118] Sinenko SA, Starkova TY, Kuzmin AA, Tomilin AN. Physiological Signaling Functions of Reactive Oxygen Species in Stem Cells: From Flies to Man. *Frontiers in cell and developmental biology*. 2021;9:714370.
[119] George S, Abrahamse H. Redox Potential of Antioxidants in Cancer Progression and Prevention. *Antioxidants* (Basel, Switzerland). 2020;9(11).
[120] Velasco-Velázquez MA, Popov VM, Lisanti MP, Pestell RG. The role of breast cancer stem cells in metastasis and therapeutic implications. *The American journal of pathology*. 2011;179(1):2-11.

[121] Bao B, Ahmad A, Azmi AS, Ali S, Sarkar FH. Overview of cancer stem cells (CSCs) and mechanisms of their regulation: implications for cancer therapy. *Current protocols in pharmacology*. 2013;Chapter 14:Unit 14.25.

[122] Gzil A, Zarębska I, Bursiewicz W, Antosik P, Grzanka D, Szylberg Ł. Markers of pancreatic cancer stem cells and their clinical and therapeutic implications. *Molecular biology reports*. 2019;46(6):6629-45.

[123] Kumari S, Badana AK, G MM, G S, Malla R. Reactive Oxygen Species: A Key Constituent in Cancer Survival. *Biomarker insights*. 2018;13:1177271918755391.

[124] Zheng J. Energy metabolism of cancer: Glycolysis versus oxidative phosphorylation (Review). *Oncology letters*. 2012;4(6):1151-7.

[125] Sullivan LB, Chandel NS. Mitochondrial reactive oxygen species and cancer. *Cancer & metabolism*. 2014;2:17.

[126] Addeo M, Di Paola G, Verma HK, Laurino S, Russi S, Zoppoli P, Falco G, Mazzone P. Gastric Cancer Stem Cells: A Glimpse on Metabolic Reprogramming. *Front Oncol*. 2021 Jun 16;11:698394.

[127] Diehn M, Cho RW, Lobo NA, Kalisky T, Dorie MJ, Kulp AN, Qian D, Lam JS, Ailles LE, Wong M, Joshua B, Kaplan MJ, Wapnir I, Dirbas FM, Somlo G, Garberoglio C, Paz B, Shen J, Lau SK, Quake SR, Brown JM, Weissman IL, Clarke MF. Association of reactive oxygen species levels and radioresistance in cancer stem cells. *Nature*. 2009 Apr 9;458(7239):780-3.

[128] Shi X, Zhang Y, Zheng J, Pan J. Reactive oxygen species in cancer stem cells. *Antioxidants & redox signaling*. 2012;16(11):1215-28.

[129] Storz P. Forkhead homeobox type O transcription factors in the responses to oxidative stress. *Antioxidants & redox signaling*. 2011;14(4):593-605.

[130] Ding S, Li C, Cheng N, Cui X, Xu X, Zhou G. Redox Regulation in Cancer Stem Cells. *Oxidative medicine and cellular longevity*. 2015;2015:750798.

[131] Suzuki S, Okada M, Shibuya K, Seino M, Sato A, Takeda H, Seino S, Yoshioka T, Kitanaka C. JNK suppression of chemotherapeutic agents-induced ROS confers chemoresistance on pancreatic cancer stem cells. *Oncotarget*. 2015;6(1):458-70.

[132] Blokhina O, Virolainen E, Fagerstedt KV. Antioxidants, oxidative damage and oxygen deprivation stress: a review. *Annals of botany*. 2003;91 Spec No(2):179-94.

[133] Thanee M, Loilome W, Techasen A, Sugihara E, Okazaki S, Abe S, Ueda S, Masuko T, Namwat N, Khuntikeo N, Titapun A, Pairojkul C, Saya H, Yongvanit P. CD44 variant-dependent redox status regulation in liver fluke-associated cholangiocarcinoma: A target for cholangiocarcinoma treatment. *Cancer Sci*. 2016 Jul;107(7):991-1000.

[134] Asai R, Tsuchiya H, Amisaki M, Makimoto K, Takenaga A, Sakabe T, Hoi S, Koyama S, Shiota G. CD44 standard isoform is involved in maintenance of cancer stem cells of a hepatocellular carcinoma cell line. *Cancer Med*. 2019;8(2):773-782.

[135] Lewerenz J, Hewett SJ, Huang Y, Lambros M, Gout PW, Kalivas PW, Massie A, Smolders I, Methner A, Pergande M, Smith SB, Ganapathy V, Maher P. The cystine/glutamate antiporter system x(c)(-) in health and disease: from molecular mechanisms to novel therapeutic opportunities. *Antioxid Redox Signal*. 2013 Feb 10;18(5):522-55.

[136] Bridges RJ, Natale NR, Patel SA. System $xc^-$ cystine/glutamate antiporter: an update on molecular pharmacology and roles within the CNS. *British journal of pharmacology*. 2012;165(1):20-34.

[137] Lu C, Ren C, Yang T, Sun Y, Qiao P, Han X, Yu Z. Fructose-1, 6-bisphosphatase 1 interacts with NF-κB p65 to regulate breast tumorigenesis via PIM2 induced phosphorylation. *Theranostics*. 2020 Jul 9;10(19):8606-8618.

[138] Climent M, Viggiani G, Chen YW, Coulis G, Castaldi A. MicroRNA and ROS Crosstalk in Cardiac and Pulmonary Diseases. *International journal of molecular sciences*. 2020;21(12).

[139] Lin YH. MicroRNA Networks Modulate Oxidative Stress in Cancer. *International journal of molecular sciences*. 2019;20(18).

[140] Graham AM, Presnell JS. Hypoxia Inducible Factor (HIF) transcription factor family expansion, diversification, divergence and selection in eukaryotes. *PloS one*. 2017;12(6):e0179545.

[141] Ziello JE, Jovin IS, Huang Y. Hypoxia-Inducible Factor (HIF)-1 regulatory pathway and its potential for therapeutic intervention in malignancy and ischemia. *The Yale journal of biology and medicine*. 2007;80(2):51-60.

[142] Qutub AA, Popel AS. Reactive oxygen species regulate hypoxia-inducible factor 1alpha differentially in cancer and ischemia. *Molecular and cellular biology*. 2008;28(16):5106-19.

[143] Heikkilä M, Pasanen A, Kivirikko KI, Myllyharju J. Roles of the human hypoxia-inducible factor (HIF)-3α variants in the hypoxia response. *Cellular and molecular life sciences: CMLS*. 2011;68(23):3885-901.

[144] Zhao T, Zhu Y, Morinibu A, Kobayashi M, Shinomiya K, Itasaka S, Yoshimura M, Guo G, Hiraoka M, Harada H. HIF-1-mediated metabolic reprogramming reduces ROS levels and facilitates the metastatic colonization of cancers in lungs. *Sci Rep*. 2014;4:3793.

[145] Labbozzetta M, Notarbartolo M, Poma P. Can NF-κB Be Considered a Valid Drug Target in Neoplastic Diseases? Our Point of View. *International journal of molecular sciences*. 2020;21(9).

[146] Rinkenbaugh AL, Baldwin AS. The NF-κB Pathway and Cancer Stem Cells. *Cells*. 2016;5(2).

[147] Myant KB, Cammareri P, McGhee EJ, Ridgway RA, Huels DJ, Cordero JB, Schwitalla S, Kalna G, Ogg EL, Athineos D, Timpson P, Vidal M, Murray GI, Greten FR, Anderson KI, Sansom OJ. ROS production and NF-κB activation triggered by RAC1 facilitate WNT-driven intestinal stem cell proliferation and colorectal cancer initiation. *Cell Stem Cell*. 2013;12(6):761-73.

[148] Cordani M, Butera G, Pacchiana R, Masetto F, Mullappilly N, Riganti C, Donadelli M. Mutant p53-Associated Molecular Mechanisms of ROS Regulation in Cancer Cells. *Biomolecules*. 2020;10(3):361.

[149] Liu D, Xu Y. p53, oxidative stress, and aging. *Antioxidants & redox signaling*. 2011;15(6):1669-78.

[150] Liu B, Chen Y, St Clair DK. ROS and p53: a versatile partnership. *Free radical biology & medicine*. 2008;44(8):1529-35.

[151] Sablina AA, Budanov AV, Ilyinskaya GV, Agapova LS, Kravchenko JE, Chumakov PM. The antioxidant function of the p53 tumor suppressor. *Nature medicine*. 2005;11(12):1306-13.

[152] Stoner CS, Pearson GD, Koç A, Merwin JR, Lopez NI, Merrill GF. Effect of thioredoxin deletion and p53 cysteine replacement on human p53 activity in wild-type and thioredoxin reductase null yeast. *Biochemistry*. 2009;48(38):9156-69.

[153] Ma Q. Role of nrf2 in oxidative stress and toxicity. *Annual review of pharmacology and toxicology*. 2013;53:401-26.

[154] He F, Ru X, Wen T. NRF2, a Transcription Factor for Stress Response and Beyond. *International journal of molecular sciences*. 2020;21(13).

[155] Thimmulappa RK, Mai KH, Srisuma S, Kensler TW, Yamamoto M, Biswal S. Identification of Nrf2-regulated genes induced by the chemopreventive agent sulforaphane by oligonucleotide microarray. *Cancer research*. 2002;62(18):5196-203.

[156] Tuy K, Rickenbacker L, Hjelmeland AB. Reactive oxygen species produced by altered tumor metabolism impacts cancer stem cell maintenance. *Redox biology*. 2021;44:101953.

[157] Wang Y, Qi H, Liu Y, Duan C, Liu X, Xia T, Chen D, Piao HL, Liu HX. The double-edged roles of ROS in cancer prevention and therapy. *Theranostics*. 2021;11(10):4839-4857.

[158] Ozben T. Oxidative stress and apoptosis: impact on cancer therapy. *Journal of pharmaceutical sciences*. 2007;96(9):2181-96.

[159] Neumann CA, Fang Q. Are peroxiredoxins tumor suppressors? *Current opinion in pharmacology*. 2007;7(4):375-80.

[160] Li X, Chen Y, Zhao J, Shi J, Wang M, Qiu S, Hu Y, Xu Y, Cui Y, Liu C, Liu C. The Specific Inhibition of SOD1 Selectively Promotes Apoptosis of Cancer Cells via Regulation of the ROS Signaling Network. *Oxid Med Cell Longev*. 2019;2019:9706792.

[161] Glasauer A, Sena LA, Diebold LP, Mazar AP, Chandel NS. Targeting SOD1 reduces experimental non–small-cell lung cancer. *The Journal of clinical investigation*. 2014;124(1):117-28.

[162] Gomez M, Germain D. Cross talk between SOD1 and the mitochondrial UPR in cancer and neurodegeneration. *Molecular and cellular neurosciences*. 2019;98:12-8.

[163] Glasauer A, Chandel NS. Targeting antioxidants for cancer therapy. *Biochemical pharmacology*. 2014;92(1):90-101.

[164] Yang H, Villani RM, Wang H, Simpson MJ, Roberts MS, Tang M, Liang X. The role of cellular reactive oxygen species in cancer chemotherapy. *J Exp Clin Cancer Res*. 2018 Nov 1;37(1):266.

[165] Ma J, Kavelaars A, Dougherty PM, Heijnen CJ. Beyond symptomatic relief for chemotherapy-induced peripheral neuropathy: Targeting the source. *Cancer*. 2018;124(11):2289-98.

[166] Somwar R, Erdjument-Bromage H, Larsson E, Shum D, Lockwood WW, Yang G, Sander, C, Ouerfelli, O, Tempst, PJ, Djaballah, H, & Varmus, HE. Superoxide dismutase 1 (SOD1) is a target for a small molecule identified in a screen for inhibitors of the growth of lung adenocarcinoma cell lines. *Proceedings of the National Academy of Sciences of the United States of America*. 2011;108(39):16375-80.

[167] Wang Y, Mandal AK, Son YO, Pratheeshkumar P, Wise JTF, Wang L, Zhang Z, Shi X, Chen Z. Roles of ROS, Nrf2, and autophagy in cadmium-carcinogenesis and its prevention by sulforaphane. *Toxicol Appl Pharmacol*. 2018 Aug 15;353:23-30.

[168] Lee JH, Khor TO, Shu L, Su ZY, Fuentes F, Kong AN. Dietary phytochemicals and cancer prevention: Nrf2 signaling, epigenetics, and cell death mechanisms in blocking cancer initiation and progression. *Pharmacology & therapeutics*. 2013;137(2):153-71.

[169] Pant K, Yadav AK, Gupta P, Islam R, Saraya A, Venugopal SK. Butyrate induces ROS-mediated apoptosis by modulating miR-22/SIRT-1 pathway in hepatic cancer cells. *Redox biology*. 2017;12:340-9.

[170] Li X, Pathi SS, Safe S. Sulindac sulfide inhibits colon cancer cell growth and downregulates specificity protein transcription factors. *BMC Cancer*. 2015;15(1):974.

[171] Jutooru I, Guthrie AS, Chadalapaka G, Pathi S, Kim K, Burghardt R, Jin, UH, & Safe, S. Mechanism of action of phenethylisothiocyanate and other reactive oxygen species-inducing anticancer agents. *Molecular and cellular biology*. 2014;34(13):2382-95.

[172] Huang CH, Lee YC, Chen YJ, Wang LJ, Shi YJ, Chang LS. Quinacrine induces the apoptosis of human leukemia U937 cells through FOXP3/miR-183/β-TrCP/SP1 axis-mediated BAX upregulation. *Toxicology and applied pharmacology*. 2017;334:35-46.

[173] Chen P, Wang H, Yang F, Chen H, He W, Wang J. Curcumin Promotes Osteosarcoma Cell Death by Activating miR-125a/ERRα Signal Pathway. *Journal of cellular biochemistry*. 2017;118(1):74-81.

[174] Yang J, Qiu B, Li X, Zhang H, Liu W. p53-p66(shc)/miR-21-Sod2 signaling is critical for the inhibitory effect of betulinic acid on hepatocellular carcinoma. *Toxicology letters*. 2015;238(3):1-10.

[175] Wu L, Liu T, Xiao Y, Li X, Zhu Y, Zhao Y, Bao J, Wu C. Polygonatum odoratum lectin induces apoptosis and autophagy by regulation of microRNA-1290 and microRNA-15a-3p in human lung adenocarcinoma A549 cells. *Int J Biol Macromol*. 2016;85:217-26.

[176] Setyawati MI, Leong DT. Mesoporous Silica Nanoparticles as an Antitumoral-Angiogenesis Strategy. *ACS Applied Materials & Interfaces*. 2017;9(8):6690-703.

[177] Yousefi AM, Safaroghli-Azar A, Fakhroueian Z, Bashash D. ZnO/CNT@Fe(3)O(4) induces ROS-mediated apoptosis in chronic myeloid leukemia (CML) cells: an emerging prospective for nanoparticles in leukemia treatment. *Artificial cells, nanomedicine, and biotechnology*. 2020; 48(1):735-45.

[178] Kou L, Sun R, Xiao S, Zheng Y, Chen Z, Cai A, Zheng H, Yao Q, Ganapathy V, Chen R. Ambidextrous Approach To Disrupt Redox Balance in Tumor Cells with Increased ROS Production and Decreased GSH Synthesis for Cancer Therapy. *ACS Appl Mater Interfaces*. 2019;11(30):26722-26730.

[179] Winkle M, El-Daly SM, Fabbri M, Calin GA. Noncoding RNA therapeutics — challenges and potential solutions. *Nature Reviews Drug Discovery*. 2021;20(8):629-51.

[180] Chen C, Jiang X, Gu S, Zhang ZJTl. *MicroRNA-155 regulates arsenite-induced malignant transformation by targeting Nrf2-mediated oxidative damage in human bronchial epithelial cells*. 2017;278:38-47.

[181] Bayraktar R, Ivan C, Bayraktar E, Kanlikilicer P, Kabil NN, Kahraman N, Mokhlis HA, Karakas D, Rodriguez-Aguayo C, Arslan A, Sheng J, Wong S, Lopez-Berestein G, Calin GA, Ozpolat B. Dual Suppressive Effect of miR-34a on the FOXM1/eEF2-Kinase Axis Regulates Triple-Negative Breast Cancer Growth and Invasion. *Clin Cancer Res*. 2018;24(17):4225-4241.

[182] Rokavec M, Öner MG, Li H, Jackstadt R, Jiang L, Lodygin D, Kaller M, Horst D, Ziegler PK, Schwitalla S, Slotta-Huspenina J, Bader FG, Greten FR, Hermeking H. IL-6R/STAT3/miR-34a

feedback loop promotes EMT-mediated colorectal cancer invasion and metastasis. *J Clin Invest.* 2014;124(4):1853-67.

[183] Zhou Y, Huang T, Siu HL, Wong CC, Dong Y, Wu F, Zhang B, Wu WK, Cheng AS, Yu J, To KF, Kang W. IGF2BP3 functions as a potential oncogene and is a crucial target of miR-34a in gastric carcinogenesis. *Mol Cancer.* 2017;16(1):77.

[184] Deng XJ, Zheng HL, Ke XQ, Deng M, Ma ZZ, Zhu Y, Cui YY. Hsa-miR-34a-5p reverses multidrug resistance in gastric cancer cells by targeting the 3'-UTR of SIRT1 and inhibiting its expression. *Cell Signal.* 2021;84:110016.

[185] Kai AK, Chan LK, Lo RC, Lee JM, Wong CC, Wong JC, Ng IO. Down-regulation of TIMP2 by HIF-1α/miR-210/HIF-3α regulatory feedback circuit enhances cancer metastasis in hepatocellular carcinoma. *Hepatology.* 2016;64(2):473-87.

[186] Ren D, Yang Q, Dai Y, Guo W, Du H, Song L, Peng X. Oncogenic miR-210-3p promotes prostate cancer cell EMT and bone metastasis via NF-κB signaling pathway. *Mol Cancer.* 2017;16(1):117.

[187] Xie Y, Wang Y, Li J, Hang Y, Jaramillo L, Wehrkamp CJ, Phillippi MA, Mohr AM, Chen Y, Talmon GA, Mott JL, Oupický D. Cholangiocarcinoma therapy with nanoparticles that combine downregulation of MicroRNA-210 with inhibition of cancer cell invasiveness. *Theranostics.* 2018;8(16):4305-4320.

[188] Wang R, Zhang S, Chen X, Li N, Li J, Jia R, Pan Y, Liang H. EIF4A3-induced circular RNA MMP9 (circMMP9) acts as a sponge of miR-124 and promotes glioblastoma multiforme cell tumorigenesis. *Mol Cancer.* 2018;17(1):166.

## Chapter 9

# Oxidative Stress: A Source of Cellular Plasticity and Tumor Heterogeneity

**Leena Arora**
**Moyna Kalia**
**Kumari Bhavya**
**and Durba Pal***
Department of Biomedical Engineering, Indian Institute of Technology Ropar, Punjab, India

## Abstract

The heterogeneity of stromal and immune cells is a modern hallmark of aggressive tumors. Advanced tumors are characterized by oxidative dysregulation, abnormal pH, high metabolism, increased vascular demands, and low oxygen tension. Oxygen-derived pro-oxidants called reactive oxygen species (ROS) produced due to homeostatic imbalance act as a game-changer in accelerating tumor heterogeneity through DNA damage and genetic modifications. Prognostically, a high ROS level is a negative factor implicated in tumor progression, resistance to cell death, genomic instability, chemoresistance, and tumor heterogeneity. Cell reprogramming, or a change in cellular plasticity within the tumor microenvironment, is a significant concern that helps in tumor heterogeneity. The aggressive heterogeneous stromal and immune cell populations like fibroblasts, macrophages, and neutrophils have an inherited ability to generate ROS, affecting tumor growth. This chapter discusses how oxidative stress governs cellular plasticity, reprogramming, and polarization of stromal and immune cells in the tumor microenvironment and the possible targeted therapeutic interventions for improving patient prognosis.

**Keywords**: oxidative stress, ROS, DNA damage, tumor microenvironment, tumor immunology, antioxidants

---

* Corresponding Author's Email: durba.pal@iitrpr.ac.in.

In: The Role of Reactive Oxygen Species in Health and Disease
Editor: Kuladip Jana
ISBN: 979-8-89113-283-2

## 1. Introduction

The generation of oxidative stress and disruption in redox balance is a well-known factor for mitochondrial and DNA mutations and enhancement of proliferative cell phenotypes in tumors [1]. It provides a platform for the neoplastic transformation of normal cells, tumor development, and progression by altering electron transport chain (ETS) and ATP synthase activity [2]. Reactive oxygen species such as free radical superoxide and hydrogen peroxide released from the tumor cells, macrophages, dendritic cells, and natural killer cells present within the tumor microenvironment (TME) modulate the stromal and immune cell phenotypes towards aggressive stages [3]. TME is a term given to the environment of a developing tumor which is comprised of cancer cells, stromal cells, immune cells, surrounding blood vessels, signaling molecules, and extracellular matrix [4]. Due to abnormal growth, high metabolic needs, and vascular demands, an oxygen-deprived region also known as a hypoxic region is created, which is one of the significant factors for ROS generation [5]. Hypoxia stabilizes the HIF-α subunits by denying the propyl hydroxylases (PHDs) mediated hydroxylation of proline residues located within oxygen-dependent death domain, repudiating von Hippel–Lindau tumor suppressor protein (VHL) (E3 ubiquitin ligase) for HIF-α subunit proteasomal degradation in the absence of oxygen with a paradoxical increase in mitochondrial ROS [6]. Therefore, tumor and stromal cells generate ROS via mitochondria, peroxisomes, or NADPH oxidase (NOX) based on metabolic stress, hypoxic stress, or pH imbalance, which is critical for cellular plasticity, phenotypic switching, or differentiation within this TME. These reactive oxygen species include free radicals, such as singlet oxygen ($^1$O2), superoxide (O2•–), hydroxyl (HO•), hydroperoxyl (HO2•), carbonate (CO3•–), peroxyl (RO2•), alkoxyl (RO•), carbon dioxide radicals (CO2•–), and nonradicals, such as hydrogen peroxide ($H_2O_2$), hypobromous acid (HOBr), hypochlorous acid (HOCl), ozone (O3), organic peroxides (ROOH), peroxynitrite (ONOO–), peroxynitrate (O2NOO–), peroxynitrous acid (ONOOH), peroxomonocarbonate (HOOCO2–), nitric oxide (NO), and hypochlorite (OCl–) [7].

Cellular plasticity in tumors can be defined as the ability of a non-cancerous/cancerous cell to differentiate into another cell type of distinct lineage because of interactions with cancer cells. Several routes can be reached. Phenotypic switching is a broader term, including epithelial-mesenchymal plasticity of tumor cells, stromal cell phenotypic and molecular changes, and immune cell reprogramming such as macrophage polarization towards M1/M2 states [8]. Differentiation is the most common route by which one pluripotent or multipotent cell gives rise to a mature cell type. One route is de-differentiation, where terminally differentiated cancer cells revert to a less differentiated stage within their lineage [9]. Trans-differentiation is another such route that takes de-differentiation a step further to a point where it delineates to other lineage and differentiates to other cell types [10]. Another way is reprogramming, where a differentiated cell reverts to its pluripotent sources, potentially adapting to any other cell type [10]. Therefore, cellular plasticity defines the framework for establishing heterogeneous cell types in the oxidative tumor micro-environment (Figure 1). ROS generation changes the molecular and metabolic profiling of several cells present within the TME, targeting essential aspects of pluripotency, self-renewal capacity, and reprogramming [11]. These changes in molecular pathways act as a positive feedback loop for ROS synthesis [12].

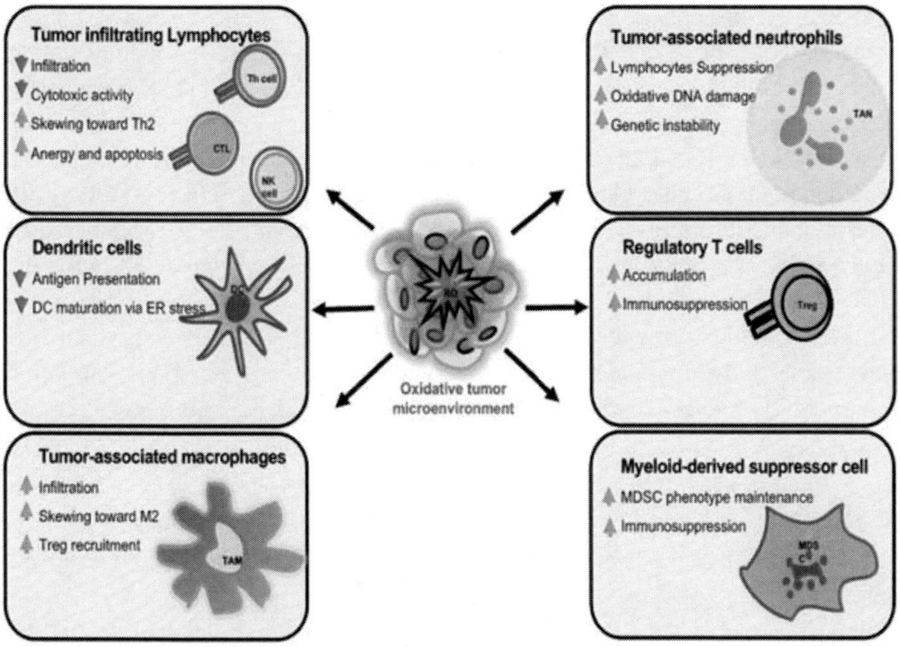


**Figure 1**. An overview of the crucial role of oxidative stress on various cell populations present in the Tumor Micro-Environment (TME). Oxidative stress promotes an immunosuppressive state by recruitment of Treg cells as well as favouring the M2 phenotype. It also decreases the Tumor Infiltrating Lymphocyte (TIL) directed killing of tumor cells.

## 2. Stromal Cell Regulation in the TME

Tumors are strikingly heterogeneous in composition primarily due to the recruitment of nearby non-cancerous stromal cells such as mesenchymal stromal cells, normal fibroblasts, adipocytes, and endothelial cells that secrete various cytokines and chemokines that facilitate stromal cell differentiation and tumor spreading [13]. Chronic oxidative stress regulates these stromal cells' differentiation and phenotype switching by regulating multiple pathways. Toullec et al., 2010 investigated the role of the AP-1 (activator protein-1) transcription factor family in regulating oxidative stress and related plasticity variations in tumor stromal cells [14]. JunD (AP-1 family protein) protects stromal cells against oxidative stress, the inactivation of which leads to persistent accumulation of $H_2O_2$. This phenomenon further leads to high HIF-1α due to a reduction in the activity of hypoxia-inducible factor (HIF)-prolyl-hydroxylases (PHDs) and enhancement of HIF-1α regulated genes [15]. These together facilitate fibroblastic and angiogenic properties leading to cancer metastasis. This section will discuss the role of oxidative stress in CAF transformation and its role in different endothelial cell phenotype formation.

### 2.1. ROS and Cancer-Associated Fibroblasts (CAFs)

Myofibroblasts are significant stromal cell populations shaping the TME. Many of the classic and emerging "hallmarks of malignancy," such as immune evasion and metabolic compartmentalization, are influenced by the presence of different functional states of CAFs [16]. Recent studies have revealed an important relationship between mature tumors, redox imbalance, fibroblast activation, and cancer aggressiveness [17]. These combined characteristics synergistically create an ecosystem that encourages the conversion of normal fibroblasts to CAFs. NOX4 and DUOX2 are two enzymes that increase ROS production and are reported to regulate cell cycle entry upon platelet-derived growth factor (PDGF) stimulation in normal human fibroblasts [18]. PDGF regulates the levels of Ras proteins and ERK1/2 activation accelerating collagen synthesis and disease progression [19]. NOX4 governs a range of cellular functions, including differentiation and senescence [20]. It is expressed in many human tumors, including bladder, colorectal, renal, and prostate cancers, where it has been linked to stromal activation and CAF-like behaviour and phenotype [21]. Hydrogen peroxide ($H_2O_2$) particularly induces oxidative stress, associated with decreased mitochondrial function and increased glucose uptake and ROS levels, leading to CAF differentiation [1]. CAFs release more $H_2O_2$ than normal cells, suggesting that extracellular $H_2O_2$ may play a role in stroma remodeling [22]. Treating NFs with CAF-conditioned media or exogenous $H_2O_2$ causes them to acquire an oxidative CAF-like state [22]. CAFs produce more $H_2O_2$ due to poor TGF signaling, which suppresses the antioxidant enzyme glutathione peroxidase 1 (GPX1) [23].

Mechanistically, the activation of various signaling pathways, including CXCL12/CXCR4, TGF- β1, NOTCH, NF-κβ, and WNT signaling either by mitochondrial ROS or NADPH oxidases, accelerates fibroblast activation (Figure 2). Chan et al., 2017, reported impaired SMAD3 and TAK1/cJUN pathways (TGF-β signaling), antioxidant enzyme glutathione peroxidase 1 (GPX1) deficiency, and high $H_2O_2$ production as a critical factor for CAF activation [23]. The p53 tumor suppressor gene also regulates Gpx-1 expression [24]. Ablation or mutation of p53 in stromal fibroblasts increased growth and metastatic spread of PC3 prostate cancer cells because of increased production of stromal-derived factor-1(SDF-1) [25]. TGF-β is known to participate in the differentiation and metabolic regulation of CAFs, by activating NOX4 and ROS production and by regulating α-SMA expression [26]. In addition, activated TGF-β1 in fibroblasts increases oxidative stress, autophagy/mitophagy, aerobic glycolysis, and down-regulation of caveolin1 (Cav1): these changes may extend to fibroblasts and promote the growth of cancer cells [27]. High amounts of reactive oxygen species (ROS) produced by cancer cells cause oxidative stress in CAFs, resulting in the formation of autophagosomes that combine with lysosomes, causing mitochondrial disruption and Cav-1 degradation [27]. Cav-1 ablation or mutation is one of the characteristics of fibroblasts found in tumor tissues. In a positive feedback loop, the downregulation of Cav-1 in CAFs causes increased ROS levels in cancer cells, which causes oxidative stress in CAFs [27]. Cav1 downregulation also causes fibroblasts to shift to catabolic metabolism and enhances the mitochondrial activity of cancer cells nearby. Cav-1 downregulation in fibroblasts is also linked to TGF signaling activation [28]. Also, epithelial-derived ROS, which inactivates p53 and p16INK4A, causes fibroblast activation and senescence in underlying CAFs in a TGF-dependent manner [29].

Zhang and co-workers have reported that normal oral fibroblasts (NOF) undergo aerobic glycolysis and secrete more ROS and L-lactate via AMPK and PGC-1α activity, leading to mitochondrial permeability transition pore (mPTP) opening, hypoxia, and mitophagy which helps acquire CAF metabolic phenotype when co-cultured with oral squamous cell carcinoma (OSSC) cells [29]. Moreover, chemokines such as CXCL12 promote fibroblast differentiation into myofibroblasts in a ROS-dependent manner. Interestingly, triple-negative breast tumors exhibit enhanced CXCL12 signaling and a related oxidative stress signature, which are correlated with the high content of myofibroblasts [14]. High levels of ROS are known to decrease senescence as well as prolong the life span of CAFs [30]. ROS has been demonstrated to transduce PDGF signaling by inactivating phosphatases, a potent activator of fibroblast growth and motility [31]. The oxidation of the catalytic site of protein tyrosine phosphatases (PTPases) has now been confirmed as a crucial regulatory mechanism in growth factor signaling [31]. The attachment of PDGF ligands to their receptors increases ROS levels, inhibiting protein tyrosine phosphatase activity, improving PDGFR signaling, and driving cell proliferation and migration [32]. PDGF has been demonstrated to promote ROS generation reciprocally in normal human fibroblasts. NOX4 and DUOX2 are regulated cell cycle entrances in normal human fibroblasts in response to PDGF stimulation [18]. Finally, while much research on PTPase oxidation has focused on pathways that regulate ROS levels, recent findings have revealed the critical function of the thioredoxin system in PTPase reactivation [33]. These findings suggest that ROS can increase PDGF signaling and, as a result, may specifically interfere with PDGFR-+ CAF activities, such as cell proliferation and migration. Hanley et al., 2018 indicated the importance of ROS in myofibroblast transdifferentiation as suppression of ROS activity by N-acetyl cysteine (NAC), GKT137831 (NOX4 inhibitor), and RNAi-mediated knockdown prevented trans-differentiation of normal fibroblasts even on exogenous TGF-β induction [26].

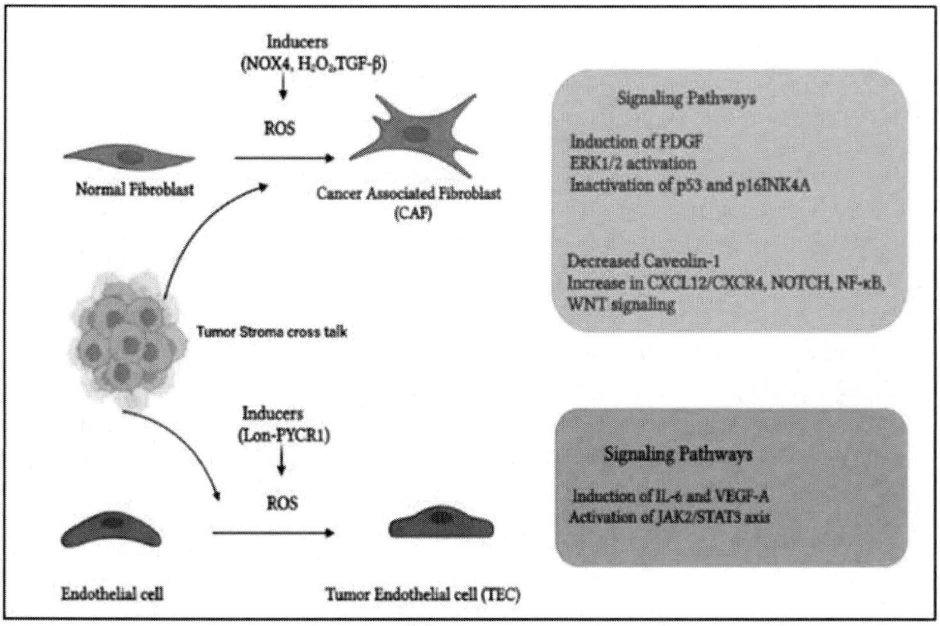


**Figure 2.** Schematic representation highlighting the effect of ROS on the transformation of normal fibroblasts into Cancer-Associated Fibroblasts (CAF), as well as endothelial cells into tumor Endothelial Cells (TEC). ROS, reactive oxygen species; PDGF, platelet-derived growth factor; NOX, nicotinamide adenine dinucleotide phosphate (NADPH) oxidase (NOX); TGFβ, transforming growth factor-β; ERKs, extracellular signal-regulated kinases; CXCL12/CXCL4, chemokine (C-X-C motif) ligand 12/4; NF-κB, nuclear factor kappa B WNT, wingless/Integrated; IL-6, interleukin 6; VEGF, Vascular endothelial growth factor; JAK/STAT, janus kinase/signal transducers and activators of transcription; Lon-PYCR1,lon-pyrroline-5-carboxylate reductase 1.

## 2.2. ROS and Tumor Endothelial Cell Heterogeneity (TECs)

Tumor endothelial cells (TECs) are markedly different from those residing within healthy normal tissues. Notably, TECs have altered morphologic and genetic phenotypes consisting of structural chromosomal modifications and mutations compared to normal endothelial cells (NECs) [34]. Oxidative stress and the resultant hypoxic surroundings of the high-grade tumors coerce the development of irregular vasculature leading to the formation of various endothelial cell subtypes [35]. Arora et al., 2021 give a comprehensive overview of different endothelial cell subtypes with representative markers present in the TME [36]. Briefly, they are divided into tip cells, stalk cells, lymphatic cells, endothelial progenitor cells, arterial & venous endothelial cells, and scavenging capillaries, etc. as revealed by single-cell RNA sequencing [36]. Mitochondrial ROS can stimulate a couple of signaling pathways [37]. Perhaps the requirement of mitochondrial ROS is imperative for the stabilization of hypoxia-inducible transcription factors (HIFs) which results in the transcription of genes vital for angiogenesis such as vascular endothelial growth factor (VEGF) [38]. Movahed et al., 2021 investigated the crucial role of oxidative stress in cancer stem cell (CSC) differentiation to TECs [39]. $H_2O_2$ release dysregulates the pentose phosphate pathway (PPP) by inhibiting the rate-limiting enzyme glucose-6-phosphate dehydrogenase (G6PD), inducing an increase in autophagy and CSC differentiation [39].

Mitochondrial oxidative stress by Lon-PYCR1 maintains an immunosuppressive tumor microenvironment that promotes cancer progression and metastasis [40]. It induces the production of IL-6 and VEGF-A from cancer cells for subsequent activation of angiogenesis by endothelial cells in the TME [40]. Lon-overexpression induces the tube formation of HUVEC and 3B11 mouse endothelial cells. The induction of tube formation by Lon was inhibited by IKK inhibitor VII (IKKi7), p38α-shRNA, p65-shRNA, and p50-shRNA, suggesting that Lon induces p38-and NF-κB-dependent inflammatory response to promote angiogenesis of endothelial cells [40]. Various angiogenic molecules are regulated by redox signaling. Upregulation of VEGF is mediated by activation of the JAK2/STAT3 signaling pathway [41]. VEGF is involved in the development of endothelial progenitor cells (EPCs) [42]. VEGFR2 is also one of the endothelial cell markers [43]. VEGFR2 expression was linked to differentiation, metastasis/recurrence, and the rate of tumorigenesis (Figure 2) [43]. Differentiation of adipose-derived mesenchymal stem cells into ECs was promoted by VEGF-A/VEGFR2 [44]. As a result, elevated VEGF could be a factor in EC differentiation from CSCs. ROS mediates VEGF/VEGFR2 expression and endothelial cell sprouting via lipid oxidation. Oxidized lipids and oxidized low-density lipoproteins strongly induce HIF-1α and VEGF expression in human macrophages and endothelial cells [45].

# 3. ROS Regulation and Myeloid Cell Networking

The myeloid lineage is a heterogeneous population including granulocytes and mononuclear phagocytes that a play critical role in anti-tumor immunity. Mononuclear phagocytes like monocytes, macrophages, and dendritic cells (DCs) also function in innate immunity by recognizing and engulfing pathogens and perform an ancillary role in adaptive immunity by presenting antigens to T cells [46]. ROS production either from immune cells or from tumor cells works dichromatically in solid tumors [47]. When in the early stages, ROS supports immune cells to disrupt/burst the cancer cells but supports the tumor heterogeneity and progression at later stages [47].

### 3.1. ROS Involvement in Macrophage Phenotype Switching

Macrophages are derived from CD34+ bone marrow immature monocytes [48] which under the chemotactic factors such as monocyte chemotactic protein-1 (MCP-1), macrophage inflammatory proteins-1 (MIP-1), and RANTES, migrate to the tumor tissues. Li et al., 2017 indicated ROS-mediated NF-κB activation, cytokine expression, disruption of tissue polarity, and monocyte infiltration in breast cancer 3D tumors [49]. Once recruited, monocytes are differentiated into the tumor-suppressive (M1) or tumor-promoting (M2/TAM) macrophage population influenced by cytokines such as colony-stimulating factor-1 (CSF1), IL4, IL13 and ROS [50]. While studying the correlation between TAM infiltration and clinicopathologic characteristics in supraglottic laryngeal carcinoma, Lin et al., 2011 found that the density of peritumoral CD68+ TAMs is significantly higher in recurrent/dead cases compared to the non-recurrent cases [49]. Also, the 5-year survival rate was lower in patients with high TAM regions. NADPH oxidases (NOXs) are well known to induce tumor inflammation, macrophage differentiation, angiogenesis, and tumor growth [51]. Xu et al., 2016 reported that NOX1 and NOX2 are critical for an increase in ROS production and the differentiation of monocytes to macrophages by the activation of MAPKs, JNK, and ERK signaling pathways in NOX1/2 double knock-out mice [51]. Zhang et al., 2019 found a positive correlation between NOX4 and CD68 or CD206 TAM macrophages and studied its potential role in macrophage polarisation and NSCLC cell growth [52]. PI3K/Akt activation and related NOX4 overexpression stimulates the release of CCL7, IL8, CSF-1, and VEGF-C production involved in macrophage migration [52].

Furthermore, NOX4-educated M2 macrophages stimulate the JNK pathway and release HB- EGF, involved in TAM-mediated tumor growth [52]. Logu et al., 2021 identified an oxidative stress sensor, transient receptor potential ankyrin 1 (TRPA1) as a target molecule in macrophage polarisation in melanomas [53]. Counterstaining of tissue samples with CD163, 4-hydroxynonenal (4-HNE), and TRPA1 indicates a positive correlation between all three and increased tumor severity [53]. Also, in macrophages, lack of the antioxidant enzyme, peroxiredoxin 5 (Prx5) led to M2 polarisation and lower levels of the M1-associated inflammatory factors iNOS, TNF, and Il-1. M2 cytokines such as IL-10, IL-6, ARG-1, HO-1, and pSTAT-3 increased as a result [53]. Stalin et al., 2019 applied NOX1 inhibitor GKT771 which reduced the tumor growth, lymph/angiogenesis, recruited proinflammatory macrophages, and natural killer T lymphocytes to the tumor microenvironment of colon carcinoma [54].

Myeloid-derived stem cells, (MDSCs) are one of the major immunosuppressive subgroups that play a key role in tumor growth and contribute to a suppressive tumor environment by creating reactive oxygen species (ROS) [55]. Furthermore, treatment of ROS inhibitors was found to totally eliminate MDSCs' suppressive impact on T cells [56]. ROS hinder recognition between the TCR and the MHC-peptide complex, reducing T cell immunological responses, but adding ROS inhibitors like catalase to the MDSCs/T cell co-culture system could diminish MDSCs' suppressive effects on T cell proliferation [57]. MDSCs isolated from NOX-2-

deficient animals produced little or no ROS and failed to decrease T cell proliferation and IFN-gamma production [58]. MDSCs have been shown to limit T cell activation by depleting cystine and cysteine, which are intimately linked to ROS generation [59]. Indeed, cystine and cysteine are required for the formation of glutathione (GSH), which neutralizes ROS. Furthermore, scavenging $H_2O_2$ using catalase causes immature myeloid cells to develop into macrophages in tumor-bearing animals, implying that ROS plays a role in maintaining the undifferentiated state of MDSC [56]. Low levels of ROS, on the other hand, might activate T cells, and anti-CD3-induced activation of the extracellular signal-regulated ERK pathway necessitated the production of $H_2O_2$ [56].

### 3.2. ROS Influenced Dendritic Cell Functionalities

Dendritic cells (DCs) are antigen-presenting cells that elicit the antigen-specific T-cell response. Various groups have discovered different DC subsets in tumors [60]. Broadly, DCs are classified into DC1, DC2, DC3, LAMP3 DC, mREG-DC, and plasmacytoid DCs based on their different gene expression profiles [60-61]. ROS can directly control DC maturation, antigen processing, and cross-activation of T cells by targeting molecular pathways such as p38-MAPK and ERK1/2 [62]. By disrupting Ca2+ homeostasis, oxidative stress can alter the ER function of dendritic cells [63]. The unfolded protein response is triggered by the build-up of misfolded proteins, and it aims to restore normal cell function by stopping translation, degrading misfolded proteins, and increasing chaperone protein expression [63]. The unfolded protein response axis PERK-eIF2-ATF4 was altered by ROS, resulting in a short-term inhibition of CD86, IL-1, and IL-12B expression in THP1 monocytes [64]. However, at later time points, these genes were elevated, indicating a pro-inflammatory response. ER stress is also caused by lipid peroxidation products such as 4-hydroxynonenal, which result from increased ROS production (4-HNE) [65]. Because of its reactivity with thiols and amine groups, this aldehyde rapidly forms protein adducts, which can induce the unfolded protein response. 4-HNE was also found to form adducts with ER-resident chaperone proteins, resulting in enhanced activation of the ER stress transcription factor XBP1 in ovarian cancer-associated DCs. This, in turn, hampered anti-tumor immunity by accumulating lipid bodies in DCs, which prevent MHC-I from reaching the cell surface [66]. DCs may potentially internalize 4-HNE from the TME because it is highly stable and can diffuse through membranes. Malondialdehyde, another prevalent lipid peroxidation product, generates protein adducts that have been found to be highly auto-immunogenic, potentially thwarting specific anti-tumor responses [67].

High levels of ROS frequently cause aberrant DC activation, probably because of ROS oxidizing high mobility group protein B1 (HMGB1), an essential molecule for DC maturation, resulting in T-cell malfunction [68]. Hypoxia further affects DCs through modulating the amounts of a variety of chemicals, including ROS [62]. In low oxygen conditions, as explained by the Warburg effect, cells will create ATP through a faster rate of anaerobic glycolysis, resulting in an increase in reactive oxygen species (ROS) in both tumor cells and immune cells [62]. A lack of energy can limit DC maturation and influence antigen presentation, while a high quantity of ROS can hinder DC maturation and influence antigen presentation via the p38-MAPK and ERK1/2 pathways [69]. Furthermore, ROS can influence the physiological state of DCs by modulating pro-inflammatory signaling pathways in DCs [62]. During cross-presentation (i.e., uptake, processing, and presentation of an extracellular antigen via the MHC I protein) to cytotoxic CD8+ T lymphocytes, ROS generation is enhanced in multiple subtypes of DCs [70]. This method entails the recruitment of NOX2 to the phagosome, which results in the alkalization of the phagosome's content via low-level ROS generation. As a result, alkalization of the phagosome preserves the structural integrity of the internalized antigen, allowing it to be presented successfully via the MHC I complex [70].

### 3.3. ROS and Neutrophil Subtypes

In human malignancies, neutrophils make up a significant component of the inflammatory cell infiltration (71). Various cell types in the TME can secrete neutrophil chemotactic chemicals. However, it plays a crucial role in tumor growth in some situations. Several enzymes (e.g., NADPH oxidase) in neutrophil phagolysosomes can convert molecular oxygen into superoxide radicals (O2) [71]. Additionally, superoxide radicals can be transformed into hydrogen peroxide. Although ROS was developed to damage bacterial membranes as host defense, it can also have a nonspecific effect on host tissue [37]. In the context of cancer, ROS produced by infiltrating neutrophils has been shown to have both genotoxic (which causes DNA damage and tumor formation) and cytotoxic (which causes tumor regression) effects [72]. NADPH oxidase in neutrophil granules alters the pH of the merged phagosome and lysosome structure, now known as the phagolysosome, resulting in the formation of reactive oxygen species (ROS) via the respiratory burst [72]. However, neutrophil-released ROS in the TME usually has a protumor effect by damaging DNA bases, resulting in mutations. Hydrogen peroxide can control many cells signaling pathways that are crucial in cellular biologies, such as the PI3K/Akt, IKK/NF-KB, and MAPK/Erk1/2 pathways [73].

Tumor-associated neutrophils decrease IL-17–producing T-cell responses, confirming that most of these cells contained the V6 TCR chain, were largely CD27 CD44+, and had deficient glutathione levels. It implies that CD27 17 T cells may have a lesser capacity to detoxify ROS than CD27+ T cells (or other T-cell subsets) [74]. Thus, in a nutshell, neutrophils are giving rise to an immunosuppressive environment favoring tumor progression through the neutrophil/ROS/γδ17 T-cell axis in the tumor microenvironment. Since these γδ T cells secrete cytokines such as IFN gamma, loss of these can hamper the protective responses against tumors [74].

## 4. ROS Involvement in Shaping Adaptive Tumor Immunity

### 4.1. B Cells

B cells are classically known to regulate inflammation and immune responses by releasing antibodies and activating T cells in the TME [75]. Single-cell transcriptomics has revealed significant heterogeneity and identified multiple subgroups of tumor-infiltrated B-cells. Hu et al., 2021 distinguished the tumor-infiltrating B cells from those in peripheral blood in being more mature, having higher clonality, more class-switching recombinations, and higher mutations [76]. Broadly, naïve B cells, regulatory B cells, follicular B cells, germinal center B cells, granzyme-secreting cells, and plasma B cells are the major subtypes residing within the tumor microenvironment. Naïve B cells are associated with good patient prognosis [77], whereas other subclasses are usually associated with poor prognosis [76]. B cell differentiation is paralleled by the strong upregulation of oxidative systems prevailing in stressed conditions [78]. Bertolotti and co-workers demonstrated the involvement of intracellular $H_2O_2$ in B cell proliferation via the Nrf2 (NF-E2-related factor-2) pathway [79]. Transcription factor PAX5A regulates B cells' early development and maturation via $H_2O_2$ Exposure of B cells to $H_2O_2$ induces the translocation of cytoplasmic redox factor APE/Ref-1 into the nucleus, increasing PAX5A binding affinity and leading to B cell development and maturation [80]. B-plasma cell differentiation requires redox remodeling, such as $H_2O_2$, which plays a significant role in generating functional B cells by influencing B cell activation and differentiation [80]. B cell-to plasma-cell differentiation is characterized by a massive expansion of the endoplasmic reticulum, finalized to sustain abundant immunoglobulin (Ig) synthesis and secretion. The increased production of disulfide-rich Ig might cause oxidative stress that could serve signaling roles in the differentiation and lifespan control of antibody- secreting cells [79]. Terminal B-cell differentiation entails redox stress, NF-E2-related factor- 2 (Nrf2) activation, and reshaping of the antioxidant responses. Endoplasmic reticulum (ER)- resident Prx4 increases dramatically during differentiation [80]. IgM secretion was not significantly affected in its absence, but more high-molecular-weight covalent complexes accumulated intracellularly [79]. These results are highly suggestive that the early intracellular production of $H_2O_2$ facilitates B-cell proliferation and reveals a role for the Nrf2 pathway in the differentiation and function of IgM-secreting cells (Figure 3).

### 4.2. T Cell

ROS at a low concentration is essential for T cell activation, expansion, and effector function in the TME [56]. High ROS production by cancer and stromal cells can reach T cells and cause oxidative stress, inducing T cell hypo-responsiveness in cancer patients. It stimulates many proximal and distal signaling pathways in T cells [37]. Mitochondrial ROS regulates T cell activation by modulating the production of IL-2 and IL-4, which is controlled by an oxidative signal generated by mitochondrial respiratory complex I [81]. Extramitochondrial ROS is most likely implicated in the Granzyme B-induced cell death process through NOX activation [82]. By preventing membrane lipid peroxidation and ferroptosis, glutathione peroxidase 4 (Gpx4) could function as a unique antioxidant enzyme to inhibit lipid peroxidation and play a critical role in CD8+ T cell homeostatic survival and both CD4+ and CD8+ T cell expansion upon TCR triggering in response to infection [83]. Reduced expression of the transcription coactivator peroxisome proliferator-activated receptor gamma, coactivator 1 alpha (PGC-1), a master regulator of genes involved in mitochondrial biogenesis, could explain the loss of mitochondrial quality control [84]. PGC-1 is activated in response to hypothermia, reactive oxygen species (ROS), and other signals. It regulates genes involved in fatty acid oxidation, glucose utilization, mitochondrial proteins, and transcription factors. Similarly, induction of PGC-1 in T cells increased T cell activity. It was linked to more prolonged survival in a B16 melanoma model, showing that poor mitochondrial quality control hinders anti-tumor immunity [85]. HIF-1 promotes the overexpression of nitrogen oxide synthase (iNOS) and

Arginase-I in the TME, allowing them to produce NO and L-Arginine and the expression of oxygen--containing molecules NOX2 and ROS [86]. These play a role in limiting T cell proliferation, activation, and differentiation and favoring effector CD8+ T cell death and immunosuppression. In a hypoxic environment, the contact between antigen and TCR is blocked. Hypoxia may also affect CD8+ T cell activity, depending on the present amount of CD8+ T cell development [87]. At multiple stages of T cell differentiation, changing oxygen levels affected T cell proliferation. These findings suggest that hypoxia is immunosuppressive, and metabolic reprogramming caused by enhanced HIF-1 activity may decrease immune response [88]. Reduced ATP synthesis due to defective OXPHOS during hypoxia could be the cause. Hypoxia is also known to enhance the formation of reactive oxygen species (ROS), which can cause apoptosis of activated T cells [88]. Activated CD8+ T cells with a partial HIF-1 impairment, on the other hand, produce more cytokines. T cells increase the expression of coinhibitory ligands, such as CTLA-4, CD244, and LAG-3, and reduce T-bet levels [89] (Figure 3).

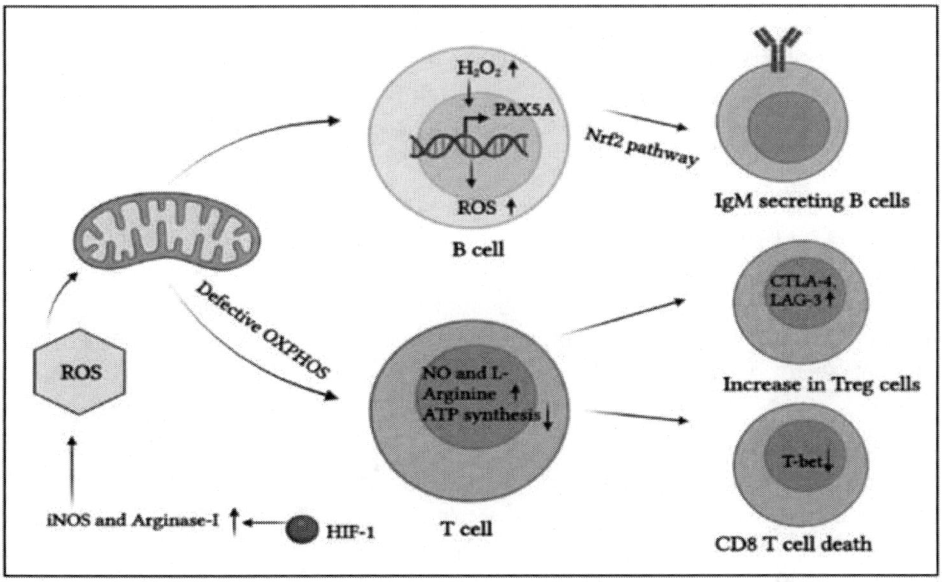


**Figure 3.** Schematic depicting how ROS promotes pro-tumor immunity by increasing the number of IgM secreting B cells and immune evasion by killing the Cytotoxic T Lymphocytes (CTLs). ROS, reactive oxygen species; OXPHOS, Oxidative phosphorylation; PAX5A, paired box 5A transcription factor; Nrf2, nuclear factor erythroid 2–related factor 2; NO, Nitric oxide; iNOS, inducible nitric oxide synthase (iNOS); Hypoxia-inducible factor; CTLA 4, cytotoxic T-lymphocyte–associated antigen 4; LAG3, lymphocyte activation gene 3; T-bet, T box expressed in T cells.

## 5. Therapeutic Implications of ROS in Tumor Microenvironment

ROS dynamically affects the TME and exerts a pivotal role in cancer angiogenesis, metastasis, and survival [1]. ROS can activate the cancer cell signaling pathways involving mitogen-activated protein kinase/extracellular signal-regulated protein kinase 1/2 (MAPK/ERK1/2), p38, C-Jun N-terminal kinase, and phosphoinositide-3-kinase/protein kinase B (PI3K/Akt)

[12-15]. These pathways can, in turn, activate the nuclear factor kappa light chain enhancer of activated B-cells (NF-KB), matrix metalloproteinase (MMPs), and vascular endothelial growth factor (VEGF) in the stromal and immune cell population. We can therapeutically target ROS by an antioxidant or pro-oxidant approach (Table 1). Table 1 describes different strategies and molecules developed to target stromal and immune cells. Briefly, CAFs can be precisely targeted by inhibiting NOX4 to both suppress and reverse CAF differentiation. Also, GKT137831 (setanaxib) is known to suppress TGFβ1 induced myofibroblast differentiation *in vivo* [90]. ROS also has a role as intracellular signaling in vascular cells. Sources of ROS include superoxide and hydrogen peroxide produced by NADPH oxidase activity in endothelial cells [37]. Studies reported that Axitinib, an inhibitor of VEGFRs currently in use as a second-line treatment for advanced renal cell carcinoma, promotes senescence of human endothelial cells *in vitro*. Mechanistically, this requires oxidative stress-dependent activation of the Ataxia Telangiectasia Mutated (ATM) kinase [91]. Various isoforms of ROS-generating NOX enzymes are expressed in the vasculature and provide distinct signaling cues through differential localization and activation. To investigate the efficacy of NOX1 inhibition as an anticancer strategy, Stalin and co-workers observed that NOX1-deficient mice treated with the novel NOX1-selective inhibitor GKT771 had a reduction in tumor growth by inhibiting angiogenesis and enhancing the recruitment of immune cells [54]. The antitumor activity of GKT771 requires an intact immune system and enhances anti-PD1 antibody activity. GKT771 can be used as a potential novel therapeutic strategy to treat colorectal cancer, particularly with checkpoint inhibition [54].

Furthermore, repolarizing TAM to M1 macrophages is an excellent strategy, and various research groups aim to achieve it. Liu et al. developed the ROS inducing polypeptide nano-vectors which utilized the galactose-functionalized zinc protoporphyrin IX (ZnPP) grafted poly(L-Lysin)-b-poly (ethylene glycol) polypeptide micelles (ZnPP PM) for TAM targeted immunopotentiator delivery, which aimed at *in vivo* repolarization of TAMs to antitumor M1 macrophages [92]. Wang et al., reported that 7S, 15R-Dihydroxy-16S, 17S-epoxy-docosapentaenoic acid (diHEP-DPA) is a novel resolvin synthesized from the substrate, docosahexaenoic acid (DHA) and its in-vitro and in-vivo studies on colorectal cancer demonstrated that diHEP-DPA reduced the expression of TAM markers by inhibiting the NF- κB signaling pathway,inhibited cancer stemness via the ROS/STAT3 signaling pathway [93]. Further, Peng et al. developed a trastuzumab-modified mannosylated liposome system for dual targeting of both the TAM-2 (M2) and HER-2 positive NSCLC cells; thus, this system was able to repolarize the pro-tumor M2 phenotype to the antitumor M1 and modulate the intracellular redox balance via ROS/NOX3/MsrA axis [94]. Zheng and the group reported that the ultra-small Cu2-x Se nanoparticle (CS NPs) could significantly polarize M0-M2 like macrophages into M1 phenotype *in vitro*, increasing antitumor efficacy and inducing robust T-cell immunity [95]. CS-NP can increase intracellular ROS in macrophages and trigger auto-ubiquitination of tumor necrosis factor receptor-associated factor 6 (TRAF 6), resulting in enhanced activation of interferon regulatory factor 5. It induces the expression of its downstream gene IL-23 and later activates M1 polarization. Therefore, ROS generated by CS-NPs can trigger M1 polarization through the novel ROS-TRAF6-IRF5-IL-23 signaling pathway [95].

**Table 1.** ROS-modulating strategies and different agents explored in cancer clinical trials

| ROS Modulating Strategies | Compounds | References |
|---|---|---|
| Antioxidant approach | | [96] |
| Intake of antioxidants | Vitamin A and E, selenium | |
| NADPH oxidase inhibition | Histamine | |
| GSH induction | Sulforaphane | |
| Nitroxide compound manipulation | Tempol | |
| Pro-oxidant approach | | |
| ROS generation | Arsenic trioxide, imexon, doxorubicin, daunorubicin, cisplatin, oxaliplatin, sunitinib, gefitinib, erlotininb, tra stuzumab, bevaccizumab | |
| GSH depletion | β-phenylethyl isotiocyanate, buthionine sulfoximine | |
| Thioredoxin inhibition | PX-12, motexanfin gadolinium | |
| Superoxide dismutase inhibition | 2-methoxyestradiol, ATN-224, disulfiram | |

| Sl. No. | Drugs | Mechanism | Application | Reference |
|---|---|---|---|---|
| 1. | GKT137831 (Setanaxib) | Dual inhibitor of NADPH oxidase (NOX4, NOX1) | Stromal cell CAFs | [90] |
| 2. | Poly I:C loaded ZnPP PM (ZnPP PM/PIC) | TAM to M1 polarisation | TAM-targeted transport of an immunologic adjuvant with ZnPP- grafted nano vectors to repolarize TAMs to M1 macrophages | [92] |
| 3. | Butylated hydroxy anisole (BHA) | Blocks the ROS mediated differentiation of M2 macrophages. | Blocks the occurrence of TAMs and potentially suppress tumorigenesis. | [97] |
| 4. | 7S,15R-Dihydroxy-16S,17S-epoxy-docosapentaenoic acid (diHEP- DPA) | Inhibit TAM by blocking NF-κB pathway, and cancer stemness via the ROS/STAT3 pathway. | TME targeted therapy against colorectal cancer | [93] |
| 5. | Trastuzumb modified mannosylated liposome system | Repolarisation of protumor M2 phenotype to antitumor M1 | The system triggers the ROS/MsrA/EGFRT790M transduction axis, thus providing a potential method for EGFRT790M-associated drug resistant NSCLC therapy. | [94] |

| Sl. No. | Drugs | Mechanism | Application | Reference |
|---|---|---|---|---|
| 6. | Ultra-small Cu2−xSe nanoparticles (CS NPs) | Increase intracellular ROS in macrophages and activates M1 polarization | CS NP can bridge innate and adaptive immunity against malignant solid tumors and act as a potential immunomodulator for antitumor immunotherapy | [95] |
| 7. | Axitinib | Promotes senescence of human endothelial cells *in vitro* via oxidative stress- dependent activation of the Ataxia Telangiectasia Mutated (ATM) kinase. | Advanced renal carcinoma | [(91] |
| 8. | GKT771 | Inhibit NOX1 and enhances recruitment of immune cells. | To treat colorectal cancer, particularly in combination with checkpoint inhibition | [54] |

## Conclusion

The tumor microenvironment plays an important role in carcinogenesis as non-tumor cells provide a niche which favors the onset of CSC that leads to heterogeneity, drug resistance, and poor prognosis. It is now known that not only tumor cells but the TME including non-cancer stromal cells such as CAFs as well as tumor-infiltrating immune cells play important roles in the crosstalk between tumor and environment to drive tumorigenesis. It is becoming increasingly clear that metabolic reprogramming and subsequent ROS generation are essential to this crosstalk. The differential effects of varying levels of ROS on biological outcomes within the cells that encompass a tumor, and its microenvironment are becoming increasingly important to our understanding of tumor initiation, growth and progression. Understanding how to exploit ROS to modulate cell-specific biological functions will not only help with the possible creation of new anticancer treatments but may also allow for increased efficacy of those treatments already in existence.

## References

[1] Liou GY, Storz P. Reactive oxygen species in cancer. *Free Radic Res.* 2010 May;44(5):479-96. doi: 10.3109/10715761003667554.

[2] Aggarwal V, Tuli HS, Varol A, Thakral F, Yerer MB, Sak K, Varol M, Jain A, Khan MA, Sethi G. Role of Reactive Oxygen Species in Cancer Progression: Molecular Mechanisms and Recent Advancements. *Biomolecules.* 2019 Nov 13; 9(11):735. doi: 10.3390/biom9110735.

[3] Aboelella NS, Brandle C, Kim T, Ding ZC, Zhou G. Oxidative Stress in the Tumor Microenvironment and Its Relevance to Cancer Immunotherapy. *Cancers* (Basel). 2021 Feb 27; 13(5):986. doi: 10.3390/cancers13050986.

[4] Whiteside TL. The tumor microenvironment and its role in promoting tumor growth. *Oncogene.* 2008 Oct 6; 27(45):5904-12. doi: 10.1038/onc.2008.271.

[5] Muz B, de la Puente P, Azab F, Azab AK. The role of hypoxia in cancer progression, angiogenesis, metastasis, and resistance to therapy. *Hypoxia* (Auckl). 2015 Dec 11;3:83-92. doi: 10.2147/HP.S93413.

[6] Weidemann A, Johnson RS. Biology of HIF-1alpha. *Cell Death Differ.* 2008 Apr;15(4):621-doi: 10.1038/cdd.2008.12.

[7] Collin F. Chemical Basis of Reactive Oxygen Species Reactivity and Involvement in Neurodegenerative Diseases. *Int J Mol Sci.* 2019 May 15; 20(10):2407. doi: 10.3390/ijms20102407.

[8] Qin S, Jiang J, Lu Y, Nice EC, Huang C, Zhang J, He W. Emerging role of tumor cell plasticity in modifying therapeutic response. *Signal Transduct Target Ther.* 2020 Oct 7; 5(1):228. doi: 10.1038/s41392-020-00313-5.

[9] Friedmann-Morvinski D, Verma IM. Dedifferentiation and reprogramming: origins of cancer stem cells. *EMBO Rep.* 2014 Mar; 15(3):244-53. doi: 10.1002/embr.201338254.

[10] Jopling C, Boue S, Izpisua Belmonte JC. Dedifferentiation, transdifferentiation and reprogramming: three routes to regeneration. *Nat Rev Mol Cell Biol.* 2011 Feb;12(2):79-89. doi: 10.1038/nrm3043.

[11] Bigarella CL, Liang R, Ghaffari S. Stem cells and the impact of ROS signaling. *Development.* 2014 Nov; 141(22):4206-18. doi: 10.1242/dev.107086.

[12] Schieber M, Chandel NS. ROS function in redox signaling and oxidative stress. *Curr Biol.* 2014 May 19; 24(10):R453-62. doi: 10.1016/j.cub.2014.03.034.

[13] Mao Y, Keller ET, Garfield DH, Shen K, Wang J. Stromal cells in tumor microenvironment and breast cancer. *Cancer Metastasis Rev.* 2013 Jun; 32(1-2):303-15. doi: 10.1007/s10555-012-9415-3.

[14] Toullec A, Gerald D, Despouy G, Bourachot B, Cardon M, Lefort S, Richardson M, Rigaill G, Parrini MC, Lucchesi C, Bellanger D, Stern MH, Dubois T, Sastre-Garau X, Delattre O, Vincent-Salomon A,

Mechta-Grigoriou F. Oxidative stress promotes myofibroblast differentiation and Tumor spreading. *EMBO Mol Med.* 2010 Jun;2(6):211-30. doi: 10.1002/emmm.201000073.

[15] Gerald D, Berra E, Frapart YM, Chan DA, Giaccia AJ, Mansuy D, Pouysségur J, Yaniv M, Mechta-Grigoriou F. JunD reduces tumor angiogenesis by protecting cells from oxidative stress. *Cell.* 2004 Sep 17; 118(6):781-94. doi: 10.1016/j.cell.2004.08.025.

[16] Xing F, Saidou J, Watabe K. Cancer associated fibroblasts (CAFs) in tumor microenvironment. *Front Biosci* (Landmark Ed). 2010 Jan 1; 15(1):166-79. doi: 10.2741/3613.

[17] Sahai E, Astsaturov I, Cukierman E, DeNardo DG, Egeblad M, Evans RM, Fearon D, Greten FR, Hingorani SR, Hunter T, Hynes RO, Jain RK, Janowitz T, Jorgensen C, Kimmelman AC, Kolonin MG, Maki RG, Powers RS, Puré E, Ramirez DC, Scherz-Shouval R, Sherman MH, Stewart S, Tlsty TD, Tuveson DA, Watt FM, Weaver V, Weeraratna AT, Werb Z. A framework for advancing our understanding of cancer-associated fibroblasts. *Nat Rev Cancer.* 2020 Mar; 20(3):174-186. doi: 10.1038/s41568-019-0238-1.

[18] Salmeen A, Park BO, Meyer T. The NADPH oxidases NOX4 and DUOX2 regulate cell cycle entry via a p53-dependent pathway. *Oncogene.* 2010 Aug 5; 29(31):4473-84. doi: 10.1038/onc.2010.200.

[19] Svegliati S, Cancello R, Sambo P, Luchetti M, Paroncini P, Orlandini G, Discepoli G, Paterno R, Santillo M, Cuozzo C, Cassano S, Avvedimento EV, Gabrielli A. Platelet-derived growth factor and reactive oxygen species (ROS) regulate Ras protein levels in primary human fibroblasts via ERK1/2. Amplification of ROS and Ras in systemic sclerosis fibroblasts. *J Biol Chem.* 2005 Oct 28; 280(43):36474-82. doi: 10.1074/jbc.M502851200.

[20] Weyemi U, Lagente-Chevallier O, Boufraqech M, Prenois F, Courtin F, Caillou B, Talbot M, Dardalhon M, Al Ghuzlan A, Bidart JM, Schlumberger M, Dupuy C. ROS-generating NADPH oxidase NOX4 is a critical mediator in oncogenic H-Ras-induced DNA damage and subsequent senescence. *Oncogene.* 2012 Mar 1; 31(9):1117-29. doi: 10.1038/onc.2011.327.

[21] Valkenburg KC, de Groot AE, Pienta KJ. Targeting the Tumor stroma to improve cancer therapy. *Nat Rev Clin Oncol.* 2018 Jun; 15(6):366-381. doi: 10.1038/s41571-018-0007-1.

[22] Arcucci A, Ruocco MR, Granato G, Sacco AM, Montagnani S. Cancer: An Oxidative Crosstalk between Solid Tumor Cells and Cancer Associated Fibroblasts. *Biomed Res Int.*; 2016:4502846. doi: 10.1155/2016/4502846.

[23] Chan JS, Tan MJ, Sng MK, Teo Z, Phua T, Choo CC, Li L, Zhu P, Tan NS. Cancer-associated fibroblasts enact field cancerization by promoting extratumoral oxidative stress. *Cell Death Dis.* 2017 Jan 19; 8(1):e2562. doi: 10.1038/cddis.2016.492.

[24] Budanov AV. The role of tumor suppressor p53 in the antioxidant defense and metabolism. *Subcell Biochem.* 2014; 85:337-58. doi: 10.1007/978-94-017-9211-0_18.

[25] Addadi Y, Moskovits N, Granot D, Lozano G, Carmi Y, Apte RN, Neeman M, Oren M. p53 status in stromal fibroblasts modulates tumor growth in an SDF1-dependent manner. *Cancer Res.* 2010 Dec 1; 70(23):9650-8. doi: 10.1158/0008-5472.CAN-10-1146.

[26] Hanley CJ, Mellone M, Ford K, Thirdborough SM, Mellows T, Frampton SJ, Smith DM, Harden E, Szyndralewiez C, Bullock M, Noble F, Moutasim KA, King EV, Vijayanand P, Mirnezami AH, Underwood TJ, Ottensmeier CH, Thomas GJ. Targeting the Myofibroblastic Cancer-Associated Fibroblast Phenotype through Inhibition of NOX4. *J Natl Cancer Inst.* 2018 Jan 1; 110(1):109–20. doi: 10.1093/jnci/djx121.

[27] Hou X, Zhang J, Wang Y, Xiong W, Mi J. TGFBR-IDH1-Cav1 axis promotes TGF-β signalling in cancer-associated fibroblast. *Oncotarget.* 2017 Sep 13; 8(48):83962-83974. doi: 10.18632/oncotarget.20861.

[28] Han M, Nwosu ZC, Piorońska W, Ebert MP, Dooley S, Meyer C. Caveolin-1 Impacts on TGF-β Regulation of Metabolic Gene Signatures in Hepatocytes. *Front Physiol.* 2020 Jan 31; 10: 1606. doi: 10.3389/fphys.2019.01606.

[29] Zhang Z, Gao Z, Rajthala S, Sapkota D, Dongre H, Parajuli H, Suliman S, Das R, Li L, Bindoff LA, Costea DE, Liang X. Metabolic reprogramming of normal oral fibroblasts correlated with increased glycolytic metabolism of oral squamous cell carcinoma and precedes their activation into carcinoma associated fibroblasts. *Cell Mol Life Sci.* 2020 Mar; 77(6):1115-1133. doi: 10.1007/s00018-019-03209-y.

[30] Yasuda T, Koiwa M, Yonemura A, Miyake K, Kariya R, Kubota S, Yokomizo-Nakano T, Yasuda-Yoshihara N, Uchihara T, Itoyama R, Bu L, Fu L, Arima K, Izumi D, Iwagami S, Eto K, Iwatsuki M, Baba Y, Yoshida N, Ohguchi H, Okada S, Matsusaki K, Sashida G, Takahashi A, Tan P, Baba H, Ishimoto T. Inflammation-driven senescence-associated secretory phenotype in cancer-associated fibroblasts enhances peritoneal dissemination. *Cell Rep*. 2021 Feb 23; 34(8):108779. doi: 10.1016/j.celrep.2021.108779.

[31] Chen KC, Zhou Y, Zhang W, Lou MF. Control of PDGF-induced reactive oxygen species (ROS) generation and signal transduction in human lens epithelial cells. *Mol Vis*. 2007 Mar 14; 13:374-87.

[32] Heldin CH. Targeting the PDGF signaling pathway in tumor treatment. *Cell Commun Signal*. 2013 Dec 20; 11:97. doi: 10.1186/1478-811X-11-97.

[33] Schwertassek U, Haque A, Krishnan N, Greiner R, Weingarten L, Dick TP, Tonks NK. Reactivation of oxidized PTP1B and PTEN by thioredoxin 1. *FEBS J*. 2014 Aug; 281(16):3545-58. doi: 10.1111/febs.12898.

[34] Nagl L, Horvath L, Pircher A, Wolf D. Tumor Endothelial Cells (TECs) as Potential Immune Directors of the Tumor Microenvironment - New Findings and Future Perspectives. *Front Cell Dev Biol*. 2020 Aug 19; 8:766. doi: 10.3389/fcell.2020.00766.

[35] Carmeliet P, Jain RK. Molecular mechanisms and clinical applications of angiogenesis. *Nature*. 2011 May 19; 473(7347):298-307. doi: 10.1038/nature10144.

[36] Arora L, Pal D. Remodeling of Stromal Cells and Immune Landscape in Microenvironment during Tumor Progression. *Front Oncol*. 2021 Mar 8; 11:596798. doi: 10.3389/fonc.2021.596798.

[37] Zorov DB, Juhaszova M, Sollott SJ. Mitochondrial reactive oxygen species (ROS) and ROS- induced ROS release. *Physiol Rev*. 2014 Jul; 94(3):909-50. doi: 10.1152/physrev.00026.2013.

[38] Thomas LW, Ashcroft M. Exploring the molecular interface between hypoxia-inducible factor signalling and mitochondria. *Cell Mol Life Sci*. 2019 May; 76(9):1759-1777. doi: 10.1007/s00018-019-03039-y.

[39] Movahed ZG, Yarani R, Mohammadi P, Mansouri K. Sustained oxidative stress instigates differentiation of cancer stem cells into tumor endothelial cells: Pentose phosphate pathway, reactive oxygen species and autophagy crosstalk. *Biomed Pharmacother*. 2021 Jul; 139: 111643. doi: 10.1016/j.biopha.2021.111643.

[40] Kuo CL, Chou HY, Chiu YC, Cheng AN, Fan CC, Chang YN, Chen CH, Jiang SS, Chen NJ, Lee AY. Mitochondrial oxidative stress by Lon-PYCR1 maintains an immunosuppressive tumor microenvironment that promotes cancer progression and metastasis. *Cancer Lett*. 2020 Apr 1; 474:138-150. doi: 10.1016/j.canlet.2020.01.019.

[41] Zhao M, Gao FH, Wang JY, Liu F, Yuan HH, Zhang WY, Jiang B. JAK2/STAT3 signaling pathway activation mediates tumor angiogenesis by upregulation of VEGF and bFGF in non-small-cell lung cancer. *Lung Cancer* 2011 Sep; 73(3):366-74. doi: 10.1016/j.lungcan.2011.01.002.

[42] Li L, Liu H, Xu C, Deng M, Song M, Yu X, Xu S, Zhao X. VEGF promotes endothelial progenitor cell differentiation and vascular repair through connexin 43. *Stem Cell Res Ther*. 2017 Oct 24; 8(1):237. doi: 10.1186/s13287-017-0684-1.

[43] Miettinen M, Rikala MS, Rys J, Lasota J, Wang ZF. Vascular endothelial growth factor receptor 2 as a marker for malignant vascular tumors and mesothelioma: an immunohistochemical study of 262 vascular endothelial and 1640 nonvascular tumors. *Am J Surg Pathol*. 2012 Apr; 36(4):629-39. doi: 10.1097/PAS.0b013e318243555b.

[44] Rautiainen S, Laaksonen T, Koivuniemi R. Angiogenic Effects and Crosstalk of Adipose- Derived Mesenchymal Stem/Stromal Cells and Their Extracellular Vesicles with Endothelial Cells. *Int J Mol Sci*. 2021 Oct 8; 22(19):10890. doi: 10.3390/ijms221910890.

[45] Inoue M, Itoh H, Tanaka T, Chun TH, Doi K, Fukunaga Y, Sawada N, Yamshita J, Masatsugu K, Saito T, Sakaguchi S, Sone M, Yamahara Ki, Yurugi T, Nakao K. Oxidized LDL regulates vascular endothelial growth factor expression in human macrophages and endothelial cells through activation of peroxisome proliferator-activated receptor- gamma. *Arterioscler Thromb Vasc Biol*. 2001 Apr; 21(4):560-6. doi: 10.1161/01.atv.21.4.560.

[46] Hirayama D, Iida T, Nakase H. The Phagocytic Function of Macrophage-Enforcing Innate Immunity and Tissue Homeostasis. *Int J Mol Sci*. 2017 Dec 29; 19(1):92. doi: 10.3390/ijms19010092.

[47] Kennel KB, Greten FR. Immune cell - produced ROS and their impact on tumor growth and metastasis. *Redox Biol.* 2021 Jun;42: 101891. doi: 10.1016/j.redox.2021.101891.

[48] Wolf AA, Yáñez A, Barman PK, Goodridge HS. The Ontogeny of Monocyte Subsets. *Front Immunol.* 2019 Jul 17;10: 1642. doi: 10.3389/fimmu.2019.01642.

[49] Li L, Chen J, Xiong G, St Clair DK, Xu W, Xu R. Increased ROS production in non-polarized mammary epithelial cells induces monocyte infiltration in 3D culture. *J Cell Sci.* 2017 Jan 1; 130(1):190-202. doi: 10.1242/jcs.186031.

[50] Ge Z, Ding S. The Crosstalk Between Tumor-Associated Macrophages (TAMs) and Tumor Cells and the Corresponding Targeted Therapy. *Front Oncol.* 2020 Nov 3;10: 590941. doi: 10.3389/ fonc.2020.590941.

[51] Xu Q, Choksi S, Qu J, Jang J, Choe M, Banfi B, Engelhardt JF, Liu ZG. NADPH Oxidases Are Essential for Macrophage Differentiation. *J Biol Chem.* 2016 Sep 16; 291(38):20030-41. doi: 10.1074/jbc.M116.731216.

[52] Zhang J, Li H, Wu Q, Chen Y, Deng Y, Yang Z, Zhang L, Liu B. Tumoral NOX4 recruits M2 tumor-associated macrophages via ROS/PI3K signaling-dependent various cytokine production to promote NSCLC growth. *Redox Biol.* 2019 Apr 22; 101116. doi: 10.1016/j.redox.2019.101116.

[53] De Logu F, Souza Monteiro de Araujo D, Ugolini F, Iannone LF, Vannucchi M, Portelli F, Landini L, Titiz M, De Giorgi V, Geppetti P, Massi D, Nassini R. The TRPA1 Channel Amplifies the Oxidative Stress Signal in Melanoma. *Cells.* 2021 Nov 11; 10(11):3131. doi: 10.3390/cells10113131.

[54] Stalin J, Garrido-Urbani S, Heitz F, Szyndralewiez C, Jemelin S, Coquoz O, Ruegg C, Imhof BA. Inhibition of host NOX1 blocks tumor growth and enhances checkpoint inhibitor- based immunotherapy. *Life Sci Alliance.* 2019 Jun 27; 2(4):e201800265. doi: 10.26508/lsa.201800265.

[55] Kumar V, Patel S, Tcyganov E, Gabrilovich DI. The Nature of Myeloid-Derived Suppressor Cells in the Tumor Microenvironment. *Trends Immunol.* 2016 Mar; 37(3):208-220. doi: 10.1016/j.it.2016. 01.004.

[56] Chen X, Song M, Zhang B, Zhang Y. Reactive Oxygen Species Regulate T Cell Immune Response in the Tumor Microenvironment. *Oxid Med Cell Longev.* 2016; 2016:1580967. doi: 10.1155/2016/1580967.

[57] Lu T, Ramakrishnan R, Altiok S, Youn JI, Cheng P, Celis E, Pisarev V, Sherman S, Sporn MB, Gabrilovich D. Tumor-infiltrating myeloid cells induce tumor cell resistance to cytotoxic T cells in mice. *J Clin Invest.* 2011 Oct; 121(10):4015-29. doi: 10.1172/JCI45862.

[58] Groth C, Hu X, Weber R, Fleming V, Altevogt P, Utikal J, Umansky V. Immunosuppression mediated by myeloid-derived suppressor cells (MDSCs) during Tumor progression. *Br J Cancer.* 2019 Jan; 120(1):16-25. doi: 10.1038/s41416-018-0333-1.

[59] Srivastava MK, Sinha P, Clements VK, Rodriguez P, Ostrand-Rosenberg S. Myeloid- derived suppressor cells inhibit T-cell activation by depleting cystine and cysteine. *Cancer Res.* 2010 Jan 1; 70(1):68-77. doi: 10.1158/0008-5472.CAN-09-2587.

[60] Collin M, Bigley V. Human dendritic cell subsets: an update. *Immunology.* 2018 May; 154(1):3-20. doi: 10.1111/imm.12888.

[61] Balan S, Saxena M, Bhardwaj N. Dendritic cell subsets and locations. *Int Rev Cell Mol Biol.* 2019; 348:1-68. doi: 10.1016/bs.ircmb.2019.07.004.

[62] Paardekooper LM, Vos W, van den Bogaart G. Oxygen in the tumor microenvironment: effects on dendritic cell function. *Oncotarget.* 2019 Jan 25; 10(8):883-896. doi: 10.18632/oncotarget.26608.

[63] Cao SS, Kaufman RJ. Endoplasmic reticulum stress and oxidative stress in cell fate decision and human disease. *Antioxid Redox Signal.* 2014 Jul 20; 21(3):396-413. doi: 10.1089/ars.2014.5851.

[64] Grootjans J, Kaser A, Kaufman RJ, Blumberg RS. The unfolded protein response in immunity and inflammation. *Nat Rev Immunol.* 2016 Aug; 16(8):469-84. doi: 10.1038/nri.2016.62.

[65] Zhong H, Yin H. Role of lipid peroxidation derived 4-hydroxynonenal (4-HNE) in cancer: focusing on mitochondria. *Redox Biol.* 2015; 4: 193-9. doi: 10.1016/j.redox.2014.12.011.

[66] Negre-Salvayre A, Coatrieux C, Ingueneau C, Salvayre R. Advanced lipid peroxidation end products in oxidative damage to proteins. Potential role in diseases and therapeutic prospects for the inhibitors. *Br J Pharmacol.* 2008 Jan; 153(1):6-20. doi: 10.1038/sj.bjp.0707395.

[67] Duryee MJ, Clemens DL, Opperman PJ, Thiele GM, Duryee LM, Garvin RP, Anderson DR. Malondialdehyde-Acetaldehyde Modified (MAA) Proteins Differentially Effect the Inflammatory Response in Macrophage, Endothelial Cells and Animal Models of Cardiovascular Disease. *Int J Mol Sci.* 2021 Nov 30; 22(23):12948. doi: 10.3390/ijms222312948.

[68] Peng X, He Y, Huang J, Tao Y, Liu S. Metabolism of Dendritic Cells in Tumor Microenvironment: For Immunotherapy. *Front Immunol.* 2021 Feb 24; 12: 613492. doi: 10.3389/fimmu.2021.613492.

[69] Han Z, Dong Y, Lu J, Yang F, Zheng Y, Yang H. Role of hypoxia in inhibiting dendritic cells by VEGF signaling in tumor microenvironments: mechanism and application. *Am J Cancer Res.* 2021 Aug 15; 11(8): 3777-3793.

[70] Gutiérrez-Martínez E, Planès R, Anselmi G, Reynolds M, Menezes S, Adiko AC, Saveanu L, Guermonprez P. Cross-Presentation of Cell-Associated Antigens by MHC Class I in Dendritic Cell Subsets. *Front Immunol.* 2015 Jul 17; 6: 363. doi: 10.3389/fimmu.2015.00363.

[71] Gregory AD, Houghton AM. Tumor-associated neutrophils: new targets for cancer therapy. *Cancer research.* 2011 Apr 1; 71(7):2411-6.

[72] Wu L, Saxena S, Singh RK. Neutrophils in the tumor microenvironment. *Tumor Microenvironment.* 2020:1-20.

[73] Wu L, Saxena S, Awaji M, Singh RK. Tumor-associated neutrophils in cancer: going pro. *Cancers.* 2019 Apr; 11(4):564.

[74] Mensurado S, Rei M, Lança T, Ioannou M, Goncalves-Sousa N, Kubo H, Malissen M, Papayannopoulos V, Serre K, Silva-Santos B. Tumor-associated neutrophils suppress pro- tumoral IL-17+ γδ T cells through induction of oxidative stress. *PLoS biology.* 2018 May 11; 16(5):e2004990.

[75] Catalán D, Mansilla MA, Ferrier A, Soto L, Oleinika K, Aguillón JC, Aravena O. Immunosuppressive Mechanisms of Regulatory B Cells. *Front Immunol.* 2021 Apr 29; 12:611795. doi: 10.3389/fimmu.2021.611795.

[76] Hu Q, Hong Y, Qi P, Lu G, Mai X, Xu S, He X, Guo Y, Gao L, Jing Z, Wang J, Cai T, Zhang Y. Atlas of breast cancer infiltrated B-lymphocytes revealed by paired single-cell RNA- sequencing and antigen receptor profiling. *Nat Commun.* 2021 Apr 12; 12(1):2186. doi: 10.1038/s41467-021-22300-2.

[77] Schaafsma E, Jiang C, Cheng C. B cell infiltration is highly associated with prognosis and an immune-infiltrated tumor microenvironment in neuroblastoma. *J Cancer Metastasis Treat.* 2021; 7(34):10.20517/2394-4722.2021.72. doi: 10.20517/2394-4722.2021.72.

[78] Price MJ, Patterson DG, Scharer CD, Boss JM. Progressive Upregulation of Oxidative Metabolism Facilitates Plasmablast Differentiation to a T-Independent Antigen. *Cell Rep.* 2018 Jun 12; 23(11):3152-3159. doi: 10.1016/j.celrep.2018.05.053.

[79] Bertolotti M, Yim SH, Garcia-Manteiga JM, Masciarelli S, Kim YJ, Kang MH, Iuchi Y, Fujii J, Vené R, Rubartelli A, Rhee SG, Sitia R. B- to plasma-cell terminal differentiation entails oxidative stress and profound reshaping of the antioxidant responses. *Antioxid Redox Signal.* 2010 Oct; 13(8):1133-44. doi: 10.1089/ars.2009.3079.

[80] Recaldin T, Fear DJ. Transcription factors regulating B cell fate in the germinal centre. *Clin Exp Immunol.* 2016 Jan; 183(1):65-75. doi: 10.1111/cei.12702.

[81] Kaminski MM, Sauer SW, Klemke CD, Süss D, Okun JG, Krammer PH, Gülow K. Mitochondrial reactive oxygen species control T cell activation by regulating IL-2 and IL-4 expression: mechanism of ciprofloxacin-mediated immunosuppression. *J Immunol.* 2010 May 1; 184(9):4827-41. doi: 10.4049/jimmunol.0901662.

[82] Jacquemin G, Margiotta D, Kasahara A, Bassoy EY, Walch M, Thiery J, Lieberman J, Martinvalet D. Granzyme B-induced mitochondrial ROS are required for apoptosis. *Cell Death Differ.* 2015 May; 22(5):862-74. doi: 10.1038/cdd.2014.180.

[83] Matsushita M, Freigang S, Schneider C, Conrad M, Bornkamm GW, Kopf M. T cell lipid peroxidation induces ferroptosis and prevents immunity to infection. *J Exp Med.* 2015 Apr 6; 212(4):555-68. doi: 10.1084/jem.20140857.

[84] Puigserver P, Spiegelman BM. Peroxisome proliferator-activated receptor-gamma coactivator 1 alpha (PGC-1 alpha): transcriptional coactivator and metabolic regulator. *Endocr Rev.* 2003 Feb; 24(1):78-90. doi: 10.1210/er.2002-0012.

[85] Sukumar M, Kishton RJ, Restifo NP. Metabolic reprograming of anti-tumor immunity. *Curr Opin Immunol.* 2017 Jun; 46:14-22. doi: 10.1016/j.coi.2017.03.011.

[86] Olson N, van der Vliet A. Interactions between nitric oxide and hypoxia-inducible factor signaling pathways in inflammatory disease. *Nitric Oxide.* 2011 Aug 1; 25(2):125-37. doi: 10.1016/j.niox.2010.12.010.

[87] Zhang Y, Ertl HC. Starved and Asphyxiated: How Can CD 8(+) T Cells within a Tumor Microenvironment Prevent Tumor Progression. *Front Immunol.* 2016 Feb 10; 7:32. doi: 10.3389/ fimmu.2016.00032.

[88] Tao JH, Barbi J, Pan F. Hypoxia-inducible factors in T lymphocyte differentiation and function. A Review in the Theme: Cellular Responses to Hypoxia. *Am J Physiol Cell Physiol.* 2015 Nov 1; 309(9):C580-9. doi: 10.1152/ajpcell.00204.2015.

[89] Okoye IS, Houghton M, Tyrrell L, Barakat K, Elahi S. Coinhibitory Receptor Expression and Immune Checkpoint Blockade: Maintaining a Balance in CD8+ T cell Responses to Chronic Viral Infections and Cancer. *Front Immunol.* 2017 Sep 29; 8:1215. doi: 10.3389/fimmu.2017.01215.

[90] Ford K, Hanley CJ, Mellone M, Szyndralewiez C, Heitz F, Wiesel P, Wood O, Machado M, Lopez MA, Ganesan AP, Wang C, Chakravarthy A, Fenton TR, King EV, Vijayanand P, Ottensmeier CH, Al-Shamkhani A, Savelyeva N, Thomas GJ. NOX4 Inhibition Potentiates Immunotherapy by Overcoming Cancer-Associated Fibroblast-Mediated CD8 T-cell Exclusion from Tumors. *Cancer Res.* 2020 May 1; 80(9):1846-1860. doi: 10.1158/0008- 5472.CAN-19-3158.

[91] Mongiardi MP, Radice G, Piras M, Stagni V, Pacioni S, Re A, Putti S, Ferrè F, Farsetti A, Pallini R, Barilà D, Levi A, Falchetti ML. Axitinib exposure triggers endothelial cells senescence through ROS accumulation and ATM activation. *Oncogene.* 2019 Jul; 38(27):5413-5424. doi: 10.1038/s41388-019-0798-2.

[92] Liu L, He H, Liang R, Yi H, Meng X, Chen Z, Pan H, Ma Y, Cai L. ROS-Inducing Micelles Sensitize Tumor-Associated Macrophages to TLR3 Stimulation for Potent Immunotherapy. *Biomacromolecules.* 2018 Jun 11; 19(6):2146-2155. doi: 10.1021/acs.biomac.8b00239.

[93] Wang L, Choi HS, Su Y, Lee B, Song JJ, Jang YS, Seo JW. 7S, 15R-Dihydroxy-16S, 17S-Epoxy-Docosapentaenoic Acid, a Novel DHA Epoxy Derivative, Inhibits Colorectal Cancer Stemness through Repolarization of Tumor-Associated Macrophage Functions and the ROS/STAT3 Signaling Pathway. *Antioxidants* (Basel). 2021 Sep 14; 10(9):1459. doi: 10.3390/antiox10091459.

[94] Peng H, Chen B, Huang W, Tang Y, Jiang Y, Zhang W, Huang Y. Reprogramming Tumor- Associated Macrophages To Reverse EGFRT790M Resistance by Dual-Targeting Codelivery of Gefitinib/Vorinostat. *Nano Lett.* 2017 Dec 13; 17(12):7684-7690. doi: 10.1021/acs.nanolett.7b03756.

[95] Zheng, Y., Han, Y., Wang, T., Liu, H., Sun, Q., Hu, S., Chen, J., Li, Z., Reprogramming Tumor-Associated Macrophages via ROS-Mediated Novel Mechanism of Ultra-Small Cu2−xSe Nanoparticles to Enhance Anti-Tumor Immunity. *Adv. Funct. Mater.* 2022, 32, 2108971. doi:10.1002/adfm.202108971.

[96] Seebacher NA, Krchniakova M, Stacy AE, Skoda J, Jansson PJ. Tumor Microenvironment Stress Promotes the Development of Drug Resistance. *Antioxidants* (Basel). 2021 Nov 11; 10(11):1801. doi: 10.3390/antiox10111801.

[97] Zhang Y, Choksi S, Chen K, Pobezinskaya Y, Linnoila I, Liu ZG. ROS play a critical role in the differentiation of alternatively activated macrophages and the occurrence of tumor-associated macrophages. *Cell Res.* 2013 Jul; 23(7):898-914. doi: 10.1038/cr.2013.75.

Chapter 10

# The Role of Bioactive Compounds and Associated Nanoparticles in the Regulation of ROS: An Innovative Strategy in Cancer Therapy

**Satish Kumar Vemuri**[1]
**Chepuri Kalyani**[2,*]
**C. D. S. L. N. Tulasi**[2]
**D. Manikantha**[2]
**G. P. V. Subbaiah**[1]
**and A. V. Gurava Reddy**[1]
[1]SMART, Sunshine Hospitals, Secunderabad, Hyderabad, Telangana State, India
[2]Jawaharlal Nehru Technological University Hyderabad, Telangana State, India

## Abstract

Reactive oxygen species (ROS), as a group referred to as highly reactive molecules, are very unstable and (are) reduced oxygen derivatives. These ROS are mostly the by-products of cellular metabolic pathways. Notably, they originate from the electron transport chain of mitochondria (ETC). In all the eukaryotic cells and tissues, conversion of molecular oxygen (small portion) into superoxide anion is a continuous process that usually takes place in the ETC. Likewise, hypochlorous acid, hydrogen peroxide, hydroxyl radical, and singlet oxygen are similar by-products produced from various cellular processes where they act as secondary messengers. Hence, low or basal levels of ROS are generally considered indispensable for the normal physiological functions within the system like sensing of oxygen, regulation of vasculature diameter, maintenance of physiological state of skeletal muscle, routine maintenance of genomic stability in terms of regulating transcription, translation and various signal transduction events. However, elevated levels of ROS production are associated with numerous disease states ranging from infection and inflammation to chronic cardiovascular, respiratory, tumorigenic, neurodegenerative, and many more disorders related to metabolisms like digestion and assimilation. Many research studies emphasize the eminent role of ROS in most cancer types. This could be due to its prominent role in promoting tumor initiation, development, and progression. Altogether,

* Corresponding Author's E-mail: Chekalyanibhaskar.tp@gmail.com.

In: The Role of Reactive Oxygen Species in Health and Disease
Editor: Kuladip Jana
ISBN: 979-8-89113-283-2

in living organisms, the intricate balance of ROS is very crucial to modulate the signaling within the cell at the base level or mutual communication among the cells to transduce the signal or to decide the fate of a cell whether to survive or die. By considering the ROS-based pathology state, Studies exhibit the potential anti-cancer properties of b-AuNPs in a phytochemical combinatorial approach that could be the future of cancer nanomedicine.it is vital/essential/necessary/imperative to adopt potential preventive or management strategies to modulate ROS and target them by a diverse range of therapeutic agents.

**Keywords**: ROS, bioactive compounds, nanoparticles, cancer, chemoprevention

## 1. Introduction

Reactive oxygen species (ROS) are a group of oxygen-containing molecules including superoxide, hydroxyl radicals, and hydrogen peroxide ($H_2O_2$) which are highly unstable. [1]. To boost their stability, these molecules scavenge electrons from completely reduced compounds. Endogenous ROS amounts are closely controlled within the cell under normal conditions by several antioxidant systems. Yet, as these reactive entities gather electrons from a variety of sources, they produce even more free radicals, leading to oxidative stress [2]. ROS can harm DNA, lipid-based membranes, and proteins in the cell when it is exposed to oxidative stress. Elevated levels of cellular ROS have indeed been connected to the development of a variety of malignancies and diseases. [3, 4].

Despite the dangers of excessive amounts of these chemicals, multiple studies have linked moderate levels of ROS to intracellular signaling pathways in a variety of cell types [5-8]. ROS plays a role in a range of metabolic pathways in the body, including the regulation of vascular diameter, oxygen sensing, genome stability, gene expression, protein expression, and skeletal muscle physiology maintenance, as well as the immune system.

### 1.1. ROS Sources and Types

**Table 1.** ROS Sources and Types

| ROS | $t_{1/2}$ | Migration distance | Mode of action | Production site | Scavenging systems |
|---|---|---|---|---|---|
| Superoxide ($O_2^{\bullet-}$) | 1-4µs | 30 nm | Reacts with Fe-S proteins Dismutates to $H_2O_2$ | Apoplast, chloroplasts, mitochondria, peroxisomes, electron transfer chains | SOD, flavonoids, ascorbate |
| Hydroxyl radical ($OH^{\bullet}$) | 1 ns | 1 nm | Extremely reactive with all biomolecules including DNA, RNA, lipids, and proteins | Iron and $H_2O_2$ (Fenton reaction) | Flavonoids, proline, sugars, ascorbate |
| Hydrogen peroxide ($H_2O_2$) | >1 ms | >1 µm | Reacts with proteins by attacking methionine and cysteine residues. Reacts with proteins of heme and DNA. | Peroxisomes, chloroplasts, mitochondria, cytosol, apoplast | APX, CAT, GPX, PER, PRX, ascorbate, glutathione |
| Singlet oxygen ($^1O_2$) | 1-4µs | 30 nm | Oxidizes proteins (His, Trp, Tyr, Met, Cysresidues), lipids, and G residues of DNA | Membranes, chloroplasts, nuclei | Carotenoids and α-tocopherol |

A minimum of one oxygen is present in ROS [9]. Hydroxy Radicals ($^{\cdot}OH$), Superoxides ($O_2^{\cdot -}$), Lipid Peroxides (LP), Hydrogen Peroxides ($H_2O_2$), and have all been identified as ROS in previous studies. The ROS types mentioned originate from the endoplasmic reticulum, electron transport chain, and other cycles. Proteins, DNA, and Lipids are the principal targets for these ROS [10-11] (Table 1).

## 1.2. Intracellular and Excellular ROS generation Sources

The compartments like cytoplasm, mitochondria, peroxisomes, and endoplasmic reticulum are the major areas of ROS production. Alteration in redox balance may impact disorders related to metabolic dysfunction, as subcellular compartments and enzymes that normally create ROS are linked to metabolic regulation. Cytosolic ROS are formed most notably through NOX (nicotinamide adenine dinucleotide phosphate [NADPH] oxidase) activity. NOX proteins produce $O_2^-$ through NADPH electron exchange. In addition to NOX-dependent ROS production, nitric oxide synthases (NOS). Endothelial and neuronal are the sources of cytosolic ROS [12]. Secondly, ROS is a natural by-product of the electron transport chain.

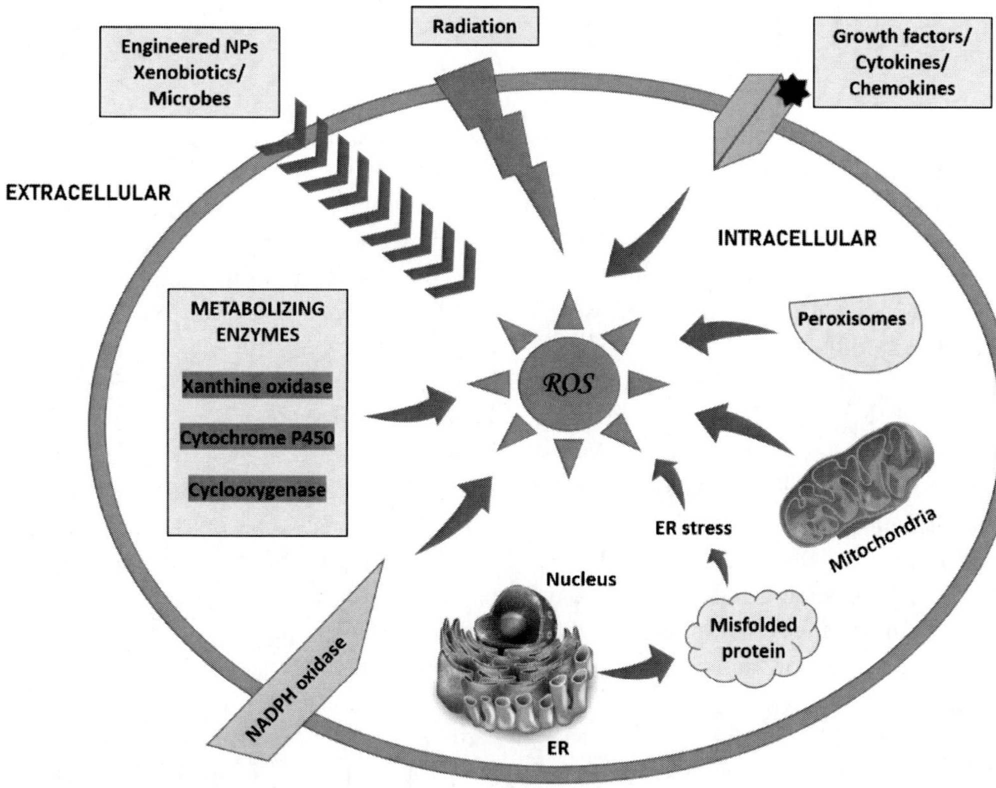


**Figure 1.** Reactive oxygen species (ROS) production sources. The external and intracellular Origins of ROS production are depicted in this diagram. Microbial infection, environmental contaminants, radiation exposure, and exposure to manufactured nanoparticles are all examples of extracellular sources of ROS (NPs). The mitochondria, endoplasmic reticulum (ER) stress, the NOX family proteins and metabolizing enzymes of cell can all produce intracellular ROS.

There are seven separate sites of mammalian mitochondrial ROS production that have been identified and widely accepted. The substrate-binding sites for the generation of ROS include pyruvate dehydrogenase (PDH), 2 oxogluterate dehydrogenase (OGDH), site IF, site IQ, Electron transferring flavoprotein ubiquinone oxidoreductase (ETFQOR), Glycerol 3 Phosphate dehydrogenase (GPDH), Site III Qo. However, the maximum capacities to produce ROS are at complex I and III [13, 14]. Peroxisomes are also a significant source of ROS. Particularly, peroxisomes produce $H_2O_2$ from an abundance of $O_2$-consuming oxidases including D-amino acid oxidase, L-α-hydroxyacid oxidase, D-aspartate oxidase, polyamine oxidase, L-pipecolic oxidases, xanthine oxidase, and ACOX (acyl-CoA oxidases) [15]. Lastly, the Endoplasmic reticulum also contributes to the production of ROS. In Endoplasmic Reticulum, the process of protein folding takes place. For successful protein folding, protein disulphide isomerase and ER oxido-reduction 1 pathways play a pivotal role. During the process, $H_2O_2$ is released. Apart from pathway of PDI/ERO1 [16], the mentioned ROS is generated through other monooxygenase systems which are associated with membrane namely cytochrome $b_5$ reductase, ER-localized NOX4 and cytochrome p450 [17]. Extracellular sources of ROS generation include ROS-inducing agents, such as radiation, pollutants, and exposure to nanomaterials (Figure 1).

## 2. Physiological Role of ROS

ROS act as signalling molecules, allowing normal physiological functioning to be regulated and maintained. Mitochondrial reactive oxygen species (ROS), particularly superoxide anion (O2·−) and hydrogen peroxide ($H_2O_2$) been shown to own a function in normal vascular physiology in reaction to shear stress. ROS mediates vasoconstriction by raising [Ca2þ]I while hydrogen peroxide functions as an endothelium-derived hyperpolarizing factor and produces vasodilation by PKG1a-mediated pathways [18, 19] and suppression of intracellular calcium mobilization [20].

Similarly, the ETC works as an $O_2$ sensor by releasing reactive oxygen species (ROS) in response to hypoxia. The generated ROS from hypoxia operates as signalling molecules, triggering a variety of functional responses, including increased synthesis and stability of hypoxia-inducible factor-1 (HIF-1) [21]. ROS are important in pathogen defence, cell communication, and mitogenic responses at low to moderate concentrations; at greater concentrations, ROS generates oxidative stress, a condition in which pro-oxidants and antioxidant components are out of balance. During oxidative stress, DNA, cellular lipids, and proteins can be damaged, resulting in reduced cell survival and impaired cell function [22, 23]. In cells, an effective anti-oxidative system composed of enzymatic and non-enzymatic antioxidants keeps ROS levels under tight control to prevent oxidative harm but does not eradicate them [24]. Superoxide dismutase, in particular, scavenges the superoxide anion and converts it to hydrogen peroxide, resulting in reduced toxicity. Peroxisomes contain catalase, which decomposes hydrogen peroxide and protects cells from highly dangerous hydroxyl radicals. Glutathione peroxidase, reductase, and transferase are important antioxidant enzymes that participate in direct removal of the reactive oxygen species (ROS) [25]. Glutathione peroxidase detoxifies most peroxides in the mitochondria or cytosol [26]. The antioxidant enzymes peroxiredoxins (Prxs) are required for redox equilibrium [27]. Ascorbate, glutathione, and vitamin E make up the non-enzymatic antioxidant defense systems in the brain [28]. These antioxidant molecules normally detoxify reactive oxygen/nitrogen species, but they do not inhibit them from forming [29]. T cells, a crucial role in adaptive immunity, require ROS for proliferation and signal transduction to activate [30].

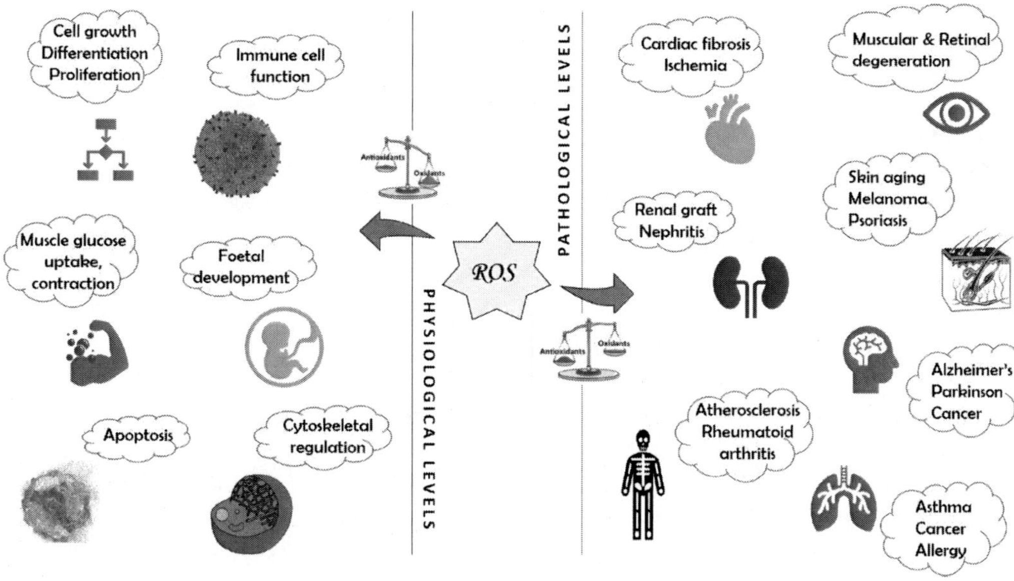


**Figure 2.** Role of Reactive oxygen species (ROS) in various physiological processes and imbalance of ROS in various pathological conditions.

ROS plays a crucial role in both the innate and acquired systems of the immunological defense mechanism. Exaggerated ROS generation as part of the oxidative burst in activated phagocytes existing in the local inflammatory milieu is one of the first lines of defense built against invasive pathogens when exposed to environmental pathogens. ROS are engaged in the acquired immune response because active phagocytes continue to create excess ROS locally, which enhances intracellular signal transduction cascades within T cells and lowers their activation threshold [31]. ROS may operate as signaling intermediates in the maintenance of skeletal muscle glucose absorption during contraction, according to the latest report. However, the findings of such studies should be regarded with caution because they vary depending on the model tested and the experimental approach [3-5]. Furthermore, the cellular redox state is seen as a new regulator of genomic stability and transcription. Several physiological processes, including cell cycle regulation, programmed cell death, DNA damage response, intermediate metabolism regulation, and autophagy are known to be regulated at the level of gene transcription. So, because deacetylase activity of sirtuins, members of the class III histone deacetylases, is dependent on cellular redox status & NAD+ availability, oxidative stress and cellular energy consumption have been identified as major transcription regulatory variables. The oxidative stress and DNA damage-responsive transcription factor E2F1, which controls cell cycle and directly binds to the promoter of sirtuin 1, the most explored member of the sirtuins family, has been found to affect the gene expression level of sirtuins. The involvement of reactive oxygen species (ROS) in transcription control is crucial under oxidative stress, and it can alter critical activities including cellular lifespan, glucose homeostasis, inflammation, and a variety of aging-related disorders like cancer [9]. As a result, ROS may play an important role in regulating homeostasis through controlling a variety of physiological activities and in the pathogenesis of diseases such as cancer, autoimmune disorders/inflammatory diseases, neurological diseases, cardiovascular diseases, and diabetes mellitus [22] (Figure 2).

## 3. ROS in Pathological Disorders

Cardiovascular disease, cancer, neurological disorders, diabetes, ischemia/reperfusion, and other diseases, as well as aging, have all been linked to oxidative stress [32-35]. There are two types of diseases: (i) The first group includes diseases characterized by pro-oxidants altering the thiol/disulfide redox state and impairing glucose tolerance—so-called "mitochondrial oxidative stress" conditions (cancer and diabetes mellitus); (ii) the second group includes diseases characterized by "inflammatory oxidative conditions" (cancer and diabetes mellitus) and increased activity of the NAD(P)H oxidase (chronic inflammation and atherosclerosis). The detrimental effects of free radical action (lipid peroxidation, DNA damage, protein oxidation) play a substantial role in the aging process [36] (Figure 2). Validated oxidative stress biomarkers provide compelling evidence for the link between oxidative/nitrosative stress and acute and chronic illnesses. Such biomarkers must be objectively examined and assessed over long periods in healthy and diseased patients. Dalle-Donne and collaborators have released an excellent study on indicators of oxidative stress in human disorders [32].

### 3.1. Neurological Disorder

#### *3.1.1. Alzheimer's Disease*

Patients with Alzheimer's disease (AD) have considerable oxidative damage in their brains, which is associated with a significant accumulation of amyloid-βpeptide (Aβ), the major constituent of senile plaques in the brain, as well as the deposition of neurofibrillary tangles and neutrophil threads [37]. In Alzheimer's disease patients, Aβ is the major component of senile plaques and cerebrovascular amyloid deposits. Aβ is a 39–43 amino acid peptide generated by proteolytic cleavage from the bigger amyloid beta (A4) precursor protein (APP). The predominant form of A is Aβ1-40, although the minor species, Aβ1-42, has a higher propensity to combine and is rich in amyloid deposits.

Although the neurochemical factors that cause this age-related degenerative process are still poorly understood, accumulating data suggests that bio-metals such as copper (Cu), iron (Fe), and zinc (Zn) play a crucial role in Aβ buildup and neuronal degeneration. Cu and Zn have high-affinity binding sites in Aβ, and APP binds these metals as well via N-terminal metal-binding domains. Cu ions bind to A monomer via three histidine residues and tyrosine or in aggregated Aβ via a bridging histidine molecule. Cu has been demonstrated to cause considerable Aβ aggregation under mildly acidic (pH 6.6) circumstances, which is similar to the microenvironment in AD neuropil.

#### *3.1.2. Parkinson's Disease*

Parkinson's disease (PD) is characterized by a loss of neurons in the substantia nigra, a region of the midbrain [35]. The cells in the substantia nigra connect with cells in the stratum via dopamine (a neurotransmitter that acts as a chemical messenger between brain and nerve cells). As a result, a decrease in nigral dopamine levels leads to a decrease in striatal dopamine, which is thought to be the source of PD symptoms [34]. In the cerebral cortex, anterior thalamus, hypothalamus, amygdala, and basal forebrain, neuronal loss and Lewy bodies, the pathological hallmarks of PD, have been found. In nerve cells, Lewy Bodies are little spherical protein deposits. They disturb the brain's regular functioning by interfering with the action of essential chemical messengers such as acetylcholine and dopamine; they are most likely generated as cells attempt to defend themselves against attack [35]. Filaments made up of α-synuclein makes up the majority of intracytoplasmic Lewy bodies. The genetic causes of Parkinson's disease are two recently discovered point mutations in α-synuclein [38].

Levodopa is a dopamine precursor and the most often used Parkinson's disease medication. Long-term usage of l-dopa likely leads to oxidation and toxicity in brain cells. If this is proven to be accurate, it will strengthen the case for using antioxidants in regular Parkinson's disease treatment. Because oxidative stress appears to be part of a cascade of biochemical changes leading to dopaminergic death, separating the effect and extent of oxidative stress from other components of the cascade that can play a primary role in the initiation of ROS and RNS is an alarming problem in understanding the pathogenesis of PD.

### 3.2. Cardiovascular Diseases

Cardiovascular tissue injury has been related to ROS-induced oxidative stress in cardiac and vascular myocytes [33]. Regardless of the lack of direct evidence linking oxidative stress to cardiovascular illness, ROS-induced oxidative stress plays a role in atherosclerosis, ischemic heart disease, hypertension, cardiomyopathies, cardiac hypertrophy, and congestive heart failure [39]. The enzymes xanthine oxidoreductase (XOR), NAD(P)H oxidase (multi subunit membrane complexes), and NOS, as well as mitochondrial cytochromes and hemoglobin, are the principal causes of oxidative stress in the cardiovascular system [40, 41]. NOSs and hemoglobin are also major sources of RNS, such as SNOs (NO-modified cysteine thiols in proteins peptides, and amino acids) & NO• which are responsible for NO• bioactivity. Increased formation of reactive oxygen species (ROS) causes peroxidation and oxidation of thiol groups in phospholipids and proteins, resulting in oxidative stress [42]. ROS attacks cause changes in membrane permeability, breakdown of the membrane lipid bilayer, and functional alteration of a variety of cellular proteins. Due to increased oxidative stress, abnormalities in myocyte function are thought to be linked to the anomalies on degradation cellular proteins and lipids, in addition to effects on sub-cellular organelles. Angiotensin II (AngII) is a multifunctional hormone that regulates vascular function by regulating a variety of cellular processes such as cell proliferation, fibrosis, death, inflammation, and migration [43, 44]. Blood pressure and fluid balance are both regulated by AngII. A growing amount of evidence suggests that the formation of reactive oxygen species (ROS) is intimately tied to AngII-induced activity.

### 3.3. Ischemic/ Reperfusion Injury

Ischemia-reperfusion injury is a clinically significant condition that occurs when blood flow is restored after a critical period of coronary blockage. Despite the low oxygen tension experienced during ischemia, moderate ROS formation is thought to occur, most likely from a mitochondrial source [45-47]. The large burst of ROS seen during reperfusion may come from a different cellular source than during ischemia, and the cause has yet to be identified conclusively. Massive formation of reactive oxygen species (ROS) during ischemia/reperfusion causes tissue injury, resulting in major difficulties in organ transplantation, stroke, and myocardial infarction [46] Reperfusion damage is caused by neutrophils, which are the main effector cells. Xanthine dehydrogenase is transformed to xanthine oxidase, which uses oxygen as a substrate, during ischemia/ reperfusion. During ischemia, excessive ATP consumption results in the accumulation of the purine catabolites hypoxanthine and xanthine, which are processed by xanthine oxidase to produce massive amounts of superoxide radical and hydrogen peroxide after reperfusion and oxygen ingress [48].

Several studies have used various pharmacological approaches to evaluate the negative effects of ischemia-reperfusion-induced oxidant generation [49]. In vivo in gene-targeted mice, the function of cellular antioxidant enzymes in the development of myocardial damage revealed that neither Cu–ZnSOD deficiency nor overexpression influenced the amount of myocardial necrosis [50]. Glutathione peroxidase overexpression does not affect the severity of myocardial damage. Overexpression of MnSOD, on the other hand, greatly reduced myocardial necrosis after myocardial injury/reperfusion. These findings suggest that MnSOD, but not Cu/ZnSOD or glutathione peroxidase, plays an important role in mice after in vivo MI/R.

### 3.4. Rheumatoid Arthritis

Rheumatoid arthritis is an autoimmune illness in which macrophages and activated T lymphocytes infiltrate the joints and surrounding tissue, causing chronic inflammation [51]. The production of free radicals at the spot of inflammation is primarily associated with the pathogenesis of the disease. Elevated levels of prostaglandins and isoprostanes in body fluids verified oxidative injury and inflammatory status in several rheumatic illnesses when compared to controls. Synovial tissue oxidative conditions are also linked to a higher rate of p53 mutations [52].

T-cells isolated from rheumatoid arthritis patients' synovial fluid reveal evidence of reduced intracellular GSH, poor phosphorylation of the adaptor protein linker for T-cell activation (LAT), and the "primed" CD45RO phenotype [53]. Changes in intracellular GSH level have been validated to cause changes in LAT subcellular localization. The aberrant expression of numerous adhesion molecules (ELAM-1, VCAM-1, ICAM-1, ICAM-2) mediates monocyte and lymphocyte migration into the rheumatoid arthritis synovium [54].

### 3.5. Diabetes

Complex I and the ubiquinone–complex III interface are the major locations of superoxide production in the mitochondrial membrane under normal conditions, where the presence of long-lived intermediates permits electrons to react with molecular dioxygen [55]. Under diabetic conditions, however, the primary sites of superoxide creation are altered, and complex II becomes the primary source of electrons that contribute to superoxide synthesis [56]. This conclusion was reached after testing a complex II inhibitor, 2-thenoyltrifluoroacetone, and an uncoupler of oxidative phosphorylation, carbonyl cyanide m- chlorophenyldihydrazone, in diverse cells exposed to elevated glucose concentrations [57].

NAD(P)H is another cause of ROS in diabetes. The enzyme is made up of two membrane-bound components, gp91phox and p22phox, which together form cytochrome b558, the complex's catalytic core [58]. NAD(P)H oxidases are a key source of glucose-induced ROS generation in the vasculature and kidney cells, according to several lines of evidence, supporting NAD(P)H as a mediator of diabetic problems [59]. Other cells' involvement hasn't been proven. Because hypertension is a common consequence of diabetes, it's probable that NAD(P)H oxidase expression is regulated similarly in both diseases. Increased angiotensin II labeling in cardiac myocytes and endothelial cells from diabetes patients causes this [59]. AngII type 1 receptor antagonists can block the high glucose-induced formation of ROS and p47phox (a cytosolic component of activated NAD(P)H), confirming a relationship between the two pathways of NAD(P)H oxidase activation. Several PKC inhibitors limit the NAD(P)H oxidase-mediated production of ROS in diabetics, implicating this family of kinases in the regulation of hyperglycemia-induced NAD(P)H oxidase activity. Because hyperglycemia-induced oxidative stress occurs in nonnucleated cells (erythrocytes) that lack mitochondria and the NAD(P)H oxidase, another mechanism of ROS generation must exist in these cells. Glucose auto-oxidation could be one explanation for such behavior [60]. In the presence of iron and copper ions, glucose and its metabolites are known to react with hydrogen peroxide to create hydroxyl radicals. In vitro tests provide evidence for this, hence in vivo investigations should be conducted.

### 3.6. Cancer

Oxidative stress causes a cellular redox inequity, detected in cancer cells as compared to normal cells; the redox imbalance may thus be linked to oncogenic activation. Mutagenesis, cancer, and aging all begin with the permanent change of genetic material caused by "oxidative damage" occurrences. DNA mutation is a vital step in carcinogenesis, and higher levels of oxidative DNA lesions have been seen in a variety of tumors, implying that such damage is involved in cancer etiology. Single/double-stranded DNA breaks, pyrimidine/purine, or deoxyribose alterations, and DNA damage are all examples of ROS-induced DNA damage. DNA damage can cause transcriptional pause or induction, signal transduction pathway induction, replication mistakes, and genomic instability, all of which are linked to cancer [61, 62]. The development of 8-OH-G is the most well-studied DNA lesion. This lesion is significant because it is simple to produce and carcinogenic, making it a possible biomarker for carcinogenesis. Damage to DN; A, mutations, and altered gene expression are all significant participants in the carcinogenesis process. The presence of oxidants appears to be a common denominator in each of these occurrences [62-64]. ROS have been shown to interfere with the expression of a variety of genes and signal transduction pathways, making them important in the carcinogenesis process [62, 65]. Because the mechanism of cell growth regulation is so complex, the participation of ROS in this process is highly dependent on the kind and concentration of the radical in question. The activation of transcription factors such as MAP-kinase/AP-1 and the NF-B pathways affects cell growth and apoptosis directly [62].

Anomalies in growth factor receptor function are linked to the development of a variety of malignancies. ROS and metals viz., nickel, beryllium, arsenic and cobalt, impact many growth factor receptors like VEGF, EGF, PDGF and so on. When both EGF & VEGF are activated, cellular Ca levels rise (II). In lung and urinary malignancies, increased expression of EGF receptors and overexpression of the EGF receptor has been found [66]. Endothelial cells, fibroblasts, and mesenchymal cells all contain PDGF, which has been reported to be overexpressed in lung and prostate cancers (Figure 3).

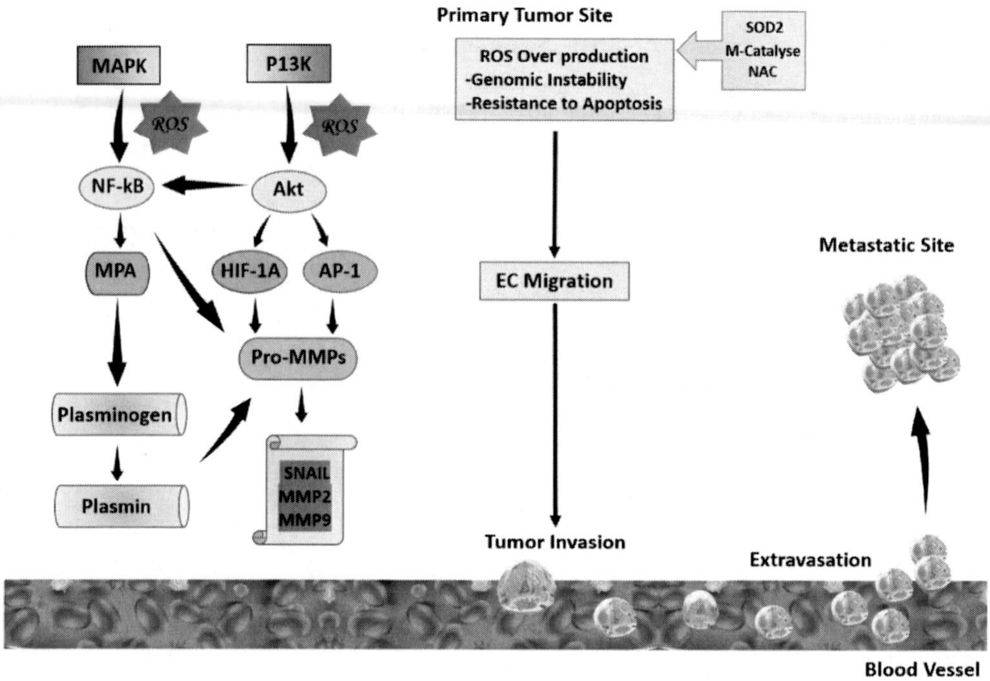


**Figure 3.** Metastasis and reactive oxygen species. Elevated levels of reactive oxygen species stimulate the PI3K/Akt/mTOR (phosphoinositide-3-kinase/AKT serine/threonine kinases/mechanistic target of rapamycin kinase), MAPK (Mitogen-activated protein kinases), and MAPK (Mitogen-activated protein kinases) signaling pathways, which activate downstream SNAIL, MMP2 (metalloproteinase 2).

## 4. Cancer Theranostics

Although enormous studies have been made in our understanding of the disease, its processes and signaling cascades, tumor biomarkers, and the pathophysiology of cancer development, treatment techniques are still in their infancy. Chemotherapeutics must be delivered to target cells without harming healthy tissues and organs, which is a huge difficulty. Non-specific delivery vehicles of chemotherapeutic drugs that target only the afflicted cells and organs without adversely impacting healthy tissues are one of the leading causes of such side effects. As a result, delivering chemotherapeutics to target cells and organs has become a major problem for modernizing therapy protocols [67]. The efflux of chemotherapeutic agents by ABC transporter pumps and the establishment of primary or acquired resistance to apoptosis, which leads to chemoresistance, are two important probable pathways for the establishment of chemoresistance in cancer cells. Most critically, it has been discovered that a subpopulation of cancer stem cells seen in tumors is responsible for multidrug resistance [68]. Furthermore, it has been observed that the majority of cancer cells escape apoptosis by generating autophagy, hypoxia, and epigenetic alterations such as DNA methylation & histone modifications, all of which are thought to have a role in cancer cell chemoresistance. Furthermore, surgical excision of cancer tissues may cause damage to healthy cells as well as dangerous neoplasms.

### 4.1. Natural Therapeutics

Plants, animals, aquatic organisms, and microbes all contribute to drugs that could be used as anti-cancer agents, making nature an appealing source of new therapeutic entities. Several bioactive chemicals derived from medicinal plants have also been demonstrated to delay metastasis and inhibit angiogenesis (Bioactive compounds). Vinca alkaloids, podophyllotoxin and its derivatives taxanes, anthracyclines, doxorubicin, etoposide, and their analogs, camptothecin (CPT) and its derivatives, and many others are naturally occurring anticancer drugs (Table 2). Even though the majority of researchers favoured natural products, the majority of natural chemicals are not developing into viable drug candidates. Insolubility, instability, and low bioavailability are few key hurdles for employing these natural chemicals, which have found them ineffective in clinical trials. In the present era, non-toxic, cost-effective, and environmentally friendly anticancer treatments are available from plant sources. As a result, several physicists and pharmacologists are working to design a more effective nanocarrier system that can target cancerous cells and deliver an ideal therapeutic dose.

**Table 2.** Phytochemicals showing *in vitro* anticancer activity via regulation of different signaling pathways in various cancer cell lines

| S.No | Phytochemical | Signaling pathway | Cell line | References |
|---|---|---|---|---|
| 1. | Allicin | STAT3 signaling pathway | HuCCT-1 cell | 69 |
| 2. | Andrographolide | PI3k/AKT/mTOR pathway | MDA-MB-231 | 70 |
| 3. | Baicalin | MAPK, ERK, and p38 pathways | HCT116 | 71 |
| 4. | Curcumin | PI3K/Akt/mTOR pathway, NF-κB pathway | LNCaP, DU145, and PC-3 | 72, 73, 74 |
| 5. | Dicumarol | Apoptotic pathway | A2780 | 75 |
| 6. | Epigallocatechin | Apoptotic & Notch signaling pathway | HuCC-T1 | 76 |
| 7. | Genistein | Apoptotic pathway | HL60 | 77 |
| 8. | Nimbolide | PI3k/AKT/mTOR and ERK signaling pathway | SCC131 and SCC4 | 78 |
| 9. | Pterostilbene | NF-κB, AKT, and ERK signaling pathways | OVCAR8, OV1063, IGROV-1, and SKOV3. | 79 |
| 10. | Reversatol | Akt/mTOR and p38-MAPK pathway | A549 and H1299 | 80, 81 |
| 11. | Urolic acid | Apoptotic pathway | MDA-MB-231, M4Beu | 82 |
| 12. | Withaferin-A | NF-κB, Prosurvival, and Apoptotic pathways | LY-3, LY-10, SudHL-6 | 83 |

### 4.2. Nanoparticles for Triggered Drug Delivery and Effective Cancer Therapy

Drug delivery using nanoparticles (NPs) has been intensively pursued to establish safer and most effective therapy techniques for a variety of essential diseases, most prominently cancer. This is due to the small size, easiness of functionalization, increased drug loading (large surface to volume proportion), easy penetration, and greater tissue retention. Aside from that, nanoparticles have high biocompatibility and biodegradability, making them an attractive method of administration in the treatment of cancer.

### *4.2.1 Bio-Inspired Nanoparticles*

Bio-inspired nanoparticles that mimic natural body components have gotten a lot of attention recently because of their ability to act as alternative and potential biocompatible drug delivery methods in cancer theranostics [84]. Because of their non-toxic, cost-effective, non-hazardous, and environmentally acceptable qualities, plant-based metal, and metal oxide NPs have gotten a lot of attention. The stability, oxidizing power, chemical inertness, hydrophilicity, as well as optical, physical, electrical, and photo-catalytic capabilities of bioengineered titanium oxide nanoparticles ($TiO_2NPs$) were all demonstrated. $TiO_2NPs$ have been proven to have crucial activities in a variety of ways, including cancer cell targeting and gene delivery. Toxic ($TiO_2$+) ions can cause oxidative stress, which can change the functioning of cancer cell components [85]. Furthermore, the metallic ions generated by metal and metal oxide NPs can interface with functional groups in proteins involved in cell membranes or enzymatic systems, such as sulfur, thiols, or carbonyls. They may also attach to nucleic acids, causing oxidation and denaturation, which results in cancer being destroyed. Super magnetism, fluorescence, surface plasmon resonance (SPR), and also light absorption and scattering ability, are all characteristics of AuNPs [86]. They can target specific locations on cancer cells and destroy them permanently. After modifying the surface phosphorus and nitrogen ligands, this targeting might be improved even more. AuNPs can also be effective in the hyperthermia technique, which is centered on producing high temperatures for the selective destruction of malignant cells by apoptosis. The heat is produced by the interaction of AuNPs with light at the SPR wavelength (808 nm), which causes excessive production of heat shock proteins and cellular metabolism disturbance. Binding AuNPs to cytokines (TNF- $\alpha$) and hormone receptors which synergize the antitumor response is also another example of selective targeting and selective damage. Similarly, due to their site-specific targeting cells and subsequent drug release, iron oxide nanoparticles (IONPs) have been found to have therapeutic potential in the treatment of cancer. Plant-mediated production of Silver NPs has a large literature collection [87]. Biosynthesized AgNPs have a wide range of applications in the pharmaceutical industry, including acting as a reducing as well as capping mediator. Manganese oxide, magnetic iron oxide, silica, and copper nanoparticles are some other examples. Manganese oxide nanoparticles ($MnO_2NPs$) can induce cancer cell death by acting as a hyperthermic agent. Copper nanoparticles (CuNPs) were also effective in inflicting selective damage on tumor cells. In addition to being optically transparent, relatively non-toxic, and biologically inert, silica nanoparticles (SiNPs) are characterized by their surface modification. These characteristics make them a perfect candidate for the incorporation of diverse dyes, such as polymethines, and the delivery of several hydrophobic drugs, such as paclitaxel and camptothecin, rather than intravenous administration of drugs that require high levels of toxic organic solvents, like dimethyl sulfoxide (DMSO). Liposomes, viral nanoparticles, lipid nanoparticles, protein nanoparticles, biosynthesized metal nanoparticles, and other bio-inspired nanomaterials are now being employed in cancer theranostics applications. The FDA has approved many nano theranostics agents, which are currently being tested in clinical trials (Table 3).

**Table 3.** Different nanocarrier systems, characteristics and dosage of photosynthesized Nanoparticles for targeted delivery – In vitro studies

| S.No | Phytochemical | Nanocarrier | Cancer cell line | Size of NP in nm | $IC_{50}$ | Reference |
|---|---|---|---|---|---|---|
| 1. | Gallic acid | Gum Arabic encapsulation | HepG2 and MCF7 | 33 to 250 | 6.61 and 8.52 μg/mL | 88 |
| 2. | Kaempferol | PLGA | OVCAR-3 | 210 | 25 μm | 89 |
| 3. | Gambogic acid | poly(ester urea urethane) nanoparticles | HCT116 | 98–267 | 6μg/mL | 90 |
| 4. | Betulinic acid | Magneto liposomes | MDA-MB-231, MCF 10A | 200 | 25 μm | 91 |
| 5. | Epigallocatechin-3-*O*-gallate | Gold Nanoparticle | MBT-2 | 20–1200 | 28.4 μm | 92 |
| 6. | Thymoquinone | Cubosomes | MCF-7 and MDA-MB-231 | 98 | 27.60 &7.80 μM | 93 |
| 7. | Gingerol | Gold Nanoparticles | MDAMB-231 | 9 | 40.3 μg/ml | 94 |
| 8. | Sulforaphane | selenium nanoparticles | HT-29 & Caco-2 | 80.2 | 0.17 & 0.38 ppm | 95 |
| 9. | Curcumin | PGMD polymer | MCF-7 & MDA-MB 231 | 218 | 43.4 & 30.5 μM | 96 |
| 10. | Withaferin-A | Nanosponges | MCF-7 | 117 | 1.57μM | 97 |

### *4.2.2. Various other Nanocarriers*

Besides metal oxides, advances in nanotechnology have been credited with the most advanced carrier systems for therapeutic agent delivery at the desired spot. The carrier protects and improves the strength of the encapsulated medicinal ingredient by protecting it from external deterioration.

Polymeric nanoparticles are one type of nanocarrier that has been developed and employed. These nanoparticles are made from natural polymers such as pectin, sodium alginate, or gelatin, as well as synthetic polymers like PLGA, PB A, and PLA [98]. Drug absorption at the cellular level would be increased, and polymeric nanoparticles would provide improved support for drug candidate distribution. Additionally, greater stability for long-term storage protects against external degradation.

Liposomes are nanocarriers that contain lipid vesicles with a lipid bilayer separating the interior aqueous core from the outside bulk. Liposomes range in size from 10 nm to 1μm or more. The internal aqueous core transports drugs, proteins, vaccines, and enzymes to their intended destination [99]. Liposomes are amphiphilic, biocompatible, and biodegradable allowing both polar and nonpolar medicines to be encapsulated. Liposome permeation and transfection significantly increases cell apoptosis.

Nanomicelles, on the other hand, are ultramicroscopic self-assembling nanosized vesicular carriers with particle diameters and morphologies less than 100 nm and a spherical shape. An amphiphilic monomer unit makes up the nanomicelles. The presence of fatty acyl chains signified the hydrophobic center, while typical hydrophilic polymers denoted the hydrophilic core [100]. This can contain both hydrophobic and hydrophilic drugs as they are amphiphilic nature. The majority of nanomicelles work by preventing active ROS formation in cells through cell cycle arrest and apoptosis.

Solid-lipid nanoparticles have a solid and lipid core and are simple to create. The particle size of SLN that has been prepared ranges from 10 to 1000 nm. The smaller particle size range could be useful in biomedical applications. They can be made using biocompatible excipients and can encapsulate lipophobic& lipophilic drugs [101]. Dendrimers are a type of nanoparticle that consists of a central core, internal branches, and external multifunctional surface groups and are nano-sized repetitively branched, star-shaped polymeric systems with a tree-like structure. Because of their multifunctional properties, dendrimers are a promising drug delivery vehicle. Multiple branches aid in the accumulation of additional molecules and serve as carriers for drug molecules, biomolecules such as proteins, and amino acids [102]. Dendrimers' monodisperse nature allows for repeatable and consistent pharmacokinetics with high absorption efficiency. Graphite, Diamonds, Graphene, Fullerene, and other allotropic forms of carbon have piqued the interest of researchers who want to learn more about the application of Carbon nanotubes (CNTs) in biological research [103]. The CNTs are made out of coiled graphene sheets with a greater surface area, longer length, and narrower diameter, making them ideal for therapeutic compounds.

## 5. Nanoparticles in Anticancer Action

### 5.1. Mechanism of Phytosynthesised NPs

When nanobiomaterials are utilized, the severe side effects of chemotherapy are prevented. They work by directly targeting cancer cells or tumors while exposing normal bodily tissues and organs to a much lesser concentration of the drug. The medicine becomes a colloid system when it is compacted in a nanobiomaterial and comes into direct touch with the biological system. It is feasible to advocate for regulated drug release into tumors, avoiding anticancer drug degradation and therefore lowering the amount of medication administered and the exposure of normal tissues to the drug concentration. Nanoparticles penetrate the apical membrane barrier of intestinal epithelial cells to reach enterocytes via transcellular transport. Exocytosis by enterocytes allows them to enter the circulation or lymphatic channels. Both clathrin and caveolae-mediated endocytosis pathways are used to ingest particles with a diameter of less than 500 nm. (Figure 4).

Several kinds of research have been conducted to assess the anticancer potential of phytosynthesized nanoparticles. Figure 4 depicts a schematic illustration of the numerous ways that phytosynthesized nanoparticles use to limit cancer cell proliferation. Because the phytochemicals are coated or bind as functional groups in the nanoparticles, phytosynthesized nanoparticles have better anticancer properties than chemically synthesized equivalents. These phytochemicals, in combination with nanoparticles, will serve as anticancer agents, increasing the intensity of apoptosis (programmed cell death) in cancer cells. Anticancer activity is exhibited by phytosynthesized nanoparticles in two ways: by binding to the surface or via cellular internalization. Because of their limited stability, certain nanoparticles dissolve into their ionic forms in the bodily fluid after being administered [104] whereas stable nanoparticles reach the cancer cells.

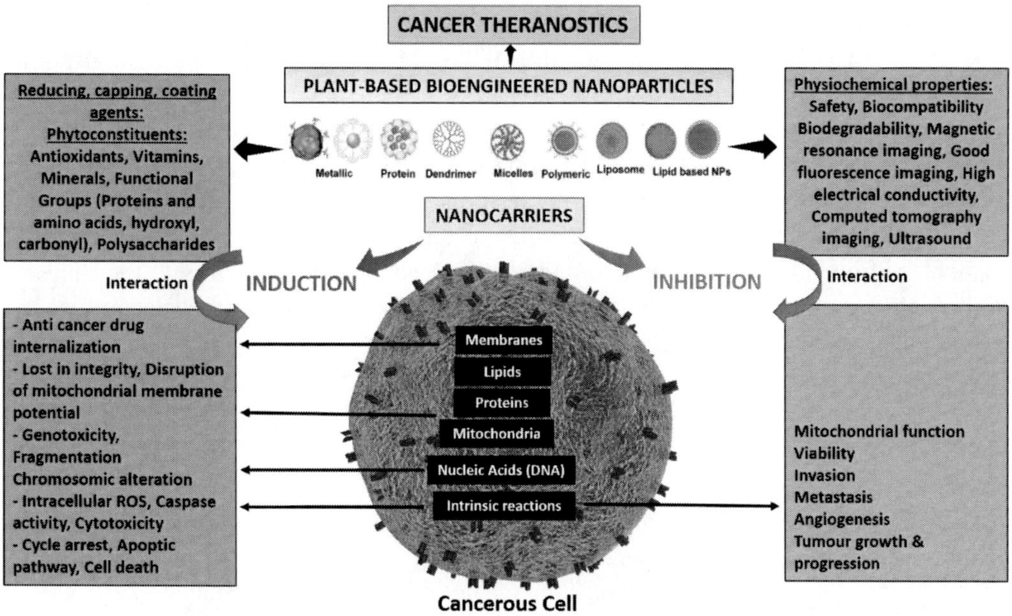


**Figure 4.** Roles of physiochemical and biological properties of bioengineered nanoparticles and their interactions with cancerous cell properties.

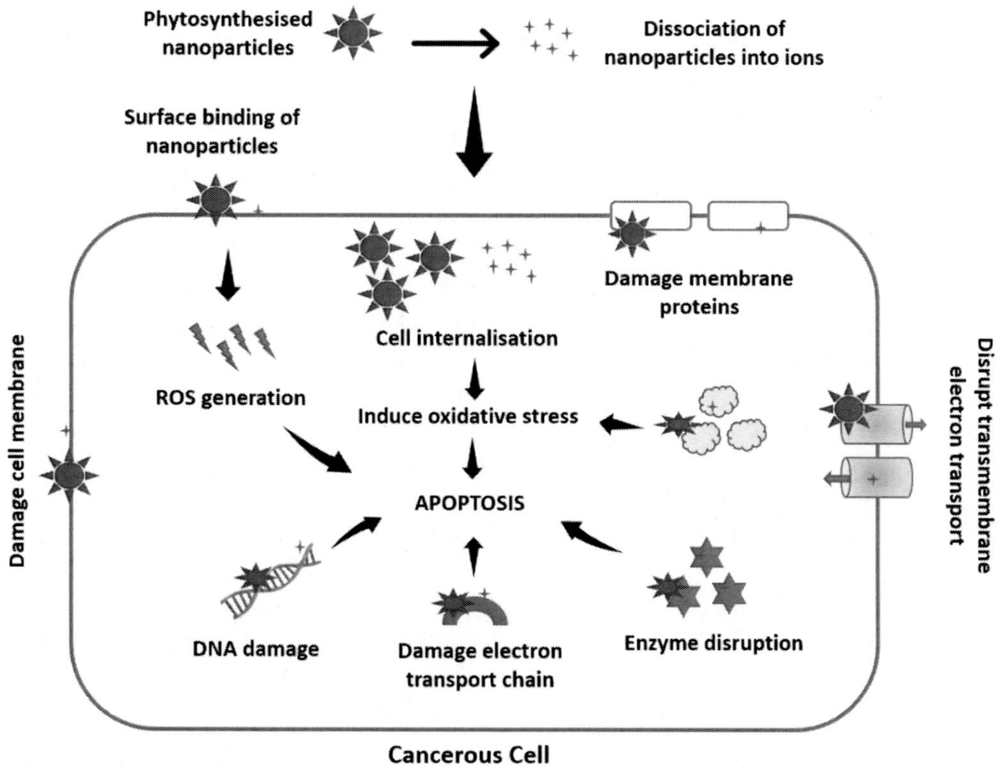


**Figure 5.** Mechanism of phytosynthesised nanoparticles.

The anticancer activity of dissociated ions in the bodily fluid is also evident, but it is less specific for cancer cells. To boost the stability of unstable nanoparticles in physiological fluids, they are nano encapsulated or nano formulated. The generation of reactive oxygen species (ROS) by nanoparticles is the main mechanism by which cancerous cells are inhibited. The activation of ROS formation at the molecular and subcellular levels can be a triggering factor for apoptosis, and therefore cancer cell eradication. ROS will form on the surface to destroy cell membranes, or nanosized particles will enter cells and generate ROS, disrupting cell membranes from within [105]. When phytosynthesized nanoparticles bind to the cancer cell surface, they may cause membrane protein degradation, transmembrane electron transport disruption, as well as cancer cell apoptosis [106]. (Figure 5)

## 5.2. NPs Inducing ROS in Cell Signaling Pathways

With revolutionary applications in biotechnology, electronics, medical drug delivery, cosmetics, material science, aerospace engineering, and biosensors, the developing field in nanotechnology has altered many divisions of the industrial field. Nanomaterials (NM) that have been manufactured have received commercial interest in several consumer items. Their new thermal, physicochemical, and electrical qualities make them easier to use in apparel, medicine, and cosmetics, increasing the likelihood of environmental and human contact [107-109]. Metal-based nanoparticles (NP) and Carbon nanotubes (CNT) have attracted the most commercial interest of entire NM because of their outstanding intrinsic features, such as excessive conductivity and tensile strength, that fulfills the need of the precise function for which these NP are designed [110, 111]. Their extensive use raises worries about unintended human exposure and the resulting negative health effects [112]. Inflammation, oxidative stress, genetic damage, and the suppression of cell division and cell death are some of the paradigms for NP-mediated toxicity [113-116]. Most of the research to date has revealed that NP toxicity is frequently associated with ROS production (which can be beneficial or harmful during biological interactions) and oxidative stress [109, 114]. Several adverse events have been linked to NP-mediated ROS reactions, including inflammation, genotoxicity, carcinogenesis and fibrosis. CNT-induced oxidative stress, for example, activates cell signaling pathways, leading to a rise in the expression of fibrotic and proinflammatory cytokines. The process for ROS synthesis differs from one NP to other, and the actual primary cellular mechanism for ROS production is still unknown.

### *5.2.1. NF-κB*

The NF-κB protein family activates genes involved in cellular stress resistance and regulates a variety of processes including inflammation, apoptosis, immunological response, and cell proliferation. The activation of NF-κB by prooxidant $H_2O_2$ via the traditional IKK-dependent route is well known. ROS such as OH•, HOCl, and $^1O_2$ as well as RNS like ONOO activate NF-κB by releasing IBs, which causes NF-B to translocate to the nucleus [117, 118]. Inside the nucleus, NF-κB causes oxidative stress and inflammation by inducing the transcription of pro-inflammatory mediators. ROS stimulate NF-κB to control the production of pro-inflammatory IL-2, IL-8, TNF, and IL-6 by lung epithelial cells and macrophages during NP-mediated lung damage [119]. Metal oxide NPs, such as Fe, Zn, Si, and Cd, cause NF-B activation in response to ROS [120-122].

#### *5.2.2. AP-1*

The transcription factor activator protein (AP)-1 generally gets activated in response to cytokines, oxidants, bacterial and viral infections, and growth factors. It regulates cell proliferation, differentiation, and death, making it a significant element in carcinogenesis [123]. Under oxidative conditions, phosphorylation of the protooncogene c-jun is thought to activate AP-1 [124]. Metal NP, like Ni, Cr, and Fe, have been found to activate AP-1 by generating reactive oxygen species (ROS) [125].

#### *5.2.3. MAPK*

MAPKs are basically serine-threonine protein kinases that regulate a variety of physiological functions such as gene expression, proliferation, mitosis, differentiation, apoptosis, and cell survival. Growth factor-regulated stress-activated MAPK, c-jun $NH_2$-terminal kinases (JNK), extracellular signal-related kinases (ERK), and p38 MAPK are all examples of MAPK. Free radicals may promote oxidative alteration of MAPK signaling proteins (e.g., MAP3 K and RTK) when ROS generation exceeds the antioxidant proteins' capacity, resulting in MAPK activation. By inhibiting and/or degrading MAPK phosphatases (MKP), ROS can activate MAPK pathways [126, 127]. Finally, the location of ROS synthesis, in addition to the kinetics of ROS production, quantity, and redox status, cellular antioxidant pools, are all likely to have a role in defining the effects of ROS on MAPK pathway activation [128]. In various cell types, Ag-NP activates JNK protein signaling and death [129] whereas $CeO_2$ NP activates p38 MAPK signaling in bronchoalveolar cells [130].

#### *5.2.4. PTP*

PTPs (protein tyrosine phosphatases) are important regulators in signal-transduction pathways that control proliferation, cell development, differentiation, and transformation. PTP's are extremely reactive cysteine residues that are vulnerable to oxidative stress caused by $H_2O_2$, free radicals, or alterations in the intracellular disulfide/thiol redox state [131, 132]. Metal NP, such as $Zn^{2+}$ and $V^{5+}$, may have a role in PTP redox control by inhibiting MAPK and EGFR [133, 134].

#### *5.2.5. Src*

Src kinases are members of the nonreceptor tyrosine kinase family, which regulates cell development. Src kinase is activated by mild oxidative stress, which then initiates a cell-signaling cascade [135]. This could explain how a modest dosage of metal NP caused lymphocyte cell death through ROS-dependent Src kinase activation [136].

## 6. Synergistic Interaction of Phytochemicals with Chemotherapeutic Drugs

The fundamental benefit of utilizing phytochemicals to make nanoparticles is that they are easier to incorporate into cancer cells. Internalization of NPs into cancer cells may cause an increase in oxidative stress and osmotic pressure, both of which can cause apoptosis.

Internalized nanoparticles may also bind to proteins, cell organelles, and DNA electron transport chain enzymes, causing them to malfunction and causing programmed cell death. It's worth noting that all of these processes could be responsible for nanoparticles' ability to inhibit cancer cells. Thus, antagonistic and synergistic mechanisms are the two types of anticancer treatments that phytosynthesized nanoparticles can demonstrate. Synergistic techniques, on the other hand, use a combination of the two or more of the above-mentioned mechanisms to use nanoparticles to suppress cancer cell development.

Furthermore, nanoparticles are attracted to cancer cells by electrostatic attraction or by disintegration ions can enter cells through ion channels. In cancer treatment, phytochemical and chemotherapeutic combinations are currently exhibiting increased therapeutic responses. Curcumin nanoparticles in combination with cisplatin reduced tumor volume and weight in rat ovarian cancer and suppressed proliferation by downregulating the PI3K/Akt and JAK/STAT3 signaling pathways [137]. Furthermore, cytotoxicity of PEGylated Gold nanoparticles of Doxorubicin in combination with Kaempferol was demonstrated against colon cancer cells (HT29) but not against normal colon cells (HTB-38) [138]. A multi-branched ionic liquid-chitosan grafted mPEG magnetic nanocarrier was created and employed for targeted multidrug delivery of doxorubicin and methotrexate as model anticancer drugs to MCF7 breast cancer cells [139, 140].

Furthermore, by combining two or more anti-tumor mechanisms, combination treatment based on nano-sized drug delivery systems has been established as a promising method. Tian et al. have reported on the utilization of Glycyrrhizic Acid-Modified DSPE-PEG-PEI Nanoparticles for Doxorubicin and Bcl-2 siRNA Co-delivery against liver cancer. Both in vitro (HepG2) and in vivo (H22-bearing mice) research demonstrated that triggering apoptosis could be useful in anti-tumor therapy [141]. In Hep-2 cells and tumor spheroids, a gold nanoparticle-mediated co-delivery of doxorubicin and polo-like kinase 1 (PLK1) siRNA revealed a synergistic impact [142].

Nanomaterials and their application in cancer theranostics have progressed in recent years, but clinical translation has been hampered by numerous difficulties. Multiple capabilities placed in a single NP make the system difficult, which is one of the greatest issues in the nano-theranostic area. Choosing the right imaging and contrast modalities for the specific clinical situation is critical to the success of cancer theranostics nanomedicine. As a result, to make the use of nanomaterials for cancer theranostics easier, an interdisciplinary approach must be used. For future clinical applications of nano-theranostics, drug encapsulation optimization, ligand conjugation efficiency, and excellent repeatability with low-cost biomaterials are required. Furthermore, extensive in vivo testing is essential and crucial. In conclusion, there is a bright perspective for bio-inspired nano-theranostics, and its clinical translation could be accomplished soon to transform cancer therapy, with numerous encouraging pieces of evidence sprouting in recent years where new drug delivery and imaging technologies are produced.

## Conclusion

Cancer stem cells and miRNAs are two newly discovered fundamentals of MDR pathways in cancer. The contribution of several pathways implicated in the development of chemotherapeutic resistance is a future challenge for MDR-related cancer therapy. As a result, the persistence of cancer stem cells even after successful chemotherapy frequently leads to disease relapse at the same or a different site. For full remission, the death of cancer stem cells is as vital as, if not more critical than, the eradication of dividing tumor cells. Resistance pathways and molecules will aid in the development of more effective and curative treatments. A focused therapy delivery, which may combine natural bioactive chemicals with additional first-line anti-tumor chemotherapeutic drugs and/or imaging, contrast, antibodies, and peptides, is necessary for better results. According to the studies, nanotechnology can effectively formulate bioactive, and this nano-formulated bioactive with a potent capacity against various cancer cells has improved efficacy and bioavailability in vivo. However, more study is necessary before this research can be implemented into clinical practice. Before introducing nano-formulations into clinical practice, it is critical to explore all elements of them in depth in research studies.

## References


[1] Jakubczyk K, Dec K, Kałduńska J, Kawczuga D, Kochman J, Janda K. Reactive oxygen species-sources, functions, oxidative damage. *Pol MerkurLekarski.* 2020;48(284):124-7.

[2] Irani K, Xia Y, Zweier JL, Sollott SJ, Der CJ, Fearon ER, Sundaresan M, Finkel T, Goldschmidt-Clermont PJ. Mitogenicsignaling mediated by oxidants in Ras-transformed fibroblasts. *Science.* 1997;275(5306):1649-52.

[3] Bauer G. HOCl-dependent singlet oxygen and hydroxyl radical generation modulate and induce apoptosis of malignant cells. *Anticancer Res.* 2013;33(9):3589-602.

[4] Bauer G. Targeting extracellular ROS signaling of tumor cells. *Anticancer Res.* 2014;34(4):1467-82.

[5] Keyer K, Imlay JA. Superoxide accelerates DNA damage by elevating free-iron levels. *ProcNatlAcadSci U S A.* 1996;93(24):13635-40.

[6] Tu BP, Weissman JS. The FAD-and O2-dependent reaction cycle of Ero1-mediated oxidative protein folding in the endoplasmic reticulum. *Mol Cell.* 2002;10(5):983-94.

[7] Bienert GP, Schjoerring JK, Jahn TP. Membrane transport of hydrogen peroxide. *Biochim Biophys Acta.* 2006;1758(8):994-1003.

[8] Sies H. Hydrogen peroxide as a central redox signaling molecule in physiological oxidative stress: oxidative eustress. *Redox Biol.* 2017;11:613-9.

[9] Bauer G. HOCl and the control of oncogenesis. *J InorgBiochem.* 2018;179:10-23.

[10] Winterbourn CC. Toxicity of iron and hydrogen peroxide: the Fenton reaction. *Toxicol Lett.* 1995;82-83:969-74.

[11] Mittler R. ROS are good. *Trends Plant Sci.* 2017;22(1):11-9.

[12] Cave AC, Brewer AC, Narayanapanicker A, Ray R, Grieve DJ, Walker S, Shah AM. NADPH oxidases in cardiovascular health and disease. *Antioxid Redox Signal.* 2006 May-Jun;8(5-6):691-728.

[13] Chen Q, Vazquez EJ, Moghaddas S, Hoppel CL, Lesnefsky EJ. Production of reactive oxygen species by mitochondria: central role of complex III. *J Biol Chem.* 2003;278(38):36027-31.

[14] Bulua AC, Simon A, Maddipati R, Pelletier M, Park H, Kim KY, Sack MN, Kastner DL, Siegel RM. Mitochondrial reactive oxygen species promote production of proinflammatory cytokines and are elevated in TNFR1-associated periodic syndrome (TRAPS). *J Exp Med.* 2011;208(3):519-33.

[15] Fransen M, Nordgren M, Wang B, Apanasets O. Role of peroxisomes in ROS/RNS-metabolism: implications for human disease. *Biochim Biophys Acta.* 2012;1822(9):1363-73.

[16] Tavender TJ, Springate JJ, Bulleid NJ. Recycling of peroxiredoxin IV provides a novel pathway for disulphide formation in the endoplasmic reticulum. *EMBO J.* 2010;29(24):4185-97.

[17] Malhotra JD, Miao H, Zhang K, Wolfson A, Pennathur S, Pipe SW, Kaufman RJ. Antioxidants reduce endoplasmic reticulum stress and improve protein secretion. *Proc Natl AcadSci* U S A. 2008;105(47):18525-30.

[18] Ellinsworth DC, Sandow SL, Shukla N, Liu Y, Jeremy JY, Gutterman DD. Endothelium-derived hyperpolarization and coronary vasodilation: diverse and integrated roles of epoxyeicosatrienoic acids, hydrogen peroxide, and gap junctions. *Microcirculation.* 2016;23(1):15-32.

[19] Burgoyne JR, Madhani M, Cuello F, Charles RL, Brennan JP, Schröder E, Browning DD, Eaton P. Cysteine redox sensor in PKGIa enables oxidant-induced activation. *Science.* 2007;317(5843):1393-7.

[20] GuSX, Stevens JW, Lentz SR. Regulation of thrombosis and vascular function by protein methionine oxidation. *Blood.* 2015;125(25):3851-9.

[21] Alfadda AA, Sallam RM. Reactive oxygen species in health and disease. *J Biomed Biotechnol.* 2012;2012:936486.

[22] Apostolova N, Victor VM. Molecular strategies for targeting antioxidants to mitochondria: therapeutic implications. *Antioxid Redox Signal.* 2015;22(8):686-729.

[23] Averill-Bates D, Grondin M, Ouellet F. Activation of apoptosis signaling pathways by reactive oxygen species. *Cryobiology.* 2018;80:170.

[24] Sharma P, Jha AÁ, Dubey RS, Pessarakli M, ROS. Oxidative damage and antioxidativedefense mechanism in plants under stressful conditions. *J Bot.* 2012;2012:1-26.

[25] Khan RA, Khan MR, Sahreen S. Brain antioxidant markers, cognitive performance and acetylcholinesterase activity of rats: efficiency of Sonchusasper. *Behav Brain Funct.* 2012;8(1):21.

[26] Lubos E, Loscalzo J, Handy DE. Glutathione peroxidase-1 in health and disease: from molecular mechanisms to therapeutic opportunities. *Antioxid Redox Signal.* 2011;15(7):1957-97.

[27] Olguín-Albuerne M, Morán J. Redox signaling mechanisms in nervous system development. *Antioxid Redox Signal.* 2018;28(18):1603-25.

[28] Ohta Y, Yashiro K, Ohashi K, Imai Y. Disruption of non-enzymatic antioxidant defense systems in the brain of rats with water-immersion restraint stress. *J ClinBiochemNutr.* 2012;51(2):136-42.

[29] Arteaga O, Álvarez A, Revuelta M, Santaolalla F, Urtasun A, Hilario E. Role of antioxidants in neonatal hypoxic–ischemic brain injury: new therapeutic approaches. *Int J Mol Sci.* 2017;18(2):265.

[30] Franchina DG, Dostert C, Brenner D. Reactive oxygen species: involvement in T cell signaling and metabolism. *Trends Immunol.* 2018;39(6):489-502.

[31] Meng TC, Fukada T, Tonks NK. Reversible oxidation and inactivation of protein tyrosine phosphatases in vivo. *Mol Cell.* 2002;9(2):387-99.

[32] Dalle-Donne I, Rossi R, Colombo R, Giustarini D, Milzani A. Biomarkers of oxidative damage in human disease. *Clin Chem.* 2006;52(4):601-23.

[33] Dhalla NS, Temsah RM, Netticadan T. Role of oxidative stress in cardiovascular diseases. *J Hypertens.* 2000;18(6):655-73.

[34] Jenner P. Oxidative stress in Parkinson's disease. *Ann Neurol.* 2003;53;Suppl 3:S26-36; discussion S36.

[35] Sayre LM, Smith MA, Perry G. Chemistry and biochemistry of oxidative stress in neurodegenerative disease. *Curr Med Chem.* 2001;8(7):721-38.

[36] Harman E. Protein oxidation in aging and age-related diseases. *Gerontology.* 1956;11:298-300.

[37] Butterfield DA, Castegna A, Lauderback CM, Drake J. Evidence that amyloid beta-peptide-induced lipid peroxidation and its sequelae in Alzheimer's disease brain contribute to neuronal death. *Neurobiol Aging.* 2002;23(5):655-64.

[38] Jin, L, Yang H. 2006. alph-Synuclein aggregation and Parkinson's disease: Factors affecting the aggregation of alpha-Synuclein;33(4):321-8.

[39] Kukreja RC, Hess ML. The oxygen free radical system: from equations through membrane-protein interactions to cardiovascular injury and protection. *Cardiovasc Res.* 1992;26(7):641-55.

[40] Berry CE, Hare JM. Xanthine oxidoreductase and cardiovascular disease: molecular mechanisms and pathophysiological implications. *J Physiol.* 2004;555(3):589-606.

[41] Hare JM, Stamler JS. NO/redox disequilibrium in the failing heart and cardiovascular system. *J Clin Invest.* 2005;115(3):509-17.

[42] Molavi B, Mehta JL. Oxidative stress in cardiovascular disease: molecular basis of its deleterious effects, its detection, and therapeutic considerations. *CurrOpinCardiol.* 2004;19(5):488-93.

[43] Romero JC, Reckelhoff JF. State-of-the-Art lecture. Role of angiotensin and oxidative stress in essential hypertension. *Hypertension.* 1999;34(4 Pt 2):943-9.

[44] Sowers JR. Hypertension, angiotensin II, and oxidative stress. *N Engl J Med.* 2002;346(25):1999-2001.

[45] Becker LB. New concepts in reactive oxygen species and cardiovascular reperfusion physiology. *Cardiovasc Res.* 2004;61(3):461-70.

[46] Kasparová S, Brezová V, Valko M, Horecký J, Mlynárik V, Liptaj T, Vancová O, Ulicná O, Dobrota D. Study of the oxidative stress in a rat model of chronic brain hypoperfusion. *Neurochem Int.* 2005;46(8):601-11.

[47] Lombardi V, Valko L, Stolc S, Valko M, Ondrejicková O, Horáková L, Placek J, Troncone A. Free radicals in rabbit spinal cord ischemia: electron spin resonance spectroscopy and correlation with SOD activity. *Cell MolNeurobiol.* 1998;18(4):399-412.

[48] Granger DN, Stokes KY, Shigematsu T, Cerwinka WH, Tailor A, Krieglstein CF. Splanchnic ischaemia–reperfusion injury: mechanistic insights provided by mutant mice. *ActaPhysiol Scand.* 2001;173(1):83-91.

[49] Chen Z, Siu B, Ho YS, Vincent R, Chua CC, Hamdy RC, Chua BH. Overexpression of MnSOD protects against myocardial ischemia/reperfusion injury in transgenic mice. *J Mol Cell Cardiol.* 1998;30(11):2281-9.

[50] Jones SP, Hoffmeyer MR, Sharp BR, Ho YS, Lefer DJ. Role of intracellular antioxidant enzymes after in vivo myocardial ischemia and reperfusion. *Am J Physiol Heart Circ Physiol. 2003*;284(1):H277-82.

[51] Bauerova K, Bezek S. Role of reactive oxygen and nitrogen species in etiopathogenesis of rheumatoid arthritis. *Gen PhysiolBiophys.* 2000;18;Spec No:15-20.

[52] Firestein GS, Echeverri F, Yeo M, Zvaifler NJ, Green DR. Somatic mutations in the p53 tumor suppressor gene in rheumatoid arthritis synovium. *ProcNatlAcadSci USA.* 1997;94(20):10895-900.

[53] Maurice MM, Nakamura H, van der Voort EA, van Vliet AI, Staal FJ, Tak PP, Breedveld FC, Verweij CL. Evidence for the role of an altered redox state in hyporesponsiveness of synovial T cells in rheumatoid arthritis. *J Immunol.* 1997;158(3):1458-65.

[54] Cunnane G, FitzGerald O, Beeton C, Cawston TE, Bresnihan B. Early joint erosions and serum levels of matrix metalloproteinase 1, matrix metalloproteinase 3, and tissue inhibitor of metalloproteinases 1 in rheumatoid arthritis. *Arthritis Rheum.* 2001;44(10):2263-74.

[55] Kwong LK, Sohal RS. Substrate and site specificity of hydrogen peroxide generation in mouse mitochondria. *Arch BiochemBiophys.* 1998;350(1):118-26.

[56] Nishikawa T, Edelstein D, Du XL, Yamagishi S, Matsumura T, Kaneda Y, Yorek MA, Beebe D, Oates PJ, Hammes HP, Giardino I, Brownlee M. Normalizing mitochondrial superoxide production blocks three pathways of hyperglycaemic damage. *Nature.* 2000;404(6779):787-90.

[57] Yamagishi SI, Edelstein D, Du XL, Brownlee M. Hyperglycemia potentiates collagen-induced platelet activation through mitochondrial superoxide overproduction. *Diabetes.* 2001;50(6):1491-4.

[58] Babior BM. NADPH oxidase: an update. *Blood.* 1999;93(5):1464-76.

[59] Li JM, Shah AM. ROS generation by nonphagocytic NADPH oxidase: potential relevance in diabetic nephropathy. *J Am SocNephrol.* 2003;14(8);Suppl 3:S221-6.

[60] Robertson RP, Harmon J, Tran PO, Tanaka Y, Takahashi H. Glucose toxicity in β-cells: type 2 diabetes, good radicals gone bad, and the glutathione connection. *Diabetes.* 2003;52(3):581-7.

[61] Marnett LJ. Oxyradicals and DNA damage. *Carcinogenesis.* 2000;21(3):361-70.

[62] Valko M, Rhodes CJB, Moncol J, Izakovic MM, Mazur M. Free radicals, metals and antioxidants in oxidative stress-induced cancer. *ChemBiol Interact.* 2006;160(1):1-40.

[63] Valko M, Morris H, Mazúr M, Rapta P, Bilton RF. Oxygen free radical generating mechanisms in the colon: do the semiquinones of vitamin K play a role in the aetiology of colon cancer? *BiochimBiophysActa.* 2001;1527(3):161-6.

[64] Valko M, Izakovic M, Mazur M, Rhodes CJ, Telser J. Role of oxygen radicals in DNA damage and cancer incidence. *Mol Cell Biochem.* 2004;266(1-2):37-56.

[65] Poli G, Leonarduzzi G, Biasi F, Chiarpotto E. Oxidative stress and cell signalling. *Curr Med Chem.* 2004;11(9):1163-82.

[66] Drevs J, Medinger M, Schmidt-Gersbach C, Weber R, Unger C. Receptor tyrosine kinases: the main targets for new anticancer therapy. *Curr Drug Targets.* 2003;4(2):113-21.

[67] Sharma A. Chemoresistance in cancer cells: exosomes as potential regulators of therapeutic tumor heterogeneity. *Nanomedicine (Lond).* 2017;12(17):2137-48.

[68] Abdullah LN, Chow EKH. Mechanisms of chemoresistance in cancer stem cells. *ClinTransl Med.* 2013;2(1):3.

[69] Chen H, Zhu B, Zhao L, Liu Y, Zhao F, Feng J, Jin Y, Sun J, Geng R, Wei Y. Allicin Inhibits Proliferation and Invasion in Vitro and in Vivo via SHP-1-Mediated STAT3 Signaling in Cholangiocarcinoma. *Cell PhysiolBiochem.* 2018;47(2):641-653.

[70] Li J, Zhang C, Jiang H, Cheng J. Andrographolide inhibits hypoxia-inducible factor-1 through phosphatidylinositol 3-kinase/AKT pathway and suppresses breast cancer growth. *Onco Targets Ther.* 2015;8:427-35.

[71] Dou J, Wang Z, Ma L, Peng B, Mao K, Li C, Su M, Zhou C, Peng G. Baicalein and baicalin inhibit colon cancer using two distinct fashions of apoptosis and senescence. *Oncotarget.* 2018;9(28):20089-20102.

[72] Shankar S, Srivastava RK. Involvement of Bcl-2 family members, phosphatidylinositol 3′-kinase/AKT and mitochondrial p53 in curcumin (diferulolylmethane)-induced apoptosis in prostate cancer. *Int J Oncol.* 2007;30(4):905-18.

[73] Qiao Q, Jiang Y, Li G. Inhibition of the PI3K/AKT-NF-κB pathway with curcumin enhanced radiation-induced apoptosis in human Burkitt's lymphoma. *J Pharmacol Sci.* 2013;121(4):247-56.

[74] Li M, Zhang Z, Hill DL, Wang H, Zhang R. Curcumin, a dietary component, has anticancer, chemosensitization, and radiosensitization effects by down-regulating the MDM2 oncogene through the PI3K/mTOR/ETS2 pathway. *Cancer Res.* 2007;67(5):1988-96.

[75] Zhang W, Su J, Xu H, Yu S, Liu Y, Zhang Y, Sun L, Yue Y, Zhou X. Dicumarol inhibits PDK1 and targets multiple malignant behaviors of ovarian cancer cells. *PLoS One.* 2017;12(6):e0179672.

[76] Kwak TW, Park SB, Kim HJ, Jeong YI, Kang DH. Anticancer activities of epigallocatechin-3-gallate against cholangiocarcinoma cells. *Onco Targets Ther.* 2017;10:137-44.

[77] Tuli HS, Tuorkey MJ, Thakral F, Sak K, Kumar M, Sharma AK, Sharma U, Jain A, Aggarwal V, Bishayee A. Molecular Mechanisms of Action of Genistein in Cancer: Recent Advances. *Front Pharmacol.* 2019;10:1336.

[78] Sophia J, Kowshik J, Dwivedi A, Bhutia SK, Manavathi B, Mishra R, Nagini S. Nimbolide, a neem limonoid inhibits cytoprotective autophagy to activate apoptosis via modulation of the PI3K/Akt/GSK-3β signalling pathway in oral cancer. *Cell Death Dis.* 2018;9(11):1087.

[79] Pei HL, Mu DM, Zhang B. Anticancer activity of pterostilbene in human ovarian cancer cell lines. *Med SciMonit.* 2017;23:3192-9.

[80] Honari M, Shafabakhsh R, Reiter RJ, Mirzaei H, Asemi Z. Resveratrol is a promising agent for colorectal cancer prevention and treatment: focus on molecular mechanisms. *Cancer Cell Int.* 2019;19(1):180.

[81] Wang J, Li J, Cao N, Li Z, Han J, Li L. Resveratrol, an activator of SIRT1, induces protective autophagy in non-small-cell lung cancer via inhibiting Akt/mTOR and activating p38-MAPK. *Onco Targets Ther.* 2018;11:7777-86.

[82] Feng XM, Su XL. Anticancer effect of ursolic acid via mitochondria dependent pathways. *OncolLett.* 2019;17(6):4761-7.

[83] McKenna MK, Gachuki BW, Alhakeem SS, Oben KN, Rangnekar VM, Gupta RC, Bondada S. Anti-cancer activity of withaferin A in B-cell lymphoma. *Cancer BiolTher.* 2015;16(7):1088-98.

[84] NgandeuNeubi GMN, Opoku-Damoah Y, Gu X, Han Y, Zhou J, Ding Y. Bio-inspired drug delivery systems: an emerging platform for targeted cancer therapy. *Biomater Sci.* 2018;6(5):958-73.

[85] Lemire JA, Harrison JJ, Turner RJ. Antimicrobial activity of metals: mechanisms, molecular targets and applications. *Nat Rev Microbiol.* 2013;11(6):371-84.

[86] Thakur NS, Bhaumik J, Kirar S, Banerjee UC. Development of gold-based phototheranosticnanoagents through a bioinspired route and their applications in photodynamic therapy. *ACS Sustain Chem Eng.* 2017:597950-7960.

[87] Rafique M, Sadaf I, Rafique MS, Tahir MB. A review on green synthesis of silver nanoparticles and their applications. *Artif Cells NanomedBiotechnol.* 2017;45(7):1272-91.

[88] Hassani A, Azarian MMS, Ibrahim WN, Hussain SA. Preparation, characterization and therapeutic properties of gum arabic-stabilized gallic acid nanoparticles. *Sci Rep.* 2020;10(1):17808.

[89] Luo H, Jiang B, Li B, Li Z, Jiang BH, Chen YC. Kaempferol nanoparticles achieve strong and selective inhibition of ovarian cancer cell viability. *Int J Nanomedicine.* 2012;7:3951-9.

[90] He M, Ro L, Liu J, Chu CC. Folate-decorated arginine-based poly (ester urea urethane) nanoparticles as carriers for gambogic acid and effect on cancer cells. *J Biomed Mater Res A.* 2017;105(2):475-90.

[91] Farcas CG, Dehelean C, Pinzaru IA, Mioc M, Socoliuc V, Moaca EA, Avram S, Ghiulai R, Coricovac D, Pavel I, Alla PK, Cretu OM, Soica C, Loghin F. Thermosensitive Betulinic Acid-Loaded Magnetoliposomes: A Promising Antitumor Potential for Highly Aggressive Human Breast Adenocarcinoma Cells Under Hyperthermic Conditions. *Int J Nanomedicine.* 2020;15:8175-8200.

[92] Hsieh DS, Wang H, Tan SW, Huang YH, Tsai CY, Yeh MK, Wu CJ. The treatment of bladder cancer in a mouse model by epigallocatechin-3-gallate-gold nanoparticles. *Biomaterials.* 2011;32(30):7633-40.

[93] Mehanna MM, Sarieddine R, Alwattar JK, Chouaib R, Gali-Muhtasib H. Anticancer activity of thymoquinone cubic phase nanoparticles against human breast cancer: formulation, cytotoxicity and subcellular localization. *Int J Nanomedicine.* 2020;15:9557-70.

[94] Alkhathlan AH, Al-Abdulkarim HA, Khan M, Khan M, Alkholief M, Alshamsan A, Almomen A, Albekairi N, Alkhathlan HZ, Siddiqui MRH. Evaluation of the Anticancer Activity of Phytomolecules Conjugated Gold Nanoparticles Synthesized by Aqueous Extracts of *Zingiberofficinale* (Ginger) and *Nigella sativa* L. Seeds (Black Cumin). *Materials (Basel).* 2021;14(12):3368.

[95] Krug P, Mielczarek L, Wiktorska K, Kaczyńska K, Wojciechowski P, Andrzejewski K, Ofiara K, Szterk A, Mazur M. Sulforaphane-conjugated selenium nanoparticles: towards a synergistic anticancer effect. *Nanotechnology.* 2019;30(6):065101.

[96] Kumari M, Sharma N, Manchanda R, Gupta N, Syed A, Bahkali AH, Nimesh S. PGMD/curcumin nanoparticles for the treatment of breast cancer. *Sci Rep.* 2021;11(1):3824.

[97] Shah HS, Nasrullah U, Zaib S, Usman F, Khan A, Gohar UF, Uddin J, Khan I, Al-Harrasi A. Preparation, Characterization, and Pharmacological Investigation of Withaferin-A Loaded Nanosponges for Cancer Therapy; In Vitro, In Vivo and Molecular Docking Studies. *Molecules.* 2021;26(22):6990.

[98] Naik JB, Pardeshi SR, Patil RP, Patil PB, Mujumdar A. Mucoadhesive micro-/Nano carriers in ophthalmic drug delivery: an overview. *BioNanoScience.* 2020;10(3):564-82.

[99] Saraf S, Jain A, Tiwari A, Verma A, Panda PK, Jain SK. Advances in liposomal drug delivery to cancer: an overview. *J Drug DelivSci Technol.* 2020;56:101549.

[100] Jo MJ, Jin IS, Park CW, Hwang BY, Chung YB, Kim JS, Shin DH. Revolutionizing technologies of nanomicelles for combinatorial anticancer drug delivery. *Arch Pharm Res.* 2020;43(1):100-109.

[101] Nasirizadeh S, Malaekeh-Nikouei B. Solid lipid nanoparticles and nanostructured lipid carriers in oral cancer drug delivery. *J Drug DelivSci Technol.* 2020;55:101458.

[102] Hwang JH, Choi CW, Chang YS, Choe YH, Park WS, Shin SM, Lee M, Lee SI. The efficacy of clinical strategies to reduce nosocomial sepsis in extremely low birth weight infants. *J Korean Med Sci.* 2005;20(2):177-81.

[103] Son KH, Hong JH, Lee JW. Carbon nanotubes as cancer therapeutic carriers and mediators. *Int J Nanomedicine.* 2016;11:5163-85.

[104] Jeevanandam J, Kulabhusan PK, Sabbih G, Akram M, Danquah MK. Phytosynthesized nanoparticles as a potential cancer therapeutic agent. *Biotech.* 2020;10(12):535.

[105] Goel A, Bhatia AK. Phytosynthesized nanoparticles for effective cancer treatment: a review. *NANOASIA.* 2019;9(4):437-43.

[106] Jadhav K, Deore S, Dhamecha D, H R R, Jagwani S, Jalalpure S, Bohara R. Phytosynthesis of Silver Nanoparticles: Characterization, Biocompatibility Studies, and Anticancer Activity. *ACS BiomaterSci Eng.* 2018;4(3):892-899.

[107] Maynard AD, Baron PA, Foley M, Shvedova AA, Kisin ER, Castranova V. Exposure to carbon nanotube material: aerosol release during the handling of unrefined single-walled carbon nanotube material. *J Toxicol Environ Health* A. 2004;67(1):87-107.

[108] Donaldson K, Murphy FA, Duffin R, Poland CA. Asbestos, carbon nanotubes and the pleural mesothelium: a review of the hypothesis regarding the role of long fibre retention in the parietal pleura, inflammation and mesothelioma. *Part Fibre Toxicol.* 2010;7(1):1-17.

[109] Nel A, Xia T, Mädler L, Li N. Toxic potential of materials at the nanolevel. *Science.* 2006;311(5761):622-7.
[110] Huang YW, Wu CH, Aronstam RS. Toxicity of transition metal oxide nanoparticles: recent insights from in vitro studies. *Materials (Basel).* 2010;3(10):4842-59.
[111] Stella GM. Carbon nanotubes and pleural damage: perspectives of nanosafety in the light of asbestos experience. *Biointerphases.* 2011;6(2):1-P17.
[112] Shvedova AA, Pietroiusti A, Fadeel B, Kagan VE. Mechanisms of carbon nanotube-induced toxicity: focus on oxidative stress. *ToxicolApplPharmacol.* 2012;261(2):121-33.
[113] Ju-Nam Y, Lead JR. Manufactured nanoparticles: an overview of their chemistry, interactions and potential environmental implications. *Sci Total Environ.* 2008;400(1-3):396-414.
[114] Li N, Xia T, Nel AE. The role of oxidative stress in ambient particulate matter-induced lung diseases and its implications in the toxicity of engineered nanoparticles. *Free RadicBiol Med.* 2008;44(9):1689-99.
[115] Stone V, Johnston H, Clift MJ. Air pollution, ultrafine and nanoparticle toxicology: cellular and molecular interactions. *IEEE Trans Nanobioscience.* 2007;6(4):331-40.
[116] Johnston HJ, Hutchison G, Christensen FM, Peters S, Hankin S, Stone V. A review of the in vivo and in vitro toxicity of silver and gold particulates: particle attributes and biological mechanisms responsible for the observed toxicity. *Crit Rev Toxicol.* 2010;40(4):328-46.
[117] Ye J, Zhang X, Young HA, Mao Y, Shi X. Chromium(VI)-induced nuclear factor-κB activation in intact cells via free radical reactions. *Carcinogenesis.* 1995;16(10):2401-5.
[118] Allen RG, Tresini M. Oxidative stress and gene regulation. *Free RadicBiol Med.* 2000;28(3):463-99.
[119] Byrne JD, Baugh JA. The significance of nanoparticles in particle-induced pulmonary fibrosis. *Mcgill J Med.* 2008;11(1):43-50.
[120] Pujalté I, Passagne I, Brouillaud B, Tréguer M, Durand E, Ohayon-Courtès C, L'Azou B. Cytotoxicity and oxidative stress induced by different metallic nanoparticles on human kidney cells. *Part Fibre Toxicol.* 2011;8:10.
[121] Murray AR, Kisin ER, Tkach AV, Yanamala N, Mercer R, Young SH, Fadeel B, Kagan VE, Shvedova AA. Factoring-in agglomeration of carbon nanotubes and nanofibers for better prediction of their toxicity versus asbestos. *Part Fibre Toxicol.* 2012;9:10.
[122] Hubbard AK, Timblin CR, Shukla A, Rincón M, Mossman BT. Activation of NF-κB-dependent gene expression by silica in lungs of luciferase reporter mice. *Am J Physiol Lung Cell Mol Physiol.* 2002;282(5):L968-75.
[123] Ding M, Shi X, Lu Y, Huang C, Leonard S, Roberts J, Antonini J, Castranova V, Vallyathan V. Induction of activator protein-1 through reactive oxygen species by crystalline silica in JB6 cells. *J Biol Chem.* 2001;276(12):9108-14.
[124] Shi H, Hudson LG, Liu KJ. Oxidative stress and apoptosis in metal ion-induced carcinogenesis. *Free RadicBiol Med.* 2004;37(5):582-93.
[125] Poljak-Blazi M, Jaganjac M, Mustapic M, Pivac N, Muck-Seler D. Acute immunomodulatory effects of iron polyisomaltosate in rats. *Immunobiology.* 2009;214(2):121-8.
[126] Guyton KZ, Liu Y, Gorospe M, Xu Q, Holbrook NJ. Activation of mitogen-activated protein kinase by H2O2: Role in cell survival following oxidant injury. *J Biol Chem.* 1996;271(8):4138-42.
[127] Tournier C, Thomas G, Pierre J, Jacquemin C, Pierre M, Saunier B. Mediation by arachidonic acid metabolites of the H2O2-induced stimulation of mitogen-activated protein kinases (extracellular-signal-regulated kinase and c-Jun NH2-terminal kinase). *Eur J Biochem.* 1997;244(2):587-95.
[128] Son Y, Cheong YK, Kim NH, Chung HT, Kang DG, Pae HO. Mitogen-activated protein kinases and reactive oxygen species: how can ROS activate MAPK pathways? *J Signal Transduct.* 2011;2011:792639.
[129] Hsin YH, Chen CF, Huang S, Shih TS, Lai PS, Chueh PJ. The apoptotic effect of nanosilver is mediated by a ROS- and JNK-dependent mechanism involving the mitochondrial pathway in NIH3T3 cells. *ToxicolLett.* 2008;179(3):130-9.
[130] Park EJ, Choi J, Park YK, Park K. Oxidative stress induced by cerium oxide nanoparticles in cultured BEAS-2B cells. *Toxicology.* 2008;245(1-2):90-100.

[131] Genestra M. Oxyl radicals, redox-sensitive signalling cascades and antioxidants. *Cell Signal.* 2007;19(9):1807-19.

[132] Barrett DM, Black SM, Todor H, Schmidt-Ullrich RK, Dawson KS, Mikkelsen RB. Inhibition of protein-tyrosine phosphatases by mild oxidative stresses is dependent on S-nitrosylation. *J Biol Chem.* 2005;280(15):14453-61.

[133] Kim YM, Reed W, Wu W, Bromberg PA, Graves LM, Samet JM. Zn2+-induced IL-8 expression involves AP-1, JNK, and ERK activities in human airway epithelial cells. *Am J Physiol Lung Cell Mol Physiol.* 2006;290(5):L1028-35.

[134] Tal TL, Graves LM, Silbajoris R, Bromberg PA, Wu W, Samet JM. Inhibition of protein tyrosine phosphatase activity mediates epidermal growth factor receptor signaling in human airway epithelial cells exposed to Zn2+. *ToxicolApplPharmacol.* 2006;214(1):16-23.

[135] Esposito F, Chirico G, Montesano Gesualdi N, Posadas I, Ammendola R, Russo T, Cirino G, Cimino F. Protein kinase B activation by reactive oxygen species is independent of tyrosine kinase receptor phosphorylation and requires SRC activity. *J Biol Chem.* 2003;278(23):20828-34.

[136] Balamurugan K, Rajaram R, Ramasami T, Narayanan S. Chromium(III)-induced apoptosis of lymphocytes: death decision by ROS and Src-family tyrosine kinases. *Free RadicBiol Med.* 2002;33(12):1622-40.

[137] Sandhiutami NMD, Arozal W, Louisa M, Rahmat D, Wuyung PE. Curcumin nanoparticle enhances the anticancer effect of cisplatin by inhibiting PI3K/AKT and JAK/STAT3 pathway in rat ovarian carcinoma induced by DMBA. *Front Pharmacol.* 2021;11:2199.

[138] Meena D, Vimala K, Kannan S. Combined delivery of DOX and kaempferol using pegylated gold nanoparticles to target colon cancer. *J Clust Sci.* 2021:1-15.

[139] Rahimi M, Shafiei-Irannejad V, D Safa KD, Salehi R. Multi-branched ionic liquid-chitosan as a smart and biocompatible nano-vehicle for combination chemotherapy with stealth and targeted properties. *CarbohydrPolym.* 2018;196:299-312.

[140] Zhang D, Ye Z, Liu H, Wang X, Hua J, Ling Y, Wei L, Xia Y, Sun S, Xiao L. Cell membrane coated smart two-dimensional supraparticle for *in vivo* homotypic cancer targeting and enhanced combinational theranostics. *Nanotheranostics.* 2021;5(3):275-287.

[141] Tian G, Pan R, Zhang B, Qu M, Lian B, Jiang H, Gao Z, Wu J. Liver-Targeted Combination Therapy Basing on Glycyrrhizic Acid-Modified DSPE-PEG-PEI Nanoparticles for Co-delivery of Doxorubicin and Bcl-2 siRNA. *Front Pharmacol.* 2019;10:4.

[142] Shrestha B, Wang L, Zhang H, Hung CY, Tang L. Gold nanoparticles mediated drug-gene combinational therapy for breast cancer treatment. *Int J Nanomedicine.* 2020;15:8109-19.

## Chapter 11

# Reactive Oxygen Species and Cancer: A Split Self Affair

**Sangramjit Mondal**
**Aiindrila Dhara**
**Debkumar Pal**
**and Nirmalya Sen***
Molecular Oncology Laboratory, Department of Biological Sciences, Bose Institute, Kolkata, India

## Abstract

Reactive Oxygen Species (ROS) are oxygen intermediates that are ubiquitously produced in almost all biological reactions. Unique balance between production and clearance of ROS protects us from various diseases including cancers. Being strong oxidizing agents, loss of ROS homeostasis often results in DNA damage, mutations and inflammatory environment which promotes oncogenic processes like metastasis, angiogenesis and proliferation. ROS also alters cancer hallmarks like tumor suppressor, immune signals and metabolic reprogramming in various cancers. Paradoxically, high ROS due to effects of cancer therapeutic drugs often results in cancer cell apoptosis. While antioxidant genes may prevent such ROS induced apoptosis and act as oncogenes in many cancers. The role of ROS in cancer has become controversial owing to its differential effect on the fates of various tumors. In this chapter, we have summarized types of ROS species, various mechanisms of ROS generation and clearance, and their regulatory effects in cancer progression. Along with this, article sums up several therapeutic regimes which are repurposed to induce ROS and subsequent cell death in cancers.

**Keywords**: ROS, free radical, DNA damage, oncogene, cancer, vitamins, antioxidants

## 1. Introduction: A Blind Date

Free radicles in biological world comprise of Reactive Oxygen Species (ROS)and Reactive Nitrogen Species (RNS) and their derivatives that act as powerful signalling molecules and regulate numerous biological processes [1]. Endogenous ROS is often generated within cells

* Corresponding Author's Emai: nirmalya@jcbose.ac.in.

In: The Role of Reactive Oxygen Species in Health and Disease
Editor: Kuladip Jana
ISBN: 979-8-89113-283-2

as a by-product of various metabolic pathways in organelles like mitochondria, peroxisomes, endoplasmic reticulum and cytosol [1, 2]. Often termed as "double edged sword," low ROS levels are associated with chronic granulomatous disease and certain autoimmune disorders [3, 4]. Again, high ROS have been reported in diabetes, neurodegenerative diseases, cardiac diseases, arthritis and cancers [1, 2, 4]. Initiation and progression of cancers due to abnormal ROS is widely studied and currently ROS related pathways are reported in almost all known human cancers. Till date, National Library of Medicine search engine; Pubmed scored for almost ~52,000 publications from the keyword 'Reactive Oxygen Species and Cancer,' indicating the importance of ROS in the disease of cancer (Figure 1). While ROS are often generated during normal cellular processes like respiration, metabolism, cell division, etc; the mechanism of ROS clearance is maintained by default antioxidant functions within cells [5]. Thus, differentiating cancer causing ROS from physiologically produced ROS is difficult. In general, low to moderate levels of ROS works as a signal transducer and can initiate oncogenic processes like cell proliferation, angiogenesis, metastasis, drug resistance, hypoxia etc [6, 7]. Nevertheless, elevated ROS levels are associated with DNA damage, gene mutations, chromosomal aberration, genomic instability, loss of immune surveillance [8-10]. This chapter will be focusing on factors deregulating ROS homeostasis that result in cancers [11].

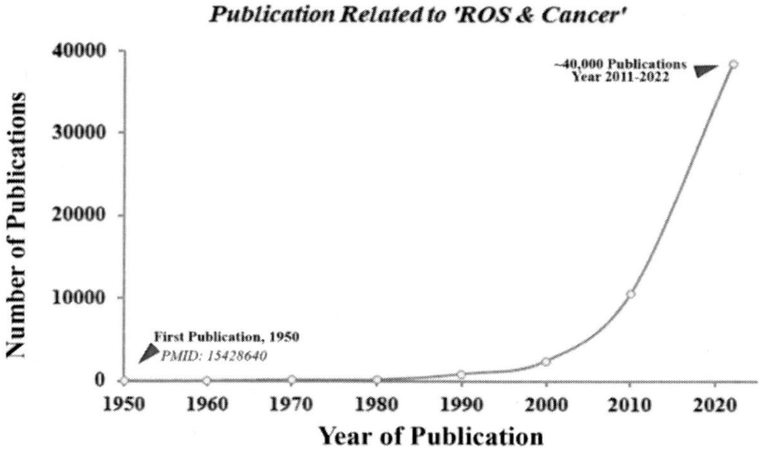


**Figure 1**. Importance of ROS in cancer research. Graph indicating number of publications per 10-year interval (till July2022) for word query 'Reactive Oxygen Species and Cancer' in Pubmed (https://pubmed.ncbi.nlm.nih.gov/).

## 2. Types of ROS- You Are Not My Type

Molecular oxygen often transitions between various free radicles/reactive states like singlet oxygen ($^1O_2$), superoxide radicle ($O_2^{\cdot -}$), perhydroxyl radicle ($HO_2.$), hydroxyl radicle (.OH), alkoxy radical ($RO^{\cdot}$) and gets stabilize via electron transfer mechanisms [1, 12]. Apart from free radicles, species capable of releasing such oxygen intermediates like hydrogen peroxide ($H_2O_2$), hypochlorous acid (HOCl), ozone ($O_3$), nitrous acid ($HNO_2$), nitrosylcation ($NO^+$), nitroxyl anion ($NO^-$) and organic peroxides (ROOH) are also involved in oncogenesis [13]. Table 1 summarizes the types of ROS, their functions and chemistry. Free radicles like superoxide anion ($O2^{-}$), singlet oxygen ($^1O_2$), and hydroxyl radical (•OH) etc are more often implicated in various cancers [14-17].

**Table 1.** Types of reactive oxygen species and their possible mechanism of generation

| ROS species | Chemistry/generation | Association with cancer | Reference |
|---|---|---|---|
| $O_2^{-*}$ | $O_2 + e^- \rightarrow O_2^{-*}$ | DNA Damage and mutation | [11, 18] |
| $H_2O_2$ | $O_2^{-*} + H^+ \rightarrow H_2O_2^*$ | Mutation, Oncogene upregulation, cell proliferation | |
| HO* | $O_2^{-*} + H_2O_2 \rightarrow O_2 + HO_2^- + HO^*$<br>(Haber-Weiss Reaction)<br>$Fe^{2+} + H_2O_2 \rightarrow Fe^{3+} + HO^- + HO^*$<br>(Fenton Reaction) | DNA Damage, Angiogenesis, metastasis | |
| Species Capable of Generating Free Radicals | | | |
| RO* | RO-H ⟹ RO-TM<br>Transition Metals<br>$RO^*$<br>Strong Oxidative Condition | Oxidizes amino acids, denature proteins. | [19] |
| HOCL | $H_2O_2 + Cl^- + H^+ \rightarrow HOCl + H_2O$<br>$HOClOCl^- + \leftrightarrow H^+$. | DNA Damage & mutation, DNA repair enzyme pathways | [20, 21] |
| $^1O_2$ | $HOCl + H_2O_2 \rightarrow {}^1O_2 + Cl^- + H^+ + H_2O$ | DNA Damage | [20] |

## 3. Regulatory Mechanism of ROS in Oncogenesis- Affair with Cancers

Persistent oxidative stress due to exogenous or endogenous ROS generation results in promotion of various cancers including breast cancer, liver cancer, lung cancer, melanoma, ovarian cancer and pancreatic cancer (22-27). Oxidative environment is a hub for mutations and cellular damage which initiates various cancer hallmarks as indicated in Figure 2 [28].

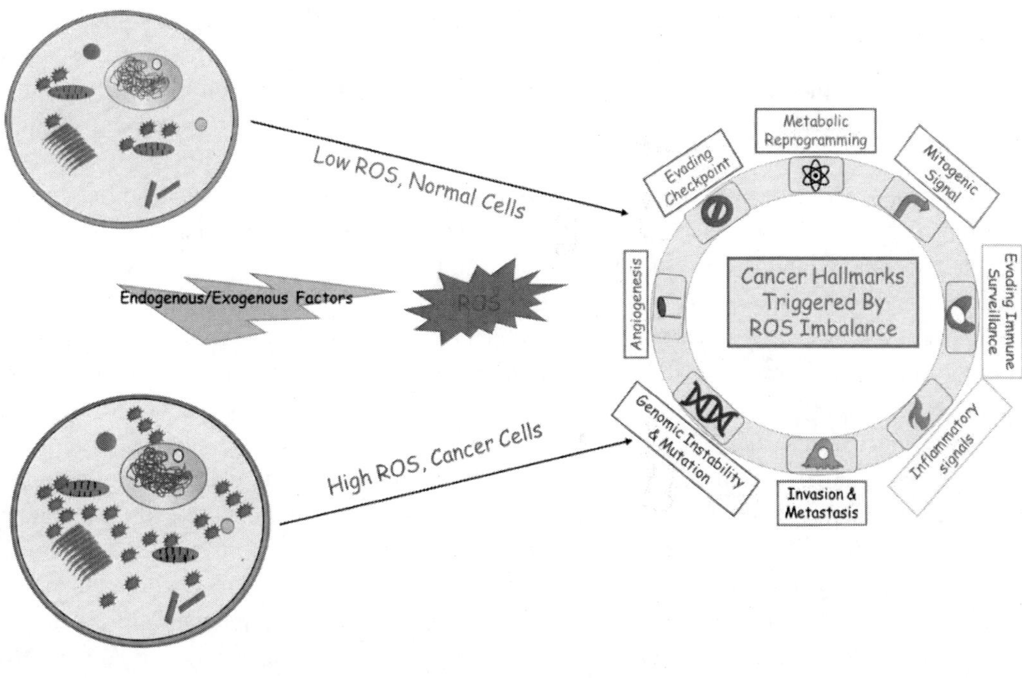


**Figure 2**. ROS can trigger oncogenesis via altering cancer Hallmarks.

Here we describe the potential alliance of ROS with the cancer hallmarks i.e; i) Mutation and genomic instability: Free radicles promote DNA damage thereby causing mutations and chromosomal aberrations. DNA damage repair functions like base excision are often affected by oxidative stress resulting in tumour initiation [29, 30]. Conversion of guanine to 8-oxo-7,8-dihydroguanine is the most usual mutation that often results in disruption of protein binding, telomere maintenance and genomic stability resulting in cancers [31-34]. Environmental carcinogens often promote DNA adducts under oxidative stress which transmits oncogenic signals in proliferating cells [33]. ii) Inflammatory signals and evasion of tumour suppressor: Continuous oxidative stress due to heightened ROS often led to inflammation, activation of oncogenes and inactivation of tumour suppressor genes [35]. For example, proto-oncogenes *c-MYC, c-FOS* are reported to be induced by oxidative stress while tumour suppressors like *Tp53* and *PTEN* are inactivated in high ROS environment [36-41]. iii) Metastasis: It is reported that high ROS mediated hyper-methylation of E-Cadherin promoter results in metastasis [42]. Interestingly, inflammatory signals like transforming growth factor β (*TGFβ*) pathway regulated via ROS promotes metastasis in pancreatic cancer [43, 44]. iv) Sustained proliferative signals: High ROS interferes with several mitogenic signals like *MAPK, PI3K/AKT* which are directly associated with cell proliferation, resistance and transformation phenotypes in breast cancer and leukemias [45-49]. v) Angiogenesis: Growing tumours are dependent on blood vessels, for nutrient and signals and ROS plays many roles in promoting the process of angiogenesis. High ROS can induce *VEGF* and *MMP3* mediated angiogenesis and genomic instability by various means [50]. Firstly, NADPH oxidase (*NOX* genes) derived ROS initiates VEGF receptor phosphorylation resulting in activation of angiogenic transcriptional program [51]. Secondly, hypoxia in tumours generates ROS which in turn induces *HIFα* that directly activates *VEGF* signalling resulting in angiogenesis. Thirdly, *VEGF* independent, ROS mediated angiogenesis can be initiated via altered lipid oxidation by-products having pro-angiogenic activities [52]. vi) Escape Immune surveillance: Diabolically, ROS are mediators of immune response like phagocytosis, antigen presentation, differentiation while exerting immunosuppressive/evasive effects on immune cells within the "tumour microenvironment" (TME) of solid tumours. For example neutrophils can reprogram mitochondrial metabolism to derive a) nutrient for tumour growth and maintain immune suppression, b) promote metastatic colonization of tumour cells by supressing NK cells and c) undergo NETosis (Neutrophil extracellular traps) utilizing plasma redox imbalance releasing a battery of inflammatory stimulus resulting in metastasis [53-56]. Macrophages on the corollary undergo polarization depending upon stimulus and can provide pro-tumorigenic cues [57, 58]. Finally, Myeloid-derived suppressor cells (MDSCs); a mixed population of immature myeloid cells were shown to suppress T cells and NK cells via ROS-dependent generation of peroxynitrite ($ONOO^-$) and promote immune suppression in tumours [59, 60]. vii) Metabolic reprogramming: Another cancer hallmark, cellular energetics and metabolic reprogramming is also associated with ROS in promoting cancers [61, 62]. High ROS levels can reduce oxidative phosphorylation by stabilization of HIF1-α, and cysteine modifications favouring aerobic glycolysis and survival [63]. Several metabolic enzymes of glycolysis, kerb cycle are modulated by ROS in cancers [49, 61, 64, 65]. Lipid oxidation and associated DNA damage due to altered metabolism triggered by ROS is known to cause cancers [66]. Owing to their allegiance with hallmarks of cancer, ROS can be associated with the initiation, progression and maintenance of the disease [67]. Figure 3 indicates the molecular/genetic events associated with increasing ROS levels that dictates the pathophysiology of cancers.

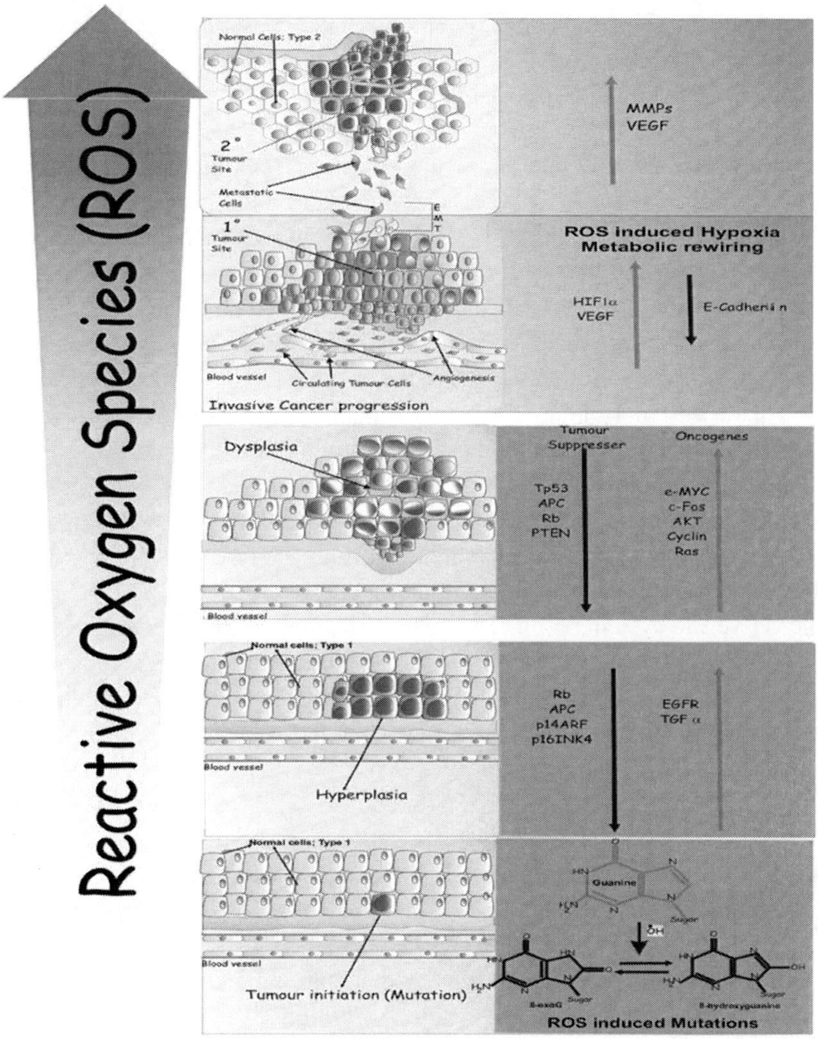


**Figure 3.** The role of ROS in cancer initiation and progression.

## 4. ROS Generation and ROS Clearance: Gifts and Breakups

There are varied signals for ROS generation which can be primarily divided into endogenous and exogenous signals. For endogenous ROS, mitochondrial metabolism acts as the primary source in mammalian system where Electron Transport Chain (ETC) produces superoxide (O2*-) at mitochondrial complex I, II & III via reduction of oxygen [68-72]. Other than ETC, endogenous catalytic enzymes such as NADPH oxidases (*NOX*), Xanthine Oxidase (*XO*), Lipoxygenases (*LOX*) and Cytochrome P-450 can generate various reactive oxygen intermediates [73-79]. Interestingly, NOX family (*NOX1-5*) of genes results in cancers as they can generate both superoxide and hydrogen peroxide [80, 81]. Enzymes Dual-oxidase 1 and 2 (*DUOX1, DUOX2*) that are capable of producing only hydrogen peroxide are also implicated in cancers [82, 83]. Apart from this, superoxide anion is also produced at endoplasmic reticulum during oxidative protein folding by protein disulfide isomerases and oxidoreductin-1enzyme

[84]. Exogenous factors like radiation, xenobiotics, and environmental pollutants often results in ROS production [85-88].

Since ROS is harmful and causes damage to cells and organs, scavenging of ROS becomes inevitable [89-91]. Several antioxidant genes like superoxide-dismutase (*SOD*), Catalase (*CAT*), peroxiredoxin (*PRX*), Glutathione peroxidase (*GPX*), etc actively clear excess ROS from the cellular milieu [51, 92, 93]. SODs are the major antioxidant defense systems against superoxide anions and consist of three isoforms in humans: the cytoplasmic Cu/ZnSOD (*SOD1*), the mitochondrial MnSOD (*SOD2*), and the extracellular Cu/ZnSOD (*SOD3*), which require catalytic transition metal (Cu, Zn or Mn) for their activation [51]. The catalysis of $O2^{\bullet -}$ to $H_2O_2$ by SOD involves alternate redox reaction against the transition metal, such as copper (Cu) and manganese (Mn) at the active site of the enzyme [94, 95]. Next, Catalase converts the hydrogen peroxide into molecular oxygen and water by a two-step reaction using Ferric ion at its active centre [92]. Peroxiredoxins are Cysteine active centre containing ROS scavenger enzymes found in all animals [96]. Mammalian cells express six isoforms of Prx as per locations; with *Prx I, II*, and *VI* being localized mainly in the cytosol; *Prx III* being restricted to mitochondria; *Prx IV* being found predominantly in the endoplasmic reticulum (ER); and *Prx V* being present in the cytosol, mitochondria, and peroxisomes [93]. Hydrogen peroxide oxidizes the active site cysteine of Prx in a redox reaction, thus neutralizing ROS. Meanwhile, oxidized Prx can be reduced by a cellular thiol, thus enabling Prx to function as a locally constrained peroxidase [93]. Additionally, glutathione (GSH) and GSH/glutathione peroxidase (GPX) enzyme system maintains an important cell redox pathway Other than neutralizing hydrogen peroxide (H2O2) and peroxynitrite (ONOO-), GSH and GPX enzymes act as buffers to protect proteins and organelles from modifications [97]. Most GPx enzymes contain active selenocysteine that uses GSH as an obligate co-substrate in the reduction of hydrogen peroxide to water. Albeit all GPxs don't use GSH, rather some of these enzymes are functionally identified as thioredoxin-dependent peroxidases containing a redox-Cystine in active centre [98].

Although essential for ROS clearance and prevention of cancers, SODs, GPx, Prx, Cat enzymes often get deregulated and causes cancers [99]. For example, SOD1 and 2 cause breast and lung cancers [100-102]. Catalase was reported to cause aggressive glioblastoma [103]. Peroxiredoxins are involved in oncogenesis, chemo-resistance and can act as predictive biomarkers in various cancers [104-106]. PrxII is known to regulate cancer stem cells and their associated cancer properties [107]. Like other ROS clearance factors, GPxs have a dichotomous role in different tumor types, acting either as a tumor suppressor or a pro-survival protein [108-111]. Table 2 summarizes the major anti-oxidant genes, ROS scavenging mechanisms and association with cancers.

Non-enzymatic cofactors like vitamins including ascorbic acids (Vit C), α-tocopherol (Vit-E), Lipoic acid, glutathione can act as scavengers of free radicles from living systems [112-116]. Glutathione has antioxidant properties since it contains reducing thiol groups that undergo redox reactions. In cells, glutathione is maintained in the reduced form by the enzyme glutathione reductase and in turn reduces other metabolites, enzymes and oxidants [117-119]. Vitamins like C, E, D and A intake has shown reduced cancer occurrence, increased cancer prevention and better prognosis in various studies indicating the importance of ROS scavenging effects of these cofactors in cancers [120-123].

**Table 2.** Various antioxidant genes/isoforms that are used by cells for ROS clearance functions

| ROS Scavenger genes | Isoforms | Mechanism | Cancers | Reference |
|---|---|---|---|---|
| *CAT* | No isoforms | Breaks down hydrogen peroxide into $H_2O$ and $O_2$. | Leukaemia, gastric cancer, colorectal cancer, prostate cancer | [124, 125] |
| *SOD* | *SOD1, SOD2* | Catabolizes Superoxide anion to $O_2$ & $H_2O_2$ | Lung cancer | [126, 127] |
| *PRX* | *PRDX1, PRDX3, PRDX4, PRDX5, PRDX6* | Neutralizes hydroperoxides to water. | Breast cancer, lung cancer, bladder cancer, prostate cancer | [128, 129] |
| *GPX* | *GPX1, GPX2* | Converts $H_2O_2$ to $H_2O$.<br>$H_2O_2$+ GSH<br>$H_2O$ + GSSG | Prostate cancer, lung Cancer, colorectal cancer, ovarian cancer | [98, 130] |
| *SESN* | *SESN1, SESN2, SESN3* | Sestrine catalyses PRX regeneration via regulating various transcription factors | Endometrial Cancer, Colorectal Cancer, Lung Cancer | [131, 132] |

## 5. Therapeutics Utilizing ROS in Killing Cancers: Menage a Trois

Variety of cancers are addicted to an environment of oxidative stress while others reprogram antioxidant axis to curtail their ROS (108-111). Table 3 summarizes ROS dependent mechanisms that promote oncogenesis. Since high levels of ROS triggers cell death, therapeutics that induces ROS are often used in treatment of cancers [3, 133-141]. Most of the therapeutics targeting cancer hallmarks, either works via increasing ROS or by decreasing antioxidants and altering redox homeostasis. Table 4 provides a list of such drugs and their usage in various cancers.

**Table 3.** ROS dependent mechanisms in various cancers

| Cancer | ROS Mechanism | Reference |
|---|---|---|
| Pancreatic ductal adenocarcinoma (PDAC) | TIGAR gene expression in PDAC controls premalignant initiation and metastasis via phenotypic switches induced by ROS. While KRAS driven PDAC initiation are supported by ROS triggered growth factor activation. | PMID: 31983610<br>PMID: 27215184 |
| Breast cancer | MYC and MCL1 driven ROS generation via mitochondrial OXPHOS results in tumor maintenance and drug resistance. | PMID: 35571115<br>PMID: 28978427 |
| Thyroid cancer | ROS generated due to activity of NADPH oxidase 1; NOX1 promotes AKT dependent cancer cell renewal | PMID: 30690057 |
| Prostate cancer | Elevated levels of NADPH oxidases 5; NOX5, increases ROS production and tumor growth | PMID: 33062666<br>PMID: 25559363 |
| Lung cancer | ROS in lung cancer promotes inflammatory signals that support tumor growth. For example, NADPH oxidase 4; NOX4 gene stimulates ROS mediated cytokine production and tumor growth. | PMID: 18176884<br>PMID: 30769285 |

**Table 4.** Therapeutics that work via alteration of ROS homeostasis in cancer

| Mode of Action | Drug | Cancer |
|---|---|---|
| Direct ROS generation | Doxorubicin | Follicular lymphoma, breast, colon cancer |
| | Cisplatin | Testicular, bladder, ovarian, cervical cancer |
| | 2-Methoxyestradiol | Renal cell carcinoma, ovarian, prostate cancer |
| Antioxidant process inhibition | Buthioninesulfoximine | Neuroblastoma, melanoma |
| | Imexon | Myeloma, melanoma, pancreatic cancer |
| Indirectly affect redox metabolism | Celecoxib: 1 | Colorectal, hepatocellulatar, breast, lung cancer |
| | Nelfinavir | Cervical cancer<br>Adenoid Cystic Cancer of the Head and Neck<br>NSCLC |
| | Bortezomib | Myeloma |
| | Fenretinide (4-hydroxyphenyl retinamide) | Breast, Bladder, Prostate, Head & Neck Carcinoma, Lung, Renal Cancers |

## 6. Discussion: Why in a Relationship? ROS a Necessary Evil

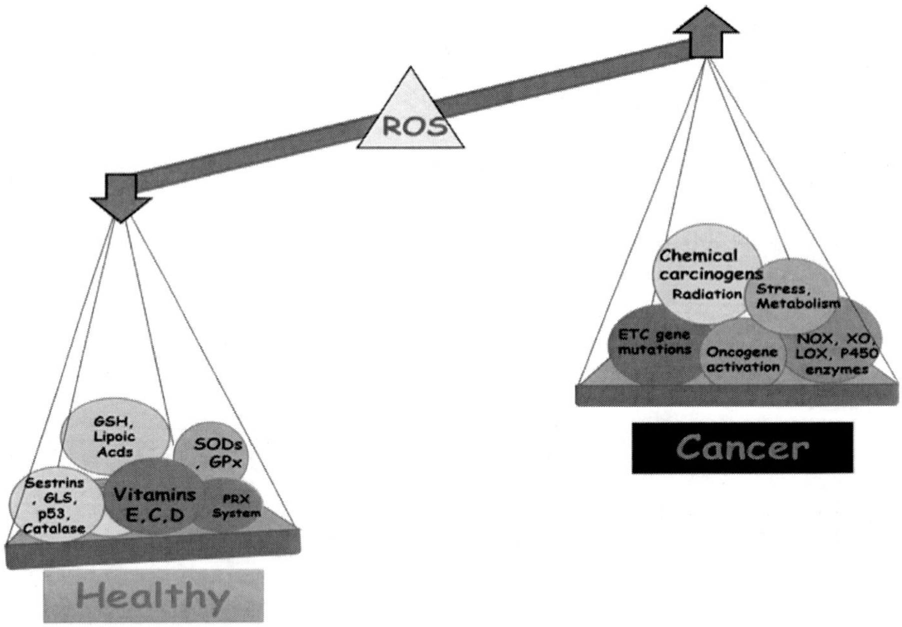


**Figure 4.** The balance of ROS predicts outcomes between Heath and Disease.

Free radicles catalyse many physiologically important reactions and hence they play crucial roles in human health [142-144]. At normal levels ROS participates in physiological functions such as metabolism, signal transduction, proliferation, immune system regulation and vascular remodelling [97, 144-152]. High ROS essentially promotes apoptosis, senescence and other cell growth inhibitory mechanisms which act as a preventive measure against diseases [3, 133-141]. On the other hand, high ROS is also responsible for initiation and progression of many diseases such as diabetes, neurodegenerative diseases, cardiac diseases, arthritis and cancers [1, 153-163]. Given this dual nature of ROS, it can be considered as ‘a necessary evil’ for living systems [28, 137, 164-166]. The interesting facet lies with the balance between useful and harmful ROS that predicts outcomes between healthy individuals and cancer patients Figure 4. Both genetic and environmental factors that generate or quench ROS are important in maintaining such balance and averting diseases. Several cohort studies on cancer patients regarding the role of food-based antioxidants like vitamins and supplements indicate a paradigm shift towards an antioxidant rich lifestyle that might prevent cancer worldwide [120-123]. In conclusion, our work here tries to connect the various aspects of cancer initiation and progression that are often triggered by reactive oxygen species and may include molecular processes like mutations, gene activation/inactivation and altered cancer hallmarks.

## Conclusion: Every End Has a Beginning

Even though ROS is involved in pathogenesis of various diseases including cancer, only one antioxidant drug "Edaravone" had been clinically approved amyotrophic lateral sclerosis (ALS) in 2017. While some cancers are addicted to high ROS and other thrive better in a non-oxidative milieu, our lifestyle, genetic makeup and environmental interactions play a major role in regulation of redox balance and prevention of cancers. Nonetheless, ROS scavengers (NCT01921426) and ROS generating small molecules (NCT00522834, NCT00592046) are currently under clinical trial for cancer treatments. These give a hope that ROS homeostasis may hold the key to successful cancer therapies.

## Acknowledgments

Apologies to those whose related publications were not cited due to space limitations. The authors acknowledge Bose Institute, India for supporting study. N. S. was funded by DBT-Ramalingaswami Fellowship (Award sanction No: BT/RLF/Re-entry/27/2016) from Department of Biotechnology, India and SERB-DST ECR Award (Award sanction No: ECR/2018/000595) from Department of Science and Technology, India. D. P was funded by scholarship from UGC-NET (Award Number211610182027). The authors have no financial conflict of interest to report.

## References

[1] Phaniendra A, Jestadi DB, Periyasamy L. Free radicals: properties, sources, targets, and their implication in various diseases. *Indian J Clin Biochem*. 2015;30(1):11-26.

[2] Snezhkina AV, Kudryavtseva AV, Kardymon OL, Savvateeva MV, Melnikova NV, Krasnov GS, ROS Generation and Antioxidant Defense Systems in Normal and Malignant Cells. *Oxid Med Cell Longev*. 2019;2019:6175804.

[3] Brieger K, Schiavone S, Miller FJ, Jr., Krause KH. Reactive oxygen species: from health to disease. *Swiss Med Wkly*. 2012;142:w13659.

[4] Martin KR, Barrett JC. Reactive oxygen species as double-edged swords in cellular processes: low-dose cell signaling versus high-dose toxicity. *Hum Exp Toxicol*. 2002;21(2):71-5.

[5] Mittler R. ROS Are Good. *Trends Plant Sci*. 2017;22(1):11-9.

[6] Kim YW, Byzova TV. Oxidative stress in angiogenesis and vascular disease. *Blood.* 2014;123(5):625-31.
[7] Xia C, Meng Q, Liu LZ, Rojanasakul Y, Wang XR, Jiang BH. Reactive oxygen species regulate angiogenesis and tumor growth through vascular endothelial growth factor. *Cancer Res.* 2007;67(22):10823-30.
[8] Dizdaroglu M, Jaruga P, Birincioglu M, Rodriguez H. Free radical-induced damage to DNA: mechanisms and measurement. *Free Radic Biol Med.* 2002;32(11):1102-15.
[9] Kennel KB, Greten FR. Immune cell - produced ROS and their impact on tumor growth and metastasis. *Redox Biol.* 2021;42:101891.
[10] Nakamura H, Takada K. Reactive oxygen species in cancer: Current findings and future directions. *Cancer Sci.* 2021;112(10):3945-52.
[11] Galadari S, Rahman A, Pallichankandy S, Thayyullathil F. Reactive oxygen species and cancer paradox: To promote or to suppress? *Free Radic Biol Med.* 2017;104:144-64.
[12] Li R, Jia Z, Trush MA. Defining ROS in Biology and Medicine. *React Oxyg Species* (Apex). 2016;1(1):9-21.
[13] Genestra M. Oxyl radicals, redox-sensitive signalling cascades and antioxidants. *Cell Signal.* 2007;19(9):1807-19.
[14] Di Mascio P, Martinez GR, Miyamoto S, Ronsein GE, Medeiros MHG, Cadet J. Singlet Molecular Oxygen Reactions with Nucleic Acids, Lipids, and Proteins. *Chem Rev.* 2019;119(3):2043-86.
[15] Schumacker PT. Reactive oxygen species in cancer cells: live by the sword, die by the sword. *Cancer Cell.* 2006;10(3):175-6.
[16] Schumacker PT. Reactive oxygen species in cancer: a dance with the devil. *Cancer Cell.* 2015;27(2):156-7.
[17] Shah MA, Rogoff HA. Implications of reactive oxygen species on cancer formation and its treatment. *Semin Oncol.* 2021;48(3):238-45.
[18] Evans JL, Goldfine ID. Alpha-lipoic acid: a multifunctional antioxidant that improves insulin sensitivity in patients with type 2 diabetes. *Diabetes Technol Ther.* 2000;2(3):401-13.
[19] Jia K, Zhang F, Huang H, Chen Y. Visible-Light-Induced Alkoxyl Radical Generation Enables Selective C(sp(3))-C(sp(3)) Bond Cleavage and Functionalizations. *J Am Chem Soc.* 2016;138(5):1514-7.
[20] Panasenko OM, Gorudko IV, Sokolov AV. Hypochlorous acid as a precursor of free radicals in living systems. *Biochemistry* (Mosc). 2013;78(13):1466-89.
[21] Pullar JM, Vissers MC, Winterbourn CC. Living with a killer: the effects of hypochlorous acid on mammalian cells. *IUBMB Life.* 2000;50(4-5):259-66.
[22] Azad N, Rojanasakul Y, Vallyathan V. Inflammation and lung cancer: roles of reactive oxygen/nitrogen species. *J Toxicol Environ Health B Crit Rev.* 2008;11(1):1-15.
[23] Brown NS, Bicknell R. Hypoxia and oxidative stress in breast cancer. Oxidative stress: its effects on the growth, metastatic potential and response to therapy of breast cancer. *Breast Cancer Res.* 2001;3(5):323-7.
[24] Calvisi DF, Ladu S, Hironaka K, Factor VM, Thorgeirsson SS. Vitamin E down-modulates iNOS and NADPH oxidase in c-Myc/TGF-alpha transgenic mouse model of liver cancer. *J Hepatol.* 2004;41(5):815-22.
[25] Chan DW, Liu VW, Tsao GS, Yao KM, Furukawa T, Chan KK. Loss of MKP3 mediated by oxidative stress enhances tumorigenicity and chemoresistance of ovarian cancer cells. *Carcinogenesis.* 2008;29(9):1742-50.
[26] Edderkaoui M, Hong P, Vaquero EC, Lee JK, Fischer L, Friess H. Extracellular matrix stimulates reactive oxygen species production and increases pancreatic cancer cell survival through 5-lipoxygenase and NADPH oxidase. *Am J Physiol Gastrointest Liver Physiol.* 2005;289(6):G1137-47.
[27] Fruehauf JP, Trapp V. Reactive oxygen species: an Achilles' heel of melanoma? *Expert Rev Anticancer Ther.* 2008;8(11):1751-7.
[28] Zablocka A, Janusz M. [The two faces of reactive oxygen species]. *Postepy Hig Med Dosw* (Online). 2008;62:118-24.

[29] Maynard S, Schurman SH, Harboe C, de Souza-Pinto NC, Bohr VA. Base excision repair of oxidative DNA damage and association with cancer and aging. *Carcinogenesis*. 2009;30(1):2-10.
[30] Melis JP, van Steeg H, Luijten M. Oxidative DNA damage and nucleotide excision repair. *Antioxid Redox Signal*. 2013;18(18):2409-19.
[31] Coluzzi E, Colamartino M, Cozzi R, Leone S, Meneghini C, O'Callaghan N. Oxidative stress induces persistent telomeric DNA damage responsible for nuclear morphology change in mammalian cells. *PLoS One*. 2014;9(10):e110963.
[32] Fouquerel E, Barnes RP, Uttam S, Watkins SC, Bruchez MP, Opresko PL. Targeted and Persistent 8-Oxoguanine Base Damage at Telomeres Promotes Telomere Loss and Crisis. *Mol Cell*. 2019;75(1):117-30 e6.
[33] Swenberg JA, Lu K, Moeller BC, Gao L, Upton PB, Nakamura J. Endogenous versus exogenous DNA adducts: their role in carcinogenesis, epidemiology, and risk assessment. *Toxicol Sci*. 2011;120 Suppl 1:S130-45.
[34] von Zglinicki T. Oxidative stress shortens telomeres. Trends Biochem Sci. 2002;27(7):339-44.
[35] Hayes JD, Dinkova-Kostova AT, Tew KD. Oxidative Stress in Cancer. *Cancer Cell*. 2020;38(2):167-97.
[36] Amstad P, Crawford D, Muehlematter D, Zbinden I, Larsson R, Cerutti P. Oxidants stress induces the proto-oncogenes, C-fos and C-myc in mouse epidermal cells. *Bull Cancer*. 1990;77(5):501-2.
[37] Cao J, Schulte J, Knight A, Leslie NR, Zagozdzon A, Bronson R. Prdx1 inhibits tumorigenesis via regulating PTEN/AKT activity. *EMBO J*. 2009;28(10):1505-17.
[38] Hussain SP, Aguilar F, Amstad P, Cerutti P. Oxy-radical induced mutagenesis of hotspot codons 248 and 249 of the human p53 gene. *Oncogene*. 1994;9(8):2277-81.
[39] Kwon D, Choi C, Lee J, Kim KO, Kim JD, Kim SJ. Hydrogen peroxide triggers the expression of Fas/FasL in astrocytoma cell lines and augments apoptosis. *J Neuroimmunol*. 2001;113(1):1-9.
[40] Kwon J, Lee SR, Yang KS, Ahn Y, Kim YJ, Stadtman ER. Reversible oxidation and inactivation of the tumor suppressor PTEN in cells stimulated with peptide growth factors. *Proc Natl Acad Sci USA*. 2004;101(47):16419-24.
[41] Yuan TL, Cantley LC. PI3K pathway alterations in cancer: variations on a theme. *Oncogene*. 2008;27(41):5497-510.
[42] Lim SO, Gu JM, Kim MS, Kim HS, Park YN, Park CK. Epigenetic changes induced by reactive oxygen species in hepatocellular carcinoma: methylation of the E-cadherin promoter. *Gastroenterology*. 2008;135(6):2128-40, 40 e1-8.
[43] Cencioni C, Comunanza V, Middonti E, Vallariello E, Bussolino F. The role of redox system in metastasis formation. *Angiogenesis*. 2021;24(3):435-50.
[44] Chang CH, Pauklin S. ROS and TGFbeta: from pancreatic tumour growth to metastasis. *J Exp Clin Cancer Res*. 2021;40(1):152.
[45] Koundouros N, Poulogiannis G. Phosphoinositide 3-Kinase/Akt Signaling and Redox Metabolism in Cancer. *Front Oncol*. 2018;8:160.
[46] Niwa K, Inanami O, Yamamori T, Ohta T, Hamasu T, Kuwabara M. Redox regulation of PI3K/Akt and p53 in bovine aortic endothelial cells exposed to hydrogen peroxide. *Antioxid Redox Signal*. 2003;5(6):713-22.
[47] Okoh VO, Felty Q, Parkash J, Poppiti R, Roy D. Reactive oxygen species via redox signaling to PI3K/AKT pathway contribute to the malignant growth of 4-hydroxy estradiol-transformed mammary epithelial cells. *PLoS One*. 2013;8(2):e54206.
[48] Silva A, Girio A, Cebola I, Santos CI, Antunes F, Barata JT. Intracellular reactive oxygen species are essential for PI3K/Akt/mTOR-dependent IL-7-mediated viability of T-cell acute lymphoblastic leukemia cells. *Leukemia*. 2011;25(6):960-7.
[49] Zhao Y, Hu X, Liu Y, Dong S, Wen Z, He W. ROS signaling under metabolic stress: cross-talk between AMPK and AKT pathway. *Mol Cancer*. 2017;16(1):79.
[50] Radisky DC, Levy DD, Littlepage LE, Liu H, Nelson CM, Fata JE. Rac1b and reactive oxygen species mediate MMP-3-induced EMT and genomic instability. *Nature*. 2005;436(7047):123-7.
[51] Ushio-Fukai M, Nakamura Y. Reactive oxygen species and angiogenesis: NADPH oxidase as target for cancer therapy. *Cancer Lett*. 2008;266(1):37-52.

[52] Shatrov VA, Sumbayev VV, Zhou J, Brune B. Oxidized low-density lipoprotein (oxLDL) triggers hypoxia-inducible factor-1alpha (HIF-1alpha) accumulation via redox-dependent mechanisms. *Blood.* 2003;101(12):4847-9.

[53] El-Hag A, Clark RA. Down-regulation of human natural killer activity against tumors by the neutrophil myeloperoxidase system and hydrogen peroxide. *J Immunol.* 1984;133(6):3291-7.

[54] Inoue M, Nakashima R, Enomoto M, Koike Y, Zhao X, Yip K. Plasma redox imbalance caused by albumin oxidation promotes lung-predominant NETosis and pulmonary cancer metastasis. *Nat Commun.* 2018;9(1):5116.

[55] Li P, Lu M, Shi J, Hua L, Gong Z, Li Q. Dual roles of neutrophils in metastatic colonization are governed by the host NK cell status. *Nat Commun.* 2020;11(1):4387.

[56] Rice CM, Davies LC, Subleski JJ, Maio N, Gonzalez-Cotto M, Andrews C. Tumour-elicited neutrophils engage mitochondrial metabolism to circumvent nutrient limitations and maintain immune suppression. *Nat Commun.* 2018;9(1):5099.

[57] Locati M, Curtale G, Mantovani A. Diversity, Mechanisms, and Significance of Macrophage Plasticity. *Annu Rev Pathol.* 2020;15:123-47.

[58] Murray PJ. Macrophage Polarization. *Annu Rev Physiol.* 2017;79:541-66.

[59] Lu T, Gabrilovich DI. Molecular pathways: tumor-infiltrating myeloid cells and reactive oxygen species in regulation of tumor microenvironment. *Clin Cancer Res.* 2012;18(18):4877-82.

[60] Talmadge JE. Pathways mediating the expansion and immunosuppressive activity of myeloid-derived suppressor cells and their relevance to cancer therapy. *Clin Cancer Res.* 2007;13(18 Pt 1):5243-8.

[61] Bhardwaj V, He J. Reactive Oxygen Species, Metabolic Plasticity, and Drug Resistance in Cancer. *Int J Mol Sci.* 2020;21(10).

[62] Kim J, Kim J, Bae JS. ROS homeostasis and metabolism: a critical liaison for cancer therapy. *Exp Mol Med.* 2016;48(11):e269.

[63] Tonks NK. Redox redux: revisiting PTPs and the control of cell signaling. *Cell.* 2005;121(5):667-70.

[64] Ghanbari Movahed Z, Rastegari-Pouyani M, Mohammadi MH, Mansouri K. Cancer cells change their glucose metabolism to overcome increased ROS: One step from cancer cell to cancer stem cell? *Biomed Pharmacother.* 2019;112:108690.

[65] Rodic S, Vincent MD. Reactive oxygen species (ROS) are a key determinant of cancer's metabolic phenotype. *Int J Cancer.* 2018;142(3):440-8.

[66] Bartsch H, Nair J. Chronic inflammation and oxidative stress in the genesis and perpetuation of cancer: role of lipid peroxidation, DNA damage, and repair. *Langenbecks Arch Surg.* 2006;391(5):499-510.

[67] Reuter S, Gupta SC, Chaturvedi MM, Aggarwal BB. Oxidative stress, inflammation, and cancer: how are they linked? *Free Radic Biol Med.* 2010;49(11):1603-16.

[68] Muller FL, Liu Y, Van Remmen H. Complex III releases superoxide to both sides of the inner mitochondrial membrane. *J Biol Chem.* 2004;279(47):49064-73.

[69] Quinlan CL, Orr AL, Perevoshchikova IV, Treberg JR, Ackrell BA, Brand MD. Mitochondrial complex II can generate reactive oxygen species at high rates in both the forward and reverse reactions. *J Biol Chem.* 2012;287(32):27255-64.

[70] Saybasili H, Yuksel M, Haklar G, Yalcin AS. Effect of mitochondrial electron transport chain inhibitors on superoxide radical generation in rat hippocampal and striatal slices. *Antioxid Redox Signal.* 2001;3(6):1099-104.

[71] Staniek K, Gille L, Kozlov AV, Nohl H. Mitochondrial superoxide radical formation is controlled by electron bifurcation to the high and low potential pathways. *Free Radic Res.* 2002;36(4):381-7.

[72] Takeshige K. [Superoxide formation and lipid peroxidation by the mitochondrial electron-transfer chain]. *Rinsho Shinkeigaku.* 1994;34(12):1269-71.

[73] Battelli MG, Polito L, Bortolotti M, Bolognesi A. Xanthine Oxidoreductase-Derived Reactive Species: Physiological and Pathological Effects. *Oxid Med Cell Longev.* 2016;2016:3527579.

[74] Cantu-Medellin N, Kelley EE. Xanthine oxidoreductase-catalyzed reactive species generation: A process in critical need of reevaluation. *Redox Biol.* 2013;1:353-8.

[75] Cho KJ, Seo JM, Kim JH. Bioactive lipoxygenase metabolites stimulation of NADPH oxidases and reactive oxygen species. *Mol Cells.* 2011;32(1):1-5.

[76] Li X, Fang P, Mai J, Choi ET, Wang H, Yang XF. Targeting mitochondrial reactive oxygen species as novel therapy for inflammatory diseases and cancers. *J Hematol Oncol*. 2013;6:19.

[77] Veith A, Moorthy B. Role of Cytochrome P450s in the Generation and Metabolism of Reactive Oxygen Species. *Curr Opin Toxicol*. 2018;7:44-51.

[78] Weidinger A, Mullebner A, Paier-Pourani J, Banerjee A, Miller I, Lauterbock L. Vicious inducible nitric oxide synthase-mitochondrial reactive oxygen species cycle accelerates inflammatory response and causes liver injury in rats. *Antioxid Redox Signal*. 2015;22(7):572-86.

[79] Zangar RC, Davydov DR, Verma S. Mechanisms that regulate production of reactive oxygen species by cytochrome P450. *Toxicol Appl Pharmacol*. 2004;199(3):316-31.

[80] Block K, Gorin Y. Aiding and abetting roles of NOX oxidases in cellular transformation. *Nat Rev Cancer*. 2012;12(9):627-37.

[81] Kamata T. Roles of Nox1 and other Nox isoforms in cancer development. *Cancer Sci.* 2009;100(8):1382-8.

[82] Faria CC, Fortunato RS. The role of dual oxidases in physiology and cancer. *Genet Mol Biol.* 2020;43(1 suppl. 1):e20190096.

[83] Little AC, Sulovari A, Danyal K, Heppner DE, Seward DJ, van der Vliet A. Paradoxical roles of dual oxidases in cancer biology. *Free Radic Biol Med*. 2017;110:117-32.

[84] Bhandary B, Marahatta A, Kim HR, Chae HJ. An involvement of oxidative stress in endoplasmic reticulum stress and its associated diseases. *Int J Mol Sci*. 2012;14(1):434-56.

[85] Henkler F, Brinkmann J, Luch A. The role of oxidative stress in carcinogenesis induced by metals and xenobiotics. *Cancers* (Basel). 2010;2(2):376-96.

[86] Lakey PS, Berkemeier T, Tong H, Arangio AM, Lucas K, Poschl U. Chemical exposure-response relationship between air pollutants and reactive oxygen species in the human respiratory tract. *Sci Rep*. 2016;6:32916.

[87] Lodovici M, Bigagli E. Oxidative stress and air pollution exposure. *J Toxicol*. 2011;2011:487074.

[88] Nuszkiewicz J, Wozniak A, Szewczyk-Golec K. Ionizing Radiation as a Source of Oxidative Stress-The Protective Role of Melatonin and Vitamin D. *Int J Mol Sci*. 2020;21(16).

[89] He L, He T, Farrar S, Ji L, Liu T, Ma X. Antioxidants Maintain Cellular Redox Homeostasis by Elimination of Reactive Oxygen Species. *Cell Physiol Biochem*. 2017;44(2):532-53.

[90] Poljsak B, Suput D, Milisav I. Achieving the balance between ROS and antioxidants: when to use the synthetic antioxidants. *Oxid Med Cell Longev*. 2013;2013:956792.

[91] Schieber M, Chandel NS. ROS function in redox signaling and oxidative stress. *Curr Biol.* 2014;24(10):R453-62.

[92] Goyal MM, Basak A. Human catalase: looking for complete identity. *Protein Cell*. 2010;1(10):888-97.

[93] Rhee SG, Woo HA, Kil IS, Bae SH. Peroxiredoxin functions as a peroxidase and a regulator and sensor of local peroxides. *J Biol Chem*. 2012;287(7):4403-10.

[94] Culotta VC, Yang M, O'Halloran TV. Activation of superoxide dismutases: putting the metal to the pedal. *Biochim Biophys Acta*. 2006;1763(7):747-58.

[95] Perry JJ, Shin DS, Getzoff ED, Tainer JA. The structural biochemistry of the superoxide dismutases. *Biochim Biophys Acta*. 2010;1804(2):245-62.

[96] Detienne G, De Haes W, Mergan L, Edwards SL, Temmerman L, Van Bael S. Beyond ROS clearance: Peroxiredoxins in stress signaling and aging. *Ageing Res Rev*. 2018;44:33-48.

[97] Di Meo S, Reed TT, Venditti P, Victor VM. Role of ROS and RNS Sources in Physiological and Pathological Conditions. *Oxid Med Cell Longev*. 2016;2016:1245049.

[98] Lubos E, Loscalzo J, Handy DE. Glutathione peroxidase-1 in health and disease: from molecular mechanisms to therapeutic opportunities. *Antioxid Redox Signal*. 2011;15(7):1957-97.

[99] Chung WH. Unraveling new functions of superoxide dismutase using yeast model system: Beyond its conventional role in superoxide radical scavenging. *J Microbiol*. 2017;55(6):409-16.

[100] Gomez ML, Shah N, Kenny TC, Jenkins EC, Jr., Germain D. SOD1 is essential for oncogene-driven mammary tumor formation but dispensable for normal development and proliferation. *Oncogene*. 2019;38(29):5751-65.

[101] Liu S, Li B, Xu J, Hu S, Zhan N, Wang H. SOD1 Promotes Cell Proliferation and Metastasis in Non-small Cell Lung Cancer via an miR-409-3p/SOD1/SETDB1 Epigenetic Regulatory Feedforward Loop. *Front Cell Dev Biol.* 2020;8:213.

[102] Papa L, Hahn M, Marsh EL, Evans BS, Germain D. SOD2 to SOD1 switch in breast cancer. *J Biol Chem.* 2014;289(9):5412-6.

[103] Flor S, Oliva CR, Ali MY, Coleman KL, Greenlee JD, Jones KA. Catalase Overexpression Drives an Aggressive Phenotype in Glioblastoma. *Antioxidants* (Basel). 2021;10(12).

[104] Lee SB, Ho JN, Yoon SH, Kang GY, Hwang SG, Um HD. Peroxiredoxin 6 promotes lung cancer cell invasion by inducing urokinase-type plasminogen activator via p38 kinase, phosphoinositide 3-kinase, and Akt. *Mol Cells.* 2009;28(6):583-8.

[105] Nicolussi A, D'Inzeo S, Capalbo C, Giannini G, Coppa A. The role of peroxiredoxins in cancer. *Mol Clin Oncol.* 2017;6(2):139-53.

[106] Zhu H, Tao X, Zhou L, Sheng B, Zhu X, Zhu X. Expression of thioredoxin 1 and peroxiredoxins in squamous cervical carcinoma and its predictive role in NACT. *BMC Cancer.* 2019;19(1):865.

[107] Chandimali N, Jeong DK, Kwon T. Peroxiredoxin II Regulates Cancer Stem Cells and Stemness-Associated Properties of Cancers. *Cancers* (Basel). 2018;10(9).

[108] Arsova-Sarafinovska Z, Matevska N, Eken A, Petrovski D, Banev S, Dzikova S. Glutathione peroxidase 1 (GPX1) genetic polymorphism, erythrocyte GPX activity, and prostate cancer risk. *Int Urol Nephrol.* 2009;41(1):63-70.

[109] Bansal A, Simon MC. Glutathione metabolism in cancer progression and treatment resistance. *J Cell Biol.* 2018;217(7):2291-8.

[110] Chang C, Worley BL, Phaeton R, Hempel N. Extracellular Glutathione Peroxidase GPx3 and Its Role in Cancer. *Cancers* (Basel). 2020;12(8).

[111] Strycharz-Dudziak M, Foltyn S, Dworzanski J, Kielczykowska M, Malm M, Drop B. Glutathione Peroxidase (GPx) and Superoxide Dismutase (SOD) in Oropharyngeal Cancer Associated with EBV and HPV Coinfection. *Viruses.* 2020;12(9).

[112] Clarke MW, Burnett JR, Croft KD. Vitamin E in human health and disease. *Crit Rev Clin Lab Sci.* 2008;45(5):417-50.

[113] Padayatty SJ, Katz A, Wang Y, Eck P, Kwon O, Lee JH. Vitamin C as an antioxidant: evaluation of its role in disease prevention. *J Am Coll Nutr.* 2003;22(1):18-35.

[114] Samoylenko A, Hossain JA, Mennerich D, Kellokumpu S, Hiltunen JK, Kietzmann T. Nutritional countermeasures targeting reactive oxygen species in cancer: from mechanisms to biomarkers and clinical evidence. *Antioxid Redox Signal.* 2013;19(17):2157-96.

[115] Traber MG, Stevens JF. Vitamins C and E: beneficial effects from a mechanistic perspective. *Free Radic Biol Med.* 2011;51(5):1000-13.

[116] Vineetha RC, Archana V, Binu P, Arathi P, Nair RH. L-Ascorbic Acid and alpha-Tocopherol Reduces Hepatotoxicity Associated with Arsenic Trioxide Chemotherapy by Modulating Nrf2 and Bcl2 Transcription Factors in Chang liver Cells. *Nutr Cancer.* 2018;70(4):684-96.

[117] Forman HJ, Zhang H, Rinna A. Glutathione: overview of its protective roles, measurement, and biosynthesis. *Mol Aspects Med.* 2009;30(1-2):1-12.

[118] Meister A. Glutathione metabolism and its selective modification. *J Biol Chem.* 1988;263(33): 17205-8.

[119] Wu G, Fang YZ, Yang S, Lupton JR, Turner ND. Glutathione metabolism and its implications for health. *J Nutr.* 2004;134(3):489-92.

[120] Chen Z, Huang Y, Cao D, Qiu S, Chen B, Li J. Vitamin C Intake and Cancers: An Umbrella Review. *Front Nutr.* 2021;8:812394.

[121] Garland CF, Garland FC, Gorham ED, Lipkin M, Newmark H, Mohr SB. The role of vitamin D in cancer prevention. *Am J Public Health.* 2006;96(2):252-61.

[122] Mamede AC, Tavares SD, Abrantes AM, Trindade J, Maia JM, Botelho MF. The role of vitamins in cancer: a review. *Nutr Cancer.* 2011;63(4):479-94.

[123] Patterson RE, White E, Kristal AR, Neuhouser ML, Potter JD. Vitamin supplements and cancer risk: the epidemiologic evidence. *Cancer Causes Control.* 1997;8(5):786-802.

[124] Galasso M, Gambino S, Romanelli MG, Donadelli M, Scupoli MT. Browsing the oldest antioxidant enzyme: catalase and its multiple regulation in cancer. *Free Radic Biol Med*. 2021;172:264-72.
[125] Goth L, Nagy T. Inherited catalase deficiency: is it benign or a factor in various age related disorders? *Mutat Res*. 2013;753(2):147-54.
[126] Somwar R, Erdjument-Bromage H, Larsson E, Shum D, Lockwood WW, Yang G. Superoxide dismutase 1 (SOD1) is a target for a small molecule identified in a screen for inhibitors of the growth of lung adenocarcinoma cell lines. *Proc Natl Acad Sci USA*. 2011;108(39):16375-80.
[127] Younus H. Therapeutic potentials of superoxide dismutase. *Int J Health Sci* (Qassim). 2018;12(3): 88-93.
[128] Kim Y, Jang HH. The Role of Peroxiredoxin Family in Cancer Signaling. *J Cancer Prev*. 2019;24(2):65-71.
[129] Poynton RA, Hampton MB. Peroxiredoxins as biomarkers of oxidative stress. *Biochim Biophys Acta*. 2014;1840(2):906-12.
[130] Masuda R, Goto K. Modeling of selenocysteine-derived reactive intermediates utilizing a nano-sized molecular cavity as a protective cradle. *Methods Enzymol*. 2022;662:331-61.
[131] Budanov AV, Sablina AA, Feinstein E, Koonin EV, Chumakov PM. Regeneration of peroxiredoxins by p53-regulated sestrins, homologs of bacterial AhpD. *Science*. 2004;304(5670):596-600.
[132] Qu J, Luo M, Zhang J, Han F, Hou N, Pan R. A paradoxical role for sestrin 2 protein in tumor suppression and tumorigenesis. *Cancer Cell Int*. 2021;21(1):606.
[133] Burdon RH. Superoxide and hydrogen peroxide in relation to mammalian cell proliferation. *Free Radic Biol Med*. 1995;18(4):775-94.
[134] Cesselli D, Aleksova A, Sponga S, Cervellin C, Di Loreto C, Tell G. Cardiac Cell Senescence and Redox Signaling. *Front Cardiovasc Med*. 2017;4:38.
[135] Colavitti R, Finkel T. Reactive oxygen species as mediators of cellular senescence. *IUBMB Life*. 2005;57(4-5):277-81.
[136] Lu T, Finkel T. Free radicals and senescence. *Exp Cell Res*. 2008;314(9):1918-22.
[137] Perillo B, Di Donato M, Pezone A, Di Zazzo E, Giovannelli P, Galasso G. ROS in cancer therapy: the bright side of the moon. *Exp Mol Med*. 2020;52(2):192-203.
[138] Pizzino G, Irrera N, Cucinotta M, Pallio G, Mannino F, Arcoraci V. Oxidative Stress: Harms and Benefits for Human Health. *Oxid Med Cell Longev*. 2017;2017:8416763.
[139] Redza-Dutordoir M, Averill-Bates DA. Activation of apoptosis signalling pathways by reactive oxygen species. *Biochim Biophys Acta*. 2016;1863(12):2977-92.
[140] Simon HU, Haj-Yehia A, Levi-Schaffer F. Role of reactive oxygen species (ROS) in apoptosis induction. *Apoptosis*. 2000;5(5):415-8.
[141] Yang S, Lian G. ROS and diseases: role in metabolism and energy supply. *Mol Cell Biochem*. 2020;467(1-2):1-12.
[142] Droge W. Free radicals in the physiological control of cell function. *Physiol Rev*. 2002;82(1):47-95.
[143] Valdez LB, Lores Arnaiz S, Bustamante J, Alvarez S, Costa LE, Boveris A. Free radical chemistry in biological systems. *Biol Res*. 2000;33(2):65-70.
[144] Yamazaki I. Free radical mechanisms in enzyme reactions. *Free Radic Biol Med*. 1987;3(6):397-404.
[145] Chen Q, Wang Q, Zhu J, Xiao Q, Zhang L. Reactive oxygen species: key regulators in vascular health and diseases. *Br J Pharmacol*. 2018;175(8):1279-92.
[146] Diebold L, Chandel NS. Mitochondrial ROS regulation of proliferating cells. *Free Radic Biol Med*. 2016;100:86-93.
[147] Du Plessis SS, Agarwal A, Halabi J, Tvrda E. Contemporary evidence on the physiological role of reactive oxygen species in human sperm function. *J Assist Reprod Genet*. 2015;32(4):509-20.
[148] Dutta S, Majzoub A, Agarwal A. Oxidative stress and sperm function: A systematic review on evaluation and management. *Arab J Urol*. 2019;17(2):87-97.
[149] Sena LA, Chandel NS. Physiological roles of mitochondrial reactive oxygen species. *Mol Cell*. 2012;48(2):158-67.
[150] Tavassolifar MJ, Vodjgani M, Salehi Z, Izad M. The Influence of Reactive Oxygen Species in the Immune System and Pathogenesis of Multiple Sclerosis. *Autoimmune Dis*. 2020;2020:5793817.

[151] Xu S, Touyz RM. Reactive oxygen species and vascular remodelling in hypertension: still alive. *Can J Cardiol*. 2006;22(11):947-51.
[152] Yang Y, Bazhin AV, Werner J, Karakhanova S. Reactive oxygen species in the immune system. *Int Rev Immunol*. 2013;32(3):249-70.
[153] Aggarwal V, Tuli HS, Varol A, Thakral F, Yerer MB, Sak K. Role of Reactive Oxygen Species in Cancer Progression: Molecular Mechanisms and Recent Advancements. *Biomolecules*. 2019;9(11).
[154] Gandhi S, Abramov AY. Mechanism of oxidative stress in neurodegeneration. *Oxid Med Cell Longev*. 2012;2012:428010.
[155] Giacco F, Brownlee M. Oxidative stress and diabetic complications. *Circ Res*. 2010;107(9):1058-70.
[156] Kaneto H, Katakami N, Matsuhisa M, Matsuoka TA. Role of reactive oxygen species in the progression of type 2 diabetes and atherosclerosis. *Mediators Inflamm*. 2010;2010:453892.
[157] Klaunig JE. Oxidative Stress and Cancer. *Curr Pharm Des*. 2018;24(40):4771-8.
[158] Liou GY, Storz P. Reactive oxygen species in cancer. *Free Radic Res*. 2010;44(5):479-96.
[159] Mateen S, Moin S, Khan AQ, Zafar A, Fatima N. Increased Reactive Oxygen Species Formation and Oxidative Stress in Rheumatoid Arthritis. *PLoS One*. 2016;11(4):e0152925.
[160] Panth N, Paudel KR, Parajuli K. Reactive Oxygen Species: A Key Hallmark of Cardiovascular Disease. *Adv Med*. 2016;2016:9152732.
[161] Peoples JN, Saraf A, Ghazal N, Pham TT, Kwong JQ. Mitochondrial dysfunction and oxidative stress in heart disease. *Exp Mol Med*. 2019;51(12):1-13.
[162] Singh A, Kukreti R, Saso L, Kukreti S. Oxidative Stress: A Key Modulator in Neurodegenerative Diseases. *Molecules*. 2019;24(8).
[163] Uttara B, Singh AV, Zamboni P, Mahajan RT. Oxidative stress and neurodegenerative diseases: a review of upstream and downstream antioxidant therapeutic options. *Curr Neuropharmacol*. 2009;7(1):65-74.
[164] Bardaweel SK, Gul M, Alzweiri M, Ishaqat A, HA AL, Bashatwah RM. Reactive Oxygen Species: the Dual Role in Physiological and Pathological Conditions of the Human Body. *Eurasian J Med*. 2018;50(3):193-201.
[165] Samimi A, Khodayar MJ, Alidadi H, Khodadi E. The Dual Role of ROS in Hematological Malignancies: Stem Cell Protection and Cancer Cell Metastasis. *Stem Cell Rev Rep*. 2020;16(2):262-75.
[166] Wang Y, Qi H, Liu Y, Duan C, Liu X, Xia T. The double-edged roles of ROS in cancer prevention and therapy. *Theranostics*. 2021;11(10):4839-57.

Chapter 12

# Interconnecting ROS and EMT: A Way Forward in Cancer Therapy

**Debarpan Mitra**
**Depanwita Saha**
**Subhabrata Guha**
**Gaurav Das**
**and Nabendu Murmu***
[1]Department of Signal Transduction and Biogenic Amines,
Chittaranjan National Cancer Institute, Kolkata, India

## Abstract

Reactive oxygen species (ROS) are found to be elevated in multiple types of cancers. Though ROS in higher levels induces cancer cell death, at lower levels it leads to activation of cell survival pathways. ROS has also recently shown induction of another important phenomenon, i.e., Epithelial Mesenchymal Transition (EMT). This is interesting considering the fact that EMT is essential for metastasis and drug resistance which is the major cause for cancer recurrence and cancer related death. In one such study, it has been shown that ROS induces EMT through activation of Snail leading to repression of E-cadherin. ROS acts as a secondary messenger in this case and regulates EMT through ECM remodelling, monitoring cellular movement, modulation of cell-cell junctions and remodelling cytoskeleton. Hence, targeting EMT through manipulation of redox may hold promise for future cancer therapy. In this book chapter, we will try to understand the interconnections between EMT and ROS, how they regulate/de-regulate each other and how their interconnections might lead to development of new therapeutic targets in cancer therapy.

**Keywords:** ROS, EMT, oxidative stress, metastasis, therapeutics

---

* Corresponding Author's Email: nabendu.murmu@cnci.ac.in.

In: The Role of Reactive Oxygen Species in Health and Disease
Editor: Kuladip Jana
ISBN: 979-8-89113-283-2

## 1. Introduction

Reactive oxygen species (ROS) are ionic radicals that contain oxygen or molecules containing single unpaired electron. They are generally highly reactive, having a very short half-life. Under a physiological state, the cellular ROS levels are kept in equilibrium with antioxidants that are biochemical in nature. Oxidative stress induced excessive ROS induction is a major player implicated in cancer initiation as well as its progression [1]. Oxidative stress is a phenomenon where the equilibrium between the oxidants and antioxidants are disrupted in the favour of oxidants, leading to the dysregulation of redox signalling and also resulting in molecular damage [2]. This is mainly mediated by their ability to damage DNA that induces mutations resulting in gain of function of oncogenes or the loss of function of the tumour suppressors [3]. Inherently, cancer cells are attributed to have high metabolic activity, mitochondrial dysfunction, oncogenic activation, aggravated cellular receptor signalling, all of which contribute to the elevation of ROS levels [4].

*ROS manifests in two ways:* 1. As unbound oxygen or radical species, namely, superoxide, sulphoxyl radicals nitric oxide, hydroxyl radical, etc. and 2. As bound oxygen or non-radical species namely, oxygen singlet, ozone, hydrogen peroxide or as highly reactive derivatives of lipids or carbohydrates [5]. The primary source of intracellular superoxide involves the oxidation of NADPH mediated by NADPH oxidase (NOXs) [6]. Other mechanisms like electron leakage during mitochondrial aerobic respiration also contribute to superoxide generation (Figure 1). Intracellularly, superoxide dismutase converts superoxide into hydrogen peroxide. This hydrogen peroxide oxidizes the cystein residues of certain proteins that results in the initiation of the redox signalling cascade. If the concentration of hydrogen peroxide rises uncontrollably, it reacts with hydroxyl radicals that irreparably damage the nucleic acids and other cellular macromolecules essential for cellular survival. Mitochondrial respiration produces low levels of ROS which can activate an array of signalling cascades, which is holistically called as redox signalling [7]. Post ROS mediated DNA damage and cell division with these damaged DNA may lead to neoplastic mutations [8]. Several redox sensitive kinases, viz. Src, JAK, PI3K, Akt, MAPK etc. are activated by the oxidative stress which in turn activates oncogenic transcription factors [9]. The two hallmarks of oncogenesis i.e., increased cellular proliferation and prevention of apoptosis are commonly mediated by the ROS activated signal transduction via the Ras-Raf-Mek pathway. This Ras mediated transformation mediated by ROS results in anchorage independent growth, morphological modulations and ultimately tumourigenesis. Also, the cytoplasmic factors involved in redox signalling, nuclear signalling factors etc. play a pivotal role in carcinogenesis. Oxidative stress induces the translocation of thioredoxin to the nucleus, which then reacts with redox factor 1 and results in the expressional modulation of AP-1, NF-κβ and Egr-1. AP-1 activation promotes signalling related to proliferation via the activation of cyclin-D and CDKs [9, 10]. Parallely, NF-κβ effects the elevation of inflammatory cytokines to promote favourable tumourogenic environment. It also modulates genes associated with extended survival, viz., cIAP1/2, Bcl-2 and Bcl-xL. NFκβ has also been implicated in VEGF and MMP-9 mediated induction of apoptosis and metastasis [11].

## 1.1. ROS and Its Implications on EMT

Metastasis is the most elusive characteristic of cancer and has proven to be a major obstacle from a therapeutic point of view. Induction of metastasis has close implications on EMT and is a process that plays significant roles in several developmental biological processes like embryogenesis, wound healing, and organ fibrosis [12, 13, 14]. EMT induces a mesenchymal phenotype on epithelial cells and is designated by the loss of epithelial phenotypeand the acquisition of mesenchymal markers. This provides the cancer cells a mode of acquiring invasive and metastatic properties. The concept of EMT has evolved over 20 years and is currently the most accepted and widely publicised mechanism of "invasion-metastasis cascade" for the cancer cells [15, 16, 17]. EMT is mostly controlled by a niche of signalling pathways that comprises ofseveral growth factors (transforming growth factor-β (TGF-β), epidermal growth factor (EGF)) and their associated downstreamtargets (Wnt, Notch, Hedgehog, nuclear-factor kappa B (NF-κB), extracellular signal-regulated kinase (ERK), and phosphatidylinositol 3-kinase (PI3K)/Akt). The phenotype plasticity of epithelial cells results in morphogenesis, repair and invasiveness. In depth understanding of the molecular mediators that control epithelial plasticity is warranted to develop viable therapeutic approaches targeting EMT. Members of the TGF-β family of growth factors can initiate and influence EMT in a variety of biological systems and pathophysiological context by activating several major signalling pathways and transcriptional regulators. These modulation takes place in response to stresses involved in tumourigenesis, including hypoxia, oncogenic or metabolic stress, inflammation, and several other physical constraints [18, 19]. Excluding the enhancing effect of the metastatic potential of cancer cells, EMT has also been implicated in promoting CSC phenotype and also in therapeutic resistance.

As the cancer cells are highly dysregulated and divide uncontrollably, they eventually face nutrient and oxygen shortage in the tumour microenvironment. Thus, the oncogenic cells face an acute hypoxic condition. Hypoxia inducible factor (HIF) signalling plays a vital role in EMT. The EMT promoter snail acts as a HIF target during hypoxia induced EMT [20]. In previous studies, hypoxia is implicated with the induction of mitochondrial ROS production and in turn mitochondrial ROS acts as a pivotal factor in hypoxia induced EMT [21, 22]. Thus, a sequential cascade of events occur wherein, ROS generation is induced by hypoxia, this ROS then induces EMT which in turn imparts stemness to the cancer cells (Figure 2).

Thus, it is evident that ROS contributes greatly in oncogenesis. Moreover, it has the potential to promote EMT, especially under hypoxic conditions of the tumour microenvironment which ultimately leads to the emergence of the most elusive chemo/radio resistant CSC population.

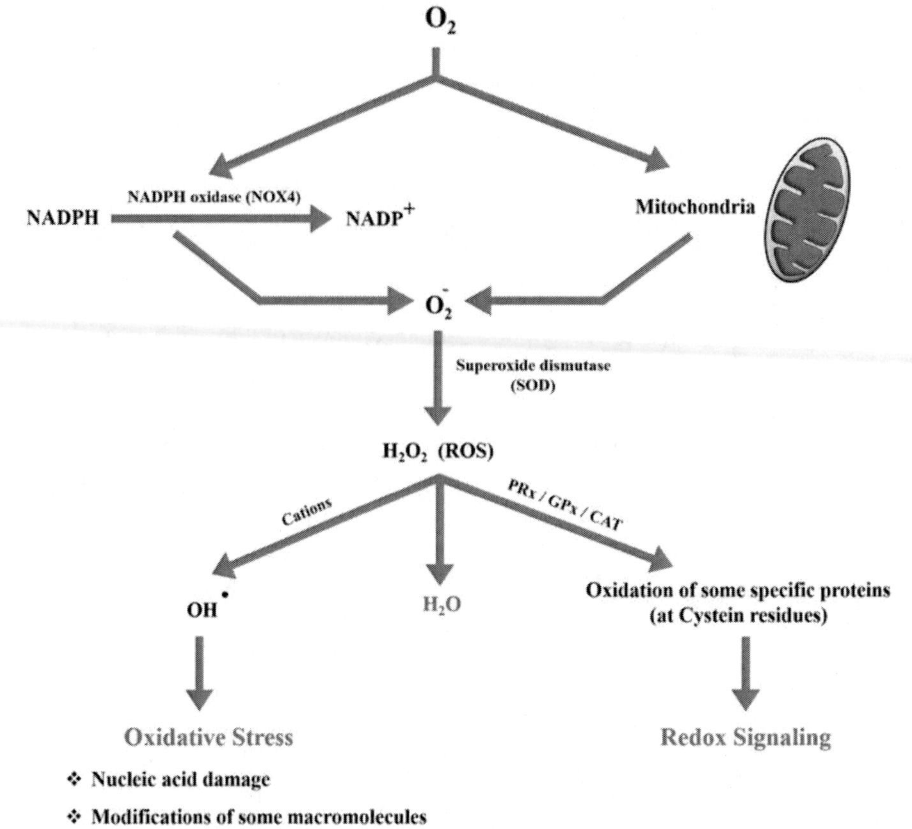


**Figure 1.** Flow diagram representing the various pathways of intracellular ROS generation.

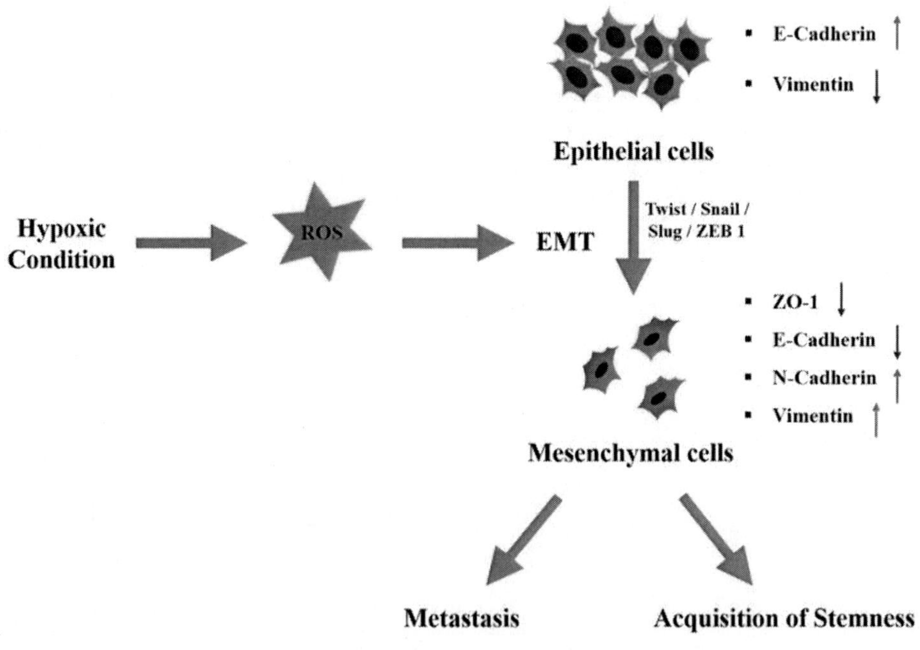


**Figure 2.** Flow diagram representing the cross-relation between ROS and EMT.

## 1.2. Signalling Mechanism and Transcription Factors Modulating ROS/EMT Interplay in Various Cancer Models

NADPH oxidase 4 (Nox4) acts as a critical redox signalling transducer in modulating TGF-β promoted EMT in pancreatic adenocarcinoma [23]. De-novo transcription of Nox4 derives elevated level of TGF-β induced ROS production, which enhances p38MAPK activation for the up-regulation of Snail and repression of E-Cadherin expression. On the other hand Nox4 derived ROS also suppresses the expression of PTP1B which plays antagonizing effect on TGF-β driven EMT phenotype. TGF-β induced NOX4/ROS also played tumorigenic role in glioblastoma through the mechanistic involvement of PI3K/AKT/HIF-1$\alpha$ pathway [24]. A recent study confirmed NF-κB/NOX4/ROS axis as a potential therapeutic target for TGF-β mediated lung cancer metastasis [25]. Also, a feedback loop is formed by TGF-β1 and ROS/HIF in order to maintain a prolonged signalling cascade initiated by either ROS/HIF or TGF-β1. During hypoxia, relatively low-level ROS is generated by the mitochondria at the early onset of hypoxia to stabilize HIF, inducing TGF-β1 gene expression at a later stage. Elevated TGF-β1 levels sustain the ROS production, maintaining a prolonged ROS/HIF/TGF-β1 signalling [26, 27]. TGF-β stimulating EMT is also regulated by the prostate transmembrane protein androgen induced-1(TMEPAI) in lung cancer [28]. TMEPAI regulated EMT and cell migration in a bidirectional mechanism. In one way, TMEPAI downregulates IRS-1 protein level, which is associated with maintaining the epithelial characteristics of cells. On the other hand TMEPAI also causes TGF-β1-induced downregulation of ferritin leading to increased ROS production which again inhibits the expression of IRS-1. Hence via increasing ROS production, TMEPAI could be a potential therapeutic target modulating destabilization of epithelial integrity as well as TGF-β induced EMT (Figure 3).

Ubiquitin specific peptidase 17 (USP-17) served as a crucial player in exerting EMT through reducing ROS production in prostate carcinoma. USP-17 has been found to be upregulated in prostate cancer cells, causing the activation of NF-κβ followed by the inhibition of mitochondrial apoptosis via reducing ROS generation with the simultaneous inhibition of MMP family (MMP2, 9, 14) to enhance EMT [29]. In oral squamous cell cancer, Aurora kinase A (AURKA) has been discovered to synchronise the ROS/EMT interaction (OSCC). One of the most prevalent cancers affecting the head and neck region is OSCC. AURKA, a serine-threonine kinase, is linked to the development of human cancers and plays a role in the creation of mitotic spindles and chromosome segregation. In patient tumour samples and OSCC cell lines, AURKA expression is up regulated. Patients with OSCC who had high levels of AURKA had a significantly worse rate of overall survival. Using cell counting kit-8 (CCK-8), EdU, colony formation, and transwell analyses *in vitro*, it was found that AURKA knockdown significantly decreased OSCC cell proliferation, migration, and invasion. The up-regulated expression of E-cadherin and the down-regulated expression of Vimentin in OSCC cells show that AURKA silencing reduced EMT. Additionally, AURKA inhibition significantly increased the expression of cleaved Caspase-3 and poly (ADP)-ribose polymerase, which in turn increased the rate of apoptosis (PARP). In AURKA-knockdown OSCC cells, the generation of reactive oxygen species (ROS) was also considerably increased. Importantly, it was discovered that inhibiting ROS production with its scavenger, n-acetylcysteine (NAC), greatly eliminated the apoptosis that was caused by AURKA silencing, together with restored proliferation and EMT. AURKA knockdown significantly slowed tumour growth *in vivo*. In order to show that AURKA inhibition may represent a unique therapeutic target for the prevention of OSCC, knockdown of AURKA inhibited cell proliferation, migration, and invasion as well as promoted apoptosis and ROS formation in OSCC [30]. ROS induced EMT has been also facilitated by Mitochondrial calcium uniporter regulator 1 (MCUR1) in hepatocellular carcinoma (HCC) [31]. Mitochondrial $Ca^{2+}$ has been reported to be the essential source of ROS in mammalian cells. Mitochondrial $Ca^{2+}$ mediated over-expression of MCUR1 which promotes the nuclear translocation of Nrf2 and activation of Notch1 signalling which ultimately promotes Snail transcription and EMT activation to enhance HCC metastasis.

DUOX2 has been reported to mediate ROS induced EMT in 5-Fluorouracil (5-FU) resistant human colon cancer [32]. 5-FU resistance increases the expression of DNA methylase TET1 which transcriptionally activates dual oxidase 2 (DUOX2) to increase the expression of ROS which again upregulates the expression of EMT associated transcription factors. So targeting TET1/DUOX2/ROS axis could be a potential therapeutic strategy against invasion and metastasis of colon cancer. In breast cancer, distal less homeobox 2(Dlx2) plays a crucial role in potentiating ROS induced EMT via activating Snail expression [33]. Interestingly it has been observed that by activating Dlx2/Snail axis, ROS causes repression of mitochondrial oxidative phosphorylation and cytochrome C oxidase inhibition in presence of oxygen and relies on glycolytic switch to potentiate EMT. Autophagy defect can also induce EMT in cervical cancer cells through the mechanistic association with Heme Oxygenase (HO-1). Autophagy defect upregulates HO-1 and increased ROS production leading to the elevated expression of Zeb-1 transcription factor followed by the increase in expression of mesenchymal and epithelial marker ration to promote EMT [34].

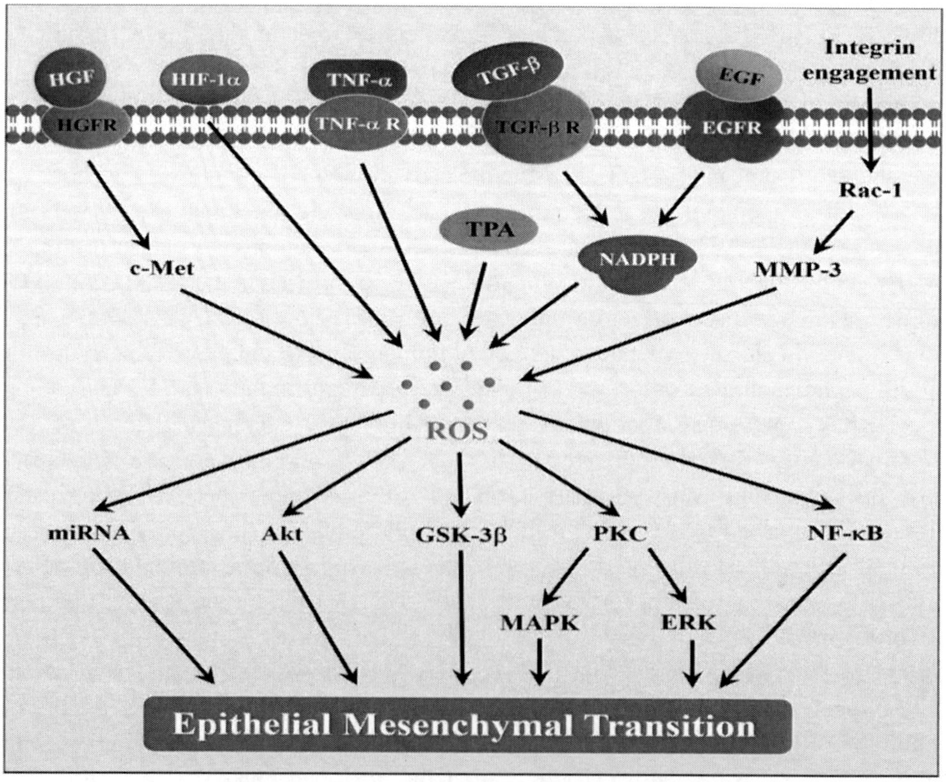


**Figure 3.** Diagram representing the various pathways implicated in ROS generation and induction of EMT.

## 1.3. Proposed Inhibitors Regulating ROS/EMT Interplay in Different Cancer Models

The natural triterpenoid Cucurbitacin B (CuB) derived from the Chinese medicinal plant Cucurbitaceae has been reported to exhibit potential anticancer effect through inducing MAPK/ERK, PI3K/Akt, and JAK/STAT signalling pathways in several cancer models including liver, lung, and breast carcinoma. Recent studies indicated that the CuB demonstrated the reversal of TGF-β induced the EMT in Gefitinib resistant non-small cell lung carcinoma cells (NSCLC) through reducing the ROS production and downregulating PI3K/Akt/mTOR signalling cascade [35]. A very recent study also demonstrated the anti-metastatic potential of small molecule Fangchinoline in NSCLC by directly targeting Nox4 causing decreased production of Nox4 derived cytosolic ROS and inhibition of the Akt-mTOR pathway [36].

TGF-β inducing the EMT has also been found to be suppressed by green tea extract and its polyphenolic compound Epigallocatechin-3-gallate (EGCG) in cervical cancer cells via decreasing ROS production with the simultaneous inhibition of phosphorylation and nuclear translocation of Smad2 and Smad 3 leading to the alteration of associated EMT markers including g E-cadherin, Vimentin, Snail, ZEB, Slug, and Twist [37].

Neem tree (*Azadirachta indica*) derived triterpene Nimbolide has been reported to exhibit the anti-proliferative and the anti-metastatic properties against pancreatic carcinoma via excessive production of the ROS through reduced PI3K/AKT/mTOR and ERK signaling, induced apoptotic cell death, leading to the disturbance in redox balance and cellular integrity [38]. Another phytochemical Matrine, extracted from Chinese herb *Sophora Flavescens Ait* inhibits EMT in pancreatic cancer cells by attenuating intracellular ROS production and downregulation of MMP2 and MMP9 through the transcriptional mediator NF-κB [39].

In breast cancer cells Ultrasound-targeted microbubble destruction (UTMD) system provided a new strategy to prevent EMT by inducing oxidative stress [40]. UTMD mediated drug/gene delivery has already shown an improved cancer therapy through reducing drug resistance and minimizing toxicity of chemotherapies. UTMD treatment increased the intracellular ROS level followed by the upregulated expression of miR-200c, functionally related with the EMT. The consecutive downregulated expression of ZEB-1, Vimentin and migratory property in UTMD treated breast cancer cells and its reversal in ROS scavenger treated cells provided the mechanistic involvement of ROS/miR-200c/ZEB-1 axis in EMT regulation of breast carcinoma. ROS mediated EMT inhibition has also been triggered by polyphenol extract of Annurca apple (APE) in breast cancer. APE exhibited promising results in ROS mediated inhibition of dual-specificity phosphatase-1 (Dusp-1) and sustained JNK activation triggering a phenotypic switch from mesenchymal to epithelial characteristics with increased E-Cadherin/Vimentin ratio and downregulated expression of matrix metalloproteases and SMAD-2/3 proteins [41]. The intracellular ROS production has been found to be reduced by 18β-glycyrrhetinic acid (18β-GA), a bioactive component of licorice root in gastric carcinoma. With the decrease of ROS production, 18β-GA showed the chemo-preventive efficacy by mechanistically inhibiting PKCα-ERK signaling along with the upregulated E-Caderin expression and downregulated vimentin expression and enzymatic activity of MMP2 and MMP9 [42].

### 1.4. Points of Target in ROS Mediated Cancer Therapy

The majority of therapies increase intracellular ROS levels and affect cancer cells' redox balance. It is commonly acknowledged that the anticancer impact of these chemotherapeutics is the production of oxidative stress and ROS-mediated cell damage in cancer. However, a number of novel treatment strategies that target intracellular ROS levels have produced conflicting results. Detecting dynamic ROS levels in tumours during and after chemotherapy in clinical settings is of increasing interest to predict ROS levels for understanding complex tumour biology during chemotherapy because it is impossible to quantitatively compute the detection of dynamic ROS levels in tumours during or after chemotherapy.

### 1.5. The Possible ROS-related Anticancer Therapeutic Strategies Are

1. The first tactic, which aims to lessen the number of converted cells by denying them of fuel, is based on lowering ROS levels to combat their function in cellular transformation.
2. The second technique takes into account the fact that cancer cells are more vulnerable than their normal counterparts to subsequent increases in ROS and are unable to achieve redox balance since their antioxidant systems have already been activated. Therefore, by causing ROS under these metabolic circumstances, a significant portion of the cells die.

ROS can be induced to cause cancer cell death by elevating the oxidative stress in the cell by using a spatiotemporal approach combined with deep and safe tissue penetration capabilities [43]. It consists of a collection of extremely reactive compounds that have developed into key signalling pathway regulators. It is now widely acknowledged that several biological processes, including gene expression, call on modest quantities of ROS. Because of their higher metabolic rate, gene mutations, and relative hypoxia, tumour cells produce more ROS than normal, which are then suppressed by their increased antioxidant enzymatic and non-enzymatic mechanisms. Due to their involvement in several signalling pathways and ability to cause DNA mutation, moderate elevations in ROS are a major factor in a number of pathologic diseases, including the promotion and development of tumours. ROS, however, can also cause planned cell death (PCD). ROS levels are slightly elevated during the early stages of cancer and as it progresses. Because they have more evolved antioxidant systems than normal cells, cancer cells therefore live on levels of ROS that are slightly higher than those in their normal counterparts. This property makes cancer cells more susceptible to outside influences that raise ROS levels. A growing variety of therapeutic methods are being developed to increase ROS levels, so they surpass the cells' redox tolerance and cause oxidative stress that is harmful to cellular viability [44-47].

Different anti-cancer strategies with ROS have been designed over time for targeting cancer cells due to the differential redox states exhibited by cancer and normal cells [48]. All the strategies are directed towards increasing the ROS production which at lower levels are important cell signalling regulators required to carry out several important cellular functions which also includes modulation of gene expression [49]. ROS is already present in an elevated state in cancer cells due to several reasons like upregulated metabolism, excess mutations in gene and hypoxic environment [6]. Thus, further increase in ROS production led to the creation of a condition known as the oxidative stress and can induce cell death or apoptosis in cancer cells [50]. Oxidative stress is an important pathophysiological marker of cancer, in fact Warburg effect enhances ROS levels in cancer cells [51]. Though ROS levels are increased in cancer cells, excessive ROS in cells lead to destruction of lipids, proteins and DNA content.

A growing body of research demonstrates that ROS operates as secondary messengers in intracellular signalling cascades that initiate and sustain the oncogenic phenotype of cancer cells, demonstrating the "two-faced" nature of ROS. However, ROS can also cause apoptosis and cellular senescence, making them an anti-tumorigenic species. Oxidative stress, which is characterised by the cumulative creation of ROS/RNS by either endogenous or external insults, is prevalent in many different types of cancer cells and is associated with alterations in the redox control of cellular signalling pathways. When compared to normal cells, a number of cancer cells have been reported to exhibit a cellular redox imbalance brought on by oxidative stress; this redox imbalance may therefore be related to oncogenic activation. Elevated amounts of oxidative DNA lesions (8-OH-G) have been found in numerous cancers, firmly implicating such damage in the aetiology of cancer. DNA mutation is a crucial stage in carcinogenesis. It seems that the initiation process is mostly connected to the DNA damage. Free radicals' structural, chemical, and biochemical characteristics, as well as their endogenous and exogenous sources, metal-mediated formation of free radicals (e.g., Fenton chemistry), DNA damage (both mitochondrial and nuclear), damage to lipids and proteins caused by free radicals, the phenomenon of oxidative stress, cancer, and the redox environment of a cell, should all receive attention. Therefore, this phenomenon might be used as a therapeutic method to fight cancer [52].

In comparison to normal cells, tumours have greater basal levels of ROS and a different redox environment. These cells may become toxic from an excessive quantity of ROS, making them more susceptible to damage from subsequent ROS shocks brought on by pharmacological agents. Drug resistance, however, may result from the increase of antioxidant capacity in response to intrinsic oxidative stress in cancer cells. Therefore, redox modulation's potential therapeutic benefits could result from the abrogation of such drug-resistant pathways. Redox-modulating drugs have been created in large quantities. Different forms of cancer have been found to be selectively killed by the redox-active system exemplified by ascorbate-driven quinone redox cycling and the group of redox-silent vitamin E analogues represented by tocopheryl succinate. These compounds work in concert to destabilise organelles like mitochondria and unleash their apoptogenic potential, which causes cancerous cells to die effectively and tumour growth to be suppressed. Clinical trials are presently being conducted to evaluate the effectiveness of novel redox-modulating medications in treating cancer patients, which is in line with this idea [53].

Alternatively, ROS can induce oxidative stress through peroxidation of lipids thus altering the lipid bilayer and generation of free radicals [54]. This is potentially dangerous as it may lead to the peroxidation of phospholipids which in turn can cause leakage of electrons in the inter- membrane space of mitochondria due to damage to the permeability pores in mitochondrial inner membrane and complexes I and III of ETC [55].

ROS also controls several signalling pathways that are associated with apoptosis and cell proliferation [56]. The main mechanism is to oxidize the cysteine or the tyrosine residue present in the active site thus creating an inter-protein bridge which affects the function of the protein [57]. Generally, ROS inhibits phosphatases whereas it may also inhibit or activate kinases, like it activates several components of the c-JNK and P38 kinase pathways which are involved in induction of apoptosis. A pleiotropic cytokine called TNF-α can either promote cell growth or cause cell death. Increased vulnerability to TNFα-induced death is brought on by continuous JNK activation, which is a significant factor in the death response. It has been proposed that ROS are necessary for sustained JNK activation in NF-κB-deficient cells, but it is unclear how ROS impact JNK activation. Mitochondrial superoxide dismutase inhibits the accumulation of TNFα-induced ROS, which in turn causes oxidation and inhibits JNK-inactivating phosphatases by changing their catalytic cysteine to sulfenic acid. This causes persistent JNK activation, which is necessary for the release of cytochrome c, the cleavage of caspase 3, and necrotic cell death. Antioxidant treatment stops $H_2O_2$ buildup, JNK phosphatase oxidation, persistent JNK activity, and both types of cell death in cells and lab animals. Without influencing liver regeneration, antioxidant therapy also halts TNFα-mediated fulminant liver failure [58].

ROS is involved in the mitogenic signal transduction as secondary messengers. $H_2O_2$ has been used to obtain the majority of experimental evidence in favour of the function of ROS as signalling molecules. Tyrosine kinase receptors (TKRs) are rapidly tyrosine phosphorylated in response to $H_2O_2$ exposure in the absence of growth factor binding, which activates downstream signalling cascades including protein kinase B (AKT). Diethylmaleate (DEM), another compound that can raise intracellular ROS levels, works by depleting the antioxidant reduced glutathione (GSH). Comparing the effects of $H_2O_2$ with DEM reveals that the latter causes redox changes that are less severe than those caused by $H_2O_2$. It was also showed that, despite their ability to activate ERKs and AKT with kinetics different from those seen after $H_2O_2$ treatment, DEM-induced redox changes are not accompanied by platelet-derived growth factor-receptor (PDGF-R) and epidermal growth factor-receptor Tyr phosphorylation. The effects of DEM are not caused by the TKR phosphorylation, as demonstrated by the fact that AG1296, a specific inhibitor of PDGF-R Tyr kinase, does not prevent the activation of these two pathways. In contrast, the Src kinase inhibitor PP2 (4-amino-5-(4-chlorophenyl)-7-(t-butyl) pyrazole (3,4-d) pyrimidine) totally blocks DEM- and H2O2-induced AKT activation while having no impact on the ERKs pathway. Last but not least, DEM-treated cells show nitration of Tyr residues in PDGF-R, indicating that ROS-induced changes other than Tyr phosphorylation can occur at the level of growth factor receptors and can be implicated in the regulation of signalling pathways. Formation of disulphide linkage between the catalytic cysteines causes Hydrogen peroxide to inactivate PTEN and activate the PI-3 kinase cascade leading to suppression of the inhibition of p38-MAPK signalling pathway [59].

ROS also induces the release of calcium ions which activates PKC and helps in cancer cell proliferation [60]. A larger increase in ROS leads to the activation of p38 mediated apoptosis.

Most of the ROS related pathways are involved in transducing their cytoplasmic signals to the nucleus where they cause modulation in the expression of transcription factors. In order to protect itself from oxidative stress, cancer cells start the expression of many antioxidant enzymes [61]. One of the leading redox sensitive transcription factors is Nrf2 which when not inhibited by Keap1 translocate to the nucleus and induces expression of many antioxidant genes like glutathione (GSH), heme-oxygenase HMOX1, which controls Fe (II) homeostasis etc.

The regulatory mechanism between Keap1 and Nrf2 is essential for protecting cells from oxidative and xenobiotic damage. Normally, the Cul3-Keap1 ubiquitin E3 ligase complex continuously ubiquitinates Nrf2 and catalyzes its rapid degradation to proteasomes. Reactive cysteine residues in Keapl undergo modification in response to electrophilic and oxidative stress, which decreases E3 ligase activity, stabilizes Nrf2, and induces a wide range of cytoprotective genes. The entire Keap1 homodimer forms a cherry-bob structure, as shown by biochemical and structural investigations, in which one molecule of Nrf2 connects with two molecules of Keap1 by means of two binding sites found within the Neh2 domain of Nrf2. For Nrf2 ubiquitination, this two-site binding appears to be essential. Missense mutations in the KEAP1 and NRF2 genes have been found in a large number of human malignancies. Due to these mutations, Nrf2 becomes constitutively activated instead of being ubiquitinated and degraded as a result of Keap1-Nrf2 complex activity. Increased expression of Nrf2 target genes benefits both healthy and cancerous cell growth and stress tolerance. Selective Nrf2 inhibitor research and development should significantly improve cancer treatment [62]. JNK which is activated upon ROS induction, activates FOXO family of transcription factors which induces expression of catalase and SODs [63]. p53 is another factor that may promote both oxidant and antioxidant gene expression depending on the level of ROS present [64]. Likewise, when ROS is present in moderate levels, it inhibits the expression of p53 while at higher levels ROS promotes its expression. P53 induces the expression sestrins which activates peroxiredoxins thus boosting the antioxidant cellular response [65]. Antioxidants are also involved in cancer progression like HMOX1 expression, involved in Fe (II) secretion and storage, and have a significant role in breast tumorigenesis [66].

ROS also leads to the expression of HIF1α and stabilizes it from rapid hydroxylation by inhibiting the activity of iron-dependent prolyl 4-hydroxylase (PHD) [67]. As a result of which HIF1α activates several important genes like VEGF and VEGFR which aid in cancer progression [68]. ROS also influences the DNA binding ability of transcription factors like it enhances the nuclear localization ability of ATM kinase involved in DNA damage repair pathway. APE1/Ref-1 is hypothesised to be a multifunctional protein involved in base excision DNA repair and redox control that is necessary for early mouse embryonic development. APE1/Ref-1 can increase the DNA-binding ability of numerous transcription factors, including NF-κB, AP-1, and p53, by reducing their essential cysteine residues. It also exhibits AP endonuclease activity. However, the precise mechanism through which APE1/Ref-1 performs its crucial tasks *in vivo* is yet unknown. APE1/Ref-1 not only assists other reducing molecules, like glutathione or thioredoxin, in decreasing the target transcription factors indirectly as well as directly. The newly discovered function of APE1/Ref-1, known as redox chaperone activity, is not reliant on either its redox activity or its AP endonuclease activity and is exerted at concentrations substantially lower than those required for its redox activity. There is proof that APE1/physical Ref-1's interaction with target transcription factors is what drives the redox chaperone activity that is crucial to NF-κ-mediated gene expression in human cells. As a result of its various metabolic actions, APE1/Ref-1 may participate in a variety of roles in a pathway that responds to oxidative stress. These discoveries also shed light on how intracellular redox control works [69]. ROS also possesses the ability to cause both activation and repression of NF-kB, an important transcription factor involved in several cellular signalling pathways [70, 71, 72, 73].

ROS also affects epigenetic regulators like the HDACs and DNMTs leading to oxidation of adenines and guanines in DNA creating one of the most potent ROS induced mutagenic lesion called 8-oxo-G [74]. ROS induces DNA hypermethylation in the promoter regions of tumor suppressor genes thus affecting their expression [75]. Oxidized Guanines create a lot of problems by causing changes in the methylation pattern in DNA and inhibition of telomerase and telomere binding proteins causing shortening of the telomere length [76, 77].

ROS also causes mutations in the mitochondrial DNA which affects the genes encoding complexes involved in the ETC chain, thus causing disruption of the whole electron transport system. The mitochondrial DNA is even more sensitive to the ROS induced mutagenesis as it is not protected by histones and mitochondrion does not possess NER repair mechanism [78].

## Conclusion

Thus, it can be concluded that the elevated ROS levels triggers the apoptosis and the antioxidant enzymes or the regulators triggered in response to the elevated ROS, actually aids in the cancer progression. Hence, for the therapeutic intervention, we need drugs that will further elevate the ROS levels in the cancer cells so as to induce cell death by apoptosis or drugs that can inhibit the antioxidant enzymes or the regulators which help in the cancer progression. Of course, therapeutic intervention can be a multipronged approach, taking place at many levels from the epigenetic modulators to molecules that alter signalling cascades to the regulators of EMT pathway, thus opening up a plethora of novel cancer therapeutics.

## References

[1] Kim S., Kim S. H., Lee C. E. SOCS1 Represses Fractionated Ionizing Radiation-Induced EMT Signaling Pathways through the Counter-Regulation of ROS-Scavenging and ROS-Generating Systems. *Cell Physiol Biochem.* 2020; 54[5]:1026-1040.

[2] Shibue T., Weinberg R. A. EMT, CSCS, and drug resistance: the mechanistic link and clinical implications. *Nat Rev Clin Oncol.* 2017; 14[10]:611-29.

[3] Møller P., Wallin H. Adduct formation, mutagenesis and nucleotide excision repair of DNA damage produced by reactive oxygen species and lipid peroxidation product. *Mutat Res.* 1998; 410[3]:271-90.

[4] Szatrowski T. P., Nathan C. F. Production of large amounts of hydrogen peroxide by human tumor cells. *Cancer Res.* 1991; 51[3]:794-98.

[5] Liou G. Y., Storz P. Reactive oxygen species in cancer. *Free Radic Res.* 2010; 44[5]:479-96.

[6] Bedard K., Krause K. H. The NOX family of ROS-generating NADPH oxidases: physiology and pathophysiology. *Physiol Rev.* 2007; 87[1]:245-313.

[7] Tian L., Cai Q., Wei H. Alterations of antioxidant enzymes and oxidative damage to macromolecules in different organs of rats during aging. *Free Radic Biol Med.* 1998; 24[9]:1477-84.

[8] Loft S., Poulsen H. E. Cancer risk and oxidative DNA damage in man. *J Mol Med (Berl).* 1996; 74[6]:297-312.

[9] Waris G., Ahsan H. Reactive oxygen species: role in the development of cancer and various chronic conditions. *J Carcinog.* 2006; 5[5]:14.

[10] Cook J. A., Gius D., Wink D. A., Krishna M. C., Russo A., Mitchell J. B. Oxidative stress, redox, and the tumor microenvironment. *Semin Radiat Oncol.* 2004; 14[3]:259-66.

[11] Xia Y., Shen S., Verma I. M. NF-KB, an active player in human cancers. *Cancer Immunol Res.* 2014; 2[9]:823-30.

[12] Lamouille S., Xu J., Derynck R. Molecular mechanisms of epithelial-mesenchymal transition. *Nat Rev Mol Cell Biol.* 2014; 15[3]:178-96.

[13] Tsai J. H., Yang J. Epithelial-mesenchymal plasticity in carcinoma metastasis. *Genes Dev.* 2013; 27[20]:2192-206.

[14] De Craene B., Berx G. Regulatory networks defining EMT during cancer initiation and progression. *Nat Rev Cancer.* 2013; 13[2]:97-110.

[15] Shi D., Guo L., Sun X., Shang M., Meng D., Zhou X., Liu X., Zhao Y., Li J. UTMD inhibit EMT of breast cancer through the ROS/MiR-200c/ZEB1 axis. *Sci Rep.* 2020; 10[1]:6657.

[16] Fidler I. J. The pathogenesis of cancer metastasis: the 'seed and soil' hypothesis revisited. *Nat Rev Cancer.* 2003; 3[6]:453-8.

[17] Thompson E. W., Newgreen D. F., Tarin D. Carcinoma invasion and metastasis: A role for epithelial-mesenchymal transition? *Cancer Res.* 2005; 65[14]:5991-95; discussion 5995.

[18] Pickup M., Novitskiy S., Moses H. L. The roles of TGFβ in the tumour microenvironment. *Nat Rev Cancer.* 2013; 13[11]:788-99.

[19] Zavadil J., Böttinger E. P. TGF-beta and epithelial-to-mesenchymal transitions. *Oncogene.* 2005; 24[37]:5764-74.

[20] Luo D., Wang J., Li J., Post M. Mouse snail is a target gene for HIF. *Mol Cancer Res.* 2011; 9[2]:234-45.

[21] Hamanaka R. B., Chandel N. S. Mitochondrial reactive oxygen species regulate hypoxic signaling. *Curr Opin Cell Biol.* 2009; 21[6]:894-9.

[22] Hiraga R., Kato M., Miyagawa S., Kamata T. Nox4-derived ROS signaling contributes to TGF-β-induced epithelial-mesenchymal transition in pancreatic cancer cells. *Anticancer Res.* 2013; 33[10]:4431-8.

[23] Su X., Yang Y., Guo C., Zhang R., Sun S., Wang Y., Qiao Q., Fu Y., Pang Q. NOX4-Derived ROS Mediates TGF-$\beta$1-Induced Metabolic Reprogramming during Epithelial-Mesenchymal Transition through the PI3K/AKT/HIF-1$\alpha$ Pathway in Glioblastoma. *Oxid Med Cell Longev.* 2021; 2021:5549047.

[24] Ma M., Shi F., Zhai R., Wang H., Li K., Xu C., Yao W., Zhou F. TGF-β promote epithelial-mesenchymal transition via NF-κB/NOX4/ROS signal pathway in lung cancer cells. *Mol Biol Rep.* 2021; 48[3]:2365-2375.

[25] McMahon S., Charbonneau M., Grandmont S., Richard D. E., Dubois C. M. Transforming growth factor Beta1 induces hypoxia-inducible factor-1 stabilization through selective inhibition of PHD2 expression. *J Biol Chem.* 2006; 281[34]:24171-81.

[26] Görlach A., Diebold I., Schini-Kerth V. B., Berchner-Pfannschmidt U., Roth U., Brandes R. P., Kietzmann T., Busse R. Thrombin activates the hypoxia-inducible factor-1 signaling pathway in vascular smooth muscle cells: Role of the p22(phox)-containing NADPH oxidase. *Circ Res.* 2001; 89[1]:47-54.

[27] Hu Y., He K., Wang D., Yuan X., Liu Y., Ji H., Song J. TMEPAI regulates EMT in lung cancer cells by modulating the ROS and IRS-1 signaling pathways. *Carcinogenesis.* 2013; 34(8):1764-72.

[28] Baohai X., Shi F., Yongqi F. Inhibition of ubiquitin specific protease 17 restrains prostate cancer proliferation by regulation of epithelial-to-mesenchymal transition (EMT) via ROS production. *Biomed Pharmacother.* 2019; 118:108946.

[29] Dawei H., Honggang D., Qian W. AURKA contributes to the progression of oral squamous cell carcinoma (OSCC) through modulating epithelial-to-mesenchymal transition (EMT) and apoptosis via the regulation of ROS. *Biochem Biophys Res Commun.* 2018; 507(1-4):83-90.

[30] Jin M., Wang J., Ji X., Cao H., Zhu J., Chen Y., Yang J., Zhao Z., Ren T., Xing J. MCUR1 facilitates epithelial-mesenchymal transition and metastasis via the mitochondrial calcium dependent ROS/Nrf2/Notch pathway in hepatocellular carcinoma. *J Exp Clin Cancer Res.* 2019; 38(1):136.

[31] Kang K. A., Ryu Y. S., Piao M. J., Shilnikova K., Kang H. K., Yi J. M., Boulanger M., Paolillo R., Bossis G., Yoon S. Y., Kim S. B., Hyun J. W. DUOX2-mediated production of reactive oxygen species induces epithelial mesenchymal transition in 5-fluorouracil resistant human colon cancer cells. *Redox Biol.* 2018; 17:224-235.

[32] Lee S. Y., Ju M. K., Jeon H. M., Lee Y. J., Kim C. H., Park H. G., Han S. I., Kang H. S. Reactive oxygen species induce epithelial-mesenchymal transition, glycolytic switch, and mitochondrial repression through the Dlx-2/Snail signaling pathways in MCF-7 cells. *Mol Med Rep.* 2019; 20[3]:2339-2346.

[33] Zhao Z., Zhao J., Xue J., Zhao X., Liu Peishu. Autophagy inhibition promotes epithelial-mesenchymal transition through ROS/HO-1 pathway in ovarian cancer cells. *Am J Cancer Res.* 2016; 6[10]:2162-77.

[34] Yuan R., Fan Q., Liang X., Han S., He J., Wang Q. Q., Gao H., Feng Y., Yang S. Cucurbitacin B inhibits TGF-β1-induced epithelial-mesenchymal transition (EMT) in NSCLC through regulating ROS and PI3K/Akt/mTOR pathways. *Chin Med.* 2022;17[1]:24.

[35] Chen B., Song Y., Zhan Y., Zhou S., Ke J., Ao W., Zhang Y., Liang Q., He M., Li S., Xie F., Huang H., Chan W. N., Cheung A. H. K., Ma B. B. Y., Kang W., To K. F., Xiao J. Fangchinoline inhibits non-small cell lung cancer metastasis by reversing epithelial-mesenchymal transition and suppressing the cytosolic ROS-related Akt-mTOR signaling pathway. *Cancer Lett.* 2022; 543:215783.

[36] Panji M., Behmard V., Zare Z., Malekpour M., Nejadbiglari H., Yavari S., Nayerpour Dizaj T., Safaeian A., Maleki N., Abbasi M., Abazari O., Shabanzadeh M., Khanicheragh P. Suppressing effects of green tea extract and Epigallocatechin-3-gallate (EGCG) on TGF-β- induced Epithelial-to-mesenchymal transition via ROS/Smad signaling in human cervical cancer cells. *Gene.* 2021;794:145774.

[37] Subramani R., Gonzalez E., Arumugam A., Nandy S., Gonzalez V., Medel J., Camacho F., Ortega A., Bonkoungou S., Narayan M., Dwivedi A. K., Lakshmanaswamy R. Nimbolide inhibits pancreatic cancer growth and metastasis through ROS-mediated apoptosis and inhibition of epithelial-to-mesenchymal transition. *Sci Rep.* 2016 Jan 25; 6:19819.

[38] Huang M., Xin W. Matrine inhibiting pancreatic cells epithelial-mesenchymal transition and invasion through ROS/NF-κB/MMPs pathway. *Life Sci.* 2018; 192:55-61.

[39] Sies H. Oxidative stress: A concept in redox biology and medicine. *Redox Biol.* 2015; 4[1]:180-83.

[40] Vuoso D. C., D'Angelo S., Ferraro R., Caserta S., Guido S., Cammarota M., Porcelli M., Cacciapuoti G. Annurca apple polyphenol extract promotes mesenchymal-to-epithelial transition and inhibits migration in triple-negative breast cancer cells through ROS/JNK signaling. *Sci Rep.* 2020;10[1]:15921.

[41] Cai H, Chen X, Zhang J, Wang J. 18β-glycyrrhetinic acid inhibits migration and invasion of human gastric cancer cells via the ROS/PKC-α/ERK pathway. *J Nat Med.* 2018; 72[1]:252-259.

[42] Yang H, Villani RM, Wang H, Simpson MJ, Roberts MS, Tang M, Liang X. The role of cellular reactive oxygen species in cancer chemotherapy. *J Exp Clin Cancer Res.* 2018;37[1]:266.

[43] Wang J, YiJ. Cancer cell killing via ROS: to increase or decrease that is the question. *Cancer Biol Ther.* 2008; 7[12]:1875-84.

[44] Halliwell B. The antioxidant paradox. *Lancet.* 2000; 355[9210]:1179-80.

[45] Ozben T. Oxidative stress and apoptosis: impact on cancer therapy. *J Pharm Sci.* 2007; 96[9]:2181-96.

[46] Trachootham D., Alexandre J., Huang P. Targeting cancer cells by ROS-mediated mechanisms: A radical therapeutic approach? *Nat Rev Drug Discov.* 2009; 8[7]:579-91.

[47] Perillo B., Di Donato M., Pezone A., Di Zazzo E., Giovannelli P., Galasso G., Castoria G., Migliaccio A. ROS in cancer therapy: the bright side of the moon. *Exp Mol Med.* 2020; 52[2]:192-203.

[48] Forman H. J., Ursini F., Maiorino M. An overview of mechanisms of redox signaling. *J Mol Cell Cardiol.* 2014; 73:2-9.

[49] Paradies G., Petrosillo G., Pistolese M., Di Venosa N., Federici A., Ruggiero F. M. Decrease in mitochondrial Complex I activity in ischemic/reperfused rat heart: involvement of reactive oxygen species and cardiolipin. *Circ Res.* 2004; 94[1]:53-9.

[50] Petrosillo G., Ruggiero F. M., Di Venosa N., Paradies G. Decreased Complex III activity in mitochondria isolated from rat heart subjected to ischemia and reperfusion: role of reactive oxygen species and cardiolipin. *F.A.S.E.B. J.* 2003; 17[6]:714-16.

[51] Valko M., Rhodes C. J., Moncol J., Izakovic M., Mazur M. Free radicals, metals and antioxidants in oxidative stress-induced cancer. *Chem Biol Interact.* 2006; 160[1]:1-40.

[52] Tomasetti M., Santarelli L., Alleva R., Dong L. F., Neuzil J. Redox-active and redox-silent compounds: synergistic therapeutics in cancer. *Curr Med Chem.* 2015; 22[5]:552-68.

[53] Huang L. E., Gu J., Schau M., Bunn H. F. Regulation of hypoxia-inducible factor 1alpha is mediated by an O2-dependent degradation domain via the ubiquitin-proteasome pathway. *Proc Natl Acad Sci U* S A. 1998; 95[14]:7987-92.

[54] Houée-Lévin C., Bobrowski K., Horakova L., Karademir B., Schöneich C., Davies M. J., Spickett C. M. Exploring oxidative modifications of tyrosine: an update on mechanisms of formation, advances in analysis and biological consequences. *Free Radic Res.* 2015; 49[4]:347-73.

[55] Thannickal V. J., Fanburg B. L. Activation of an H2O2-generating NADH oxidase in human lung fibroblasts by transforming growth factor beta 1. *J Biol Chem.* 1995; 270[51]:30334-38.

[56] Meng T. C., Fukada T. F., Tonks N. K. Reversible oxidation and inactivation of protein tyrosine phosphatases *in vivo*. *Mol Cell.* 2002; 9[2]:387-99.
[57] Kamata H., Honda S., Maeda S., Chang L., Hirata H., Karin M. Reactive oxygen species promote TNFalpha-induced death and sustained JNK activation by inhibiting MAP kinase phosphatases. *Cell.* 2005; 120[5]:649-61.
[58] Esposito F., Chirico G., Montesano Gesualdi N., Posadas I., Ammendola R., Russo T., Cirino G., Cimino F. Protein kinase B activation by reactive oxygen species is independent of tyrosine kinase receptor phosphorylation and requires SRC activity. *J Biol Chem.* 2003; 278[23]:20828-34.
[59] Corcoran A., Cotter T. G. Redox regulation of protein kinases. *FEBS J.* 2013; 280[9]:1944-65.
[60] Gopalakrishna R., Jaken S. Protein kinase C signaling and oxidative stress. *Free RadicBiol Med.* 2000; 28[9]:1349-61.
[61] Taguchi K., Motohashi H., Yamamoto M. Molecular mechanisms of the Keap1–Nrf2 pathway in stress response and cancer evolution. *Genes Cells.* 2011; 16[2]:123-40.
[62] Meister A. Selective modification of glutathione metabolism. *Science.* 1983; 220[4596]:472-77.
[63] Melnik B. C. P53: key conductor of all anti-acne therapies. *J Transl Med.* 2017; 15[1]:195.
[64] Latimer H. R., Veal E. A. Peroxiredoxins in regulation of MAPK signalling pathways; sensors and barriers to signal transduction. *Mol Cells.* 2016; 39[1]:40-5.
[65] Weinberg E. D. The role of iron in cancer. *Eur J Cancer Prev.* 1996; 5[1]:19-36.
[66] Abid M. R., Spokes K. C., Shih S. C., Aird W. C. NADPH oxidase activity selectively modulates vascular endothelial growth factor signaling pathways. *J Biol Chem.* 2007; 282[48]:35373-85.
[67] Guo Z., Kozlov S., Lavin M. F., Person M. D., Paull T. T. ATM activation by oxidative stress. *Science.* 2010; 330[6003]:517-21.
[68] Ando K., Hirao S., Kabe Y., Ogura Y., Sato I., Yamaguchi Y., Wada T., Handa H. A new APE1/Ref-1-dependent pathway leading to reduction of NF-kappaB and AP-1, and activation of their DNA-binding activity. *Nucleic Acids Res.* 2008; 36[13]:4327-36.
[69] Toledano M. B., Leonard W. J. Modulation of transcription factor NF-kappa B binding activity by oxidation-reduction *in vitro*. *Proc Natl Acad Sci U S A.* 1991; 88[10]:4328-32.
[70] Matthews J. R., Kaszubska W., Turcatti G., Wells T. N., Hay R. T. Role of cysteine62 in DNA recognition by the P50 subunit of NF-kappa B. *Nucleic Acids Res.* 1993;21[8]:1727-34.
[71] Lingappan K. NF-κB in Oxidative Stress. *CurrOpinToxicol.* 2018; 7:81-86.
[72] Matsushima S., Kuroda J., Ago T., Zhai P., Park J. Y., Xie L. H., Tian B., Sadoshima J. Increased oxidative stress in the nucleus caused by Nox4 mediates oxidation of HDAC4 and cardiac hypertrophy. *Circ Res.* 2013 Feb 15; 112[4]:651-63.
[73] Ushijima T. Detection and interpretation of altered methylation patterns in cancer cells. *Nat Rev Cancer.* 2005; 5[3]:223-31.
[74] Grollman A. P., Moriya M. Mutagenesis by 8-oxoguanine: an enemy within. *Trends Genet.* 1993; 9[7]:246-9.
[75] Opresko P. L., Fan J., Danzy S., Wilson D. M., Bohr V. A. Oxidative damage in telomeric DNA disrupts recognition by TRF1 and TRF2. *Nucleic Acids Res.* 2005; 33[4]:1230-39.
[76] Hanahan D., Weinberg R. A. Hallmarks of cancer: the next generation. *Cell.* 2011;144[5]:646-74.
[77] Nishikawa M. Reactive oxygen species in tumor metastasis. *Cancer Lett.* 2008;266[1]:53-9.

## Chapter 13

# The Dynamic Role of Reactive Oxygen Species in Relation to Ovarian Cancer

**Aparajita Pal***
**Sunita Jhulki**
**and Sukanya Kundu**
Diamond Harbour Women's University, Sarisha, West Bengal, India

## Abstract

Reactive oxygen species (ROS) are generated when molecular oxygen reacts with electrons. ROS is multifunctional in nature, having both positive and negative roles in many physical and physiological processes in our body. ROS and antioxidants remain in a perfect balance in healthy cells but hypoxic condition, shifts the redox balance and activates different redox sensitive signaling pathways. Altered level of ROS sometime exerts positive effects upon various cellular operations like autophagy, signal transduction, apoptosis, tumorigenesis inhibition whereas in some cases it implicates negative effects like mitochondrial and DNA damage, organ failure, aging, reproductive infertility, biomolecule degradation, etc. and induces several irreversible deleterious diseases. ROS plays a crucial role in the progression of various diseases including cardiovascular disease, chronic autoimmune diseases, neurodegenerative disorder, etc. ROS is also responsible for carcinogenesis and induces malignant tumor growth. ROS is involved in various stages of cancer development and plays a very delicate role in tumor development and progression. Among various types of cancer, ovarian cancer is one of the most deleterious gynecological malignancies and fifth principal cause of cancer-related death in women. Involvement of ROS in ovarian cancer is widely established by various groups of researchers. Antioxidant therapy and ROS inhibition can be a probable way out to deal with ROS mediated sufferings. On the other hand, various anticancer therapeutics greatly rely on cellular damage via ROS. Here we have analyzed the overall functions of ROS in health and diseases, its role in progression of cancer and more specifically towards ovarian cancer along with the therapeutic strategies involving ROS against ovarian cancer.

**Keywords:** ROS, oxidative stress, ovarian cancer, signal transduction, tumorigenesis, anticancer therapeutics

---

* Corresponding Author's Email: aparajita.zoology@gmail.com.

In: The Role of Reactive Oxygen Species in Health and Disease
Editor: Kuladip Jana
ISBN: 979-8-89113-283-2

## 1. Introduction

Reactive oxygen species (ROS) is a group of unstable oxygen-containing small molecules that are generated by cells in both normal and pathological condition of body. Generally, ROS are short-lived metabolic by-products of different cellular organelles that increase the oxidative stress. All types of ROS, such as superoxide ($O2^{\bullet -}$), peroxyl ($ROO^{\bullet}$), hydroxyl ($OH^{\bullet}$), hypochlorous acid (HOCl), singlet oxygen ($_1O^2$), hydrogen peroxide ($H_2O_2$), nitric oxide ($NO^{\bullet}$), etc. are eminently reactive due to their unpaired valence electrons [1, 2]. At normal level, ROS can significantly regulate the various kinds of cellular activities including different signaling pathways, gene transcription and associated with immunological defense system of body against pathogenic attack [3]. Although, at higher concentration, ROS performs as reactive damaging molecules to cells that have been capable of destroying DNA, proteins, lipids, after reacting with them. Overproduction of ROS causes several harmful or irreversible diseases including neurological disorders, chronic auto-immune diseases and cardiovascular, sensory disorders, etc. ROS can also contribute to the appearance of carcinogenesis [3]. Antioxidant associated defense mechanisms of body usually balance the production and depletion of ROS; mainly the antioxidant enzymes including glutathione peroxidase, superoxide dismutase, catalase, glutathione, glutathione reductase and other non-enzymatic antioxidant like β-carotene, α-tocopherol play a crucial role in protecting the body from free radicals [4].

ROS plays a very important role in tumor development and progression via dysregulating the redox homeostasis with reduced mechanical activity of antioxidant enzymes, thus leading to malignant conversion by targeting several molecular factors/signaling pathways [1]. ROS has been widely involved in the pathogenesis of cancer. On the one hand, ROS and/or Reactive nitrogen species (RNS) can promote molecular genetic changes leading to tumor initiation, growth, and development as well as to therapeutic resistance. On the other hand, long-term elevated ROS/RNS levels have cytotoxic effects and can induce the activation of apoptotic pathways [5].

Ovarian cancer is the most common malignant tumor of the female reproductive tract. Involvement of ROS in progression and genetic instability of ovarian cancer and its chemoresistance against different chemotherapeutics has been established by different research groups [6]. A plethora of studies have been conducted on the role of ROS in the development and its strong association with ovarian cancer. This chapter will update the relevant literature and focuses on the sources and effects of ROS in diseases with a special emphasis on ovarian cancer from different perspectives.

## 2. Sources of ROS

In the cell, ROS are commonly produced endogenously via the electron transport chain (ETC) or derived from metabolic by-products of different intracellular organelles (mitochondria, peroxisomes, etc.) and by the enzymes such as NADPH oxidases (NOX), nitric oxide synthase (NOS) and xanthine oxidase (XO). Some pollutants, drugs and ionizing radiation are also involved in ROS generation [2, 7].

## 2.1. ROS Production from Mitochondria

Mitochondrial respiratory chain is one of the leading resources of ROS in cell; where multiple enzyme complexes are involved directly in the process of emanating electrons. These involves NADH (Nicotinamide adenine dinucleotide) or $FADH_2$ (flavin adenine dinucleotide) and through the process of partial reduction reaction with oxygen (act as electron acceptor) these electrons produce superoxide ($O2^{\bullet-}$) free radical [8].

The mitochondrial structure comprises of an outer membrane, an intermediate space, an inner membrane and a matrix. The inner mitochondria system consisting of five complexes, among them complex I to complex III considerably participate in mtROS (mitochondrial ROS) production. Complex I known as NADH-ubiquinone oxidoreductase that are associated with ubisemiquinone, FMN, NAD, etc. [1, 8]. In ETC, non-enzymatic reaction is responsible for the formation of superoxide anions; these superoxide anions become hydrogen peroxide ($H_2O_2$) via superoxide dismutase (SOD) enzymes. SOD1 and SOD2 are two types of SOD enzymes mainly located in mitochondrial intermembrane space and mitochondrial matrix. Furthermore, in mitochondria a radical-reaction between the superoxide ($O2^{\bullet-}$) and nitric oxide (NO) can significantly produce peroxynitrite ($ONOO^-$) which is a type of reactive oxidant molecules. These ROS can also cause irreversible DNA damage and alteration of mitochondrial protein [8].

There is an interconnection between complex II (oxidoreductase) and TCA cycle. As an intermediate product, Succinate is formed by TCA cycle that is involved in production of ROS. Additionally, in complex III of ETC, p66SHC (Src homologous–collagen homologue adaptor protein) can produce ROS via oxidative reaction of cytochrome c [1].

## 2.2. ROS Production from NADPH Oxidase and Xanthine Oxidase Activity

As previously mentioned, NADPH oxidase (NOX) is the direct source of cellular ROS generation and is capable of producing ROS without the help of another intracellular organelle. NOX is composed of five isoforms (NOX1 to NOX5) that are expressed in endothelial cells. They act as electron transferring agents that help in superoxide formation. It has been reported that NOX also regulates different kinds of physiological processes including cell growth, apoptosis, migration, contraction, etc. [3].

NOX1 is especially present in epithelial cells of colon, lungs alveolar and NOX3 is mainly exhibited in auditory and sensory system of the inner ear. Research exhibits that NOX2 is considerably associated with immune system related to defense mechanism of body against various kinds of pathogenic (e.g., bacteria) attack and this type of NOX isoform is significantly expressed in phagocytes [3]. NOX5, eminently expressed in testis and lymphoid tissue; whereas high level of NOX4 is remarkably found in kidney and in different types of vascular cells, neurons, etc. According to study reports NOX-generated ROS significantly degrades the cofactor of enzyme nitric oxide synthase (NOS) and by this way it can increase the level of superoxide production [3, 7].

Xanthine oxidase, another important source of cellular ROS in body, consists of both dehydrogenase and oxidase form. Oxidase form of XO generates ROS by utilizing only $O_2$ but as an electron acceptor both $NAD^+$ and $O_2$ are used by dehydrogenase form of XO that are involved in ROS production [7].

## 3. ROS in Relation to Physiological Processes

### 3.1. ROS in Reproduction

During fertilization, spermatozoa generate ROS for their maturation and fusion with oocyte membrane. Over production of ROS in spermatozoa causes mitochondrial dysfunction as ROS decreases mitochondrial membrane potential. It damages sperm DNA and reduce sperm quality leading to male infertility [9]. Lower amount of ROS is needed for capacitation process, but excess amount of ROS exerts negative effect in male reproductive health. Oxidative stress can reduce sperm membrane fluidity by affecting polyunsaturated fatty acids in the membrane [10]. In case of female reproduction, ROS is involved in corpus luteum functioning. At the last stage of corpus luteum, ROS induce apoptotic cell death and regression of the luteal layer. It has been reported that ROS is also involved in age related declination of testosterone secretion from Leydig cells [11].

### 3.2. ROS in Signal Transduction

Experimental evidence suggested that ROS plays a crucial role in major cell signaling pathways. ROS can modify some "redox-sensitive" factors into signaling molecules. It can react with cysteine residue of enzymes. Some growth factors, hormones namely angiotensin II and insulin employ ROS partially in their intracellular messaging system. In this way, ROS can mimic the effects of ligand and leads to ligand independent signaling cascades [12].

Some important signaling pathways involving ROS are:

#### *3.2.1. MAPK Pathway*

Three main groups of MAPK signaling are extracellular signal-regulated protein kinase (ERK1-2), c-Jun NH2-terminal kinase (JNK1-3) and p38. ROS can trigger the activation of ERK and JNK cascades. On the other hand, other components in MAPK signaling alter their effect depending upon intracellular ROS level. ROS can also activate p38 signaling pathway [12].

#### *3.2.2. PI3K/Akt (Phosphatidylinositol 3-Kinase/Protein Kinase 3) Pathway*

Oxidative stress can activate this pathway in different cell types through phosphorylation. Akt pathway can induce metastasis, angiogenesis and tumor inhibition. ROS triggers the activation of Akt pathway in various cell lines [12].

#### *3.2.3. PKC (Protein kinase C) Pathway*

PKC has different isoforms. These isoforms are involved in several cellular activities like, differentiation, proliferation, apoptosis, etc. Cysteine rich regions of PKC alpha are ROS sensitive. Various isoforms of PKC can be activated by ROS [12].

#### *3.2.4. NF-κβ Pathway*

NF-κβ transcription can modulate enormous gene activities involved in immune response, cell growth, cell differentiation, apoptosis, etc. ROS can interact with NF-κβ signal transduction pathways in multiple ways and regulate its activities. ROS phosphorylate NF-κβ and some other participants in this pathway. ROS stimulates this pathway in the cytoplasm but inhibits NF-κβ activity in the nucleus [12].

#### *3.2.5. Nrf2 (Nuclear Factor (Erythroid-Derived 2) Factor 2) Pathway*

Nrf2 binds with cytoskeletal protein Keap1 and remains inactivated in normal condition of a cell. But under oxidative stress Nrf2 gets phosphorylated and activated and translocated into the nucleus. It then regulates other gene functions [12].

#### *3.2.6. Fork Head Box O (FoxO) Pathway*

This pathway is involved in differentiation, cell proliferation, apoptosis, DNA damage, etc. ROS oxidizes cysteine residue of FoxO and modulates the activity of this factor [12].

#### *3.2.7. HIF-1 α Pathway*

ROS can destabilize this transcription factor in hypoxic condition [12].

### 3.3. Immune Defense & ROS

ROS has an important cytotoxic effect against some harmful pathogen like protozoa, parasitic fungi, helminths, mycobacteria, etc. ROS plays an important role in the innate immunity as defense mechanisms and in essential cell types involved in adaptive immune response [9, 1]

### 3.4. ROS in Relation to Damage of Biomolecule

Toxic ROS molecules are harmful to almost all types of biomolecules like protein, lipid, nucleic acids, carbohydrates, etc. Therefore, they can cause lipid peroxidation, DNA damage, genetic mutation, carbohydrate damage, cellular membrane damage and carcinogenesis also [9]

#### *3.4.1. Lipid Peroxidation*

Polyunsaturated fatty acids (PUFA) are more prone to oxidative degradation. PUFA is present in cell membrane and oxidized into lipid peroxyl and lipid alkoxyl free radicals [9].

#### *3.4.2. Protein Oxidation*

ROS can alter protein structure by peptide bond breakage, protein-protein crosslinking and side chain modification. Protein carbonyl and peroxides are formed because of these and in turn they can alter protein functions. Protein carbonyls are irreversible and get accumulated in the cell, implicating age-related diseases [13]. **Figure 1** briefly depicts the sources of ROS and its physiological function in our body.

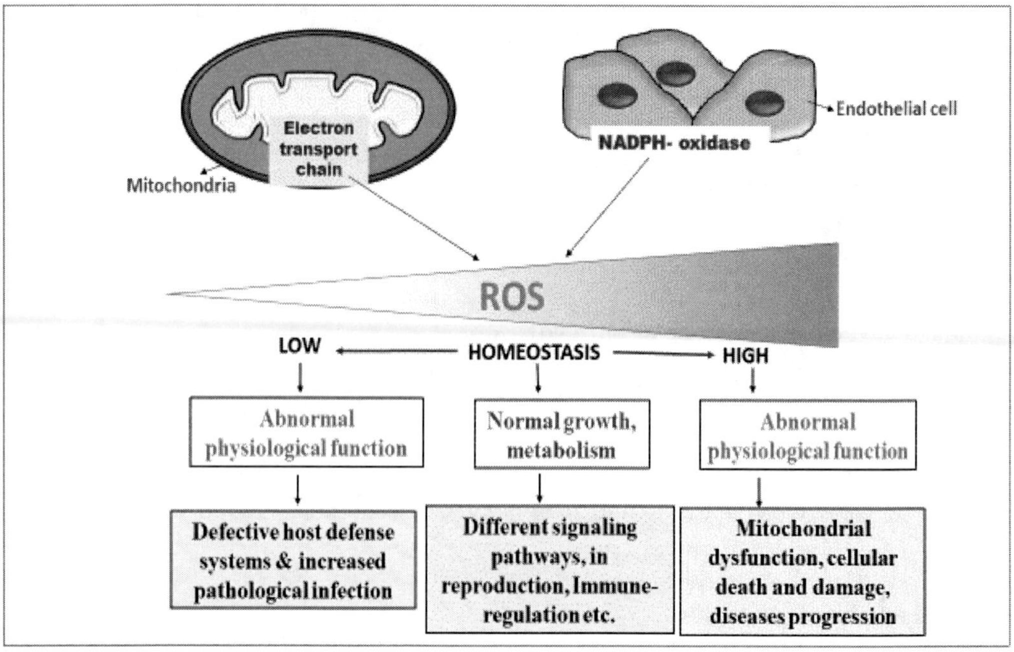


**Figure 1.** Sources of ROS and its physiological function in our body.

## 4. Involvement of ROS in Diseases

Generally, ROS is present in all aerobic cells. There is a balance between ROS and biochemical antioxidants within a healthy cell. Sometimes due to some physiological causes like antioxidant depletion, ROS overproduction, this balance got disturbed. This phenomenon is called oxidative stress. Oxidative stress can damage cellular metabolism and other important activities within a cell that leads to some serious health hazards among organisms. To counterbalance this situation, a cell must restore that redox balance [14]. ROS plays an important role in the progression of various diseases including cardiovascular disease, chronic autoimmune diseases, neurodegenerative disorder, etc. Moreover, ROS is also responsible for carcinogenesis by inducing malignant tumor growth. Recently, researchers have been engaged in investigating the paradoxical effects of ROS in cancer [1].

### 4.1. Interplay of ROS in Different Kind of Cancers

ROS are involved in all stages of cancer development. ROS somehow plays a very delicate role in tumor development and progression [1]. On the contrary, novel therapeutic strategies against cancer often employ the fine tuning of intracellular ROS signaling to deprive cells effectively from ROS-induced tumor promoting events. High level of ROS mainly dysregulates the redox homeostasis with reducing the mechanical activity of antioxidant enzymes, thus leading to malignant conversion by targeting several molecular factors/signaling pathways including NF-κB, Nrf2, etc. ROS-induced signaling pathways/transcription factors, promote neoplasm malignancy by enhancing cancer cell proliferation, migration, anti-apoptosis and angiogenesis properties [15]. ROS also creates a strong interlink between chronic-inflammation and cancer development by modulating the expression of different inflammatory factors, including pro-inflammatory cytokines, AP-1 (activator protein 1), HIF-1α (hypoxia-inducible factor-1 alpha), growth factors, PPAR-γ (peroxisome proliferator-activated receptor gamma), etc. [15]. However, the inflammatory cells that are residing in tumor micro-environment also generates ROS via elevating the activity of NOX, myeloperoxidase, etc. Besides, these cells can significantly alter the function of biological/chemical factors by increasing the expression of COX-2 (cyclooxygenase 2) and lipoxygenase pathway. ROS- induced overexpression of NF-κB play a crucial role in chemotherapeutic drug (5-fluorouracil, paclitaxel, doxorubicin, etc.) resistance by upregulating the expression of survivin, Akt, B-cell lymphoma 2 (Bcl-2), cyclin D1, COX-2 at the level of transcription [15].

EMT (Epithelial to mesenchymal transition) is a key process for initiation of cancer cell metastasis. ROS increases the EMT by upregulating the activity of EMT-responsible factors such as, TGF β1, cathepsin, matrix metalloproteinase, etc. [15]. It has been reported that ROS plays a vital role in metastasis by ameliorating TGF-β1-mediated Urokinase type Plasminogen Activator (uPA) and MMP-9 expression [16]. Another report showed that, in hepatocellular carcinoma, over activity of mitochondrial Calcium-Uniporter Regulator-1 (MCUR1) is observed, where MCUR1 acts as EMT inducing factors through ROS mediated Notch1/Nrf2 axis. Therefore, reducing the activity of MCUR1 relatively decreased the metastasis of hepatocellular carcinoma [15].

Additionally, NOX-derived ROS also have been reported to alter the function of NK (natural killer) cells, which in turn increase the metastasis of melanoma [17].

Aquaporin-5 is commonly known as peroxiporin and is responsible for high levels of $H_2O_2$ influx in pancreatic cancer cells (BxPC3 cell line), which subsequently stimulates anti-apoptosis and proliferation of signaling complex [18]. Angiogenesis, a process of blood vessels sprouting, is crucial for nourishment of tumor. ROS is associated with tumor angiogenesis and increases the activation of different angiogenic factors, and it also upregulates the expression of NF-κB pathway in different cancers [15]. Mutated p53 considerably increases the ROS-induced expression of HIF-1 and vascular endothelial growth factor and hypoxia-inducible factor (VEGF-A) in colon carcinoma (human HCT116 cell line), thus enhancing cancer progression [19]. ROS has been shown to upregulate GPER (G-protein estrogen receptor) HIF-1α /VEGF by increasing EGFR and ERK signaling in breast cancer cells [15].

#### *4.1.1. Potential Role of ROS in Ovarian Cancer*

Ovarian cancer is the third most frequent gynecological malignancy and fifth principal cause of cancer-related death which is found in women [5]. Annually 290,000 new cases of ovarian carcinoma and nearly 180,000 ovarian cancer-related deaths are reported throughout the world in the previous few years [5]. According to the origin, ovarian cancer is broadly divided into 3-types: Epithelial cell, germ cell and stromal cell cancer. Among them epithelial ovarian cancer (EOC) is more common, approximately 90% ovarian cancers are epithelial in origin [20]. Based on the factors of various histopathological or clinical and genetic factors, ovarian cancer is divided into two types- type I and type II tumor. Type I tumor exhibit less death rate (10%) and it usually grow from extraovarian lesions (benign or noncancerous) which resides in the ovary. Whereas type II tumors are more aggressive in nature, and they are serous-tubule intra-epithelial carcinoma (STIC) in origin [5]. Ovarian cancer is generally prevalent in postmenopausal female with the highest incidence among the patients 50 years' age or above. Ovarian cancer is not easily diagnosed until it reached at advanced stages and it's silent and obscure symptoms make it hard to diagnose and cure appropriately [20] Gastro-intestinal dysfunction, abdominal enlargement, pelvic and abdominal pain are common symptoms found in patients with last stage of ovarian cancer [21]. According to study reports 70% of ovarian cancers are diagnosed at stage III or IV and the survival rate is 47.4% of all people who have been suffering from ovarian cancer for 5 years [21].

Endogenous factors like stress, inflammation and exogenous factors such as radiation, pollution, etc. are responsible for excessive production of reactive oxygen species (ROS) in ovarian cells. The high level of oxidant molecules disrupts the antioxidant defense mechanism and significantly increases the oxidative stress-mediated damage of different biomolecules (nucleic acid, lipid, protein, etc.) [5]. The ovarian carcinoma cells manifested an increased expression of iNOS, MPO, NAD(P)H oxidase, as well as enhanced NO levels which in turn correlated with expression in iNOS. Moreover, these carcinoma cells showed lower apoptosis, which was markedly induced by the inhibition of iNOS thus indicating a strong link between apoptosis and the NO/iNOS pathways in these cells [6]. An overview about the involvement of ROS in case of ovarian cancer is discussed in the following section.

#### *4.1.1.1. Involvement of ROS in Genetic Instability of Ovarian Cancer*

Oxidative stress plays a crucial role in alternation of genetic stability and DNA damage. Hydroxyl radicals and hydrogen peroxide are two reactive molecules mainly generated by Fenton reaction and these reactive molecules can significantly enhance transferrin Receptor 1(TfR1) axis [5]. TfR1, important for iron uptake in cells and its overexpression is associated with malignant tumorigenesis. TfR1 enhances the development of ovarian cancer by inducing double stranded DNA breakage in fallopian tube epithelial cells [5, 22].

8-hydroxy-2′-deoxyguanosine (8-OHdG), is a well-known DNA damage biomarker. It is a product of guanine oxidation, generally used to assess risk related to cancer advancement. Furthermore, 8-OHdG is also responsible for transformation of nucleotide-bases (GC to AT) that leads to mutation (somatic) of ovarian cancer [5, 17]. Mitochondrial ETC is one of the major sources for ROS production, so the chances of mutation in mtDNA is higher than nuclear DNA. There is no DNA repair system that recovers the damage of mtDNA. According to research reports, mutated form of mtDNA was found in ovarian cancer. Study showed that there is strong interconnection between mtDNA mutation and therapeutic resistance of ovarian cancer cells [7, 5]. Additionally, in mutated-mtDNA HGSC (High -grade serous carcinoma) the chances of platinum resistance are relatively high compared to normal mtDNA HGSC model making them chemoresistant to certain anticancer drugs [23].

Tumor suppressor gene p53, is involved in important cellular processes like DNA repair, apoptosis, cell growth and it also reverses genomic instability. Study revealed that approximately 96% of High -grade serous carcinoma (HGSC) has mutant p53. However, mutation of p53 results in generation of high levels of ROS by activation of different redox signaling/enzymes. Mutated p53 can increase tumor growth by sustaining the Warburg effect in cancer cells [5].

#### *4.1.1.2. Involvement of ROS in Signaling Pathways of Ovarian Cancer*

mTOR signaling and PI3K/AKT axis are key regulators that influence the progression and therapeutic resistance of ovarian cancer. Oxidative stress increases the invasiveness of ovarian cancer by upregulating the activity of PI3K/AKT and mTOR pathway [24]. Other reports exhibited that ROS mediated activation of PI3K/AKT pathway increases the phosphorylation of glycogen synthase kinase 3β (GSK3β). Phosphorylated-GSK3β has been observed in many ovarian cancers [25]. This phosphorylated form of GSK3β is incapable of deactivating the activity of β-catenin. Wnt/ β catenin axis is crucial for ovarian cancer progression and poor prognosis [25, 5]. In addition, Wnt/β-catenin is documented to increase expression of ABCB1/MDR-1 (genes for multidrug resistance) and trigger anti-apoptosis of ovarian cancer cells [25]. ROS (derived from NOX1) also promotes the Wnt/β-catenin signaling pathway by blocking the activity of nucleoredoxin. Here, nucleoredoxin is a negative regulator of Wnt pathway. This Wnt/β-catenin significantly increased cell proliferation and is associated with ovarian cancer pathogenesis. Reports suggested that Wnt/β-catenin enhanced the metastasis and it has the capability to sustain the stem- cell like properties of ovarian cancer cells [5].

Involvement of ROS was also found in angiogenesis of ovarian tumor. It has been reported that $H_2O_2$–induced EGFR (Epidermal growth factor receptor) overexpression plays an important role in the angiogenesis of human ovarian cancer, via upregulating the activity of AKT/p70S6K1 axis. In these reports, ovarian cancer cells also elevated VEGF (angiogenesis factor) expression that is induced by overactivated-EGFR [24]. At hypoxic condition, a paradoxical event exhibited by ROS on ovarian cancer metastasis. ROS can induce the high expression of HIF-1α and EMT responsible for gene expression [5]. ROS have been considered to debilitate the expression of anti-metastasis factor, such as E-cadherin through increasing the activity of lysl oxidase in epithelial ovarian carcinoma (EOC) demonstrating a negative correlation between Lysl oxidase and E-cadherin. It has been demonstrated that suppression of E-cadherin expression subsequently increases metastasis of ovarian cancer cells [26].

#### *4.1.1.3. Involvement of ROS in Ovarian Cancer Related Inflammatory Factors*

ROS can potentially increase the expression of inflammation associated factors including cytokine, growth factors like TGFβ, etc. and are responsible for growth and poor prognosis of ovarian cancer [5]. TAMs (tumor-associated macrophages), a crucial inflammatory component which are mainly located in tumor microenvironment. Studies have demonstrated that ROS is a key factor in the release of pro-inflammatory cytokines, including IL1-α, TNF-α, IL-10, etc. from TAMs, which induce the progression of ovarian cancer [5]. Interaction of tumor cells and TAMs creates a metabolic product known as itaconic acid. In peritoneal macrophages, the production of itaconic acid is significantly stimulated in ovarian cancer. Itaconic acid subsequently increases ROS production and promotes ROS-induced tumor growth [5]. In addition, neutrophils of ovarian cancer patients exhibit high levels of ROS, thereby resulting in increased growth and metastasis of tumor [27].

Ascites is generally seen in the ovarian tumor microenvironment and is responsible for disseminating cancer cells to areas (peritoneal or pelvic) close to the ovary [28]. This phenomenon is strongly associated with oxidative stress [29]. As reported by Pakula et al., serous ovarian tumor-generated ascites increases oxidative stress in HPMCs (human peritoneal mesothelial cells) enhancing the activity of oxidative phosphorylation associated enzymes and initiates ovarian cancer cell expansion and migration [30]. The interrelationship of ROS with ovarian cancer is presented in the **Figure 2.**

## 5. Contribution of ROS in Anticancer Therapeutics against Ovarian Cancer

As discussed, in normal cells ROS may act as a signal transducing agent and are involved in other biochemical processes. But when the redox balance breaks down in a normal cell, it turns into an abnormal condition. In this situation ROS concentration in a cell rises and breaks cellular proteins, DNAs, lipids and pushes a cell into senescence, degradation or in a diseased condition. Beside these, there are multiple reports in the literature showing ROS can serve as a potential tumor suppressor and can induce cell apoptosis, necrosis, autophagy and other cell damaging processes. Multiple drugs and agents employ these features to target a diseased cell leading to cell death. Simply, ROS exhibits prominent characteristics of two-edged sword. This complex and crucial role of ROS has been proved as both advantageous and disadvantageous. Now it's an understandable approach to the researchers to find out which one is more applicable, where and why. More specifically, researchers are aiming to develop therapeutic strategies based on modulating ROS levels to treat cancer. In this context, here we will discuss ROS regulated cell death pathways with a special emphasis to the treatment of ovarian cancer [31].

### 5.1. Role of ROS in Cancer Therapeutics

The most common way by which ROS kills transformed cells is the activation of programmed cell death. Radiotherapy as well as chemotherapy causes an increase in intracellular ROS leading to apoptosis via extrinsic or intrinsic pathways. ROS also triggers apoptosis by inactivating or increasing the ubiquitination of the anti-apoptotic protein Bcl-2 and by increasing the intracellular levels of Bax and Bad [32]. ROS are also able to induce necrosis, which was originally considered an unregulated form of cell death which is now recognized as necroptosis, another type of programmed cell death [33]. Moreover, many recent studies indicate the therapeutic approach to kill cancer cells by ROS-induced autophagy. Enhanced ROS levels can impair the multidrug resistance of cancer cells thereby increasing the efficacy of anticancer therapeutics [32]. Following section focuses on the involvement of ROS in different kind of programmed cell death.

#### *5.1.1. Involvement of ROS in Apoptosis*

ROS play significant role in death receptor mediated apoptosis and mitochondria mediated apoptosis. During Fas activation in death receptor mediated apoptosis, Fas ligand (FasL) induced rapid ROS production by involving NADPH oxidase. In Fas-tyrosine phosphorylation event, epidermal growth factor receptor (EGFR) and Fas interaction requires FasL mediated ROS response. This event gives an indication to recruit Fas-associated death domain and caspase 8 in apoptosis induction process. Beside this, FasL-induced ROS resolves ubiquitination and deterioration by proteasome of FLICE (Fas-associated death domain-like interleukin-1beta-converting enzyme)-inhibitory protein (FLIP). This event further demands Fas activation [31].

In mitochondria mediated apoptosis, a pore complex is formed, through which cytochrome c is released. Here the role of ROS is to give stability to the pore complex. ROS exploits both the cell signaling cascade and oxidative moderation of the pore complex components. c-Jun,

NH-2 terminal Kinase (JNK) are known as stress activated protein kinases (SAPKs). These signaling molecules are involved in ROS-mediated pore complex opening and aid in apoptosis [31].

A wide array of drugs used in cancer therapeutics induce oxidative stress. Procarbazine, used for the treatment of Hodgkin's lymphoma and brain cancers, induces apoptosis by oxidative DNA damage. 5-fluorouracil, a pyrimidine analog, produces ROS through p53-dependent pathways and induces apoptosis in colon and rectal cancer cells [32].

Moreover, Doxorubicin-dependent cytotoxicity is linked to the stimulation of a Fenton reaction that generates hydroxyl radicals successfully used for the treatment of Kaposi's sarcoma, breast and bladder cancer and acute lymphocytic leukemia (33). Arabinocytosine, hampered DNA replication, followed by anthracyclines to increase ROS, has been shown to drive programmed cell death, with beneficial effects for patients with acute myeloid leukemia (AML) [34].

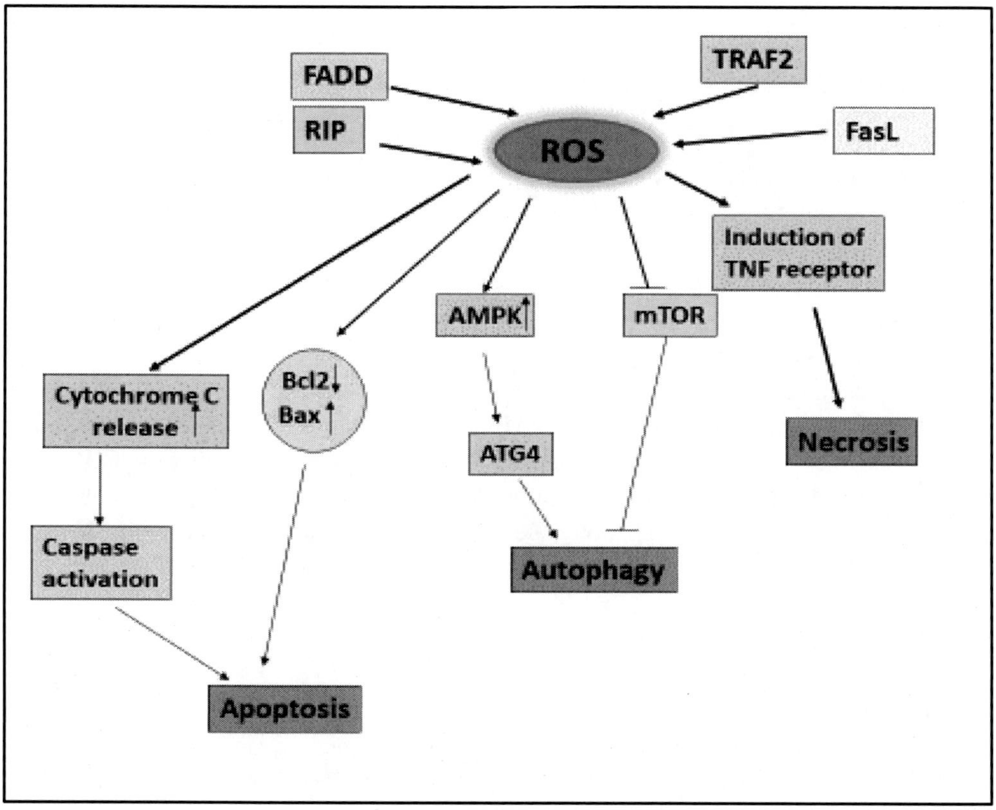


**Figure 2.** ROS mediated activation of different signaling pathways, transcription factors and inflammatory factors that are responsible for progression of ovarian cancer.

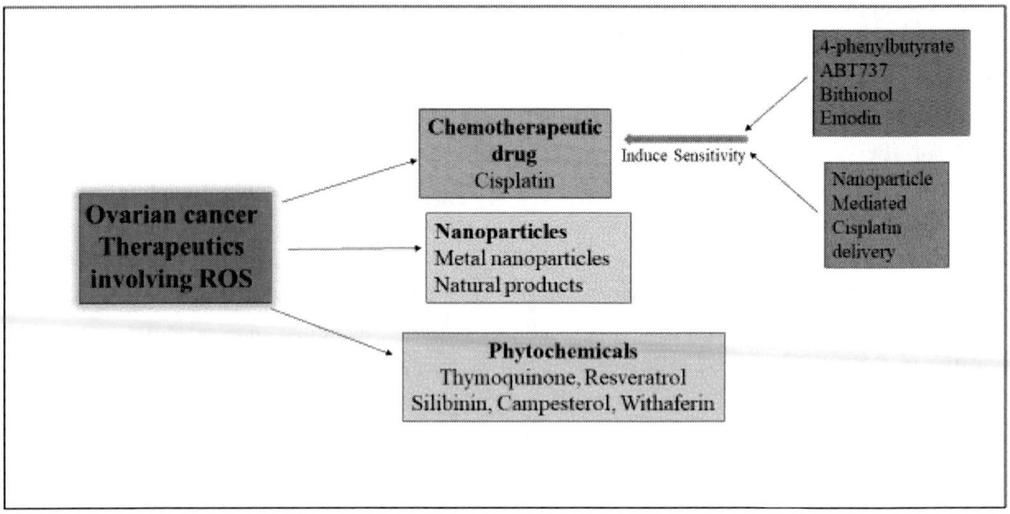


**Figure 3.** Interrelationship of ROS in different cellular death processes.

Oxygen free radicals stimulate apoptosis through VDAC (voltage-dependent anion channel)-dependent permeability of mitochondrial outer membrane. ANT (adenine-nucleotide translocase) undergoes conformational changes and inhibits nucleotide binding by ROS stimulation and promotes calcium entry. Calcium surges hypothesize cyclophilin D-ANT complex formation leading to the opening of pore and apoptosis [31].

### *5.1.2. ROS and Necrosis*

Studies suggested that necrotic cell death also involves ROS. Both the mitochondria and NADPH oxidase generated ROS, is involved in necrotic cell death. TNF induced necrotic cell death results from the accumulation of ROS by receptor-interacting protein (RIP), tumor necrosis factor receptor (TNFR)-associated factor-2 (TRAF2) and FADD [35]. Fiers W., et al. proposed that mitochondria is supposed to be the main source of ROS in necrosis (36). Kim et al. speculated Nox1 is responsible for TNFα-induced ROS generation in mouse fibroblasts [37].

### *5.1.3. Role of ROS in Autophagy*

Autophagy is a physiological process by which a eukaryotic cell breaks down and its organelles and macromolecules undergo recycling. Like apoptosis and necrosis, autophagy can also exploit oxidative stress in cellular response. ROS acts as a signaling molecule in survival-prone autophagy. ROS-mediated autophagy is a new discovery in biological science. In radiotherapy and conventional chemotherapeutics, malignant tumors become resistant to proapoptotic therapies and here lies the importance of autophagic cell death to degrade those kind of malignancies [31].

Specifically, it has been reported that $H_2O_2$-dependent inactivation of ATG4 (autophagy-related gene-4) increases LC3-associated autophagosomes and the ATM (ataxia-telangiectasia mutated (ATM) protein kinase)-mediated oxidation of AMP-activated protein kinase (AMPK) inhibits mammalian target of rapamycin 1(mTORC1) which is a pivotal negative regulator of autophagy. In this connection, it has been reported that $H_2O_2$ induces autophagic cell death in glioma cells after treatment with sanguinarine (a polycyclic ammonium ion) which increases electron leakage from mitochondria and induces NOXs. Rapamycin, administered in combination with inhibitors of HSP90, causes mitochondrial damage with accompanying oxidative stress and autophagy and reduces tumor growth in RAS-dependent tumors [32]. **Figure 3** depicts the role of ROS associated with various cellular death processes.

## 5.2. Current Therapeutic Status of Ovarian Cancer Involving ROS

Nowadays cytoreduction and the combination of paclitaxel and carboplatin-based chemotherapeutics are among the most exploited ovarian cancer treatment strategies. Angiogenesis inhibitors like bevacizumab and poly (ADP-ribose) polymerase (PARP) inhibitors like Olaparib and niraparib are more efficient and safer in ovarian cancer treatment. Patients having BRCA1/2 mutation or homologous recombination deficiency are better treated with these PARP inhibitors [38].

Different polyphenols and flavonoids, such as curcumin, genistein, quercetin and resveratrol can maintain the balance of antioxidant systems by scavenging ROS. These antioxidant phytochemicals inhibit human cancers, including ovarian cancer, by their anticancer activities, including anti-angiogenic, anti-proliferative, anti-inflammatory, and pro-apoptotic properties [39]. Reports have shown that inhibiting ROS production by pharmacological inhibitors, antagonists, or scavengers can inhibit the proliferation of ovarian cancer cells. Diphenylene iodinenium, a ROS inhibitor, promotes apoptosis in ovarian cancer cells by inhibiting NADPH oxidase (NOX). Lysophosphatidic acid (LPA), a bioactive lipid mediator, induced cell proliferation by stimulating NOX-dependent ROS generation in ovarian cancer [40].

## 5.3. Treatments of Ovarian Cancer and ROS Participation

There are several ovarian cancer treatment strategies associated with the involvement of ROS. Among them a few are discussed in the subsequent section.

### *5.3.1. Cisplatin Induced Chemotherapeutic Strategy*

Cisplatin is a platinum based anti-cancer drug having antimicrobial activity and used in the treatment of various carcinomas including ovarian cancer. It has been used in various cancer treatments including ovarian cancer. But cisplatin usually becomes chemoresistant because it has the tendency to internalize cellular organelles and stimulate autophagy, unfolded protein response (UPR) and other defensive mechanisms. Therefore, enhancing cisplatin sensitivity to cancer cells may be the feasible way-out to treat cancer [38].

*In vitro* experiments suggested that treatment of 4-phenylbutyrate with cisplatin can increase cisplatin sensitivity to ovarian cancer cells. This can increase ROS accumulation in A2780 cell line and inhibit histone deacetylase activity and promote apoptosis and DNA damage (40). Compounds like ABT737, Bithionol and Emodin also act synergistically to enhance the activity of cisplatin by the involvement of ROS [41].

Reports suggested that Cisplatin can also induce ROS generation which in turn results in the upregulation of Sirtuin 2 (SIRT2). This SIRT2 can resensitize human ovarian cancer cell line to cisplatin, enhancing cisplatin-induced cell death [42].

Ongoing studies are focusing on the employment of different nanoparticles which can increase ROS accumulation and sensitize ovarian cancer cells to cisplatin sensitivity [42]

Chemoresistance is a major challenge for current ovarian cancer therapy. Previous studies indicated that epithelial-mesenchymal transition (EMT) and cancer stem cell (CSC) are the culprits of the chemoresistance, metastasis, tumor relapse and death in ovarian cancer patients. A very recent report suggested that cisplatin in combination with a dual PI3K/mTOR inhibitor (BEZ235) induced higher ROS level and more apoptosis in chemoresistant epithelial ovarian cancer cells. Moreover, this treatment combination effectively inhibited PI3K/Akt/mTOR signaling pathway, reversed EMT and decreased CSC marker expression in chemoresistant ovarian cancer cells [43].

### *5.3.2. Celastrol Loaded Nanoparticle in Ovarian Cancer Treatment*

Celastrol is a natural product, isolated from traditional medicine and has been proven to be curative for inflammation and cancers. This drug was used to treat ovarian cancer in preclinical models. Due to the low specificity and low solubility of celastrol loaded nanoparticles have been made to check its efficacy against ovarian cancer. It is reported that celastrol loaded nanoparticles stimulated the production of ROS and can amplify the treatment efficacy in SKOV3 cell line. Folic acid groups on the nanoparticle surface led them to the targeted delivery to the cell or tissues. Celastrol induced apoptosis by activating glycogen synthase kinase 3β which in turn activated TNF- α-induced NF-κB signaling pathway and impaired tumorigenesis [44].

### *5.3.3. Gold Nanoparticles Involving ROS Pathway in Treatment of Ovarian Cancer*

Researchers have experimentally established that peanut-shaped gold nanoparticles (AuP-NPs) can be a potential anti-cancer agent in treating ovarian cancer. The cytotoxic activity and mechanism of action are investigated on SKOV3 cell line. It has been established that gold nanoparticles can induce apoptosis and autophagy through ROS-mediated pathways [45].

### *5.3.4. Phytochemicals in Ovarian Cancer Treatments*

In addition, with other conventional drugs, phytochemicals also have potential effects upon ovarian cancer cells. Flavonoids like silibinin, campesterol, thymoquinone, resveratrol, withaferin are few important flavonoids having anti-cancer effects against ovarian cancer. Ovarian cancer cell lines like A2780, OVCAR-3 and SKOV-3 were tested, and it was postulated that flavonoids can decrease the viability of these cancer cells. Flavonoids exploit extrinsic apoptotic pathway in presence of ROS scavenging on the other hand it will prefer intracellular ROS production to activate intrinsic apoptotic pathway [46].

Silibinin, a natural polyphenolic flavonoid, can increase ROS accumulation and downregulate ERK and Akt pathway which in turn inhibits tumor growth and cell survival in case of ovarian carcinomas. It has been experimentally established that both *in vitro* (A2780 cell line) and *in vivo*, administration of Silibinin is effective against ovarian cancer [47].

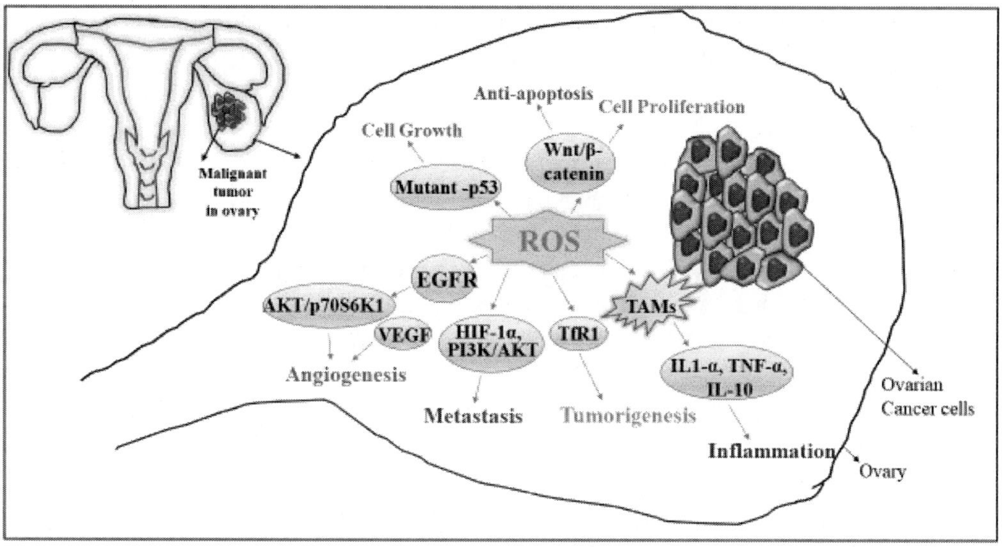


**Figure 4.** Summary of Ovarian cancer therapies involving ROS.

Campesterol, an important phytosterols, induced cell death signals and cell death in human ovarian cancer cells by the generation of reactive oxygen species (ROS), and calcium concentrations [48].

Thymoquinone, the active ingredient of *Nigella sativa*, triggered ROS-mediated cytotoxicity against ovarian cancer cell line (Caov-3 cells) and induced its anti-cancer effect by intrinsic pathway of apoptosis and cellular oxidative stress [49].

Resveratrol, a natural polyphenol compound, has been displayed its anticancer effects against human ovarian cancer cell lines (A2780 and SKOV3). Enhanced ROS production was involved in the cell death via ROS-dependent decrease of Notch1 signaling [50]. Moreover, resveratrol effectively can eliminate ovarian cancer stem cells *in vitro* through induction of cell death and inhibition of self-renewal capacity. Resveratrol-induces the loss of the self-renewal capacity of ovarian CSCs in which ROS played an important role [51].

Normally cancer treatment via doxorubicin is constricted because of its serious side effects. But combination treatment of doxorubicin with withaferin, a phytochemical, can increase the efficacy of the drug synergistically, thereby reducing its side effects. *In vitro* experiments with A2780, A2780/CP70 and CaOV3 cell lines showed apoptosis, cell proliferation inhibition upon treatment with withaferin-doxorubicin combination. This combination treatment resulted in enhanced ROS production. Excess ROS caused DNA damage, autophagy and exerted deleterious effects on ovarian cancer cells (52). **Figure 4** describes the therapeutic role of ROS in ovarian cancer.

## 6. Conclusion

Since last century, ROS has gained attention due to its multifactorial action in the body. ROS regulates several pathways as an intermediate factor and is closely linked with different pathological conditions mainly in its uncontrolled production. But how ROS modulates the diseasing causing pathways remains unclear. So, further research should continue to elucidate detailed mechanisms of action of ROS in different diseases. The multiplex interconnection between ROS levels and cancer is essentially based on accurate fine-tuning between ROS production and scavenging. Cancer initiation and progression exploit the slight increases in ROS levels. Various lines of evidence support the role of oxidative stress in the development and progression of ovarian cancer. Ovarian cancer being a detrimental one requires more attention towards its treatment and management. This overview may open the way for the evaluation of therapeutic approaches targeting oxidative imbalance: the oxidative stress status may represent the key to treat and, eventually, to prevent ovarian cancer. Further extensive research and clinical trials are required to investigate the possible involvement of ROS in anticancer therapeutics.

## Acknowledgments

I would like to acknowledge the publisher, editor, and support staff at Nova Science Publishers, for their help in the various stages of developing and producing the chapter. Authors would also like to acknowledge Diamond Harbour Women's University for its support.

## Abbreviations

| | |
|---|---|
| Akt | Ak strain transforming |
| EGFR | Epidermal growth factor receptor |
| VEGF | Vascular endothelial growth factor |
| HIF-1 α | Hypoxia-inducible factor 1 alpha |
| TfR1 | Transferrin receptor 1 |
| IL-1α | Interleukin-1 alpha |
| TAMs | Tumor-associated macrophages |
| TNF-α | Tumor necrosis factor alpha |
| IL-10 | Interleukin 10 |
| Bcl2 | B-cell lymphoma 2 |
| Bax | BCL2-Associated X Protein |
| AMPK | AMP-activated protein kinase |
| ATG4 | Autophagy-related gene 4 |
| mTOR | mammalian target of rapamycin |
| RIP | Receptor interacting protein |
| TRAF-2 | TNF receptor-associated factor 2 |
| FADD | Fas-associated death domain protein |
| Fas L | Fas ligand. |

## References

[1] Yang, S., and Lian, G. (2020). ROS and diseases: role in metabolism and energy supply. *Mol Cell Biochem.*, 467 (1-2), 1–12. Available from: doi: 10.1007/s11010-019-03667-9.

[2] Prasad, S., Gupta, S. C., and Tyagi, A. K. (2017). Reactive oxygen species (ROS) and cancer: role of antioxidative nutraceuticals. *Cancer Lett.*, 387, 95-105. Available from: doi: 10.1016/j.canlet.2016.03.042.

[3] Briegera, K., Schiavonea, S., Miller, Jr. F. J., and Krause, K. H. (2012). Reactive oxygen species: from health to disease., *Swiss Med Wkly.*, 142, w13659. Available from: doi: 10.4414/smw.2012.13659.

[4] Kim, G. H., Kim, J. E., Rhie, S. J., and Yoon, S. (2015). The role of oxidative stress in neurodegenerative diseases., *Exp Neurobiol.*, 24 (4), 325-340. Available from: doi: 10.5607/en.2015.24.4.325.

[5] Ding, D. N., Xie, L. Z., Shen, Y., Li, J., Guo, Y., Fu, Y., Liu, F. Y., and Han, F. J. (2021). Insights into the Role of Oxidative Stress in Ovarian Cancer. *Oxid Med Cell Longev.*, 2021, 1-20. Available from: doi: 10.1155/2021/8388258.

[6] Saed, G. M., Diamond, M. P., and Fletcher, N. M. (2017). Updates of the role of oxidative stress in the pathogenesis of ovarian cancer. *Gynecol Oncol.*, 45(3), 595-602. Available from: doi: 10.1016/j.ygyno.2017.02.033.

[7] Xu, T., Ding, W., Ji, X., Ao, X., Liu, Y., Yu, W., and Wang, J. (2019). Oxidative stress in cell death and cardiovascular diseases. *Oxid Med Cell Longev.*, 2019, 1-11. Available from: doi: 10.1155/2019/9030563.

[8] Li, X., Fang, P., Mai, J., Choi, E. T., Wang, H., and Yang, X. F. (2013). Targeting mitochondrial reactive oxygen species as novel therapy for inflammatory diseases and cancers. Available from: *J Hematol Oncol.*, 6, 19. doi: 10.1186/1756-8722-6-19.

[9] Datta, K., Singha, S., and Chattopadhyay, P. (2000). Reactive oxygen species in health and disease. *Natl Med J India.*, 13(6), 303-310.

[10] Sanocka, D., and Kurpisz, M. (2004). Reactive oxygen species and sperm cells. *Reprod Biol Endocrinol.*, 2, 12. Available from: doi: 10.1186/1477-7827-2-12.

[11] Hanukoglu, I. (2006). Antioxidant protective mechanisms against reactive oxygen species (ros) generated by mitochondrial p450 systems in steroidogenic cells. *Drug Metab Rev.*, 38 (1-2), 171–196. Available from: doi: 10.1080/03602530600570040.

[12] Drobot, L. B., Samoylenko, A. A., Vorotnikov, A. V., Tyurin-Kuzmin, P. A., Bazalii, A. V., Kietzmann, T., Tkachuk, V. A., and Komisarenko, S. V. (2013). Reactive oxygen species in signal transduction. *Biochem Biotechnol Mod Med.*, 85(6), 209-217. Available from: doi:10.15407/ubj85.06.209.

[13] Santos, A. L., Sinha, S., and Lindner, A. B. (2018). The good, the bad, and the ugly of ros: new insights on aging and aging-related diseases from eukaryotic and prokaryotic model organisms. *Oxid Med Cell Longev.*, 2018, 1941285. Available from: doi: 10.1155/2018/1941285.

[14] Waris, G., and Ahsan, H. (2006). Reactive oxygen species: role in the development of cancer and various chronic conditions. *J Carcinog.*, 5, 14. Available from: doi: 10.1186/1477-3163-5-14.

[15] Aggarwal, V., Tuli, H. S., Varol, A., Falak, T., Yerer, M. B., Sak, K., Varol, M., Jain, A., Khan, M. A., and Sethi, G. (2019). Role of reactive oxygen species in cancer progression: molecular mechanisms and recent advancements. *Biomolecules.*, 9, 735. Available from: doi: 10.3390/biom9110735.

[16] Liao, Z., Chua, D., and Tan, N. S. (2019). Reactive oxygen species: a volatile driver of field cancerization and metastasis. *Mol Cancer.*, 18, 65. Available from: doi: 10.1186/s12943-019-0961-y.

[17] Aydin, E., Johansson, J., Nazir, F. H., Hellstrand, K., and Martner, A. (2017). Role of nox2-derived reactive oxygen species in nk cell–mediated control of murine melanoma metastasis. *Cancer Immunol Res.*, 5, 804–811. Available from: doi: 10.1158/2326-6066.CIR-16-0382.

[18] Rodrigues, C., Pimpao, C., Mosca, A. F., Coxixo, A. S., Lopes, D., Silva, I. V., Pedersen, P. A., Antunes, F., and Soveral, G. (2019). Human aquaporin-5 facilitates hydrogen peroxide permeation affecting adaptation to oxidative stress and cancer cell migration. *Cancers (Basel).*, 11(7), 932. Available from: doi: 10.3390/cancers11070932.

[19] Khromova, N. V., Kopnin, P. B., Stepanov, E. V., Agapova, L. S., and Kopnin, B. P. (2009). p53 hot-spot mutants increase tumor vascularization via ROS-mediated activation of the HIF1/VEGF-A pathway. *Cancer Lett.*, 276(2), 143–151. Available from: doi: 10.1016/j.canlet.2008.10.049.

[20] Roett, M. A., and Evans, P. (2009). Ovarian cancer: an overview. *Am Fam Physician.*, 80(6), 609-616.

[21] Stewart, C., Ralyea, C., and Lockwood, S. (2019). Ovarian cancer: An integrated review. *Semin Oncol Nurs.*, 35(2), 151-56. Available from: doi: 10.1016/j.soncn.2019.02.001.

[22] Candelaria, P. V., Leoh, L. S., Penichet, M. L., and Daniels-Wells, T. R. (2021). Antibodies targeting the transferrin receptor 1 (tfr1) as direct anti-cancer agents. *Front immunol.*, 12, 607692. Available from: doi: 10.3389/fimmu.2021.607692.

[23] Wang, C., Han, S., Chen, X., and Guo, W. (2020). Pathogenic heteroplasmic somatic mitochondrial dna mutation confers platinum-resistance and recurrence of high-grade serous ovarian cancer. *Cancer manag res.*, 12, 11085–11093. Available from: doi: 10.2147/CMAR.S277724.

[24] Liu, L. Z., Hu, X. W., Xia, C., He, J., Zhou, Q., Shi, X., Fang, J., and Jiang, B. H. (2006). Reactive oxygen species regulate epidermal growth factor-induced vascular endothelial growth factor and hypoxia-inducible factor-1α expression through activation of AKT and P70S6K1 in human ovarian cancer cells. *Free Radic Biol Med.*, 41(10), 1521–1533. Available from: doi: 10.1016/j.freeradbiomed.2006.08.003.

[25] Arend, R. C., Londono-Joshi, A. I., Straughn, Jr. J. M., and Buchsbaum, D. J. (2013). The Wnt/β-catenin pathway in ovarian cancer: A review. *Gynecol Oncol.*, 131(3), 772–779. Available from: doi: 10.1016/j.ygyno.2013.09.034.

[26] Savant, S. S., Sriramkumar, S., and Hagan, H. M. O. (2018). The role of inflammation and inflammatory mediators in the development, progression, metastasis, and chemoresistance of epithelial ovarian cancer. *Cancers.*, 10(8), 251. Available from: doi: 10.3390/cancers10080251.

[27] Klink, M., Jastrzembska, K., and Nowak, M. (2008). Ovarian cancer cells modulate human blood neutrophils response to activation *in vitro*. *Scand J Immunol.*, 68(3), 328–336. Available from: doi: 10.1111/j.1365-3083.2008.02139. x.

[28] Worzfeld, T., Pogge von Strandmann, E., Huber, M., Reinartz, S., and Muller, R. (2017). The unique molecular and cellular microenvironment of ovarian cancer. *Front Oncol.*, 7, 24. Available from: doi: 10.3389/fonc.2017.00024

[29] Mikuła-Pietrasik, J., Uruski, P., Szubert, S., Sajdak, S., Tykarski, A., and Książek, K. (2017). Malignant ascites determine the transmesothelial invasion of ovarian cancer cells. *Int J Biochem Cell Biol.*, 92, 6–13. Available from: doi: 10.1016/j.biocel.2017.09.002.

[30] Pakuła, M., Mikuła-Pietrasik, J., Stryczynski, L., Szuber, S.., Moszynski, R., Szpurek, D., Sajdak, S., Tykarski, A., and Książek, K. (2018). Mitochondria-related oxidative stress contributes to ovarian cancer-promoting activity of mesothelial cells subjected to malignant ascites. *Int J Biochem Cell Biol.*, 98, 82–88. Available from: doi: 10.1016/j.biocel.2018.03.011.

[31] Wang, J., and Yi, J. (2008). Cancer cell killing via ros: To increase or decrease that is the question. *Cancer Biol Ther.*, 7, 1875-1884. Available from: doi: 10.4161/cbt.7.12.7067.

[32] Hwang, P. M., Bunz, F., Yu, J., Rago, C., Chan, T. A., Murphy, M. P., Kelso, G. F., Smith, R. A., Kinzler, K. W., and Vogelstein, B. (2001). Ferrodoxin reductase affects p53-dependent, 5-fluorouracil-induced apoptosis in colorectal cancer cells. *Nat Med.*, 7(10), 1111-1117. Available from: doi: 10.1038/nm1001-1111.

[33] Perillo, B., Donato, M. D., Pezone, A., Zazzo, E. D., Giovannelli, P., Galasso, G., Castoria, G., and Migliaccio, A. (2020). ROS in cancer therapy: the bright side of the moon. *Exp Mol Med.*, 52(2), 192-203. Available from: doi: 10.1038/s12276-020-0384-2.

[34] Prieto-Bermejo, R., Romo-Gonzalez, M., Perez-Fernandez, A., Ijurko, C., and Hernandez- Hernandez, A. (2018). Reactive oxygen species in hematopoiesis: leukemic cells take a walk on the wild side. *J Exp Clin Cancer Res.*, 37(1), 125. Available from: doi: 10.1186/s13046-018-0797-0.

[35] Galluzzi, L., and Kroemer, G. (2008). Necroptosis: a specialized pathway of programmed necrosis. *Cell.*, 135(7), 1161-1163. Available from: doi:10.1016/j.cell.2008.12.004.

[36] Fiers, W., Beyaert, R., Declercq, W., and Vandenabeele, P. (1999). More than one way to die: apoptosis, necrosis and reactive oxygen damage. *Oncogene.*, 18(54), 7719-7730. Available from: doi:10.1038/sj.onc.1203249.

[37] Kim, Y. S., Morgan, M. J., Choksi, S., and Liu, Z. G. (2007). Tnf-induced activation of the nox1 NADPH oxidase and its role in the induction of necrotic cell death. *Mol Cell.*, 26(5), 675-687. Available from: doi:10.1016/j.molcel.2007.04.021.

[38] Mirzaei, S., Hushmandi, K., Zabolian, A., Saleki, H., Torabi, S. M. R., Ranjbar, A., SeyedSale, S. H., Sharifzadeh, S. O., Khan, H., Ashrafizadeh, M., Zarrabi, A., and Ahn, K. S. (2021). Elucidating role of

reactive oxygen species (ros) in cisplatin chemotherapy: a focus on molecular pathways and possible therapeutic strategies. *Molecules.*, 26(8), 2382. Available from: doi: 10.3390/molecules26082382.

[39] Landis-Piwowar, K. R., Milacic, V., Chen, D., Yang, H., Zhao, Y., Chan, T. H., Yan, B., and Dou, Q. P. (2006). The proteasome as a potential target for novel anticancer drugs and chemosensitizers. *Drug Resist Updat.*, 9(6), 263-273. Available from: doi: 10.1016/j.drup.2006.11.001.

[40] Kobayashi, H., Imanaka, S., and Shigetomi, H. (2022). Revisiting therapeutic strategies for ovarian cancer by focusing on redox homeostasis (review). *Oncol Lett.*, 23(3), 80. Available from: doi:10.3892/ol.2022.13200.

[41] Almotairy, A. R. Z., Montagner, D., Morrison, L., Devereux, M., Howe, O., and Erxleben, A. (2020). Pt (IV) pro-drugs with an axial hdac inhibitor demonstrate multimodal mechanisms involving dna damage and apoptosis independent of cisplatin resistance in A2780/A2780cis cells. *J Inorg Biochem.*, 210:111125. Available from: doi:10.1016/j.jinorgbio.2020.111125.

[42] Wang, W., Im, J., Kim, S., Jang, S., Han, Y., Yang, K. M., Kim, S. J., Dhanasekaran, D. N., and Song., Y. S. (2020). ROS-induced sirt2 upregulation contributes to cisplatin sensitivity in ovarian cancer. Antioxidants (Basel)., 9(11), 1137. Available from: doi: 10.3390/antiox9111137.

[43] Deng, J., Bai, X., Feng, X., Ni, J., Beretov, J., Graham, P., and Li, Y. (2019). Inhibition of PI3K/Akt/mTOR signaling pathway alleviates ovarian cancer chemoresistance through reversing epithelial-mesenchymal transition and decreasing cancer stem cell marker expression. *BMC Cancer.*, 19(1), 618. Available from: doi: 10.1186/s12885-019-5824-9.

[44] Niu, W., Wang, J., Wang, Q., and Shen, J. (2020). Celastrol loaded nanoparticles with ros-response and ros-inducer for the treatment of ovarian cancer. *Front Chem.*, 8, 574614. Available from: doi:10.3389/fchem.2020.574614.

[45] Piktel, E., OScilowska, I., Suprewicz, L., Depciuch, J., Marcinczyk, N., Chabielska, E., Wolak, P., Wollny, T., Janion, M., Parlinska-Wojtan, M., and Bucki, R. (2021). ROS-mediated apoptosis and autophagy in ovarian cancer cells treated with peanut-shaped gold nanoparticles. *Int J Nanomedicine.*, 16, 1993-2011. Available from: doi:10.2147/IJN.S277014.

[46] Tavsana, Z., and Kayali, H. A. (2019). Flavonoids showed anticancer effects on the ovarian cancer cells: involvement of reactive oxygen species, apoptosis, cell cycle and invasion. *Biomed Pharmacother.*, 116, 109004. Available from: doi:10.1016/j.biopha.2019.109004.

[47] Cho, H. J., Suh, D. S., Moon, S. H., Song, Y. J., Yoon, M. S., Park, D. Y., Choi, K. U., Kim, Y. K., and Kim, K. H. (2013). Silibinin inhibits tumor growth through downregulation of extracellular signal-regulated kinase and akt *in vitro* and *in vivo* in human ovarian cancer cells. *J Agric Food Chem.*, 61(17), 4089-4096. Available from: doi: 10.1021/jf400192v.

[48] Bae, H., Park, S., Yang, C., Song, G., and Lim, W. (2021). Disruption of endoplasmic reticulum and ros production in human ovarian cancer by campesterol. *Antioxidants (Basel).*, 10(3), 379. Available from: doi: 10.3390/antiox10030379.

[49] Taha, M. M. E., Sheikh, B. Y., Salim, L. Z. A., Mohan., S., Khan, A, Kamalidehghan, B., Ahmadipour, F., and Abdelwahab, S. I. (2016). Thymoquinone induces apoptosis and increase ros in ovarian cancer cell line. *Cell Mol Biol.*, 62(6), 97-101.

[50] Kim, T. H., Park, J. H., and Woo, J. S. (2019). Resveratrol induces cell death through ROS-dependent downregulation of notch1/pten/akt signaling in ovarian cancer cells. *Mol Med Rep.*, 19(4), 3353-3360. Available from: doi:10.3892/mmr.2019.9962.

[51] Seino, M., Okada, M., Shibuyam, K., Seino, S., Suzuki, S., Takeda, H., Ohta, T., Kurachi, H., and Kitanaka, C. (2015). Differential contribution of Ros to resveratrol-induced cell death and loss of self-renewal capacity of ovarian cancer stem cells. *Anticancer Research.*, 35(1), 85-96.

[52] Fong, M. Y., Jin, S., Rane, M., Singh, R. K., Gupta, R., and Kakar, S. S. (2012 Jul 30). Withaferin A synergizes the therapeutic effect of doxorubicin through ros-mediated autophagy in ovarian cancer., Edited by Yunli Zhou. *PLoS ONE*, 7(7), e42265. Available from: doi: 10.1371/journal.pone.0042265.

## Chapter 14

# Therapeutic Strategies Targeting Oxidative Stress in Diabetes

**Abhishek Sehrawat[1]**
**Jayapriya Mishra[1]**
**Satinder Kaur[1]**
**Inderpal Singh Sidhu[2]**
**Bhawana Thukral[3]**
**Gurjit Kaur Bhatti[4]**
**Sarabjit Singh Mastana[5]**
**P. Hemachandra Reddy[6]**
**and Jasvinder Singh Bhatti[1,*]**

[1]Department of Human Genetics and Molecular Medicine, School of Health Sciences, Central University of Punjab, Bathinda India
[2]Department of Zoology, Sri Guru Gobind Singh College, Sector 26, Chandigarh India
[3]Department of Nutrition and Dietetics, University Institute of Applied Health Sciences, Chandigarh University, Mohali India
[4]Department of Medical Lab Technology, University Institute of Applied Health Sciences, Chandigarh University, Mohali India
[5]School of Sport, Exercise and Health Sciences, Loughborough University, Loughborough, UK
[6]Department of Internal Medicine, Texas Tech University Health Sciences Center, Lubbock, Texas, USA; Department of Pharmacology and Neuroscience, Texas Tech University Health Sciences Center, Lubbock, Texas, USA; Department of Public Health, Graduate School of Biomedical Sciences, Texas Tech University Health Sciences Center, Lubbock, Texa, USA; Department of Neurology, Texas Tech University Health Sciences Center, Lubbock, Texas, USA; Department of Speech, Language, and Hearing Sciences, Texas Tech University Health Sciences Center, Lubbock, Texas, USA

## Abstract

Diabetes mellitus, a commonly known composite metabolic disorder affecting multiple organs such as kidney, heart, nerve, and eye is the most common cause of mortality and

[*] Corresponding Author's E-mail: jasvinder.bhatti@cup.edu.in.

In: The Role of Reactive Oxygen Species in Health and Disease
Editor: Kuladip Jana
ISBN: 979-8-89113-283-2

morbidity globally. Several clinical and experimental studies indicate that hyperglycemia-induced overproduction of reactive free radicals promotes the advancement of diabetes and its complications. Usually, in response to enhanced reactive oxygen species (ROS) production, our body has endogenous antioxidant defense mechanisms, or it can uptake exogenous antioxidants from the food/diet, which neutralizes ROS and thus maintains bodily homeostasis. Any impairment of balance between antioxidants and ROS production will give rise to a condition known as "oxidative stress." Evidence suggests that oxidative stress also impairs glucose metabolism, induces structural and functional damage to cellular macromolecules such as DNA, proteins, and lipids, and alters the body's enzymatic system. It also interferes with different cellular signaling pathways and thus might induce microvascular and macrovascular complications in diabetes. As our understanding related to the molecular mechanisms of ROS generation and its detoxification evolves, it clearly indicates that therapies targeting ROS (either preventing their generation or directly scavenging them) in combination with traditional clinical therapeutics may prove more beneficial in treating diabetes. In this chapter, we intend to summarize the basics of diabetes, contribution of oxidative stress in its etiology, and discuss currently available therapeutics or novel strategies targeting oxidative stress in diabetes

**Keywords**: hyperglycemia, diabetes, reactive oxygen species, oxidative stress, diabetic complications, antioxidants

## 1. Introduction

Diabetes mellitus is a multifactorial and persistent metabolic disorder delineated by steady hyperglycemia. The term "mellitus" was first coined by John Rollo in 1798 and literally means "sweet as sugar" [1]. According to recent estimates, diabetes affects 537 million adults globally and an upsurge of approximately 783 million is expected in the next 25 years (2021). The global rise in prevalence of diabetes at such a huge rate appears to cause a huge socioeconomic burden soon, especially in the developing countries. In most cases, diabetes remains undiagnosed; approximately 1 in 2 people are not familiar with the disease, representing a lack of awareness about such dreadful disease [2]. If diabetes remains undiagnosed for prolonged periods, it will result in serious complications such as nephropathy, retinopathy, cardiomyopathy and neuropathy, and ultimately total organ damage. These microvascular and macrovascular complications drastically enhance the mortality and morbidity rates among diabetics [3].

Broadly, type 1 diabetes mellitus (T1DM) and type 2 diabetes mellitus (T2DM) are the two basic forms of diabetes mellitus; some other types such as gestational diabetes mellitus (GDM), monogenic diabetes also exists. The β-cell dysfunction is a common key element in pathophysiology of both T1DM and T2DM, however the mechanisms of β-cell impairment differs [4]. Besides that, reduced insulin sensitivity in the target tissues (primarily in skeletal muscles and liver), a condition commonly termed as insulin resistance (IR) also represents the T2DM. Mainly, the incompetency of β-cells to produce insulin in sufficient amounts to regulate the glucose levels seems integral for pre-diabetes as well as overt diabetes [5, 6]. The mechanisms behind impaired β-cell function are complex and yet unclear, however, several genome-wide studies have revealed diversity of risky gene forms which enhances the risk of developing the disease [7, 8]. Accumulating evidence also indicates the role of oxidative stress, mitochondrial dysfunction and ER stress in the damage to β-cells [9]. The levels of inflammatory cytokines such as TNF-α, IL-6 were also reported to be elevated in IR and it appears that interaction between these signaling pathways may exacerbate the glucolipotoxicity stress in β-cells [10, 11]. Indeed, the crosstalk between the genetic background and the environment acts as a key driver in the pathophysiology of diabetes. Environmental factors might trigger the activation of autoimmune response such as in the T1DM and this process varies between individuals depending on the genetic background/make-up of an individual [12].

There is additional evidence that oxidative stress has a role in the etiology of numerous metabolic disorders such as hyperglycemia, heart disease, obesity, and cancer. In fact, the high glucose environment in diabetes also promotes reactive oxygen species (ROS) overproduction via different pathways such as hexosamine pathway, advanced glycation end products (AGEs) formation, DAG/PKC pathway and others [13, 14]. Moreover, the tissues or organs with high metabolic activity such as pancreatic islets are highly vulnerable to deterioration or destruction caused by excessive ROS production and oxidative stress as they completely rely on mitochondrial oxidative phosphorylation for ATP needs (particularly in hyperglycemia); where ROS production is unavoidable [15, 16]. In addition, the antioxidant defense system also operates at low levels in β-cells making them more vulnerable to ROS induced stress damage.

Furthermore, several investigations have been carried out to determine the efficacy of possible pharmacological agents targeting the oxidative stress. Traditionally, therapies involving conventional drugs have been used to treat this metabolic disease despite their clinical side effects on prolonged usage. It entails the use of anti-hyperglycemic drugs such as metformin, drugs enhancing the insulin secretion like sulfonylureas and meglitinides, improving the insulin sensitivity using thiazolidinediones; limiting the absorption of carbohydrates by α-glucosidase-2 inhibitors and preventing the reabsorption of glucose in kidneys via sodium-glucose transporter inhibitors (SGLT-2) inhibitors [17, 18]. Recently, the use of organic compounds derived from plants, commonly known as phytochemicals or bioactive compounds including polyphenols, alkaloids, flavonoids etc. has been emerged as a novel treatment approach for diabetes due to their anti-hyperglycemic effects [19, 20, 21]. Researchers from all over the world have discovered a plethora of plants-derived bioactive compounds that regulate the progression of diabetes. For instance, Eurocristatine, an alkaloid, reduces fasting blood glucose (FBG) and alleviates the insulin resistance (IR) by PI3K/Akt signaling pathway activation [22]. Similarly, treatment of high-fat diet (HFD) fed SD mice with curcumin for 8 weeks also demonstrated anti-diabetic effects via up-regulation of PI3K/Akt pathway [23].

In order to study the pathogenesis of diabetes and its complications, several *in-vivo* and *in-vitro* models have been used extensively. The *in-vitro* models such as hepatoma cell line HepG2 (an *in-vitro* alternative to primary human hepatocytes) or insulin-secreting β-cell lines such as INS-1, RIN, MIN etc. have empowered researchers to directly test a variety of anti-diabetic compounds [24, 25]. Additionally, different types of animal models such as Goto-Katazaki (GK) rat [26], Zucker Diabetic Fatty (ZDF) rats [27], KK-rats, TallyHo mouse [28], Wistar rats, C57BL/6J mice, *db/db* mice and others have been developed to further confirm the *in-vitro* results [29, 30, 18]. These models include the genetically modified or chemically induced models by Streptozotocin (STZ), alloxan or monosodium glutamate (MSG) [31]. We seek to summarize the role of oxidative stress in diabetes and the currently available therapeutic approaches/strategies for treating diabetes by targeting oxidative stress in this article, beginning with the basics and progressing to the pathogenic mechanism of diabetes.

## 2. Diabètes : Background and its Classification

Diabetes is a persistent metabolic disorder wherein the body either doesn't produce enough insulin or it is unable to use the secreted insulin; resulting in high blood-glucose levels or hyperglycemia [32]. Under normal physiological conditions, insulin is released by the pancreatic β-cells in response to elevating glucose levels, which further promotes its uptake and subsequent metabolism in the cells thereby releasing energy. However, IR, a condition defined by reduced insulin-sensitivity in organs like skeletal muscles and liver and the failure of β-cells to release insulin leads to abnormally high levels of glucose in the body due to impaired glucose metabolism. As a result of IR, higher concentrations of insulin are required for normalizing the glucose levels; consequently, giving rise to a condition known as "hyperinsulinemia". The IR is the primitive detectable anomaly of diabetes followed by β-cell dysfunction in its background resulting in overt-diabetes [33, 34]. Prolonged hyperglycemia results in long-term manifestations such as diabetic nephropathy (DN)- ultimately leading to end-stage renal failure; diabetic retinopathy (DR)- causing significant loss of vision; diabetic cardiomyopathy (DCM)- resulting in heart failure, stroke and diabetic neuropathy (DNeu)- damage to nerves that may cause foot ulcers, feet amputations or the damage to autonomic nervous system [35, 36]. Other forms of complications include dementia or cognitive impairment, sexual dysfunction and depression [37, 38, 39].

Recent research has mostly focused on clarifying the mechanism behind both IR and -cell dysfunction in diabetes, but there is still some uncertainty. However, it has been found that oxidative stress, mitochondrial dysfunction and ER stress are the major mechanisms playing a central role [40]. It has also been demonstrated that aging, glucotoxicity, and lipotoxicity majorly induces β-cell failure [41, 42]. It has also been indicated that alteration of various signaling pathways in response to hyperglycemia is responsible for generation of reactive free radicals and oxidative stress which exacerbates the diabetic complications, but the specific mechanisms are not yet clear [43].

*Risk factors and symptoms:* As per the International Diabetes Federation (IDF), a diverse range of risk factors might promote the development of diabetes including genetics, family history, obesity, dietary factors, sedentary lifestyle, physical inactivity, high blood pressure, environmental stress [44]. The common symptoms of diabetes include polydipsia, polyuria, polyphagia, blurred vision, unexplained weight loss and tingling sensations especially in lower extremities [45]. In attempt to normalize the blood glucose levels, kidney produces more urine leading to more frequent urination (polyuria), consequently thirst increases (polydipsia) in response to normalize body fluids. Furthermore, reduced cellular uptake of glucose promotes polyphagia.

*Diagnosis of diabetes:* Currently, diabetes is diagnosed according to the guidelines recommended by the WHO and ADA. The preferred diagnostic tests involve-random blood glucose test, fasting plasma glucose (FPG) test, 2-hour plasma glucose (2PG) test after a 75g oral glucose Tolerance Test (OGTT), HbA1c. People with random plasma glucose ≥ 11.1 mmol/L (200 mg/dl), FPG values ≥ 7.0 mmol/L (126 mg/dl), HbA1c ≥ 6.5% (48 mmol/mol) are considered to have diabetes and 2PG values ≥ 11.1 mmol/L (200mg/dl) and [46]. If any of the conditions/criteria are met, then a person is considered as diabetic.

*Classification of Diabetes:* The American Diabetes Association (ADA) classifies diabetes into four general categories namely type-1 diabetes (T1DM), type-2 diabetes (T2DM), gestational diabetes mellitus (GDM) and other forms of diabetes due to other causes such as maturity-onset diabetes of young adults (MODY), monogenic diabetes, and drug- or chemical-induced diabetes [47]. However, the majority of cases fall under the category of T1DM and T2DM. Figure 1 illustrates a comparison between healthy, T1DM and T2DM state (Figure 1).

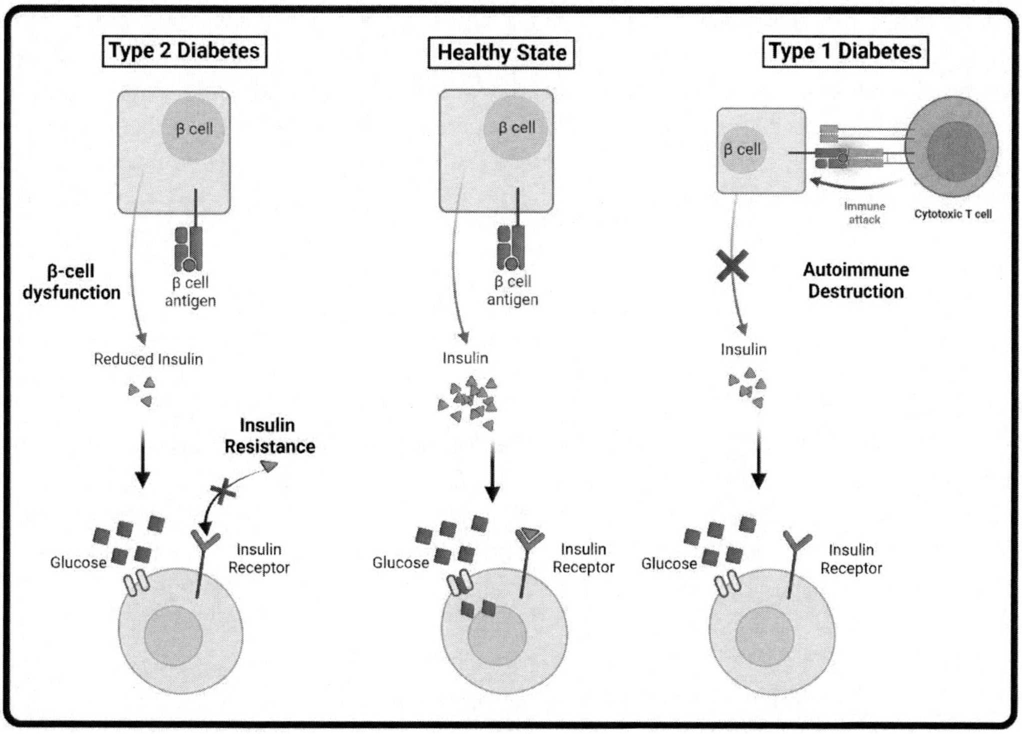


**Figure 1.** A comparison between healthy, type 1 diabetes and type 2 diabetes mellitus.

The T1DM and T2DM represent a state of hyperglycemia having β-cell dysfunction in common. The activated immune response against β-cells in T1DM renders them incapable of producing insulin, whereas β-cell impairment in T2DM reduces their insulin secretion ability. In addition, due to the lowered insulin sensitivity in target tissues such as skeletal muscles, adipose tissue and liver insulin resistance occurs in T2DM.

The juvenile onset diabetes or T1DM or insulin dependent diabetes mellitus is an autoimmune disease involving β-cell destruction and absolute insulin deficiency [48]. It has been suggested that any deformities in the or human leukocyte antigen region (HLA) proteins (the proteins which help in distinguishing self/non-self cells) results in the autoimmune reaction against β-cells. Various genetic (such as HLA, insulin-VNTR) as well as environmental factors (including diet, gut microbiota and viral infections) are responsible for enhanced susceptibility of the T1DM [49]. The patients suffering from T1DM needs inulin shots to survive.

The non-insulin dependent diabetes mellitus or T2DM is the most prevalent type involving development of resistance to insulin in combination with the aberrant insulin secretion. It accounts for approximately 90-95% of diabetes and mostly it was found associated with the lifestyle [50]. Generally, lack of physical activity, intake of calorie rich foods, obesity in children, smoking, aging are the most crucial risk factors driving the T2DM pathogenesis [51].

In this form of diabetes, mainly β-cell dysfunction and IR are the two main factors linked to its progression [52].

Recently, a new form of diabetes, namely the type 3 diabetes (T3DM), has been proposed by scientists. T3DM denotes a condition of strong association between T2DM and consequent cognitive impairment; since neurons are unable to take up glucose because of brain IR. In other words, T3DM can be defined as a metabolic syndrome like T1DM and T2DM, starting from brain IR through hyperglycemia, neuronal alterations, neuronal stress, accumulation of neurotoxins and subsequently leading to neurodegeneration [53]. Studies have also suggested a cross-link between pathogenesis of neurodegenerative diseases such as Alzheimer's (AD), hyperglycemia and IR [54]. The role of impaired insulin signaling in the etiology of AD has also been demonstrated. Because of the shared molecular and pathological features among T1DM and T2DM, the term T3DM has been proposed [55]. The etiology of T3DM is not yet fully revealed, but it has been proposed that apart from Aβ-accumulation, tau-hyperphosphorylation and neuroinflammation; accumulation of ROS and oxidative stress are the major/key components associated with the neurodegeneration [56, 54].

The GDM is also one of the most common and serious medical complications in pregnancy characterized by hyperglycemia that occurs most likely after 24 weeks and disappears after pregnancy. Consequently, it enhances the chances of the newborn and mother developing obesity, T2DM, and cardiovascular disease in the future [57]. Additionally, MODY which accounts for 1-2 percent of all diabetes, is also a rare type of diabetes, representing a heterogenous group of monogenic, autosomal dominant diseases [58, 59]. The mutations in different types of genes are such as those encoding the insulin promoting factor-1, glucokinase, hepatocyte nuclear factor 1-α, insulin promoting factor-1 etc. result in most of the cases of MODY [60].

Another commonly known form that one should be aware of is "pre-diabetes", defined as an asymptomatic, intermediate state of hyperglycemia prior to diabetes, where the levels of glucose are elevated than normal but below the threshold levels. Observational evidence suggested that pre-diabetic individuals are more likely to develop overt-diabetes and various complications at later stages [61]. However, this condition does not appear benign, but it is still an at-risk state whose diagnosis is necessary to avoid future complications.

## 3. Epidemiology of Diabetes

Diabetes is a global epidemic, prevalence of which has been increasing worldwide, especially the developing nations with a huge force of working-age adults such as China, India and USA are becoming epicenters/hotspots. According to recent IDF estimates, India ranks second after China with 74.2 million diabetics (20-79 years) and this estimate is projected to rise to approximately 124.9 million by 2045. The major drivers of this current global epidemic are obesity, unhealthy diet, energy-rich diets, family history, physical inactivity, population ageing and sedentary lifestyle [62, 63, 64, 65, 51]. Most of the premature deaths are caused by diabetes recurrent complications. Cardiovascular disorders contribute most to mortality of diabetic patients [63]. About 10% of patients die because of renal failure, and around half of the patients die of cardiovascular disease, and [64]. The prevalence and incidence of T1DM is also on a rise with a 2-5% increase in incidence rates/year globally and a prevalence of 1in 300 in the US [66]. In a US based study, similar rising trends in T2DM were observed [67].

Emerging evidence indicates that demographic or socio-economic pattern of T2DM is changing, with shift from older onset to young stages with a more aggressive disease phenotype enhancing the risk of premature progression of chronic complications [68]. This will have unfavorable long-term outcomes, adversely affecting the quality of life in future, and therefore, posing higher threat of current and future burden of disease or serious health challenge in next few decades [62, 68]. In UK, a study estimated that the T2DM prevalence increased from 3.21/10,000 persons in 2004 to 5.26/10,000 persons in 2014 with a significant increase in patients aged 16-34 years. They also related the rise in prevalence to male gender from deprived areas compared to females. The mortality rates were higher in patients with T2DM compared to patients without diabetes [69]. The global burden of disease study in China based on prevalence and mortality rates revealed that prevalence of diabetes had increased from 3.7% to 6.6% between 1990 to 2016 and similarly the mortality rates increased by 63.5%. Additionally, both prevalence and mortality rates were significantly higher in young adults between 15-49 years [70]. Another age-related diabetes study revealed that an inverse or a negative correlation between age of diabetes emergence and diabetes-induced mortality and morbidity exists, with younger patients having higher mortality and morbidity rates compared to usual onset of T2DM [71]. The prevalence of diabetes in India is also shifting from elderly to youth, metropolitan cities to rural areas and prosperous to impoverished strata and hence posing a greater economic burden on patients, their families as well as on the society [72].

Several studies have revealed differences in the incidence of T2DM according to race, ethnicity, socioeconomic positions but the mechanisms and cause are not well understood [73, 74, 75]. A study by Agardh and colleagues on incidence of T2DM and socio-economic position measured in terms of education level, occupation and income over the time period between 1966 and 2010 indicated that there is an increased risk of T2DM in middle- and low-income countries having low education level and low occupation that is lower socio-economic positions (SEP) compared with higher SEP groups [74]. Bakker et al. revealed that the risk of developing T2DMrelated complications such as retinopathy, nephropathy, coronary artery and cerebrovascular disease in South Asians is higher as compared to White population [75]. A recent study by Khan et al. on global epidemiology of T2DM based on Global Burden of Disease (1990-2017) data revealed the prevalence rate of 6059 cases /100,000 in 2017 and it is projected to increase to 7079 individuals per 100,000 by 2030. Such a high prevalence of T2DM is rising the burden of human suffering in lower-income countries [76].

There is mounting evidence that diabetes is on the rise, especially in developing nations but on the contrary, some recent studies indicate a stabilizing or declining incidence of diabetes. Study by Magliano and colleagues over three time periods revealed that the incidence of diabetes in high income nations increased consistently from 1990s to mid-2000s, while a declining pattern emerged since then. They suggested that effective prophylactic strategies, awareness campaigns and education on public health might be the factors contributing to such decline [77]. A similar contrasting multi-country study of drift in T2DM incidence in middle-income and high-income countries also revealed a stabilizing or declining trend but the reasons for such declines in incidence need further investigation [78].

Overall, it can be stated that diabetes as a global pandemic is evidencing a frequent increase in young adults in middle- and lower-income countries, and unhealthy diet, obesity, changing lifestyle, population ageing being the major contributors of the disease (Figure 2). The increasing prevalence and incidence of T2DM and its complications worldwide consequently lead to rise in socio-economic burden. The early diagnosis, change in diet and lifestyle modification might be a good solution for the disease.

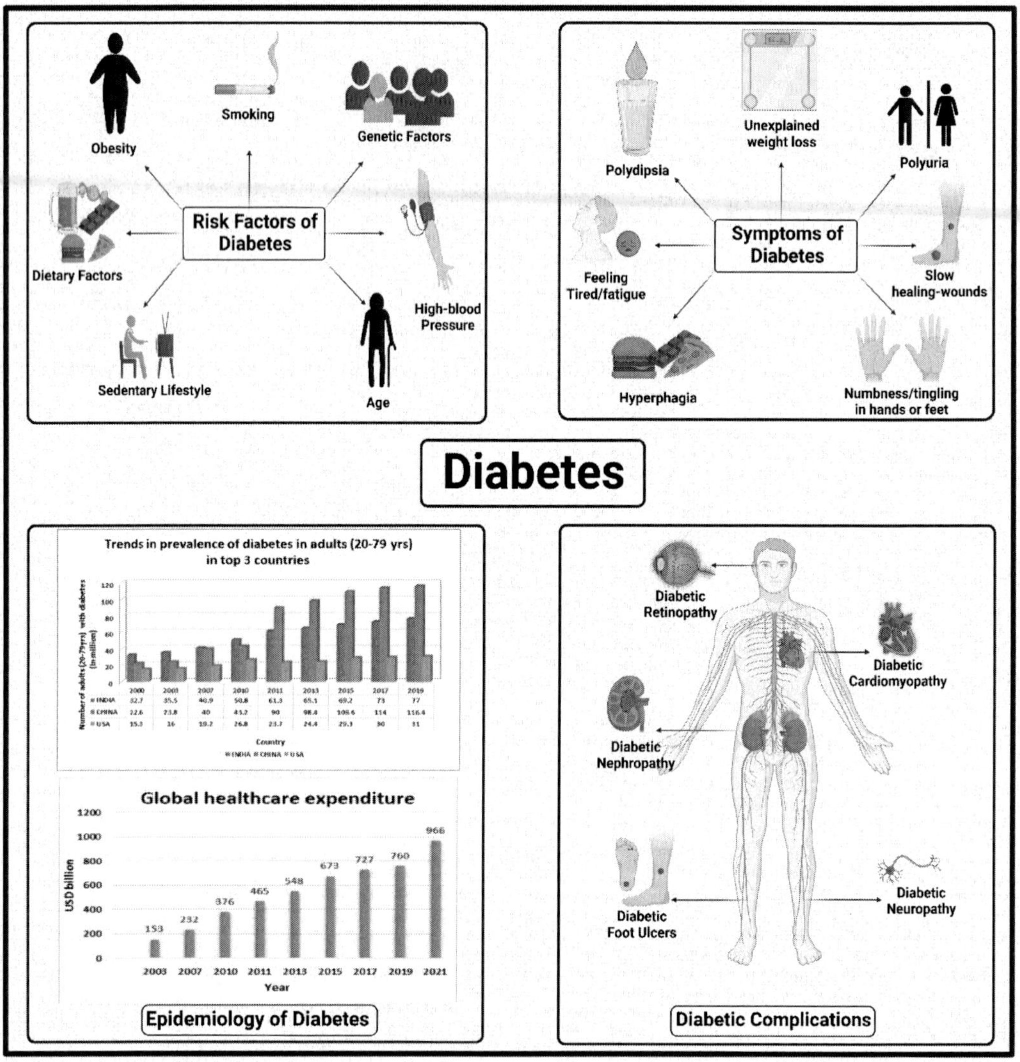


**Figure 2.** Epidemiology of diabetes.

This figure depicts the major risk factors, symptoms of diabetes along with the global epidemiological data and long-term diabetic complications. Diabetes is characterized by symptoms such as excessive thirst, hunger, urge to urinate more frequently, lack of energy, slow healing of wounds and sometimes blurred vision. The lack of physical activities, obese lifestyle, and intake of calorie rich diet besides some genetic factors represents the common causal factors of diabetes. Notably, India, China and US are top countries having a large population of diabetics. This ultimately leads to a rise in healthcare costs. If left untreated or undiagnosed for longer periods, various diabetic complications like diabetic neuropathy, retinopathy, nephropathy and cardiomyopathy might appear.

## 4. Diabetic Complications

If left undiagnosed for longer periods, diabetes may lead to several life-threatening complications. Broadly, these consequences of diabetes can be categorized into acute and chronic complications. The acute or short-term complications include diabetic ketoacidosis, whereas the chronic or long-term complications can be classified into microvascular and macrovascular. Macrovascular complications involve damage to large blood vessels of circulatory system and causes cerebrovascular disease (influencing brain), cardiovascular disease (affecting heart), and peripheral vascular disease (impacting peripheral arteries) while microvascular complications affect the small blood vessels of the circulatory system and contributes to DN (kidney disease), DR (eye disease) and DNeu (nerve disease) [72]. Studies related to diabetes complications suggests that cardiovascular disease is the primary cause of death in diabetic patients [79]. Studies had revealed multiple mechanisms leading to pathological changes in these complications including increased formation of AGEs, oxidative stress, activation of aldose reductase (polyol) pathway, activation of protein kinase C (PKC), altered expression and action of growth factors [80]. Some major diabetic complications are described below.

### 4.1. Diabetic Nephropathy

One of the most common complications of diabetes that ultimately leads to kidney failure or end-stage renal disease (ESRD) is DN. Approximately 40% of diabetic patients (both T1DM and T2DM) are affected by this complication [81]. The important pathological features of DN are accumulation of ECM proteins such as collagen (Collagen I and IV) and fibronectin in proximal tubular cells (PTCs), mesangial expansion, glomerular basement membrane (GBM) thickening, glomerular hypertrophy, renal fibrosis, inflammation and podocyte injury involving apoptosis and damage to podocyte adhesive capacity [82].

DN progresses in stages and eventually leads to renal failure. Clinically, it is characterized by polyuria, microalbuminuria, proteinuria, GFR reduction (at rate of 2-20ml/minute/year) and increase in blood pressure [83]. There are various profibrotic and proinflammatory factors (mainly IL-6, IL-1 and TNF-$\alpha$) that drive renal fibrosis and inflammation respectively. The molecular pathways provoking DN remain unknown, however, recent studies have reported that accumulation of AGEs and oxidative stress in addition to genetic predisposition, environmental and epigenetic variables (such as long non-coding RNAs, microRNAs) collectively regulate the aetiology of DN [84].

### 4.2. Diabetic Retinopathy

DR is a common progressive complication of small blood vessels which consequently leads to blindness or permanent vision loss. The proliferative DR (PDR) and non-proliferative DR (NPDR) are the two types of DR. NPDR is the early stage of DR marked by formation of microaneurysms, arising because of capillary dilations and loss of pericytes. It further progresses to the late-stage PDR, a condition represented by neovascularization and may lead to severe vision impairment due to vitreous hemorrhage [85]. Endothelial cell dysfunction, retinal neovascularization, development of acellular capillaries, formation of microaneurysms, and basement membrane thickening, all represent the complicated pathogenesis of DR [86, 87]. Also, retina cells in DR show significant differences in their morphology compared to normal cells [88]. Furthermore, in-vitro studies have revealed the impairment of blood-retinal vascular barrier (BRB) permeability in DR [89]. The alterations in BRB permeability result in the leakage of plasma proteins and lipids promoting the formation of hard exudates in the retina. Mechanically, the hyperglycemia-induced oxidative stress also promotes the pathogenesis of DR. The activation of PKC pathway, polyol pathway, hexosamine pathway and accumulation of AGEs are the key mechanisms implicated in the pathology of DR [90]. Therefore, the maintenance of glucose levels to optimal levels is the key to prevent the progression of DR.

### 4.3. Diabetic Neuropathy

Neuropathy is a frequent complication in diabetes causing nerve impairment throughout the body resulting in alteration of autonomic, motor and sensory functions [91]. DNeu is the major cause of disability and death among diabetics leading to reduced quality of life. It affects approximately 60-70% of patients with diabetes [92]. However, a recent study estimated that approximately 50% of diabetic adults suffer from DNeu in their lifetime indicating the huge economic health burden of the disease [93]. The two major forms of DNeu known to co-exist in diabetic patients are DPN and DAN. Distal symmetric polyneuropathy (DSPN) also defined as diabetic peripheral neuropathy (DPN) is the commonly occurring mode of DNeu caused by hyperglycemia [94]. It involves impairment of sensory function that begins distally in the fingers, toes or the lower extremities and the patients suffer from pain causing "painful DPN" [95, 96] and consequently develop advanced neuropathic conditions involving cranial nerve, limbs [91]. Following this, patients lose the ability to feel temperature and painful stimuli; altered foot morphology (asymptomatic phase) could result in foot ulceration which further enhances the risk of limb loss or feet amputations in diabetic patients [97]. Diabetic autonomic neuropathy (DAN) is a potentially life-threatening disorder of autonomic nervous system in diabetes that affects the cardiovascular, urogenital, gastrointestinal, pupillomotor and sudomotor systems [98, 96].

Basically, the pathophysiology of DNeu is complex and the precise processes causing the disease are not known. It is known that hyperglycemia induced metabolic and morphological changes affect the nervous system. Several recent studies have investigated mechanisms of pathogenesis of DPN and revealed that it is multifactorial, that is, it involves activation of multiple molecular pathways such as AGEs pathway, polyol pathway, MAPK pathway, NF-kB signaling, Wnt-pathway, GSK3 signaling apart from hyperglycemia induced oxidative stress, low-grade inflammation, impaired nerve perfusion, dyslipidemia, reduced nerve growth factors that cause nerve dysfunction [99, 100].

### 4.4. Diabetic Cardiomyopathy

The individuals suffering from diabetes are at a higher risk of developing DCM. DCM came into existence in 1972 when Rubler et al. described his observation of 4 adult diabetic individuals suffering from congestive heart failure (CHF) independent of other common causes of heart failure such as congenital heart disease, coronary artery disease, alcoholism, or hypertension [101]. This observation laid the basis of this new disease and subsequent studies thereafter supported the existence of DCM. Studies defined DCM as a diabetic complication in the non-existence of dyslipidemia, coronary artery disease, hypertension which consequently increases the risk of heart failure [102]. A clinical study by Kannel and colleagues revealed that diabetic subjects are at higher risk (four to five-fold) of heart failure and suggested that cardiomyopathy is associated with diabetes and heart disease [103]. Furthermore, several clinical and experimental studies have demonstrated that DCM is related to structural and functional changes in the heart such as, systolic, diastolic dysfunction and left ventricular (LV) hypertrophy [104, 105, 106]. The exact mechanism of this disease is not known, but several pathophysiological mechanisms have been identified, among them small vessel disease, autonomic dysfunction, myocardial fibrosis and metabolic derangements/disturbances are the major mechanisms [107]. The other potential mechanisms implicated in DCM are hyperglycemia, oxidative stress, myocardial lipotoxicity, mitochondrial dysfunction and over-activation of renin-angiotensin system [108, 109].

### 4.5. Molecular Pathogenesis of Diabetes

Under normal physiological conditions, the feedback loops or interplay between pancreatic insulin secretion and insulin action (in insulin-sensitive tissues such as skeletal muscle, liver and adipose) maintains the normoglycemic state in body. Insulin, a hormone produced by pancreatic β-cells, is critical in regulating blood glucose levels. It is first synthesized as pre-proinsulin, processed to proinsulin and further converted to insulin and stored in secretory granules till use [110]. Besides the rising glucose levels, various nutrients and hormones also promote release of insulin [111]. Insulin directs the cells to uptake glucose where it can be metabolized to generate ATP/energy. Besides this, insulin also inhibits gluconeogenesis and glycogenolysis in the liver. In contrast, when this feedback loop is disturbed either due to β-cell dysfunction or IR, a state of hyperglycemia is induced [50]. In other words, impaired tissue-sensitivity to insulin (IR) and β-cell failure to maintain a sufficiently high levels of insulin to overcome IR (impaired β-cell function) contributes to the pathogenesis of T2DM [112]. The first noticeable abnormality, prior to onset of overt diabetes, is IR, accompanied by a compensatory rise in glucose-stimulated insulin secretion, or GSIS, and is a predictive factor for the development of T2DM [34]. The pathogenic mechanisms underlying the major forms of diabetes are outlined below.

## 5. Pathogenesis of T1DM

As discussed earlier, T1DM is an autoimmune disorder represented by β-cells loss or reduction in mass of β-cells and insulitis. It involves a T-cell mediated hypersensitivity reaction whereby the aberrant T-cells recognize and damage the β-cells consequently resulting in a hypo-insulinemic and hyperglycemic state in the body [113]. T1DM can develop at any stage of life, but it is more frequent in childhood. The pathophysiology of T1DM is not fully clear, however, some autoimmune mechanisms are well known. In this process, autoantibodies released against β-cell antigens are responsible for the destruction of insulin secreting cells. Some novel targets (autoantigens) of autoantibodies generated in T1DM involve insulin, the SLC30A zinc transporter 8 (ZnT8), the protein tyrosine phosphatase–related molecules IA-2 and glutamic acid decarboxylase 65KDa isoform (GAD65) [114, 115, 116]. Majorly, the environmental factors (chemicals, dietary factors, gut microbiota and bacterial/viral infections) as well as various genetic factors (HLA genes and non-HLA genes) in combination serves a critical role in generation of such autoimmune response against β-cells [117].

Several lines of evidence highlight the importance of major susceptibility genes for T1DM. The HLA are the most potential genetic factors/determinants known, accounting for approximately 40-50 percent of T1DM [118]. Recent data indicates that individuals with specific HLA class II haplotypes such as HLA-DR4-DQ8 and HLA-DR3-DQ2 are at a high risk of developing autoantibodies targeting β-cells and T1DM [119, 120]. Usually, autoantibodies IA-2 and ZnT8 appear first in the process [121]. Recent studies have demonstrated that risk of T1DM is significantly increased, if a child with a family history carries any of the above-mentioned HLA risk genotypes [122, 123]. These genetic risk factors also influence the sort of autoantibody that appears first. Besides this, other non-HLA genes including some single nucleotide polymorphisms such as *INS, SLC30A8, PTPN22 and BACH2* also affect the development and progression of T1DM. These variants are known to regulate the initiation of the auto-antibodies production/autoimmunity as well as they can alter the rate of destruction of β-cells at later stages [124, 125, 126, 127].

Environmental factors also play an integral part in the pathophysiology of T1DM. Recent studies have focused more on the microbial communities residing inside the gut or gastrointestinal tract (GIT) in healthy individuals including the impact of dietary components on the composition of the gut biota. Since these microbes are responsible for synthesizing several essential vitamins, minerals and enhances the immunity of the individual, therefore any alteration in this microbial ecosystem either due to diet, age, lifestyle factors or drugs could lead to the pathogenesis of variety of diseases, among them one is T1DM [128]. The presence of pathogens including bacteria and viruses (especially enteroviruses) [129] may promote the pathogenesis of T1DM due to their inherent ability of mimicking autoantigens of human β-cells as well as they can also cause β-cell damage via directly infecting and replicating inside β-cells [130, 131, 132]. Similarly, dietary intake such as a diet rich in sugars and carbohydrates also act as a causative factor, promoting T1DM progression [133]. In addition, insulitis (inflammation of the islets) is also a major hallmark of T1DM. This condition refers to a state of inflammation-mediated invasion of mononuclear cells in the islets leading to β-cell loss [134]. The direct contact of β-cells with T-cells, activated macrophages and cytokines such as TNF-α and IL-1β are some of the possible reasons behind pancreatic β-cell apoptosis [135]. Collectively, the T1DM etiology is a highly multifactorial, heterogenous and complicated where genetic risk factors, environmental factors and their association plays a major role.

## 6. Pathogenesis of T2DM

It is evident from literature that IR and β-cell dysfunction are the two primary characteristics in the T2DM pathophysiology. The functional impairment of β-cells leads to reduced insulin secretion whereas IR indicates defective insulin action; consequently, both lead to hyperglycemia.

*Insulin resistance (IR):* IR is the primary characteristic of T2DM that may exist for a longer time periods prior to development of overt diabetes with visible symptoms. The diminished ability of insulin-sensitive tissues to counter normal insulin levels defines condition of IR. The molecular mechanisms underlying this are not yet well established however, several *in-vivo* studies demonstrated that impairment or alteration in the major proteins involved in the intracellular insulin signaling such as the transcription factor forkhead (Foxo1a), insulin receptor substrate (IRS), protein kinase B-β (PKB-β) isoforms are involved in IR [136, 137, 138].

It is well known that activation of IRS-1 tyrosine kinase activity is necessary for insulin-mediated signaling, therefore, any alteration in IRS-1 kinase activity or PI3K pathway say via mutation can hinder the action of insulin on target tissues, leading to IR [139]. In addition, adipose tissues via regulating the levels of free fatty acids (FFAs) and adipokines/ adipocytokines such as resistin, adiponectin, visfatin, leptin, IL-6 and TNF-α plays crucial role in induction of IR as explained in detail [140]. In obese individuals, the rise in adipose tissues might up-regulatethe above-mentioned substances and thus impair the insulin signaling. The elevated levels of FFAs also impair the tissue sensitivity towards insulin by reducing the IRS-1 phosphorylation at tyrosine, inhibiting the activity of IRS-1 associated PI3K and disrupting the insulin signaling pathways [141]. TNF-α also acts in a similar manner by suppressing the insulin-induced IRS-1 phosphorylation at tyrosine as well as down-regulating the expression of GLUT-4 [142, 143]. Moreover, the upregulation of different types of inflammatory cytokines results in chronic inflammation, which further promotes the pathogenesis of IR.

Another putative mechanism promoting IR is mitochondrial dysfunction. The imperative free-radicals produced from the electron transport chain (ETC) are known to enhance the insulin-sensitivity, whereas if generated in uncontrolled levels due to mitochondrial impairment/dysfunction might impair the insulin signaling and induce IR [144]. So, the abnormal accumulation of adipose tissue, FFAs, inflammatory cytokines, adipokines in combination with mitochondrial dysfunction can impair insulin signaling and its response in insulin-target tissues.

*β-cell dysfunction:* The second most important key factor involved in pathophysiology of T2DM. Since insulin secretion is dependent on the mass of β-cells and its function, therefore, reduction of β-cell mass or functional abnormality can directly affect the insulin levels. Several types of metabolic abnormalities are related to loss of β-cell including glucotoxicity, lipotoxicity, inflammation and oxidative stress [4, 145]. The term "toxicity" here refers to damage to pancreatic β-cells as well as glucose homeostasis that might lead to the pathophysiology of T2DM; and if the damage occurs due to prolonged exposure to glucose and lipids, then it is termed as "glucotoxicity" and "lipotoxicity" respectively. The mechanisms underlying glucotoxicity and lipotoxicity appear unclear; however, a line of evidence indicates that mitochondrial dysfunction, oxidative stress, inflammation and ER-stress might play an important role [146, 147, 148, 145, 149]. FFAs are known to impair the protein folding capacity of ER, cause aberrant post-translational modifications such as palmitoylation and may result in misfolded proteins accumulating, thereby leading to ER stress [150, 151]. FFAs also induces the islet amyloid polypeptide (IAPP) accumulation in β-cells and contributes to inflammation. Moreover, excess glucose and lipids promotes the production of ROS including superoxide and thus leads to oxidative stress which further alters the structure, functions of mitochondria; and consequently, harms the β-cells [146, 152, 153]. It is also well known that intake of excess calories, lack of physical activity, sedentary lifestyle promotes obesity which is linked with development of T2DM. The excess production of adipose derived factors and cytokines such as IL-1β, TNF-α, adiponectin, leptin and resistin in obese individuals also plays a crucial role in impairment of pancreatic β-cell function [154].

A study by Talchai et al. contradicts the notion that β-cell apoptosis is the primary cause of β-cell loss; they proposed that pancreatic β-cell dedifferentiation (the reversion of β-cell into progenitor-like cells) can be the mechanism underlying β-cell dysfunction [155]. Besides these factors, autophagy, a process recognized to maintain cellular homeostasis via balancing the formation, breakdown and recycling of cellular materials through lysosomal pathway also plays a significant role in maintaining β-cell homeostasis [156]. The deficiency of autophagy in β-cell leads to abnormal functioning of mitochondria, hinders the unfolded protein response (UPR), elevates ER-stress, altered the antioxidant genes expression (such as SOD, GPx, HO-1 and CAT impaired the physiology of pancreatic β-cells [157, 158, 159].

Another unusual but interesting aspect of β-cell dysfunction has been revealed by Almaca and colleagues. They proposed that the cellular microenvironment inside pancreatic islets might have a potential role in β-cell dysfunction. Pancreatic β-cells are surrounded by other types of cells submerged in extracellular matrix and the local interactions among them actuate the functioning of pancreatic cells. The correlation between various types of acellular and cellular components in pancreas defines the physiology and pathology of the β-cells. Therefore, any chronic alterations in this islet niche might act as a potent mechanism promoting β-cell dysfunction [160].

## 7. Metabolic Homeostasis and Oxidative Stress

For their survival, living creatures require a constant supply of energy from their surroundings to sustain life, which relies on a series of metabolic processes. To keep the metabolism in check, energy storage and consumption are closely regulated; any disruption in metabolic homeostasis will lead to a diseased state. The synthesis of biomolecules, including enzymes and hormones, must be regulated alongside the catabolic process to maintain metabolic homeostasis, and metabolic regulation through intracellular respiration and oxidative phosphorylation is essential to maintain metabolic equilibrium [161]. Several elements of research have investigated the potential significance of oxidative stress, as it is an etiological factor in diabetes characterized be abnormal in the creation of the free radical generation and antioxidant defense mechanisms. The higher kinetics of biochemical reactions attained in elemental single-electron reactions are favorable and serve as a source of radical species that are either undesirable by-products or exceed the stated parameters [162]. Oxygen-possessing molecules with an uneven number of electrons produced from such reactions are called free radicals, and they can easily react with other molecules. Because free radicals react so quickly with other molecules, they can create enormous chain chemical reactions in the body. However, these oxygen free radicals, also known as ROS, tend to accumulate in cells and tissues, and failure to counteract their negative effects leads to increased oxidative stress in the biological system [163]. Although ROS production is a fundamental phenomenon, many biological activities of the body are fully reliant on it, such as cellular signaling pathways for cellular differentiation and apoptotic cell death, immune system molecules, and protein phosphorylation. Indeed, ROS are required only at low levels to maintain a physiologically active state, whereas rising ROS levels and oxidative stress will have detrimental effects such as derailed homeostasis, oxidative phosphorylation imbalance, macromolecular damage such as DNA, proteins, and lipids, and finally cellular dysfunctions [164].

## 8. Sources of Free Radicals

Superoxide and hydroxyl radicals are the two major oxygen free radicals. They're made from molecular oxygen that's been reduced chemically. Increased levels of these free radicals can lead to cell death and apoptosis, which can lead to stroke, neurodegeneration, cancer, cardiovascular disease, diabetes, and other significant problems. Internally, such free radicals are formed as a result of cellular respiration and exhausted antioxidant defenses, or externally, due to radiation, exposure to xenobiotics, anthropogenic activities or cigarette smoking [162]. There are different sources of ROS production such as peroxisomes, mitochondria and endoplasmic reticulum (ER); many metabolic changes, including lipid peroxidation also promote ROS generation. Another organ that has received little study is the glucosochondria, which is known to generate ROS because of cellular respiration and is an unavoidable source of ROS generation and is speculated to exacerbate intracellular oxidative stress. It has at least nine known superoxide anion-producing sites, as well as numerous ROS defense systems [165]. In the process of oxidative phosphorylation, approximately 0.2-5% singlet electrons leak during their transfer between complexes, reducing molecular oxygen and forming $O2^{\bullet-}$, $H_2O_2$ and $HO^{\bullet}$. In addition, the membrane-bound NADPH oxidase (NOX), and nitric oxide synthases isoforms (NOS1–3) are the main endogenous sites of cellular redox-reactive species generation, including ROS and reactive nitrogen species (RNS) [166, 167]. The magnitude or severity of ROS-induced impairment varies depending on the components and nature of radical intermediates. Highly energetic molecules like $HO^{\bullet}$ and $ONOO^{-}$, interact with other species near the point of generation, whereas less energetic molecules, like $^1O_2$ and $O2^{\bullet-}$, can disperse over greater range and make a significant impact in more precise regions. These might even interact against each other to produce greater reactive species, for example when $O2^{\bullet-}$ reacts with NO to make $ONOO^{-}$ [162].

### 8.1. Mitochondria

As discussed earlier that, mitochondria are a primary source of free radicals; particularly, complex I and III are the two key locations in the ETC where superoxide radicals are produced. Because superoxide generation is non-enzymatic, the higher the metabolic rate, the more ROS is produced. The superoxide radicals are converted to $H_2O_2$ with the help of a mitochondrial enzyme called mitochondrial superoxide dismutase (MnSOD). In addition, two more enzymes namely Catalase (CAT) and Glutathione peroxidase (GPx) are involved in the detoxification of $H_2O_2$. Many other mitochondrial enzymes like monoamine oxidase, glycerol phosphate dehydrogenase and α-ketoglutarate dehydrogenase, electron transfer flavoprotein (ETF), ETF quinone dehydrogenase, aconitase and p66shc are also involved in formation of free radicals.

ETF-QO is an end point for electrons flowing from nine flavoprotein acyl-CoA dehydrogenases and two N-methyl dehydrogenases that comprise flavin (FAD) [168], making it sensitive to ROS. However, p66shc is a pro-apoptotic protein that can modulate mitochondrial ROS generation and aging [169, 170]. When exposed to oxidative stress, p66Shc localizes to the mitochondrial intermembrane space, wherein it interacts with cytochrome-c, resulting in the production of ROS.

## 8.2. Peroxisomes

Peroxisomes are abundant sub cellular single membrane bound structures without DNA that primarily metabolize oxidatively and are assumed to be the principal source of intracellular $H_2O_2$. Several enzymatic reactions including GOX (photorespiratory glycolate oxidase- an enzyme of both α-oxidation and β-oxidation), acyl-CoA oxidase, flavin oxidase and dismutase $O2^{\bullet-}$ free radicals are liable for production of $H_2O_2$ in peroxisomes. A significant number of ROS and RNS are evolved in peroxisomes which contribute to oxidative stress including $H_2O_2$, $O2^{\bullet-}$, $^{\bullet}OH$, $NO^{\bullet}$ and $ONOO^{-}$. Peroxisomes, like mitochondria and chloroplasts, also generate superoxide anion ($O2^{\bullet-}$) during normal metabolic processes. The organellar matrix and the peroxisomal membrane are two key sites implicated in the generation of $O2^{\bullet-}$. The enzymes xanthine oxidoreductase (XOR) and urate oxidase (UO) are known to be involved in peroxisomal catalytic processes in the matrix and are also responsible for the production of $O2^{\bullet-}$ and H2O2. It has also been observed that ETC in the peroxisomal membrane is also involved in the formation of free radicals such as $O2^{\bullet-}$ through the complex formed of cyt-b, the flavoprotein NADH and ferricyanide reductase [171, 172].

## 8.3. Endoplasmic Reticulum

Many cellular activities, such as protein folding, synthesis, and transport of proteins to their target organelles after post-translational modifications, as well as calcium storage and assistance in maintaining homeostasis and lipid metabolism, are carried out alone by the ER. In particular, ER stress has been seen as a result of alteration in protein folding pathways. This results in accumulation of misfolded proteins which impedes the cellular homeostasis by inducing the UPR, which eventually stimulates ROS formation and elevates ER stress. Furthermore, in the ER lumen, oxidative denaturation of proteins provides a strong oxidative environment [171].

The generation of ROS is also aided by ER enzymes like CYP-450 (cytochrome p-450), b5 enzymes, and diamine oxidase. The ER-electron coupling system, which includes the protein disulphide isomerase (PDI) and endoplasmic reticulum oxidoreductin-1 (ERO1), the intra endoplasmic reticulum-glutathione disulphide (GSSG)/glutathione (GSH), and the NADPH-p450 reductase (NPR) system, is also thought to be a source of ROS. Other endogenous supply of ROS includes prostaglandin synthesis, decreased riboflavin, adrenalin autooxidation, FMNH2, FADH2 and CYP-450 [173, 174]. Besides these major sources of ROS, some authors have reported that thymidine catabolism, polyamine catabolism, oxidase activity of some enzymes like XOR, cyclooxygenases, NADPH oxidase family (NOX) and many other enzymes, are also involved in ROS production [171].

## 9. Antioxidant Mechanisms in the Body

Normally antioxidants inside the body are found in enzymatic form that catalyzes reaction which tends to decrease reactivity of free radical. Production of free radicals occurs continuously in the body but in pathologies like diabetes their production gets exaggerated to a level where the antioxidant mechanism of body is insufficient to dispose of them. Major key players include SOD CAT, GPx and paraoxonase (PON) and thioredoxin (Trx). The process of ATP generation in ATP contributes to formation of ROS. Basically, the endogenous antioxidant system consists of enzymatic and non-enzymatic antioxidants. The SOD, CAT, GPx and many other enzymes are among the enzymatic antioxidants that effectively neutralize the deleterious effects of $O2^{\bullet -}$, $H_2O_2$, and $^{\bullet}OH$ like potential oxidants. While the nonenzymatic antioxidants consists of melatonin, L-arginine, glutathione, vitamins, metals and omega-3; derived from metabolic and nutrient sources also contribute significantly to counteract the free radicals [175].

The enzyme SOD-1 is found at three locations in different forms 1) SOD-1 or Cu-Zn SOD in cytosolic location; 2) mitochondrial SOD-2 or Mn-SOD in mitochondrial location; 3) Extracellular SOD or Cu-Zn SOD in extracellular locations [176]. GPx's are found in 4 isoforms GPx-1, GPx-2, GPx-3 and GPx-4 out of which GPx-1 is found in sufficient amounts. GPx-1 form specifically occurs in cytosolic, mitochondrial and to some extent in peroxisomal locations in certain cells. GPx-1 form is much more efficient in removing intracellularly accumulated $H_2O_2$ [177]. GPx is a selenium containing enzymes but are occupied by cysteine at their active site. GPx is found in various cellular compartments mainly GPx1 in cytosol and mitochondria, GPx-2 in cytosol and nucleus and GPx-3 in plasma and GPx-4 is associated with membrane to prevent them from damage. Trx antioxidant system is made of NADPH, thioredoxin reductase and Trx [178].

Trx contains dithiol and disulphide in their active site. The Trx system in communion with glutathione and glutaredoxin sustains the reducing conditions of cell which is helpful in elimination of ROS. In cytoplasm and nucleus of Trx-1 form is found and in mitochondria which is a sensitive site for producing ROS. Thioredoxin reductase-1(TRX1) and Thioredoxin reductase-2(TRX2) are residents of cytosol and mitochondria requiring selenium for their activation [179]. CAT is mainly located in peroxisome, and it serves to destroy $H_2O_2$ that could prove hazardous to cells when produced in a chaotic manner [180]. Approximately 6 million molecules of $H_2O_2$ are converted to water and oxygen in just a single minute owing to higher catalytic efficiency of CAT. The CAT shows peak of their activity in liver, the major detoxification organ and erythrocyte cells [181]. PON are found in 3 isoforms and calcium is required for their activity. Oxidized LDL cholesterol undergoes hydrolysis by HDL-associated esterase PON-1. They reduce the peroxidation adducts generated by reaction of ROS with lipid peroxides thus exerting antioxidant effect [182]. Iron cluster containing Lipoic acid synthase is an enzyme found in mitochondria that produces lipoic acid inside body [183].

According to recent research, antioxidants can manage autoxidation by halting the free radicals proliferation or restricting free radical formation, reducing oxidative stress, improving immunological response, and extending healthy lifespan [184]. An antioxidant becomes oxidized when it encounters a free radical. As a result, the body's antioxidant supply must be replenished on a regular basis. Autoxidation is prevented by additives such as radical-trapping antioxidants, which intervene directly in the chain process. When evaluating antioxidant efficacy against autoxidation, the absorption of oxygen, the formation of peroxide, and the identification of the breakdown product should all be taken into account [185]. $O2^{\bullet -}$, the most common cellular free radical, causes a plethora of negative alterations by generating stronger radical species like $^{\bullet}OH$ and $ONOO^{-}$, many of which are attributed to an increase in peroxidative activities and a deficiency in antioxidants. An electron or an active hydrogen atom, such as one in an OH group, is required for these free radical scavenging reactions. In general, antioxidants containing active OH groups, such as vitamins C and E, flavanols, and polyphenols are excellent radical traps [186]. With the help of this protective mechanism, it is possible to regulate homeostasis while also delaying the process of aging and limiting the frequency of disease onset. Failure to maintain the balance between the free radical generation and neutralization by the antioxidant system results in the development of numerous metabolic disorders.

## 10. The Crosstalk between Oxidative Stress and Pathogenesis of Diabetes

When aberrant chemical interactions occur in the body and impair the normal metabolism, the patient is at risk of development of metabolic disease. Diabetes is one of the most prevalent metabolic and inflammatory disease that is dominating a large number of populations of the world including India. The high production of ROS in diabetes is indeed an outcome of hyperglycemia, which seems to have a potential role in the formation of diabetes complications, according to recent research [168]. The mitochondrial ETC enzymes, lipoxygenases, xanthine oxidases, nitric oxide synthases, cyclooxygenases, and peroxidases are all increased by hyperglycemia in diabetic individuals [187]. Due to the formation of ROS, oxidative stress and cellular death; numerous signaling pathways that are primarily initiated by hyperglycemia seem to make a significant contribution in diabetes and its complications. Figure 3 describes the cellular pathways involved in ROS-mediated pathogenesis of diabetes. Hyperglycemia activates several regulatory signaling pathways, which result in inflammation, cellular degradation, release of cytokines and as a result, diabetes consequences such as nephropathy, neuropathy, retinopathy, and cardiovascular disease [187]. The degradation of antioxidants such as CAT, SOD, and GSH-Px may result in a rise in ROS generation in diabetics due to a reduction in antioxidant production and, as a result, an increase in oxidative stress susceptibility [188]. In T2DM, pathway-selective insulin resistance (SEIRR) elevates mitochondrial ROS generation from FFA and inactivates anti-atherosclerotic enzymes, contributes to atherosclerosis and cardiomyopathy. The susceptibility of diabetics to ROS varies substantially. Inflammatory impairment in T1DM is driven in part by islet ROS, but in T2DM, increased food flux and subsequent ROS generation are likely to underlie loss of β-cell function [189, 190] (Figure 3). Some of the major contributing factors that have established a link between ROS generation and involvement in diabetic complications such as AGE, Aldose reductase (AR), PKC and Hexosamine are briefly discussed here.

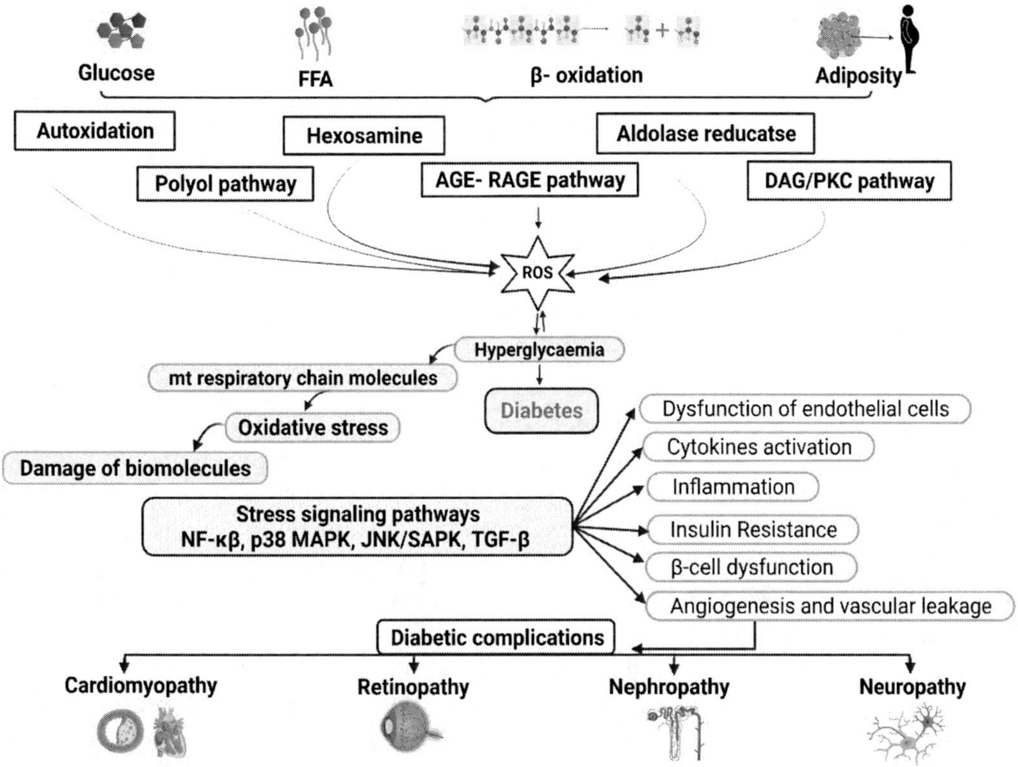


**Figure 3.** ROS modulating pathways leading to Diabetes conditions.

Normally, the increasing levels of glucose concentrations as well as free fatty acids, β-oxidation and adiposity are going to disturb several metabolic pathways including autoxidation, polyol pathway, hexosamine pathway, AGE- RAGE pathway, aldolase reductase pathway and DAG/PKC pathway. It results in elevation of mitochondrial ROS generation and exerts their harmful effects on macromolecules like DNA, protein, and lipids. Also, the overproduction of ROS will induce hyperglycemia in the body directly leads to diabetes. While damaged biomolecules activate the stress signaling pathways and involve abnormal metabolism that affects different organs very badly. The cardiovascular system, renal system, nervous system is severely damaged through this and termed as diabetic complications.

## 11. Impairment of Aldose Reductase and Free Radical Generation

The polyol pathway has been demonstrated to be a dominant contributor of diabetes-induced oxidative stress in previous investigations utilizing AR impaired animals [191, 192]. The complex I of ETC uses NADH as a source of electrons, resulting in enhanced superoxide radical production in the mitochondria. Because of the increased flux of electron donors into the ETC oxidative phosphorylation in mitochondria is boosted in diabetes. The inner mitochondrial membrane (IMM) potential rises, preventing electron transport within complex III. As a result, electrons return to coenzyme Q, they are transported one after another to molecular oxygen, resulting in production of superoxide [193]. The polyol pathway transforms glucose to fructose, which is then metabolized into fructose-3-phosphate and 3-deoxyglucosone, both are more strong and effective non-enzymatic glycation agents than glucose. As a result, increasing glucose flow through the polyol pathway may significantly raise AGE formation, resulting in ROS production. Accumulating evidence also supports the detrimental role of AR not only in diabetes and cardiovascular disease but also in diabetes related complications such as retinopathy, nephropathy, and neuropathy. The AR-dependent polyol pathway also contributes to diabetic cataractogenesis [192].

## 12. Advanced Glycation End Products (AGEs) Formation

Free radicals are thought to make a significant contribution in the development and advancement of late-diabetic complications because of their potential to effectively oxidization and degradation of biomolecules like nucleic acids, proteins, and lipids [194]. Biomolecules like proteins or lipids that become glycated as a consequence of sugar consumption are known as AGEs. The carbonyl group of a reducing sugar combines with the amino group of a lipid, protein, or DNA to produce Schiff bases, which then rearrange to Amadori products that are highly unstable and contribute to irreversible AGE development. This process is known as Maillard reaction where autoxidation of biomolecules occurs and results in formation of AGEs. The production of AGEs and their interactions with their receptors, receptors for AGE (RAGE), leads to ROS generation, which is also responsible for the inactivation of antioxidants. Because of the high levels of circulating glucose, AGE precursors, and oxidative stress in diabetics, AGE production is enhanced. Raised volumes of AGEs, such as N ε-(carboxymethyl) lysine (CML), glucosepane, pentosidine, or MG-derived hydroimidazolone, have been identified in the serum and tissues of individuals with T1DM and T2DM. Additionally, AGE build-up in diabetic tissue has also been linked to diabetes related complications [195, 196]. Even during diabetes, retinal pericytes accumulate AGEs, which are thought to impair pericyte survival and function. Furthermore, AGEs increase mRNA levels of RAGE in pericytes via internal ROS production. As a result, these positive feedback mechanisms transform AGE signals again, further alleviating the cytotoxic effects of AGEs on pericytes (dysfunctional pericytes) in DR. The consequences include vascular inflammation in retina and angiogenesis. Likewise, through increase in angiogenesis and endothelial cell dysfunction (faulty endothelium-dependent vasodilation in diabetes) develops DCM. Apart from that, DN is also another complication arising due to accumulation of AGEs where disruption of proximal tubular cells, mesangial cells, and glomerulosclerosis occurs. Also, DNeu is also observed with signs including demyelination of neurons, interruption in axonal transport [197].

## 13. Role of DAG/PKC Pathway

In cells where glucose transport is mostly independent of insulin, the metabolic repercussions of hyperglycemia can be seen. The activation of the PKC pathway, which results in intracellular hyperglycemia, has also been linked to the pathophysiology of diabetes. Increased production of diacylglycerol (DAG), a key intrinsic cellular co-factor for activation of PKC, activates PKC via glycolytic products such as dihydroxy-acetone phosphate (DHAP) and glyceraldehyde-3-

phosphate (G3P). The increased expression of PKC and the prolonged elevation of DAG in the vasculature can be maintained, which stimulates transforming growth factor-β (TGF- β). It regulates the gene expression of proteoglycans and collagen and develops abnormalities in diabetic tissues via elevating TGF- β levels, which is believed to promote thickening of capillary basement membrane [198]. PKC-dependent regulation of nicotinamide adenine dinucleotide phosphate (NADPH) oxidase is one of the major components linked to enhanced glucose-induced ROS generation. PKC isoforms play a role in endothelial dysfunction by altering NO production and inflammatory responses in the presence of elevated cholesterol and glucose levels. It can be said that ROS production leads to PKC activation and vice versa, implying that they are both implicated in the accumulation of oxidative stress in hyperglycemic individuals. Undoubtedly, many other complications like atherosclerosis and cardiovascular diseases appear due to endothelial dysfunction via ROS/RNS production. Another result of PKC activation is a disruption in lipid control, which influences the expression level of the low-density lipoprotein receptor (LDLR), which may be another cause of atherosclerosis (199].

## 14. Hexosamine Pathway

There is substantial evidence that the hexosamine signaling pathway and some other stress-sensitive signaling pathways like NF-κβ, p38 MAPK, and JNK/SAPK are activated by hyperglycemia and ROS. Increased intracellular glucose may be diverted into the hexosamine pathway, which could promote diabetic macrovascular disease. Only a small percentage of glucose (about 1–3%) is metabolized via the hexosamine route in a normal physiological state. Higher flux across the hexosamine pathway can be caused by excessive intracellular glucose levels along with FFA [198, 200]. The glucose metabolism via hexosamine biosynthesis pathway causes formation of the O-linked sugar moiety molecule, -N-acetylglucosamine (O-GlcNAc) to proteins, which affects their biological function. Temporary activation of HBP/O-GlcNAc pathways is a dynamic, beneficial strategy for enhancing cell viability in normal physiology; therapies that abruptly block this route reduce stress tolerance. Prolonged dysregulation of HBP/O-GlcNAc mechanisms, on the other hand, is seen to be harmful in pathophysiological contexts such as diabetes and its complications by developing IR [201, 202,203]. However, there is significant evidence that the hexosamine pathway has a role in the pathophysiology of diabetic vascular and renal problems. The underlying mechanism seems to be transcriptional control, which is likely influenced by transcription factor O-GlcNAc alteration involves in development of IR ultimately cause both T1DM and T2DM [204, 205].

## 15. Oxidative Stress and β-Cell Dysfunction

Moreover, the β-cell is another target of oxidative stress, response to a glucose stimulus, cells are responsible for sensing and secreting the right amount of insulin present in the body. Several studies have demonstrated that prolonged exposure to excessive glucose, higher FFA levels, or both cause β-cell dysfunction. Chronic hyperglycemia in T2DM patients is linked to reduced β-cell activity and tends to increase the risk of development of the disease [203]. Apparently, ROS appears to play a role in β-cell failure in regard to inflammation. Recent research, however, suggests that ROS may play a role in glucose-stimulated insulin secretion (GSIS). Another intracellular source of ROS is NADPH oxidases (NOXs) that are associated to plasma or subcellular membrane. Islets and β-cells have been found to contain several members of the NOX family [206]. The β-cells are vulnerable to ROS and RNS released by phagocytotic cells. Furthermore, immune cells generate cytokines that exacerbate β-cell oxidative damage. This pathway is known to be important in chronic autoimmune illness T1DM [207]. Because of their low antioxidant defence mechanism, both secretory ability and survival rate are affected by oxidative stress in β-cells, and both characteristics lead to β-cell failure. The negative consequences of persistently increased blood glucose concentrations (along with the generation of ROS/RNS) are referred to as glucose toxicity. In T2DM, glucotoxicity, along with lipotoxicity, is crucial because β-cells often live for longer periods and are exposed to significant glucose and FFA levels, which lead to β-cell function deterioration [208]. It can be stated that increased level of oxidative stress will cause β-cell inflammation, limiting the secretory activity of the cell and contributing to the advancement of both T1DM and T2DM respectively.

## 16. Clinically Available Therapeutics for Diabetes

Some of the more commonly used pharmacological interventions involving us of drugs such as metformin, sulfonylurea, meglitinides, thiazolidinediones, GLP-1 agonists, DPP-4 inhibitors and SGLT2 inhibitors are discussed here. These interventions along with their mechanism of action are depicted in Figure 4.

The illustration represents mechanisms of action of some commonly used drugs for treatment of diabetes. a) Sulfonylurea/Meglitinides acts at the level of pancreas and elevates the insulin secretion via binding to the sulfonylurea receptor-1 (SUR1) receptors and blocking the $K^+$ movement outside the cell, thereby promoting inward flux of $Ca^{++}$. b) Metformin decreases hyperglycemia, insulin resistance and lipotoxicity via downregulating glucose production and fatty acid synthesis respectively in the liver. c) The intestine secretes a glucagon-like peptide 1 (GLP-1) hormone, with potential to reduce the levels of blood sugar. However, GLP-1 has a short half-life since it is degraded by the dipeptidyl peptidase-4 (DPP-4). So, the DPP-4 inhibitors and GLP-1 agonists are used to maintain the normal levels of GLP-1. d) Similarly, the normal process of glucose reabsorption by kidney is elevated in diabetes, therefore, SGLT-2 inhibitors are used to downregulate the glucose reabsorption thereby maintaining normoglycemia.

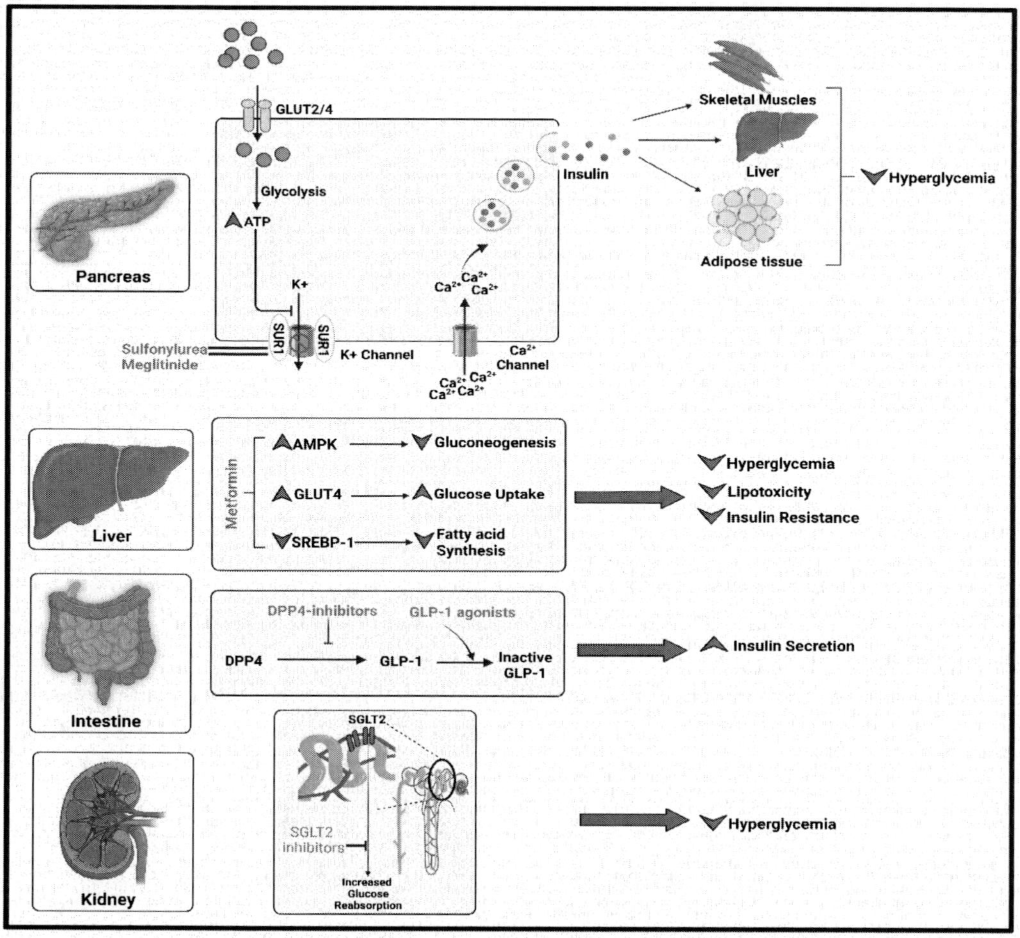


**Figure 4.** Mode of action of various pharmacological interventions.

## 16.1. Metformin

The first line of drug used for the treatment of T2DM is metformin. It activates AMPK pathway and decreases liver glucose synthesis, improves tissue sensitivity towards insulin by increasing glucose uptake from peripheral tissue [209]. In order to enhance glucose sensitivity, it increases insulin receptor expression and influences protein kinase activity. It does not directly affect production of insulin endogenously but rather exerts its effects indirectly by adjusting the level of incretin hormones from gut. The gut mediated mechanism of metformin involves an interplay between gut, brain and liver as per in-vivo experiments conducted in rats which demonstrates that when intestinal cells are exposed to metformin hepatic glucose synthesis occurs via nervous vagus efferent and nucleus tractussolitarius through activation of AMPK and GLP-1 receptors [210]. Metformin depletes cellular store of ATP resulting in deficient flow of ATP required which are normally used to carry energy consuming hepatic gluconeogenesis. It also performs its effects independent of ATP production by mitochondrial ETC through influencing enzymes like pyruvate decarboxylase and other enzymes and thus decreases hepatic gluconeogenesis in ATP independent manner [211]. Metformin is a complex-I inhibitor of mitochondria [212] and is less toxic because it only inhibits mitochondrial complex-I when mitochondria is in activated state thus sparing other mitochondria and hence helps to maintain a significant pool of mitochondria by avoiding unnecessary mitochondrial inhibition. Metformin downregulates the levels of sterol regulatory element binding protein-1 (SREBP-1) and exhibits antilipidemic effects; furthermore, via AMPK activation it also shows antihyperglycemic effects [213]. Metformin decreases ATP levels and increases AMP levels by inhibiting enzyme AMP deaminase enzyme. Gluconeogenesis consumes 4 molecules of ATP and 2 molecules of GTP for synthesis of one molecule of glucose from pyruvate or lactate. Hence by decreasing ATP concentration metformin reduces gluconeogenesis. Metformin inhibits mitochondrial glycerol-3 phosphate dehydrogenase enzyme that impedes the transport of NADH from cytosol to mitochondria. The increase in NADH levels prevents conversion of lactate to pyruvate as this conversion requires $NAD^+$ [214].

### 16.2. Sulfonylureas

In contrast to Metformin, Sulfonylureas work by increasing the concentration of insulin produced inside the body. The receptor for binding sulfonylureas is present on pancreatic β-cells and by stopping influx of $K^+$ through ATP-dependent channel the intracellular concentration of $K^+$ drops severely causing membrane depolarization which leads to enhanced movement of calcium inside β-cells causing increased contraction of actomyosin complex leading to increased exocytosis of insulin thus increasing insulin level [215]. Sulfonylureas not only bind to membranes to sulfonylureas receptors (SURs) on membranes of β-cells but SURs on cardiac and endothelial cells hence extending positive effects on cardiovascular complications [216]. These drugs are categorized in 2 classes including first class agents including chlorpropamide, tolazamide, tolbutamide, and second-generation agents including glipizide, glimepiride and glyburide. Second class agents are preferred for patients as they have high potency [217]. The second-generation drugs differ from first generation drug in having higher number of substitutions at either end of compound or thus are generated from modifications in first generation SUs. These substitutions help to prolong their action and improve their pharmacological and pharmacokinetic properties. Glimepiride as compared to glyburide and glipizide have less chances of inducing ventricular arrythmias, cardiac changes and has no effect on blood pressure thus making it a preferable option. Interaction of sarcolemma ATP dependent $K^+$ channels of cardiac myocytes specifically with glimepiride rather than directly binding ATP channel in mitochondria is a key to prevent tissue damage during repeated ischemic stroke and thus prolong the time needed for myocardial infarct needed to cause cellular damage [218]. Glimepiride is sometimes categorized under class third generations' sulfonylureas drug. They also produce their effects by binding to Epac-2. The binding of Epac-2 with Rap-1 protein is known to enhance the fusion of insulin vesicles with plasmalemma, thereby elevating insulin secretion. Sulfonylureas imitate the effects of c-AMP in activating Epac-2 and increasing accessibility of insulin [219]. Sulfonylureas produce their effects. They also inhibit glucagon secretion from pancreatic islet as well as enhance glycogen synthesis in liver thus greatly reducing glucose levels in body up to a greater extent [220]. They reduce the rate of clearance of insulin from liver. These drugs enhance the ability of peripheral tissues to take glucose [221].

## 16.3. Thiazolidinediones

A new class of drugs having ability to bind with very high affinity to adipocytic PPAR-γ (peroxisome proliferator activated receptor-γ) to increase adipogenesis and fatty acid uptake. PPARs are part of nuclear hormone receptors. PPARs can control gene transcription in both ligand dependent and ligand independent ways [222]. PPAR-γ targets the fatty acid oxidation and transportation, glucose metabolism, cholesterol biosynthesis and transportation, apoptosis and anti-inflammatory response genes. PPAR-γ is inherently expressed all over the kidney principally in collecting ducts and distal medullary and at a depressed levels in renal vasculature and proximal tubules [223]. PPAR-γ ligands can directly influence vascular functions as they are expressed in endothelial cells and vascular smooth muscle cells [224]. PPAR-γ is the main controller of adipocyte differentiation [225]. Examples of thiazolidinediones include pioglitazone, rosiglitazone, troglitazone. Besides this mTOR modulator TZDS includes MSDC-0106 and MSDC-0602 are pioglitazone derivatives with classical PPAR-γ agonist activity but have protective effects on mitochondria are currently under clinical trials. Balaglitazone is a partial resemblance to PPAR-γ and thus increases insulin sensitivity [226]. However, heart attacks and oedema are the complications that are commonly associated with pioglitazone and rosiglitazone. Its administration is also associated with increased chances of bone fractures [227]. PGC-1α is implicated in biogenesis of mitochondria is positively stimulated by Pioglitazone and as PGC-1α is expressed in renal portion it gives desirable effects against DN which is a complication that aggravates diabetes even more [150]. PGC-1α acts as a coactivator of transcription for PPAR-γ. When tissues are exposed to high demands of energy then expression of PGC-1α from mitochondria of these tissues comes into play which controls biogenesis of mitochondria [228].

## 16.4. Meglitinide Analogues

These are a class of oral drugs that enhances insulin secretion from pancreas [229]. Meglitinides bind to SUR-1 receptor on β-cells with lower affinity than sulfonylureas but stimulate insulin secretion in same way. By stopping activity of $K^+$ dependent ATP channel of pancreatic β-cells they tend to enhance glucose secretion where blockage of ATP dependent $K^+$ channel leads to membrane depolarization and increased level of calcium inside β-cells boost release of insulin. Nateglinide and repaglinide are meglitinide analogues that are currently prescribed for diabetes medication [230]. Receptor of both sulphonylureas and meglitinide both bind to SUR receptor on but at different position as meglitinides bind to a high affinity receptor and that to very loosely than sulphonylureas hence meglitinides have lesser time to exert their action and require much more glucose levels than sulphonylureas to release insulin in blood. Because they share similar mechanisms of action, they are never prescribed together [231]. These are more effective for patients with irregular schedule of meals [232]. A healthy kidney is not an essential prerequisite for excretion of repaglinide as it has a shorter duration of action and hence can be given to patients with kidney problems which generally arise as a side effect of diabetes [233]. The Nateglinide shows specificity in binding SUR-1 receptor than SUR-2 receptor and its much shorter half-life enables it to get eliminated easily before it reaches a toxic concentration inside tissues. Its potency is directly proportional to glucose levels in blood. Moreover, insulin secretion under the action of nateglinide is dependent on the time at which this drug is taken.

To increase its effectiveness, it is prescribed before having meal as it increases insulin secretion in a much higher way before meal than in fasting state or having it after meal. Repaglinide never accumulates in plasma at a higher level as it is easily excreted via bile [230].

### 16.5. Alpha-Glucosidase-2 Inhibitors

These are class of drugs known to hinder the carbohydrates absorption from gut. The conversion of non-absorbable form of carbohydrates in absorbable form is impeded by the AGIs. Normally after a meal, there is a spike in glucose levels as enzymes digesting carbohydrates get activated in gut after a meal. Hence AGIs (Alpha-glucosidase-2 inhibitors) inhibit enzymes glucoamylase, sucrase, maltase and isomaltase. These enzymes are found in brush bordered epithelium of small intestine. AGIs including miglitol, voglibose and acarbose are used to regulate diabetes, to decrease release of glucose after a meal [234]. Hence these drugs do not directly or indirectly effect insulin secretion from pancreas but slow down absorption of carbohydrates from gut which generally spikes after a meal. Besides diabetes, they produce positive effects by reducing hypertension and decreasing susceptibility to cardiovascular disease which are common complications that arise as a side effect of diabetes. Intestinal membrane bound enzyme alpha-glucosidase which hydrolyses oligosaccharides and disaccharides to glucose and other monosaccharides thus decreasing sugar levels is reversibly inhibited by Miglitol [235]. Sulfonylureas are more effective in reducing blood glucose concentrations than AGIs inhibitors. AGI-2 inhibitors can bind the carbohydrate binding region of alpha-glucosidase in competition to oligosaccharides thus prolonging cleavage of oligosaccharides to monosaccharides [236]. AGI-2 inhibitors prolong the time taken for absorption of sugar from gut by inhibiting alpha glucosidase-2 enzymes. These drugs bind reversibly to enzyme to inhibit their action. Besides these drugs also affects colonic starch fermentation by exerting metabolic action on them [237]. Since these drugs inhibit carbohydrate digestion in small intestine by mimicking carbohydrates and binding alpha glucosidases these carbohydrates are sent into colon for further digestion giving rise to side effects such as flatulence [238]. Acarbose exerts its effects at three levels. It exerts action at gut by modulating level of gut hormones, and interaction with gut microbiota. At intestine level, unabsorbed acarbose also affects immunity of intestine by decreasing proinflammatory cytokine expression on gut cells. AGI drugs make patients prone to lung cancer slightly and prone to gastrovascular cancer [239]. Side effects of these drugs include diarrhea, flatulence and abdominal pain which are dependent on dose level of drugs [240].

### 16.6. Dipeptidyl Peptidase-4 Inhibitors

Normally, the enzyme Dipeptidyl peptidase-4(DPP-4) shows its effects indirectly by acting on incretin hormones including gastric inhibitory peptides (GIP) and Glucagon like peptide-1 (GLP-1). The enteroendocrine L-cells of small intestine following post translational breakage of proglucagon by prohormone convertase (PC)1/3 secretes GLP-1. GLP-1 and GLP-1-$NH_2$ are the functional forms of GLP-1 released in circulation and they are further broken into GLP-1and GLP-1-$NH_2$ respectively [241]. PC1/3 mainly resides in K-cells of proximal jejunum and duodenum releases GIP from preproGIP post translational splitting [242]. Both these hormones are released just after meal ingestion to decrease sugar level that generally rises following a meal, but these hormones have shorter half-lives hence are degraded by DPP-4 readily. GIP and GLP-1 have half-life of 5 and 2 minutes [243]. DPP-4 enzymes belong to the family of enzymes called serine proteases. DPP-4 enzymes are found in kidney, liver, intestine, placenta, lymphocytes and endothelial cells. Saxagliptin, vildagliptin are some of examples of DPP-4 inhibitors [244]. GLP-1 enhances insulin release along with hindrance to glucagon hence exerting its effects rigidly in a glucose-dependent manner. The human gene DPP-4 promoter carries consensus sequence for transcription factors like hepatocyte nuclear factor-1 and nuclear factor-κβ. DPP-4 cleaves many peptides comprising chemokines, neuropeptides and regulatory peptides. They show higher propensity of breaking peptides at position-2 proline but can also degrade after hydroxyproline, dehydroproline, alanine, glycine, threonine, valine or leucine [245]. PC-2 residing in α-cells of pancreas releases GIP which induces insulin secretion in a manner similar to that caused by GIP [246]. The membrane bound form of DPP-4 is known as CD-26 while soluble form which is known for cleaving preproglucagon is known as DPP-4 simply. The co-stimulation migration, memory development, thymic maturation of T-cells also falls under control of membrane bound form of DPP-4 or CD26 hence DPP-4 inhibitors can also prove useful in treatment of type-1 diabetes. T-cell activation, proliferation and its immune functions are controlled by CD-26. Thus DPP-4 inhibitors can prove useful in treatment of type-1 diabetes also as their membrane bound form is intimately connected to regulation of immune system [247].

## 16.7. Sodium-Glucose Transporter Inhibitors

The SGLT-2 transporters are a large family of membrane proteins involved in the transport of amino acids, glucose, osmolytes, vitamins and ions. SGLT-2 inhibitors comprise empagliflozin, dapagliflozin and canagliflozin. SGLT-1 is found in intestinal mucosal cells of small intestine and SGT-2 is found in the first and third segment of PCT where it participates in reabsorption of almost 90% of filtered glucose. Reabsorption of glucose is elevated in patients of diabetes by increasing expression of SGLT-2in kidneys in PCT region particularly which is responsible for maximum reabsorption of glucose [248]. Hepatocyte nuclear factor1α (HNF-1α) regulates SGLT-2 expression at transcriptional level by binding to promoter of SGT-2 gene [249]. SGLT-2 specific inhibitors decrease blood glucose level by promoting removal of glucose via urine by causing hindrance to process of reabsorption in PCT. This in turn alleviates patients from hypertension caused by enhanced blood glucose levels as their action is exerted at kidney level, they are specifically used in treatment of DNeu also. SGLT-1 exerts their action at the location of gut and at distal S3 portion of proximal tubules where SGLT-1 inhibitors help to cope up with increased blood glucose level after ingestion of meals. The time after ingestion of meal is described to be a sensitive time during which glucose level spike in a much-exaggerated manner than in normal conditions due to activation of several carbohydrate digesting enzymes that increase blood glucose levels. LX4211 is a SGLT-1 and SGLT-2 inhibitor that targets both the transporter are found in apical membrane of PCT S1 and S3 portions of PCT working in communion basolateral membrane glucose transporters GLUT-1 and GLUT-2 that are responsible for transcellular transport of glucose [250].

The main difference in both transporter lies in their affinity where SGLT-2 and GLUT-2 working in S1 segment of nephron shows less affinity, but SGLT-1 and GLUT-2 shows high affinity. ISIS 388626 is an antisense SGLT-2 oligonucleotide composed of 12 nucleotides that decreases mRNA of SGLT-2 in proximal tubules of rodents, dogs and non-human primates by 70 – 80% and decreases blood glucose level by increasing elimination of glucose via urine [251]. The plasma glucose lowering effects along with increased elimination of glucose via urine coupled with protective β-cell regeneration in rat animal model has proved the potential and efficacy of SGLT inhibitors in treating T1DM also [252].

## 17. Emerging Therapeutic Approaches

### 17.1. Antioxidant Gene Regulation by FOXO (Forkhead Box O)

FOXO are transcription factors comprising mainly FOXO-1, FOXO3 and FOXO4. Increased cellular stress serves to activate FOXO transcription factors. An increase in intracellular concentration of ROS causes localization of FOXO to nucleus, where it is transcriptionally active. In presence of insulin, nerve growth factor, insulin like growth factor-1(IGF-1), phosphorylation of FOXO protein happens to be a game changing process that changes its site of action [253]. Phosphorylation of FOXO by insulin and IGF-1(insulin growth factor-1) causes FOXO to switch its location from nucleus to cytoplasm where it is degraded by ubiquitin dependent pathway [254]. A master regulator FOXO activity is its different post translational modifications like phosphorylation, acetylation and ubiquitination to FOXO. Cell cycle arrest, DNA repair, apoptosis, autophagy and glucose homeostasis are some of essential cellular processes that are regulated by FOXO. The site of action of FOXO is decided by the modifications that happen to take place in FOXO. For example, phosphorylation of FOXO proteins by AKT/protein kinase- B and SGL (serum and glucocorticoid induced kinases (SGKs) renders FOXO protein in inactive state by holding them in cytoplasm because FOXO can be activated only when it reaches nucleus. Similarly, ERK, JNK and p38 also phosphorylate the FOXO proteins. In contrast the phosphorylation performed by AMPK and CDK-2 (cyclin dependent kinase-2) stimulates FOXO activity by changing its location from cytoplasm to nucleus. Iκβ also phosphorylates FOXO and causes its destruction by ubiquitin- proteasomal machinery. Similarly, deacetylases also regulate FOXO activity. Deacetylases like Sirt-1 are double edged swords because depending on target gene promoter they increase or decrease FOXO binding with DNA. Phosphorylation and ubiquitination mediated degradation are complementary to each other as phosphorylation performed by IK-β and Akt marks protein for ubiquitination thus establishing a link between two. However, acetylation in most cases causes FOXO to head for nucleus under effect of oxidative stress to activate genes responsible for autophagy [255]. If β-catenin partners with FOXO its activity receives a boost [256]. For example, stress induced kinase JNK phosphorylates FOXO-4 that promotes its entry into nucleus and activates its transcriptional activity [257]. The acetylation of FOXO by c-AMP responsive element binding proteins (CREB) binding protein (CBP/P300) positively influences FOXO transcriptional activity throughout oxidative stress while deacetylases like sirtuins (SIRT-1) performs deacetylation activity on FOXO proteins that negatively effects transcriptional activity of FOXO proteins [258]. FOXO acetylation depends on HDACs (histone deacetylases), SIRT-1 and SIRT-2(silent information regulators), CBP/p300 co-activator. CBP/p300 could activate gene transcription of FOXO by adding acetyl residues to unmask chromatin that is transcriptionally active. FOXO can interact with SIRT-1 only in nucleus under influence of ROS exposure. Unlike SIRT-1, SIRT-2 can deacetylate FOXO both in nucleus and cytoplasm. SIRT-1 targets genes that are involved in cell survival while SIRT-2 is responsible for $p27^{kip}$ and Mn dismutase thus helps cell machinery to fight against threat imposed by ROS in form of impostor. Unlike polyubiquitination, monoubiquitylation performed in MDM-2 pushes FOXO in nucleus while deubiquitylation performed by USP-7 decreases transcription of FOXO [259]. Resveratrol is known to stimulate activity of sirtuins which deacetylates liver kinase-B1 which serves to activate AMPK and send cell to fate of apoptosis by caspase-12 signaling thus removes endoplasmic stress induced oxidative stress also [260, 261].

Other reports present contradictory views regarding acetylation and deacetylation where acetylation suppresses, and deacetylation promotes FOXO transcriptional activity [262]. The expression of genes that cause autophagy and Ubiquitin proteosome system and are responsible for decreasing ROS levels are increased by FOXO transcription factors. A specific form of autophagy is lipophagy which releases lipids from cytosolic lipid droplets under positive regulation of FOXO [263]. Epigallocatechin gallate (ECGC) in green tea and quercetin found in onion serves to increase levels of FOXO transcription factor [264, 265]. By phosphorylating FOXO proteins with help of phosphoinositide -3 kinase (PI-3kinase) and serine threonine kinase Akt just like insulin ECGC imitates insulin signaling thus tricking cellular machinery by stimulating insulin dependent activity even in absence of insulin [266]. SGLT-1 transporter provides a gateway for direct absorption of quercetin glucosidase or MDR-2(multidrug transporter protein-2) these are excreted into lumen. Carrier mediated transport facilitates their uptake after which these are hydhydrolyzed intracellular β-glucosidase [267].

## 17.2. NRF-2 Activators (Nuclear Factor Erythroid Related Factor-2)

The Nrf-2 is a master controller of antioxidant response. Expression of about 100 genes are under control of NRF-2KEAP/ARE pathway. These genes recognize DNA damage, regulate proteosome function, detoxification enzymes, direct antioxidant protein and free radical metabolism [268]. it is principally regulated by four E3 ubiquitin ligase complexes mediated ubiquitylation and proteasomal degradation: KEAP-1Cullin (CUL)3 RING Box protein 1(Rbx-1), βTrCP-S phase kinase associated protein (SKP-1)-CUL-1RBX-1, WD repeat protein,3-CUL-4 damaged DNA binding protein (DDB-1) under various conditions. CUL3-RBX1 is a ubiquitin ligase complex for which KEAP-1 is an adaptor. KEAP-1 binds with NRF-2 causing ubiquitinoylation by 26S proteosome in cytoplasm [269].

Sensor cysteines of KEAP-1 react with ROS as the ROS levels increase that permits NRF-2 to escape from KEAP-1 mediated degradation as KEAP-1 marks NRF-2 for degradation. Hence under conditions of ROS accumulation NRF-2, gathers in nucleus and activate expression of cell protecting genes [270]. Driven by cellular stress, NRF-2 stimulates expression of genes containing ARE (antioxidant responsive element binding protein). Expression of glucose transporter-1(GLUT-1) is increased is increased under NRF-2 stimulation that increases glucose supply in glycolysis. Phosphoglucomutase (PGM-5) converts glucose-6phosphate to glucose-1phosphate. For glycogen synthesis NRF-2 stimulates expression of PGM5, 1,4alpha glucan branching enzymes, phosphorylase kinase regulatory subunit alpha-1, glucosidase alpha acid thus affects glycogen metabolism. Free radicals' formation is encouraged by iron [271]. Curcumin is known to be potent activator of NRF-2 hence is recognized as an antioxidant that could alleviate the stress caused by ROS [272]. Resveratrol also is an NRF-2 activator responsible for reducing stress caused by ROS [273].

### 17.3. Enzyme Antioxidants Imitators

Some of drugs have been designed that tends to imitate effects of the antioxidant enzymes produced inside body thus reinforcing the effects of antioxidant machinery of body. Examples include SOD mimetic out of which Mn SOD are specifically imitated as they exert their effects directly on mitochondria. Cyclic polyamines, salen, porphyrin, nitroxides are some of SOD imitators. Cyclic polyamines include M40403 that has shown positive results at cytosol and mitochondrial level [274]. Mn (III) *meso*-tetrakis (*N*-ethylpyridinium-2-yl) porphyrin, $Mn^{III}TE\text{-}2\text{-}PyP^{5+}$ (AEOL-10113) has a potent antioxidant property owing to presence of ortho cationic pyridyl nitrogen [275]. SOD, CAT and GPx antioxidant activity is shown by synthetic salen-manganese compound EUK-8. EUK-8 proved specifically helpful in treatment of diabetes type-1 patients [276]. SOD mimicking activity of nitroxides is dependent on conditions of pH. Nitroxides have poor SOD activity in their original form. Decreased pH increases their SOD activity indicating role of environment in nitroxides activity. Besides this tempol [4-hydroxy 2,2,6,6, tetramethylpiperidine -N-oxyl also gives desirable results [277].

### 17.4. Phytochemicals as Antioxidant

Flavonoid Fisetin (3,3,4,7-tetrahydroxy-flavone) found in onion and strawberry exhibit considerable antioxidant activity [278]. Flavonoid kaempferol found in Indian gooseberry exert significant antioxidant effects indirectly targeting hyperglycemia which is regarded as main culprit for generating ROS in diabetes patients. it targets hyperglycemia and also improves IR by affecting NF-κβ pathway. It is also effective against cardiovascular complication that arise as a side effect of diabetes by affecting MAPK pathway. Serine threonine kinases ERK-1 and 2, JNK and p38(stress activated kinases) are part of MAPK pathway. Activation or phosphorylation of ERK-1 and 2 is linked with reduced severity of myocardial ischemia reperfusion injury [279]. Pomegranate exerts antioxidant and inflammatory effect on account of flavonoids, anthocyanins, ellagitannins, proanthocyanins found in it which either activates or inhibit PPAR-γ and NF-κβ which prevent oxidation of lipids by attenuating ROS generation [280]. White mulberry performs antioxidant effects by inhibition of polyol pathway and apoptosis that is induced by hyperglycemia [281]. Plant goji berry found in China contain several phytochemicals that alleviates oxidative stress and retinal damage caused by diabetes. It contains zeaxanthin, betaine, carotene, β-sitosterol and polysaccharides. A non-essential amino acid called taurine found in goji berry spreads its action along three directions. Firstly, it inhibits the activity of enzyme caspase-3 thus providing an antiapoptotic response. Secondly it also targets PPAR-γ receptors which is a major target of many diabetes therapeutics. Thirdly it decreases m-RNA levels of VEGF providing antiangiogenic properties to it. It protects retina from damage caused by excessive ROS by preventing excitotoxicity caused by glutamate by decreasing levels of glutamate and NMDA (n-methyl-D-aspartate) receptor [282].

*Abeliophyllum distichium*; Aeglemarmelos; *Agrimoniapilosa* ledeb; Dendrobiumchrysotoxum; *Zea* mays, *Glycyrrhiza uralensi* and *Perilla frutescens* are some of plants that contain chemicals that target polyol pathway responsible for ROS generation. *Osteomeles schwerinae, Zingiber zerumbet and Cnidium officinale* are some of plants that contain phytochemicals for targeting hexosamine pathway. The AGE and their receptors are targeted by *Aster Koraiensis*; *Camellia nitidissima*; *Carpobrotus edulis; Litchi chinenesis; Melissa officinalis*; *Polygonatum odoratum*; *P. cuspidatum*; *P. multiflorum* and *Prunella vulgaris* [283].

Table 1 depicts some phytoactive compounds that exhibit considerable anti-diabetic and antioxidant property via targeting various pathway to reduce ROS species and associated complications that result from type-2 diabetes. 8-phenylnaringenin and xanthohumol decreases levels of protein galectin 3, nitro tyrosine, AGE in liver thus alleviating oxidative stress. Didyamin decreases levels of aldolase reductase and α-glucosidase while sudachitin directly controls mitochondrial generation by affecting NRF levels and increases mRNA of GLUT-4. Hesperidin produces its effect by principally aiming enzymes like SOD, CAT and GPx. GLU-T4 levels are increased by tangeretin. Nobiletin stimulates MAPK-ERK pathway along with decrease in LDL receptor expression on liver cells. Cosmosiin produces its affects via affecting GLU-T4 movement. For example, the levels of enzyme glucose-6 phosphorylase, glycogen phosphorylase, fructose-1,6diphosphatases are brought to normal levels by naringin. Naringin also produces its effects on stomach by decreasing absorption of sugars and enhances insulin secretion from pancreatic-β cells [284]. Specifically, 3,4dimethoxy benzoic anhydride; benzo[*b*])-1, 4-diazabicyclo[2.2.2]octane, and cinnamic acid extracted from Bael plant exhibit antioxidant effects against retinopathy caused by diabetes [285]. Similarly, agrimonia Pilosaledeb found in China contains afzelin, luteolin, and luteolin-7-glucoside and agrimoniin. These phytochemicals perform inhibitory action on aldolase reductase that is responsible for ROS generation. Most notable chemical found in agrimonia Pilosaledeb is a triterpenoid that PPAR-γ receptor along with activation of mRNA of GLUT-4 and adiponectin thus spreading its action along various pathways and ultimately targeting ROS generation [286].

## 17.5. Vitamins and Other Antioxidants

Vitamin A, vitamin C, vitamin E, carotenoids and lipoic acid are well studied antioxidants that are very effective for treating diabetes. Vitamin A and vitamin E are known to possess higher antioxidant property. Ubiquinol, Uric acid also serves as antioxidant produced inside body [287]. Rats fed on HFD decrease liver diglyceride and triglyceride levels [71]. Coenzyme Q10 has greater ability to dispose of radicals than vitamin E. Vitamin E (α-tocopherol) is a leading antioxidant because it constitutes the structure of membranes which is a location prone to lipid peroxidation. Specifically, glycation modification of proteins that result from ROS exposure are targeted by vitamin E [288]. Nicotinamide is a form of vitamin B3 that protects pancreatic β-cells from death by stabilizing sufficient pool of $NADP^+$ and interfering with process of DNA repair by negatively effecting enzyme poly ADP ribose polymerase (PARP) thus sending cell to ultimate destination of death. Receptors of vitamin D are found in pancreatic β-cells hence vitamin D plays role of good antioxidant in treatment of diabetes [289]. The therapeutic effects of insulin sensitizers such as protein kinase-C inhibitors are increased by coenzyme Q10. Coenzyme Q10 administration is especially helpful in treating cardiovascular complications that emerge as a side effect of diabetes [290].

**Table 1.** Mode of action of various bioactive compounds or plant-based therapeutics

| Phytochemicals | Mechanism of action | Reference |
|---|---|---|
| 8-phenylnaringenin | Decreases nitro tyrosine, galectin-3 and AGE. | [291] |
| Didyamin | Decreases aldolase reductase and alpha glucosidase | [284] |
| Xanthohumol | Decreases galectin-3, nitro tyrosine and AGE | [284] |
| Sudachitin | Increases mRNA of GLU-T4 | [284] |
| Hesperidin | Increases enzymes SOD, CAT and GPx | |
| Tangeretin | Increases SOD, CAT and GPx | [292] |
| Nobiletin | Targets MAPK-ERK pathway and decreases LDL receptor expression on liver cells. | [284] |
| Cosmosiin | Increases GLUT-4 levels | [293] |
| Naringin | Targets enzymes glucose-6 phosphorylase, glycogen phosphorylase and fructose1,6 diphosphatases | [284] |
| Taurine | Targets PPAR-γ pathway and cause increase in levels of caspase-3 and decrease in levels of VEGF. | [282] |
| Kaempferol | Targets MAPK pathway and NF-κβ pathway | [279] |
| Resveratrol | Activates NRF-2 transcription factor, activate sirtuins to deacetylate liver kinase-B1 which further activates AMPK leading to activation of caspase-12 | [273] |
| Curcumin | Activates NRF-2 transcription factor | [272] |
| ECGC | Activates FOXO transcription factors | |
| Quercetin | Activates FOXO transcription factors | |

## 17.6. Other Novel Strategies

Recent studies have also demonstrated some novel forms of therapeutics directly targeting the mitochondria and ER oxidative stress (Table 2). Some of drugs have been designed to enter mitochondria dependent on the action potential of mitochondria to enter it includes CoQ (Mito-Q), TEMPO (Mito TEMPO) or plastoquinone owing to presence of triphenyl phosphonium ion joined attached to these lipophilic antioxidant molecules [294]. While entry of these drugs is dependent on action potential of mitochondria but entry of special tetrapeptides with +3 charge called Szeto-Schiller peptides is independent of action potential of mitochondria [295]. These peptides are known to interact with cardiolipin which is very susceptible to damage caused by ROS. These peptides specifically interact with cardiolipins to stabilize the cristae containing respiratory complexes. ROS oxidizes cardiolipin thus decreasing its affinity to bind cytochrome-c thus releasing cytochrome-c from clutches of inner mitochondrial membrane [296]. A lethal form of cytochrome-c is generated by converting it into oxygenase with help of oxidized cardiolipin by breaking $Met^{80}$ -Fe ligation on heme of oxidized cardiolipin. This oxygenase is catastrophic as it doesn't even require ROS to further oxidize cardiolipin [297].

**Table 2.** Novel treatment strategies targeting mitochondrial and ER oxidative stress in diabetes

| Drugs | Mechanism of action | Reference |
|---|---|---|
| Mito-Q | Interacts with complex-1 and 3 of mitochondria | [294] |
| DS-1150b | Increases GLUT-4 levels | [298] |
| Mito-TEMPO | Interacts with complex-1 and 3 of mitochondria | [294] |
| Dantrolene | Targets ryanodine receptors on ER of β-cells | [299] |
| Phenyl butyric acid | Targets ER stress (corrects folding of proteins) | [299] |

| Drugs | Mechanism of action | Reference |
|---|---|---|
| Metallic Nonantioxidant | Cycles between different oxidation states to reduce ROS species | [300] |
| Szeto-Schiller peptides | Targets cardiolipin containing respiratory complexes | [295] |
| Tauroursodeoxycholic acid | Targets ER stress (corrects folding of proteins) | [299] |
| MLR-1023 (tolimidone) | Activates Lyn kinase | [298] |
| OPK88003 | Binds GLP-1 and GIP receptors | [301] |
| LY2405319 | Mimics FGF-21 | [302] |
| RO0281675, RO438920, LY2121260, PSN-GK1 | Activates glucokinase | [303, 304] |
| AMG837, GW-9508, TUG-424, AM-1638, AM-5262, GW-9508, TUG-424, AM-1638 AM-5262 | Mimics GPCR-40 ligands | [303, 304] |
| Mitochonic acid | Interacts with MINOS and specifically with mitofilin, ATP synthase and Hsp-70 | [305, 306] |

Mitochonic acid is a newer class of drug that serves to increase ATP production independent of ETC [305]. Liposomes are used for targeting mitochondria because of their unique ability to enclose both hydrophilic and hydrophobic drugs and protecting them for degradation. After altering surface of liposomes with dequalinium and 4-carboxybutyltriphenylphosphonium bromide so that they can carry negatively charged drugs into mitochondria. Resveratrol enclosing nanoparticles are being suggested to treat oxidative stress resulting from ROS [307]. In course of transfer of healthy mitochondria from mesenchymal stem cells to injured cell with dysfunctional mitochondria tunnelling nanotubes with 50-100nm are formed. Hence, mesenchymal stem cell could prove to be an effective therapeutic option [308]. A specific micro-RNA called mi-R708 is enhanced in response to stress caused by ER and by increased exposure of glucose to pancreatic β-cells. The secretion of insulin is balanced by neuronatin. Neuronatin selectively targets this micro-R708 to restore insulin secretion stimulated by glucose. Cerium oxide nanoparticles have shown better results in terms of mi-RNA delivery [309]. The drop of oxidative stress in neuronal cells and optimization of the pancreatic β-cell isolation by targeting inner mitochondrial membrane by novel peptide D-Arg 2'-6'dimethyltyrosine-Lys-Phe-$NH_2$ which significantly improved function after transplantation [310]. Cerium oxide nanoparticles are similar to CAT enzyme in protecting cell from oxidative stress caused by ROS. After preparing large pore size Mesoporous silica particles loaded with cross linked glutathione oxidases and cerium nanoparticles (MCF) these MCF are encapsulated with a model protein inside chitosan microgels sensitive to pH [311]. Novel mechanisms drugs have been categorized in 4 types based on their mechanism of action. First class of drugs are glucokinase activators that directly acts on β-cells to stimulate insulin secretion by targeting selectively pancreas; targeting liver and pancreas both and G-protein coupled receptor-40 agonist (GPCR-40 agonist). Examples of it RO0281675, RO438920, LY2121260 and PSN-GK1 are glucokinase activators while AMG837, GW-9508, TUG-424, AM-1638, AM-5262, GW-9508, TUG-424, AM-1638 and AM-5262 are GPCR-40 agonists. Out of these some of the drugs are still under clinical trials include TAK-875, LY2881835, AMG837[303, 304]. The second class of drugs include oxyntomodulin analogues. The post translational modification of pre-proglucagon results in release of oxyntomodulin from enteroendocrine cells of small intestine as a reaction to nutrient ingested. OPK88003 is an oxyntomodulin analogue [301]. Both the GLP-1 receptors and glucagon receptors are occupied by oxyntomodulin hence enlarging its area of action [312]. LY2405319 is an FGF-21 analogue which is used to increase insulin sensitivity [302]. 5

receptors of FGF (fibroblast growth factor) include FGF-1, FGF-2, FGF-3, FGF-4 and FGF-5 receptors. These FGF receptors belong to the same family. FGF-2 is involved in pancreatic development while FGF-4 is involved in cholesterol metabolism [313]. Other drugs include DS-1150b that stimulate GLUT-4 transport and MLR-1023 (tolimidone) that is a selective allosteric activator of Lyn kinase thus helps to increase insulin sensitivity [298]. Nanoparticles synthesized using cerium oxide (nanoceria), iron oxide, cobalt oxide, copper oxide, manganese oxide, vanadium pentoxide and from noble metals like gold, silver and platinum have shown significant antioxidant activity and thus have been given terminology of nano antioxidants. The antioxidant ability of these metallic nanoparticles is attributed to their ability of cycling between different oxidation sites and hence they can dispose of ROS easily. For example, nanoceria are characterized by several oxygen vacancies and can switch between $Ce3^{+}$ and $Ce4^{+}$[300]. 2 chaperones namely phenyl butyric acid and tauroursodeoxycholic acid are 2 chaperones that have proved to be a boon in alleviating ER stress responsible for Type-1 diabetes. Similarly, dantrolene targets ryanodine receptors of ER of β-cells have also proved helpful [299]. Erythropoietin (EPO) and its receptor EPO-R are activated by hypoxia and by increased exposure to ROS. Survival of β-cells is increased by EPO by influencing Wnt activity or sirtuins activity. Wnt activity specifically reduces the cardiovascular complications associated with diabetes [314].

## Conclusion

Given the fact that diabetes is the most common of all metabolic illnesses, there are undoubtedly multiple diabetogenic factors in everyday life. This is a metabolic disorder distinguished by some deformities and damages that arises due to hyperglycemia, IR, and many other complications like cardiac disease, blindness due to retinal damage, stroke, and neuronal damage. However, like other major diseases ROS and oxidative stress are also playing the role of causative factor for all types of diabetes also. And most importantly, ROS is generating inside the body which is greatly influenced by the elevated glucose levels that is hyperglycemia, and vice versa. Increasing levels of ROS/RNS are leading to disturbance in normal metabolism through lipid peroxidation, autoxidation, protein oxidation and DNA damage. To counteract the harmful effects of free radicals, antioxidants already have demonstrated their therapeutic potential of diabetes. Improvement of medical sciences provide some potential pharmaceutical drugs, but nothing to prevent the disease and progression of related complications. Researchers are continuously putting their effort into finding effective antioxidants to nullify the negative impact of oxidative stress generates inside the body and delayed the aging process. However, further studies are needed to limit excess ROS production and to protect the pancreatic β-cell from its detrimental effects and restore its normal function to avoid the disease and its consequences.

## Acknowledgements

This research has been supported by Science and Engineering Research Board (SERB), a statutory body of the Department of Science and Technology (DST), Government of India

(Grant No. EEQ/2020/000188). BioRender software was used to create the figures appended in this manuscript.

## References


[1] Lakhtakia R. The history of diabetes mellitus. *Sultan Qaboos University Medical Journal.* 2013 Aug; 13(3):368.

[2] Ogurtsova K, Guariguata L, Barengo NC, Ruiz PL, Sacre JW, Karuranga S, Sun H, Boyko EJ, Magliano DJ. IDF diabetes Atlas: Global estimates of undiagnosed diabetes in adults for 2021. *Diabetes Research and Clinical Practice.* 2022 Jan 1; 183:109118.

[3] Ling W, Huang Y, Huang YM, Fan RR, Sui Y, Zhao HL. Global trend of diabetes mortality attributed to vascular complications, 2000–2016. *Cardiovascular Diabetology.* 2020 Dec; 19:1-2.

[4] Cnop M, Welsh N, Jonas JC, Jorns A, Lenzen S, Eizirik DL. Mechanisms of pancreatic β-cell death in type 1 and type 2 diabetes: many differences, few similarities. *Diabetes.* 2005 Dec 1; 54(suppl_2):S97-107.

[5] Rutter GA, Pullen TJ, Hodson DJ. Martinez− Sanchez A, Martinez-Sanchez A. Martinez− Sanchez A. Pancreatic β-cell identity, glucose sensing and the control of insulin secretion. *Biochem J.* 2015; 466:203-18.

[6] Kahn SE, Cooper ME, Del Prato S. Pathophysiology and treatment of type 2 diabetes: perspectives on the past, present, and future. *The Lancet.* 2014 Mar 22; 383(9922):1068-83.

[7] Hakonarson H, Grant SF, Bradfield JP, Marchand L, Kim CE, Glessner JT, Grabs R, Casalunovo T, Taback SP, Frackelton EC, Lawson ML. A genome-wide association study identifies KIAA0350 as a type 1 diabetes gene. *Nature.* 2007 Aug 2; 448(7153):591-4.

[8] Sladek R, Rocheleau G, Rung J, Dina C, Shen L, Serre D, Boutin P, Vincent D, Belisle A, Hadjadj S, Balkau B. A genome-wide association study identifies novel risk loci for type 2 diabetes. *Nature.* 2007 Feb 22; 445(7130):881-5.

[9] Zaccardi F, Webb DR, Yates T, Davies MJ. Pathophysiology of type 1 and type 2 diabetes mellitus: a 90-year perspective. *Postgraduate Medical Journal.* 2016 Feb 1; 92(1084):63-9.

[10] King GL. The role of inflammatory cytokines in diabetes and its complications. *Journal of Periodontology.* 2008 Aug; 79:1527-34.

[11] Lytrivi M, Castell AL, Poitout V, Cnop M. Recent insights into mechanisms of β-cell lipo-and glucolipotoxicity in type 2 diabetes. *Journal of Molecular Biology.* 2020 Mar 6; 432(5):1514-34.

[12] Åkerblom HK, Knip M, Hyöty H, Reijonen H, Virtanen S, Savilahti E, Ilonen J. Interaction of genetic and environmental factors in the pathogenesis of insulin-dependent diabetes mellitus. *Clinica Chimica Acta.* 1997 Jan 17; 257(2):143-56.

[13] Papachristoforou E, Lambadiari V, Maratou E, Makrilakis K. Association of glycemic indices (hyperglycemia, glucose variability, and hypoglycemia) with oxidative stress and diabetic complications. *Journal of Diabetes Research.* 2020 Oct 12; 2020.

[14] Brownlee M. Biochemistry and molecular cell biology of diabetic complications. *Nature.* 2001 Dec 13; 414(6865):813-20.

[15] Leloup C, Tourrel-Cuzin C, Magnan C, Karaca M, Castel J, Carneiro L, Colombani AL, Ktorza A, Casteilla L, Pénicaud L. Mitochondrial reactive oxygen species are obligatory signals for glucose-induced insulin secretion. *Diabetes.* 2009 Mar 1; 58(3):673-81.

[16] Lenzen S, Drinkgern J, Tiedge M. Low antioxidant enzyme gene expression in pancreatic islets compared with various other mouse tissues. *Free Radical Biology and Medicine.* 1996 Jan 1; 20(3):463-6.

[17] Kahn SE, Cooper ME, Del Prato S. Pathophysiology and treatment of type 2 diabetes: perspectives on the past, present, and future. *The Lancet.* 2014 Mar 22; 383(9922):1068-83.

[18] King AJ. The use of animal models in diabetes research. *British Journal of Pharmacology.* 2012 Jun; 166(3):877-94.

[19] Li Y, Liu Y, Liang J, Wang T, Sun M, Zhang Z. Gymnemic acid ameliorates hyperglycemia through PI3K/AKT-and AMPK-mediated signaling pathways in type 2 diabetes mellitus rats. *Journal of Agricultural and Food Chemistry.* 2019 Oct 14;67(47):13051-60.

[20] Gao Y, Zhang M, Wu T, Xu M, Cai H, Zhang Z. Effects of D-pinitol on insulin resistance through the PI3K/Akt signaling pathway in type 2 diabetes mellitus rats. *Journal of Agricultural and Food Chemistry.* 2015 Jul 8;63(26):6019-26.

[21] Lee M, Li H, Zhao H, Suo M, Liu D. Effects of hydroxysafflor yellow A on the PI3K/AKT pathway and apoptosis of pancreatic β-cells in type 2 diabetes mellitus rats. *Diabetes, Metabolic Syndrome and Obesity: Targets and Therapy.* 2020; 13:1097.

[22] Zhang H, Hui J, Yang J, Deng J, Fan D. Eurocristatine, a plant alkaloid from Eurotium cristatum, alleviates insulin resistance in db/db diabetic mice via activation of PI3K/AKT signaling pathway. *European Journal of Pharmacology.* 2020 Nov 15; 887:173557.

[23] Xia ZH, Zhang SY, Chen YS, Li K, Chen WB, Liu YQ. Curcumin anti-diabetic effect mainly correlates with its anti-apoptotic actions and PI3K/Akt signal pathway regulation in the liver. *Food and Chemical Toxicology.* 2020 Dec 1; 146:111803.

[24] Donato MT, Tolosa L, Gómez-Lechón MJ. Culture and functional characterization of human hepatoma HepG2 cells. *Protocols in in vitro hepatocyte research.* 2015:77-93.

[25] Skelin M, Rupnik M, Cencič A. Pancreatic beta cell lines and their applications in diabetes mellitus research. *ALTEX-Alternatives to Animal Experimentation.* 2010 May 1; 27(2):105-13.

[26] Guest PC. Characterization of the Goto-Kakizaki (GK) rat model of type 2 diabetes. *Pre-Clinical Models: Techniques and Protocols.* 2019:203-11.

[27] Clark JB, Palmer CJ, Shaw WN. The diabetic Zucker fatty rat. *Proceedings of the society for experimental Biology and Medicine.* 1983 May; 173(1):68-75.

[28] Kim JH, Stewart TP, Zhang W, Kim HY, Nishina PM, Naggert JK. Type 2 diabetes mouse model TallyHo carries an obesity gene on chromosome 6 that exaggerates dietary obesity. *Physiological Genomics.* 2005 Jul 14; 22(2):171-81.

[29] Cefalu WT. Animal models of type 2 diabetes: clinical presentation and pathophysiological relevance to the human condition. *ILAR Journal.* 2006 Jan 1; 47(3):186-98.

[30] Rees DA, Alcolado JC. Animal models of diabetes mellitus. *Diabetic Medicine.* 2005 Apr; 22(4):359-70.

[31] Islam MS. Experimental rodent models of type 2 diabetes: a review. *Methods and Findings in Experimental and Clinical Pharmacology.* 2009 May 1; 31(4):249-61.

[32] Tamarai K, Bhatti JS, Reddy PH. Molecular and cellular bases of diabetes: Focus on type 2 diabetes mouse model-TallyHo. *Biochimica et Biophysica Acta (BBA)-Molecular Basis of Disease.* 2019 Sep 1; 1865(9):2276-84.

[33] Mercurio V, Carlomagno G, Fazio V, Fazio S. Insulin resistance: is it time for primary prevention? *World Journal of Cardiology.* 2012 Jan 1; 4(1):1.

[34] DeFronzo RA. Pathogenesis of type 2 diabetes mellitus. *Medical Clinics.* 2004 Jul 1; 88(4):787-835.

[35] Forbes JM, Cooper ME. Mechanisms of diabetic complications. *Physiological Reviews.* 2013 Jan; 93(1):137-88.

[36] Padhi S, Nayak AK, Behera A. Type II diabetes mellitus: a review on recent drug based therapeutics. *Biomedicine & Pharmacotherapy.* 2020 Nov 1; 131:110708.

[37] Hamed SA. Brain injury with diabetes mellitus: evidence, mechanisms and treatment implications. *Expert Review of Clinical Pharmacology.* 2017 Apr 3; 10(4):409-28.

[38] Shivavedi N, Kumar M, Tej GN, Nayak PK. Metformin and ascorbic acid combination therapy ameliorates type 2 diabetes mellitus and comorbid depression in rats. *Brain Research.* 2017 Nov 1; 1674:1-9.

[39] Yuan P, Ma D, Gao X, Wang J, Li R, Liu Z, Wang T, Wang S, Liu J, Liu X. Liraglutide ameliorates erectile dysfunction via regulating oxidative stress, the RhoA/ROCK pathway and autophagy in diabetes mellitus. *Frontiers in Pharmacology.* 2020 Aug 12; 11:1257.

[40] Eguchi N, Vaziri ND, Dafoe DC, Ichii H. The role of oxidative stress in pancreatic β cell dysfunction in diabetes. *International Journal of Molecular Sciences.* 2021 Feb 3; 22(4):1509.

[41] Baidwan S, Chekuri A, Hynds DL, Kowluru A. Glucotoxicity promotes aberrant activation and mislocalization of Ras-related C3 botulinum toxin substrate 1 [Rac1] and metabolic dysfunction in pancreatic islet β-cells: reversal of such metabolic defects by metformin. *Apoptosis.* 2017 Nov; 22:1380-93.

[42] Vilas-Boas EA, Almeida DC, Roma LP, Ortis F, Carpinelli AR. Lipotoxicity and β-cell failure in type 2 diabetes: Oxidative stress linked to NADPH oxidase and ER stress. *Cells.* 2021 Nov 26; 10(12):3328.

[43] Volpe CM, Villar-Delfino PH, Dos Anjos PM, Nogueira-Machado JA. Cellular death, reactive oxygen species (ROS) and diabetic complications. *Cell Death &Disease.* 2018 Jan 25; 9(2):119.

[44] Federation ID. IDF Diabetes Atlas, tenth. *International Diabetes.* 2021.

[45] American Diabetes Association. Diagnosis and classification of diabetes mellitus. *Diabetes Care.* 2013 Jan; 36(Suppl 1):S67.

[46] World Health Organization. *Classification of diabetes mellitus.*

[47] American Diabetes Association. Classification and Diagnosis of Diabetes: Standards of Medical Care in Diabetes—2021. *Diabetes Care.* 2021; 44:S15-33.

[48] Kaul K, Tarr JM, Ahmad SI, Kohner EM, Chibber R. Introduction to diabetes mellitus. *Diabetes: An Old Disease,* A New Insight. 2013:1-1.

[49] Paschou SA, Papadopoulou-Marketou N, Chrousos GP, Kanaka-Gantenbein C. On type 1 diabetes mellitus pathogenesis. *Endocrine Connections.* 2018 Jan; 7(1):R38.

[50] Stumvoll M, Goldstein BJ, Van Haeften TW. Type 2 diabetes: principles of pathogenesis and therapy. *The Lancet.* 2005 Apr 9; 365(9467):1333-46.

[51] Chatterjee S, Khunti K, Davies MJ. Type 2 diabetes. *The Lancet.* 2017 Jun 3; 389(10085):2239-51.

[52] Roden M, Shulman GI. The integrative biology of type 2 diabetes. *Nature.* 2019 Dec 5; 576(7785):51-60.

[53] Caberlotto L, Nguyen T, Lauria M, Priami C, Rimondini R, Maioli S, Cedazo-Minguez A, Sita G, Morroni F, Corsi M, Carboni L. Cross-disease analysis of Alzheimer's disease and type-2 Diabetes highlights the role of autophagy in the pathophysiology of two highly comorbid diseases. *Scientific Reports.* 2019 Mar 8;9(1):1-3.

[54] Nguyen TT, Ta QT, Nguyen TT, Le TT, Vo VG. Role of insulin resistance in the Alzheimer's disease progression. *Neurochemical Research.* 2020 Jul; 45:1481-91.

[55] Kandimalla R, Thirumala V, Reddy PH. Is Alzheimer's disease a type 3 diabetes? A critical appraisal. *Biochimica et Biophysica Acta (BBA)-Molecular Basis of Disease.* 2017 May 1; 1863(5):1078-89.

[56] Butterfield DA, Halliwell B. Oxidative stress, dysfunctional glucose metabolism and Alzheimer disease. *Nature Reviews Neuroscience.* 2019 Mar; 20(3):148-60.

[57] Plows JF, Stanley JL, Baker PN, Reynolds CM, Vickers MH. The pathophysiology of gestational diabetes mellitus. *International Journal of Molecular Sciences.* 2018 Oct 26; 19(11):3342.

[58] Gaál Z, Balogh I. Monogenic Forms of Diabetes Mellitus. *Exp Suppl.* 2019; 111:385-416.

[59] Bishay RH, Greenfield JR. A review of maturity onset diabetes of the young (MODY) and challenges in the management of glucokinase-MODY. *Medical Journal of Australia.* 2016 Nov; 205(10):480-5.

[60] Velho G, Robert JJ. Maturity-onset diabetes of the young (MODY): genetic and clinical characteristics. *Hormone Research in Paediatrics.* 2002; 57(Suppl. 1):29-33.

[61] Bansal N. Prediabetes diagnosis and treatment: A review. *World J Diabetes* 2015; 6 (2): 296-303

[62] Lascar N, Brown J, Pattison H, Barnett AH, Bailey CJ, Bellary S. Type 2 diabetes in adolescents and young adults. *The Lancet Diabetes &Endocrinology.* 2018 Jan 1; 6(1):69-80.

[63] Zheng Y, Ley SH, Hu FB. Global aetiology and epidemiology of type 2 diabetes mellitus and its complications. *Nature Reviews Endocrinology.* 2018 Feb; 14(2):88-98.

[64] Susan van D, Beulens JW, Yvonne T. van der S, Grobbee DE, Nealb B. The global burden of diabetes and its complications: an emerging pandemic. *European Journal of Cardiovascular Prevention & Rehabilitation. 2010* May 1; 17(1_suppl):s3-8.

[65] Virally M, Blicklé JF, Girard J, Halimi S, Simon D, Guillausseau PJ. Type 2 diabetes mellitus: epidemiology, pathophysiology, unmet needs and therapeutical perspectives. *Diabetes &Metabolism.* 2007 Sep 1; 33(4):231-44.

[66] Maahs DM, West NA, Lawrence JM, Mayer-Davis EJ. Epidemiology of type 1 diabetes. *Endocrinology and Metabolism Clinics.* 2010 Sep 1; 39(3):481-97.

[67] Mayer-Davis EJ, Lawrence JM, Dabelea D, Divers J, Isom S, Dolan L, Imperatore G, Linder B, Marcovina S, Pettitt DJ, Pihoker C, Saydah S, Wagenknecht L; SEARCH for Diabetes in Youth Study. Incidence Trends of Type 1 and Type 2 Diabetes among Youths, 2002-2012. *New England Journal of Medicine.* 2017 Apr 13; 376(15):1419-1429.

[68] Song SH. Emerging type 2 diabetes in young adults. *Adv Exp Med Biol.* 2012; 771:51-61.

[69] Zghebi SS, Steinke DT, Carr MJ, Rutter MK, Emsley RA, Ashcroft DM. Examining trends in type 2 diabetes incidence, prevalence and mortality in the UK between 2004 and 2014. *Diabetes, Obesity and Metabolism.* 2017 Nov; 19(11):1537-45.

[70] Liu M, Liu SW, Wang LJ, Bai YM, Zeng XY, Guo HB, Liu YN, Jiang YY, Dong WL, He GX, Zhou MG. Burden of diabetes, hyperglycaemia in China from to 2016: findings from the 1990 to 2016, global burden of disease study. *Diabetes &Metabolism.* 2019 Jun 1; 45(3):286-93.

[71] Al-Saeed AH, Constantino MI, Molyneaux L, D'Souza M, Limacher-Gisler F, Luo C, Wu T, Twigg SM, Yue DK, Wong J. An inverse relationship between age of type 2 diabetes onset and complication risk and mortality: the impact of youth-onset type 2 diabetes. *Diabetes Care.* 2016 May 1; 39(5):823-9.

[72] Pradeepa R, Mohan V. Prevalence of type 2 diabetes and its complications in India and economic costs to the nation. *European Journal of Clinical Nutrition.* 2017 Jul; 71(7):816-24.

[73] Golden SH, Yajnik C, Phatak S, Hanson RL, Knowler WC. Racial/ethnic differences in the burden of type 2 diabetes over the life course: a focus on the USA and India. *Diabetologia.* 2019 Oct; 62:1751-60.

[74] Agardh E, Allebeck P, Hallqvist J, Moradi T, Sidorchuk A. Type 2 diabetes incidence and socio-economic position: a systematic review and meta-analysis. *International Journal of Epidemiology.* 2011 Jun 1; 40(3):804-18.

[75] Bakker LE, Sleddering MA, Schoones JW, Meinders AE, Jazet IM. Mechanisms in endocrinology: pathogenesis of type 2 diabetes in South Asians. *European Journal of Endocrinology.* 2013 Nov; 169(5):R99-114.

[76] Khan MA, Hashim MJ, King JK, Govender RD, Mustafa H, Al Kaabi J. Epidemiology of type 2 diabetes–global burden of disease and forecasted trends. *Journal of Epidemiology and Global Health.* 2020 Mar; 10(1):107.

[77] Magliano DJ, Islam RM, Barr EL, Gregg EW, Pavkov ME, Harding JL, Tabesh M, Koye DN, Shaw JE. Trends in incidence of total or type 2 diabetes: systematic review. *BMJ.* 2019 Sep 11; 366.

[78] Magliano DJ, Chen L, Islam RM, Carstensen B, Gregg EW, Pavkov ME, Andes LJ, Balicer R, Baviera M, Boersma-van Dam E, Booth GL. Trends in the incidence of diagnosed diabetes: a multicountry analysis of aggregate data from 22 million diagnoses in high-income and middle-income settings. *The Lancet Diabetes &Endocrinology.* 2021 Apr 1; 9(4):203-11.

[79] Viigimaa M, Sachinidis A, Toumpourleka M, Koutsampasopoulos K, Alliksoo S, Titma T. Macrovascular complications of type 2 diabetes mellitus. *Current Vascular Pharmacology.* 2020 Mar 1; 18(2):110-6.

[80] He Z, King GL. Microvascular complications of diabetes. *Endocrinology and Metabolism Clinics.* 2004 Mar 1; 33(1):215-38.

[81] Gross JL, De Azevedo MJ, Silveiro SP, Canani LH, Caramori ML, Zelmanovitz T. Diabetic nephropathy: diagnosis, prevention, and treatment. *Diabetes Care.* 2005 Jan 1; 28(1):164-76.

[82] Tang J, Yao D, Yan H, Chen X, Wang L, Zhan H. The role of MicroRNAs in the pathogenesis of diabetic nephropathy. *International Journal of Endocrinology.* 2019 Dec 1; 2019.

[83] Persson F, Rossing P. Diagnosis of diabetic kidney disease: state of the art and future perspective. *Kidney international supplements.* 2018 Jan 1; 8(1):2-7.

[84] Papadopoulou-Marketou N, Paschou SA, Marketos N, Adamidi S, Adamidis S, Kanaka-Gantenbein C. Diabetic nephropathy in type 1 diabetes. *Minerva Medica.* 2017 Dec 4; 109(3):218-28.

[85] Wang W, Lo AC. Diabetic retinopathy: pathophysiology and treatments. *International Journal of Molecular Sciences.* 2018 Jun 20; 19(6):1816.

[86] Durham JT, Herman IM. Microvascular modifications in diabetic retinopathy. *Current Diabetes Reports.* 2011 Aug; 11:253-64.

[87] Lechner J, O'Leary OE, Stitt AW. The pathology associated with diabetic retinopathy. *Vision Research.* 2017 Oct 1; 139:7-14.

[88] Liu X, Zhang Y, Liang H, Zhang Y, Xu Y. MicroRNA-499-3p inhibits proliferation and promotes apoptosis of retinal cells in diabetic retinopathy through activation of the TLR4 signaling pathway by targeting IFNA2. *Gene.* 2020 May 30; 741:144539.

[89] Ji Q, Han J, Wang L, Liu J, Dong Y, Zhu K, Shi L. MicroRNA-34a promotes apoptosis of retinal vascular endothelial cells by targeting SIRT1 in rats with diabetic retinopathy. *Cell Cycle.* 2020 Nov 1; 19(21):2886-96.

[90] Stitt AW, Curtis TM, Chen M, Medina RJ, McKay GJ, Jenkins A, Gardiner TA, Lyons TJ, Hammes HP, Simo R, Lois N. The progress in understanding and treatment of diabetic retinopathy. *Progress in Retinal and Eye Research.* 2016 Mar 1; 51:156-86.

[91] Said G. Diabetic neuropathy—a review. *Nature clinical practice Neurology.* 2007 Jun; 3(6):331-40.

[92] Charnogursky G, Lee H, Lopez N. Diabetic neuropathy. *Handbook of Clinical Neurology.* 2014 Jan 1; 120:773-85.

[93] Hicks CW, Selvin E. Epidemiology of peripheral neuropathy and lower extremity disease in diabetes. *Current Diabetes Reports.* 2019 Oct; 19:1-8.

[94] Vinik AI, Mehrabyan A. Diabetic neuropathies. *Medical Clinics.* 2004 Jul 1; 88(4):947-99.

[95] Feldman EL, Callaghan BC, Pop-Busui R, Zochodne DW, Wright DE, Bennett DL, Bril V, Russell JW, Viswanathan V. Diabetic neuropathy. *Nature Rev Dis Primers.* 2019 Jun 13; 5(1):41.

[96] Tesfaye S, Boulton AJ, Dyck PJ, Freeman R, Horowitz M, Kempler P, Lauria G, Malik RA, Spallone V, Vinik A, Bernardi L. Diabetic neuropathies: update on definitions, diagnostic criteria, estimation of severity, and treatments. *Diabetes Care.* 2010 Oct; 33(10):2285.

[97] Duby JJ, Campbell RK, Setter SM, White JR, Rasmussen KA. Diabetic neuropathy: an intensive review. *American Journal of Health-System Pharmacy.* 2004 Jan 15; 61(2):160-73.

[98] Freeman R. Diabetic autonomic neuropathy. *Handbook of Clinical Neurology.* 2014 Jan 1; 126:63-79.

[99] Albers JW, Pop-Busui R. Diabetic neuropathy: mechanisms, emerging treatments, and subtypes. *Current Neurology and Neuroscience Reports.* 2014 Aug; 14:1-1.

[100] Dewanjee S, Das S, Das AK, Bhattacharjee N, Dihingia A, Dua TK, Kalita J, Manna P. Molecular mechanism of diabetic neuropathy and its pharmacotherapeutic targets. *European Journal of Pharmacology.* 2018 Aug 15; 833:472-523.

[101] Rubler S, Dlugash J, Yuceoglu YZ, Kumral T, Branwood AW, Grishman A. New type of cardiomyopathy associated with diabetic glomerulosclerosis. *The American Journal of Cardiology.* 1972 Nov 8; 30(6):595-602.

[102] Acar E, Ural D, Bildirici U, Sahin T, Yılmaz I. Diabetic cardiomyopathy. *Anadolu Kardiyol Derg.* 2011 Dec; 11(8):732-7.

[103] Kannel WB, Hjortland M, Castelli WP. Role of diabetes in congestive heart failure: the Framingham study. *The American Journal of Cardiology.* 1974 Jul 1; 34(1):29-34.

[104] Yilmaz S, Canpolat U, Aydogdu S, Abboud HE. Diabetic cardiomyopathy; summary of 41 years. *Korean Circulation Journal.* 2015 Jul 1; 45(4):266-72.

[105] Murarka S, Movahed MR. Diabetic cardiomyopathy. *Journal of Cardiac Failure.* 2010 Dec 1; 16(12):971-9.

[106] Jia G, Whaley-Connell A, Sowers JR. Diabetic cardiomyopathy: a hyperglycaemia-and insulin-resistance-induced heart disease. *Diabetologia.* 2018 Jan; 61(1):21-8.

[107] Goyal BR, Mehta AA. Diabetic cardiomyopathy: pathophysiological mechanisms and cardiac dysfunction. *Human & Experimental Toxicology.* 2013 Jun; 32(6):571-90.

[108] Mizamtsidi M, Paschou SA, Grapsa J, Vryonidou A. Diabetic cardiomyopathy: a clinical entity or a cluster of molecular heart changes? *European Journal of Clinical Investigation.* 2016 Nov; 46(11):947-53.

[109] Bugger H, Abel ED. Molecular mechanisms of diabetic cardiomyopathy. *Diabetologia.* 2014 Apr; 57:660-71.

[110] Tokarz VL, MacDonald PE, Klip A. The cell biology of systemic insulin function. *Journal of Cell Biology.* 2018 Jul 2; 217(7):2273-89.

[111] Fu Z, R Gilbert E, Liu D. Regulation of insulin synthesis and secretion and pancreatic Beta-cell dysfunction in diabetes. *Current Diabetes Reviews.* 2013 Jan 1; 9(1):25-53.
[112] DeFronzo RA, Bonadonna RC, Ferrannini E. Pathogenesis of NIDDM: a balanced overview. *Diabetes Care.* 1992 Mar 1; 15(3):318-68.
[113] Atkinson MA, Eisenbarth GS, Michels AW. Type 1 diabetes. *The Lancet.* 2014 Jan 4; 383(9911):69-82.
[114] Ilonen J, Hammais A, Laine AP, Lempainen J, Vaarala O, Veijola R, Simell O, Knip M. Patterns of β-cell autoantibody appearance and genetic associations during the first years of life. *Diabetes.* 2013 Oct 1; 62(10):3636-40.
[115] Ziegler AG, Hummel M, Schenker M, Bonifacio E. Autoantibody appearance and risk for development of childhood diabetes in offspring of parents with type 1 diabetes: the 2-year analysis of the German BABYDIAB Study. *Diabetes.* 1999 Mar 1; 48(3):460-8.
[116] Wenzlau JM, Juhl K, Yu L, Moua O, Sarkar SA, Gottlieb P, Rewers M, Eisenbarth GS, Jensen J, Davidson HW, Hutton JC. The cation efflux transporter ZnT8 (Slc30A8) is a major autoantigen in human type 1 diabetes. *Proceedings of the National Academy of Sciences.* 2007 Oct 23; 104(43):17040-5.
[117] Bauer W, Gyenesei A, Krętowski A. The multifactorial progression from the islet autoimmunity to type 1 diabetes in children. *International Journal of Molecular Sciences.* 2021 Jul 13; 22(14):7493.
[118] Xie Z, Chang C, Zhou Z. Molecular mechanisms in autoimmune type 1 diabetes: a critical review. *Clinical Reviews in Allergy & Immunology.* 2014 Oct; 47:174-92.
[119] Ziegler AG, Rewers M, Simell O, Simell T, Lempainen J, Steck A, Winkler C, Ilonen J, Veijola R, Knip M, Bonifacio E. Seroconversion to multiple islet autoantibodies and risk of progression to diabetes in children. *JAMA* 2013 Jun 19; 309(23):2473-9.
[120] Krischer JP, Lynch KF, Schatz DA, Ilonen J, Lernmark Å, Hagopian WA, Rewers MJ, She JX, Simell OG, Toppari J, Ziegler AG. The 6 year incidence of diabetes-associated autoantibodies in genetically at-risk children: the TEDDY study. *Diabetologia.* 2015 May; 58:980-7.
[121] Ilonen J, Lempainen J, Hammais A, Laine AP, Härkönen T, Toppari J, Veijola R, Knip M, Finnish Pediatric Diabetes Register. Primary islet autoantibody at initial seroconversion and autoantibodies at diagnosis of type 1 diabetes as markers of disease heterogeneity. *Pediatric Diabetes.* 2018 Mar; 19(2):284-92.
[122] Steck AK, Rewers MJ. Genetics of type 1 diabetes. *Clinical Chemistry.* 2011 Feb 1; 57(2):176-85.
[123] Redondo MJ, Steck AK, Pugliese A. Genetics of type 1 diabetes. *Pediatric Diabetes.* 2018 May; 19(3):346-53.
[124] Lempainen J, Laine AP, Hammais A, Toppari J, Simell O, Veijola R, Knip M, Ilonen J. Non-HLA gene effects on the disease process of type 1 diabetes: from HLA susceptibility to overt disease. *Journal of Autoimmunity.* 2015 Jul 1; 61:45-53.
[125] Lempainen J, Hermann R, Veijola R, Simell O, Knip M, Ilonen J. Effect of the PTPN22 and INS risk genotypes on the progression to clinical type 1 diabetes after the initiation of β-cell autoimmunity. *Diabetes.* 2012 Apr 1; 61(4):963-6.
[126] Kirchhoff K, Machicao F, Haupt A, Schäfer SA, Tschritter O, Staiger H, Stefan N, Häring HU, Fritsche A. Polymorphisms in the TCF7L2, CDKAL1 and SLC30A8 genes are associated with impaired proinsulin conversion. *Diabetologia.* 2008 Apr; 51:597-601.
[127] Achenbach P, Lampasona V, Landherr U, Koczwara K, Krause S, Grallert H, Winkler C, Pflüger M, Illig T, Bonifacio E, Ziegler AG. Autoantibodies to zinc transporter 8 and SLC30A8 genotype stratify type 1 diabetes risk. *Diabetologia.* 2009 Sep; 52:1881-8.
[128] Saffouri GB, Shields-Cutler RR, Chen J, Yang Y, Lekatz HR, Hale VL, Cho JM, Battaglioli EJ, Bhattarai Y, Thompson KJ, Kalari KK. Small intestinal microbial dysbiosis underlies symptoms associated with functional gastrointestinal disorders. *Nature Communications.* 2019 May 1; 10(1):2012.
[129] Yeung WC, Rawlinson WD, Craig ME. Enterovirus infection and type 1 diabetes mellitus: systematic review and meta-analysis of observational molecular studies. *BMJ.* 2011 Feb 3; 342.
[130] Ylipaasto P, Klingel K, Lindberg AM, Otonkoski T, Kandolf R, Hovi T, Roivainen M. Enterovirus infection in human pancreatic islet cells, islet tropism in vivo and receptor involvement in cultured islet beta cells. *Diabetologia.* 2004 Feb; 47:225-39.

[131] Krogvold L, Edwin B, Buanes T, Frisk G, Skog O, Anagandula M, Korsgren O, Undlien D, Eike MC, Richardson SJ, Leete P. Detection of a low-grade enteroviral infection in the islets of langerhans of living patients newly diagnosed with type 1 diabetes. *Diabetes.* 2015 May 1; 64(5):1682-7.

[132] Yoon JW. The role of viruses and environmental factors in the induction of diabetes. *Current Topics in Microbiology and Immunology. 1990* Jan 1; 164:95-123.

[133] Lamb MM, Frederiksen B, Seifert JA, Kroehl M, Rewers M, Norris JM. Sugar intake is associated with progression from islet autoimmunity to type 1 diabetes: the Diabetes Autoimmunity Study in the Young. *Diabetologia.* 2015 Sep; 58:2027-34.

[134] In't Veld P. Insulitis in human type 1 diabetes: the quest for an elusive lesion. *Islets.* 2011 Jul 1;3(4):131-8.

[135] Eizirik DL. Mandrup-Poulsen T. A choice of death—the signal-transduction of immune-mediated beta-cell apoptosis. *Diabetologia.* 2001; 44:2115-33.

[136] Yamauchi T, Tobe K, Tamemoto H, Ueki K, Kaburagi Y, Yamamoto-Honda R, Takahashi Y, Yoshizawa F, Aizawa S, Akanuma Y, Sonenberg N. Insulin signalling and insulin actions in the muscles and livers of insulin-resistant, insulin receptor substrate 1-deficient mice. *Molecular and Cellular Biology.* 1996 Jun; 16(6):3074-84.

[137] Nakae J, Biggs WH, Kitamura T, Cavenee WK, Wright CV, Arden KC, Accili D. Regulation of insulin action and pancreatic β-cell function by mutated alleles of the gene encoding forkhead transcription factor Foxo1. *Nature Genetics.* 2002 Oct; 32(2):245-53.

[138] Cho H, Mu J, Kim JK, Thorvaldsen JL, Chu Q, Crenshaw III EB, Kaestner KH, Bartolomei MS, Shulman GI, Birnbaum MJ. Insulin resistance and a diabetes mellitus-like syndrome in mice lacking the protein kinase Akt2 (PKBβ). *Science.* 2001 Jun 1; 292(5522):1728-31.

[139] Abdul-Ghani MA, DeFronzo RA. Pathogenesis of insulin resistance in skeletal muscle. *Journal of Biomedical Biotechnology.* 2010; 2010:476279.

[140] Jaganathan R, Ravindran R, Dhanasekaran S. Emerging role of adipocytokines in type 2 diabetes as mediators of insulin resistance and cardiovascular disease. *Canadian Journal of Diabetes.* 2018 Aug 1; 42(4):446-56.

[141] Yu C, Chen Y, Cline GW, Zhang D, Zong H, Wang Y, Bergeron R, Kim JK, Cushman SW, Cooney GJ, Atcheson B. Mechanism by which fatty acids inhibit insulin activation of insulin receptor substrate-1 (IRS-1)-associated phosphatidylinositol 3-kinase activity in muscle. *Journal of Biological Chemistry.* 2002 Dec 27; 277(52):50230-6.

[142] Feinstein R, Kanety H, Papa MZ, Lunenfeld B, Karasik A. Tumor necrosis factor-alpha suppresses insulin-induced tyrosine phosphorylation of insulin receptor and its substrates. *Journal of Biological Chemistry.* 1993 Dec 15; 268(35):26055-8.

[143] Peraldi P, Spiegelman B. TNF-α and insulin resistance: summary and future prospects. *Molecular and Cellular Biochemistry.* 1998 May; 182:169-75.

[144] Cheng Z, Tseng Y, White MF. Insulin signaling meets mitochondria in metabolism. *Trends in Endocrinology & Metabolism.* 2010 Oct 1; 21(10):589-98.

[145] Wang C, Guan Y, Yang J. Cytokines in the progression of pancreatic β-cell dysfunction. *International Journal of Endocrinology.* 2010 Oct; 2010.

[146] Morgan D, Oliveira-Emilio HR, Keane D, Hirata AE, Santos da Rocha M, Bordin S, Curi R, Newsholme P, Carpinelli AR. Glucose, palmitate and pro-inflammatory cytokines modulate production and activity of a phagocyte-like NADPH oxidase in rat pancreatic islets and a clonal beta cell line. *Diabetologia.* 2007 Feb; 50:359-69.

[147] Qi D, Cai K, Wang O, Li Z, Chen J, Deng B, Qian L, Le Y. Fatty acids induce amylin expression and secretion by pancreatic β-cells. *American Journal of Physiology-Endocrinology and Metabolism.* 2010 Jan; 298(1):E99-107.

[148] Krizhanovskii C, Fred RG, Oskarsson ME, Westermark GT, Welsh N. Addition of exogenous sodium palmitate increases the IAPP/insulin mRNA ratio via GPR40 in human EndoC-βH1 cells. *Upsala Journal of Medical Sciences.* 2017 Jul 3; 122(3):149-59.

[149] Robertson RP. Chronic oxidative stress as a central mechanism for glucose toxicity in pancreatic islet beta cells in diabetes. *Journal of Biological Chemistry.* 2004 Oct 8; 279(41):42351-4.

[150] Gwiazda KS, Yang TL, Lin Y, Johnson JD. Effects of palmitate on ER and cytosolic Ca2+ homeostasis in β-cells. *American Journal of Physiology-Endocrinology and Metabolism.* 2009 Apr; 296(4):E690-701.

[151] Baldwin AC, Green CD, Olson LK, Moxley MA, Corbett JA. A role for aberrant protein palmitoylation in FFA-induced ER stress and β-cell death. *American Journal of Physiology-Endocrinology and Metabolism.* 2012 Jun 1; 302(11):E1390-8.

[152] Kajimoto Y, Kaneto H. Role of oxidative stress in pancreatic β-cell dysfunction. *Mitochondrial Pathogenesis: From Genes and Apoptosis to Aging and Disease.* 2004:168-76.

[153] Molina AJ, Wikstrom JD, Stiles L, Las G, Mohamed H, Elorza A, Walzer G, Twig G, Katz S, Corkey BE, Shirihai OS. Mitochondrial networking protects β-cells from nutrient-induced apoptosis. *Diabetes.* 2009 Oct 1; 58(10):2303-15.

[154] Zhao YF, Feng DD, Chen C. Contribution of adipocyte-derived factors to beta-cell dysfunction in diabetes. *The International Journal of Biochemistry & Cell Biology.* 2006 May 1; 38(5-6):804-19.

[155] Talchai C, Xuan S, Lin HV, Sussel L, Accili D. Pancreatic β cell dedifferentiation as a mechanism of diabetic β cell failure. *Cell.* 2012 Sep 14; 150(6):1223-34.

[156] Ebato C, Uchida T, Arakawa M, Komatsu M, Ueno T, Komiya K, Azuma K, Hirose T, Tanaka K, Kominami E, Kawamori R. Autophagy is important in islet homeostasis and compensatory increase of beta cell mass in response to high-fat diet. *Cell Metabolism.* 2008 Oct 8; 8(4):325-32.

[157] Jung HS, Chung KW, Kim JW, Kim J, Komatsu M, Tanaka K, Nguyen YH, Kang TM, Yoon KH, Kim JW, Jeong YT. Loss of autophagy diminishes pancreatic β cell mass and function with resultant hyperglycemia. *Cell Metabolism.* 2008 Oct 8;8(4):318-24.

[158] Quan W, Hur KY, Lim Y, Oh SH, Lee JC, Kim KH, Kim GH, Kim SW, Kim HL, Lee MK, Kim KW. Autophagy deficiency in beta cells leads to compromised unfolded protein response and progression from obesity to diabetes in mice. *Diabetologia.* 2012 Feb; 55:392-403.

[159] Cunha DA, Hekerman P, Ladrière L, Bazarra-Castro A, Ortis F, Wakeham MC, Moore F, Rasschaert J, Cardozo AK, Bellomo E, Overbergh L. Initiation and execution of lipotoxic ER stress in pancreatic β-cells. *Journal of Cell Science.* 2008 Jul 15; 121(14):2308-18.

[160] Almaça J, Caicedo A, Landsman L. Beta cell dysfunction in diabetes: the islet microenvironment as an unusual suspect. *Diabetologia.* 2020 Oct; 63:2076-85.

[161] Pang G, Xie J, Chen Q, Hu Z. Energy intake, metabolic homeostasis, and human health. *Food Science and Human Wellness.* 2014 Sep 1; 3(3-4):89-103.

[162] Niedzielska E, Smaga I, Gawlik M, Moniczewski A, Stankowicz P, Pera J, Filip M. Oxidative stress in neurodegenerative diseases. *Molecular Neurobiology.* 2016 Aug; 53:4094-125.

[163] Le Lay S, Simard G, Martinez MC, Andriantsitohaina R. Oxidative stress and metabolic pathologies: from an adipocentric point of view. *Oxidative medicine and cellular longevity.* 2014 Oct; 2014.

[164] Pizzino G, Irrera N, Cucinotta M, Pallio G, Mannino F, Arcoraci V, Squadrito F, Altavilla D, Bitto A. Oxidative stress: harms and benefits for human health. *Oxidative Medicine and Cellular Longevity.* 2017 Oct;2017.

[165] Andreyev AY. Kushnareva YE, Starkov AA. Mitochondrial metabolism of reactive oxygen species. *Biochemistry* (Mosc). 2005; 70:200-14.

[166] Rodriguez R, Redman R. Balancing the generation and elimination of reactive oxygen species. *Proceedings of the National Academy of Sciences.* 2005 Mar 1; 102(9):3175-6.

[167] Sharifi-Rad M, Anil Kumar NV, Zucca P, Varoni EM, Dini L, Panzarini E, Rajkovic J, Tsouh Fokou PV, Azzini E, Peluso I, Prakash Mishra A. Lifestyle, oxidative stress, and antioxidants: back and forth in the pathophysiology of chronic diseases. *Frontiers in Physiology.* 2020 Jul 2; 11:694.

[168] Ha H, Hwang IA, Park JH, Lee HB. Role of reactive oxygen species in the pathogenesis of diabetic nephropathy. *Diabetes Research and Clinical Practice.* 2008 Nov 13; 82 Suppl 1:S42-5.

[169] Lone A, Harris RA, Singh O, Betts DH, Cumming RC. p66Shc activation promotes increased oxidative phosphorylation and renders CNS cells more vulnerable to amyloid beta toxicity. *Scientific Reports.* 2018 Nov 20; 8(1):1-7.

[170] Zorov DB, Juhaszova M, Sollott SJ. Mitochondrial reactive oxygen species (ROS) and ROS-induced ROS release. *Physiological Reviews.* 2014 Jul; 94(3):909-50.

[171] Snezhkina AV, Kudryavtseva AV, Kardymon OL, Savvateeva MV, Melnikova NV, Krasnov GS, Dmitriev AA. ROS generation and antioxidant defense systems in normal and malignant cells. *Oxidative Medicine and Cellular Longevity*. 2019 Aug 5; 2019.

[172] Del Río LA, López-Huertas E. ROS generation in peroxisomes and its role in cell signaling. *Plant and Cell Physiology*. 2016 Jul 1; 57(7):1364-76.

[173] Zeeshan HM, Lee GH, Kim HR, Chae HJ. Endoplasmic reticulum stress and associated ROS. *International Journal of Molecular Sciences*. 2016 Mar 2; 17(3):327.

[174] Phaniendra A, Jestadi DB, Periyasamy L. Free radicals: properties, sources, targets, and their implication in various diseases. *Indian Journal of Clinical Biochemistry*. 2015 Jan; 30:11-26.

[175] Pham-Huy LA, He H, Pham-Huy C. Free radicals, antioxidants in disease and health. *International Journal of Biomedical Science*. 2008 Jun; 4(2):89.

[176] Fukai T, Ushio-Fukai M. Superoxide dismutases: role in redox signaling, vascular function, and diseases. *Antioxidants & Redox Signaling*. 2011 Sep 15; 15(6):1583-606.

[177] Lubos E, Loscalzo J, Handy DE. Glutathione peroxidase-1 in health and disease: from molecular mechanisms to therapeutic opportunities. *Antioxidants and Redox Signaling*. 2011 Oct 1; 15(7):1957-97.

[178] He L, He T, Farrar S, Ji L, Liu T, Ma X. Antioxidants maintain cellular redox homeostasis by elimination of reactive oxygen species. *Cellular Physiology and Biochemistry*. 2017; 44(2):532-53.

[179] Dunn LL, Buckle AM, Cooke JP, Ng MK. The emerging role of the thioredoxin system in angiogenesis. *Arteriosclerosis, Thrombosis, and Vascular Biology*. 2010 Nov 1; 30(11):2089-98.

[180] Glorieux C, Calderon PB. Catalase, a remarkable enzyme: targeting the oldest antioxidant enzyme to find a new cancer treatment approach. *Biological Chemistry*. 2017 Oct 1; 398(10):1095-108.

[181] Sies H. Oxidative stress: a concept in redox biology and medicine. *Redox biology*. 2015 Apr 1; 4:180-3.

[182] Shunmoogam N, Naidoo P, Chilton R. Paraoxonase (PON)-1: a brief overview on genetics, structure, polymorphisms and clinical relevance. *Vascular Health and Risk Management*. 2018 Jun 18:137-43.

[183] Padmalayam I. Targeting mitochondrial oxidative stress through lipoic acid synthase: a novel strategy to manage diabetic cardiovascular disease. *Cardiovascular & Hematological Agents in Medicinal Chemistry (Formerly Current Medicinal Chemistry-Cardiovascular & Hematological Agents)*. 2012 Sep 1; 10(3):223-33.

[184] Tan BL, Norhaizan ME, Liew WP, Sulaiman Rahman H. Antioxidant and oxidative stress: a mutual interplay in age-related diseases. *Frontiers in Pharmacology*. 2018 Oct 16; 9:1162.

[185] Yagi K. A rapid method for evaluation of autoxidation and antioxidants. *Agricultural and Biological Chemistry*. 1970 Jan 1; 34(1):142-5.

[186] Lü JM, Lin PH, Yao Q, Chen C. Chemical and molecular mechanisms of antioxidants: experimental approaches and model systems. *Journal of Cellular and Molecular Medicine*. 2010 Apr; 14(4):840-60.

[187] Volpe CM, Villar-Delfino PH, Dos Anjos PM, Nogueira-Machado JA. Cellular death, reactive oxygen species (ROS) and diabetic complications. *Cell Death &Disease*. 2018 Jan 25; 9(2):119.

[188] Asmat U, Abad K, Ismail K. Diabetes mellitus and oxidative stress-A concise review. *Saudi Pharm J*. 2016 Sep; 24(5):547-553.

[189] Wu X, Chen K, Williams KJ. The role of pathway-selective insulin resistance and responsiveness in diabetic dyslipoproteinemia. *Current Opinion in Lipidology*. 2012 Aug 1; 23(4):334-44.

[190] Tiwari BK, Pandey KB, Abidi AB, Rizvi SI. Markers of oxidative stress during diabetes mellitus. *Journal of Biomarkers*. 2013; 2013.

[191] Obrosova IG, Pacher P, Szabó C, Zsengeller Z, Hirooka H, Stevens MJ, Yorek MA. Aldose reductase inhibition counteracts oxidative-nitrosative stress and poly (ADP-ribose) polymerase activation in tissue sites for diabetes complications. *Diabetes*. 2005 Jan 1; 54(1):234-42.

[192] Tang WH, Martin KA, Hwa J. Aldose reductase, oxidative stress, and diabetic mellitus. *Frontiers in Pharmacology. 2012* May 9; 3:87.

[193] Yang XH, Trumpower BL. Protonmotive Q cycle pathway of electron transfer and energy transduction in the three-subunit ubiquinol-cytochrome c oxidoreductase complex of Paracoccus denitrificans. *Journal of Biological Chemistry*. 1988 Aug 25; 263(24):11962-70.

[194] Rösen P, Nawroth PP, King G, Möller W, Tritschler HJ, Packer L. The role of oxidative stress in the onset and progression of diabetes and its complications: asummary of a Congress Series sponsored byUNESCO-MCBN, the American Diabetes Association and the German Diabetes Society. *Diabetes Metabolism Research and Reviews.* 2001 May; 17(3):189-212.

[195] Nowotny K, Jung T, Höhn A, Weber D, Grune T. Advanced glycation end products and oxidative stress in type 2 diabetes mellitus. *Biomolecules.* 2015; 5 (1): 194–222.

[196] Monnier VM, Sell DR, Genuth SA. Glycation products as markers and predictors of the progression of diabetic complications. *Annals of the New York Academy of Sciences.* 2005 Jun; 1043(1):567-81.

[197] Yamagishi SI, Nakamura K, Imaizumi T. Advanced glycation end products (AGEs) and diabetic vascular complications. *Current Diabetes Reviews.* 2005 Jan 1; 1(1):93-106.

[198] Aronson D. Hyperglycemia and the pathobiology of diabetic complications. *Cardiovascular diabetology: Clinical, metabolic and inflammatory facets.* 2008; 45:1-6.

[199] Lien CF, Chen SJ, Tsai MC, Lin CS. Potential role of protein kinase C in the pathophysiology of diabetes-associated atherosclerosis. *Frontiers in Pharmacology.* 2021 Jul 2; 12:716332.

[200] Kitada M, Zhang Z, Mima A, King GL. Molecular mechanisms of diabetic vascular complications. *Journal of diabetes investigation.* 2010 Jun; 1(3):77-89.

[201] Qin CX, Sleaby R, Davidoff AJ, Bell JR, De Blasio MJ, Delbridge LM, Chatham JC, Ritchie RH. Insights into the role of maladaptive hexosamine biosynthesis and O-GlcNAcylation in development of diabetic cardiac complications. *Pharmacological Research.* 2017 Feb 1; 116:45-56.

[202] Du XL, Edelstein D, Rossetti L, Fantus IG, Goldberg H, Ziyadeh F, Wu J, Brownlee M. Hyperglycemia-induced mitochondrial superoxide overproduction activates the hexosamine pathway and induces plasminogen activator inhibitor-1 expression by increasing Sp1 glycosylation. *Proceedings of the National Academy of Sciences.* 2000 Oct 24; 97(22):12222-6.

[203] Evans JL, Goldfine ID, Maddux BA, Grodsky GM. Are oxidative stress− activated signaling pathways mediators of insulin resistance and β-cell dysfunction? *Diabetes.* 2003 Jan 1; 52(1):1-8.

[204] Yki-Järvinen H, Helve E, Koivisto VA. Hyperglycemia decreases glucose uptake in type I diabetes. *Diabetes.* 1987 Aug 1; 36(8):892-6.

[205] Buse MG. Hexosamines, insulin resistance, and the complications of diabetes: current status. *American Journal of Physiology-Endocrinology and Metabolism.* 2006.

[206] Block K, Gorin Y, Abboud HE. Subcellular localization of Nox4 and regulation in diabetes. *Proceedings of the National Academy of Sciences.* 2009 Aug 25; 106(34):14385-90.

[207] Eizirik DL, Colli ML, Ortis F. The role of inflammation in insulitis and β-cell loss in type 1 diabetes. *Nature Reviews Endocrinology.* 2009 Apr; 5(4):219-26.

[208] Drews G, Krippeit-Drews P, Düfer M. Oxidative stress and beta-cell dysfunction. *Pflügers Archiv-European Journal of Physiology.* 2010 Sep; 460:703-18.

[209] Teodoro JS, Nunes S, Rolo AP, Reis F, Palmeira CM. Therapeutic options targeting oxidative stress, mitochondrial dysfunction and inflammation to hinder the progression of vascular complications of diabetes. *Frontiers in Physiology.* 2019 Jan 17; 9:1857.

[210] Apostolova N, Iannantuoni F, Gruevska A, Muntane J, Rocha M, Victor VM. Mechanisms of action of metformin in type 2 diabetes: Effects on mitochondria and leukocyte-endothelium interactions. *Redox Biology.* 2020 Jul 1; 34:101517.

[211] Argaud D, Roth H, Wiernsperger N, Leverve XM. Metformin decreases gluconeogenesis by enhancing the pyruvate kinase flux in isolated rat hepatocytes. *European Journal of Biochemistry.* 1993 May; 213(3):1341-8.

[212] Owen MR, Doran E, Halestrap AP. Evidence that metformin exerts its anti-diabetic effects through inhibition of complex 1 of the mitochondrial respiratory chain. *Biochemical Journal.* 2000 Jun 15; 348(3):607-14.

[213] Yap F, Craddock L, Yang J. Mechanism of AMPK suppression of LXR-dependent Srebp-1c transcription. *International Journal of Biological Sciences.* 2011; 7(5):645.

[214] An H, He L. Current understanding of metformin effect on the control of hyperglycemia in diabetes. *The Journal of Endocrinology.* 2016 Mar; 228(3):R97.

[215] Sola D, Rossi L, Schianca GP, Maffioli P, Bigliocca M, Mella R, Corlianò F, Fra GP, Bartoli E, Derosa G. Sulfonylureas and their use in clinical practice. *Arch Med Sci* 2015; 11 (4): 840-8.

[216] Singh AK. Deciding oral drugs after metformin in type 2 diabetes: an evidence-based approach. *Indian Journal of Endocrinology and Metabolism.* 2014 Sep; 18(5):617.

[217] Chaudhury A, Duvoor C, Reddy Dendi VS, Kraleti S, Chada A, Ravilla R, Marco A, Shekhawat NS, Montales MT, Kuriakose K, Sasapu A. Clinical review of antidiabetic drugs: implications for type 2 diabetes mellitus management. *Frontiers in Endocrinology.* 2017 Jan 24; 8:6.

[218] Basit A, Riaz M, Fawwad A. Glimepiride: evidence-based facts, trends, and observations. *Vascular Health and Risk Management.* 2012 Aug 15:463-72.

[219] Thulé PM, Umpierrez G. Sulfonylureas: a new look at old therapy. *Current diabetes reports.* 2014 Apr; 14:1-8.

[220] Kalra S, Bahendeka S, Sahay R, Ghosh S, Md F, Orabi A, Ramaiya K, Al Shammari S, Shrestha D, Shaikh K, Abhayaratna S. Consensus recommendations on sulfonylurea and sulfonylurea combinations in the management of Type 2 diabetes mellitus–International Task Force. *Indian Journal of Endocrinology and Metabolism.* 2018 Jan 1; 22(1):132-57.

[221] Lv W, Wang X, Xu Q, Lu W. Mechanisms and characteristics of sulfonylureas and glinides. *Current Topics in Medicinal Chemistry.* 2020 Jan 1; 20(1):37-56.

[222] Hauner H. The mode of action of thiazolidinediones. *Diabetes Metabolism Research and Reviews.* 2002 Mar; 18(S2):S10-5.

[223] Zhou TB, Drummen GP, Jiang ZP, Long YB, Qin YH. Association of peroxisome proliferator-activated receptors/retinoic acid receptors with renal diseases. *Journal of Receptors and Signal Transduction.* 2013 Dec 1; 33(6):349-52.

[224] Marx N, Schönbeck U, Lazar MA, Libby P, Plutzky J. Peroxisome proliferator-activated receptor gamma activators inhibit gene expression and migration in human vascular smooth muscle cells. *Circulation Research.* 1998 Nov 30; 83(11):1097-103.

[225] Spiegelman BM. PPAR-gamma: adipogenic regulator and thiazolidinedione receptor. *Diabetes.* 1998 Apr 1; 47(4):507-14.

[226] Soccio RE, Chen ER, Lazar MA. Thiazolidinediones and the promise of insulin sensitization in type 2 diabetes. *Cell Metabolism.* 2014 Oct 7; 20(4):573-91.

[227] Lebovitz HE. Thiazolidinediones: the forgotten diabetes medications. *Current diabetes reports.* 2019 Dec; 19:1-3.

[228] Yun CW, Han YS, Lee SH. PGC-1α controls mitochondrial biogenesis in drug-resistant colorectal cancer cells by regulating endoplasmic reticulum stress. *International Journal of Molecular Sciences.* 2019 Apr 5; 20(7):1707.

[229] Black C, Donnelly P, McIntyre L, Royle PL, Shepherd JP, Thomas S. Meglitinide analogues for type 2 diabetes mellitus. *Cochrane Database Syst Rev.* 2007 Apr 18; 2007(2):CD004654.

[230] Guardado-Mendoza R, Prioletta A, Jiménez-Ceja LM, Sosale A, Folli F. The role of nateglinide and repaglinide, derivatives of meglitinide, in the treatment of type 2 diabetes mellitus. *Arch Med Sci.* 2013 Oct 31;9(5):936-43.

[231] Grant JS, Graven LJ. Progressing from metformin to sulfonylureas or meglitinides. *Workplace Health & Safety.* 2016 Sep; 64(9):433-9.

[232] Apovian CM, Okemah J, O'Neil PM. Body weight considerations in the management of type 2 diabetes. *Advances in Therapy.* 2019 Jan; 36:44-58.

[233] Hasslacher C, Multinational Repaglinide Renal Study Group. Safety and efficacy of repaglinide in type 2 diabetic patients with and without impaired renal function. *Diabetes Care.* 2003 Mar 1; 26(3):886-91.

[234] Coleman RL, Scott CA, Lang Z, Bethel MA, Tuomilehto J, Holman RR. Meta-analysis of the impact of alpha-glucosidase inhibitors on incident diabetes and cardiovascular outcomes. *Cardiovascular Diabetology.* 2019 Dec; 18(1):1-7.

[235] Scott LJ, Spencer CM. Miglitol: a review of its therapeutic potential in type 2 diabetes mellitus. *Drugs.* 2000 Mar; 59(3):521-49.

[236] Yen FS, Wei JC, Lin MC, Hsu CC, Hwu CM. Long-term outcomes of adding alpha-glucosidase inhibitors in insulin-treated patients with type 2 diabetes. *BMC Endocrine Disorders.* 2021 Dec; 21(1):1-0.

[237] Van de Laar FA. Alpha-glucosidase inhibitors in the early treatment of type 2 diabetes. *Vascular health and risk management.* 2008 Dec 1; 4(6):1189-95.

[238] Derosa G, Maffioli P. α-Glucosidase inhibitors and their use in clinical practice. *Archives of Medical Science: AMS.* 2012 Nov 11; 8(5):899.

[239] Zhao Y, Wang Y, Lou H, Shan L. Alpha-glucosidase inhibitors and risk of cancer in patients with diabetes mellitus: a systematic review and meta-analysis. *Oncotarget.* 2017 Oct 10; 8(46):81027.

[240] Min SH, Yoon JH, Hahn S, Cho YM. Efficacy and safety of combination therapy with an α-glucosidase inhibitor and a dipeptidyl peptidase-4 inhibitor in patients with type 2 diabetes mellitus: A systematic review with meta-analysis. *Journal of Diabetes Investigation.* 2018 Jul; 9(4):893-902.

[241] Deacon CF, Johnsen AH, Holst JJ. Degradation of glucagon-like peptide-1 by human plasma in vitro yields an N-terminally truncated peptide that is a major endogenous metabolite in vivo. *The Journal of Clinical Endocrinology & Metabolism.* 1995 Mar 1; 80(3):952-7.

[242] Campbell E. J, Drucker J, D. Pharmacology, Physiology, and Mechanisms of Incretin Hormone Action. *Cell Metabolism.* 2013; 17(6):819-37.

[243] Seino Y, Fukushima M, Yabe D. GIP and GLP-1, the two incretin hormones: similarities and differences. *Journal of Diabetes Investigation.* 2010 Feb; 1(1-2):8-23.

[244] Singh AK, Jatwa R, Purohit A, Ram H. Synthetic and phytocompounds based dipeptidyl peptidase-IV (DPP-IV) inhibitors for therapeutics of diabetes. *Journal of Asian Natural Products Research.* 2017 Oct 3; 19(10):1036-45.

[245] Thornberry NA, Gallwitz B. Mechanism of action of inhibitors of dipeptidyl-peptidase-4 (DPP-4). *Best Practice & Research Clinical Endocrinology & Metabolism.* 2009 Aug 1; 23(4):479-86.

[246] Fujita Y, Wideman RD, Asadi A, Yang GK, Baker R, Webber T, Zhang T, Wang R, Ao Z, Warnock GL, Kwok YN. Glucose-dependent insulinotropic polypeptide is expressed in pancreatic islet α-cells and promotes insulin secretion. *Gastroenterology.* 2010 May 1; 138(5):1966-75.

[247] Gurgel Penaforte-Saboia J, Couri CE, Vasconcelos Albuquerque N, Lauanna Lima Silva V, Bitar da Cunha Olegario N, Oliveira Fernandes V, Montenegro Junior RM. Emerging roles of dipeptidyl peptidase-4 inhibitors in delaying the progression of type 1 diabetes mellitus. *Diabetes, Metabolic Syndrome and Obesity: Targets and Therapy.* 2021 Feb 10:565-73.

[248] Rieg T, Masuda T, Gerasimova M, Mayoux E, Platt K, Powell DR, Thomson SC, Koepsell H, Vallon V. Increase in SGLT1-mediated transport explains renal glucose reabsorption during genetic and pharmacological SGLT2 inhibition in euglycemia. *American Journal of Physiology-Renal Physiology.* 2014 Jan 15; 306(2):F188-93.

[249] Hirsch JR. Loo DD, and Wright EM. Regulation of Na+/glucose cotransporter expression by protein kinases in Xenopus laevis oocytes. *Journal of Biological Chemistry.* 1996; 271:14740-6.

[250] Zambrowicz B, Lapuerta P, Strumph P, Banks P, Wilson A, Ogbaa I, Sands A, Powell D. LX4211 therapy reduces postprandial glucose levels in patients with type 2 diabetes mellitus and renal impairment despite low urinary glucose excretion. *Clinical Therapeutics.* 2015 Jan 1; 37(1):71-82.

[251] Poulsen SB, Fenton RA, Rieg T. Sodium-glucose cotransport. *Current Opinion in Nephrology and Hypertension.* 2015 Sep; 24(5):463.

[252] Chen J, Fan F, Wang JY, Long Y, Gao CL, Stanton RC, Xu Y. The efficacy and safety of SGLT2 inhibitors for adjunctive treatment of type 1 diabetes: a systematic review and meta-analysis. *Scientific reports.* 2017 Mar 9; 7(1):1-9.

[253] Kloet DE, Burgering BM. The PKB/FOXO switch in aging and cancer. *Biochimica et Biophysica Acta (BBA)-Molecular Cell Research.* 2011 Nov 1; 1813(11):1926-37.

[254] O'Neill BT, Lee KY, Klaus K, Softic S, Krumpoch MT, Fentz J, Stanford KI, Robinson MM, Cai W, Kleinridders A, Pereira RO. Insulin and IGF-1 receptors regulate FoxO-mediated signaling in muscle proteostasis. *The Journal of Clinical Investigation.* 2016 Sep 1; 126(9):3433-46.

[255] Zhao Y, Wang Y, Zhu WG. Applications of post-translational modifications of FoxO family proteins in biological functions. *Journal of Molecular Cell Biology.* 2011 Oct 1; 3(5):276-82.

[256] Greer EL, Brunet A. FOXO transcription factors at the interface between longevity and tumor suppression. *Oncogene.* 2005 Nov; 24(50):7410-25.

[257] Essers MA, Weijzen S, de Vries-Smits AM, Saarloos I, de Ruiter ND, Bos JL, Burgering BM. FOXO transcription factor activation by oxidative stress mediated by the small GTPase Ral and JNK. *The EMBO Journal.* 2004 Dec 8; 23(24):4802-12.

[258] Motta MC, Divecha N, Lemieux M, Kamel C, Chen D, Gu W, Bultsma Y, McBurney M, Guarente L. Mammalian SIRT1 represses forkhead transcription factors. *Cell.* 2004 Feb 20; 116(4):551-63.

[259] Boccitto M, G Kalb R. Regulation of Foxo-dependent transcription by post-translational modifications. *Current Drug Targets.* 2011 Aug 1; 12(9):1303-10.

[260] Saraswat K, Rizvi SI. Novel strategies for anti-aging drug discovery. *Expert Opinion on Drug Discovery.* 2017 Sep 2; 12(9):955-66.

[261] Liu H, Yang J, Li L, Shi W, Yuan X, Wu L. The natural occurring compounds targeting endoplasmic reticulum stress. *Evidence-Based Complementary and Alternative Medicine.* 2016 Oct; 2016.

[262] van der Horst A, Tertoolen LG, de Vries-Smits LM, Frye RA, Medema RH, Burgering BM. FOXO4 is acetylated upon peroxide stress and deacetylated by the longevity protein hSir2SIRT1. *Journal of Biological Chemistry.* 2004 Jul 9; 279(28):28873-9.

[263] Webb AE, Brunet A. FOXO transcription factors: key regulators of cellular quality control. *Trends in Biochemical Sciences.* 2014 Apr 1; 39(4):159-69.

[264] Belguise K, Guo S, Sonenshein GE. Activation of FOXO3a by the green tea polyphenol epigallocatechin-3-gallate induces estrogen receptor α expression reversing invasive phenotype of breast cancer cells. *Cancer Research.* 2007 Jun 15; 67(12):5763-70.

[265] Huang CY, Chan CY, Chou IT, Lien CH, Hung HC, Lee MF. Quercetin induces growth arrest through activation of FOXO1 transcription factor in EGFR-overexpressing oral cancer cells. *The Journal of Nutritional Biochemistry.* 2013 Sep 1; 24(9):1596-603.

[266] Bartholome A, Kampkötter A, Tanner S, Sies H, Klotz LO. Epigallocatechin gallate-induced modulation of FoxO signaling in mammalian cells and C. elegans: FoxO stimulation is masked via PI3K/Akt activation by hydrogen peroxide formed in cell culture. *Archives of Biochemistry and Biophysics.* 2010 Sep 1; 501(1):58-64.

[267] Srivastava RK, Tang SN, Zhu W, Meeker D, Shankar S. Sulforaphane synergizes with quercetin to inhibit self-renewal capacity of pancreatic cancer stem cells. *Frontiers in Bioscience* (Elite edition). 2011; 3:515.

[268] David JA, Rifkin WJ, Rabbani PS, Ceradini DJ. The Nrf2/Keap1/ARE pathway and oxidative stress as a therapeutic target in type II diabetes mellitus. *Journal of diabetes research.* 2017 Aug 20; 2017.

[269] Itoh K, Wakabayashi N, Katoh Y, Ishii T, O'Connor T, Yamamoto M. Keap1 regulates both cytoplasmic-nuclear shuttling and degradation of Nrf2 in response to electrophiles. *Genes to Cells.* 2003 Apr; 8(4):379-91.

[270] Baird L, Llères D, Swift S, Dinkova-Kostova AT. Regulatory flexibility in the Nrf2-mediated stress response is conferred by conformational cycling of the Keap1-Nrf2 protein complex. *Proceedings of the National Academy of Sciences.* 2013 Sep 17; 110(38):15259-64.

[271] He F, Ru X, Wen T. NRF2, a transcription factor for stress response and beyond. *International Journal of Molecular Sciences.* 2020 Jul 6; 21(13):4777.

[272] Lin X, Bai D, Wei Z, Zhang Y, Huang Y, Deng H, Huang X. Curcumin attenuates oxidative stress in RAW264. 7 cells by increasing the activity of antioxidant enzymes and activating the Nrf2-Keap1 pathway. *PloS one.* 2019 May 21; 14(5):e0216711.

[273] Kim EN, Lim JH, Kim MY, Ban TH, Jang IA, Yoon HE, Park CW, Chang YS, Choi BS. Resveratrol, an Nrf2 activator, ameliorates aging-related progressive renal injury. *Aging* (Albany NY). 2018 Jan; 10(1):83.

[274] Filograna R, Godena VK, Sanchez-Martinez A, Ferrari E, Casella L, Beltramini M, Bubacco L, Whitworth AJ, Bisaglia M. Superoxide dismutase (SOD)-mimetic M40403 is protective in cell and fly models of paraquat toxicity: implications for Parkinson disease. *Journal of Biological Chemistry.* 2016 Apr 22; 291(17):9257-67.

[275] Spasojević I, Chen Y, Noel TJ, Yu Y, Cole MP, Zhang L, Zhao Y, Clair DK, Batinić-Haberle I. Mn porphyrin-based superoxide dismutase (SOD) mimic, MnIIITE-2-PyP5+, targets mouse heart mitochondria. *Free Radical Biology and Medicine.* 2007 Apr 15; 42(8):1193-200.

[276] Olcott AP, Tocco G, Tian J, Zekzer D, Fukuto J, Ignarro L, Kaufman DL. A salen-manganese catalytic free radical scavenger inhibits type 1 diabetes and islet allograft rejection. *Diabetes.* 2004 Oct 1; 53(10):2574-80.

[277] Ranjbar A, Kheiripour N, Ghasemi H, Rabiei MA, Dadras F, Khoshjou F. Antioxidative effects of tempol on mitochondrial dysfunction in diabetic nephropathy. *Iranian journal of kidney diseases.* 2018 Mar 1; 12(2):84.

[278] Thakur P, Kumar A, Kumar A. Targeting oxidative stress through antioxidants in diabetes mellitus. *Journal of Drug Targeting.* 2018 Oct 21; 26(9):766-76.

[279] Suchal K, Malik S, Khan SI, Malhotra RK, Goyal SN, Bhatia J, Ojha S, Arya DS. Molecular pathways involved in the amelioration of myocardial injury in diabetic rats by kaempferol. *International Journal of Molecular Sciences.* 2017 May 15; 18(5):1001.

[280] Islam M, Alam F, Solayman M, Khalil M, Kamal MA, Gan SH. Dietary phytochemicals: natural swords combating inflammation and oxidation-mediated degenerative diseases. *Oxidative Medicine and Cellular Longevity. 2016* Oct; 2016.

[281] Mahmoud AM, Abd El-Twab SM, Abdel-Reheim ES. Consumption of polyphenol-rich Morus alba leaves extract attenuates early diabetic retinopathy: the underlying mechanism. *European Journal of Nutrition.* 2017 Jun; 56:1671-84.

[282] Yu X, Xu Z, Mi M, Xu H, Zhu J, Wei N, Chen K, Zhang Q, Zeng K, Wang J, Chen F. Dietary taurine supplementation ameliorates diabetic retinopathy via anti-excitotoxicity of glutamate in streptozotocin-induced Sprague-Dawley rats. *Neurochemical Research.* 2008 Mar; 33:500-7.

[283] Parveen A, Kim JH, Oh BG, Subedi L, Khan Z, Kim SY. Phytochemicals: Target-based therapeutic strategies for diabetic retinopathy. *Molecules.* 2018 Jun 23; 23(7):1519.

[284] Gandhi GR, Vasconcelos AB, Wu DT, Li HB, Antony PJ, Li H, Geng F, Gurgel RQ, Narain N, Gan RY. Citrus flavonoids as promising phytochemicals targeting diabetes and related complications: A systematic review of in vitro and in vivo studies. *Nutrients.* 2020 Sep 23; 12(10):2907.

[285] Sankeshi V, Kumar PA, Naik RR, Sridhar G, Kumar MP, Gopal VH, Raju TN. Inhibition of aldose reductase by Aegle marmelos and its protective role in diabetic cataract. *Journal of Ethnopharmacology.* 2013 Aug 26; 149(1):215-21.

[286] Guo T, Zhu L, Tan J, Zhou X, Xiao L, Liu X, Wang B. Promoting effect of triterpenoid compound from Agrimonia pilosa Ledeb on preadipocytes differentiation via up-regulation of PPARγ expression. *Pharmacognosy magazine.* 2015 Jan; 11(41):219.

[287] Bajaj S, Khan A. Antioxidants and diabetes. *Indian Journal of Endocrinology and Metabolism.* 2012 Dec; 16(Suppl 2):S267.

[288] Pazdro R, Burgess JR. The role of vitamin E and oxidative stress in diabetes complications. *Mechanisms of ageing and development.* 2010 Apr 1; 131(4):276-86.

[289] Zhang P, Li T, Wu X, Nice EC, Huang C, Zhang Y. Oxidative stress and diabetes: antioxidative strategies. *Frontiers of Medicine.* 2020 Oct; 14:583-600.

[290] Greenberg S, Frishman WH. Co-enzyme Q10: a new drug for cardiovascular disease. *The Journal of Clinical Pharmacology.* 1990 Jul; 30(7):596-608.

[291] Luís C, Costa R, Rodrigues I, Castela Â, Coelho P, Guerreiro S, Gomes J, Reis C, Soares R. Xanthohumol and 8-prenylnaringenin reduce type 2 diabetes–associated oxidative stress by downregulating galectin-3. *Porto Biomedical Journal.* 2019 Jan; 4(1).

[292] Sundaram R, Shanthi P, Sachdanandam P. Effect of tangeretin, a polymethoxylated flavone on glucose metabolism in streptozotocin-induced diabetic rats. *Phytomedicine.* 2014 May 15; 21(6):793-9.

[293] Rao YK, Lee MJ, Chen K, Lee YC, Wu WS, Tzeng YM. Insulin-mimetic action of rhoifolin and cosmosiin isolated from Citrus grandis (L.) Osbeck leaves: enhanced adiponectin secretion and insulin receptor phosphorylation in 3T3-L1 cells. *Evidence-Based Complementary and Alternative Medicine.* 2011 Jan 1; 2011.

[294] Murphy MP, Smith RA. Targeting antioxidants to mitochondria by conjugation to lipophilic cations. *Annu. Rev. Pharmacol. Toxicol.* 2007 Feb 10; 47:629-56.

[295] Zhao K, Zhao GM, Wu D, Soong Y, Birk AV, Schiller PW, Szeto HH. Cell-permeable peptide antioxidants targeted to inner mitochondrial membrane inhibit mitochondrial swelling, oxidative cell death, and reperfusion injury. *Journal of Biological Chemistry.* 2004 Aug 13; 279(33):34682-90.

[296] Orrenius S, Zhivotovsky B. Cardiolipin oxidation sets cytochrome c free. *Nature Chemical Biology.* 2005 Sep 1; 1(4):188-9.

[297] Kagan VE, Tyurin VA, Jiang J, Tyurina YY, Ritov VB, Amoscato AA, Osipov AN, Belikova NA, Kapralov AA, Kini V, Vlasova II. Cytochrome c acts as a cardiolipin oxygenase required for release of proapoptotic factors. *Nature Chemical Biology.* 2005 Sep 1; 1(4):223-32.

[298] Perreault L, Skyler JS, Rosenstock J. Novel therapies with precision mechanisms for type 2 diabetes mellitus. *Nature Reviews Endocrinology.* 2021 Jun; 17(6):364-77.

[299] Clark AL, Urano F. Endoplasmic reticulum stress in beta cells and autoimmune diabetes. *Current Opinion in Immunology.* 2016 Dec 1; 43:60-6.

[300] Lushchak O, Zayachkivska A, Vaiserman A. Metallic nanoantioxidants as potential therapeutics for type 2 diabetes: A hypothetical background and translational perspectives. *Oxidative Medicine and Cellular Longevity.* 2018 Jun 27; 2018.

[301] De Block CE, Dirinck E, Verhaegen A, Van Gaal LF. Efficacy and safety of high-dose glucagon-like peptide-1, glucagon-like peptide-1/glucose-dependent insulinotropic peptide, and glucagon-like peptide-1/glucagon receptor agonists in type 2 diabetes. *Diabetes, Obesity and Metabolism.* 2022 May; 24(5):788-805.

[302] Gaich G, Chien JY, Fu H, Glass LC, Deeg MA, Holland WL, Kharitonenkov A, Bumol T, Schilske HK, Moller DE. The effects of LY2405319, an FGF21 analog, in obese human subjects with type 2 diabetes. *Cell Metabolism.* 2013 Sep 3; 18(3):333-40.

[303] Ghislain J, Poitout V. Targeting lipid GPCRs to treat type 2 diabetes mellitus—Progress and challenges. *Nature Reviews Endocrinology.* 2021 Mar; 17(3):162-75.

[304] Ren Y, Li L, Wan L, Huang Y, Cao S. Glucokinase as an emerging anti-diabetes target and recent progress in the development of its agonists. *Journal of Enzyme Inhibition and Medicinal Chemistry.* 2022 Dec 31; 37(1):606-15.

[305] Matsuhashi T, Sato T, Kanno SI, Suzuki T, Matsuo A, Oba Y, Kikusato M, Ogasawara E, Kudo T, Suzuki K, Ohara O. Mitochonic acid 5 (MA-5) facilitates ATP synthase oligomerization and cell survival in various mitochondrial diseases. *EBioMedicine.* 2017 Jun 1; 20:27-38.

[306] Suzuki T, Yamaguchi H, Kikusato M, Hashizume O, Nagatoishi S, Matsuo A, Sato T, Kudo T, Matsuhashi T, Murayama K, Ohba Y. Mitochonic acid 5 binds mitochondria and ameliorates renal tubular and cardiac myocyte damage. *Journal of the American Society of Nephrology.* 2016 Jul 1; 27(7):1925-32.

[307] Ashrafizadeh M, Javanmardi S, Moradi-Ozarlou M, Mohammadinejad R, Farkhondeh T, Samarghandian S, Garg M. Natural products and phytochemical nanoformulations targeting mitochondria in oncotherapy: an updated review on resveratrol. *Bioscience Reports.* 2020 Apr 30; 40(4).

[308] Wang J, Li H, Yao Y, Zhao T, Chen YY, Shen YL, Wang LL, Zhu Y. Stem cell-derived mitochondria transplantation: a novel strategy and the challenges for the treatment of tissue injury. *Stem Cell Research & Therapy.* 2018 Dec; 9(1):1-0.

[309] Grieco GE, Brusco N, Licata G, Nigi L, Formichi C, Dotta F, Sebastiani G. Targeting microRNAs as a therapeutic strategy to reduce oxidative stress in diabetes. *International Journal of Molecular Sciences.* 2019 Dec 17; 20(24):6358.

[310] Thomas DA, Stauffer C, Zhao K, Yang H, Sharma VK, Szeto HH, Suthanthiran M. Mitochondrial targeting with antioxidant peptide SS-31 prevents mitochondrial depolarization, reduces islet cell apoptosis, increases islet cell yield, and improves posttransplantation function. *Journal of the American Society of Nephrology.* 2007 Jan 1; 18(1):213-22.

[311] Kim MY, Kim J. Chitosan microgels embedded with catalase nanozyme-loaded mesocellular silica foam for glucose-responsive drug delivery. *ACS Biomaterials Science & Engineering.* 2017 Apr 10; 3(4):572-8.

[312] Pocai A. Action and therapeutic potential of oxyntomodulin. *Molecular Metabolism.* 2014 Jun 1; 3(3):241-51.

[313] Kharitonenkov A, Shiyanova TL, Koester A, Ford AM, Micanovic R, Galbreath EJ, Sandusky GE, Hammond LJ, Moyers JS, Owens RA, Gromada J. FGF-21 as a novel metabolic regulator. *The Journal of Clinical Investigation.* 2005 Jun 1; 115(6):1627-35.

[314] Maiese K, Chong ZZ, Shang YC, Wang S. Novel directions for diabetes mellitus drug discovery. *Expert Opinion on Drug Discovery.* 2013 Jan 1; 8(1):35-48.

# Chapter 15

# Reactive Oxygen Species in the Development and Progression of Type 2 Diabetes

**Dipanjan Banerjee**
**and Suman Dasgupta***
Metabolic Disease Biology Laboratory, Department of Molecular Biology and Biotechnology, Tezpur University, Tezpur, Assam, India

## Abstract

Reactive Oxygen Species (ROS) have been implicated as one of the critical regulators involved in the onset and progression of different metabolic diseases including type 2 diabetes (T2D). ROS can be generated from pathways associated with mitochondrial oxidative phosphorylation, ER stress, and NADPH oxidase (NOX). Several studies with transgenic mice have revealed NOX as the central in mediating insulin resistance in T2D pathogenesis. ROS-mediated activation of NOX signaling markedly impairs insulin synthesis from the pancreatic β-cells and attenuates insulin action on the target cells, such as skeletal muscle cells, adipocytes, and hepatocytes. Accumulating evidence suggests that NOX activation plays a pivotal role in the attenuation of PI3K/Akt phosphorylation along with the induction of inhibitory IRS1/2 phosphorylation associated with the impairment of the insulin signaling pathway. Increased ROS production has also been associated with the macrophage-adipocyte crosstalk implementing adipose tissue and vascular inflammation involved in T2D development and complications. In this chapter, we summarize the knowledge of the contribution of ROS in type 2 diabetes.

**Keywords:** ROS, T2D, insulin, adipogenesis, lipogenesis, lipolysis, therapeutics

## 1. Introduction

### 1.1. ROS Generation in type 2 diabetes (T2D)

Oxidative stress and oxidative damage to cells and tissues are very common endpoints of chronic diseases, like atherosclerosis, and diabetes [1-3]. Hyperglycemia, or high blood glucose

* Corresponding Author's E-mail: suman.dsut@gmail.com.

In: The Role of Reactive Oxygen Species in Health and Disease
Editor: Kuladip Jana
ISBN: 979-8-89113-283-2

levels, is a symptom that characterizes diabetes. Moreover, chronic or intermittent hyperglycemia is linked to the development of diabetes complications [4]. Hyperglycemia in diverse organs can affect metabolic molecules that include diacylglycerol (DAG), protein kinase C (PKC), and NADPH-oxidase, ending in the production of reactive oxygen species (ROS) [5-6], which has previously been referred to as the "dangerous metabolic route in diabetes". This signaling pathway has been studied for its role in controlling angiogenesis, oxidative stress, and cellular death. Hyperglycemia in diabetic patients causes the increased production of ROS by mitochondrial respiratory chain enzymes, xanthine oxidases, lipoxygenases, cyclooxygenases, nitric oxide synthases, and peroxidases. The production of advanced glycation end products (AGEs) [6], the release of pro-inflammatory cytokines, and cellular death (pathological autophagy and/or apoptosis) are also observed in hyperglycaemic states.

Type 2 diabetes (T2D) is a metabolic disorder in which mitochondria play an important role. Mitochondria are the primary generator of reactive oxygen species (ROS) which involves in redox equilibrium as well as other processes including apoptosis, Ca2+ metabolism, and ATP (adenosine triphosphate) or heat synthesis [7-8]. ROS can operate as signaling molecules, but when they are overproduced, they cause mitochondrial malfunction and a reduction in ATP synthesis. Mitochondrial dysfunction is causally linked with the development and progression of T2D [9]. Under diabetes circumstances, increased ROS production occurs in a variety of tissues and cells. The nonenzymatic glycosylation process, the electron transport chain in mitochondria, and membrane-bound NADPH oxidase are all producers of ROS in cells. The electron transport chain is associated with increased levels of ROS generation in diabetic patients. However, it has been discovered that membrane-bound NADPH oxidase is a major contributor to the generation of ROS [10-11]. The membrane-bound subunits gp91 phox (Nox2)/Nox1/Nox4 and p22 phox, as well as the oxidase's catalytic site and cytosolic components p47 phox and p67 phox, make up NADPH oxidase [12-14]. AGEs, insulin, and other stimuli can also activate NADPH oxidase.

### 1.2. ROS in Insulin Synthesis and Insulin Action

#### *β-Cell Dysfunction*

Loss of β-cell function has been considered as a novel mechanism underpinning β-cell failure [15-17], which is important to the onset and progression of diabetes mellitus. Increased production of ROS, mitochondrial dysfunction, ER stress, and induction of autophagy are critically linked to the development of T2D and are important for β-cell dysfunction [18-20]. The β-cells primary function is to maintain the blood glucose homeostasis in the body by secreting insulin in response to high blood glucose levels. Glucose once enters the β-cell through GLUT2 (glucose transporter-2) [21] is metabolized by glycolytic pathway to generate pyruvate, NADH, and ATP. Pyruvate enters the mitochondrial tricarboxylic acid (TCA) cycle oxidation, resulting in the synthesis of additional NADH [22]. Through the mitochondrial electron transport chain (ETC), NADH is converted into ATP. The mitochondrial ETC is the major source of cellular ROS in pancreatic β-cells and complexes I and III, which are found within the inner mitochondrial membrane, produce extremely reactive superoxide ($O_2^{\cdot-}$) ions by reducing molecular oxygen with a single electron [23-24]. ROS are also produced by cytosolic and plasma membrane oxidoreductases, which oxidize NAD(P)H and directly produce ROS by reducing molecular oxygen. Higher levels of ROS production in the β-cell of diabetes patients are implicated in β-cell damage. β-cells display a highly efficient mechanism of glucose uptake through GLTU2, a high-Km glucose transporter when exposed to a high glucose concentration [25-26]. When glucose transportation is impeded due to peripheral insulin resistance, the constant rise in glycolytic flux may increase ROS generation in the β-cells, potentially causing pathological consequences [27-28]. Increase in intracellular $Ca^{2+}$ can boost mitochondrial ROS production and activate protein kinase C (PKC), resulting in NADPH oxidase-dependent superoxide and other species that can lead to β-cell damage [29-30].

The generation of ROS is induced by high levels of free fatty acids (FFA) in several tissues, including the pancreatic β-cells. Greater levels of circulating FFA result in its increased absorption by β-cells, which then be utilized as a metabolic fuel by the mitochondria via oxidation. The resulting flux to acetyl-CoA stimulates the TCA cycle to create additional NADH and FADH2 along with increasing levels of $O_2^{\cdot-}$ [31]. Palmitate, a saturated FFA, is known to activate NOX in β-cells and islets, resulting in enhanced $O_2^{\cdot-}$ generation, and NOX activation can also potentiate glucose-stimulated insulin secretion under acute palmitate exposure [32]. Thus, NOX is an important source of intracellular $O_2^{\cdot-}$ since it may be triggered by nutrients like glucose or saturated fatty acids, as well as crucial stimulus-secretion coupling factors including $Ca^{2+}$ influx and protein kinase C activation [14].

β-cells are metabolically active, but their antioxidant defenses are less than those of other cells and tissues. In this regard, it has been discovered that the activity of free radical detoxifying enzymes and redox-regulating enzymes in pancreatic islets is low [33-34]. Under the diabetic condition, the expression of ROS markers 8-hydroxy-deoxyguanosine (8-OHdG) and 4-hydroxy-2, 3-nonenal (4-HNE) was shown to be higher in islets [35]. Furthermore, because antioxidant enzymes like catalase and glutathione peroxidase are expressed at low levels in β-cells, they are particularly sensitive to increased ROS levels. As a result, ROS are likely to be implicated in the deterioration of β-cells noticed in diabetic conditions.

### 1.3. Impairment of Insulin Signaling

Insulin resistance is the most prevalent result of impaired insulin signaling [36] and plays a major role in developing T2D. Insulin resistance develops when normal insulin levels are insufficient to cause a functional insulin response in adipose, liver, or muscle cells [37]. Insulin signaling is mainly disrupted by NOX-mediated cellular oxidative stress, leading to peripheral insulin resistance (Figure 1) [38]. Membrane-bound enzyme complexes of the NOX family transport electrons from NADPH to oxygen, resulting in the production of superoxide. There are seven isoforms of the NOX family, NOX1, NOX2, NOX3, NOX4, NOX5, dual oxidase 1 (DUOX1), and DUOX2 which are highly conserved in mammals and extensively expressed in diverse organs [39-40]. NOX1–3 requires cytosolic activators (p47phox, p67phox, NADPH oxidase organizer 1), but NOX4 is constitutively active and does not require an activator protein. NOX4 has been found to attenuate membrane translocation of glucose transporter 4 (GLUT4) [40-41]. GLUT4 is the primary glucose transporter in peripheral insulin-sensitive tissues such as skeletal muscle and adipose tissue, and it has been the subject of intense research in the field of insulin resistance [42-43]. Insulin stimulation promotes GLUT4 expression and its migration to the plasma membrane that facilitates cellular glucose uptake from the bloodstream. NOX4 is a potent enzyme that generates substantial levels of cellular ROS [44] which has been activated by the PI3-kinase dependent Rac phosphorylation [45-46]. Moreover, casein kinase 2 (CK2) is activated by increased ROS which in turn activates retromer that sends a signal to the trans-Golgi network, which transports GLUT4 to lysosomes for its destruction rather than migration to plasma membrane. As a result, in an oxidative environment, intravascular glucose levels increased. Chronic oxidative stress triggers a number of stress-sensitive signalling pathways, including NF-κB, JNK/SAPK, and p38 MAPK. When exposed to an oxidative environment, a variety of serine/threonine protein kinases such as IKK, JNK, ERK, and p38 MAPK are activated. Among these, transcription factor NF-κB plays a critical role in regulating inflammatory responses [47-48]. The NF-κB pathway is triggered when an active serine/threonine kinase, IKK, phosphorylates the inhibitory subunit IκB and thereby facilitates its degradation and release of NF-κB which then migrated to the nucleus and regulates inflammatory gene expression [49-50]. Insulin signalling has been found to be inhibited by the IKK and JNK. Inhibitory phosphorylation of IRS-1, a key component of insulin signalling, at serine 302, 307, and 612 residues by JNK and IKK prevents IRS-1 binding to the insulin receptor and thus promoting IRS-1 degradation and insulin signalling inhibition [51-53]. The capacity of IRS or IR proteins to recruit and activate downstream SH-2 domain containing signalling molecules are impaired by serine/threonine phosphorylation, as IRS protein is responsible for signal transduction connecting from the insulin receptor [36,54]. ROS can also directly activate several serine/threonine kinases implicated in insulin signalling inhibition such as PKC, PKB, mTOR, and GSK3 that desensitize the insulin signalling cascade by phosphorylating IR or IRSs on certain serine/threonine residues [36, 38].

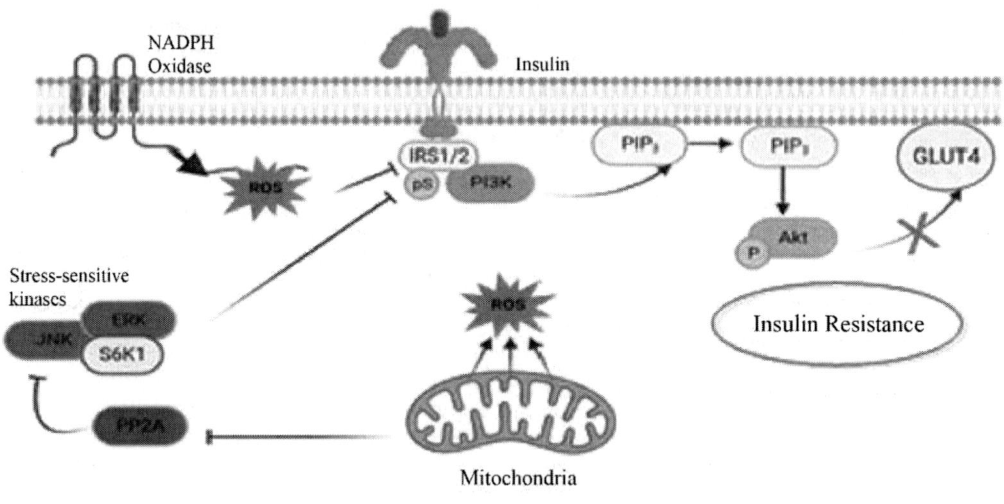


**Figure 1.** Reactive Oxygen Species (ROS) induces insulin resistance through the inhibition of insulin signaling cascade and activation of stress-sensitive kinases.

## 2. Adipose Tissue Function Impairment by ROS

Adipose tissue is one of the major cell types that play a critical role in maintaining lipid and energy homeostasis. Imbalance of lipid homeostasis in obesity is mainly responsible for the onset and progression of T2D pathophysiology. A link between excess level of ROS in adipose tissue dysfunction and T2D progression has been revealed recently. Adipose tissue functions are mainly governed by adipogenesis (conversion of adipocytes from pre-adipocytes), lipogenesis (triglyceride synthesis), and lipolysis (triacylglycerols hydrolysis) and recent studies implicated the involvement of ROS in adipocyte dysfunction.

### *Adipogenesis*

Adipose tissue health is dependent on the constant regeneration of its cells, particularly adipocytes. Adipocyte differentiation or adipogenesis is a multi-step process in which adipocytes differentiated from the multipotent mesenchymal stem cells (MSCs) in white adipose tissue (WAT) [55-56]. The transcription factors CCAAT/enhancer binding protein (C/EBP)β, followed by peroxisome proliferator-activated receptor (PPAR)γ and C/EBPα, are key executors of the terminal differentiation programme in adipogenesis, inducing the expression of adipocyte-specific genes like adiponectin and adipocyte protein 2 [56]. To maintain the differentiated state and functioning of adult adipocytes, sustained expression of PPARγ is essential. During adipogenesis, signalling pathways triggered by insulin/IGF1 receptor tyrosine kinase induces differentiation by activating the Akt kinase and its targets protein kinase C (PKC) isoforms and mammalian target of rapamycin complex-1 (mTORC1), as well as PPARγ and C/EBPα protein expression [57-58]. NOX4 plays an important role in adipogenesis. NOX4-mediated production of $H_2O_2$ has been identified as a key regulator of adipogenesis and adipose tissue function in both health and illness [59-60]. NOX4 generated $H_2O_2$ activates MAP kinase phosphatase-1 (MKP-1) which inhibits ERK1/2 [61]. Inhibition of ERK1/2 promotes an inhibitory serine phosphorylation of IRS1, which blocks IRS1 tyrosine phosphorylation in response to insulin receptor stimulation [62] necessary for phosphatidylinositol 3-kinase activation (PI3K) and thus impairs adipogenic differentiation.

### *Lipolysis*

Lipid mobilization in adipose tissues, which includes lipogenesis and lipolysis, is critical for systemic energy metabolism regulation [63]. Adipose tissue serves as the principal energy storage depot. Lipolysis hydrolyses triacylglycerol (TAG) reserves in adipose tissue to release glycerol and fatty acid (FA) and so their supply meets energy demands during periods of negative energy balance [64]. Under anabolic circumstances, the adipose tissue stores energy in the form of lipids like FA and TAG, a process known as lipogenesis. The control of FA trafficking into and out of adipocytes (lipolysis and lipogenesis) comprises metabolic as well as endo-, para-, and autocrine processes that are partially dependent on redox signalling [65]. In adipocytes, adipose tissue triglyceride lipase (ATGL) and hormone-sensitive lipase (HSL) play major role in lipolysis by classical or inflammatory lipolytic pathways [66-67]. The classical lipolytic pathway involved in the activation of β-adrenergic receptor which in turn stimulate adenylyl cyclase that generates cAMP [66] and it's associated signalling through different protein kinases (Figure 2). On the other hand, inflammatory lipolysis is triggered by toll like receptor 4 (TLR4) activation and IL-6 cytokine expression [68]. Through a cascade of proteins phosphorylation involving different protein kinases (PKA, PKC, PKG), lipolytic signals activates neutral lipases (ATGL, HSL) [69-70] that stimulates lipolysis with the help of ATGL co-activator CGI-58, and perilipin 1 (PLIN1; lipid droplet coating). ROS, especially $H_2O_2$ stimulates lipolysis through the activation of β-adrenergic receptor and adenylyl cyclase [71-72]. Adenylyl cyclase produces cAMP, which binds to PKA and releases its catalytic component, causing phosphorylation of its targets such as HSL, PLIN1, and CGI-58 that activates lipolysis. PKC can be activated by $H_2O_2$ in two ways, $H_2O_2$ either releases DAG from cellular membrane and activates PKC or increases intracellular concentrations of $Ca^{2+}$ and therefore favours PKC activation [73-74]. On the other hand, $O_2$•− and $H_2O_2$ phosphorylates HSL at Ser522 and induces its translocation from cytosol to lipid droplet that facilitates triglyceride hydrolysis.

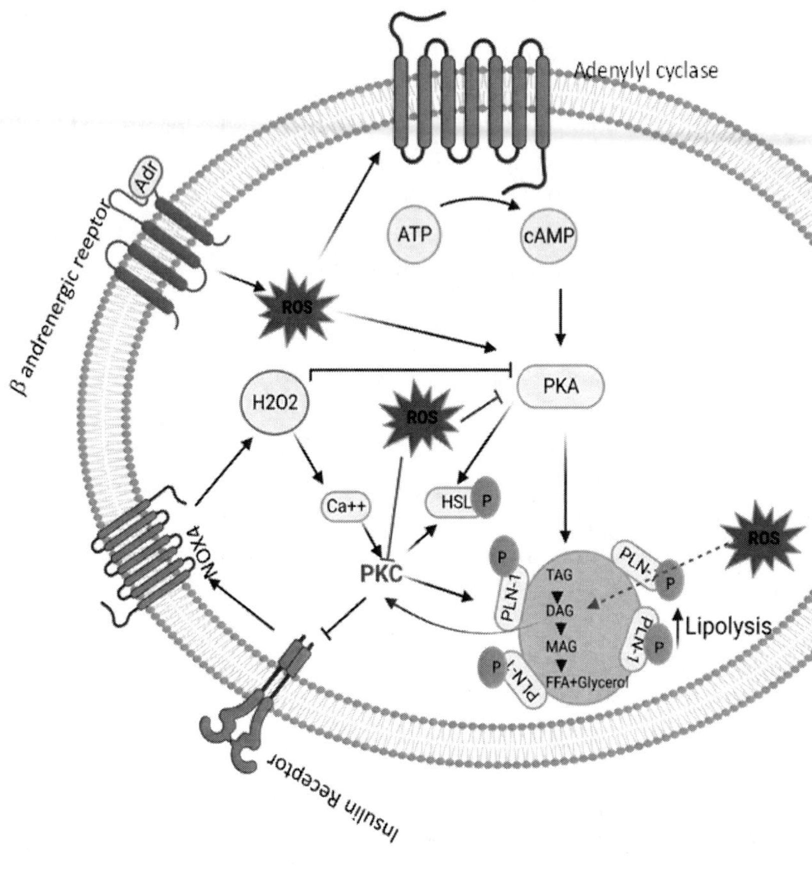


**Figure 2.** The classical lipolytic pathway associated activation of β-adrenergic receptor stimulate the adenylyl cyclase generating cAMP and governing lipolysis through different protein kinases.

### *Lipogenesis*

The lipogenic pathway involving triglyceride synthesis is critically regulated by the cellular redox state. The production of FA and TAG in the liver and adipose tissue is referred to as lipogenesis. In adipose tissue, TAG can be hydrolyzed by lipoprotein lipase (LPL) to release FA, which then enters to adipocyte through CD-36 and fatty acid transport protein-1 (FATP1) and get esterified to TAG and stored [75].

Circulating carbohydrates are converted into FA, which are subsequently utilized by the cells to synthesize TAG or other lipid molecules in de novo lipogenesis [76]. Some of the enzymes that are involved in lipogenesis are fatty acid synthase (Fasn), diacylglycerol O-acyltransferase1 (Dgat1), stearoyl-CoA desaturase-1 (Scd1), and acetyl-CoA carboxylase (Acc) [77-78]. $H_2O_2$ has long been thought to play a key function in cellular processes such as glucose transport and absorption [79]. In ROS-eliminated mice, it was noticed that lipogenic genes (Acly, Scd1, Fasn, Acaca, and Srebf1) expressions were downregulated [80-81]. In rat adipocytes, ROS-induced lipogenic gene downregulation appears to be mediated through the reduction of sterol-regulatory element-binding transcription factor 1 transcriptional activity [82].

## 3. ROS in T2D Complications

Oxidative stress is considered to be the major cause of type 2 diabetes (T2D), a metabolic condition characterized by hyperglycemia and insulin resistance [83]. As a result of an excess of glucose and oxidative phosphorylation, mitochondria experiences increased accumulation of ROS and associated oxidative stress [84]. Since the endoplasmic reticulum (ER) is a generator of ROS, it also plays a significant part in oxidative stress [85]. The ROS produced in mitochondria contributes to ER stress because of the close link between the two organelles provided by mitochondrial-associated membranes (MAMs). As a result of this vicious cycle, stress and mitochondrial malfunction are major factors in cellular redox balance. Pancreatic β-cells are particularly vulnerable to oxidative stress and mitochondrial dysfunction as mitochondria play a key role in insulin release and their exposure to hyperglycemia [86-87]. In fact, crosstalk between both mechanisms is connected to changes in glucose homeostasis and can result in the insulin-resistance status associated with diabetes. Diabetes-related acute or long-term elevation of blood glucose levels increases ROS generation and triggers apoptosis in β-cells [86]. The development of diabetic complications involves both apoptosis and necroptosis, and they may result in tissue damage to the heart, retina, kidneys, and neurological system [88]. The advancement of diabetes related problems may be established by the imbalance between autophagy and apoptosis wherein autophagy inhibits the process of apoptosis [89-90]. It has been shown that activation of PKC and mTORC1 pathway are both involved in the regulation of cellular autophagy [91-92].

## 4. Targeting ROS in T2D Therapeutics

Type 2 diabetes (T2D) is treated with a variety of medications that lower intracellular ROS levels. The majority of them either strengthen the antioxidant defence system or focus on the inhibition of ROS generation. The medications that inhibit aldose reductase are Tolrestat, Lidorestat, Zenarestat, Ponalrestat, Zopolrestat, and Epalrestat. The Fidarestat, and Ranirestat prevent the production of AGEs are pyridoxamine and aminoguanidine [93-94]. PKC412 (midostaurin), ruboxistaurinmesylate, and ruboxistaurin (RBX) are medications with the ability to block the PKC pathway [95-96]. Metformin, a biguanide functions as a hypoglycemic drug by scavenging oxygenation radicals [97-98]. By lowering both NADPH oxidase and mitochondrial ROS, it lowers intracellular ROS levels [97-99]. Additionally, it also prevents AGEs formation. A neuro-hormonal antagonist of the third generation is carvingilol which protects body's natural antioxidant mechanisms, reducing oxidative stress. Other drugs used in treating T2D are meglitinides, dipeptidyl peptidase-4 (DPP-4) inhibitors, sulfonylureas, sodium/glucose cotransporter-2 (SGLT-2) inhibitors, incretin mimetics (GLP-1 receptor agonists), amylin mimetics, and insulin analogues [100]. Commercially available drugs and their uses are listed in Table 1. Despite the fact that numerous medications exist that control type 2 diabetes by lowering ROS, the majority of them failed in clinical trials as a result of their considerable side effects. Extended use of these medications causes several side effects, including edema, severe liver damage, renal toxicity, diarrhoea, and hepatic dysfunction. Moreover, some of the medications are also poorly absorbed by the body [101].

**Table 1.** Pharmacological interventions for treatment of type 2 diabetes and its complications

| Therapeutic strategies | Mode of action | Effects |
|---|---|---|
| Biguanides | Increasing AMPK and lowering of AGEs production, and activation of NF-κB and NADPH oxidase | Reduction of the hepatic glucose production, inhibits ROS production and inflammation |
| Sulfonylureas | It closes the ATP-sensitive K+ channels by binding with SUR-1 | Increase the sensitivity of β–cells to glucose and promote insulin secretion |
| Thiazolidinediones | PPAR-γ activation and reduction of AGEs generation | Improvement of glucose uptake and reduction of inflammation |
| SGLT-2 inhibitors | AGEs generation ↓Nox4 inhibition, targeting Nrf2/HO-1 pathway, AMPK-mTOR and AMPK/PI3K/AKT/eNOS signalling cascade | Decrease the production of free-radicals and alleviates inflammation |

**Table 2.** Biochemical interventions for treatment of type 2 diabetes and its complications

| Therapeutic strategies | Mode of action | Effects |
|---|---|---|
| Proanthocyanidins from grape seed, *Abromaaugusta* leaves, *Alchorneacordifolia* leaves, *Bauhinia variegata* bark | Reduction of ER stress, inhibition of hexosamine pathway | Improve β-cell functionality |
| Amorfrutin from *Glycyrrhiza foetida* and *Amorpha fruticosa*, triterpene from *Bumelia sartorum*, phenolic from *Xanthium* | Inhibition of PKC and other stress sensitive signaling pathway | Improve insulin sensitivity |
| Curcumin from *Curcuma longa* | Inhibits PKC/MAPK Signaling pathway and Keap1/Nrf2/ARE/HO-1 signaling pathway; inhibiting IKKβ phosphorylation and preventing NF-κB activation | Boosts antioxidant enzyme levels, lowers oxidative stress and inflammation, lowers hyperglycemia, and lowers oxidative stress, all of which lessen diabetes and its consequences. |
| Resveratrol | Targetes AMPK/FoxO1 pathway and Keap1/Nrf2/ ARE signalling cascade; NOX1,2,4 and JAK/STAT signalling pathway suppression | Improves ROS scavenging capacity, antiapoptotic and anti-inflammatory role |

Traditional drugs made from plant sources are often used as folk medicine in the treatment of type 2 diabetes. The therapeutic action of medicinal plants and plant-derived compounds are primarily examined for their ability to regulate the blood glucose levels and oxidative stress. Phytochemicals exert their effects by targeting different metabolic pathways. Ginseng, a plant-derived ginsenoside, prevents cytokine-induced apoptosis of β-cells by decreasing the generation of nitric oxide (NO), reactive oxygen species (ROS), inhibiting the expression of the genes p53 and p21, and inhibiting caspases and PARP. Berberine isolated from Coptidis rhizoma (CR) and Cortex phellodendri, when administered intragastrically to STZ-induced diabetic rats, demonstrated improvement of blood glucose levels via lowering oxidative stress and blocking the polyol pathway. Fenugreek leaf powder attenuates STZ-induced oxidative stress in diabetic rats wherein it lowers blood glucose through enhancing the antioxidant system [102-105]. The uses of other phytochemicals in T2D are summarized in Table 2. A growing body of research indicates that obesity, and sedentary lifestyle favour T2D pathogenesis, therefore, a healthy lifestyle is necessary for a successful treatment strategy. The lifestyle interventions are non-pharmacological ways to treat T2D that focus on a person's lifestyle, such as their eating habits, drinking, or physical activity [106-108] (Table 3).

**Table 3.** Lifestyle interventions for treatment of type 2 diabetes and its complications

| Therapeutic strategies | Mode of action | Effects |
|---|---|---|
| Dietary intervention | AMPK/SIRT1/PGC-1α pathway inhibition, targeting FOXO/SIRT3/SOD2 and Nrf2 pathway | Antioxidant, anti-inflammatory, and antifibrotic qualities; enhances glycemic management, lipid profile, and endothelial function; prevents atherosclerosis and lowers cardiovascular risk |
| Physical activity | Keap1/Nrf2/ARE pathway stimulation, GLUT4 translocation | Antioxidant levels; insulin sensitivity |

## 5. Future Perspectives

It is evident that a variety of factors critically regulate the onset and progression of type 2 diabetes. In this context, it is well known that oxidative stress plays a significant role in the progression of the illness and in the development of diabetes complications. Overproduction of ROS that are not neutralized by the scavenging or detoxifying pathways leads to oxidative stress, which creates a redox imbalance. Vascular dysfunction would be significantly influenced by the imbalance between ROS formation and clearance. Type 2 diabetes and its associated complications such as hypertension and atherosclerosis would eventually result from the EC and VSMC damage that was gradually becoming worse. In hyperglycemia, mitochondrial and cytosolic ROS generation and increased ROS production are due to the ER stress in β-cells dysfunction. Persistent hyperglycemia associated with enhanced ROS production in insulin targets cells causing the onset of insulin resistance. Higher levels of circulatory FFA lead to the lipotoxicity and that strongly correlated with mitochondrial metabolism and ER activity resulting generation of massive amount of ROS. Recently, it has become clear that lipotoxicity plays a significant part in the dysfunction and demise of β-cells along with the impairment of adipose tissue function and thus governing insulin section and its action. Moreover, hyperglycemia increases DAG formation, PKC, and NADPH-oxidase activation which results in the production of ROS and oxidative stress in type 2 diabetes. The inflammatory responses are caused by the abnormal immune activation and the cellular death that generates AGEs also exacerbate its effects upon binding with the receptor RAGE. As a result, the gold standard treatment for stopping the onset and progression of type 2 diabetes may involve development of novel therapeutics targeting ROS.

## Conclusion

It is now clear that oxidative stress not only contributes to the development of Type 2 diabetes but also is a contributing factor. Moreover, antioxidant therapy against diabetic vascular problems merits investigation, and substances from food and medicinal plants are extensively researched in this area. Therefore, it is important to maintain the cell's ROS generation in order to keep the blood glucose levels under control. Using conventional pharmacological agents would be of great help for the treatment of type 2 diabetes and its complications. Besides these pharmacological drugs, physical activity and weight control are also beneficial to maintain the redox state. In diabetic conditions, dietary intake of antioxidants can assist in maintaining the cellular redox state. An adjuvant therapy for the treatment of diabetic vascular problems can include therapeutic foods or plants. Therefore, to examine the effectiveness in the human body, preclinical and clinical studies of phytochemicals for treatment against type 2 diabetes and its complications are of urgent need. Phytochemicals with stronger antioxidants potency can limit the oxidative stress and the risk of numerous issues related to type 2 diabetes. However, a healthy lifestyle and eating habits are also critical for type 2 diabetes and associated complications.

## References

[1] Houstis N, Rosen ED, Lander ES. Reactive oxygen species have a causal role in multiple forms of insulin resistance. *Nature.* 2006 Apr 13; 440(7086):944-8.

[2] West IC. Radicals and oxidative stress in diabetes. *Diabetic Medicine.* 2000 Mar; 17(3):171-80.

[3] Tangvarasittichai S. Oxidative stress, insulin resistance, dyslipidemia and type 2 diabetes mellitus. *World Journal of Diabetes.* 2015 Apr 4; 6(3):456.

[4] Inzucchi SE, Bergenstal RM, Buse JB, Diamant M, Ferrannini E, Nauck M, Peters AL, Tsapas A, Wender R, Matthews DR. Management of hyperglycemia in type 2 diabetes, a patient-centered approach: update to a position statement of the American Diabetes Association and the European Association for the Study of Diabetes. *Diabetes Care.* 2015 Jan; 38(1):140-9.

[5] Giacco F, Brownlee M. Oxidative stress and diabetic complications. *Circulation research.* 2010 Oct 29;107(9):1058-70.

[6] Vanessa Fiorentino T, Prioletta A, Zuo P, Folli F. Hyperglycemia-induced oxidative stress and its role in diabetes mellitus related cardiovascular diseases. *Current Pharmaceutical Design.* 2013 Oct 1; 19(32):5695-703.

[7] Di Meo S, Reed TT, Venditti P, Victor VM. Role of ROS and RNS Sources in Physiological and Pathological Conditions. *Oxidative Medicine and Cellular Longevity.* 2016; 2016:1245049.

[8] Zorov DB, Juhaszova M, Sollott SJ. Mitochondrial reactive oxygen species (ROS) and ROS-induced ROS release. *Physiological Reviews.* 2014 Jul; 94(3):909-50.

[9] Galicia-Garcia U, Benito-Vicente A, Jebari S, Larrea-Sebal A, Siddiqi H, Uribe KB, Ostolaza H, Martín C. Pathophysiology of type 2 diabetes mellitus. *International Journal of Molecular Sciences.* 2020 Jan; 21(17):6275.

[10] Fu XJ, Peng YB, Hu YP, Shi YZ, Yao M, Zhang X. NADPH oxidase 1 and its derived reactive oxygen species mediated tissue injury and repair. *Oxidative Medicine and Cellular Longevity.* 2014 Jan 19; 2014.

[11] Forrester SJ, Kikuchi DS, Hernandes MS, Xu Q, Griendling KK. Reactive oxygen species in metabolic and inflammatory signaling. *Circulation Research.* 2018 Mar 16; 122(6):877-902.

[12] Panday A, Sahoo MK, Osorio D, Batra S. NADPH oxidases: an overview from structure to innate immunity-associated pathologies. *Cellular &Molecular Immunology.* 2015 Jan; 12(1):5-23.

[13] Altenhöfer S, Radermacher KA, Kleikers PW, Wingler K, Schmidt HH. Evolution of NADPH oxidase inhibitors: selectivity and mechanisms for target engagement. *Antioxidants &Redox Signaling.* 2015 Aug 10; 23(5):406-27.

[14] Rastogi R, Geng X, Li F, Ding Y. NOX activation by subunit interaction and underlying mechanisms in disease. *Frontiers in Cellular Neuroscience.* 2017 Jan 10; 10:301.

[15] Accili D, Talchai SC, Kim-Muller J, Cinti F, Ishida E, Ordelheide AM, Kuo T, Fan J, Son J. When β-cells fail: lessons from dedifferentiation. *Diabetes, Obesity and Metabolism.* 2016 Sep; 18:117-22.

[16] Halban PA, Polonsky KS, Bowden DW, Hawkins MA, Ling C, Mather KJ, Powers AC, Rhodes CJ, Sussel L, Weir GC. β-cell failure in type 2 diabetes: postulated mechanisms and prospects for prevention and treatment. *The Journal of Clinical Endocrinology & Metabolism.* 2014 Jun 1; 99(6):1983-92.

[17] Swisa A, Glaser B, Dor Y. Metabolic stress and compromised identity of pancreatic beta cells. *Frontiers in Genetics. 2017* Feb 21; 8:21.

[18] Rocha M, Apostolova N, Diaz-Rua R, Muntane J, Victor VM. Mitochondria and T2D: role of autophagy, ER stress, and inflammasome. *Trends in Endocrinology & Metabolism.* 2020 Oct 1; 31(10):725-41.

[19] Kim YK, Sussel L, Davidson HW. Inherent beta cell dysfunction contributes to autoimmune susceptibility. *Biomolecules.* 2021 Mar 30; 11(4):512.

[20] Burgos-Morón E, Abad-Jiménez Z, Martínez de Marañón A, Iannantuoni F, Escribano-López I, López-Domènech S, Salom C, Jover A, Mora V, Roldan I, Solá E. Relationship between oxidative stress, ER stress, and inflammation in type 2 diabetes: the battle continues. *Journal of Clinical Medicine.* 2019 Sep 4; 8(9):1385.

[21] Adeva-Andany MM, Pérez-Felpete N, Fernández-Fernández C, Donapetry-García C, Pazos-García C. Liver glucose metabolism in humans. *Bioscience reports.* 2016 Dec 1; 36(6).

[22] Oeggl R, Neumann T, Gätgens J, Romano D, Noack S, Rother D. Citrate as cost-efficient NADPH regenerating agent. *Frontiers in Bioengineering and Biotechnology.* 2018 Dec 21; 6:196.

[23] Zhao RZ, Jiang S, Zhang L, Yu ZB. Mitochondrial electron transport chain, ROS generation and uncoupling. *International Journal of Molecular Medicine.* 2019 Jul 1; 44(1):3-15.

[24] Sha W, Hu F, Bu S. Mitochondrial dysfunction and pancreatic islet β-cell failure. *Experimental and Therapeutic Medicine. 2020* Dec 1; 20(6):1-.

[25] Thorens B. GLUT2, glucose sensing and glucose homeostasis. *Diabetologia.* 2015 Feb; 58(2):221-32.

[26] Berger C, Zdzieblo D. Glucose transporters in pancreatic islets. *PflügersArchiv-European Journal of Physiology.* 2020 Sep; 472(9):1249-72.

[27] Fu Z, Gilbert ER, Liu D. Regulation of insulin synthesis and secretion and pancreatic Beta-cell dysfunction in diabetes. *Current diabetes reviews.* 2013 Jan 1; 9(1):25-53.

[28] Fu Z, Gilbert ER, Liu D. Regulation of insulin synthesis and secretion and pancreatic Beta-cell dysfunction in diabetes. *Current Diabetes Reviews.* 2013 Jan 1; 9(1):25-53.

[29] Dikalov S. Cross talk between mitochondria and NADPH oxidases. *Free Radical Biology and Medicine.* 2011 Oct 1; 51(7):1289-301.

[30] Görlach A, Bertram K, Hudecova S, Krizanova O. Calcium and ROS: A mutual interplay. *Redox Biology.* 2015 Dec 1; 6:260-71.

[31] Martínez-Reyes I, Chandel NS. Mitochondrial TCA cycle metabolites control physiology and disease. *Nature Communications.* 2020 Jan 3; 11(1):102.

[32] Sato Y, Fujimoto S, Mukai E, Sato H, Tahara Y, Ogura K, Yamano G, Ogura M, Nagashima K, Inagaki N. Palmitate induces reactive oxygen species production and β-cell dysfunction by activating nicotinamide adenine dinucleotide phosphate oxidase through S rcsignaling. *Journal of Diabetes Investigation.* 2014 Jan; 5(1):19-26.

[33] Stancill JS, Broniowska KA, Oleson BJ, Naatz A, Corbett JA. Pancreatic β-cells detoxify H2O2 through the peroxiredoxin/thioredoxin antioxidant system. *Journal of Biological Chemistry.* 2019 Mar 29; 294(13):4843-53.

[34] Benáková Š, Holendová B, Plecitá-Hlavatá L. Redox homeostasis in pancreatic β-cells: from development to failure. *Antioxidants.* 2021 Mar 27; 10(4):526.

[35] Pan G, Deshpande M, Pang H, Stemmer PM, Carruthers NJ, Shearn CT, Backos DS, Palaniyandi SS. 4-Hydroxy-2-nonenal attenuates 8-oxoguanine DNA glycosylase 1 activity. *Journal of Cellular Biochemistry.* 2020 Dec; 121(12):4887-97.

[36] Boucher J, Kleinridders A, Kahn CR. Insulin receptor signaling in normal and insulin-resistant states. Cold Spring Harbor perspectives in biology. 2014 Jan 1; 6(1):a009191.

[37] Wilcox G. Insulin and insulin resistance. *Clinical Biochemist Reviews.* 2005 May;26(2):19.

[38] Rains JL, Jain SK. Oxidative stress, insulin signaling, and diabetes. *Free Radical Biology and Medicine.* 2011 Mar 1; 50(5):567-75.

[39] Tarafdar A, Pula G. The role of NADPH oxidases and oxidative stress in neurodegenerative disorders. *International Journal of Molecular Sciences.* 2018 Nov 30;19(12):3824.

[40] Bedard K, Krause KH. The NOX family of ROS-generating NADPH oxidases: physiology and pathophysiology. *Physiological Reviews.* 2007 Jan; 87(1):245-313.

[41] Youn JY, Gao L, Cai H. The p47 phox-and NADPH oxidase organiser 1 (NOXO1)-dependent activation of NADPH oxidase 1 (NOX1) mediates endothelial nitric oxide synthase (eNOS) uncoupling and endothelial dysfunction in a streptozotocin-induced murine model of diabetes. *Diabetologia. 2012* Jul; 55:2069-79.

[42] Vargas E, Podder V, Carrillo Sepulveda MA. *Physiology, Glucose Transporter Type 4.* 2023 May 1. In: StatPearls [Internet]. Treasure Island (FL): StatPearls Publishing; 2023 Jan–. PMID: 30726007.

[43] Chadt A, Al-Hasani H. Glucose transporters in adipose tissue, liver, and skeletal muscle in metabolic health and disease. *PflügersArchiv-European Journal of Physiology.* 2020 Sep; 472:1273-98.

[44] Buck T, Hack CT, Berg D, Berg U, Kunz L, Mayerhofer A. The NADPH oxidase 4 is a major source of hydrogen peroxide in human granulosa-lutein and granulosatumor cells. *Scientific Reports.* 2019 Mar 5; 9(1):3585.

[45] Nakanishi A, Wada Y, Kitagishi Y, Matsuda S. Link between PI3K/AKT/PTEN pathway and NOX proteinin diseases. *Aging and Disease.* 2014 Jun; 5(3):203.

[46] Benchoula K, Parhar IS, Wong EH. The crosstalk of hedgehog, PI3K and Wnt pathways in diabetes. *Archives of Biochemistry and Biophysics.* 2021 Feb 15; 698:108743.

[47] Papaconstantinou J. The role of signaling pathways of inflammation and oxidative stress in development of senescence and aging phenotypes in cardiovascular disease. *Cells.* 2019 Nov 4; 8(11):1383.

[48] Evans JL, Goldfine ID, Maddux BA, Grodsky GM. Oxidative stress and stress-activated signaling pathways: a unifying hypothesis of type 2 diabetes. *Endocrine Reviews.* 2002 Oct 1; 23(5):599-622.

[49] Israël A. The IKK complex, a central regulator of NF-κB activation. *Cold Spring HarborPerspectives in Biology. 2010* Mar 1; 2(3):a000158.

[50] Hinz M, Scheidereit C. The IκB kinase complex in NF-κB regulation and beyond. *EMBO Reports.* 2014 Jan; 15(1):46-61.

[51] Rajan MR, Fagerholm S, Jönsson C, Kjølhede P, Turkina MV, Strålfors P. Phosphorylation of IRS1 at serine 307 in response to insulin in human adipocytes is not likely to be catalyzed by p70 ribosomal S6 kinase. *PloS one.* 2013 Apr 2; 8(4):e59725.

[52] Herschkovitz A, Liu YF, Ilan E, Ronen D, Boura-Halfon S, Zick Y. Common inhibitory serine sites phosphorylated by IRS-1 kinases, triggered by insulin and inducers of insulin resistance. *Journal of Biological Chemistry.* 2007 Jun 22; 282(25):18018-27.

[53] Boura-Halfon S, Zick Y. Phosphorylation of IRS proteins, insulin action, and insulin resistance. *American Journal of Physiology-Endocrinology and Metabolism.* 2009 Apr; 296(4):E581-91.

[54] Copps KD, White MF. Regulation of insulin sensitivity by serine/threonine phosphorylation of insulin receptor substrate proteins IRS1 and IRS2. *Diabetologia.* 2012 Oct; 55:2565-82.

[55] Zhang K, Yang X, Zhao Q, Li Z, Fu F, Zhang H, Zheng M, Zhang S. Molecular mechanism of stem cell differentiation into adipocytes and adipocyte differentiation of malignant tumor. *Stem Cells International.* 2020 Oct; 2020.

[56] Ambele MA, Dhanraj P, Giles R, Pepper MS. Adipogenesis: a complex interplay of multiple molecular determinants and pathways. *International Journal of Molecular Sciences.* 2020 Jun 16; 21(12):4283.

[57] Rozengurt E, Sinnett-Smith J, Kisfalvi K. Crosstalk between insulin/IGF-1 and GPCR signaling systems: a novel target for the anti-diabetic drug metformin in pancreatic cancer. Clinical cancer research: *An official journal of the American Association for Cancer Research.* 2010 May 5; 16(9):2505.

[58] Boucher J, Kleinridders A, Kahn CR. Insulin receptor signaling in normal and insulin-resistant states. *Cold Spring Harbor Perspectives in Biology.* 2014 Jan 1; 6(1):a009191.

[59] De Vallance E, Li Y, Jurczak MJ, Cifuentes-Pagano E, Pagano PJ. The role of NADPH oxidases in the etiology of obesity and metabolic syndrome: contribution of individual isoforms and cell biology. *Antioxidants & Redox Signaling.* 2019 Oct 1; 31(10):687-709.

[60] Chen F, Haigh S, Barman S, Fulton DJ. From form to function: the role of Nox4 in the cardiovascular system. *Frontiers in Physiology.* 2012 Nov 1; 3:412.

[61] Schröder K, Wandzioch K, Helmcke I, Brandes RP. Nox4 acts as a switch between differentiation and proliferation in preadipocytes. *Arteriosclerosis, Thrombosis, and Vascular Biology.* 2009 Feb 1; 29(2):239-45.

[62] Fritsche L, Neukamm SS, Lehmann R, Kremmer E, Hennige AM, Hunder-Gugel A, Schenk M, Häring HU, Schleicher ED, Weigert C. Insulin-induced serine phosphorylation of IRS-2 via ERK1/2 and mTOR: studies on the function of Ser675 and Ser907. *American Journal of Physiology-Endocrinology and Metabolism.* 2011 May; 300(5):E824-36.

[63] Duncan RE, Ahmadian M, Jaworski K, Sarkadi-Nagy E, Sul HS. Regulation of lipolysis in adipocytes. *Annul Revew of Nutrition.* 2007 Aug 21; 27:79-101.

[64] Ahmadian M, Duncan RE, Jaworski K, Sarkadi-Nagy E, SookSul H. Triacylglycerol metabolism in adipose tissue. *Future Lipidology.* 2007 Apr 1; 2(2):229-37.

[65] Abou-Rjeileh U, Contreras GA. Redox regulation of lipid mobilization in adipose tissues. *Antioxidants.* 2021 Jul 7; 10(7):1090.

[66] Yang A, Mottillo EP. Adipocyte lipolysis: from molecular mechanisms of regulation to disease and therapeutics. *Biochemical Journal.* 2020 Mar 13; 477(5):985-1008.

[67] Chaves VE, Frasson D, Kawashita NH. Several agents and pathways regulate lipolysis in adipocytes. *Biochimie.* 2011 Oct 1; 93(10):1631-40.

[68] McKernan K, Varghese M, Patel R, Singer K. Role of TLR4 in the induction of inflammatory changes in adipocytes and macrophages. *Adipocyte.* 2020 Jan 1; 9(1):212-22.

[69] Li Y, Li Z, Ngandiri DA, Llerins Perez M, Wolf A, Wang Y. The molecular brakes of adipose tissue lipolysis. *Frontiers in Physiology*. 2022 Feb 24; 13:826314.

[70] Fricke K, Heitland A, Maronde E. Cooperative activation of lipolysis by protein kinase A and protein kinase C pathways in 3T3-L1 adipocytes. *Endocrinology.* 2004 Nov 1; 145(11):4940-7.

[71] Persad S, Panagia V, Dhalla NS. Role of H 2 O 2 in changing β-adrenoceptor and adenylyl cyclase in ischemia-reperfused hearts. *Molecular and Cellular Biochemistry.* 1998 Sep; 186:99-106.

[72] Little SA, de Haen C. Effects of hydrogen peroxide on basal and hormone-stimulated lipolysis in perifused rat fat cells in relation to the mechanism of action of insulin. *Journal of Biological Chemistry.* 1980 Nov 25; 255(22):10888-95.

[73] Ringvold HC, Khalil RA. Protein kinase C as regulator of vascular smooth muscle function and potential target in vascular disorders. *Advances in Pharmacology.* 2017 Jan 1; 78:203-301.

[74] Siflinger-Birnboim A, Goligorsky MS, Del Vecchio PJ, Malik AB. Activation of protein kinase C pathway contributes to hydrogen peroxide-induced increase in endothelial permeability. Laboratory Investigation; *Journal of Technical Methods and Pathology.* 1992 Jul 1; 67(1):24-30.

[75] Czech MP, Tencerova M, Pedersen DJ, Aouadi M. Insulin signalling mechanisms for triacylglycerol storage. *Diabetologia.* 2013 May; 56:949-64.

[76] Ameer F, Scandiuzzi L, Hasnain S, Kalbacher H, Zaidi N. De novo lipogenesis in health and disease. *Metabolism.* 2014 Jul 1; 63(7):895-902.

[77] Oppi-Williams C, Suagee JK, Corl BA. Regulation of lipid synthesis by liver X receptor α and sterol regulatory element-binding protein 1 in mammary epithelial cells. *Journal of Dairy Science.* 2013 Jan 1; 96(1):112-21.

[78] Conte G, Mele M, Chessa S, Castiglioni B, Serra A, Pagnacco G, Secchiari P. Diacylglycerol-acyltransferase 1, stearoyl-CoA desaturase 1, and sterol regulatory element binding protein 1 gene polymorphisms and milk fatty acid composition in Italian Brown cattle. *Journal of Dairy Science.* 2010 Feb 1; 93(2):753-63.

[79] Abou-Rjeileh U, Contreras GA. Redox regulation of lipid mobilization in adipose tissues. *Antioxidants.* 2021 Jul 7; 10(7):1090.

[80] Okuno Y, Fukuhara A, Hashimoto E, Kobayashi H, Kobayashi S, Otsuki M, Shimomura I. Oxidative stress inhibits healthy adipose expansion through suppression of SREBF1-mediated lipogenic pathway. *Diabetes.* 2018 Jun 1; 67(6):1113-27.

[81] Latorre J, Ortega F, Oliveras-Cañellas N, Comas F, Lluch A, Gavaldà-Navarro A, Morón-Ros S, Ricart W, Villarroya F, Giralt M, Fernández-Real JM. Specific adipose tissue Lbp gene knockdown prevents diet-induced body weight gain, impacting fat accretion-related gene and protein expression. *Molecular Therapy-Nucleic Acids.* 2022 Mar 8; 27:870-9.

[82] Crewe C, Zhu Y, Paschoal VA, Joffin N, Ghaben AL, Gordillo R, Oh DY, Liang G, Horton JD, Scherer PE. SREBP-regulated adipocyte lipogenesis is dependent on substrate availability and redox modulation of mTORC1. *JCI Insight.* 2019 Jul 16; 5(15):e129397.

[83] Tangvarasittichai S. Oxidative stress, insulin resistance, dyslipidemia and type 2 diabetes mellitus. *World Journal of Diabetes.* 2015 Apr 15; 6(3):456-80.

[84] Guo C, Sun L, Chen X, Zhang D. Oxidative stress, mitochondrial damage and neurodegenerative diseases. *Neural Regeneration Research.* 2013 Jul 25; 8(21):2003-14.

[85] Cao SS, Kaufman RJ. Endoplasmic reticulum stress and oxidative stress in cell fate decision and human disease. *Antioxidants &Redox Signaling.* 2014 Jul 20; 21(3):396-413.

[86] Eguchi N, Vaziri ND, Dafoe DC, Ichii H. The role of oxidative stress in pancreatic β cell dysfunction in diabetes. *International Journal of Molecular Sciences.* 2021 Feb 3; 22(4):1509.

[87] Cerf ME. Beta cell dysfunction and insulin resistance. *Frontiers in Edocrinology.* 2013 Mar 27; 4:37.

[88] Volpe CM, Villar-Delfino PH, Dos Anjos PM, Nogueira-Machado JA. Cellular death, reactive oxygen species (ROS) and diabetic complications. *Cell Death &Disease.* 2018 Jan 25; 9(2):119.

[89] Zou Z, Tao T, Li H, Zhu X. mTORsignaling pathway and mTOR inhibitors in cancer: Progress and challenges. *Cell & Bioscience.* 2020 Dec; 10(1):1-1.

[90] Meng Y, Wang W, Kang J, Wang X, Sun L. Role of the PI3K/AKT signalling pathway in apoptotic cell death in the cerebral cortex of streptozotocin-induced diabetic rats. *Experimental and Therapeutic Medicine*. 2017 May 1; 13(5):2417-22.

[91] Kaleli HN, Ozer E, Kaya VO, Kutlu O. Protein kinase C isozymes and autophagy during neurodegenerative disease progression. *Cells.* 2020 Feb 27; 9(3):553.

[92] Basu A. Regulation of autophagy by protein kinase C-ε in breast cancer cells. *International Journal of Molecular Sciences.* 2020 Jan; 21(12):4247.

[93] Nawaz MI, Abouammoh M, Khan HA, Alhomida AS, Alfaran MF, Ola MS. Novel drugs and their targets in the potential treatment of diabetic retinopathy. *Medical science monitor: International Medical Journal of Experimental and Clinical Research.* 2013; 19:300.

[94] Srivastava SK, Ramana KV, Bhatnagar A. Role of aldose reductase and oxidative damage in diabetes and the consequent potential for therapeutic options. *Endocrine Reviews.* 2005 May 1; 26(3):380-92.

[95] Mochly-Rosen D, Das K, Grimes KV. Protein kinase C, an elusive therapeutic target? *Nature reviews Drug Discovery.* 2012 Dec; 11(12):937-57.

[96] Mehta NN, Sheetz M, Price K, Comiskey L, Amrutia S, Iqbal N, Mohler ER, Reilly MP. Selective PKC beta inhibition with ruboxistaurin and endothelial function in type-2 diabetes mellitus. *Cardiovascular Drugs and Therapy.* 2009 Feb; 23:17-24.

[97] Erlich DR, Slawson DC, Shaughnessy A. Diabetes update: new drugs to manage type 2 diabetes. *FP Essentials.* 2013 May 1; 408:20-4.

[98] Hundal RS, Inzucchi SE. Metformin: new understandings, new uses. *Drugs.* 2003 Sep; 63:1879-94.

[99] Bhatti JS, Sehrawat A, Mishra J, Sidhu IS, Navik U, Khullar N, Kumar S, Bhatti GK, Reddy PH. Oxidative stress in the pathophysiology of type 2 diabetes and related complications: Current therapeutics strategies and future perspectives. *Free Radical Biology and Medicine.* 2022 Apr 7.

[100] Ordelheide AM, Hrabě de Angelis M, Häring HU, Staiger H. Pharmacogenetics of oral antidiabetic therapy. *Pharmacogenomics.* 2018 Apr; 19(6):577-87.

[101] Panigrahy SK, Bhatt R, Kumar A. Reactive oxygen species: sources, consequences and targeted therapy in type 2 diabetes. *Journal of Drug Targeting.* 2017 Feb 7; 25(2):93-101.

[102] Srividya AR, Dhanabal SP, Satish Kumar MN, Parth Kumar HB. Antioxidant and antidiabetic activity of Alpiniagalanga. *International Journal of Pharmacognosy and Phytochemical Research.* 2010; 3(1):6-12.

[103] Xiang YZ, Shang HC, Gao XM, Zhang BL. A comparison of the ancient use of ginseng in traditional Chinese medicine with modern pharmacological experiments and clinical trials. *Phytotherapy Research.* 2008 Jul; 22(7):851-8.

[104] Liu WH, Hei ZQ, Nie H, Tang FT, Huang HQ, Li XJ, Deng YH, Chen SR, Guo FF, Huang WG, Chen FY. Berberine ameliorates renal injury in streptozotocin-induced diabetic rats by suppression of both oxidative stress and aldose reductase. *Chinese Medical Journal.* 2008 Apr 20; 121(08):706-12.

[105] Annida B, Prince PS. Supplementation of fenugreek leaves reduces oxidative stress in streptozotocin-induced diabetic rats. *Journal of Medicinal Food.* 2005 Sep 1;8(3):382-5.

[106] Bhatti GK, Reddy AP, Reddy PH, Bhatti JS. Lifestyle modifications and nutritional interventions in aging-associated cognitive decline and Alzheimer's disease. *Frontiers in Aging Neuroscience.* 2020 Jan 10; 11:369.

[107] Huang XL, Pan JH, Chen D, Chen J, Chen F, Hu TT. Efficacy of lifestyle interventions in patients with type 2 diabetes: a systematic review and meta-analysis. *European Journal of Internal Medicine.* 2016 Jan 1; 27:37-47.

[108] Minor RK, Allard JS, Younts CM, Ward TM, de Cabo R. Dietary interventions to extend life span and health span based on calorie restriction. *Journals of Gerontology Series A: Biomedical Sciences and Medical Sciences.* 2010 Jul 1; 65(7):695-703.

# Chapter 16

# Physiological and Immunoregulatory Roles of Reactive Oxygen Species Mediated Signalling Pathways in Lymphatic Filariasis

**Pritha Chakraborty**[1]
**Nabarun Chandra Das**[1]
**and Suprabhat Mukherjee**[1*]
[1]Integrative Biochemistry and Immunology Laboratory, Department of Animal Science, Kazi Nazrul University, Asansol, West Bengal, India

## Abstract

Reactive oxygen species (ROS) are produced as by-products of aerobic energy metabolism for both the host and the filarial nematodes. Host immune system triggers ROS generation as a "weapon" against the parasite and mediates intracellular signalling cascades to mitigate immunopathological consequences of lymphatic filariasis (LF). With onset of LF, higher concentration of host ROS targets the pathogen and disrupts the normal redox biology of the cells, hence causing oxidative stress. It triggers the TLRs, RLRs and NLRs towards inflammatory responses causing immunopathology of LF. In India, LF poses a myriad threat to the mankind. It blocks the lymphatic system resulting in clinical manifestations including lymphedema, lymphangitis and elephantiasis. However, parasites can endure the higher levels of ROS to reside within the host for years using filarial antioxidants, namely, glutathione S-transferase (GST), thioredoxin reductase (TrxR) etc. and have crosstalk with the host immune cells including NF-κB axis to escape the host immune surveillance. Beside ROS, other reactive signalling molecules such as NO interacts with macrophages and releases arginase-1 necessary for release of anti-inflammatory cytokines and tissue remodelling. Taking this as a clue targeting the redox regulatory system can be an alternative therapeutic approach for intervention of LF. In this chapter, we are discussing about the different ROS mediated signalling cascade that have physiological and immunoregulatory impact on LF. Herein, we also emphasize the importance of regulatory redox system in developing therapeutics to target the lymph-dwelling parasites.

* Corresponding Author's E-mails: babaimbc@gmail.com; suprabhat.mukherjee@knu.ac.in.

In: The Role of Reactive Oxygen Species in Health and Disease
Editor: Kuladip Jana
ISBN: 979-8-89113-283-2

**Keywords:** Redox biology, lymphatic filariasis, *W. bancrofti*, glutathione S-transferase (GST), thioredoxin reductase (TrxR)

## Abbreviations

| | |
|---|---|
| ROS: | Reactive oxygen species |
| LF: | Lymphatic filariasis |
| TLR: | Toll-like receptors; |
| GST: | glutathione S-transferase |
| TrxR: | thioredoxin reductase |
| NO: | nitric oxide |
| WHO: | World Health Organization |
| MDA: | mass drug administration |
| RS: | reactive species |
| GR: | glutathione reductase |
| Cys: | cysteine |
| SOD: | superoxide dismutases |
| CED: | cell death abnormal |
| EGL-1: | egg-laying defective |
| $O2^-$: | superoxide anion |
| $H_2O_2$: | hydrogen peroxide |
| $OH^-$: | hydroxyl radical |
| $^1O_2$: | singlet oxygen |
| OXPHOS: | oxidative phosphorylation |
| Mn-SOD: | manganese superoxide dismutase |
| Cu/Zn-SOD: | Cu/Zn-superoxide dismutase |
| GPx: | glutathione peroxidase |
| RNS: | Reactive nitrogen species |
| PARP: | poly (ADP-ribose) polymerase |
| FoxO3: | forkhead box O-3 |
| HIF-1α: | hypoxia-inducible factor-1α |
| NRF2: | nuclear factor erythroid 2–related factor 2 |
| PI3K: | phosphatidylinositol 3 kinase |
| TOR: | target of rapamycin |
| ATG4: | Autophagy-regulating protease |
| FAD: | Flavin adenine dinucleotide |
| NADPH: | Nicotinamide adenine dinucleotide phosphate |
| TLR4: | Toll like receptor-4 |
| NF-κβ: | Nuclear factor kappa-light-chain-enhancer of activated B cells |
| WbGST: | *W bancrofti* GST |
| WbTRX: | *B. malayi* have thioredoxin |
| IFN: | interferon |
| MHC: | Major histocompatibility complex |
| BmTPX: | *B. malayi* thioredoxin peroxidase |

## 1. Introduction

Lymphatic filariasis (LF) has been known for its leading cause of parasitic disabilities affecting millions of individuals globally. It is caused by the parasites *Wuchereria bancrofti*, *Brugia malayi*, and *Brugia timori*. India among several other tropical and subtropical countries shows a prevalence of bancroftian filariasis that affects over 120 million people globally [1]. Following similar life cycles, all three parasites are transmitted by the bites of the infective mosquitoes. The adult worms live in the lymphatic vessels while the microfilariae circulate in the peripheral blood. LF is responsible for considerable suffering, deformity, and disability but is usually not fatal [2]. According to WHO, 863 million people in 47 countries are still threatened out of which 90% of the total LF cases are caused by *W. bancrofti,* and the rest are caused by the Brugian parasites [3, 4]. Infections with these nematodes are associated with chronic disease [5]; however, a proportion of individuals develop severe immunopathologies like lymphostasis, lymphedema, lymphangitis, hydrocele, chyluria, and finally elephantiasis [6]. Secondary infections may also be caused by bacteria, fungi, and protozoa which cause organ dysfunction [7]. Pathology of filariasis is due to host-parasite interactions and is associated with various inflammatory conditions [8], characterized by high levels of circulating immune complexes and increased production of proinflammatory cytokines [9] predominantly by macrophages [8]. Amongst different cells of the immune system, macrophages have been recognized as the most important cells associated with filarial infection [7].

Bancroftian filariasis impairs 36 million individuals with chronic manifestations as per the reports of WHO 2022 [3]. In India, National Vector Borne Disease Control Programme (Directorate General of Health Services, Ministry of Health & Family Welfare) have reported about 21 states being affected with cases of LF among which, Maharashtra, Orissa, Bihar, Tamil Nadu and Uttar Pradesh have the highest numbers. The Government of India have a signatory to eliminate LF from India and in pursuit of their goals, they have started mass drug administration (MDA) nationwide as preventive medication under the scheme of Hathipaon Mukt Bharat [10].

Antihelmintics like Albendazole, Ivermectin, Diethylcarbamazine are currently administered to the individuals affected by LF for the management of morbidity. Long-term treatment regimen follows with limited efficacies on adult filarids due to constant repetitive doses and prolonged treatment duration. Among other side effects, the emergence of resistance to the drugs has been posing a serious threat to the available antihelmintics since the possibility of alteration in the mode of pathogenesis gives rise to an alarming situation. Invasion of parasites poses a serious threat in the form of oxidative challenge. Parasites have developed their own survival mechanism by enduring higher levels of reactive oxygen species (ROS) to escape the oxidative stress posed by the host. It is a multi-component system containing functionally active diverse biomolecules. Therefore, exploring the redox biology of LF and targeting the highly versatile antioxidant machinery is the need of the hour in developing an alternative strategy to eliminate both microfilariae and adult filarids. In this context, herein we will be discussing different immunoregulatory roles of ROS-mediated signalling cascades that can be explored to develop a therapeutic target againstLF .

## 2. Concept of ROS

Host cell produces different reactive species (RS) which are direct oxidants having oxygen or oxygen like electronegative elements with highly active moieties during cellular metabolism and pathological conditions [11]. Free radicals are independent species consisting of unpaired electrons that are generally unstable but chemically reactive [12]. It holds the potential to oxidize cellular biomolecules including proteins and nucleic acids [13]. Out of these free radicals, some are superoxide radical, hydroxyl radical and some are non-radical like hydrogen peroxide as listed in Table 1 [14-27].

**Table 1.** Free radicals and oxidants present within the cell

| Sl. No | Reactive Oxidants | Examples |
|---|---|---|
| 1 | Reactive oxygen species (ROS) | Superoxide ($O_2^{\cdot-}$) |
| | | Hydroxyl radical ($OH^{\cdot}$) |
| | | Ozone ($O_3$) |
| | | Singlet oxygen ($^1O_2$) |
| | | Hydrogen peroxide ($H_2O_2$) |
| | | Alkoxyl radical ($RO^{\cdot}$) |
| | | Peroxyl radical ($RO2^{\cdot}$) |
| | | Hydroperoxyl radical ($HO_2^{\cdot}$) |
| | | Lipid alkoxyl ($LO^{\cdot}$) |
| | | Lipid peroxide (LOOH) |
| 2 | Reactive nitrogen species (RNS) | Nitric oxide($NO^{\cdot}$) |
| | | Nitrosoniumcation ($NO^+$) |
| | | Nitrous acid ($HNO_2$) |
| | | Nitrogen dioxide ($NO_2$) |
| | | Nitrosyl anion ($NO^-$) |
| | | Peroxynitrite ($ONOO^-$) |
| | | Alkyl peroxynitrite (ROONO) |
| 3 | Reactive chlorine species (RCS) | Nitryl chloride ($NO_2Cl$) |
| | | Hypochlorite ion ($OCl^-$) |
| 4 | Reactive sulphur species (RSS) | Thiyl radical ($R\text{-}S^{\cdot}$) |
| | | Perthiyl radical ($RSS^{\cdot}$) |

ROS or Reactive Oxygen Species is now used as an umbrella term since in medicine and biology, there are different types of "reactive species" which are named as per the nature of the reactive atom. By the action of oxidation and reduction, also known as redox reaction, oxygen nitrogen and/or sulphur is released which forms the reactive oxygen species, reactive nitrogen species and reactive sulphur species respectively [28]. There are various sources through which ROS is generated at different physiological concentrations. In human cells, there are various enzymatic sources to generate ROS, mediating through thiol-based modification of target proteins [29, 30]. Under normal physiological conditions, molecular oxygen is partially reduced and thus ROS in the form of superoxide anion ($O_2^{\cdot-}$), hydrogen peroxide ($H_2O_2$) hydroxyl radical ($OH^-$) and singlet oxygen ($^1O_2$) are generated as a product of different enzymatic reactions [31]. Reduced form of oxygen gives rise to superoxide which gets dismutated into hydrogen peroxide and can readily penetrate the cell membrane to generate hydroxyl radical via Fenton's reaction [32, 33].

$$(H_2O_2 + Fe^{3+} \rightarrow Fe^{3+} + OH^- + OH^{\cdot})$$

Lower levels of ROS production are an essential step to maintain physiological functions which include signal transduction, cell proliferation, host defence and gene expression [32, 34]. There is a cellular balance between ROS generation and clearance. It is mainly generated via the mitochondrial electron transport chain through a single-electron leak at respiratory complexes I and III of the oxidative phosphorylation (OXPHOS) pathway [34, 35]. Different types of intracellular antioxidant enzymes are present showing antioxidative defence mechanisms viz manganese superoxide dismutase (Mn-SOD) that are found in mitochondrial matrix; Cu/Zn-superoxide dismutase (Cu/Zn-SOD) found in cytosol, catalase, glutathione reductase (GR) and glutathione peroxidase (GPx). Oxygen and hydrogen peroxide is dismuted by the SODs while the GPx and catalase convert the hydrogen peroxide into $H_2O$ and $O_2$ [11, 36].

## 3. Filarial Redox Regulatory System

Counteracting the oxidative stress, and developing therapeutic strategies, the need of the hour is to understand the redox biology of filarial nematodes. Filarial nematodes are well known to survive within the host over a considerably longer period of time hereby establishing chronic inflammation and modulating the host immune body [37]. The question arises how the parasite can survive over many years without getting eliminated from the host body? The answer to this question lies in the redox regulatory circuit developed by the parasite which they have adapted for long-term survivability inside the host body and handle the ROS challenges. There are different enzymatic and non-enzymatic components regulating the redox circuit as discussed in this section.

Free radicals and radical-free systems, maintain the redox system and balance the oxidative stress. Different protein complexes are formed by thiol-oxidoreductases containing cysteine (Cys) [38]. For filarial parasites, Thioredoxin (Trx) and Glutathione (GSH) are the detoxifying systems responsible for counteracting the effect of ROS generated within the host. Trx are small ubiquitously formed proteins that are critical for redox regulation of protein functioning [39]. It carries electrons to antioxidant enzymes and biosynthetic catalytic cycles. Furthermore, cytosolic proteins are protected via the formation of intra-molecular disulphides [40]. Trx can potentiate cytokine release, regulates DNA synthesis, and cell growth, inhibit the actions of gene transcription, and induce apoptosis [41–44]. Also, Trx can control the DNA binding activity of transcription factors like NF-κB and AP-1 [45]. Glutathione S-transferase (GST) is identified as the most effective antioxidant containing glycine, cysteine, and glutamic acid [46, 47]. It is a sulfhydryl water-soluble antioxidant and is ubiquitously synthesized within different cells of the living system [48]. It holds the potential to remove the ROS, RNS, toxic metabolites, and electrophiles that are generated in the host body and restore homeostasis. Also, Morris et al., 2014 have also stated that GSH can detoxify xenobiotics and, hence can be a drug and/or vaccine candidate against drug resistance and pathology [49]. Depletion of total cell glutathione in a healthy cell can trigger apoptosis and results in cell death [48, 50].

Filarial nematodes have a thiol-dependent redox system involving both GSH and Trx. Studies conducted by Kunchithapautham et al., 2003 have shown the presence of Trx within *B.*

*malayi* as well as other parasites in abundance in all the life stages [39, 51–53]. As per his studies, Trx reduces the presence of toxic hydroxyl radicals [39]. Following the experiment, it was postulated that these antioxidant systems play an important role in the immune evasion of filarial nematodes and prolong survivability [54]. Bhargava et al., 1983 have reported reduced levels of endogenous GST among adult filaria resulting in a drop in microfilarial output and death of the parasites suggestive of interference of GST metabolism within the filarial parasites [55]. Sommer et al., 2001 have implicated that sigma-class GST released by *O. volvulus* acts as prostaglandin D2 synthase and is involved in biotransformation and biosynthesis of drosopterin [56]. Also, omega-class GSTs of *O. volvulus* are involved in metabolism, repair, and detoxification of the filarial parasites [53, 57, 58]. Apart from these, filarial nematodes produces GPx, superoxide dismutases (SODs) to protect themselves from the generated ROS and toxic contents released by the immune cells for the elimination of the parasites. GPx acts as hydroperoxidase and is abundantly found in adults. It holds the potential to remove host immunoregulatory lipids and protect the adults from the generated ROS [52]. SODs are involved in catalysis of superoxide anion into hydrogen peroxide. They are also known as metalloenzymes which protect the parasites from endogenous oxidative stress [59]. Studies have reported Cu/Zn SODs have direct role in the production of $H_2O_2$ and stimulation of cell proliferation and angiogenesis [53].

## 4. Host Regulated Anti-Filarial Redox Signalling Pathways

Generally, ROS are considered as one of the host defence molecules that trigger cellular mechanism for destructing pathogens including filarids. Overproduction of ROS in the presence of invading pathogen, specifically filarial parasites increase oxidative stress, resulting in ROS-mediated killing of the parasites [60–67]. Recent studies and their mechanistic explorations figured out two ROS-induced pathways, apoptosis, and autophagy, highly responsible for parasitic death. Among apoptosis, the cell death abnormal (CED) protein-mediated route is highly specific for nematodes. H. R. Horvitz and his colleagues first introduced the egg-laying defective (EGL-1) and CED-3, CED-4 and CED-9 proteins with their contributions to mediating apoptosis [68, 69]. ROS with their inductive properties signal increased expressions of pro-apoptotic proteins EGL-1, CED-3, and CED-4 while an excess amount of ROS may downregulate expressions of anti-apoptotic protein CED-9 [60, 61, 63, 70]. In normal homeostat conditions CED-9 blocks activity of CED-4 and CED-3. In an altered situation, oxidative stress directs EGL-1 to regulate the activity of CED-9 and allows CED-4 to activate CED-3, the ultimate contributor that executes apoptosis. Evidences from last few years expresses that beside CED proteins, the involvement of specific mammalian-specific Caspase proteins is also highly influential in mediating ROS-generated apoptosis in filarial parasites [54, 61–64, 66, 67, 71]. Extrinsic as well as intrinsic, routes of cell death have been found in most studies. Caspase, is a cysteine-aspartic proteases family protein that normally resides in pro-inactive forms. In oxidative stress, ROS initiates p53 expression by inducing DNA damage and signals mitochondria to release cytochrome c, leading to the activation of Caspase proteins by cleaving at a specific sequence next to aspartate [54, 72]. Similarly, endoplasmic reticulum (ER) stress regulates the activation of caspase proteins and ultimately promotes apoptosis in filarial parasites [73]. Cleaving and activation of procaspase-8 in extrinsic and procaspase-9 in intrinsic pathways are the most crucial steps in initiating apoptosis. Both the active mediator Caspase-8/9 then interacts with procaspase-3 to finally target poly (ADP-ribose) polymerase (PARP) and inhibit the filarial DNA repair mechanism [54, 60-64, 66, 67]. A schematic representation is depicted in Figure 1.

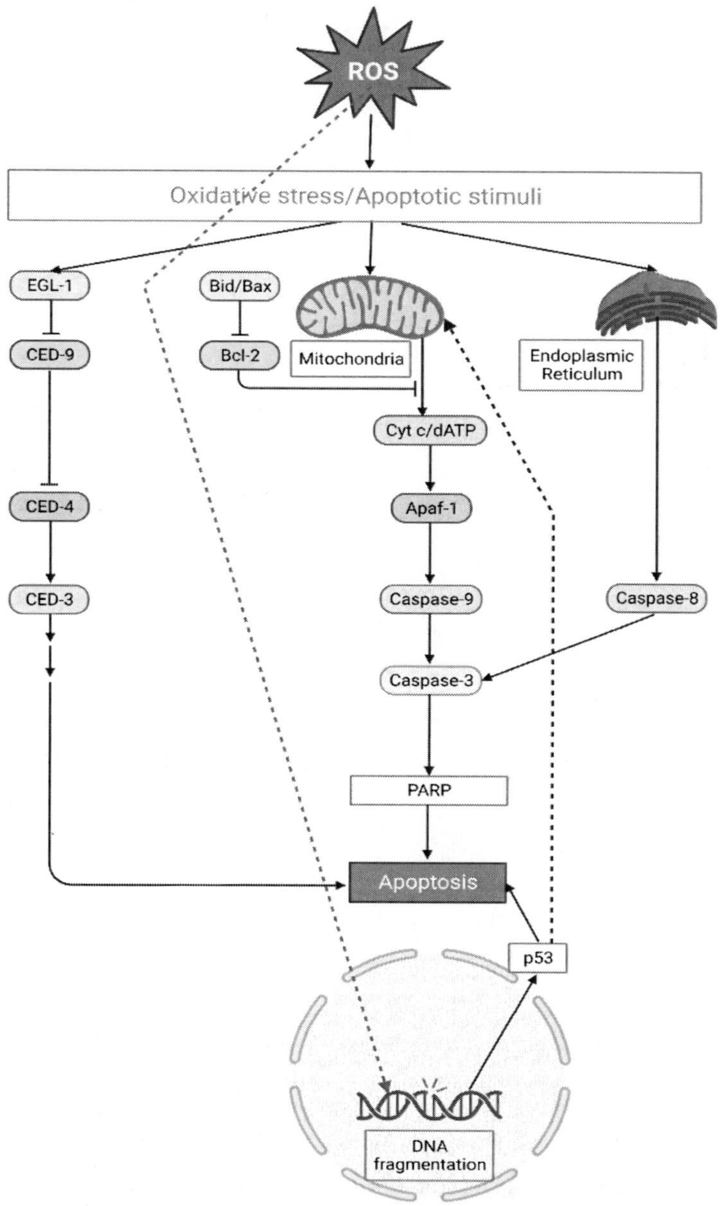


**Figure1.** ROS triggered, signalling cascade of cell death pathways in filarial parasites. Host body in the presence of filarial pathogen generates ROS, which in turn signals to oxidative stress, In the stressed condition nematode specific cell death pathway, i.e., EGL-1/CED-9/CED-4/CED-3 pathway are activated within the parasites and promote apoptotic death. Evidence from current studies demonstrated ROS has the ability to initiate mammalian-specific extrinsic pathway (Caspase-9/Caspase-3/PARP) of cell death in filarial parasites. Moreover, ROS also initiate p53 activation by damaging the genomic DNA and triggering intrinsic pathway (Cytochrome-c/Caspase-9/Caspase-3/PARP).

Surprisingly, recent studies also discovered oxidative stress due to the excess ROS activating another death signalling, autophagy in filarial parasites. In stressed condition ROS enhance the expression of autophagy-related proteins like Beclin 1, LC3, and Sequestosome-1 by activating the forkhead box O-3 (FoxO3), hypoxia-inducible factor-1α (HIF-1α), nuclear factor erythroid 2–related factor 2 (NRF2) and p53 transcriptional factors and promotes the autophagy [74–76]. Further, by inhibiting the signaling cascade of phosphatidylinositol 3 kinases (PI3K)-Protein kinase B (AKT)- the target of rapamycin (TOR), ROS induces the autophagy [77, 78]. Similarly, ROS with its oxidizing property inactivating the Autophagy-regulating protease (ATG4) promotes the formation of autophagosome as shown in Figure 2 [79].

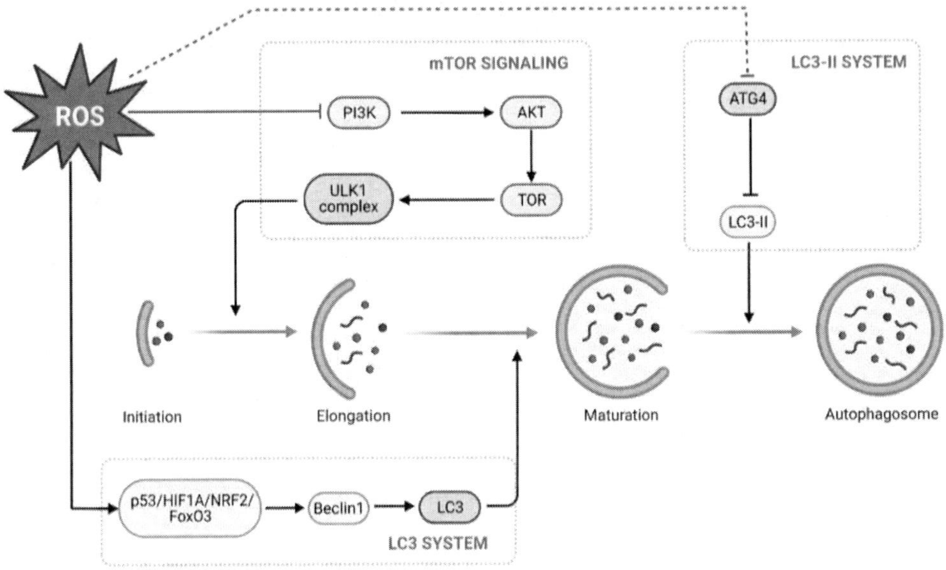


**Figure2.** Autophagy in filarial parasites can be triggered by the induction of ROS. Initially, ROS blocks the activity of PI3K and activates the TOR signalling pathway to begin the initiation of autophagosome formation. Later by inducing transcription factors, like FoxO3, HIF-1α, NRF2 and p53, ROS drives the expression of the Beclin 1 and LC3 and promotes the maturation process. ROS also oxidizes and inactivates ATG4 to allow LC3-II to continue the autophagosome formation.

## 5. Filarial Antioxidants and Their Cross-Talk with the host Redox System

Traditionally, these antioxidants maintain the redox homeostasis of biological systems. Herein, filarial parasites for their prolonged survival express these antioxidants against the host redox system and evade the ROS mediated immune system resulting in chronic filariasis.

GST is considered as both a prokaryotic and eukaryotic metabolic enzyme, having crucial role in detoxification. Results from different studies concluded that functionally filarial GST and mammalian GST are same. They follow a similar pattern in counteracting oxidative stress but differ in structural homology [80]. Recently, Ahmad et al. (2008) isolated and characterized GST from bovine filarial parasite, *Setaria cervi* which is considered a homolog of *W. bancrofti* [81]. GST provides extreme protection to filarial parasites not only from external hazards like free radicals posted by the host, granule contents of immune cells, and several therapeutic effects, also from endogenous toxins and xenotoxins like, malondialdehyde [82]. These multi-functional dimeric proteins also have heme-binding properties. In particular, Nu-class GST with its ligandin activity detoxify high lethal concentration of heme and promote larval development as well as survival of the parasites [83]. Most surprisingly, GST also has a unique property to accumulate resistance against therapeutic drugs and antibiotics [84, 85].

TrxR is a reductase enzyme, that plays a crucial role in inhibiting ROS-induced apoptosis and proves its importance in the prolonged survival of the parasites within the host. A study of Joardar et al. (2017) recently exhibited the presence of TrxR throughout the life cycle of *S. cervi* [86]. Furthermore, studies also revealed in comparison to human TrxR from brugian parasites and onchocerciasis-causing parasites have similar activity against filarial thioredoxins [87, 88]. Isolation followed by characterization explored each monomer contains a Flavin adenine dinucleotide (FAD) group, Nicotinamide adenine dinucleotide phosphate (NADPH) domain, and a disulfide active site [89]. Filarial TrxR is crucial in inhibiting toll-like receptor 4 - Nuclear factor kappa-light-chain-enhancer of activated B cells (TLR4-NF-κB) signalling and downregulates inflammation due to the active infection [86, 90]. Besides TrxR, GR is also an important disulfide reductase, plays a crucial role in maintaining intracellular redox balance. Isolation and sequencing of the GR from *Caenorhabditis. elegans*, *O. volvulus*, and *S. digitata* concluded filarial GR is highly similar to the mammalian GR [91]. GR in filarial infection maintains a high reduced glutathione/oxidized glutathione (GSH/GSSG) ratio, providing protection against oxidative stress [92].

Another potential filarial antioxidant SOD, has profound role in lowering ROS mediating the killing of the parasites. Studies revealed SOD is very common in filarial parasites and resides in both larval and adult forms of the *S. cervi*, *Litomosoides carinii*, *Dirofilaria immitis* and *B. malayi* [54]. Filarial SODs exist in two forms, Mn SODs localized only in mitochondria, and CuZn SODs, found in peroxisomes and cytosol. The catalytic activity of these enzymes protects filarial cells by converting $O_2^-$ to $H_2O_2$ [93].

Thiol-specific filarial antioxidant TPx and GPx both have enormous biological activity to protect microfilariae as well as the adults from the free radical of oxygen [94]. According to recent studies, GPx is localized in *L. carinii* and *S. cervi* whereas, *B. malayi*, *O. volvulus* and *C. elegans* are the most common for the TPx [54]. Catalase, the least antioxidant enzyme is found to be present in both the life stages of *B. malayi* [95]. This heme enzyme provides immense protection to the parasites by converting the toxic hydrogen peroxide to water and oxygen [96].

## 6. Current Trends in Therapeutic Approaches

Targeting the adult filariae can be an option to eliminate LF which can be achieved by disrupting the worm's metabolic pathways. Physiological and immunoregulatory studies over the years have given the clue to potentiate a winning strategy in developing novel anti-filarial therapy by targeting the thiol-based redox system of parasites [86, 91, 97].

## Phytocompounds Targeting Filarial Antioxidants

As earlier mentioned, parasite-generated antioxidants have the potency to combat the oxidative challenges within the host. Medicinal plants have active compounds derived from natural sources and act against infectious agents. Thus, exploring anti-filarial activities of potential phytocompounds are discussed here. Mishra et al., 2022 have reported the potential of bioactive compounds extracted from *Ocimum sanctum* which have reduced the motility of adult filarids, and postulated that due to higher amounts of oxidative stress generated by *O. sanctum* there is a significant motility reduction in the filarial parasites [98]. Furthermore, Ray et al., 2018 have reported potential antifilarial activity induced by polyphenol-enriched ethanolic extract of *Cajanu scarabaeoides* (L.) Thouars. As per the reports, ethanolic extract has posed lethality against the oocytes, mf as well as the adult stage bovine parasite is considered as WHO recommended model of *W. bancrofti* [99]. Saini et al., 2013 used an n-butanol extract of *Diospyros perigrena* and reported decreased levels of GSH in the adult *S. cervi* at different dose levels suggestive of apoptosis induced due to lower levels of intracellular GSH [100]. Aqueous extracts of *Azadirachtin indica* have indicated an alteration in internal redox system in the parasites with the significant increase in SOD levels. Furthermore, alternation in the GSH levels indicates misbalance in the primary internal redox regulator has been reported leading to reduced motility in parasites [101].

## Immunotherapeutic Approaches to Target Filarial Redox System

Apart from phytocompounds, there is another possibility to inhibit the survivability of parasites within the host by developing immunization or vaccination [80]. Other than filarial parasites, helminth parasites have been targeted using GSTs as vaccines [47, 80, 102]. The underlying mechanism suggests inactivation of GST enzymatic activity [103]. *In silico* studies on sequence homology, the 3D structure of *W. bancrofti* GST (WbGST) has been prepared by Nathan et al., 2005 and reported the potential new GST inhibitors for the development of drugs using molecular docking studies [104]. Also, *B. malayi* has thioredoxin (WbTRX) belonging to the TRX family which can be used as a multiple-mode vaccination with secretory capability [39]. Sampa et al., 2021 have designed a multi-epitope peptide-based vaccine against the filarial parasites viz targeting thioredoxin protein [105]. As per her reports, among different TLRs, TLR5 interacts with the vaccine candidate via TLR-mediated immune activation. TLRs can trigger several signalling cascades that initiates release to interferon (INF-$\gamma$) crucial for macrophage activation and increased expression of MHC-I and MHC-II molecules [106]. Anand et al., 2012 evaluated the protective efficiency of *B. malayi* thioredoxin peroxidase (BmTPX) in mice model in which anti-BmTPX antibodies were evaluated and higher levels of cellular proliferative responses were reported [107].

## Artificially Synthesized Bio-Active Nanoparticle as Potent Anti-Filarial Mediator

Among various research strategies developed against LF, the synthesis of bioactive nanomaterials has been an optimal choice in the nano-medicinal approach. AuNPs have given promising results in pharmaceutical applications. Roy et al., 2018 have examined biopolymer-inspired biogenic AuNPs against *W. bancrofti* and *S. cervi* and reported increase in the production of ROS in mf of *W. bancrofti* in a dose-dependent manner. H2DCFDA staining supported the increase in the production of ROS in both *W. bancrofti* and *S. cervi* [108]. Yadav et al., 2022 have reported increased levels of ROS and induced oxidative stress using green silver nanoparticles synthesized using *Andrographis paniculate* [109]. Nano-pharmaceuticals hence have proved to be useful in developing antifilarial drugs [90].

## 7. Prospects and Challenges

In LF, filarial parasites can modulate the host immune responses for their own survival [6, 7]. Although, the host can generate ROS to eliminate the parasites whereas parasites can counteract the effect of generated ROS using their own redox system as already discussed in this chapter. Herein, understanding the protective immune responses of the host will provide insights in the identification of vaccine candidate and adjuvants to induce protective responses in the host [110]. Furthermore, genome wide screening is also opted to identify different vaccine candidates [111–113]. Research has been focussing on eliminating the infective stage larvae (L3) and preventing its establishment in the host, hence studies are currently undergoing exploring the transcriptome and secretome of larval stages of filarial parasites which can be targeted to develop immunization against LF [114, 115]. Most of the investigations reported in this chapter emphasize on modification of the filarial redox regulatory pathways. Several trials have been running globally targeting TrxR system since the TrxR is present in all the stages of the filarial parasites [86]. Apart from targeting the TrxR directly, there are different signalling molecules associated with TrxR which can be altered that can reverse the current scenario and eliminate the prolonged surviving parasites. Herein, thiol-based antioxidant systems, GSH and Trx are considered as a potential target in anti-filarial research, but there are some challenges which are posing hurdles in the road to elimination of LF globally. Within the laboratory, development of prophylactic vaccine for the affected population as well as endemic population is highly expensive. Furthermore, there are still under developed knowledge about the protective immunity within the host, hence developing novel therapeutics targeting the anti-oxidant system of the filarial parasites. Availability of *W. bancrofti* and maintaining them into the laboratory is still a challenge for continuing the research work. However, developing the filarial model in rodent system has been a challenge throughout the globe. Although, *in vivo* data is available for other species as compared to *W. bancrofti*, hence, the study and real time data on bancroftian LF remains poor [54].

## Conclusion

863 million people in 47 nations still remains to be at risk from LF infection. A few million people harbour the infection without displaying any clinical symptoms, whereas about 40 million people experience the disease's chronic side effects. The parasite relies on these asymptomatic people to propagate the illness and sustain their life cycle in the environment. In order to halt the transmission cycle, WHO has recommended MDA within a community to

target the asymptomatic patients and break the lifecycle of filarial parasites. Although the goal of global elimination is still far away, there is an urgent need to develop alternative strategies due to resistance and immune evasion strategies developed by the parasites. Filarial antioxidants and/or redox regulatory proteins maintain the redox homeostasis and protect the biomolecules like protein, carbohydrates, lipids, nucleic acids, and other necessary compounds required for survival and growth from the oxidative damage of free radicals. Continuous exposure to living cells also leads to dysfunctionality and death. Therefore, with hope researchers now targeting these redox regulatory proteins to develop anti-filarial therapeutics. Indeed, several studies have experimentally shown therapeutic inhibition of redox regulatory proteins triggering oxidative stress-induced apoptosis in parasites. Last but not least, establishing a global timeframe for LF eradication is not absolutely necessary; this can be done regionally based on the needs. A very sophisticated strategy is needed to combat a vector-transmitted infection that could spread zoonotic diseases within a community. Planning and carrying out the following stages in the control of LF in endemic areas, where the virus still exists and is spreading, must take this into account.

## Acknowledgements

SM acknowledges University Grants Commission (UGC) (Ref no. F.2-12/2019(STRIDE-1) and KNU-UGC-STRIDE (Ref no. KNU/R/STRIDE/1077/21; Ref no. KNU/R/LETTER/ 1190/22) and Department of Science and Technology-Science & Engineering Research Board (DST-SERB) (Ref no.- SRG/2021/002605) for supporting his research activities and laboratory through awarding research grants. PC thanks DST, Govt. of India for the award of DST-INSPIRE fellowship (IF190998). The authors gratefully acknowledge BioRender.com used for drawing the figures.

## References

[1] Lupenza ET, Gasarasi DB, Minzi OM. Lymphatic filariasis elimination status: *Wuchereriabancrofti* infections in human populations and factors contributing to continued transmission after seven rounds of mass drug administration in Masasi District, Tanzania. *PLoS One.* 2022; 17(1):e0262693.

[2] Mukherjee S, Joardar N, Sinha Babu SP. Exploring the homolog of a novel proinflammatory microfilarial sheath protein (MfP) of *Wuchereriabancrofti* in the adult-stage bovine filarial parasite *Setariacervi.* J Helminthol. 2020 Nov 27; 94:e15.

[3] Lymphatic filariasis. [Dec 22]. Available from: https://www.who.int/news-room/fact-sheets/detail/lymphatic-filariasis

[4] Babu S, Nutman TB. Immunology of lymphatic filariasis. *Parasite Immunol.* 2014 Aug;36(8):338–46.

[5] Kwarteng A, Ahuno ST. Immunity in filarial infections: lessons from animal models and human studies. *Scand J Immunol.* 2017;85(4):251–7.

[6] Mukherjee S, Mukherjee S, Maiti TK, Bhattacharya S, Babu SPS. A novel ligand of Toll-like receptor 4 from the sheath of *Wuchereriabancrofti* Microfilaria induces proinflammatory response in macrophages. *Journal of Infectious Diseases.* 2017; 215(6):954–65.

[7] Mukherjee S, Mukherjee S, Bhattacharya S, Sinha Babu SP. Surface proteins of Setariacervi induce inflammation in macrophage through Toll-like receptor 4 (TLR4)-mediated signalling pathway. *Parasite Immunol.* 2017;39(1):e12389.

[8] Taylor MJ, Bandi C, Hoerauf AM, Lazdins J. Wolbachia bacteria of filarial nematodes: a target for control? *Parasitology today.* 2000;16(5):179–80.

[9] Babu S, Anuradha R, Kumar NP, George PJ, Kumaraswami V, Nutman TB. Filarial lymphatic pathology reflects augmented toll-like receptor-mediated, mitogen-activated protein kinase-mediated proinflammatory cytokine production. *Infect Immun.* 2011;79(11):4600–8.

[10] Filaria: *National Center for Vector Borne Diseases Control* (NCVBDC) [cited 2022 Jul 25].

[11] Nandi A, Yan LJ, Jana CK, Das N. Role of Catalase in Oxidative Stress- and Age-Associated Degenerative Diseases. *Oxid Med Cell Longev.* 2019 Nov 11; 2019:1–19.

[12] Phaniendra A, Jestadi DB, Periyasamy L. Free Radicals: Properties, Sources, Targets, and Their Implication in Various Diseases. *Indian Journal of Clinical Biochemistry.* 2015; 30(1):11–26.

[13] Guo C, Ding P, Xie C, Ye C, Ye M, Pan C, Cao X, Zhang S, Zheng S. Potential application of the oxidative nucleic acid damage biomarkers in detection of diseases. *Oncotarget.* 2017; 8(43):75767-75777.

[14] Ray PD, Huang BW, Tsuji Y. Reactive oxygen species (ROS) homeostasis and redox regulation in cellular signaling. *Cell Signal.* 2012; 24(5):981–90.

[15] Kodavanti PRS, Valdez M, Richards JE, Agina-Obu DI, Phillips PM, Jarema KA, Kodavanti UP. Ozone-induced changes in oxidative stress parameters in brain regions of adult, middle-age, and senescent Brown Norway rats. *ToxicolApplPharmacol.* 2021; 410:115351.

[16] Liang P, Kolodieznyi D, Creeger Y, Ballou B and Bruchez MP (2020) Subcellular Singlet Oxygen and Cell Death: Location Matters. *Front. Chem.* 8:592941.

[17] Riley PA. Free Radicals in Biology: Oxidative Stress and the Effects of Ionizing Radiation. *Int J Radiat Biol.* 1994; 65(1):27–33.

[18] Lu Y, Wang A, Shi P, Zhang H, Li Z. Quantum Chemical Study on the Antioxidation Mechanism of Piceatannol and Isorhapontigenin toward Hydroxyl and Hydroperoxyl Radicals. *PLoS One.* 2015; 10(7):e0133259.

[19] Barrera G. *Oxidative Stress and Lipid Peroxidation Products in Cancer Progression and Therapy.* ISRN Oncol. 2012; 2012:1–21.

[20] Violi F, Marino R, Milite MT, Loffredo L. Nitric oxide and its role in lipid peroxidation. *Diabetes Metab Res Rev.* 1999; 15(4):283–8.

[21] Martínez MC, Andriantsitohaina R. Reactive Nitrogen Species: Molecular Mechanisms and Potential Significance in Health and Disease. *Antioxid Redox Signal.* 2009; 11(3):669–702.

[22] Weidinger A, Kozlov A. Biological Activities of Reactive Oxygen and Nitrogen Species: Oxidative Stress versus Signal Transduction. *Biomolecules.* 2015; 5(2):472–84.

[23] O'Neill CA, van der Vliet A, Eiserich JP, Last JA, Halliwell B, Cross CE. Oxidative damage by ozone and nitrogen dioxide: synergistic toxicity in vivo but no evidence of synergistic oxidative damage in an extracellular fluid. *Biochemical Society Symposia.* 1995; 61:139–52.

[24] Salgo MG, Squadrito GL, Pryor WA. Peroxynitrite Causes Apoptosis in Rat Thymocytes. *BiochemBiophys Res Commun.* 1995; 215(3):1111–8.

[25] Gray MJ, Wholey WY, Jakob U. Bacterial Responses to Reactive Chlorine Species. *Annu Rev Microbiol.* 2013; 67(1):141–60.

[26] Panasenko OM, Evgina SA, Driomina ES, Sharov VS, Sergienko VI, Vladimirov YA. Hypochlorite induces lipid peroxidation in blood lipoproteins and phospholipid liposomes. *Free RadicBiol Med.* 1995; 19(2):133–40.

[27] Olson KR. Are Reactive Sulfur Species the New Reactive Oxygen Species? *Antioxid Redox Signal.* 2020 Dec 1;33(16):1125–42.

[28] Sies H, Jones DP. Reactive oxygen species (ROS) as pleiotropic physiological signalling agents. *Nat Rev Mol Cell Biol.* 202015; 21(7):363–83.

[29] Zeida A, Trujillo M, Ferrer-Sueta G, Denicola A, Estrin DA, Radi R. Catalysis of Peroxide Reduction by Fast Reacting Protein Thiols. *Chem Rev.* 2019; 119(19):10829–55.

[30] Poole LB. The basics of thiols and cysteines in redox biology and chemistry. *Free RadicBiol Med.* 2015; 80:148–57.

[31] Nita M, Grzybowski A. The Role of the Reactive Oxygen Species and Oxidative Stress in the Pathomechanism of the Age-Related Ocular Diseases and Other Pathologies of the Anterior and Posterior Eye Segments in Adults. *Oxid Med Cell Longev.* 2016; 1–23.

[32] Chen HY. Why the Reactive Oxygen Species of the Fenton Reaction Switches from Oxoiron(IV) Species to Hydroxyl Radical in Phosphate Buffer Solutions? A Computational Rationale. *ACS Omega.* 2019; 4(9):14105–13.

[33] Sun H, Zhang C, Cao S, Sheng T, Dong N, Xu Y. Fenton reactions drive nucleotide and ATP syntheses in cancer. J Mol Cell Biol. 2018; 10(5):448–59.

[34] Dröge W. Free Radicals in the Physiological Control of Cell Function. *Physiol Rev.* 2002; 82(1):47–95.

[35] Brand MD. The sites and topology of mitochondrial superoxide production. *ExpGerontol.* 2010; 45(7–8):466–72.

[36] von Ossowski I, Hausner G, Loewen PC. Molecular evolutionary analysis based on the amino acid sequence of catalase. *J MolEvol.* 1993; 37(1):71–6.

[37] Perry RN, Wharton DA. *Molecular and physiological basis of nematode survival.* CABI; 2011.

[38] Fomenko DE, Gladyshev VN. Comparative genomics of thiol oxidoreductases reveals widespread and essential functions of thiol-based redox control of cellular processes. *Antioxid Redox Signal.* 2012;16(3):193–201.

[39] Kunchithapautham K, Padmavathi B, Narayanan RB, Kaliraj P, Scott AL. Thioredoxin from Brugiamalayi: defining a 16-kilodalton class of thioredoxins from nematodes. *Infect Immun.* 2003; 71(7):4119–26.

[40] Arnér ESJ, Holmgren A. Physiological functions of thioredoxin and thioredoxin reductase. *Eur J Biochem.* 2000; 267(20):6102–9.

[41] Andrade RM, Reed SL. New drug target in protozoan parasites: the role of thioredoxin reductase. *Front Microbiol.* 2015; 6:975.

[42] Bertini R, Howard OM, Dong HF, Oppenheim JJ, Bizzarri C, Sergi R, Caselli G, Pagliei S, Romines B, Wilshire JA, Mengozzi M, Nakamura H, Yodoi J, Pekkari K, Gurunath R, Holmgren A, Herzenberg LA, Herzenberg LA, Ghezzi P. Thioredoxin, a redox enzyme released in infection and inflammation, is a unique chemoattractant for neutrophils, monocytes, and T cells. *J Exp Med.* 1999; 189(11):1783-9.

[43] Silberstein DS, McDonough S, Minkoff MS, Balcewicz-Sablinska MK. Human eosinophil cytotoxicity-enhancing factor. Eosinophil-stimulating and dithiol reductase activities of biosynthetic (recombinant) species with COOH-terminal deletions. *J Biol Chem.* 1993; 268(12):9138–42.

[44] Schenk H, Vogt M, Dröge W, Schulze-Osthoff K. Thioredoxin as a potent costimulus of cytokine expression. *J Immunol.* 1996; 156(2):765–71.

[45] Tanaka T, Nakamura H, Nishiyama A, Hosoi F, Masutani H, Wada H, Yodoi J. Redox regulation by thioredoxin superfamily; protection against oxidative stress and aging. *Free Radic Res.* 2000; 33(6):851-5.

[46] Saha SK, Roy P, Mondal MK, Roy D, Gayen P, Chowdhury P, Babu SPS. Development of chitosan based gold nanomaterial as an efficient antifilarial agent: A mechanistic approach. *CarbohydrPolym.* 2017; 157:1666-1676

[47] Gupta S, Bhandari YP, Reddy MV, Harinath BC, Rathaur S. Setariacervi: immunoprophylactic potential of glutathione-S-transferase against filarial parasite Brugiamalayi. *ExpParasitol.* 2005;109(4):252–5.

[48] Lomaestro BM, Malone M. Glutathione in health and disease: pharmacotherapeutic issues. *Annals of Pharmacotherapy.* 1995;29(12):1263–73.

[49] Morris G, Anderson G, Dean O, Berk M, Galecki P, Martin-Subero M, Maes M. The glutathione system: a new drug target in neuroimmune disorders. *MolNeurobiol.* 2014;50(3):1059-84.

[50] Duke RC, Ojcius DM, Young JDE. Cell suicide in health and disease. Sci Am. 1996;275(6):80–7.

[51] Ghosh I, Eisinger SW, Raghavan N, Scott AL. Thioredoxin peroxidases from *Brugiamalayi. MolBiochemParasitol.* 1998; 91(2):207–20.

[52] Ajonina-Ekoti I, Ndjonka D, Tanyi MK, Wilbertz M, Younis AE, Boursou D, Kurosinski MA, Eberle R, Lüersen K, Perbandt M, Breloer M, Brattig NW, Liebau E. Functional characterization and immune

recognition of the extracellular superoxide dismutase from the human pathogenic parasite Onchocerca volvulus (OvEC-SOD). *Acta Trop.* 2012; 124(1):15-26.

[53] Hotterbeekx A, Perneel J, Vieri MK, Colebunders R, Kumar-Singh S. The secretome of filarial nematodes and its role in host-parasite interactions and pathogenicity in onchocerciasis-associated epilepsy. *Front Cell Infect Microbiol.* 2021; 11:662766.

[54] Mukherjee S, Joardar N, Sinha Babu SP. Redox Regulatory Circuits as Targets for Therapeutic Intervention of Bancroftian Filariasis: Biochemical, Molecular, and Pharmacological Perspectives. In: Chakraborti S, Chakraborti T, Chattopadhyay D, Shaha C, editors. *Oxidative Stress in Microbial Diseases. Singapore*: Springer Singapore; 2019. pp. 185–208.

[55] Bhargava KK, le Trang N, Cerami A, Eaton JW. Effect of arsenical drugs on glutathione metabolism of Litomosoidescarinii. *MolBiochemParasitol.* 1983;9(1):29–35.

[56] Sommer A, Nimtz M, Conradt HS, Brattig N, Boettcher K, Fischer P, Walter RD, Liebau E. Structural analysis and antibody response to the extracellular glutathione S-transferases from Onchocerca volvulus. *Infect Immun.* 2001; 69(12):7718-28.

[57] Gupta S, Singh A, Yadav M, Singh K, Rathaur S. MALDI mass sequencing and characterization of filarialglutathione-S-transferase. *BiochemBiophys Res Commun.* 2007;356(2):381–5.

[58] Liebau E, Höppner J, Mühlmeister M, Burmeister C, Lüersen K, Perbandt M, Schmetz C, Büttner D, Brattig N. The secretory omega-class glutathione transferase OvGST3 from the human pathogenic parasite Onchocerca volvulus. *FEBS J.* 2008; 275(13):3438-53.

[59] Rathaur S, Sharma S, Singh RN, Henkle K, Selkirk ME. Antibody responses of Wuchereriabancrofti patients to recombinant *Brugiapahangi* superoxide dismutase. *Indian J Exp Biol.* 200; 39(1):35-40.

[60] Mukherjee N, Parida PK, Santra A, Ghosh T, Dutta A, Jana K, Misra AK, Sinha Babu SP. Oxidative stress plays major role in mediating apoptosis in filarial nematode Setariacervi in the presence of trans-stilbene derivatives. *Free Radic Biol Med.* 2016; 93:130-44.

[61] Saini P, Gayen P, Kumar D, Nayak A, Mukherjee N, Mukherjee S, Pal BC, Babu SP. Antifilarial effect of ursolic acid from Nyctanthesarbortristis: molecular and biochemical evidences. *Parasitol Int.* 2014;63(5):717-28.

[62] Ahmad R, Srivastava AK. Biochemical composition and metabolic pathways of filarial worms Setariacervi: search for new antifilarial agents. *J Helminthol.* 2007; 81(3):261-80.

[63] Mukherjee N, Joardar N, Babu SPS. Antifilarial activity of azadirachtin fuelled through reactive oxygen species induced apoptosis: a thorough molecular study on *Setariacervi*. *J Helminthol.* 2019;93(5):519–28.

[64] Nayak A, Gayen P, Saini P, Mukherjee N, Sinha Babu SP. Molecular evidence of curcumin-induced apoptosis in the filarial worm *Setariacervi*. *Parasitol Res.* 2012;111(3):1173–86.

[65] Mukherjee N, Saini P, Mukherjee S, Roy P, Sinha Babu SP. In vitro antifilarial activity of Azadirachta indica aqueous extract through reactive oxygen species enhancement. *Asian Pac J Trop Med.*2014; 7(11):841–8.

[66] Mukherjee S, Joardar N, Mondal S, Schiefer A, Hoerauf A, Pfarr K, et al. Quinolone-fused cyclic sulfonamide as a novel benign antifilarial agent. *Sci Rep.* 2018; 8(1):12073.

[67] Mukherjee N, Saini P, Mukherjee S, Roy P, Gayen P, Babu SPS. Ethanolic extract of Azadirachtaindica (A. Juss.) causing apoptosis by ROS upregulation in Dirofilariaimmitis microfilaria. *Res Vet Sci.* 2014;97(2):309–17.

[68] Hengartner MO, Robert Horvitz H. Programmed cell death in Caenorhabditis elegans. *Curr Opin Genet Dev.* 1994; 4(4):581–6.

[69] Ellis HM, Horvitz HR. Genetic control of programmed cell death in the nematode C. elegans. *Cell.* 1986; 44(6):817–29.

[70] Mukherjee N, Mukherjee S, Saini P, Roy P, Sinha Babu SP. Antifilarial effects of polyphenol rich ethanolic extract from the leaves of Azadirachtaindica through molecular and biochemical approaches describing reactive oxygen species (ROS) mediated apoptosis of Setariacervi. *ExpParasitol.* 2014; 136:41–58.

[71] Senathilake KS, Karunanayake EH, Samarakoon SR, Tennekoon KH, de Silva ED, Adhikari A. Oleanolic acid from antifilarialtriterpenesaponins of Dipterocarpuszeylanicus induces oxidative stress and apoptosis in filarial parasite *Setariadigitatain vitro*. *ExpParasitol.* 2017; 177:13-21.

[72] Redza-Dutordoir M, Averill-Bates DA. Activation of apoptosis signalling pathways by reactive oxygen species. *Biochimicaet Biophysica Acta (BBA) - Molecular Cell Research.* 2016; 1863(12):2977–92.

[73] Dolai S, Pal S, Yadav RK, Adak S. Endoplasmic reticulum stress-induced apoptosis in Leishmania through Ca2+-dependent and caspase-independent mechanism. *J Biol Chem.* 2011; 286(15):13638–46.

[74] Chang KC, Liu PF, Chang CH, Lin YC, Chen YJ, Shu CW. The interplay of autophagy and oxidative stress in the pathogenesis and therapy of retinal degenerative diseases. *Cell Biosci.* 2022; 12(1):1.

[75] Roberts DJ, Tan-Sah VP, Ding EY, Smith JM, Miyamoto S. Hexokinase-II Positively Regulates Glucose Starvation-Induced Autophagy through TORC1 Inhibition. *Mol Cell.* 2014; 53(4):521–33.

[76] Kimball SR, Gordon BS, Moyer JE, Dennis MD, Jefferson LS. Leucine induced dephosphorylation of Sestrin2 promotes mTORC1 activation. *Cell Signal.* 2016; 28(8):896–906.

[77] Zhang L, Wang H, Xu J, Zhu J, Ding K. Inhibition of cathepsin S induces autophagy and apoptosis in human glioblastoma cell lines through ROS-mediated PI3K/AKT/mTOR/p70S6K and JNK signaling pathways. *ToxicolLett.* 2014; 228(3):248–59.

[78] Zhang J, Kim J, Alexander A, Cai S, Tripathi DN, Dere R, Tee AR, Tait-Mulder J, Di Nardo A, Han JM, Kwiatkowski E, Dunlop EA, Dodd KM, Folkerth RD, Faust PL, Kastan MB, Sahin M, Walker CL. A tuberous sclerosis complex signalling node at the peroxisome regulates mTORC1 and autophagy in response to ROS. *Nat Cell Biol.* 2013; 15(10):1186-96

[79] Scherz-Shouval R, Shvets E, Fass E, Shorer H, Gil L, Elazar Z. Reactive oxygen species are essential for autophagy and specifically regulate the activity of Atg4. *EMBO J.* 2007; 26(7):1749–60.

[80] Veerapathran A, Dakshinamoorthy G, Gnanasekar M, Reddy MVR, Kalyanasundaram R. Evaluation of *Wuchereriabancrofti* GST as a Vaccine Candidate for Lymphatic Filariasis. *PLoSNegl Trop Dis.* 2009; 3(6):e457.

[81] Ahmad R, Srivastava AK, Walter RD. Purification and biochemical characterization of cytosolic glutathione-S-transferase from filarial worms *Setariacervi*. *Comp BiochemPhysiol B Biochem Mol Biol. 2008*; 151(3):237–45.

[82] van Rossum AJ, Brophy PM, Tait A, Barrett J, Jefferies JR. Proteomic identification of glutathione S-transferases from the model nematode Caenorhabditis elegans. *Proteomics.* 2001; 1(11):1463–8.

[83] Perally S, Lacourse EJ, Campbell AM, Brophy PM. Heme transport and detoxification in nematodes: subproteomics evidence of differential role of glutathione transferases. *J Proteome Res.* 2008; 7(10):4557–65.

[84] Liebau E, Eckelt VH, Wildenburg G, Teesdale-Spittle P, Brophy PM, Walter RD, Henkle-Dührsen K. Structural and functional analysis of a glutathione S-transferase from Ascarissuum. *Biochem J.* 1997; 324 (Pt 2):659-66.

[85] Liebau E, Eschbach ML, Tawe W, Sommer A, Fischer P, Walter RD, et al. Identification of a stress-responsive Onchocerca volvulus glutathione S-transferase (Ov-GST-3) by RT-PCR differential display. *MolBiochemParasitol.* 2000; 109(2):101–10.

[86] Joardar N, Sinha Babu SP. A review on the druggability of a thiol-based enzymatic antioxidant thioredoxin reductase for treating filariasis and other parasitic infections. *Int J BiolMacromol.* 2020 Jan 1; 142:125–41.

[87] Bulman CA, Bidlow CM, Lustigman S, Cho-Ngwa F, Williams D, Rascón AA. Repurposing auranofin as a lead candidate for treatment of lymphatic filariasis and onchocerciasis. *PLoSNegl Trop Dis.* 2015; 9(2):e0003534.

[88] Ngwewondo A, Scandale I, Specht S. Onchocerciasis drug development: from preclinical models to humans. *Parasitol Res.* 2021; 120(12):3939–64.

[89] Buey RM, Fernández-Justel D, de Pereda JM, Revuelta JL, Schürmann P, Buchanan BB, Balsera M. Ferredoxin-linked flavoenzyme defines a family of pyridine nucleotide-independent thioredoxinreductases. *ProcNatlAcadSci* U S A. 2018; 115(51):12967-12972.

[90] Joardar N, Mukherjee N, Sinha Babu SP. Nanopharmaceuticals to target antifilarials: Administration of old age drugs in a novel way. In: Formiga FR, Inamuddin, Severino P, editors. *Applications of Nanobiotechnology for Neglected Tropical Diseases*. Academic Press; 2021. pp. 329–56.

[91] Fata F, Gencheva R, Cheng Q, Lullo R, Ardini M, Silvestri I, Gabriele F, Ippoliti R, Bulman CA, Sakanari JA, Williams DL, Arnér ESJ, Angelucci F. Biochemical and structural characterizations of

thioredoxinreductaseselenoproteins of the parasitic filarial nematodes Brugiamalayi and Onchocerca volvulus. *Redox Biol.* 2022; 51:102278.
[92] Ren X, Zou L, Zhang X, Branco V, Wang J, Carvalho C, Holmgren A, Lu J. Redox Signaling Mediated by Thioredoxin and Glutathione Systems in the Central Nervous System. *Antioxid Redox Signal.* 2017; 27(13):989-1010.
[93] Ou XU, Tang LA, Mccrossan M, Henkleduhrsen K, Selkirk ME. Brugiamalayi: Localisation and Differential Expression of Extracellular and Cytoplasmic CuZn Superoxide Dismutases in Adults and Microfilariae. *ExpParasitol.* 1995; 80(3):515–29.
[94] Maizels RM, Blaxter ML, Scott AL. Immunological genomics of Brugiamalayi: filarial genes implicated in immune evasion and protective immunity. *Parasite Immunol.* 2001; 23(7):327–44.
[95] Henkle-Dührsen K, Kampkötter A. Antioxidant enzyme families in parasitic nematodes. *MolBiochemParasitol.* 2001; 114(2):129–42.
[96] Nandi A, Yan LJ, Jana CK, Das N. Role of Catalase in Oxidative Stress- and Age-Associated Degenerative Diseases. *Oxid Med Cell Longev.* 2019; 2019:9613090.
[97] Joardar N, Guevara-Flores A, Martínez-González J de J, Sinha Babu SP. Thiol antioxidant thioredoxin reductase: A prospective biochemical crossroads between anticancer and antiparasitic treatments of the modern era. *Int J BiolMacromol.* 2020; 165(Pt A):249–67.
[98] Mishra A, Kumar V, Singh A. Deciphering the anti-filarial potential of bioactive compounds from *Ocimum sanctum* : a combined experimental and computational study. *Pharm Biol.* 2022; 60(1):2237–52.
[99] Ray AS, Joardar N, Mukherjee S, Rahaman CH, Sinha Babu SP. Polyphenol enriched ethanolic extract of Cajanusscarabaeoides (L.) Thouars exerts potential antifilarial activity by inducing oxidative stress and programmed cell death. *PLoS One.* 2018; 13(12):e0208201.
[100] Saini P, Mukherjee N, Mukherjee S, Roy P, Gayen P, Kumar D, Pal BC, Babu SP. Diospyrosperigrena bark extract induced apoptosis in filarial parasite *Setariacervi* through generation of reactive oxygen species. *Pharm Biol.* 2015; 53(6):813-23.
[101] Mukherjee N, Saini P, Mukherjee S, Roy P, Sinha Babu SP. In vitro antifilarial activity of Azadirachta indica aqueous extract through reactive oxygen species enhancement. *Asian Pac J Trop Med.* 2014; 7(11):841–8.
[102] Morrison CA, Colin T, Sexton JL, Bowen F, Wicker J, Friedel T, Spithill TW. Protection of cattle against Fasciola hepatica infection by vaccination with glutathione S-transferase. *Vaccine.* 1996; 14(17-18):1603-12.
[103] Grzych JM, Grezel D, Xu CB, Neyrinck JL, Capron M, Ouma JH, Butterworth AE, Capron A. IgA antibodies to a protective antigen in human Schistosomiasismansoni. *J Immunol.* 1993; 150(2):527-35.
[104] Nathan ST, Mathew N, Kalyanasundaram M, Balaraman K. Structure of glutathione S-transferase of the filarial parasite Wuchereriabancrofti: a target for drug development against adult worm. *J Mol Model.* 2005;11(3):194–9.
[105] Gorai S, Das NC, Gupta PSS, Panda SK, Rana MK, Mukherjee S. Designing efficient multi-epitope peptide-based vaccine by targeting the antioxidant thioredoxin of bancroftian filarial parasite. *Infect Genet Evol.* 2022; 98:105237.
[106] Hajishengallis G. Toll gates to periodontal host modulation and vaccine therapy. *Periodontol* 2000. 2009; 51:181-207.
[107] Anand SB, Rajagopal V, Kaliraj P. Brugiamalayithioredoxin peroxidase as a potential vaccine candidate antigen for lymphatic filariasis. *ApplBiochemBiotechnol.* 2012;167(5):1351–64.
[108] Roy P, Saha SK, Gayen P, Chowdhury P, Sinha Babu SP. Exploration of antifilarial activity of gold nanoparticle against human and bovine filarial parasites: A nanomedicinal mechanistic approach. *Colloids Surf B Biointerfaces.* 2018; 161:236–43.
[109] Yadav S, Ahmad F, Rathaur S. Antifilarial efficacy of andrographolide: Ex vivo studies on bovine filarial parasite Setariacervi. *Comparative Biochemistry and Physiology Part C: Toxicology & Pharmacology.* 2022; 261:109442.
[110] Kalyanasundaram R. Lymphatic filariasis: current status of elimination using chemotherapy and the need for a vaccine. *Communicable Diseases of the Developing World.* 2016;97–124.

[111] Schussek S, Trieu A, Doolan DL. Genome-and proteome-wide screening strategies for antigen discovery and immunogen design. *Biotechnol Adv.* 2014;32(2):403–14.
[112] Versteeg L, Almutairi MM, Hotez PJ, Pollet J. Enlisting the mRNA vaccine platform to combat parasitic infections. *Vaccines* (Basel). 2019;7(4):122.
[113] Kalyanasundaram R, Khatri V, Chauhan N. Advances in Vaccine Development for Human Lymphatic Filariasis. *Trends Parasitol.* 2020; 36(2):195–205.
[114] Klei TR, Rajan T v. The filaria. Vol. 5. Springer Science & Business Media; 2002.
[115] Gnanasekar M, Rao KVN, He YX, Mishra PK, Nutman TB, Kaliraj P, et al. Novel phage display-based subtractive screening to identify vaccine candidates of Brugiamalayi. *Infect Immun.* 2004;72(8):4707–15.

# Chapter 17

# The Impact of Oxidative Stress on Male Infertility Induction and Its Management by Nutraceuticals

**Debidas Ghosh**[1,*]
**AdrijaTripathy**[1]
**Dipanwita Mitra**[1]
**and Puja Das**[2]
[1]Department of Biomedical Laboratory Science and Management
[2]Centre for Life Sciences,
Vidyasagar University, Midnapore, West Bengal, India

## Abstract

Infertility is a condition associated with major medical and social preoccupation. Male etiology is responsible for nearly half of the infertility cases and is caused by alterations in sperm concentration, motility, and/ or morphology. While environmental, physiological, and genetic influences are recognized in this concern, oxidative stress (OS) appears to be a dominant factor for impairing sperm functions and interfering pregnancy. It has become apparent over the past 20 years that the physiological levels of reactive oxygen species (ROS) have an impact on several signaling pathways that regulate biological and physiological redox sensitive processes. Reactive oxygen species modulate sperm chromatin condensation by adjusting the number of germ cells and inducing apoptosis or proliferation of spermatozoa. ROS are also involved in the processes of capacitation, acrosome reaction, mitochondrial stability, and sperm motility. ROS can function as intracellular messenger by modulating the NADPH oxidase enzyme complex in the cell membrane and intervening the respiratory chain. In spermatozoa, superoxide anion metabolism is regulated by the NADH oxidoreductase enzyme, which works in close conjunction with the mitochondrial respiratory chain and xanthine oxidase found in sperm. Seminal leukocytes are another source of ROS, producing 1000 times more of these molecules than sperm cells under physiological conditions. This is because seminal leukocytes represent the first line of defense against offending infectious agents, using primarily oxidative and inflammatory mechanisms. However, ROS can become a double-edged sword that results in cellular injury. Indeed, different ROS are generated to

[*] Corresponding Author's Email: debidasghosh999@gmail.com.

In: The Role of Reactive Oxygen Species in Health and Disease
Editor: Kuladip Jana
ISBN: 979-8-89113-283-2

counteract infectious agents can also damage host cells, which can result in the disintegration of sperm cell membrane or sperm DNA damage. Sperm cells have abundant lipids in their plasma membrane mostly in the form of polyunsaturated fatty acids (PUFAs), especially docosa hexaenoic acid. Increased ROS production induces PUFA peroxidation within the sperm cell membrane, which induces cell dysfunction due to loss of membrane fluidity and integrity which are required for successful sperm-oocyte fusion after the capacitation and acrosome reaction biochemical cascades. Byproducts of lipid peroxidation bind to and disrupt mitochondrial proteins of the electron transport chain, which leads to electron leakage and consequently decreases mitochondrial membrane potential, decreases ATP production as well as sperm motility led to male infertility.

Assessment of sperm ROS levels have been developed which can be classified into direct (Chemo luminescence assay, Flow cytometry, Electron spin resonance, Cytochrome c reduction, Thiobarbituric acid assay, Xylenol orange-based assay, FITC-labelled lectins) and indirect assays (Lipid peroxidation levels assay, Total antioxidant capacity assay, Comet assay). Currently, there are no infertility guidelines that recommend routine ROS measurement and there is still an ongoing debate on which type of patients have to be tested for the oxidative burden. Hyperviscosityof semen has also been suggestive of increased ROS because it is attributed to increased malondialdehyde levels. Increased leukocytes or round cells which are one of the principal sources of ROS may suggest further testing for ROS in semen. Abnormal sperm morphology also correlates with high levels of ROS. The hypo-osmotic swelling test suggests sperm membrane damage due to lipid peroxidation and this also focuses on higher levels of ROS in semen.

**Keywords:** oxidative stress, ROS, male infertility, germ cell apoptosis, antioxidants, nutraceuticals

# 1. Introduction

## 1.1. Oxidative Stress

Oxidative stress is an imbalance of free radicals and antioxidants in the body, which can lead to cell and tissue damage. Oxidative stress may occur naturally and plays a role in the aging process. The body's natural immune response can also trigger oxidative stress temporarily. This type of oxidative stress causes mild inflammation that goes away after the immune system fights off an infection or repairs an injury.

Oxidative stress plays an important role in the pathogenesis of chronic diseases such as cardiovascular diseases, diabetes, neurodegenerative diseases, and cancer. Long term exposure of oxidative stress increased the levels of pro-oxidant factors that can cause structural defects at mitochondrial and nuclear DNA levels, as well as functional alteration of several enzymes and cellular structures.

Oxidative stress has more harmful properties than helpful ones. It can break down cells and tissue homeostasis which can also result in inflammation. These factors can lead to lifelong diseases like diabetes, cardiovascular diseases, cancer, in some cases [1, 2].

There are several factors which help in the production of oxidative stress and excess free radicals.

- Food style
- Lifestyle
- Certain conditions
- Environmental factors such as pollution and radiation

#### *1.1.1. Food Style*

High intakes of macronutrients can promote oxidative stress and subsequently contribute to inflammation via nuclear factor-kappa B- *(NF-κB-)* mediated cell signaling pathways. Dietary carbohydrates, animal-based proteins, and fats are important to highlight here because they may contribute to the long-term consequences of nutritionally mediated inflammation [2]. Oxidative stress is a central player of metabolic ailments associated with high-carbohydrate and animal-based protein diets and excessive fat consumption. Fast food, junk food, processed food and preserved food also favor oxidative stress imposition.

#### *1.1.2. Lifestyle*

Smoking, alcohol consumption, adequate or inappropriate diet, exercise, training or untrained condition contribute to oxidative stress. Similarly static postural long-term work, continuous strenuous work, unfavorable environment of workstation is unhealthy work style that also impact oxidative stress.

#### *1.1.3. Certain Conditions*

Oxidative stress is sometimes induced by mild inflammation, diabetes, psychological distress etc.

#### *1.1.4. Environmental Factors Such as Pollution and Radiation*

Environmental factors such as salinity, drought, metal toxicity, extreme temperature, air pollutants, ultraviolet-B (UV-B) radiation, pesticides, and pathogen infection leads to stress [3, 4]. The oxidative stress is caused either by the direct effects of environmental stress or by indirect reactive oxygen species (ROS) generation and accumulation, which damage a cell before elimination [5].

## 2. ROS and RNS

### 2.1. ROS

Reactive Oxygen species (ROS) are highly reactive chemicals formed from diatomic oxygen ($O_2$). ROS are byproducts of the normal metabolism of oxygen. ROS has roles in cell signaling and homeostasis [6, 7]. ROS can cause irreversible damage to DNA as they oxidize and modify some cellular components and prevent them from performing their original functions. Mainly ROS represents a dual role; whether they will act as harmful, protective or signaling factors depends on the balance between ROS production and disposal at the right time and place. In other hand, oxygen toxicity can arise both from uncontrolled production and from the inefficient elimination of ROS by the antioxidant system. Some examples of ROS are [8]:

Hydroxyl radical (HO•)
Hydroxide ion (HO−)
Triplet oxygen ($O_2^{2\bullet}$)
Superoxide anion ($O_2^{\bullet-}$)
Peroxide ion ($O_2$−2) (Figure 1)

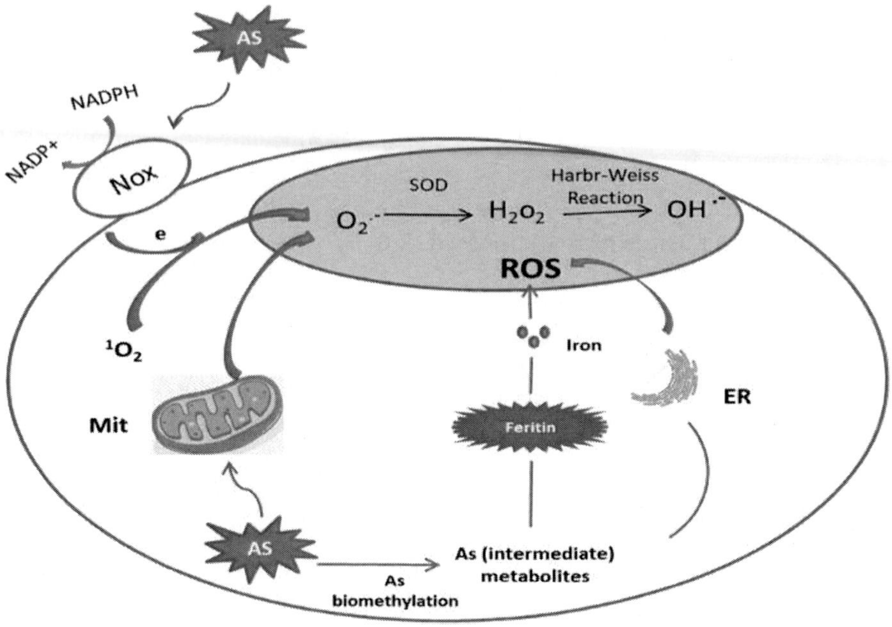


**Figure 1.** Generation of ROS.

## 2.2. RNS

Reactive nitrogen species (RNS) are a family of antimicrobial molecules derived from nitric oxide (•NO) and superoxide ($O_2^{\bullet-}$) produced via the enzymatic activity of inducible nitric oxide synthase 2 (NOS2) and NADPH oxidase respectively. NOS2 is expressed primarily in macrophages after induction by cytokines and microbial products, notably interferon-gamma (IFN-γ) and lipopolysaccharide.

## 2.3. Generation of RNS

RNS are produced in animals starting with the reaction of nitric oxide (•NO) with superoxide ($O_2^{\bullet-}$) to form peroxynitrite ($ONOO^-$).

NO (nitric oxide) + $O_2^{\bullet-}$ (superoxide) → $ONOO^-$ (peroxynitrite)

Reactive nitrogen species (RNS) are various nitric oxide–derived compounds, including nitroxyl anion, nitrosoniumcation, higher oxides of nitrogen, S-nitrosothiols, and dinitrosyl iron complexes. RNS have been recognized as playing a crucial role in the physiologic regulation of many, if not all, living cells, such as smooth muscle cells, cardiomyocytes, platelets, and nervous and juxtaglomerular cells. They possess pleiotropic properties on cellular targets after both posttranslational modifications and interactions with reactive oxygen species. Elevated levels of RNS have been implicated in cell injury and death by inducing nitrosative stress [9].

Superoxide anion ($O_2^-$) is a reactive oxygen species that reacts quickly with nitric oxide (NO) in the vasculature. The reaction produces peroxynitrite and depletes the bioactivity of NO. This is important because NO is a key mediator in many important vascular functions including regulation of smooth muscle tone and blood pressure, platelet activation, and vascular cell signaling [10].

Peroxynitrite itself is a highly reactive species which can directly react with various biological targets and components of the cell including lipids, thiols, amino acid residues, DNA bases, and low-molecular weight antioxidants. However, these reactions happen at a relatively slow rate. This slow reaction rate allows it to react more selectively throughout the cell. Peroxynitrite is able to get across cell membranes to some extent through anion channels. Additionally, peroxynitrite can react with other molecules to form additional types of RNS including nitrogen dioxide ($\bullet NO_2$) and dinitrogen trioxide ($N_2O_3$) as well as other types of chemically reactive free radicals. Important reactions involving RNS include:

- $ONOO^- + H^+ \rightarrow$ ONOOH (peroxynitrous acid) $\rightarrow \bullet NO_2$ (nitrogen dioxide) + $\bullet OH$ (hydroxyl radical)
- $ONOO^- + CO_2$ (carbon dioxide) $\rightarrow ONOOCO_2^-$ (nitrosoperoxycarbonate)
- $ONOOCO_2^- \rightarrow \bullet NO_2$ (nitrogen dioxide) + $O{=}C(O\bullet)O^-$ (carbonate radical)
- $\bullet NO + \bullet NO_2 \rightleftharpoons N_2O_3$ (dinitrogen trioxide)

### 2.4. Generation of ROS through Different Pathway

ROS is generated in cells, during the combustion of carbohydrate, protein and fat (Aerobic metabolism) as a byproduct.

- Infection and inflammation triggers ROS generation. It has been noted that spermatic leukocytes contribute 1000-fold more ROS generation than aerobic metabolism.
- The people who are suffering from MAI (Male Accessory gland Infections) produces higher amount of ROS than others.
- Varicocele that is the enlargement of vein within the scrotum is another factor that increased the ROS generation as it increased scrotal temperature which is associated with ROS generation.
- Not only that patient, specifically with hyperglycemia / diabetes mellitus increased the oxidative stress and ultimately ROS generation.
- 48% seminal leucocytes and 107% seminal ROS level is increased due to smoking.
- Acetaldehyde is a byproduct of ethanol metabolism. Due to excessive alcohol consumption this byproduct interludes with proteins and lipids leading to an increase in ROS production.

# 3. Free Radical

## 3.1. Free Radical Generation

A free radical is termed a molecular species which can contain an unpaired electron in its atomic orbit and can exist independently. Most organic radicals are quite unstable and very reactive. There are two kinds of radicals, neutral radicals and charged radicals as shown below.

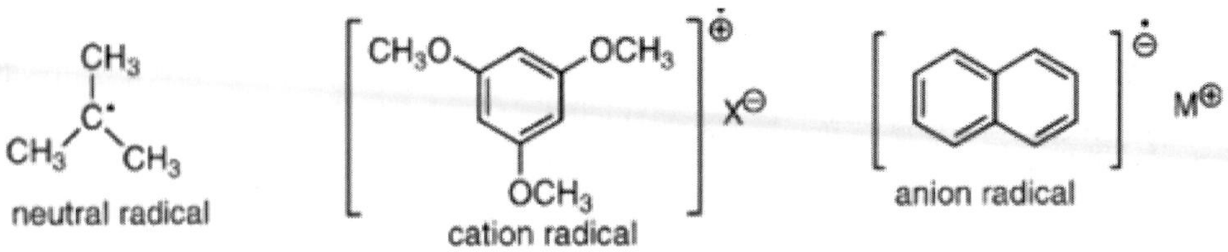


The most common well known free radicals which consists of oxygen are:

- Hydroxyl radicals
- Superoxide anion
- Hydrogen peroxide radical
- Hypochlorite radical
- Nitric oxide radical
- Peroxyntrite radical

Free radicals are generated internally from the sources mainly mitochondria, and peroxisomes.

## 3.2. Mechanism of Action of Free Radicals in the Body

Free radicals can cause chain reactions that ultimately damage cells. For example, a superoxide molecule may react with a fatty acid and capture one of its electrons. The fatty acid then becomes a free radical that can react with another fatty acid nearby. As this chain reaction continues, the permeability and fluidity of cell membranes changes, decreased activity of protein present in cell membrane, and receptor proteins undergo changes in structure that either alter or stop their function. If receptor proteins designed to react to insulin levels undergo a structural change it can negatively affect glucose uptake.

This reaction of free radical can continue unchecked unless stopped by a defense mechanism of the body. The body's two major defense systems are- free radical detoxifying enzymes and antioxidant chemicals. Free radical detoxifying enzyme (superoxide dismutase, catalase, glutathione peroxidase) systems are responsible for protecting the insides of cells from free radical damage. An antioxidant is any molecule that can block free radicals from stealing electrons; antioxidants act both inside and outside of cells.

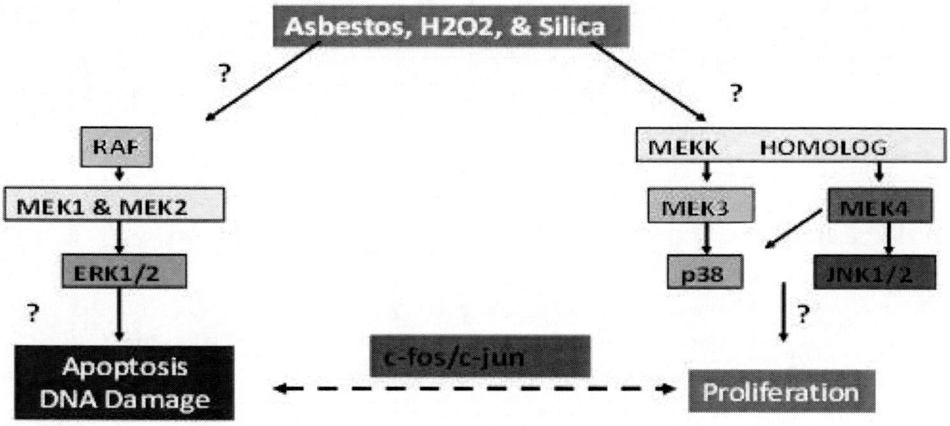


**Figure 2.** ROS – induced Mitogen-Activated Protein Kinase (MAPK) cascade.

Reactive oxygen species signaling occurs through the enzyme/gene cascades in which aging-regulating genes *p66shc*, *Sirtuin*, *FOXO3a*, and *Klotho* play a critical role together with some protein kinases particularly protein kinase Akt/PKB. Akt kinase was frequently considered to be a surviving factor. It also said that Akt kinase with other mitogen activated protein kinases (MAPKs) can participate in damaging ROS signaling processes [11](Figure 2).

### 3.3. Signal Transduction by Free Radicals

Reactive oxygen species are sometimes produced by mitochondria as part of aerobic respiration and that phagocytic cells could produce oxidants as part of a host defense mechanism, the role of oxidants within cells began to shift with the observation that the production of hydrogen peroxide was essential for normal growth factor signaling.

A number of cellular stimuli that induce ROS production also in parallel can activate MAPK pathways in multiple cell types. Direct exposure of cells to exogenous $H_2O_2$, to mimic oxidative stress, leads to activation of MAPK pathways. ROS can alter protein structure and function by modifying critical amino acid residues of proteins [12], the oxidative modification of signaling proteins by ROS may be one of the plausible mechanisms for the activation of MAPK pathways [13].

### 3.4. Free Radicals and Gene Expression

NF-kB is involved in the induced expression of the interleukin-2 gene and in a wide variety of biological responses. In particular, it is implicated in inflammatory reactions, growth control, and apoptosis [14] and is the first eukaryotic transcription factor shown to respond directly to oxidative stress in certain types of cells. NF-kB is activated by micromolar concentrations of hydrogen peroxide [15]. A moderate pro-oxidative shift in the glutathione redox state was also found to enhance NF-kB activation in Molt-4 and Jurkat T cell lines [16]. However, NF-kB activation is not strictly dependent on ROS. Several signal cascades have recently been identified that can activate NF-kB without involvement of ROS.

In T lymphocytes AP-1 is involved in the expression of the interleukin-2 gene and other immunologically relevant genes [17]. Many different oxidative stress-inducing stimuli, e.g., relatively low concentrations of hydrogen peroxide, UV light, g-irradiation, and interleukin-1, lead to AP-1 activation by different mechanisms.

## 4. Linkage Between Oxidative Stress and Male Reproduction

In several research, it has been shown that male reproduction capacity is affected by oxidative stress (OS). The one of the most important proposed caused of male infertility is OS and ROS.

Increased ROS along with decreased antioxidant defense results in redox imbalance, reduced sperm motility and sperm DNA damage. Spermatozoa are highly susceptible to the deleterious effects of ROS due to the large amounts of unsaturated fatty acids found in their cell membranes. Reactive oxygen species promote peroxidation of lipids, resulting in intracellular oxidative burden. The sequence of events involves lipid peroxidation, loss of membrane integrity with increased permeability, reduced sperm motility, structural DNA damage, and apoptosis. Several intrinsic and extrinsic factors have been associated with increased OS in the male reproductive system [18].

## 5. Oxidative Stress and Reproductive Endocrinology

An imbalance between the creation and buildup of reactive oxygen species (ROS) in cells and tissues and the body's capacity to detoxify these reactive byproducts results in the situation known as oxidative stress (OS). Oxygen-containing molecules with an unbalanced electron number are known as free radicals or ROS. They can easily interact with other molecules because of their unequal quantity and cause longer chemical reaction known as oxidation. Through redox dependent signaling, ROS influences a variety of cellular activities at low concentrations, such as proliferation, differentiation, and immunological modulation. But when different exogenous (smoking, radiation, environmental toxins, drugs etc.) and endogenous factors (high metabolic rate, by products of aerobic metabolism, diseases etc.) produce excessive ROS that can hamper the normal homeostasis of physiological system. Along with enzymatic i.e., superoxide dismutase, catalases, and peroxidases and non-enzymatic i.e., vitamins, minerals, and steroids, etc. pathways, antioxidants protect against excessive production of ROS levels. High levels of ROS can increase the chance of infertility both directly by causing OS and indirectly by influencing the hypothalamic axis that controls the release of hormones. Infertility is brought on by ROS, which lower male sex hormone levels and disrupt the endocrine balance that controls male reproductive activities. Therefore, ROS is known as "endocrine disruptors." They disturb the hormonal regulation of male reproduction, crosstalk between the hypothalamo-pituitary-testicular axis (HPTA) and other hypothalamo hormonal axes like hypothalamo pituitary adrenal axis (HPAA) in addition to interfering with endocrine pathways [19, 20].

### 5.1. Role of Gonadotropins

Luteinizing hormone (LH) and follicle-stimulating hormone (FSH), known as gonadotropins, which govern testicular activities, are released and produced by the anterior pituitary in response to the gonadotropin releasing hormone (GnRH) secreted by the hypothalamus. Leydig-cell activity and spermatogenesis are both influenced by these gonadotrophic hormones. LH and FSH receptors are found on the Leydig cells and the membrane of Sertoli cells respectively. They work together to promote healthy spermatogenesis, sperm density, and testosterone synthesis. Pituitary LH is also known as interstitial cell stimulating hormone (ICSH). LH is in charge of promoting both *in vivo* [21] and *in vitro* [22] testosterone synthesis. The activation of various mitochondrial enzymes involved in cholesterol side chain cleavage as well as other enzymes involved in the pathway of steroidogenesis is under the acute effect of LH and synthesis of various enzymes responsible for the conversion of pregnanolone to testosterone are as a chronic LH impact.

Testicular steroids are produced and secreted under the effect of FSH. FSH is responsible for the production and secretion of testicular steroids and also enhances the activity of LH on androgen production. FSH also increases the activity of two vital androgenic enzymes, i.e., $\Delta 5$, $3\beta$-HSD, $17\beta$-HSD. By enhancing the number and sensitivity of LH receptors in the testis, FSH can work synergistically with LH for the synthesis of testosterone [23].

### 5.2. Effect of ROS on HPGA and HPAA

The primary male reproductive hormones act directly or via cross-linked among multiple endocrine axes to regulate reproductive processes in addition to the central regulation through the HPGA. So, any interruption to these networks could have a negative impact on male fertility (Figure 3).

The male reproductive endocrinology may be negatively impacted by ROS due to the oxidative stress they cause by interfering with the HPA either directly or indirectly through cross-talks. Stress considerably lowers the levels of testosterone, follicle-stimulating hormone (FSH), luteinizing hormone (LH), and gonadotropin-releasing hormone (GnRH) in males. The quality of sperm may be harmed by such consequences [24].

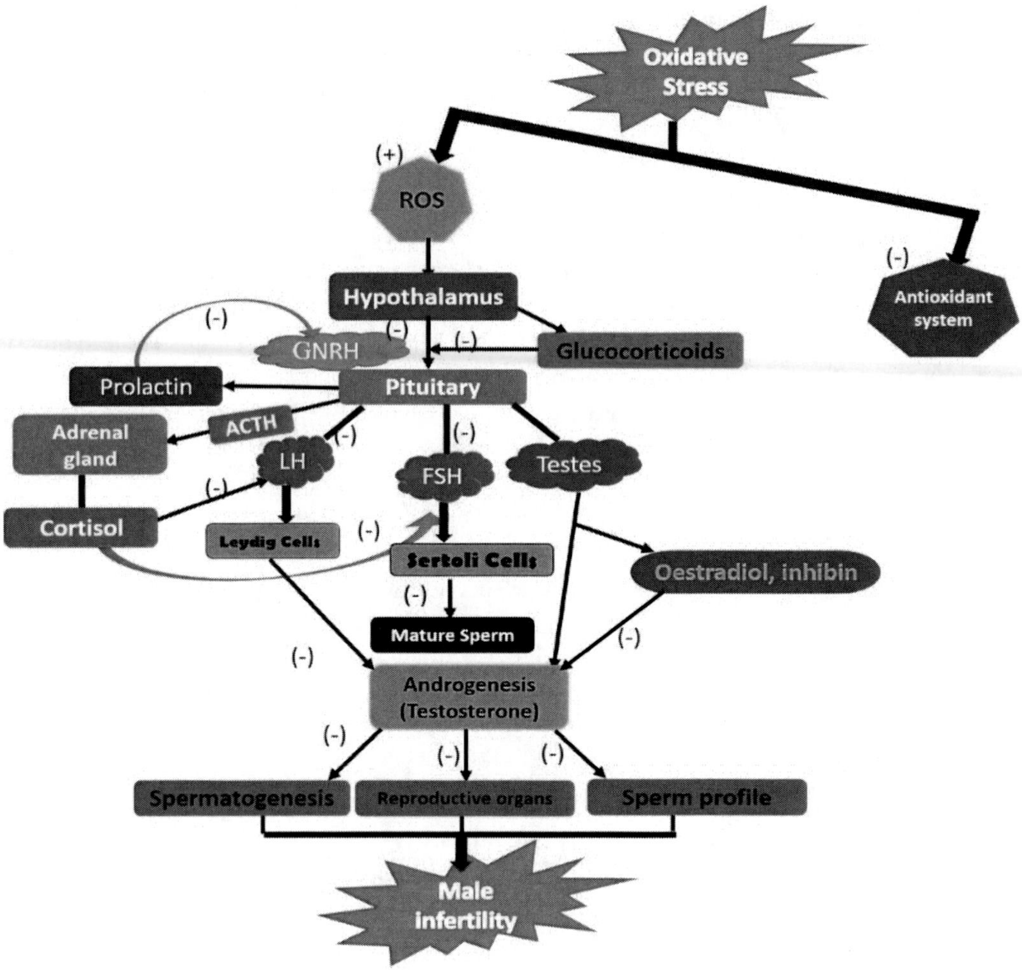


**Figure 3.** Effect of oxidative stress on male infertility through hypothalamo-pituitary-testicular axis (HPTA) and hypothalamo pituitary adrenal axis (HPAA) regulation.

Followings are the mechanism of actions of ROS generated through oxidative stress and their effects on key reproductive hormones:

| ROS generation through oxidative stress | Effects on male reproductive endocrinology |
|---|---|
| Elevating stress hormone- cortisol levels and turning on the immune-inflammatory system | Diminishes the LH and serum testosterone level by inhibiting the production of androgens and causing Leydig cells to die. |
| HPTA is impacted by glucocorticoids | Decrease the release of GnRH from hypothalamus and thereby lowers the release of LH and FSH from anterior pituitary, reducing the expression of steroidogenic enzymes, decreasing the testicular sensitivity to gonadotropins, or inducing Leydig and germ cell apoptosis. |
| Reducing the activities of antioxidant enzymes and upregulating the NADPH oxidase activity affecting mitochondrial homeostasis. | Diminishes testosterone and LH levels and interferes with Sertoli cell functioning. |
| Releasing catecholamines like adrenaline | Reduces testosterone levels by inhibiting the HPTA. |
| Lowering the natural antioxidants and free radical scavengers found in food | Lowers the production of testosterone, release of LH, and the androgen profile |

| ROS generation through oxidative stress | Effects on male reproductive endocrinology |
|---|---|
| Activating NOX and XO enzymes in the mitochondria | Reduces LH, FSH, and testosterone levels |
| Elevating the liver's cytochrome P450s enzyme activity, changing the amounts of essential metals in body, and decreasing antioxidant levels | Decreases serum levels of testosterone, LH, and FSH, accelerates apoptosis in Sertoli cells and Leydig cells, |
| Reducing the antioxidant enzymes activities and alter the membrane potential of mitochondrial | Reduces LH sensitivity by weakening LH receptors, accelerates lipid peroxidation in Leydig cells, and decreases the rate of steroidogenesis, testosterone production, and release. |

## 6. Oxidative Stress and Spermatogenesis

Spermatogenesis, an important function of the testes and seminiferous tubules in testis play a pivotal role in spermatogenesis process. During spermatogenesis process, at first, two diploid primary spermatocytes are produced from diploid spermatogonium by mitotic division. By the action of meiosis, each primary spermatocyte split into two haploid secondary spermatocytes. These focus the end of meiosis and produce four spermatids. Spermatids differentiated into mature sperm by the process of spermiogenesis.

Cellular damage results from oxidative stress, when the generation of ROS surpasses the antioxidant defense mechanisms. ROS are free radicals with a single unpaired electron, commonly known as free radicals. They are the substances produced as byproducts of oxygen metabolism. Due to the lone pair of electrons in the outer orbit, they produce extremely reactive compounds.

Prior research has revealed that reactive oxygen species (ROS) can affect human spermatozoa in two different ways. In the context of physiology, ROS is required in specific concentration for capacitation, acrosome reaction and DNA condensation. The sperm and seminal plasma itself are rich sources of antioxidant defenses. Antioxidants in seminal plasma are evidently beneficial in reducing sperm oxidative damage after ejaculation. However, sperm must depend on the epididymal/testicular antioxidants and their own intrinsic antioxidant capacity for protection because they are not in contact with seminal plasma antioxidants throughout spermatogenesis and epididymal storage. Therefore, during epididymal transit, sperm is particularly susceptible to oxidative damage, especially in cases of epididymal inflammation. High levels of DNA damage, antioxidant depletion, and subsequent disruption of spermatogenesis are all results of oxidative stress in the testes. Oxidative stress can be very detrimental to spermatozoa and affects the quality of sperm.

Numerous cell structures, particularly the phospholipids in cellular membranes, are damaged because of an imbalance between the generation and consumption of ROS. On the other hand, at high concentrations, ROS cause sperm dysfunctions such as decreased motility, loss of mitochondrial membrane potential, DNA damage, poor sperm-zona pellucida binding, and sperm-oocyte, as well as apoptosis [25]. Lipid peroxidation, a form of oxidative damage to lipids, triggers inflammatory signaling cascades that in turn create a burden on the antioxidant defense system of the body. Lipid peroxidation causes reduced sperm motility, loss of membrane integrity with increased permeability, structural DNA damage, and apoptosis [18].

## 6.1. The Physiological Role of Reactive Oxygen Species (ROS) on Spermatozoa

Reactive oxygen species are considered regulators of several intracellular pathways, modulating the activation of different transcription factors. It stimulates cyclic adenosine monophosphate (cAMP) in sperms, promoting tyrosine phosphorylation by tyrosine phosphatase inhibition [26]. Several studies reported that higher ROS levels stimulate sperm capacitation, hyperactivation, motility, acrosome reaction, and chemotaxis and chromatin compaction in maturing spermatozoa.

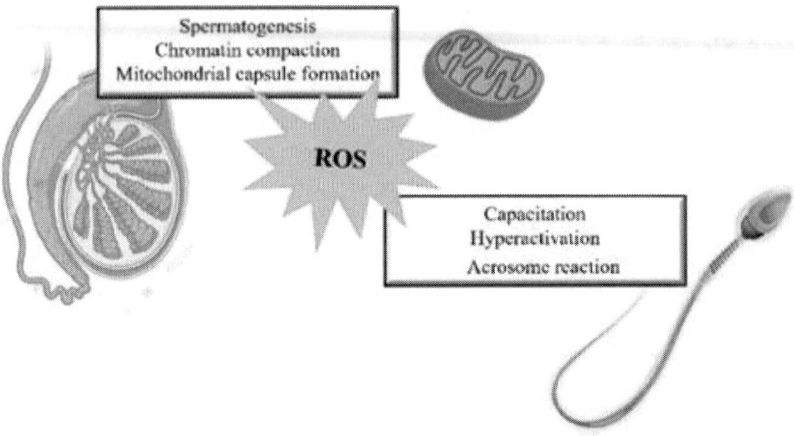


**Figure 4.** The physiological role of ROS on sperm function.

Appropriate quantities of ROS are necessary for cysteine-thiol group oxidation, which is necessary for chromatin compaction and stabilization in spermatozoa, during spermatogenesis and epididymal transit [27]. Furthermore, protein thiol oxidation, which is catalyzed by glutathione peroxidase and phospholipid hydroperoxide, is implicated in the creation of mitochondrial capsules by ROS (Figure 1). This enzyme is then converted to a cross-linked structural protein, comprising a significant component of the mitochondrial sheath of mature spermatozoa (Figure 4).

Hyperactivation is a sperm motility pattern that enhances the ability of the sperm to penetrate the cumulus cell layer surrounding the oocyte. When the incubation medium was supplied with a sufficient concentration of $H_2O_2$, a discernible rise in the number of spermatozoa exhibiting hyperactivated motility was seen [26].

Capacitation is a crucial physiological process of sperm maturation that occurs during its passage through the female reproductive tract to enable sperm-oocyte interaction. During this process, ROS act as an essential intracellular messenger to activate downstream signaling molecules mediating significant changes in the membrane architecture (i.e., membrane destabilization, permeability, and lipid redistribution), which are prerequisites for successful fertilization [30]. The acrosome reaction is an irreversible process that involves the release of acrosomal hydrolytic enzymes to attain sperm oocyte fusion. To activate tyrosine phosphorylation of putative target proteins, particularly on the pre-equatorial area of the sperm head surface, the production of both $O_2$ and $H_2O_2$ within physiological limits is necessary [27].

## 6.2. The Pathological Effect of ROS on Spermatozoa

Significantly elevated ROS levels have been found in 25–40% of infertile patients' semen samples [28]. High concentrations and prolonged exposure to ROS cause extensive damage to various integral cellular biomolecules (Figure 5), including proteins, lipids, and nucleic acids, which ultimately hampers multiple cellular functions.

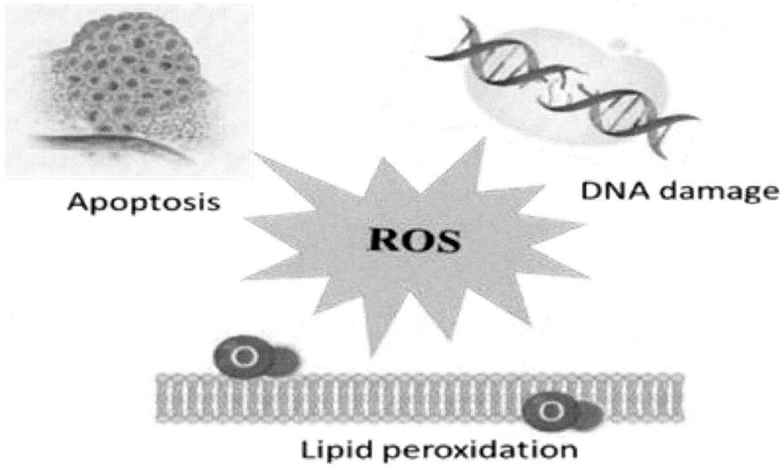


**Figure 5.** The pathological effect of ROS on sperm function.

Docosahexaenoic acids are found in greater concentrations in the plasma membrane of sperm and are necessary for maintaining the ideal fluidity required for multiple membrane fusion events. However, these fatty acids act as potential substrates for peroxidation as they comprise methylene groups with highly reactive hydrogen atoms, thereby enhancing the susceptibility of sperm to oxidative damage [29]. When a hydrogen atom from a fatty acid combines with an oxygen atom from ROS to form a lipid radical, a chain process known as lipid peroxidation is started. These unstable radicals react rapidly with oxygen molecules to form peroxy fatty acid radicals, which are thereafter converted into lipid peroxides. In the presence of a transitional metal ion, lipid peroxide is catalyzed into OH• [30] that have the ability to attract electrons from other PUFAs to generate new radicals and thereby propagate the lipid peroxidation chain reaction. When the radicals interact with one another to form the non-reactive substance known as MDA, the chain reaction is stopped. This by-product is frequently employed as a biomarker to gauge the degree of spermatozoa peroxidation damage [31].

The sustained lipid peroxidation chain reaction results in the progressive loss of membrane fatty acids, with a consequent decrease in membrane fluidity, increased non-specific permeability to ions, and inhibition of membrane-bound receptors and enzymes.

Nitric oxide (NO•), oxygen dioxide ($O_2$), and hydrogen peroxide (OH•) have been shown to attach to sperm DNA molecules, causing excessive modifications and deletions in their nucleotide bases and strand breaks, along with other numerous genotoxic effects. Nitric oxide (NO•), $O_2$, and OH• have been shown to mediate lipid peroxidation. Sperm DNA is especially prone to oxidative damage owing to its poor chromatin compaction and incomplete termination. Additionally, ROS has been linked to an apoptotic stimulus that causes the release of certain signaling molecules from the mitochondria, which are necessary for the activation of programmed cell death [32].

### *6.2.1. Sperm Motility*

Male gametes are no different from other living cells in that they need energy to grow and function. Adenosine triphosphate (ATP) is a compound that is specially used by mammalian spermatozoa to keep the intracellular environment stable and for cellular functions like motility, capacitation, hyperactivation, and the acrosome reaction, all of which are essential for successful fertilization. The most crucial aspect of semen samples is thought to be sperm motility for evaluating fertility. WHO states that 40% is the lower reference limit for overall motility [33]. Spermatozoa are highly differentiated cells that have the head, mid-piece, and tail as their three unique parts. The paternal genome is very compressed and is housed in the sperm head's nucleus. The mid-piece, which extends posteriorly from the head, is followed by the flagellum [34] (Figure 6). The fibrous sheath is located inside the flagellum, which contains the primary portion of the spermatozoa. Glycolysis happens in the head and the main component of the flagellum, and the oxidative phosphorylation occurs in the mitochondria, are two metabolic routes that the spermatozoa use to create adenosine triphosphate [35].

Adenosine diphosphate (ADP) is the precursor to ATP, hence adding and removing phosphate groups from ADP molecules is the fundamental step in all biological activities. A spermatozoon may survive only on glycolytic energy, but in order to differentiate and mature, it needs oxidative phosphorylation.

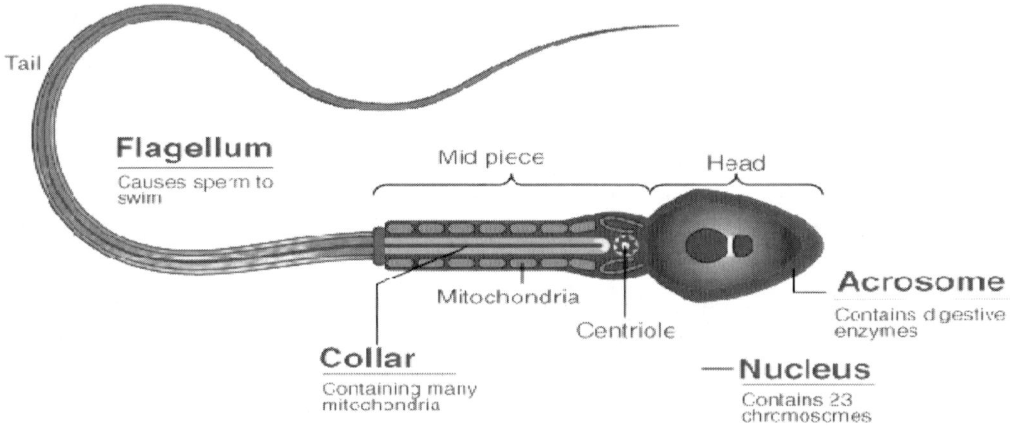


**Figure 6.** Structure of mature sperm.

For the purpose of producing energy, spermatozoa select a very species-specific metabolic pathway. Studies revealed that mice, rats, hamsters and human spermatozoa produce energy through glycolysis for fertilization [36].

In spermatozoa, mitochondria that use ATP from intracellular stores to produce energy cover the axosome and related dense fibers of the middle parts. It is generally known that the presence of ROS can cause axonemal and mitochondrial damage, which immobilizes spermatozoa. It has been established that decreased motility is caused by the ROS-induced peroxidation of lipids in the spermatozoa membrane, which reduces flexibility and suppression of motility mechanisms [37]. Additionally, ROS-induced mitochondrial DNA damage impairs spermatozoa motility by reducing ATP and energy availability, activating caspases, and finally causing apoptosis.

The vulnerability of spermatozoa to oxidative stress is associated with mitochondrial dysfunction leads to low ATP levels, and distortion of the functional Na+-K+ gradient, which permits water inflow into the cell [38]. The generation of cytotoxic aldehydes, which disrupt the electron transport chain and diminish sperm motility, may also result from the oxidative destruction of sperm membrane lipids by ROS [39]. Therefore, high superoxide anion, the most common type of ROS in sperm, has generally been regarded as a sign of low-quality sperm. This concept has been validated by a plethora of research studies on human sperm, which mostly obtain their energy from glycolysis [40]. Therefore, sperm motility is significant in male infertility.

### *6.2.2. Sperm Viability*

Sperm viability, which is frequently used as a sperm quality indicator, is the percentage of alive, membrane-intact spermatozoa in a semen sample. Sperm viability testing is used to detect the percentage of dead or alive sperm in situations where there is low sperm motility, which prevents the sperm from moving swiftly and efficiently toward the egg. As per WHO guidelines, 58% is the lower reference criterion for viable (spermatozoa with intact membranes) sperm [38]. Many sperm that are not moving are still alive but are unable to swim because of metabolic or structural abnormalities. A viability test is performed to evaluate the permeability of the sperm membrane.

Polyunsaturated fatty acids (PUFAs) are abundant in spermatozoa membranes, making them particularly vulnerable to lipid peroxidation. Spermatozoa lack the essential repair systems and are unable to reverse the oxidative damage, making them very sensitive to oxidative stress. Axonemal damage, a decline in sperm viability, and an increase in midpiece sperm morphological abnormalities are all due to oxidative stress. oxidative stress induced ROS production appear to have negative effects on spermatozoa, altering their morphology and dynamic cellular characteristics, which ultimately reduces the percentage of live or viable sperm that negatively impact on fertilization capacity. Lipid peroxidation is specifically linked to changes in membrane fluidity and permeability, supporting the role of oxidative stress in abnormalities in the motility and morphology of sperm [41].

Therefore, we can say, this stress induced ROS generation may negatively impact on membrane permeability of sperm which can easily be tested by popular eosin-nigrosine staining method by which viable and non-viable sperm seen under microscope. Eosin-nigrosine staining measures the ability of eosin to pass through the membranes that indicates the membranes are morphologically altered due to cell injury or death (non-viable). On the other hand, due to the intact permeability of plasma membrane of live sperm, that does not retain eosin that reflects viable sperm. The processes of fertilization require the intact membrane for successful fusion of spermatozoa with egg whereas membrane disruption is one of the important factors for male infertility.

### *6.2.3. Sperm Acrosome*

The anterior region of the sperm nucleus is covered by a specialized, cap-like structure called the acrosome. Despite the fact that the acrosome varies in size and shape between mammalian species, they have the same basic structural characteristics. Just prior to fertilization, the acrosome of mammalian sperm passes through a Ca2+-dependent exocytotic process. Sperm must undergo this process, known as the acrosome reaction, in order to pass through the zona pellucida which is the plasma membrane of female egg cell. The outer acrosomal membrane and the sperm plasma membrane repeatedly fuse during the acrosomal reaction, which was originally observed in sea urchins and starfish [42, 43]. This enables the acrosomal contents to exit the sperm cell. The glycoprotein coat of the oocyte is dissociated by lytic acrosomal materials, creating a hole through which the sperm head enters into the surface of the oocyte before fusing with it. Acrosome reactions must be better understood in order to distinguish between healthy and abnormal human fertilization. Sperm plasma and the outer acrosomal membrane undergo a specific occurrence of fusing and fenestration in this process. Different hydrolytic enzymes and proteolytic enzymes are released after sperm capacitation, mainly acrosin trypsin-like serine protease that hydrolyzes the egg membrane to enable the spermatozoa to navigate the interstices of the corona radiata and hyaluronidase, to digest the zona pellucida in facilitation of successful sperm- zona pellucida penetration in early fertilization process [44]. The acrosome is in an intact state if its enzymes are active. When the female genital tract undergoes the acrosome reaction before to fertilization, an intact acrosome avoids the loss of acrosomal enzymatic activity.

Spermatozoa that have undergone an acrosome reaction only can penetrate the zona pellucida, and fuse with the oocyte. Male infertility may be brought on by an early or premature acrosome response. The morphology of the sperm represents usual sperm development in the testis, indicating proper spermatogenesis. Defective sperm morphology noted in men who have sperm maturation problems, and there is an increased risk male infertility. In various infertile situations, low sperm acrosome integrity has been noted [45]. The degree of sperm acrosin activity has been hypothesized to be a good indicator of sperm reproductive potential. Acrosomes have a unique structure made of proteins and membranes that makes them vulnerable to oxidative stress and high levels of ROS. Sperm proteins and membranes are affected by oxidative stress. ROS are implicated for the pathogenesis of diseases by their oxidative alteration of proteins. Free radicals are oxidized by proteins, which results in various modifications or degradations of the constituent amino acids. Morphology and functioning of acrosome are adversely impacted lipid peroxidation and protein oxidation [46, 47]. The ability of sperm to penetrate oocytes with intact zona is decreased as a result of oxidative stress, which can also result in an early acrosomal reaction.

Physiologically, low level of ROS in semen plays a functional role in acrosomal reaction, but high ROS concentration has a detrimental effect on acrosomal reaction. Based on these findings, we hypothesized that elevated ROS and oxidative stress can create detrimental effects acrosomal reaction in many ways that ultimate goes up towards the male infertility.

#### *6.2.4. Sperm Membrane Integrity*

Several sperm features are assessed while diagnosing male infertility. The functional integrity of the sperm membrane has, however, received very little consideration. In addition to being crucial for sperm metabolism, the integrity of the membrane is also necessary for the spermatozoa's capacity to fertilize, namely the acrosome reaction, and the attachment of the spermatozoon to the egg surface. Therefore, evaluation of membrane function may be a valuable predictor of the spermatozoa's capacity to fertilize, and the integrity and functional activity of the sperm membrane are of vital relevance in the fertilization process. The cell membrane's capacity to allow for selective molecular transport is one of its characteristics. The hypo-osmotic swelling (HOS) test was frequently employed to assess the integrity of the plasma membrane [48] during a long time ago. It has been proposed that the HOS test offers more accurate data regarding the spermatozoa's ability to fertilize than the traditional parameters of semen analysis [49]. Given the significance of the sperm membrane in conception, spermiogram examination pays close attention to the membrane's integrity. Even though the spermatozoa's membrane is structurally intact, accurate fertilization with the egg won't occur if the spermatozoa's membrane is biochemically damaged; for this reason, the HOS test is a better indicator than the eosin-nigrosine stain for fertilization.

When exposed to hypo-osmotic environments, spermatozoa will take up water in an attempt to attain osmotic equilibrium. This water intake will increase sperm volume and cause the plasma membrane to inflate, providing the lowest surface to volume ratio. These hypo-osmotic circumstances seem to be particularly harmful to the sperm tail. This is known as swelling. The capacity of sperm tails to expand in the presence of a hypoosmotic solution is thought to be an indication that normal water transfer across the membrane occurs, i.e., a marker of membrane integrity and normal functional activity. An abnormal HOS test revealed that sperm membrane abnormalities could cause some qualitative defects in embryos, resulting in normal structure but physiologically dysfunctional embryos [50]. Moreover, extremely low implantation rates after IVF and embryo transfer were also connected with abnormally low HOS test findings [51]. Low levels of swelling in the HOS test have been linked to both implantation difficulties and greater risks of miscarriage [52, 53]. Due to the high concentration of polyunsaturated fatty acids (PUFA) in the plasma membrane of the human sperm and the relative deficiency of scavenging antioxidant enzyme in the cytoplasm, the sperm membrane is constantly at risk of oxidative damage. Plasma membrane integrity is impaired by lipid peroxidative damage. Changes in membrane structure and function are carried on due to the lipid peroxidation of the spermatozoa membrane. Male infertility may be greatly impacted by impaired sperm activity caused by oxidative stress-induced damage to the sperm membrane.

## 7. Oxidative Stress and Androgenesis

Androgenesis is also known as steroidogenesis, a major function of testes. There are two major steroids hormones, testosterone and androstenedione, collectively known as androgens, which are mainly synthesized in testes. Testosterone is the vital androgen produced in the interstitial compartment of Leydig cells in the testis and is responsible for maintaining spermatogenesis, sex development, sexual function and male fertility. Testosterone is made from cholesterol via a series of enzymatic reactions. Two major pathways of testosterone formation from pregnanolone, one is through the formation of progesterone, 17α hydroxysteroid progesterone and androstenedione which is known as Δ4 pathway. The other is through the formation of 17α hydroxypregnanolone, dihydroepiandrosterone and androstenediol which is known as Δ5 pathway [54]. At first, cholesterol is transited by the steroidogenic acute regulatory protein (StAR) to the interior mitochondrial membrane, where cholesterol is converted into pregnenolone by a different side-chain cleavage enzyme.

In the classical delta 5 pathway, pregnenolone is first converted into 17 α- hydroxy-pregnenolone by the enzyme 17α-hydroxylase activity and then dehydroepiandrosterone

(DHEA) by the action of C17-C20 lyase. After that, DHEA is turned into androstenedione catalyzed by Δ5, 3β-hydroxysteroid dehydrogenase (Δ5, 3β -HSD) or androstenediol by 17 β- hydroxysteroiddehydrogenase (17β- HSD). Both this androstenedione and androstenediol converted into testosterone by Δ5, 3β - HSD and 17β- HSD, respectively [55]. Sometimes testosterone may be changed into dihydrotestosterone which has greater affinity to the androgen receptor. In humans, there is another pathway which is delta 4 pathways, by which there is a minute conversion of progesterone into 17 α- hydroxyprogesterone and then androstenedione by the poor activity of C17-C20 lyase activity on the substrate 17 hydroxyprogesterone compared to 17 hydroxy-pregnenolone.

Oxidative stress induced ROS can hamper testicular androgenesis or steroidogenesis due to the diminution of antioxidant enzyme defense system. The steroidogenic acute regulatory protein (StAR) and 3-hydroxysteroid dehydrogenase isomerase (3-HSD) activities were significantly reduced as a result of the oxidative stress imposed by lipopolysaccharide in Leydig cell membranes. Furthermore, the inhibition of StAR-mediated cholesterol transfer activity and the alteration of Leydig cell mitochondrial function were linked to these effects [56]. Activities of androgenic key enzymes i.e., 3β - HSD and 17β- HSD of cultured Leydig cell were decreased. Reverse transcription study confirms that, the diminution may be due to the low level of StAR protein. In oxidative stress there is an imbalance redox homeostasis that leads to the elevation of oxidants concentrations that inhibit the gene expressions of steroidogenic enzymes thereby altering androgenesis was noted (Figure 7).

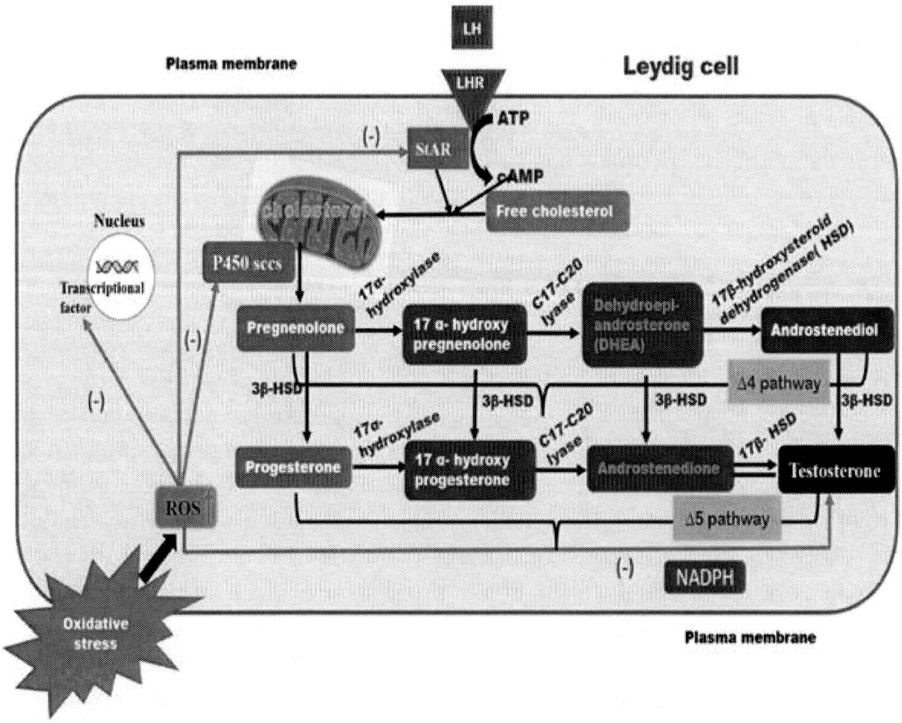


**Figure 7.** Effect of oxidative stress on androgenesis.

## 8. Oxidative Stress and Germ Cell Apoptosis

Apoptosis, or "programmed cell death," is a physiological mechanism that enables cells to self-destruct and results in the regulated death of cells. In order to control the proportion of Sertoli cells to germ cells, apoptosis is a crucial component of proper spermatogenesis. A large wave of apoptosis occurs in the testes between two and four weeks after birth during the first cycle of spermatogenesis in order to maintain the delicate balance between Sertoli cells and germ cells. [57]. By preventing the overproduction of germ cells from seminiferous tubules, apoptosis plays a critical function in maturation in selectively killing the premeiotic spermatogonia during the initial stage of spermatogenesis. Caspases, a particular family of cysteine proteases, regulate molecular mechanisms of apoptosis. Mammalian caspases can be divided into two functional groups on the basis of the entry position into the apoptosis cascade: initiator caspases (caspase 2, 8, 9, and 10) and executioner caspases (caspase 3, 6, and 7). Initiator caspases initiate the apoptosis signal and activate the executioner caspase while the executioner caspases conducted the mass proteolysis for apoptosis. Apoptotic caspases are initiated after the signal transmitted either through the extrinsic (death receptor mediated pathway) or an intrinsic (mitochondrial pathway) pathway. When a ligand binds to cell surface death receptors, it causes Pro-caspase-8 and -10 to cleave and initiate the extrinsic pathway. These receptors include tumor necrosis factor receptor 1 (TNFR 1), TNF-related apoptosis-inducing ligand (TRAIL), FasR etc. Alteration of mitochondrial membrane permeability (MMP) trigger the release of Cytochrome C and the creation of the apoptosome that activates Caspase-9, thereby intrinsic pathway is initiated. Activated Caspase-9 then cleaves Caspase-3, which drives subsequent cell death-related events.

Anti-apoptotic proteins, including Bcl-2, Bcl-XL, and Bcl-w, which are found in the outer mitochondrial membrane, regulate or prevent the release of cytochrome C. In order to initiate apoptosis, pro-apoptotic BaX, Bid, and BaK must translocate to mitochondria by creating mitochondrial pores and competing with these anti-apoptotic proteins [58]. Numerous situations, including oxidative stress, cause abnormal testicular apoptosis. The testis is vulnerable to several stresses, including inflammation, radiation, and contact with various substances that cause germ cell apoptosis [59]. Oxidative stress induced ROS generation alters the mitochondrial membrane permeability, expression of germ cell pro-apoptotic proteins namely, Bax, Bad and anti-apoptotic protein, Bcl2. The intrinsic apoptotic signaling pathway is connected to mitochondrial activities, such as ROS production. DNA damage and cellular apoptosis are greatly influenced by increased ROS. Elevated ROS production causes the release of soluble proteins from the intermembrane gap that activate caspases in the cytosol, such as cytochrome c, that influences the receptivity to apoptotic signals [60]. However, for germ cell apoptosis, intrinsic pathway mainly responsible over extrinsic pathway and there are very limited studies for extrinsic pathway induced germ cell apoptosis (Figure 8).

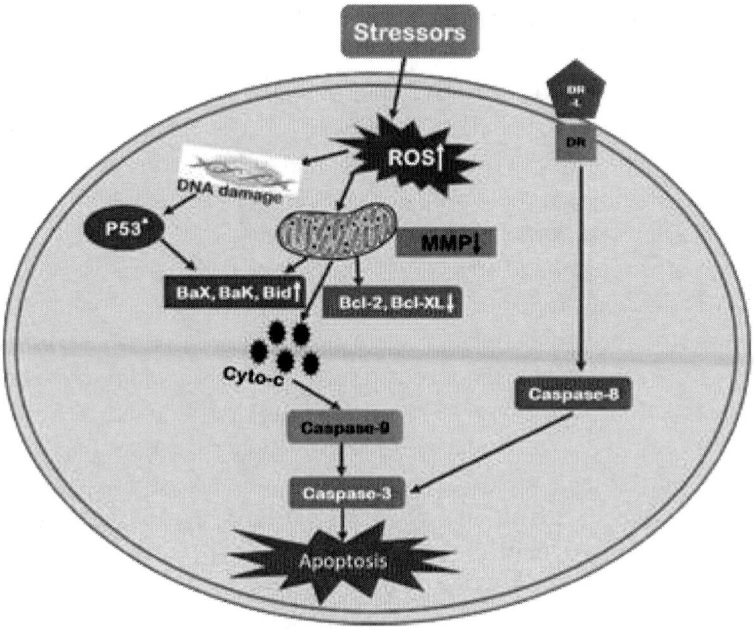


**Figure 8.** Effect of oxidative stress on germ cell apoptosis.

## 9. Oxidative Stress and Male Infertility

Infertility is generally understood to be the inability to conceive after at least one year of unprotected intercourse. Unfortunately, 15% of couples struggle to conceive and require medical guidance to increase their chances of a healthy pregnancy [61]. As per data, 48.5 million couples suffer from infertility worldwide [62]. Male infertility problems accounts for around half of all occurrences of infertility and this problem experience one in 20 males in the reproductive age range (between puberty and 40 years of age). Between 2.5% to 15% of men worldwide are infertile, which correlates to at approximately 30 million men globally. In general, female factor contribute to 50% cases of infertility, 20-30% of the fertility related problems are due to male, and other 20-30% is due to the mixed factors of both male and female [63]. It is now known that 30-80% of cases may have a significant contributory pathology caused by reactive oxygen species (ROS) that causes damage to sperm [64]. According to World Health Organization recommendations, the male factor is one of the factors that directly contributes to low sperm concentration and motility. The prevailing thought is that oxidative stress (OS) is a contributing component and one of the likely causes of idiopathic male infertility. As per statistics from the United States, male infertility is caused by oxidative stress and spermatozoa are more vulnerable to free radicals. Supplementing with antioxidants can sometimes repair the impairment. As a result, OS has recently been a major focus for medical professionals and researchers due to its direct impact on fertility.

## 10. Antioxidant as a Free Radical Quencher : An Overview

Oxygen is highly reactive atom that is an important part of highly potential damaging molecules known as "free radicals." Free radicals attack the healthy cells of the body resulting in loss of structure and function of those molecules. Humans have evolved a highly complex antioxidant protection system in the body to protect the cells and organ system against ROS. It has both endogenous and exogenous origin that function interactively and synergistically to neutralize free radicals [65]. The component of the antioxidant include:

- *Nutrient-derived antioxidant* includes carotenoids, ascorbic acid (vitamin C), tocopherols and tocotrienols (Vitamin E), other low molecular weight compounds such as glutathione and lipoic acid.
- *Antioxidant enzymes* such as catalase, superoxide dismutase, glutathione peroxidase and glutathione reductase catalyze free radical quenching reactions.
- *Metal binding proteins,* such as ferritin, lactoferin, albumin, and ceruloplsmin that sequester free iron and copper ions can catalyzes oxidative reaction.

### 10.1. Antioxidant Blocks Oxidation: Possible Ways

Antioxidant block oxidation process by neutralizing free radicals. For this purpose, antioxidants themselves become oxidized. This phenomenon takes place in two possible pathways:

#### *10.1.1. Chain – Breaking*

An individual radical may form when a free radical releases or abstracts an electron. The newly formed second molecule then does the same thing to a third molecule and like this more unstable product are generated. The process can be terminated either by the stabilization of the radical by a chain- breaking antioxidant such as β-carotene, vitamin C and E or its simply degraded into a harmless product [66].

#### *10.1.2. Preventive*

Simultaneously, antioxidants stop oxidation by minimizing the chain initiation rate. This occurs either by scavenging radicals or preventing oxidation by stabilizing transition metal radicals i.e., copper and iron.

#### *10.1.2. Mechanism of Free Radical Destruction by Antioxidant*

The rate of chemical reactions of lipid oxidation can be and or slow down if a compound inhibits the formation of free radicals in the initiation step interrupts the free radical chain propagation. The initiation of free radical formation can be delayed by metal peroxide stabilizers, chelating agents and singlet oxygen inhibitors. The propagation of free radical chain reaction can be minimized by the donation of hydrogen from the antioxidants and the metal chelating agents (Figure 9).

(1) R:H + O::O + Initiator $\longrightarrow$ R* + HOO*
(2) R* + O::O $\longrightarrow$ ROO*
(3) ROO* + R:H $\longrightarrow$ ROOH + R*
(4) RO:OH $\longrightarrow$ RO* + HO*
(5) R::R + *OH $\longrightarrow$ R:R-O*
(6) R* +R* $\longrightarrow$ R:R
(7) R* +ROO* $\longrightarrow$ ROOR
(8) ROO* + ROO* $\longrightarrow$ ROOR + $O_2$
(9) ROO* + AH $\longrightarrow$ ROOH + A*
(10)ROO* + A* $\longrightarrow$ ROOA.

**Figure 9.** Mechanism of action of antioxidants for the free radical destruction.

Antioxidants have a direct role on defense systems at different levels such as preventive, radical scavenging, repair etc.

## 10.2. Role of Different Antioxidants as a Defense System at Different Levels

- *The first line of defense* is the preventive antioxidants as it suppresses free radicals' formation. The exact mechanism and the site of free radical formation within the cell are not well elucidated yet but hydroperoxides and hydrogen peroxide which are originated from metal-induced decompositions is one of the important sources of free radical generation. Glutathione peroxidase, phospholipid hydroperoxide glutathione peroxidase (PHGPX), glutathione-s-transferase and peroxidase are known to destroy lipid hydroperoxides to corresponding alcohols.
- *The second line defense* is the antioxidants that scavenge the active radicals to suppress chain initiation and/or break the chain propagation reactions. Various endogenous radical-scavenging antioxidants are known, some are hydrophilic, and others are lipophilic. Vitamin C, uric acid, bilirubin, albumin and thiols are hydrophilic, radical scavenging antioxidants where vitamin E is considered as the most potent radical-scavenging lipophilic antioxidant.
- *The third line of defense* is known as repair and *de novo* antioxidants. This category of antioxidants only comes into play after free radical damage has occurred. These *de novo* enzymes help to repair the damage to biomolecules caused by free radicals and reconstitute the damaged cell membrane. The DNA repair system also plays an important role in the total defense system against oxidative damage. Various kinds of enzymes such as glycosylases and nucleases, which repair damaged DNA are under third line defense [67].

There is another important function called adaptation where the signal for the production and reactions of free radicals induces formation and transport of the appropriate antioxidant to the right site.

Two principal mechanisms of action have been proposed to establish antioxidant activity. The first is a chain-breaking mechanism by which the primary antioxidant donates an electron to the free radical present in the systems.

The second mechanism involves removal of ROS/reactive nitrogen species initiators (secondary antioxidants) by reducing the chain-initiating catalyst. Antioxidants may exert their effect on biological system by different mechanism including electron donation, metal ion chelation, co-antioxidants or by gene expression regulation etc.

### 10.3. Types of Antioxidants

Antioxidant system includes antioxidant enzymes, nutrient-derived antioxidants, metal binding proteins and numerous other antioxidant phyto-nutrients present in a wide variety of plant foods [68].

#### *10.3.1. Natural Antioxidant*

Natural antioxidants are widely distributed in food and medicinal plants. These natural antioxidants, especially polyphenols and carotenoids, exhibit a wide range of biological effects, including anti-inflammatory, anti-aging, anti-atherosclerosis and anti-cancer.

#### *10.3.2. Enzymatic Antioxidants:*

Catalase, peroxidase, superoxide dismutase, glutathione-s-transferase, glutathione reductase etc.

#### *10.3.3. Non-Enzymatic Antioxidants*

Ascorbic acid, alpha-tocopherol, carotene, lycopene, lutein etc.

The balance between oxidants and antioxidants in favour of oxidants is termed "oxidative stress." Regulation of reducing and oxidizing (redox) state is critical for cell viability, activation, proliferation. The various antioxidants scavenge the superoxide and free radicals or prevent the cellular oxidation of oxidizable substance resulting in decreased free radicals' formation and thus in prevention of many pathophysiological conditions. Certain quantities of ROS are necessary for our physiological functions and / or normal cell functions but it is injurious in larger amounts for body's natural antioxidant defence system that cause different diseases. Now a day oxidative stress is associated with a huge number of pathophysiological conditions including infertility, diabetes, cardiovascular disease, neuropathies and cancer etc.

## 11. Impact of Dietary Antioxidant in Male Infertility Management

The word "Nutraceutical" was coming into focus in the 1989 from "nutritional" and "pharmaceutical" terms and recently it is used to indicate food, or part of the food, which provides important benefits for human health, by enhancing treatment and preventing disease. Examples of some nutraceuticals having direct role in male reproductive system is as follows:

### 11.1. Myo-Inositol

It is the stereoisomer of inositol which is a carbocyclic polyol. It is generally found in cereals, nuts, fruits as well as in human cells. Myo-inositol is synthesized from glucose-6-phosphate and in the male reproductive system is mainly produced by follicle- stimulating hormone (FSH)-responsive Sertoli cells. It is involved in many processes such as capacitation, acrosome reaction and regulation of sperm motility. It restores mitochondrial cristae morphology and improves the mitochondrial membrane potential.

### 11.2. L-carnitine

It is highly concentrated in epididymis and seminal fluid. It is involved in the mitochondrial β-oxidation of long chain free fatty acids. Low level of L-carnitine reduces fatty acid within mitochondria reducing energy production and sperm motility. L-carnitine administration results in an improvement in seminal parameters, increasing sperm quality and motility [69].

### 11.3. Selenium

It is an essential trace element mainly found in grains, meat, fish, poultry, and eggs. It is a constituent of selenoproteins that serves as structural components of mature spermatozoa and is involved in some reproductive functions such as testosterone metabolism. Sperm capsular selenoprotein has an important structural role in spermatozoa in the form of glutathione peroxidase, which protect against peroxidation.

### 11.4. Folic Acid

It is one of the most important vitamins because it is involved in cell development and is essential for DNA, transfer RNA and protein synthesis. Alteration of folic acid metabolism has been linked to male subfertility. In a murine model it has been shown that a paternal insufficient folate intake alters the mouse sperm epigenome negativity impacting on pregnancy outcomes.

### 11.5. Zinc

It is a trace mineral essential for normal functioning of the male reproduction system. Zinc deficiency has been linked with male sterility and subfertility, playing a crucial role in normal testicular development and spermatogenesis. It affects the stability of sperm chromatin and in general biological membranes as it influences the fluidity of lipids. Exposure to stress, cigarette, smoke, pollution, and alcohol can deplete zinc and its deficiency has been linked also to low sperm counts and testosterone levels [70].

### 11.6. Green Tea

Tea (*Camellia sinensis*) is one of the most popular beverages in many countries. The main reason for this interest is the awareness of people about the therapeutic properties of this drink. Tea contains, group of catechins that includes catechin, epicatechin, gallocatechin, epigallocatechin, catechingallete, epicatechingallate, gallocatechingallate, and epigallocatechingallate and vitamins, minerals polyphenols, flavonols (theaflavins, thearubigins) adenine, theobromine, theophylline, xanthine, gallotannin, small amounts of aminophylline and a yellow volatile oil which is solid at room temperatures and has strong aromatic smell and flavor. A significant decrease of ROS was observed after treatment with green tea [71].

### 11.7. Garlic

Garlic *(Allium sativum)* has been used as medicine worldwide since ancient times. Diallyltetrasulphide is a bioactive organo-sulphur component of garlic, which has an antioxidant effect. Garlic contains a wide variety of phytochemical and micro-components such as trace elements, vitamins, fructans, flavonoids and sulphur compounds, which can effectively scavenge free radicals. Some studies have reported that garlic harms the function of testes and has spermicidal effect on spermatozoa, but some others demonstrated the opposite effect. Rats fed with 10%, 15% or 30% *Allium sativum* (AS) containing diet, showed a significant increase in the number of empty seminiferous tubules in testes and a decrease in the level of testosterone. Male rat fed with crude garlic showed damaged testicular function and spermatogenesis. Also decreased prostate weight was associated with a reduction in citric acid contained in rats fed with 30% garlic [72].

### 11.8. Lycopene

Lycopene is a beta carotenoid, well available in tomato, gac fruit, watermelon, carrot etc. It is a tetraterpin assembled from 8-isoprene unit. Lycopene is a potent antioxidant. The free radical quenching ability of lycopene is well established. Lycopene can restore the infertile condition in male by different ways. Lycopene can be able to minimize the free radical by enhancing the gene expression of the antioxidant enzymes like testicular catalase, glutathione-s-transferase, peroxidase, superoxide dismutase etc. On the other way lycopene can be able to minimize the infertility by reducing the testicular apoptosis [73].

## References


[1] Jung KJ, Lee EK, Kim JY, Zou Y, Sung B, Heo HS, Kim MK, Lee J, Kim ND, Yu BP, Chung HY. Effect of short term calorie restriction on pro-inflammatory NF-kB and AP-1 in aged rat kidney. *Inflammation Research*. 2009 Mar; 58:143-50.

[2] Limón Pacheco JH, Carballo MA, Gonsebatt ME. Antioxidants against environmental factor-induced oxidative stress. *Nutritional Antioxidant Therapies: Treatments and Perspectives*. 2017:189-215.

[3] Xie X, He Z, Chen N, Tang Z, Wang Q, Cai Y. The roles of environmental factors in regulation of oxidative stress in plant. *BioMed research international.* 2019 Oct; 2019.

[4] Waszczak C, Carmody M, Kangasjärvi J. Reactive oxygen species in plant signaling. *Annual review of plant biology*. 2018 Apr 29; 69:209-36.

[5] Edreva A. Generation and scavenging of reactive oxygen species in chloroplasts: a submolecular approach. *Agriculture, Ecosystems & Environment.* 2005 Apr 2; 106(2-3):119-33.

[6] Hayyan M, Hashim MA, AlNashef IM. Superoxide ion: generation and chemical implications. *Chemical reviews.* 2016 Mar 9; 116(5):3029-85.

[7] Martinez MC, Andriantsitohaina R. Reactive nitrogen species: molecular mechanisms and potential significance in health and disease. *Antioxidants & redox signaling.* 2009 Mar 1; 11(3):669-702.

[8] Pacher P, Beckman JS, Liaudet L. Nitric oxide and peroxynitrite in health and disease. *Physiological reviews.* 2007 Jan; 87(1):315-424.

[9] Wu M, Desai DH, Kakarla SK, Katta A, Paturi S, Gutta AK, Rice KM, Walker Jr EM, Blough ER. Acetaminophen prevents aging-associated hyperglycemia in aged rats: effect of aging-associated hyperactivation of p38-MAPK and ERK1/2. *Diabetes/metabolism Research and Reviews.* 2009 Mar; 25(3):279-86.

[10] Thannickal VJ, Fanburg BL. Reactive oxygen species in cell signaling. *American Journal of Physiology-Lung Cellular and Molecular Physiology.* 2000 Dec 1; 279(6):L1005-28.

[11] Son Y, Cheong YK, Kim NH, Chung HT, Kang DG, Pae HO. Mitogen-activated protein kinases and reactive oxygen species: how can ROS activate MAPK pathways? *Journal of Signal Transduction.* 2011; 2011:792639.

[12] Baeuerle Pa and Baltimore D (1996). Nf-Kb: Ten Years After. *Cell* 87: 13–20.

[13] Manna SK, Zhang HJ, Yan T, Oberley LW, Aggarwal BB. Overexpression of manganese superoxide dismutase suppresses tumor necrosis factor-induced apoptosis and activation of nuclear transcription factor-κB and activated protein-1. *Journal of Biological Chemistry.* 1998 May 22; 273(21):13245-54.

[14] Galter D, Mihm S, DRÖGE W. Distinct effects of glutathione disulphide on the nuclear transcription factors κB and the activator protein-1. *European Journal of Biochemistry.* 1994 Apr; 221(2):639-48.

[15] Matsuda S, Moriguchi T, Koyasu S, Nishida E. T lymphocyte activation signals for interleukin-2 production involve activation of MKK6-p38 and MKK7-SAPK/JNK signaling pathways sensitive to cyclosporin A. *Journal of Biological Chemistry.* 1998 May 15; 273(20):12378-82.

[16] Alahmar AT. Role of oxidative stress in male infertility: an updated review. *Journal of human reproductive sciences.* 2019 Jan; 12(1):4.

[17] Baker HW. Relative incidence of etiological disorders in male infertility. *Male reproductive dysfunction.* 1986:341-72.

[18] Santen RJ, Paulsen CA. Hypogonadotropic eunuchoidism. I. Clinical study of the mode of inheritance. *The Journal of Clinical Endocrinology & Metabolism.* 1973 Jan 1; 36(1):47-54.

[19] Hedger MP, de Kretser DM. Leydig cell function and its regulation. *The genetic basis of male infertility.* 2000:69-110.

[20] Hall PF, Eik-nes KB. The influence of gonadotropins *in vivo* upon the biosynthesis of androgens by homogenate of rat testis. *Biochimica et biophysica acta.* 1963 Jan 1; 71:438-47.

[21] Aitken RJ. Reactive oxygen species as mediators of sperm capacitation and pathological damage. *Molecular reproduction and development.* 2017 Oct; 84(10):1039-52.

[22] Wagner H, Cheng JW, Ko EY. Role of reactive oxygen species in male infertility: An updated review of literature. *Arab journal of urology.* 2018 Mar 1; 16(1):35-43.

[23] Kothari S, Thompson A, Agarwal A, du Plessis SS. Free radicals: their beneficial and detrimental effects on sperm function. *Indian J Exp Biol* 2010 May; 48(5):425-35.

[24] Griveau JE, Renard P, Lannou DL. An *in vitro* promoting role for hydrogen peroxide in human sperm capacitation. *International journal of andrology.* 1994 Dec; 17(6):300-7.

[25] O'Flaherty C, de Lamirande E, Gagnon C. Reactive oxygen species modulate independent protein phosphorylation pathways during human sperm capacitation. *Free Radical Biology and Medicine.* 2006 Mar 15; 40(6):1045-55.

[26] van der Horst G, du Plessis SS. Not just the marriage of Figaro: but the marriage of WHO/ESHRE semen analysis criteria with sperm functionality. *Adv Androl Online.* 2017; 4:6-21.

[27] Pahune PP, Choudhari AR, Muley PA. The total antioxidant power of semen and its correlation with the fertility potential of human male subjects. *Journal of clinical and diagnostic research: JCDR*. 2013 Jun 1; 7(6):991-5.

[28] Oborna I, Fingerova H, Novotny J, Brezinova J, Svobodova M, Aziz N. Reactive oxygen species in human semen in relation to leukocyte contamination. *Biomedical Papers of the Medical Faculty of Palacky University in Olomouc*. 2009 Mar 1; 153(1).

[29] Ayala A, Muñoz MF, Argüelles S. Lipid peroxidation: production, metabolism, and signaling mechanisms of malondialdehyde and 4-hydroxy-2-nonenal. *Oxid Med Cell Longev*. 2014; 2014:360438.

[30] Collodel G, Moretti E, Micheli L, Menchiari A, Moltoni L, Cerretani D. Semen characteristics and malondialdehyde levels in men with different reproductive problems. *Andrology*. 2015 Mar; 3(2): 280-6.

[31] Mahfouz RZ, du Plessis SS, Aziz N, Sharma R, Sabanegh E, Agarwal A. Sperm viability, apoptosis, and intracellular reactive oxygen species levels in human spermatozoa before and after induction of oxidative stress. *Fertility and Sterility*. 2010 Feb 1; 93(3):814-21.

[32] World Health Organisation. WHO laboratory manual for the examination of human semen and sperm-cervical mucus interaction. Cambridge university press; 1999 May 13.

[33] Malo AF, Gomendio M, Garde J, Lang-Lenton B, Soler AJ, Roldan ER. Sperm design and sperm function. *Biology Letters*. 2006 Jun 22; 2(2):246-9.

[34] Fouquet JP, Kann ML, Edde B, Wolff A, Desbruyeres E, Denoulet P. Differential distribution of glutamylated tubulin during spermatogenesis in mammalian testis. *Cell Motility and the Cytoskeleton*. 1994; 27(1):49-58.

[35] Miki K, Qu W, Goulding EH, Willis WD, Bunch DO, Strader LF, Perreault SD, Eddy EM, O'Brien DA. Glyceraldehyde 3-phosphate dehydrogenase-S, a sperm-specific glycolytic enzyme, is required for sperm motility and male fertility. *Proceedings of the National Academy of Sciences*. 2004 Nov 23; 101(47):16501-6.

[36] Armstrong JS, Rajasekaran M, Chamulitrat W, Gatti P, Hellstrom WJ, Sikka SC. Characterization of reactive oxygen species induced effects on human spermatozoa movement and energy metabolism. *Free Radical Biology and Medicine*. 1999 Apr 1; 26(7-8):869-80.

[37] Woo AL, James PF, Lingrel JB. Sperm motility is dependent on a unique isoform of the Na, K-ATPase. *Journal of Biological Chemistry*. 2000 Jul 7; 275(27):20693-9.

[38] Chen JJ, Yu BP. Alterations in mitochondrial membrane fluidity by lipid peroxidation products. *Free Radical Biology and Medicine*. 1994 Nov 1; 17(5):411-8.

[39] Mazzilli F, Rossi T, Marchesini M, Ronconi C, Dondero F. Superoxide anion in human semen related to seminal parameters and clinical aspects. *Fertility and Sterility*. 1994 Oct 1; 62(4):862-8.

[40] Nowicka-Bauer K, Nixon B. Molecular changes induced by oxidative stress that impair human sperm motility. *Antioxidants*. 2020 Feb 4; 9(2):134.

[41] Dan JC. Studies on the acrosome. I. Reaction to egg-water and other stimuli. *The Biological Bulletin*. 1952 Aug; 103(1):54-66.

[42] Dan JC, Kitahara A, Kohri T. Studies on the acrosome. II. Acrosome reaction in starfish spermatozoa. *The Biological Bulletin. 1954* Oct; 107(2):203-18.

[43] Zalata AA, Ahmed AH, Allamaneni SS, Comhaire FH, Agarwal A. Relationship between acrosin activity of human spermatozoa and oxidative stress. *Asian Journal of Andrology*. 2004 Dec 1; 6(4):313-8.

[44] Tzvetkova P, ZHU WJ, Jing LI, Tzvetkov D. Evaluation on sperm acrosome integrity of infertile men with varicocele. *Journal of Reproduction and Contraception*. 2007 Mar 1; 18(1):19-24.

[45] deLamirande E, O'Flaherty C. Sperm activation: role of reactive oxygen species and kinases. Biochimica et Biophysica Acta (BBA)-Proteins and Proteomics. 2008 Jan 1; 1784(1):106-15.

[46] Chen SJ, Allam JP, Duan YG, Haidl G. Influence of reactive oxygen species on human sperm functions and fertilizing capacity including therapeutical approaches. *Archives of gynecology and obstetrics*. 2013 Jul; 288:191-9.

[47] Curry MR, Watson PF. Osmotic effects on ram and human sperm membranes in relation to thawing injury. *Cryobiology*. 1994 Feb 1; 31(1):39-46.

[48] Jeyendran RS, Van der Ven HH, Perez-Pelaez M, Crabo BG, Zaneveld LJ. Development of an assay to assess the functional integrity of the human sperm membrane and its relationship to other semen characteristics. *Reproduction.* 1984 Jan 1; 70(1):219-28.
[49] Kiefer D, Check JH, Katsoff D. The value of motile density, strict morphology, and the hypoosmotic swelling test in *in vitro* fertilization-embryo transfer. *Archives of andrology.* 1996 Jan 1; 37(1):57-60.
[50] Biljan MM, Buckett WM, Taylor CT, Luckas M, Aird I, Kingsland CR, Lewis-Jones DI. Effect of abnormal hypo-osmotic swelling test on fertilization rate and pregnancy outcome in *in vitro* fertilization cycles. *Fertility and sterility.* 1996 Sep 1; 66(3):412-6.
[51] Buckett WM, Luckas MJ, Aird IA, Farquharson RG, Kingsland CR, Lewis-Jones DI. The hypo-osmotic swelling test in recurrent miscarriage. *Fertility and Sterility.* 1997 Sep 1; 68(3):506-9.
[52] Check JH, Check ML, Katsoff D. Prognosis for sperm fertilizability: analysis of different variables in men. *Archives of Andrology.* 2002 Jan 1; 48(1):73-83.
[53] Stanczik FZ. (1997). Steroid hormone In: *Infertility, contraception, and reproductive endocrinology.* Rojerio AL, Daniel RM, Richard JP, Donna S (Eds.). pp. 35-46. Massachusetts, Blackwell science.
[54] Sharpe RM. Regulation of spermatogenesis. *The physiology of reproduction.* 1994.
[55] Allen JA, Diemer T, Janus P, Hales KH, Hales DB. Bacterial endotoxin lipopolysaccharide and reactive oxygen species inhibit Leydig cell steroidogenesis via perturbation of mitochondria. *Endocrine.* 2004 Dec; 25:265-75.
[56] Rodriguez I, Ody C, Araki K, Garcia I, Vassalli P. An early and massive wave of germinal cell apoptosis is required for the development of functional spermatogenesis. *The EMBO journal.* 1997 May 1; 16(9):2262-70.
[57] Baines CP. The mitochondrial permeability transition pore and ischemia-reperfusion injury. *Basic research in cardiology.* 2009 Mar; 104:181-8.
[58] Kasahara E, Sato EF, Miyoshi M, Konaka R, Hiramoto K, Sasaki J, Tokuda M, Nakano Y, Inoue M. Role of oxidative stress in germ cell apoptosis induced by di (2-ethylhexyl) phthalate. *Biochemical Journal.* 2002 Aug 1; 365(3):849-56.
[59] Fan XY, Chen XY, Liu YJ, Zhong HM, Jiang FL, Liu Y. Oxidative stress-mediated intrinsic apoptosis in human promyelocytic leukemia HL-60 cells induced by organic arsenicals. *Scientific Reports. 2016* Jul 19; 6(1):29865.
[60] Agarwal A, Mulgund A, Hamada A, Chyatte MR. A unique view on male infertility around the globe. *Reproductive Biology and Endocrinology.* 2015 Dec; 13(1):1-9.
[61] Martinez G, Daniels K, Chandra A. *Fertility of men and women aged 15-44 years in the United States: National Survey of Family Growth,* 2006-2010. Department of Health and Human Services, Centers for Disease Control and Prevention, National Center for Health Statistics; 2012.
[62] Sharlip ID, Jarow JP, Belker AM, Lipshultz LI, Sigman M, Thomas AJ, Schlegel PN, Howards SS, Nehra A, Damewood MD, Overstreet JW. Best practice policies for male infertility. *Fertility and Sterility.* 2002 May 1; 77(5):873-82.
[63] Tremellen K. Oxidative stress and male infertility—a clinical perspective. *Human reproduction update.* 2008 May 1; 14(3):243-58.
[64] Jacob RA. The integrated antioxidant system. *Nutrition Research.* 1995 May 1; 15(5):755-66.
[65] Atta EM, Mohamed NH, Silaev AA. Antioxidants: An overview on the natural and synthetic types. *European Chemical Bulletin.* 2017; 6(8):365-75.
[66] Lobo V, Patil A, Phatak A, Chandra N. Free radicals, antioxidants and functional foods: Impact on human health. *Pharmacognosy reviews.* 2010 Jul; 4(8):118.
[67] Krishnamurthy P, Wadhwani A. Antioxidant enzymes and human health. *Antioxidant Enzyme.* 2012 Oct 3; 1:3-18.
[68] De Luca MN, Colone M, Gambioli R, Stringaro A, Unfer V. Oxidative stress and male fertility: role of antioxidants and inositols. *Antioxidants.* 2021 Aug 13;10(8):1283.
[69] Garolla A, Petre GC, Francini-Pesenti F, De Toni L, Vitagliano A, Di Nisio A, Foresta C. Dietary supplements for male infertility: a critical evaluation of their composition. *Nutrients.* 2020 May 19; 12(5):1472.
[70] Roychoudhury S, Agarwal A, Virk G, Cho CL. Potential role of green tea catechins in the management of oxidative stress-associated infertility. *Reproductive biomedicine online.* 2017 May 1; 34(5):487-98.

[71] Musavi H, Tabnak M, Sheini FA, Bezvan MH, Amidi F, Abbasi M. Effect of garlic (Allium sativum) on male fertility: a systematic review. *Journal of Herbmed Pharmacology*. 2018 Oct 7; 7(4):306-12.

[72] Tripathy A, Ara F, Ghosh P, Ghosh D. Effect of lycopene on testicular androgenic and anti-oxidative enzymes in cyproterone acetate-induced male infertile Wistar strain albino rats: An *in vitro* study. *Andrologia*. 2020 Apr; 52(3):e13494.

[73] Squadrito GL, Pryor WA. Oxidative chemistry of nitric oxide: the roles of superoxide, peroxynitrite, and carbon dioxide. *Free Radical Biology and Medicine*. 1998 Sep 1; 25(4-5):392-403.

## Chapter 18

# Fluoride and Arsenic Mediated Oxidative Damage Causing Disorders in the Reproductive System

**Priyankar Pal**
**Sagnik Biswas**
**Rubia Mondal**
**and Prabir Kumar Mukhopadhyay***
Department of Life Sciences, Presidency University, Kolkata, India

### Abstract

A number of exogenous agents and endogenous processes generate reactive oxygen species (ROS) and/or reactive nitrogen species (RNS) within living systems. ROS and RNS provide cellular protection, maintain cell activities, and act as signal transduction molecules at optimal concentration but their excess production can lead to irreversible oxidation of proteins, lipids, and nucleic acids. A state of imbalance between the prooxidants (ROS and RNS) and antioxidant defense produces oxidative stress and disorders of different physiological systems. Groundwater contaminated with fluoride and arsenic acting as exogenous agents is responsible for causing several system disorders in animals and human beings. Male and female reproductive systems are vulnerable to oxidative damage caused by these two contaminants. Fluoride exhibits an adverse impact on the male reproductive system in terms of structural alterations of the testis, epididymis, and spermatozoa along with reduced steroidogenesis and fertility. The abundance of polyunsaturated fatty acids and decreased antioxidant defense make spermatozoa more vulnerable to fluoride-mediated oxidative damage. Altered spermatogenesis, decreased serum testosterone, and structural abnormalities of reproductive organs are the key features of arsenic-mediated male reprotoxicities. Structural anomalies of the ovary and uterus particularly follicular dysplasia, altered endometrial receptivity, adverse embryo implantation, and decreased litter size, have been well characterized due to overexposed fluoride. Suppressed ovarian steroidogenesis, degeneration of ovarian follicles and uterine cells, and prolonged diestrus are the adverse outcomes of arsenic exposure. Apart from animal experimentations, epidemiological data also demonstrate the reprotoxic effects of these two environmental pollutants due to enhanced oxidative stress in the respective reproductive organs.

* Corresponding Author's Email: prabir.dbs@presiuniv.ac.in.

In: The Role of Reactive Oxygen Species in Health and Disease
Editor: Kuladip Jana
ISBN: 979-8-89113-283-2

**Keywords:** Fluoride, arsenic, ROS, reproductive toxicity, sperm, ovum, antioxidants

## Introduction

Oxygen plays a fundamental role in respiration and energy metabolism within most living creatures. In biological systems, it exerts both beneficial effects along with damaging consequences and this self-contradictory nature of oxygen is usually termed as oxygen paradox. Within the reducing environment of the intracellular milieu, oxygen may undergo univalent partial reduction and generate different reactive oxygen species (ROS) which are the products of aerobic environment and thus generate oxygen toxicity [1]. Singlet oxygen ($^1O_2$), superoxide radicals ($O_2^{\bullet-}$), hydroxyl radicals (•OH) and hydrogen peroxide ($H_2O_2$) are commonly defined as ROS which are the metabolic by-products, and any living system is under constant attack from such free radicals. Interestingly, complex antioxidant defense mechanisms are present in living systems which can combat ROS mediated cellular attack. A proper balance between ROS generation and antioxidant defensive ability needs to be required for normal cellular function, while perturbation of such equilibrium leads to generation of oxidative stress. This stress is considered as a potential reason for pathophysiology causing different system disorders.

Several endogenous and exogenous agents are liable for the generation of prooxidants which in turn cause oxidative stress. Cellular metabolism, drug metabolism, endogenous metabolism, ion influx, and pathophysiology are the major probable sources of the generation of endogenous prooxidants; while dietary ingredients, pathogens, drugs, and toxicants are the major sources of exogenous agents [2]. Environmental toxicants/pollutants from several natural and anthropogenic sources have led to serious health hazards and make the different life forms on earth more unsafe. These toxicants include various biochemically synthetic compounds, toxic by-products from several industrial discharges, metals, metalloids, and different organic compounds [3]. Based on vertebrate studies it has been reported that environmental toxicants are associated with oxidative stress-related pathological conditions, especially exogenous pollutants. Among environmental pollutants, fluoride and arsenic contamination in groundwater have become a global concern in both developing as well as developed countries. Concurrent contagions with both of these toxicants are endemic in India, Bangladesh, China, Argentina, and Mexico [4]. Both natural and anthropogenic sources are responsible for environmental contamination with fluoride and arsenic. Occupationally several industries such as pesticide, mining, pharmaceuticals, food, beverage, microelectronics, and glass are liable for exposure of both these toxicants to the environment [4]. Contaminated drinking water is the primary mode through which humans are exposed to these toxicants.

Numerous studies have reported the toxic effects of fluoride and arsenic in regard to male and female reproduction which are responsible for impairment of fertility in male and female. Here we have summarized the existing information on fluoride and arsenic toxicities from available literature highlighting the reproductive aspects as collected from both human as well as animal studies.

## Fluoride

Fluorine is very reactive in nature due to its highest electro-negativity, and it has the second highest electron affinity. At the room temperature it exists as gaseous form $F_2$ which can react explosively with the other elements. Due to very low stability, it can never be found as isolated form in nature but as an ionized form i.e., fluoride or as complex form with other elements which are the constituents of minerals in rocks.

### Source

- The sources of fluoride are both geogenic as well as anthropogenic [5]. Among geogenic sources, the highest fluoride levels are observed in granites, syenites, quartz monzonites, granodiorites, and alkaline volcanic. The most abundant fluoride-rich minerals found in rocks and sediments are fluorapatite [$Ca_5(PO_4)_3F$], fluorite (CaF2), topaz [$Al_2(SiO_4)F_2$], villiaumite (NaF), amphiboles, and micas. Granitic rocks that contain fluorinated minerals including pegmatites, amphibolites, muscovite, hornblende, and biotite micas supply fluoride to groundwater and soil by soil-forming processes and weathering [6]. Coal contains about 295 pm fluorine [7]. Precipitation and atmospheric air are also geogenic sources containing 0.00001–0.0004 mg/L fluoride [8].
- The major anthropogenic way of ground water fluoride contamination is the use of phosphatic fertilizers in an intuitive manner which is very common in India [9]. Brick industries, coal-based power stations, smelting, aluminum and glass are another anthropogenic source of fluoride [10]. Irrigation of fluoride enriched water also contaminates the ground water [11]. Clay, which is used in brick manufacturing, contains several hundred ppm of fluoride [12].

### Global Distribution

- Ground water fluoride shows both temporal and spatial variation. Fluoride contamination in ground water has become a grave question in several countries such as China, India, Kenya, Argentina and Ethiopia while Mexico is the moderately affected country [5].
- South-East Asia can be considered as the epicenter of ground water fluoride contamination where Pakistan, India and Sri Lanka are highly distressed countries. In India, about 62 million people covering 17 states are suffering from dental and skeletal fluorosis. About 75% of groundwater of East Punjab region in Pakistan exceeds WHO permissible limit of fluoride where the higher concentration has been noted in shallow water aquifer as compared to deep water aquifer [13]. In Sri Lanka, high fluoride concentrations are observed mainly in the eastern and southeastern parts. China is the most contaminated country of East Asia and here the most affected areas are belonging from high temperature and low rain fall.

- The main source of fresh water supply in the rural communities of Africa is ground water and high fluoride concentration in this ground water has been reported in East African rift valley area. This rift valley is the part of Great Rift Valley of Africa which extends from Jordan valley down through Ethiopia, Uganda, Sudan, and Kenya to Tanzania. Apart from these, Malawi, Cameroon, South Africa, Nigeria and Ghana are also affected.
- Groundwater fluoride concentration in most parts of Europe is within the permissible limits; hence the research work is limited to European subcontinents.
- Estonia and Germany have also been reported due to having fluoride contamination. In Germany up to 8.8 mg/L of fluoride has been reported from private wells. Mexico, USA and Canada are highly affected countries of North America. In Northern Mexico, fluoride has been found to be in the range of 0.5–3.7 mg/L. In the USA, groundwater fluoride ranges from less than 0.2 to 3.58 mg/L. In Canada, up to 15.1 mg/L of fluoride has been reported in Lake St. Martin region of Manitoba [14]. Argentina is the most affected country in South America.

## Metabolism

After entering the gastrointestinal tract metabolism of fluoride takes place in three different steps which are absorption, distribution and renal excretion.

### *Absorption*

Fluoride is absorbed through the stomach and small intestine in a pH dependent and independent manner respectively. Absorption of the gastric fluoride (about 20-25%) is positively associated with gastric acidity and within gastric lumen ionic fluoride is converted into hydrogen fluoride which can cross the gastric mucosa [15]. About 75-80% fluoride is absorbed from proximal small intestine as ionic form in pH independent manner which can cross the intestinal epithelium [16]. Though it has been reported that 'Fluc' protein is involved in fluoride absorption but its role in mammalian cells is still incomprehensible [17]. CLC anion transporter and FEX exporter are involved in fluoride absorption in bacteria and fungi respectively [18, 19].

### *Distribution*

After absorption the level of plasma fluoride reaches the peak level within 20-60 minutes irrespective of quantity of ingested fluoride. Then the level declines to base after 3-11 hours. Plasma is the key compartment from which fluoride is distributed to hard and soft tissues [15]. About 99% of fluoride is retained in the mineralized tissues, specifically in bone and a small extent in teeth (dentin and enamel). Distribution of fluoride in bone does not occur uniformly rather than cancellous bones exhibit higher fluoride concentration as compared to compact bones due to large surface area [20]. Concentration of enamel fluoride depends on the extent of fluoride exposure during the formation of enamel. The severity of dental fluorosis is radically associated with dentine fluoride concentration rather than enamel [21]. Distribution of fluoride to different soft tissues depends on the rate of blood flow to different organs. The ratio of fluoride concentration in tissue/plasma varies in different tissues ranging from 0.4-0.9 where kidney exerts the greatest ratio compared to other organs [22].

### *Renal Excretion*

The principal pathway through which fluoride is removed from the body is urinary excretion. In case of healthy physiological circumstances, about 60% of the absorbed fluoride is excreted in adult individuals while 45% in children [23]. Reabsorption of fluoride through renal tubule is pH dependent. In case of higher urinary pH, the concentration of hydrogen fluoride, which is the most permeable form of fluoride, becomes lower and the concentration of ionic form becomes abundant. In this situation, fluoride persists as ionic form within renal tubule and excreted out. When urinary pH is lower, then the abundant form of fluoride is hydrogen fluoride which transmits through renal tubule and returns to the systemic circulation.

## Commencement of Fluoride Toxicity

Interestingly fluoride exerts biphasic task depending on its concentration, i.e., up to permissible limit (1.5 ppm) it exhibits mitogenic activities and beyond such limit it shows the signs of different physiological anomalies.

In some regions, artificial fluoridation of water was performed to get better dental health by application of sodium fluoride (NaF) and silicofluoride compounds (sodium silicofluoride and hydrofluosilicic acid). Approximately 100 national and international organizations have recognized the public health benefits of community water fluoridation to prevent dental caries. Fluoride is liable for both teeth and bone integrity and about 99% of the fluoride in the body is in the hard tissues. Consumption of fluoride at optimal level decreases teeth demineralization, provokes dental remineralization and reduces dentin hypersensitivity via its cariostatic activities. Fluoride also promotes bone cell proliferation, increases the mineral deposition in cancellous bone and decreases the solubility of bone crystal [24].

Nevertheless, toxic consequences of fluoride can be classified as acute and chronic toxicity. Acute fluoride toxicity includes diarrhea, vomiting, nausea, acidosis and cardiac arrhythmias. However, excessive intake of fluoride for prolonged period of time (chronic toxicity) results in serious health complications which is otherwise called fluorosis. Fluorosis is of three types-skeletal fluorosis, dental fluorosis and non-skeletal fluorosis. Skeletal fluorosis includes osteoporosis, osteosclerosis, crippling deformities, spinal compressions and restricted movements of joints. Dental fluorosis is hypomineralization of the teeth enamel which results from excessive fluoride ingestion during enamel development. In the case of non-skeletal fluorosis different soft tissues and systems are affected including hepatic, renal, cardiac, neuro, immune and also reproductive.

## Mechanism of Fluoride Toxicity

Fluoride puts forth diverse cellular effects with dose, time and cell-type-dependent manner. The main noxious effect played by fluoride in cells is its interaction with enzymes and hence it acts as a potent inhibitor of enzyme. At millimolar concentrations fluoride inhibits several enzymes, including phosphatases, both *in vivo* and *in vitro* [25]. The modes of action of fluoride are as follows:

## Generation of Oxidative Stress

Exposure of fluoride augments the production of superoxide anions which activate its downstream consequences like hydroxyl radicals, hydrogen peroxide and peroxynitrite. Fluoride also enhances the generation of nitric oxide (NO) which reacts with superoxide to form peroxynitrite, and also with thiols as well as metal centers in proteins to form nitrosyl adducts. Fluoride interferes with the formation of disulfide bond which results in accumulation of misfolded proteins within endoplasmic reticulum and results endoplasmic reticulum stress. Apart from amplification of free radicals, fluoride also quenches the activities of antioxidant enzymes like glutathione peroxidase, superoxide dismutase and catalase. Excess free radicals generated after fluoride exposure attack membrane phospholipids and cause lipid peroxidation which may hamper the activities of membrane lipids [26]. Fluoride activates stress mediated signaling cascade which involves MAP kinase. Ras/MEKK/MEK-mediated pathway activates extracellular signal-regulated protein kinase (ERK) where RAS is a direct target of fluoride [27]. Fluoride also activates the stress response genes binding protein (Bip) and heat shock protein Hsp70 [26] which also results in provoked ROS generation.

## Necrosis

High concentration of fluoride is responsible cell death which follows the pathway of necrosis and previous report on mice hepatocytes support such statement [28].

## Apoptosis

Surplus oxidative strain generally leads to cell death by means of apoptosis. Fluoride causes apoptosis by oxidative stress mediated lipid peroxidation, therefore results mitochondrial dysfunctions and activation of downstream signaling cascades [29]. Fluoride alters the expressions of pro and anti-apoptotic genes. It down-regulates the expression of anti-apoptotic gene Bcl-2 while enhancing the release of cytochrome C and activation of pro-apoptotic genes like caspase-3, -8 and -9 [30].

## Inhibition of Enzymatic Activities

Fluoride inhibits anaerobic production of energy from glucose by inhibiting glycolysis through diminishing the activity of enolase. Enolase is a metal activated dimeric enzyme which uses two magnesium ions per subunit. Fluoride inhibits this enzymatic activity by competing with magnesium ion which has been reported in several cells and tissues like liver, kidney, muscle, red blood cells, intestine, brain, retina, testis and embryonic cells [31]. Apart from enolase, fluoride also inhibits the activities of acid phosphatase, aconitase, adenylyl cyclase, acetyl cholinesterase, amylase, arginase, Ca-ATPase, catalase, cytochrome-c-oxidase, enolase, glucose-6-phosphatase, glutathione peroxidase, glycogen synthase, lactate dehydrogenase, lipase, Na+/K+ ATPase, protein phosphatase, pyrophosphatase, pyruvate kinase and succinate dehydrogenase [31].

## Effects on Male Reproductive System

Several animal experimentations and clinical investigations have reported that fluoride exhibits adverse impacts on male reproductive system which include structural and functional anomalies of primary and accessory reproductive organs, functional defects of spermatozoa, decreased sperm concentration, altered sperm potentials, decreased hormonal levels and as well as impaired fertility. Fluoride can cross the blood-testis barrier (BTB) which is a physical barrier, restricts the diffusion of several exogenous as well as endogenous toxicants. Fluoride damages the structural entity of BTB by altering the expressions of BTB associated proteins such as interleukin-1 alpha (IL-1a) and actin relative protein 3 which in turn decreases the expression of filamentous actin and impairs the BTB. Thus, fluoride can enter the reproductive system and exert detrimental outcomes.

The extent of toxic consequences of fluoride in male reproductive system can be categorized in several aspects and illustrated in Figures 1, 2 and 3 which are as follows:

### Structural Alterations

Healthy structural integrity of reproductive organs is prerequisite for proper functioning, but fluoride is responsible for structural anomalies of both primary and accessory reproductive organs.

- *Effects on testis*
  - Histological features of healthy testis can be characterized by compact orientation of oval or round shaped seminiferous tubules, presence of intact Leydig cells between interstitial spaces of seminiferous tubules, abundance of different developmental stages of germ cells including mature spermatozoa within seminiferous tubules. Fluoride causes messiness of seminiferous tubules, disorganization of Leydig cells, reduction of germ cells along with spermatozoa [32].
  - Disrupted spermatogenic cells, appeared as limp within seminiferous tubules is another abnormality obtained by scanning electron microscopy [33].
  - Marked infiltration in the interstitial area of seminiferous tubules, lack of matured spermatocytes, necrotic changes in the seminiferous tubular layers along with their defoliation and vacuolization in Leydig cells are also damaging outcomes due to fluoride exposure [34].

- *Effects on epididymis*
  - Epididymis is a unique accessory reproductive organ which is made up of coiled and convoluted tubule restraining nutritious luminal fluid essential for attainment of spermatozoal fertility. Structural orientation has sub-divided the epididymis into three different regions such as caput, corpus and cauda. The cauda region of epididymis stores the mature spermatozoa and plays an important role in spermatozoal nutrition and transportation.
  - Histological features of healthy caput epididymis can be characterized by the presence of many tubules lined with pseudostratified epithelium having stereocilia. Intertubular interstitium is present within the tubules. Tubular lumen contains bundles of spermatozoa. Plenty secretory granules are present in epithelial cells. On the contrary, fluoride causes damaging consequences which include less dense or absent of stereocilia on pseudostratified epithelial cells, presence of scanty secretary granules in epithelial cells and lumen of epididymis, decreased height of pseudostratified columnar epithelium, dilated lumen and fragmented spermatozoa [35]. Nuclear pyknosis of epithelial cells is another detrimental feature due to fluoride threat [36].
  - Structural organization of cauda epididymis from normal animals can be characterized by presence of tubules having larger diameter as compared to caput, presence of abundant stereocilia in pseudostratified epithelium, presence of Intertubular interstitium, presence of secretory granules in the lumens along with bundles of spermatozoa having well defined heads and tails. Fluoride treatment causes structural detoriation of cauda epididymis which include larger tubules with denudation of the cells in lumens, decreased height of pseudostratified epithelial cell, absence or less dense stereocilia at several points of epithelial lining, reduction in the number of secretory granules as well as decreased sperm concentration in lumens [36].
- *Effects on Vas deferens*
  - Vas deferens of normal animals can be characterized by the presence of three muscle layers- outer longitudinal, middle circular and inner longitudinal. Lamina propiaremains between inner longitudinal muscle layer and pseudostratified epithelium which contains stereocilia. Epithelial cell layer is folded to form stellate lumen within which sperm bundles are present. Fluoride causes several degenerative changes including nuclear pyknosis in the epithelium, clumped stereocilia, presence of debris within lumen, increased thickness of lamina propia and presence of fewer spermatozoa in the lumen [33].
- *Effects on seminal vesicle and prostate gland*
  - Healthy seminal vesicle is characterized by presence of glandular epithelium and presence of mucosal folds extending into the lumen. Prostate gland is lined by simple and pseudostratified columnar epithelium. Intervening fibro-muscular stroma is also present.
  - Fluoride intoxication is not responsible for structural alterations of seminal vesicle and prostate gland [36].

## Redox Imbalance

Fluoride mediated augmented oxidative stress due to altered redox homeostasis affects testis, epididymis and spermatozoa. Fluoride is responsible for testicular and epididymal oxidative strain which is evidenced by decreased antioxidant strata and enhanced lipid peroxidation. Spermatozoa are vulnerable for oxidative injuries due to profusion of polyunsaturated fatty acids in their plasma membrane and inadequate antioxidant aptitude. Mechanism of oxidative stress as well as the molecular approach of oxidative damages can be described as follows:

- *Effects on testis*
  - Antioxidant defense system which depends on the activities of several enzymatic and non-enzymatic markers becomes altered due to overexposed fluoride. Superoxide dismutase (SOD) is an enzyme which converts the superoxide radicals to hydrogen peroxide. Catalase is another marker enzyme which subsequently degrades hydrogen peroxide to water and oxygen. Glutathione peroxides (GPx) convert hydrogen peroxide into water in liaison with reduced glutathione (GSH), which is considered as an internal antioxidant. Finally, Glutathione S-transferase (GST) plays an important role in the conjugation of the electrophilic compounds (formed from xenobiotics metabolism) with GSH for excretion from the body.
  - ROS induced cellular disturbances also involves lipid peroxidation and formation of malondialdehyde (MDA). Carbonylation of amino acids which is also named as protein carbonylation (PC) is another marker of oxidative stress.
  - Fluoride decreases the activities of such enzymatic antioxidant markers such as SOD, catalase, GST and GPx [37]. Fluoride also enhances the levels of MDA and PC which reflects the disparaging consequences of fluoride in testicular tissues.
  - When a cell is under stress condition, then nuclear factor-erythroid factor 2-related factor 2 (Nrf2) dissociates from Kelch-like ECH-associated protein 1 (Keap1), which is a regulator of Nrf2 and acts as sensor for oxidative stress. Then it translocates to nucleus and binds to many components like heme oxygenase-1 (HO-1), g-GGL, and NAD(P)H:quinoneoxido-reductase (NQO) which are involved in antioxidant defense. Fluoride down-regulates the expression of Nrf2, HO-1, g-GGL, NQO while up-regulates Keap1 expression [38]. Therefore, it can be concluded that fluoride mediated testicular oxidative damage follows Nrf2 signaling cascade.
- *Effects on epididymis*
  - Fluoride up-regulates the level of MDA and down-regulates the protein and mRNA expressions of GSH and GSH related enzymes such as GPx, GSH reductase (GR) and GST [39]. Hence it can be stated that epididymal antioxidant defense is affected by fluoride through glutathione related enzymes. Fluoride also diminishes the mRNA expression of epididymal SOD1 and catalase and quenches the antioxidant activities [40].
- *Effects on spermatozoa*
  - Fluoride mediated augmented lipid peroxidation and diminished activities of antioxidant enzymes leads to loss of integrity of spermatozoal plasma membrane which surrounds the acrosome, mitochondria and tail. This loss of plasma

membrane integrity and fluidity interferes the fusion of sperm-oocyte. Surplus oxidative damage leads to abundance of abnormal spermatozoa [41].

## Altered Steroidogenesis and Hormonal Imbalance

Fluoride disrupts steroidogenesis and alters hormonal levels involved in male reproduction. Fluoride alters the structural entity and enzymatic activity of Leydig cells as well as interferes with the hypothalamus-hypophysis-testis axis. Therefore, testosterone synthesis is hampered due to fluoride intoxication. Mechanism of fluoride mediated decreased steroidogenesis can be explained as follows:

- Leydig cells need normal level of expression as well as function of epidermal growth factor receptor (EGFR), androgen receptor (AR) and G protein for synthesis of testosterone. Fluoride reduces the expression of EGFR as well as AR and impedes the G protein in the Leydig cells. These events exert a negative impact on normal testosterone level. Fluoride is also responsible for decreasing the diameter of Leydig cell along with extensive degenerative alterations which alter the ability to synthesize testosterone.

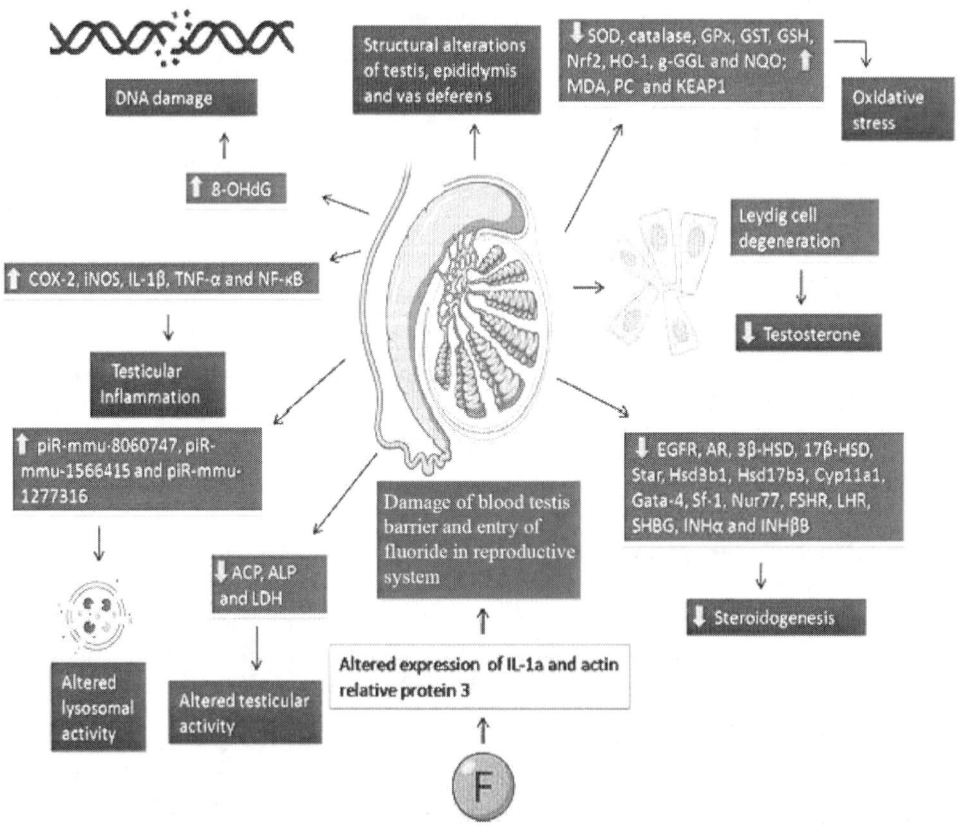


**Figure 1.** Effects of fluoride on alterations of testicular structure, redox balance, DNA status, steroidogenesis, lysosomal activities and functions.

- Synthesis of testosterone is under the control of two rate limiting enzymes which are 3β-hydroxysteroid dehydrogenase (3β-HSD) and 17β-hydroxysteroid dehydrogenase (17β-HSD). Fluoride declines the activities of enzymes and decreases the testosterone level [42].
- The regulatory enzymes involved in steroidogenesis are steroidogenic acute regulatory protein (Star), cholesterol side-chain cleavage enzyme (Cyp11a1), 3β-hydroxy dehydrogenase type I (Hsd3b1) and 17β-hydroxy dehydrogenase type III (Hsd17b3). Transcription factors which are involved in regulation of these genes are steroidogenic factor-1 (Sf-1), GATA binding protein-4 (Gata-4) and nerve growth factor IB (Nur77). Fluoride diminishes the expressions of Star, Hsd3b1, Hsd17b3, Cyp11a1, Gata-4, Sf-1 and Nur77 in Leydig cells and affects steroidogenesis [43].
- Sex hormone-binding globulin (SHBG) and follicle-stimulating hormone receptor (FSHR) are expressed in Sertoli cells while luteinizing hormone receptor (LHR) is present in Leydig cells. FSH secretion is also under the control of inhibin. Fluoride inhibits the genomic expression of FSHR, LHR, SHBG, INHα and INHβB and down-regulates testicular steroidogenic activities [44] (Figure 1).

- *DNA damage*
  - ROS mediated oxidative stress leads to DNA damage in each human cell. Main product of the oxidative stress induced DNA damage is 8- Hydroxy deoxyguanosine (8-OHdG) which is oxidized form of nitrogen base guanine and undergoes point mutation (transversion) due to its binding ability with adenine and cytosine. Therefore, presence of 8-OhdG in tissues, urine or serum is considered an indicator of DNA damage. Fluoride causes enhanced expression of such markers in testicular tissues [45]. Another report speculated that overexposure of fluoride causes fragmentation of testicular genomic DNA [32]. Apart from testis spermatozoal DNA is also soft target of fluoride [32].
- *Testicular inflammation*
  - Spermatogenesis as well as steroidogenesis are dependent on several inflammatory markers which include cyclooxygenase-2 (COX-2), iNOS, interleukin-1β (IL-1β), TNF-α, and NF-κB. Expressions of all such markers become up-regulated in case of inflammation and affect spermatogenesis. Fluoride enhances the expression of such markers and promotes testicular inflammation [46].
- *Altered lysosomal activity*
  - PIWI-interacting RNAs (piRNAs) are involved in gametogenesis and also regulate the genomic expression involved in gametogenesis. Fluoride enhances the expressions of some piRNAs among 28 different piRNAs which include piR-mmu-8060747, piR-mmu-1566415 and piR-mmu-1277316. Altered expressions such piRNAs enhance the levels of downstream signaling genes such as Ap4e1 and Ap1s3 (involved in acidic hydrolysis of the lysosomes); Gga2 (engaged in protein transport in lysosome and Golgi network) and Gla (associated with lysosomal signaling pathway). Atypical expressions of piRNAs and their target genes alter lysosomal degradation pathway which results testicular damages [47].
- *Altered expression of testicular functional marker enzymes*

- Acid phosphatase (ACP) and alkaline phosphatase (ALP) are vital biomarkers of primordial germ cells. ALP is responsible for spermatogenic growth, glucose transportation to the spermatogenic cells and testosterone biosynthesis. Lactate dehydrogenase (LDH) is another enzyme responsible for energy metabolism of spermatozoa. Fluoride diminishes the activities of ACP, ALP and LDH which results in reduced energy supply to spermatozoa, altered steroidogenesis and cessation of spermatogenesis [32].

- *Mitophagy*
  - Mitophagy is a kind of autophagy which has a pivotal role in the protection of eukaryotic cells against excessive accumulation of mitochondria. PINK1/Parkin pathway is engaged in elimination of damaged mitochondria and therefore mitophagy. PHB2 is an inner mitochondrial membrane protein also involved in mitochondrial membrane protein. Fluoride up-regulates the mRNA and protein expressions of PINK1 and PHB2 in testis and resulting in mitophagy [48] (Figure 2).
- *Apoptosis*
  - Fluoride is liable for enhanced expression of pro-apoptotic genes like caspase-3, caspase-9 and bax while decreased expression of anti-apoptotic gene bcl-2 in Leydig cells and therefore results apoptosis [49].

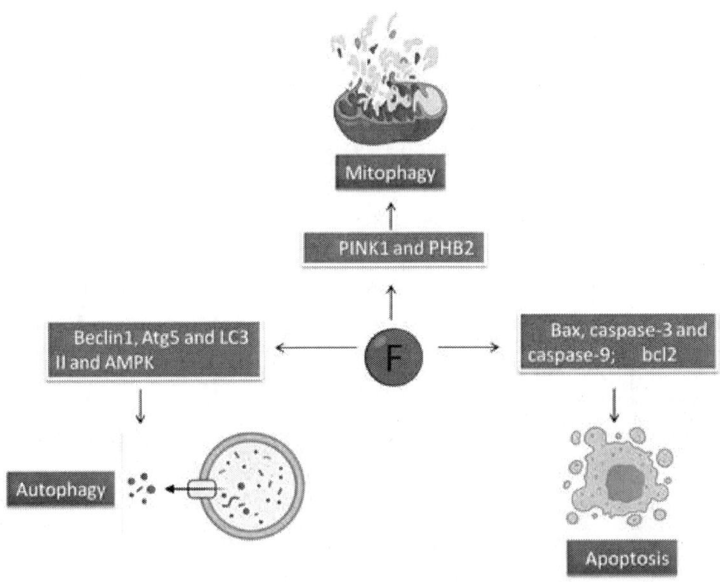


**Figure 2.** Effects of fluoride on testicular mitophagy, autophagy and apoptosis.

- *Autophagy*
  - Proteins responsible for initiation of autophagy and formation of autophagosome-lysosome fusion are Beclin1, Atg5 and LC3 II. One experimental report divulged that fluoride enhances the mRNA and proteins expressions of such markers in Leydig cells and therefore promotes autophagy [50].

- Mammalian targets of rapamycin (mTOR) promote some signaling events which inhibit autophagy. AKT and AMPK are the regulators of mTOR. Activity of mTOR is also suppressed due to dephosphorylation. Fluoride inhibits phosphorylation of mTOR and up-regulates the expression of AMPK in Leydig cells [50]. Therefore, mTOR signaling pathway is also involved in fluoride mediated testicular autophagy.

## Spermatozoal Abnormalities

- *Altered sperm potential*
  - Healthy spermatozoal features are prerequisite for successful fertilization but fluoride alters sperm potentials by affecting spermatozoal count, motility, viability and plasma functional integrity [37]. Development and functional life span of spermatozoa are dependent on testicular zinc concentration and angiotensin converting enzyme, but the levels of these factors declined due to overexposed fluoride which in turn affects spermatozoal count.
  - Spermatogenesis depends on gonadotropin-regulated testicular RNA helicase (GRTH/DDX25), which is present in nucleus and cytoplasm of meiotic spermatocytes and round spermatid. The expression of some genes such as phosphoglycerate kinase 2 (PGK2), transitional protein-2 (TP2) and HMG2 are under the control of DDX25. Fluoride down-regulates the mRNA and protein expressions of such markers and therefore impairs spermatogenesis [51].
  - In the male reproductive system cytochrome 450 expressions is found in spermatogonium, spermatocyte and Leydig cell. Cytochrome P450 is involved in metabolism of xenobiotic compounds and detoxification. Fluoride decreases the expression of such genes in testis and hampers spermatogenesis [52].
  - Balance between germ cell differentiation and apoptosis depends on cAMP responsive element modulator (CREM) and activator of CREM (ACT) but fluoride up-regulates the expressions of both the proteins and obstructs spermatogenesis [52].
  - Seminal vesicle fructose content and concentration of androgen carrier protein are two imperative factors responsible for spermatozoal motility, but fluoride causes depletion of both these factors which in turn results decreased spermatozoal motility [32].
  - One important subunit of Ubiquinol cytochrome c reductase in complex III of electron transport chain is mt-Cytb. COX is the terminal protein of electron transport chain catalyzes the production of water from oxygen. Fluoride inhibits the genomic expression of mt-COX2 and mt-Cytb resulting arrest of spermatozoal motility by decreasing ATP production via electron transport chain [53] (Figure 3).

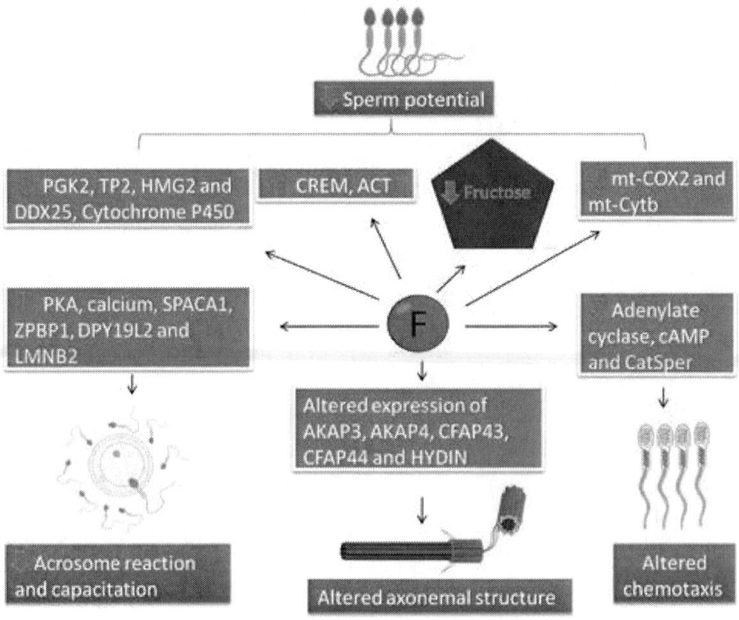


**Figure 3.** Effects of fluoride on spermatozoal features and associated molecular events.

- *Abnormalities of acrosome reaction and capacitation*
  - The late stage of capacitation comprises a number of signaling events which include tyrosine phosphorylation, protein kinase A (PKA) activation and intracellular alkalization. Fluoride down-regulates tyrosine phosphorylation, PKA activation and diminishes intracellular calcium concentration which hinders acrosome reaction and capacitation [54].
  - There are some key regulatory proteins responsible for acrosome biogenesis which are sperm acrosome associated 1 (SPACA1), zona-pellucida binding protein 1 (ZPBP1) and dpy-19like 2 (DPY19L2). Lamin B2 (LMNB2) is another protein which participates in nuclear lamina formation. Fluoride down-regulates the expressions of such markers and perturbs the structural entity of both acrosome and nuclear lamina; therefore, affects spermatogenesis [55].
  - One trypsin-like serine protease is ACR protein which is present on sperm acrosome and plays a pivotal role in spermatozoal progression through cumulus cells. One epididymal protein is SPAM1 which facilitates spermatozoa to break cumulus cells. Another serine protease is PRSS1 responsible for spermatozoal maturation within epididymis. Fluoride alters the expressions of such markers and affects acrosome reaction as well as altering fertilizing ability of spermatozoa [56].
- *Altered Chemotaxis*
  - Chemotaxis is the process in which sperm propels towards oocyte due to secretion of chemo-attractant by oocyte. Chemo-attractant binds with sperm plasma membrane receptor and activates intracellular adenylate cyclase, cAMP. Intracellular calcium concentration also becomes increased. Cation channel of sperm (CatSper) is one channel protein of sperm flagellum which permits calcium to enter into cell. Fluoride affects the expression of all such genes and hinders chemotaxis of spermatozoa [57].

- *Altered axonemal structure*
  - The structure of spermatozoal flagella depends on the intact arrangement of 9+2 microtubule. There are eleven regulatory proteins which are involved in maintenance of flagellar structure. Fluoride alters the expression of five regulatory genes which are A-kinase anchoring protein 3 (AKAP3), A-kinase anchor protein 4 (AKAP4), Cilia and Flagella Associated Protein 43 (CFAP43), Cilia and Flagella Associated Protein 44 (CFAP44) and HYDIN. As a result, axonemal structure and fibrous sheath of spermatozoa are affected [58].

## Effects on Female Reproductive System

Epidemiological report reveals that fluoride causes adverse impacts in female reproductive systems of individuals living in the fluorosis endemic areas. There is a close link between increased fluoride levels and decreased fertility rates [59]. Fluoride can cross the placental barrier and directly damage the mammalian fetuses [59]. Fluoride also enhances pregnancy complication, rate of miscarriage as well as reduction of birth rate [60]. Most of the available reports divulge that fluoride impedes female reproductive system by affecting the structural entity, redox homeostasis, reproductive hormones, fertility as well as developmental outcomes [61]. The detail mechanism of fluoride mediated female reprotoxicities is described here and also mentioned in Table 1.

### Structural Alterations

- *Effects on ovary*
  - Histological features of normal ovary can be characterized as normal arrangement of growing and graafian follicles, intact zona-pellucida layer with proliferating granulosa cells. Fluoride intoxication causes a decreased number of different types of follicles. Follicular atresia is another degenerative outcome due to fluoride threat [62].
- *Effects on uterus*
  - Histological characterization indicates the presence of three layers. The outermost layer is serosa which is formed by epithelial cells having single nuclei. The thick middle layer is myometrium which contains smooth muscles. The inner layer is endometrium and contains regularly arranged columnar epithelial cells having single nuclei.
  - Fluoride causes degenerative changes of uterus which can be characterized as-squeezing of endometrial cells, presence of space in endometrial layer and the endometrial cells become larger, endometrial glands become hypertrophic, altered shapes and sizes in the blood vessels of myometrium, cellular irregularities of myometrium [61].

**Table 1.** Fluoride induced female reproductive disorders and associated mechanism

| Applied chemical | Model organism | Dose, duration and mode of treatment | Observation | Reference |
|---|---|---|---|---|
| NaF | Wistar rats | 200 ppm orally for 28 days | Follicular atresia along with structural degenerations of ovary | [62] |
| NaF | Sprague-Dawley (SD) rats | 100, 150, and 200 ppm via drinking water for 6 months | Structural alterations of endometrium and cellular irregularities of myometrium | [61] |
| NaF | Wistar-albino rats | 100 mg/L orally for 60 days | Endometrial deterioration | [63] |
| Fluoride | Mice | 100 mg $F^-$ /L via drinking water for 90 days | Decreased mRNA expressions of antioxidant markers while up-regulation of pro-oxidants markers in ovary. | [64] |
| NaF | Swiss-strain albino mice | Oral administration at 10 mg/kg bw/day for 30 days | Reduced glutathione content of ovary | [65] |
| NaF | Wistar-albino rats | 100 mg/L orally for 60 days | Decreased uterine antioxidant strata | [63] |
| NaF | Sprague-Dawley (SD) rats | (100 and 200 mg/L in the drinking water for 12 weeks) | Down-regulation of FSH, LH, estrogen and progesterone and FSHR | [71] |
| Fluoride | Mice | 25, 50 and 100 mg/L via drinking water fluoride for 90 days | Arrests the follicular growth as well as oocyte development by up-regulating NDUFV2, SDHA and CYC | [66] |
| NaF | Mice of the KM strain | 50, 100, 150, 200mg/L via water for 5 weeks | Arrest of oocyte maturation by altering genomic expressions | [67] |
| Fluoride | Mice | g 25, 50 and 100 mg/L F by drinking water for 90 days | Apoptosis of granulosa cells | [68, 69] |
| Fluoride | Kunming mice | 100 mg $F^-$ /L by drinking water for 90 days | Apoptosis by up-regulation of Bax, Caspase-3 and Caspase-9 | [64] |
| NaF | Mice | 50, 100, 150 mg F− /L in drinking water for 90 days | Decreased litter size by up-regulation of MMP-9 and TIMP-1 | [70] |

- Endometrial deterioration can also be characterized by diffuse apoptosis in the surface epithelium and also stromal glandular epithelium and eosinophil as well as lymphocyte infiltration in the stroma [63].

## Redox Imbalance

Altered redox balance and concomitant oxidative stress is one of the putative mechanisms of fluoride mediated female reproductive anomalies. Overexposed fluoride causes excess ROS generation like superoxide anions, peroxynitrite which in turn causes cytotoxicity.

- *Effects on ovary*
  - The activities of ovarian antioxidant enzymes such as SOD, GPx and catalase become decreased in ovarian tissues due to fluoride threat. Not only the enzymatic activities, but also the mRNA expressions of such markers (SOD 1, GSH-Px1 and CAT) become down-regulated. However, total content of ovarian ROS, nitric oxide, MDA and iNOS become increased upon fluoride intoxication [64].
  - Reduced glutathione content is another damaging effect in ovarian tissues [65].
- *Effects on uterus*
  - Uterine SOD, Catalase and GPx activities become diminished, and level of MDA become up-regulated due to fluoride threat [63].

## Altered Steroidogenesis and Hormonal Imbalance

- Steroid hormones like estrogen and progesterone play vital roles in growth and differentiation of reproductive tissues as well as maintain fertility. Estrogen provokes the sensitivity of the granulosa cells to FSH and LH, therefore enhances the biosynthesis of progesterone by the granulosa cells. Estrogen also controls steroidogenesis, endorses the proliferation of granulosa cells and maintains the development of follicles [61]. Secretion of the reproductive hormones from ovary is under the control of FSH and LH secretion. FSH stimulates ovarian growth as well as follicular development. LH participates in ovulation, follicular maturation, and development of corpus luteum and synthesis of steroid hormones.
- Fluoride down-regulates the synthesis of FSH, LH, estrogen and progesterone, therefore alters the steroidogenesis [61].
- Fluoride inhibits the expression of follicle-stimulating hormone receptor (FSHR), which is a G protein-coupled receptor (GPCR) and alters the binding of hormones with receptor and affects the downstream signaling cascades [61].
- Interestingly, fluoride up-regulates the expressions of the receptors of estrogen and progesterone which in turn affect embryo implantation and endometrial receptivity [61].
- *Arrest of follicular development*
  - Some important components of mitochondrial respiratory chain complexes are NDUFV2, SDHA and CYC which take part in oxidative phosphorylation and ATP

synthesis. Fluoride up-regulates the expressions of such markers which in turn damages the mitochondrial respiratory chain and causes mitochondrial dysfunction. This detrimental effect ultimately arrests follicular growth as well as oocyte development [66].

- *Affect in oocyte maturation*
  - There are some germline specific genes which are involved in maturation of oocytes in ovary. The genes are Stra8 (helps in progression of the meiosis), DAZL (involved in proliferation and differentiation of primordial germ cells), Nobox (promotes folliculogenesis) and Sohlh1 (helps oocyte differentiation). Fluoride inhibits the expressions of all the genes and arrests the formation of mature oocyte [67].
  - Oocyte growth as well as acrosome reaction is associated with a few genes which are Gdf-9 (promotes growth and differentiation of the granulosa cells), Bmp15 (endorse oocyte maturation), H1oo (helps in oocyte maturation) and ZP2 (helps in acrosome reaction). Fluoride down-regulates the expressions of these genes and affects oocyte maturation [67].
  - One important marker which is accountable for oocyte maturation by glucose transportation in the oocyte is NNAT; however, methylation of such genes causes its inactivation. Fluoride is responsible for hypermethylation of this gene and inhibits the expression of this gene which in turn arrests oocyte maturation by disturbing glucose transport [68].
- *Apoptosis*
  - Fluoride provokes the expression of PCNA which is cell proliferation marker. Increased expression of this gene causes apoptosis of granulose cells [69].
  - Fluoride up-regulates the expressions of pro-apoptotic genes in ovary such as Bax, Caspase-3 and Caspase-9 and results apoptosis [64].
- *Birth abnormalities*
  - Pregnancy is associated with successful embryo implantation which is controlled by MMP-9/TIMP-1 system. MMP-9 is the member of matrix metalloproteinases (MMP), and TIMP-1 belongs to TIMPs. Combined activities of these two control the invasion of trophoblast as well as angiogenesis during embryonic implantation. Fluoride up-regulates the expressions of both MMP-9 and TIMP-1 which leads to decreased litter size [70].

## Arsenic

Arsenic is a metalloid having an atomic number of 33 and an atomic weight of 74.92 g/mol. It belongs to the nitrogen group of elements (group V in the Periodic table). Arsenic is a naturally occurring element widely distributed in the earth's crust. It exists in two solid modifications: yellow, and grey or metallic. It is found in nature in a number of minerals including realgar (AsdS), orpiment ($As_2$), arsenolite (As~O~a) as well as in iron minerals such as arsenopyrite (FeAsS) and loellingite ($FeAs_2$). In soils, arsenic is present in two valence states: +3 as arsenite and +5 as arsenate. Arsenic, a known carcinogen present in ground water, has been considered a toxicant to people throughout the world [71, 72]. Arsenical compounds are environmental toxins with multiple effects in animals and human populations [73, 74].

## Source

Most of the researchers agree that the main origin of arsenic in groundwater is geological as compared to others.

- *Geophysical source*
  - Theoretically it may be seemed that arsenic containing sulfide minerals (arsenopyrite), or their other products had been carried out in the geologic past from some earliest volcanic belt such as the foothills of Himalayas and deposited with the alluvium in the delta part of Ganga-Brahmaputra Basin which lies in West Bengal and Bangladesh. These buried arsenic minerals under the alluvium of Bengal delta are triggered to be responsible for water arsenic contamination [75, 76]
  - Excessive abstraction of groundwater facilitates to invade the environmental oxygen into the well which enhances the oxidation of arsenopyrite beneath the aquifer resulting in leaching the arsenic into water [77].
- *Organo-arsenic source*
  - Arsenosugar, arsenocholine, arsenobetaine tetra methyl arsonium salt and arsenic containing lipid are the main example of organo arsenic compounds which are abundantly found in marine bios as well as terrestrial species also.
  - Inorganic arsenic is converted as an organo-arsenic by the biomethylation process which easily incorporated in the biomass by the ingestion of arsenic contained food and aquifer [78].
- *Pesticide source*
  - The use of arsenical pesticides e.g., Monosodium methyl arsenate (MSMAsV), Disodium methyl arsenate (DSMAsV), Dimethyl arsinic acid (DMAsV) or cacodylic acid (CCA) causes a potentially significant human health concern [79].
- *Industrial source*
  - The uses of arsenic rich coals in thermal power plants of India are responsible for the arsenic pollution in air, soil and water [80].
  - Copper melting factories where arsenic is obtained as a byproduct during the copper production [81].
  - In gold mining activity arsenic pollution occurs due to coexistence of arsenic in gold bearing minerals [82].

## Global Distribution

- The natural contamination of as in groundwater has been reported worldwide, and the majority of these belong to South Asian and South American regions [83]. The severely affected countries include Bangladesh, India, China, Nepal, Cambodia, Vietnam, Myanmar, Laos, Indonesia and USA [84].

- In addition, countries like Argentina, Chile, Hungary, Canada, Pakistan, Mexico, and South Africa are also affected [85]. However, the South and Southeast Asian Belt is considered as the most arsenic polluted areas including India, Bangladesh, Nepal, Vietnam and China [85]. The developed countries, like USA and Canada are also experienced widespread levels of arsenic contamination in groundwater although the concentrations are characteristically lower in comparison with the Asian countries [86].

## Metabolism

Arsenic metabolism includes different steps like absorption, biomethylation, distribution and excretion.

- *Absorption*
  - Arsenic is present in four forms as arsenate, arsenite, arsenic and arsinic. Inorganic arsenic is more intoxicant than organoarsenic which is found in trivalent and pentavalent state in nature. Both soluble forms of arsenic compounds are hurriedly and extensively absorbed from the GI tract [87].
- *Biomethylation*
  - After absorption of arsenic, biomethylation occurs in the liver by enzymatic and non-enzymatic methylation processes. It has been explored that the inorganic form of arsenic is detoxified by methylation and methylated by the attendance of methyl donor SAM (S-adenosyl methionine) along with glutathione (GSH) by help of arsenic methyltransferase (AS3MT) and resulted monomethylated [e.g., monomethyl arsonous acid (MMAIII) monomethylarsonic acid (MMAV)] and dimethylated arsenical products [e.g., dimethylarsinous acid (DMAIII), dimethylarsinic acid (DMAV)], which readily eliminated by micturition [88].
  - $As^{5+}$ (Arsenate) is converted into $AS^{3+}$ (arsenite) by the arsenate reductase concomitant GSH (glutathione) and GSSG (glutathione disulfide). The enzyme arsenic-3-methyltransferase (AS3MT) converts $As^{3+}$ into $MMA^{5+}$ (Pentavalent monomethyl arsonic acid), with S-adenosyl methionine (SAM) acts as the methyl donor and, hydrolyzed into S-adenosyl homocysteine (SAH). Following the reduction of $MMA^{5+}$ to $MMA^{3+}$, further methylation and reduction process $DMA^{5+}$ is reduced into $DMA^{3+}$ [88]. The final products of ingested arsenic are MMA and DMA which are readily eliminated from the body during micturition [88]. DMA is capable of forming ROS by reacting with molecular $O_2$ (oxygen) resulting in the formation of arsenical metabolites such as dimethylarsenic radical and the dimethylarsenic peroxyl radical [89]. Inorganic arsenic has been revealed to inhibit the numerous antioxidant systems in the body, such as glutathione, glutathione peroxidase, thioredoxin reductase, and superoxide dismutase. The accumulation of ROS, hydroxyl radicals, superoxide radicals and $H_2O_2$ cause abnormal gene expression at lower concentrations and denaturation of proteins,

lipids, and DNA in greater concentrations which ultimately progress to cellular death [90].

- *Distribution*
  - The tissue distribution of arsenic in body accompanied with extent of ingestion and the type of arsenical compounds. Arsenic is deposited in liver, heart, kidneys and lungs abundantly whereas the least amount is found in muscular and nerve tissue. This metalloid is also expansively deposited in hair and nails due to profound thiol content (-SH) in keratin. Due to the parity of chemical reactivity of arsenic with phosphorus, it is accumulated in bone and teeth also and retained there for extensive duration. Arsenic has also an ability to cross the placental barrier and consequently quantity of arsenic in umbilical cord is alike to the maternal blood. The frequent intake of aresinous oxide promotes the accumulation of arsenic at notable concentration in the tissue of brain and also spinal cord [91].
- *Excretion*
  - Arsenic is excreted from the body through urine, feces, milk, sweat, hair, skin and lungs. Urinary route is the main route for excretion of arsenic from humans. Dimethyl arsenic acid ($DMA^{+5}$) is the chief form of arsenic which is eliminated via urine by the multistep methylation process of arsenical trivalent and pentavalent compounds. Half-life of arsenic is 72-120 hrs. Organic arsenicals are readily excreted hurriedly in respect of inorganic forms [92].

## Commencement of Arsenic Toxicity

- Arsenicosis may be developed insidiously after several months to a few years depending on the quantity of arsenic intake. Humans can be exposed to arsenic through drinking water or inhalation of arsenic dust particles. Though, there are medicinal values to combat against the numerous diseases, but recurrent uses pesticides, herbicides, insecticides, rodenticides, and food preservatives and as consequence use of arsenic contained fossil fuels are adequate to victimize in the aquatic environment in addition to human health [93].
- Arsenic ingestion can lead to acute and chronic poisoning. However, there are no explicit symptoms for acute arsenic poisoning, but some non-specific symptoms of chronic poisoning include weariness, colitis, loss of reflexes, weight loss, weakness, anorexia, gastritis, hair loss, and anorexia. Several experimental reports have implicated long term arsenic poisoning in the pathogenesis of diseases such as cardiovascular diseases, hyperkeratosis, disturbance in the nervous and peripheral vascular system, hyper-pigmentation, circulatory disorders, hepatic and kidney disorder [94].
- The antioxidant defense system of humans and other animals are capable of detoxifying inorganic pentavalent and trivalent arsenic compounds to methylated form and then excrete them via urine [95]. Genotoxicity of arsenic prevents DNA damage repair thus contributing to its carcinogenic property resulting in skin cancer, renal cancer, bladder cancer, liver cancer etc. [73]. Likewise experimental studies have also

indicated that arsenic is also responsible for epigenetic alterations in humans and animals [96].

## Mechanism of Arsenic Toxicity

The chemistry of arsenic is an important aspect in case of mode of action. In the late 19th century, it was hypothesized that arsenic reacted with sulphur and phosphates inactivating several enzymatic reactions which were essential for normal functioning of an individual [97]. The formation of reactive oxygen and nitrogen species by arsenic is one of the most extensively studied modes of action for commencement of arsenic toxicity [97].

- *Generation of oxidative stress*
  - Several *in-vivo* and *in-vitro* studies reported that arsenic toxicity is majorly mediated through the induction of oxidative stress [98]. For instance, an *in-vitro* study using various cell line models revealed that arsenic was involved in the activation of NF-κB, a pro-inflammatory factor via induction of oxidative stress [99].
  - Furthermore, data obtained from previous *in-vivo* studies also indicated that exposure of animals to arsenic led to an increase in malondialdehyde (MDA) concentration, a marker of lipid peroxidation, protein carbonylation, oxidative stress, as well as glutathione disulfide concentration. The exposure was also reported to have also caused a marked reduction in the concentration of glutathione and activities of antioxidant enzymes [100].
  - The mitochondrion is the main site for arsenic-induced oxidative stress [101]. The structural mitochondria integrity is usually altered following exposure to arsenic with concomitant deterioration of mitochondrial membrane potential [98].
  - Furthermore, the alteration of mitochondrial membrane potential is expected to lead to an uncontrolled generation of reactive oxygen species (ROS) as a result of inefficient reduction of molecular oxygen during the electron transport chain (ETC), a biochemical process that takes place within the inner mitochondrial membrane [101].
  - In a similar study, exposure of rats to sodium arsenite reduced the protein expression of mitochondria SOD. It also downregulated the activities of mitochondrial SOD, CAT, and complexes I, II, and IV of the electron transport chain [102].
- *Epigenetic modulations and genotoxicity*
  - Epigenetics is the study of how cells control gene activity without altering the DNA sequence.
  - It was reported that arsenic causes hypermethylation of the of the p53 gene and thus gained the attention of the researchers regarding the epigenetic effects of arsenic [97].
  - Exposure to inorganic arsenic is found to be associated with epigenetic modifications like gene specific DNA methylation, histone acetylation,

phosphorylation, methylation and altered expression of miRNAs, which may induce carcinogenesis or other diseases [97].

- The genotoxicity of arsenic cause deoxyribonucleic acid modification such as chromosomal aberrations, mutation, micronuclei formation, deletion, sister chromatid exchange [103].
- One of the mechanisms of arsenic destroys to DNA is base adjustment in particular 8-oxoguanine is one of the most frequently formed DNA nuclease modifications which are a mutagenic miscoding lesion that led to G: C to T: A transverse [103].
- Recent studies documented cytogenetic monitoring by using chromosomal aberration and micronuclei assay in order to observe genotoxic effects of arsenic in human and animal population [103].
- Inhibition of DNA repair is considered one of the most important effects of genotoxicity of arsenic. Nucleotide excision repair and base excision repair are the two processes of DNA repair which are inhibited by reactive oxygen species of arsenic [97].
- Consequently, alterations in DNA repair capacity are associated to the presence of polymorphisms in DNA repair genes which are related to risk of developing disturbance induced by arsenic [103].

- *Apoptosis*
  - $As_2O_3$ is cytotoxic and carcinogenic both *in vitro* and *in vivo* condition. Accumulation of $H_2O_2$, release of cytochrome c from mitochondria, reduction of mitochondrial membrane potential and caspase activation are common pathways of arsenic enhanced cellular death or apoptosis [98].
  - Arsenic influenced ROS showed a relation with Akt mediated cell survival pathway inhibition in acute promyelocytic leukemia U937 cells and activated caspase 3 [98].
  - Recent research has implied that JNK kinase is an important component in $As_2O_3$ influenced apoptosis. $As_2O_3$ is also able to up regulate the pro-apoptotic proteins like bax, bad, bmf, bim and down-regulate bcl-2, bcl-xl, mcl-1 [98].
  - It was documented that $As_2O_3$ increased the cancer cells sensitivity to death receptor induced apoptosis. Evidence showed that $As_2O_3$ sensitized human glioma cells through DR5 up-regulation by TRAIL induced apoptosis [98].
  - TNFR family members are also called as death receptors and induce apoptosis. Adaptor proteins act as linkers between caspases and death receptors. Death receptors usually have a death effectors domain (DED), a death domain (DD) and a caspase recruited domain (CARD) and by the involvement of these domains both initiator and effectors caspases become activated and trigger apoptosis [98].
  - The p53 has a pivotal involvement in arsenic induced apoptosis but the mechanism is not clear [103].

## Effects on Male Reproductive System

The basic constituent of male reproductive system includes a pair of testes, epididymis, vas deferens, ejaculatory duct along with accessory sex glands like seminal vesicle, prostate and bulbourethral gland. Spermatozoa are generated by the seminiferous tubules of testes and then transported via reproductive ducts to epididymis where they are stored and attain maturity. The entire system is under the control of hormones from hypothalamus, pituitary, and gonads. All these organs are vulnerable to toxicants from the environment, drugs, hormonal disruptors, etc. Arsenic is one of such environmental toxicants which affects the male reproductive system at different levels responsible for structural and functional alterations. Arsenic impairs the gene expression of Cldn11 and Ocln which codes for claudin 11 and occludin proteins. These proteins are highly responsible for brain testes barrier (BTB) maintenance [103].

## Structural Alterations of Testis and Accessory Reproductive Organs

- *Weight of reproductive organs*
  - Arsenic intoxication affects the male gonads extensively [104]. According to Khan et al., [105] male rats treated with sodium arsenite caused marked reduction in the weight of testes, epididymis and accessory sex glands. The decrease in weight may be due to the decrease in the size of seminiferous tubules, arrest of spermatogenesis and inhibition of synthesis of testosterone [106]. Arsenic trioxide intoxication also caused significant reduction in weight of testes, epididymis, prostate and seminal vesicles glands [100].
- *Histological alterations of testis and epididymis*
  - Pant et al. [107] reported the effect of sodium arsenite in mice altered the structure of testes. The mean volume of seminiferous tubules was reduced, and germinal epithelium was not intact. The spermatogenic cells had become vacuolated with less compactness.
  - Maharjan et al. [108] reported that arsenic causes the shrinkage of seminiferous tubules along with the reduction in germ cells.
  - Sodium meta-arsenite caused the dose-dependent changes in morphology of seminiferous tubules in Swiss albino mice [109]. The mice exhibited the reduction in the diameter of seminiferous tubules and cell populations of various gametes. These changes were more pronounced with the increase in dose and duration. The definite effect of arsenic on spermatogenesis was indicated by the atrophy of Leydig cells in testes.
  - A significant decrease of spermatogenesis at stage VII within the seminiferous epithelium is observed due to the effect of arsenic [110].
  - Sodium arsenite caused the necrotic changes along with the depletion of spermatozoa from the seminiferous tubules in male rats [105]. The necrotic changes are characterized by loss of germinal cells along with shrunken and irregular seminiferous tubules.
  - The exposure to environmental arsenic induces histological changes in the epididymis of Black Bengal goats [104]. The histological changes are manifested in the form of increased thickness of covering and trabeculae in epididymis. The diameter of ductus deferens was narrow in arsenic-exposed goats.
- *Structural alterations of spermatozoa*
  - The morphology and viability of sperm are considered as the useful parameters to differentiate between fertility and infertility conditions of males [83].

- The exposure of arsenic causes sperm morphological abnormalities in rats [110, 111]. The production of reactive oxygen species due to arsenic causes morphological abnormalities [83].
- Arsenic trioxide caused spermatozoal anomalies in rats where spermatozoal head deformities were also noted [100].

## Functional Alterations

- *Spermatozoal features*
  - Semen evaluation parameters such as motility, viability, sperm count, sperm membrane functional integrity and sperm DNA integrity are considered as the key indices of male reproductive system [112]. Reduction in count of sperm due to arsenic is reported in mouse [83].
  - The motility of spermatozoa plays an important role in reaching the sperm at the site of fertilization [112]. Sodium arsenite caused the reduction in motility of spermatozoa [113]. Binding of arsenic with thiol proteins results in the reduction in sperms motility [112].
  - Arsenic trioxide resulted in low spermatozoal count, decreased viability, low motility and deformed plasma membrane integrity of [100].
  - The intact functional membrane of sperm is considered as the prerequisite for the sperm fertilizing ability to egg surface [112]. Reddy et al. [114] reported the reduction in membrane integrity of sperm of mice exposed to arsenic during the transplacental and lactational stages. The exposure of arsenic trioxide to male rabbit bucks causes the reduction in sperm plasma membrane integrity [83].
  - DEAD-Box Helicase 3, Y-Linked protein (Ddx3y) is a Y-linked gene which plays a vital role in spermatogenesis and maturation of sperm. Exposure to arsenic trioxide decreased Ddx3y gene expression and protein level in the testicular tissue of mice, resulting in defective spermatogenesis and development of sperm [112].
- *Oxidative stress*
  - Arsenic caused the reduction in antioxidant enzymes [83] by increasing oxidative stress. The action of antioxidant defense system is also reduced due to sodium arsenite [28].
  - Arsenic trioxide resulted in increased lipid peroxidation (MDA), increased protein oxidation (protein carbonyl formation) and decreased GSH content along with attenuated activities of SOD, catalase and GST indicating a state of redox imbalance in testis of adult rats [100].
- *Steroidogenesis*
  - The reduction in the level of testosterone due to arsenic toxicity was reported by different researchers [100, 110].
  - Arsenic toxicity leads to atrophy of Leydig cells, subsequently decreasing the level of testosterone [115] as these cells are the main source for the synthesis of testosterone.
  - Testicular steroidogenesis is governed by two enzymes: 3β-HSD and 17β-HSD [110]. The reduction in the level of testosterone may be due to reduction in these

enzymes in arsenic toxicity [111]. Moreover, arsenic may bind with these enzymes directly and hinder their activities [112].

- Arsenic caused testicular toxicities via inhibition of gonadotropins like FSH and LH [110]. This reduction may be due to high level of cortisol which had antigonadal effects [83] (Figure 4).

- *Androgen receptors*
  - Androgen receptor (AR) plays a vital role in the process of spermatogenesis. AR is one of the members of the steroid hormone receptor family. Testosterone binds with AR which activates, interacts with DNA and mediates androgen receptor target genes [103].
  - Treatment of TM4 cells with sodium arsenite completely blocked androgen-mediated AR transcriptional activity. This further blocks the amino-carboxyl terminal interaction of the androgen receptor upon arsenic exposure [103].
  - Exposure of PC3, LNCaP and LAPC4 cells to arsenic trioxide results suppressed activities of AR through AR protein down-regulation, decreased hormone binding to AR, AR nuclear translocation disruption, and alteration in inhibition of AR binding with DNA [103].

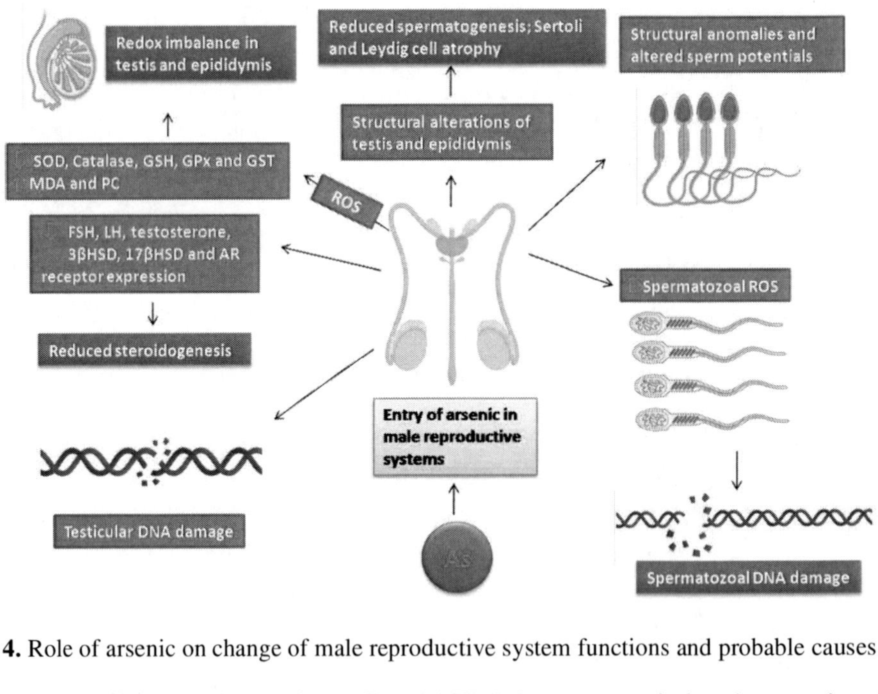


**Figure 4.** Role of arsenic on change of male reproductive system functions and probable causes.

  - Scaffold attachment factor B1 (SAFB1) has a transcriptional repression domain, and it binds with the hormone receptor and inhibits the activity of a hormonal receptor. SAFB1 null male mice were be found to be infertile due to elevated apoptosis in germ cells, hyperplasia in Leydig cells, and decreased testosterone level associated with insulin growth factor 1(IGF1) [103].
  - Male rats exposed to sodium arsenite displayed decreased SAFB1 in testis indicating apoptosis of germ cells, hyperplasia in Leydig cells, decreased

testosterone level associated with IGF1, and repression of the hormonal receptor. The suppressed androgen receptor activity might be due to defective SAFB1 [103]

- *DNA damage*
  - The integrity of sperm DNA is very important for fertilization and different toxicants contribute to DNA damage, thereby affecting its fertilization ability to oocyte [112].
  - Arsenic alters the functions of cells by binding to thiol groups of proteins and damaging the DNA [116].
  - Arsenic trioxide at a dose of 3mg/kg body weight resulted in an increased number of DNA damaged spermatozoa [100].
  - Exposure of mice to sodium arsenite in drinking water for 3 months increased DNA strand breaks in testicular tissue of mice [83].
  - The genotoxic effect to sodium arsenite was determined through comet assay in testicular cells and bone marrow. Comet assay showed increased DNA damage through the high tail length, tail DNA % and olive tail moments. These changes highlight the genotoxic effect of sodium arsenite on testicular tissue as one of the causes of arsenic-induced male reproductive toxicity [83].
- *Apoptosis*
  - Arsenic mediates cell death through apoptosis of germ cells in testes and somatic Sertoli cells *in vitro.* Arsenic trioxide induces oxidative stress in TM4 Sertoli cells, reactive oxygen species (ROS) mediated cytotoxicity, apoptosis at the mitochondrial level, and cell death through both caspase-dependent and caspase-independent pathways [103].
  - Exposure to sodium arsenite caused an elevation of apoptotic markers such as cytochrome p450 family 1 subfamily A polypeptide 1(CYP1A1), p53 and caspase 2, in male mice. This high toxicity leads to cellular damage and DNA damage in testicular tissue [103].
  - Exposure of arsenic trioxide in TM4 cell line showed increased activation of caspase 3, caspase 8 and caspase 9 and which plays an important role in the process of apoptosis. This released the apoptosis-inducing factor (AIF) through the generation of ROS [103].
- *Signalling pathways*
  - Mitogen-activated protein kinases (MAPKs) are regulatory proteins which are involved in cell signalling mediated by external stimuli. Extracellular signal-regulated kinases (ERK) pathway is involved in various male reproductive functions such as spermatogenesis, and function of Sertoli cells [103].
  - Encouragement of ERK1/2 inhibits the function of Sertoli cells and elevates apoptosis in testes [103]. Protein kinase B (AKT) regulates oxidative stress through cell growth regulation, survival of cells, the proliferation of cells, and inflammation with an immune reaction. Both MAPK and AKT are directly involved in phosphorylation of nuclear factor kappa B (NF-κB), enhancing binding of NF-κB to target DNAs, and elevating expression/activity of genes under the control of NF-κB [98]. Spermatogenesis and function of Sertoli cells in testes are controlled by NF-κB. Activated NF-κB causes a defect in spermatogenesis both in human and mice [103].

- Arsenic exposure initiated ERK/AKT/NF-κB signalling pathways in different cells [98]. Exposure to sodium arsenite elevated the expression of ERK1/2, IKKγ, PI3K, AKT and NFκB with increased phosphorylation of ERK/AKT level in testes of rats. This leads to reproductive toxicity through ERK/AKT/NF-kB signaling [103].
- *In vivo* studies with a different model, systems show that arsenic induces inflammation in testes [103]. Exposure to sodium arsenite in rats increased the inflammatory markers such as iNOS, TNF-α, COX-2, vascular endothelial growth factor (VEGF), NF κB, and caspase-3 [103].
- Exposure of 1-day-old chicken to arsenic trioxide increased inflammatory markers NF-κB, iNOS, COX-2, TNF-α and PTGEs in a dose-dependent manner [98].

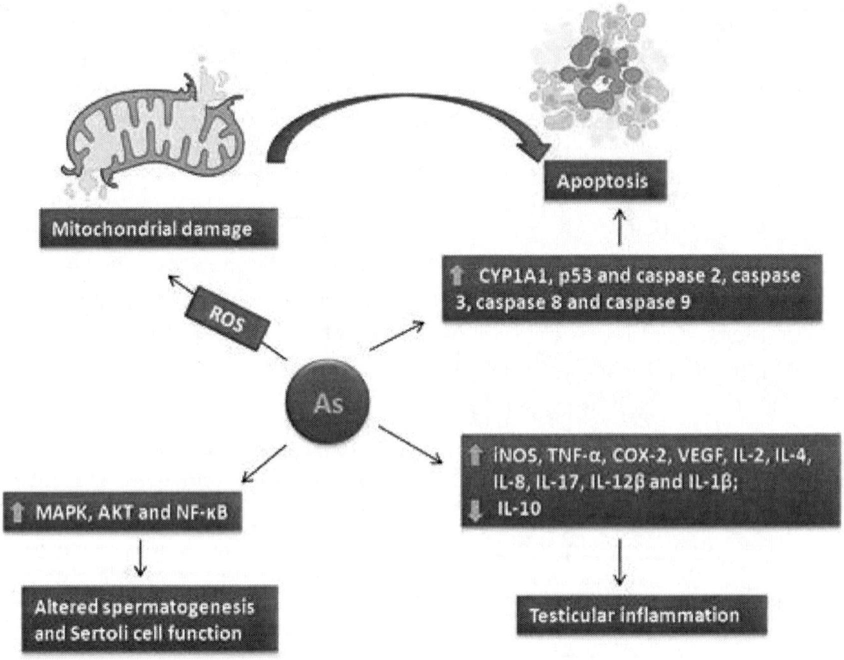


**Figure 5.** Effects of arsenic on testicular apoptosis and inflammation due to altered expressions of different marker proteins.

- The induced iNOS regulates interleukins (IL) such as interleukin-2 (IL-2), interleukin-4 (IL-4), interleukin-8 (IL-8), interleukin-17 (IL-17), interleukin-12β (IL-12β) and interleukin-1β (IL-1β), which play important roles in immune response and inflammation [98]. Arsenic up-regulates such inflammatory markers while decreases the expression of anti-inflammatory marker cytokine IL-10 which leads to testicular inflammation (Figure 5).

## Effects on Female Reproductive System

The devastating effect of arsenic (inorganic, organic and gas) on female reproductive system turned the reproductive organs' normal structure-function into malicious one. Diverse animal experimental models have been developed throughout the decades to check the effect and mode of action of arsenic and arsenicals on major reproductive organs like ovaries and uterus including the secondary cell line [117, 118]. Major findings indicated towards ovarian follicular growth arrest, altered steroidogenesis, hormonal imbalance including gross structural alterations of ovarian and uterine tissues [119, 120]. Uterine luminal epithelium damage, decrease number of endometrial glands with shortened microvilli were common findings [121, 122]. Disruption in steroidogenesis, hormonal insufficiency could lead to increased risk of infertility. Women from high level of arsenic (26.1 μg/L- 31.6 μg/L) affected area of Bangladesh reported a significant increase in spontaneous abortion/miscarriage than that of arsenic free area [123]. Arsenic wields its toxicity by targeting different reproductive organs using different pathways as follows.

### Mechanism of Toxicity

- Arsenic is familiar for targeting the reproductive organ, as in case of female reproductive system. It causes hindrance in activities of steroidogenic enzymes and encumber folliculogenesis [124]. The effect of arsenic on the estrus cycle was characterized by prolonged diestrus and low levels of plasma and serum estradiol and progesterone [124, 125]. Similarly, it also upsets the plasma gonadotropins by reducing the level of LH and FSH [126].
- It also indicated that direct adverse impact of arsenic on midbrain norepinephrine and serotonin level; corticosterone level; hypothalamic-pituitary-gonadal axis caused in compromise of gonadotropins. Arsenic mediated point mutations within the DNA binding domain (DBD) lead to confirmation of DBD-glucocorticoid receptor (GR) binding which further enhances the inhibitory response [127, 128, 129].
- Even at very low dose (1–50 μmol/kg arsenite) arsenic could also disrupt estrogen receptor (ER) followed by ER-dependent gene transcription of the 17β-estradiol (E2)–inducible vitellogenin II gene. In-vitro studies conducted using human breast cancer MCF-7 cells also unveiled significant inhibition of ER-regulated reporter gene and the native ER-regulated GREB1 gene activation [130].
- The altered follicular growth pattern, granulosa cell degeneration, follicular atresia, uterine endometrial degeneration, luminal epithelium thinning and presence of apoptotic bodies in both ovary and uterus could be the major outcome of arsenic mediated deregulation of steroid receptor and receptor mediated response.
- Another major outcome of arsenic is generation of oxidative stress in different systems including reproductive [131]. Data generated from a series of scientific studies reported inflation of ovarian and uterine lipid peroxidation (MDA/TBARS level) and protein carbonylation [132]. Diminution of superoxide dismutase (SOD), catalase activities have always been a common finding in arsenic treated ovaries and uteri [120, 122]. Furthermore, diminishing pool of GSH and GPx in arsenic treated model might

be another potential reason behind excessive ROS generation in reproductive tissues [133, 134]. As a common fate of oxidative stress, DNA damage and damage mediated cell death could take place [135]. In this regard, it is also reported that arsenic induced ovarian DNA damage by measuring the comet tail length. Whether this DNA damage could cause apoptosis in ovary and uterus, apoptotic markers expression was studied. Increased expression of apoptotic death markers; procaspase-3, cleaved caspase-3, Bax and p53 gene were found in arsenic treated uterus and ovary followed by a concomitant suppression of anti-apoptotic Bcl-2 gene expression [122, 135, 136, 137].

- Recent study had found that ovarian oxidative trauma, mitochondrial damage, and autophagy could play essential roles in the pathogenesis of arsenic trioxide-induced reproductive toxicity. Diminished expression of the related mRNA (Bcl-2 and mTOR) and protein (mTOR) in adult F1-females exposed to higher arsenic at a dose of 2 and 20 ppm elicit the autophagy pathway [138].
- Apart from these, arsenic could develop transgenerational genotoxicity and changes in global DNA methylation in F0-F4 generation of rats treated with 1mg/L solution.

The major effect of arsenic on the ovaries and uterus of different animals and its probable pathways has been detailed in Table 2.

## Developmental Toxicity

- Arsenic in drinking water and food chain caused severe adverse pregnancy outcomes like spontaneous abortion, stillbirth, foetal death, and low birth weight [150, 151]. Severe developmental toxicities like birth defects, neural tube disorders, delayed maturations were also found in experimental rat and mouse model [152, 153].
- Study conducted at Matlab region, Bangladesh reported increased tendency of spatial pattern of foetal loss and infant death due to presence of arsenic in drinking water and food chain. The pregnant women from higher arsenic (319 μg/L) zone were the prime victim [154]. Study from Lanyag Basin, Taiwan suggested that babies born in arsenic affected (<0.15 ppb to 3.59 ppm) area had less birth weight (30g lighter) than control region. Higher rate of preterm delivery was another consequence of arsenic contamination [155]. Population based cohort study from Bangladesh observed arsenic exposure during pregnancy increased the risk of foetal lower respiratory tract infection and diarrhea by 69% and 20% respectively [156].
- Increased neonatal and post-neonatal mortality was found in an area of Chile with high arsenic content in the supplied water [157]. A cohort study including 200 Mexican pregnant women revealed that excessive arsenic in drinking water is associated with adverse birth outcome and also the increased amount of urinary MMA is negatively correlated with new-born birth weight and gestational age [158].

More information regarding arsenic mediated developmental toxicity and effect on pregnancy has been abridged in Table 3.

**Table 2.** Arsenic mediated reproductive toxicities and associated mechanism

| Applied chemical | Model organism | Dose, duration and mode of treatment | Observation | Reference |
|---|---|---|---|---|
| Arsenic (III) oxide | Fish (Colisa fasciatus) | 2.0 mg/L and 14.0 mg/L exposure for 15 and 30 days respectively | Ill effect observed at 14 mg/L exposure for 30 days. Development of oocyte (II and III stage) ceased. Atretic follicle number increased. | [139] |
| Arsenite | Chinese hamster ovary K1 cells | 40μm of arsenite treatment | After 8 hrs of treatment an appreciable number of fractionated cells in sub-G1 population was detected. Increased intracellular peroxide level leads to elevation of ROS. | [118] |
| Sodium arsenite | Rat | 0.4 ppm/100 g bwt/rat/day orally for 28 days | Decreased ovarian and uterine weights, ovarian Δ5-3β-HSD and 17β-HSD activities, plasma level gonadotropins, midbrain norepinephrine activities were suppressed. Peroxidase activity in both the ovary and uterus was attenuated. Midbrain serotonin level was increased. | [127] |
| Sodium arsenite | Rat | 0.4 ppm/100 g bwt/rat/day via drinking water for period of 28 days | Prolonged diestrus phase, reduced level of plasma LH and FSH. Decreased activity of delta five, 3 beta-hydroxysteroid dehydrogenase (Δ5, 3β-HSD), and 17 beta-hydroxysteroid dehydrogenase (17β-HSD) followed by a decrement in ovarian and uterine peroxidase activities. Decreased ovarian and uterine weight with high accumulation of arsenic in plasma and organs. | p140] |
| Sodium arsenite | Rat | 50, 100, and 200 ppm in drinking water for 28 days | Increased in ovarian DNA damage followed by gross decrease in ovarian weight | [136] |
| Sodium arsenite | Rat | 4 μg/ml for 28 days was gavaged | Treatment caused a decrease in estradiol level along with LH and FSH. Degeneration in endometrial gland and luminal epithelial cells were seen. Down-regulation of estrogen receptor (ER alpha), the estrogen-responsive gene - vascular endothelial growth factor (VEGF), and G1 cell cycle proteins, cyclin D1 and CDK4, was also found. | [141] |
| Sodium arsenite | Rat | 3 ppm from mating to gestation period via drinking water. Weaned pups were administered same dose as mother for 1-3months | Arrest in follicular maturation with prolonged diestrus and low estradiol was observed. DNA damage confirmed by presence of pycnotic chromatin, karyorrhexis, significant cytoplasmic vacuolization and also vasculature damage. | [142] |
| Sodium arsenite | Rat | 0.4 ppm, 28 days by drinking water | Prolonged diestrus, decrease in number of endometrial pinopodal cells and gland pits were seen. Height reduction of uterine surface epithelium along with focal lesion and shedding of endometrium. Collagenous-irregular endometrial stromal cell number increased. Atrophied glands were present. | [143] |

**Table 2.** (Continued)

| Applied chemical | Model organism | Dose, duration and mode of treatment | Observation | Reference |
|---|---|---|---|---|
| Arsenic | Goat | Environmental exposure | Thickening of uterine horn, decrease in uterine gland number and diameter were observed. Thickening of cervical wall with a shortening in cervical villi was observed. | [144] |
| Arsenic trioxide | Rat | 0.05,0.10,0.20,0.40 mg/L for 31 weeks through drinking water | Expression of ovarian and uterine estrogen receptor-binding fragment-associated gene 9 (Ebag9) decreased with a simultaneous increase in estrogen-responsive finger protein (efp) mRNA expression. | [145] |
| Sodium arsenite | Mice | 8 mg/kg bwt for 16 days by intraperitoneal injection | Elevated ovarian ROS and MDA level with diminishing pool of SOD and GPx was observed. Oxidative stress mediated up regulation of p66Shc protein expression leading to a fate of increased follicular death. | [132] |
| Sodium arsenite | Rat | 10 mg/kg bwt for 8 days orally | Attenuation in activities of uterine superoxide dismutase (SOD), catalase and glutathione peroxidase (GPx). Ovarian and uterine structural alterations followed by DNA breakage, necrosis and disrupted steroidogenesis. | [133] |
| Inorganic arsenic (III) | Mice | 0, 0.15, 1.5 and 15 mg/L in drinking water throughout the gestation period | Early onset of puberty in female off-springs.<br>Advanced vaginal opening time with an increase in LH hormone level.<br>Increased expression of hypothalamic Kisspeptin, GnRH1, Oct2 and Ttf1 in pubertal rats. Decreased expression of Oct2 and Ttf1 in adult rats. | [146] |
| Arsenic trioxide | Rat | 3 mg/kg bwt for 28 days orally | Shrunken uterine endometrial gland with thinning of microvilli density.<br>Inflation of ROS generation. Increased expression of apoptotic death markers; procaspase-3, cleaved caspase-3, and Bax. | [122] |
| Sodium arsenite | Rat | 10 mg/kg bwt orally for 8 days | Up-regulation of inflammatory markers like tumor necrosis factor (TNF $\alpha$) and nuclear factor $\kappa$B (NF $\kappa$B). Pro-apoptotic Bax gene and p53 gene expressions were also raised followed by suppression of anti-apoptotic Bcl-2 gene expression | [137] |
| Naturally occurred arsenic | Human | Exposure through drinking water and food | Women from arsenic affected area displayed significant increase in urinary arsenic excretion which is positively correlated with primary ovarian insufficiency development | [147] |

| Applied chemical | Model organism | Dose, duration and mode of treatment | Observation | Reference |
|---|---|---|---|---|
| Arsenic trioxide | Mice | 0, 0.2, 2 and 20 ppm to the parents from 35 days before breeding and is continued until weaning. The female offspring received the same treatment until maturity. | induces autophagy in the ovaries of mature F1-female mice via activation of autophagic genes (PDK1, TSC2, P13K, P62, ATG13, AMPK, ULK1, ATG12, ATG5, LC3, Beclin1, ATG3, ATG7 and p62) and proteins (Beclin1, LC3-I, II and mTOR), resulting in decreased female gametes number as compared to the control. | [138] |
| Arsenic trioxide | Rat | 1 mg /L | Transgenerational genotoxicity and changes in global DNA methylation were observed in the F0-F4 generation. Aberrant morphology in ovary was observed. | [148] |
| Sodium arsenate | Mice | 6 mg/kg, 12 mg/kg given orally throughout the gestation period | Diminution in the fetuses' weight, hind limb, fore limb, tail and snout length, crown rump and head circumferences were noticed. Malformations in tail, feet, arms, legs, skin and eyes were also noted. Increased DNA damage of fetuses was found. | [149] |
| Sodium arsenite | Rat | 1,3,5 mg/kg bwt, orally for 30 days | Decrease in ovarian and uterine weight along with gross body weight. Growing follicular growth arrest in dose dependent fashion. Excessive ROS productions in both ovary and uterus. | [134] |

**Table 3.** Arsenic mediated developmental toxicities

| Model organism | Treatment/exposure | Observations | Reference |
|---|---|---|---|
| Mice | Sodium arsenite | Loss of fecundity. Arsenic exposed placenta displayed severe dysplasia. | [153] |
| Mice | Sodium arsenate at a dose of 4.8, 9.6, or 14.4 mg/kg | Mean foetal weight reduced than control. Fetus born from arsenic exposed mother exhibited neural tube disorder featuring vertebral and calvarial abnormalities. | [152] |
| Rat | Sodium arsenite at a dose of 1.0, 1.5 and 2.0mg/kg body weight | Delayed maturation of Purkinje cells was observed on post-natal day 11. | [159] |
| Human | Exposed via drinking water. Country: USA | Strong positive correlation was found between high level of arsenic in drinking water and rate of spontaneous abortion. | [160] |
| Human | Exposed by drinking water arsenic with a range of $\geq$0.10 mg/L. Country: Bangladesh | Spontaneous abortion, stillbirth, and preterm birth rates were substantially higher in the exposed group than non- exposed group. | [161] |
| Human | Exposed via drinking water. Country: Bangladesh | Total still birth rate (3.4%) increased with increment of ground water arsenic contamination.<br>Arsenic at <10 μg/L, 10-50 μg/L and >50 μg/L showed 2.96%, 3.79% and 4.43% of still birth. | [162] |
| Human | Exposed via drinking water. Country: Thailand | Infants exposed to arsenic in utero indicate robust activation of an integrated network of NF-kB, inflammation, cell proliferation, stress, and apoptosis pathway. | [163] |
| Human | Exposed via drinking water. Country: Chile | Prominent and significant association between arsenic exposure and late foetal mortality and neonatal death was observed. | [157] |
| Human | Exposed via drinking water. Country: Bangladesh | Statistically significant association between arsenic exposure and birth-defects was observed with an odds ratio=1.005, 95% confidence interval 1.001-1.010. | [164] |
| Human | Exposed via drinking water. Country: Bangladesh | Pregnant women exposed to >50μg/L significantly increased the risks of foetal loss (relative risk = 1.14, 95% confidence interval: 1.04, 1.25) and infant death (relative risk = 1.17, 95% confidence interval: 1.03, 1.32). | [165] |
| Human | Exposed via drinking water. Country: India | High concentrations of arsenic (> or =200 μg/L) in drinking water during pregnancy was coupled with a six-fold increased risk of stillbirth. | [166] |

## Conclusion

It is evident from preceding discussion that fluoride and arsenic are equally responsible for causing oxidative damages in male and female reproductive systems leading to a number of cellular complications including DNA damage, mitochondrial dysfunction, autophagy and apoptosis. These impediments ultimately reduce the fertilizing ability. Oxidative stress being the prime offender of fluoride and arsenic mediated reproductive afflictions; use of antioxidant might be a relevant approach to combat such jeopardy. This literature offers a better understanding on 'infertility war' caused by fluoride and arsenic contaminated food and water available globally.

## References

[1] Pruchniak MP, Araźna M, Demkc U. Biochemistry of oxidative stress. In: *Advances in Experimental Medicine and Biology*. 2016;878:9-19.

[2] Rahal A, Kumar A, Singh V, Yadav B, Tiwari R, Chakraborty S, Dhama K. Oxidative stress, prooxidants, and antioxidants: the interplay. *Biomed Res Int*. 2014;2014:761264.

[3] Ahmad S. Oxidative stress from environmental pollutants. *Arch Insect Biochem Physiol*. 1995;29(2):135-57.

[4] Chouhan S, Flora SJS. Arsenic and fluoride: Two major ground water pollutants. *Indian Journal of Experimental Biology*. 2010;48(7):666-78.

[5] Ali S, Thakur SK, Sarkar A, Shekhar S. Worldwide contamination of water by fluoride. *Environmental Chemistry Letters*. 2016;14(3):291-315.

[6] Vithanage M, Bhattacharya P. Fluoride in the environment: sources, distribution and defluoridation. Vol. 13, *Environmental Chemistry Letters*. 2015;13(2):131-47.

[7] Churchill HV, Rowley RJ, Martin LN. Flourine Content of Certain Vegetation in Western Pennsylvania Area. *Anal Chem*. 1948;20(1):69-71.

[8] Gupta SK, Deshpande RD, Agarwal M, Raval BR. Origin of high fluoride in groundwater in the North Gujarat-Cambay region, India. *Hydrogeol J*. 2005;13(4):596-605.

[9] Kundu MC, Mandal B. Assessment of potential hazards of fluoride contamination in drinking groundwater of an intensively cultivated district in West Bengal, India. *Environ Monit Assess*. 2009;152(1–4):97-103.

[10] Pickering WF. The mobility of soluble fluoride in soils. *Environ Pollution Ser B, Chem Phys*. 1985;9(4):281-308.

[11] Pettenati M, Perrin J, Pauwels H, Ahmed S. Simulating fluoride evolution in groundwater using a reactive multicomponent transient transport model: Application to a crystalline aquifer of Southern India. *Appl Geochemistry*. 2013;29:102-16.

[12] MacDonald H. Fluoride as air pollutant. *Fluoride Q Rep*. 1969;2(1):4–12.

[13] Farooqi A, Masuda H, Kusakabe M, Naseem M, Firdous N. Distribution of highly arsenic and fluoride contaminated groundwater from east Punjab, Pakistan, and the controlling role of anthropogenic pollutants in the natural hydrological cycle. *Geochem J*. 2007;41(4):213-34.

[14] Desbarats AJ. On elevated fluoride and boron concentrations in groundwaters associated with the Lake Saint-Martin impact structure, Manitoba. *Appl Geochemistry*. 2009;24(5):915-27.

[15] Whitford GM. The metabolism and toxicity of fluoride. *Monographs in oral science*. 1996:16(2):1-153.

[16] Nopakun J, Messer HH, Voller V. Fluoride absorption from the gastrointestinal tract of rats. *J Nutr*. 1989;119(10):1411-7.

[17] Stockbridge RB, Robertson JL, Kolmakova-Partensky L, Miller C. A family of fluoride-specific ion channels with dual-topology architecture. *Elife*. 2013;2:e01084.

[18] Stockbridge RB, Lim HH, Otten R, Williams C, Shane T, Weinberg Z, Miller C. Fluoride resistance and transport by riboswitch-controlled CLC antiporters. *Proc Natl Acad Sci USA*. 2012 Sep 18;109(38):15289-94.

[19] Li S, Smith KD, Davis JH, Gordon PB, Breaker RR, Strobel SA. Eukaryotic resistance to fluoride toxicity mediated by a widespread family of fluoride export proteins. *Proc Natl Acad Sci USA*. 2013;110(47):19018-23.

[20] Weidmann SM, Weatherell JA. The uptake and distribution of fluorine in bones. *J Pathol Bacteriol.* 1959;78:243-55.

[21] Vieira APGF, Hancock R, Dumitriu M, Limeback H, Grynpas MD. Fluoride's effect on human dentin ultrasound velocity (elastic modulus) and tubule size. *Eur J Oral Sci.* 2006;114(1):83-8.

[22] Whitford GM, Pashley DH, Reynolds KE. Fluoride tissue distribution: Short-term kinetics. *Am J Physiol - Ren Fluid Electrolyte Physiol.* 1979;5(2):141-8.

[23] Villa A, Anabalon M, Zohouri V, Maguire A, Franco AM, Rugg-Gunn A. Relationships between fluoride intake, urinary fluoride excretion and fluoride retention in children and adults: An analysis of available data. *Caries Res*. 2010;44(1):60-8.

[24] Palmer CA, Anderson JJB. Position of the American Dietetic Association: The impact of fluoride on health. *J Am Diet Assoc*. 2000;100(10):1208-13.

[25] Mendoza-Schulz A, Solano-Agama C, Arreola-Mendoza L, Reyes-Márquez B, Barbier O, Del Razo LM. The effects of fluoride on cell migration, cell proliferation, and cell metabolism in GH4C1 pituitary tumour cells. *Toxicol Lett*. 2009;190(2):179-86.

[26] Barbier O, Arreola-Mendoza L, Del Razo LM. Molecular mechanisms of fluoride toxicity. *Chemico-Biological Interactions*. 2010:188(2):319-33.

[27] Wittinghofer A, Gremer L, Gilsbach B, Reza Ahmadian M. Fluoride complexes of oncogenic Ras mutants to study the Ras-RasGAP interaction. *Biol Chem.* 2008;389(9):1163-71.

[28] Ghosh J, Das J, Manna P, Sil PC. Cytoprotective effect of arjunolic acid in response to sodium fluoride mediated oxidative stress and cell death via necrotic pathway. *Toxicol Vitr.* 2008;22(8):1918-26.

[29] Karube H, Nishitai G, Inageda K, Kurosu H, Matsuoka M. NaF activates MAPKs and induces apoptosis in odontoblast-like cells. *J Dent Res*. 2009;88(5):461-5.

[30] Urbansky ET. Fate of fluorosilicate drinking water additives. *Chem Rev*. 2002;102(8):2837-54.

[31] Strunecka A, Strunecky O. Mechanisms of fluoride toxicity: From enzymes to underlying integrative networks. *Applied Sciences*. 2020;10(20):7100.

[32] Pal P, De A, Roychowdhury T, Mukhopadhyay PK. Vitamin C and E supplementation can ameliorate NaF mediated testicular and spermatozoal DNA damages in adult Wistar rats. *Biomarkers.* 2022;27(4):361-74.

[33] Susheela AK, Kumar A. A study of the effect of high concentrations of fluoride on the reproductive organs of male rabbits, using light and scanning electron microscopy. *J Reprod Fertil.* 1991;92(2):353–60.

[34] Giri DK, Ghosh RC, Kashyap DK, DewanganG, Maiti S. Hae-mato-Biochemical Alterations in Subacute Oral Toxicity of Sodium Fluoride in Wistar Rats. *J Anim Res*. 2015;5(3):595–8.

[35] Kumar A, Susheela AK. Effects of chronic fluoride toxicity on the morphology of ductus epididymis and the maturation of spermatozoa of rabbit. *Int J Exp Pathol*. 1995;76(1):1–11.

[36] Chinoy NJ, Sequeira E. Effects of fluoride on the histoarchitecture of reproductive organs of the male mouse. *Reprod Toxicol* [Internet]. 1989;3(4):261–7.

[37] Pal P, Mukhopadhyay PK. Fluoride induced testicular toxicities in adult Wistar rats. *Toxicol Mech Methods.* 2021;31(5):383–92.

[38] Thangapandiyan S, Miltonprabu S. Epigallocatechin gallate exacerbates fluoride-induced oxidative stress mediated testicular toxicity in rats through the activation of Nrf2 signaling pathway. *Asian Pacific J Reprod*. 2015;4(4):272–87.

[39] Sun Z, Li S, Yu Y, Chen H, Ommati MM, Manthari RK, Niu R, Wang J. Alterations in epididymal proteomics and antioxidant activity of mice exposed to fluoride. *Arch Toxicol*. 2018 Jan;92(1):169-180.

[40] Sun Z, Li S, Guo Z, Li R, Wang J, Niu R, Wang J. Effects of Fluoride on SOD and CAT in Testis and Epididymis of Mice. *Biol Trace Elem Res*. 2018;184(1):148-153.

[41] Long Hu, Jin Ying, Lin Mu, Sun Yu, Zhang Liang, Clinch Carole. Fluoride toxicity in the male reproductive system. *Fluoride*. 2009;42(4):260-76.
[42] Ghosh D, Das S, Maiti R, Jana D, Das UB. Testicular toxicity in sodium fluoride treated rats: Association with oxidative stress. *Reprod Toxicol*. 2002;16(4):385–90.
[43] Orta Yilmaz B, Korkut A, Erkan M. Sodium fluoride disrupts testosterone biosynthesis by affecting the steroidogenic pathway in TM3 Leydig cells. *Chemosphere*. 2018;212:447-55.
[44] Han H, Sun Z, Luo G, Wang C, Wei R, Wang J. Fluoride exposure changed the structure and the expressions of reproductive related genes in the hypothalamus-pituitary-testicular axis of male mice. *Chemosphere*. 2015;135:297–303.
[45] Feng D, Huang H, Yang Y, Yan T, Jin Y, Cheng X, Cui L. Ameliorative effects of N-acetylcysteine on fluoride-induced oxidative stress and DNA damage in male rats' testis. *Mutat Res Genet Toxicol Environ Mutagen*. 2015;792:35-45.
[46] Zhang S, Jiang C, Liu H, Guan Z, Zeng Q, Zhang C, Lei R, Xia T, Gao H, Yang L, Chen Y, Wu X, Zhang X, Cui Y, Yu L, Wang Z, Wang A. Fluoride-elicited developmental testicular toxicity in rats: roles of endoplasmic reticulum stress and inflammatory response. *Toxicol Appl Pharmacol*. 2013 Sep 1;271(2):206-15.
[47] Li Y, Min cheng, Zhao Y, Wang J, Wang J. Effects of fluoride on PIWI-interacting RNA expression profiling in testis of mice. *Chemosphere*. 2021;269:128727.
[48] Liang C, He Y, Liu Y, Gao Y, Han Y, Li X, Zhao Y, Wang J, Zhang J. Fluoride exposure alters the ultra-structure of sperm flagellum via reducing key protein expressions in testis. *Chemosphere*. 2020 May;246:125772.
[49] Song Gh, Wang RL, Chen ZY, Zhang B, Wang HL, Liu ML, Gao JP, Yan XY. Toxic effects of sodium fluoride on cell proliferation and apoptosis of Leydig cells from young mice. *J Physiol Biochem*. 2014;70(3):761-8.
[50] Zhang J, Zhu Y, Shi Y, Han Y, Liang C, Feng Z, Zheng H, Eng M, Wang J. Fluoride-Induced Autophagy via the Regulation of Phosphorylation of Mammalian Targets of Rapamycin in Mice Leydig Cells. *J Agric Food Chem*. 2017;65(40):8966-8976.
[51] Yang R, Tsai-Morris CH, Kang JH, Dufau ML. Elucidation of RNA binding regions of gonadotropin-regulated testicular RNA helicase (GRTH/DDX25) to transcripts of a chromatin remodeling protein essential for spermatogenesis. *Horm Mol Biol Clin Investig*. 2015;22(3):119-30.
[52] Zhang J, Zhu Y, Liang C, Qie M, Niu R, Sun Z, Wang J, Wang J. Effects of Fluoride on Expression of P450, CREM and ACT Proteins in Rat Testes. *Biol Trace Elem Res*. 2017;175(1):156-160.
[53] Sun Z, Zhang W, Xue X, Zhang Y, Niu R, Li X, Li B, Wang X, Wang J. Fluoride decreased the sperm ATP of mice through inhabiting mitochondrial respiration. *Chemosphere*. 2016 Feb;144: 1012-7.
[54] Kim J, Kwon WS, Rahman MS, Lee JS, Yoon SJ, Park YJ, You YA, Pang MG. Effect of sodium fluoride on male mouse fertility. *Andrology*. 2015;3(3):544-51.
[55] Jiang S, Liang C, Gao Y, Liu Y, Han Y, Wang J, Zhang J. Fluoride exposure arrests the acrosome formation during spermatogenesis via down-regulated Zpbp1, Spaca1 and Dpy19l2 expression in rat testes. *Chemosphere*. 2019;226:874-882.
[56] Liu Y, Liang C, Gao Y, Jiang S, He Y, Han Y, Olfati A, Manthari RK, Wang J, Zhang J. Fluoride Interferes with the Sperm Fertilizing Ability via Downregulated SPAM1, ACR, and PRSS21 Expression in Rat Epididymis. *J Agric Food Chem*. 2019;67(18):5240-5249.
[57] Lu Z, Wang S, Sun Z, Niu R, Wang J. *In vivo* influence of sodium fluoride on sperm chemotaxis in male mice. *Arch Toxicol*. 2014;88(2):533-9.
[58] Liang C, Gao Y, He Y, Han Y, Manthari RK, Tikka C, Chen C, Wang J, Zhang J. Fluoride induced mitochondrial impairment and PINK1-mediated mitophagy in Leydig cells of mice: *In vivo* and *in vitro* studies. *Environ Pollut*. 2020;256:113438.
[59] Liang S, Zhao MH, Ock SA, Kim NH, Cui XS. Fluoride impairs oocyte maturation and subsequent embryonic development in mice. *Environ Toxicol*. 2016;31(11):1486-95.
[60] Zhang Z, GZ Z, Liu X, Sun G, Guan G. Effects of organic fluoride exposure on the development their offspring. *Fluoride*. 1993;26(5):223-4.

[61] Zhou Y, Zhang H, He J, Chen X, Ding Y, Wang Y, Liu X. Effects of sodium fluoride on reproductive function in female rats. *Food Chem Toxicol*. 2013;56:297-303.

[62] Maity PP, Dash M, Dey A, Maity M, Chattopadhyay S. Vitamin C and E in Fluoride Induced Uterine Oxidative Stress and Apoptotic Gene Expression in Wistar Rats. *Proc Zool Soc*. 2021;74(2)189-97.

[63] Guney M, Oral B, Demirin H, Karahan N, Mungan T, Delibas N. Protective effects of vitamins C and E against endometrial damage and oxidative stress in fluoride intoxication. *Clin Exp Pharmacol Physiol*. 2007;34(5–6):467-74.

[64] Wang HW, Zhao WP, Liu J, Tan PP, Zhang C, Zhou BH. Fluoride-induced oxidative stress and apoptosis are involved in the reducing of oocytes development potential in mice. *Chemosphere*. 2017;186:911-8.

[65] Chawla SL, Rao M V. Protective effect of melatonin against fluorideinduced oxidative stress in the mouse ovary. *Fluoride*. 2012;45(2):125–32.

[66] Zhao W peng, Wang H wei, Liu J, Zhang Z hao, Zhu S quan, Zhou B hua. Mitochondrial respiratory chain complex abnormal expressions and fusion disorder are involved in fluoride-induced mitochondrial dysfunction in ovarian granulosa cells. *Chemosphere*. 2019;215:619-25.

[67] Yin S, Song C, Wu H, Chen X, Zhang Y. Adverse effects of high concentrations of fluoride on characteristics of the ovary and mature oocyte of mouse. *PLoS One*. 2015;10(6): e0129594.

[68] Liu X, Nie ZW, Gao YY, Chen L, Yin SY, Zhang X, Hao C, Miao YL. Sodium fluoride disturbs DNA methylation of NNAT and declines oocyte quality by impairing glucose transport in porcine oocytes. *Environ Mol Mutagen*. 2018;59(3):223-233.

[69] Zhao WP, Wang HW, Liu J, Tan PP, Lin L, Zhou BH. JNK/STAT signalling pathway is involved in fluoride-induced follicular developmental dysplasia in female mice. *Chemosphere*. 2018;209:88-95.

[70] Wang H wei, Zhao W peng, Tan P pan, Liu J, Zhao J, Zhou B. The MMP-9/TIMP-1 System is Involved in Fluoride-Induced Reproductive Dysfunctions in Female Mice. *Biol Trace Elem Res*. 2017;178(2):253-60.

[71] Zhou Y, Qiu Y, He J, Chen X, Ding Y, Wang Y, Liu X. The toxicity mechanism of sodium fluoride on fertility in female rats. *Food Chem Toxicol*. 2013 Dec;62:566-72.

[72] Pant N, Kumar R, Murthy RC, Srivastava SP. Male reproductive effect of arsenic in mice. *Biometals*. 2001;14(2):113-7.

[73] Liu SX, Athar M, Lippai I, Waldren C, Hei TK. Induction of oxyradicals by arsenic: implication for mechanism of genotoxicity. *Proc Natl Acad Sci USA*. 2001;98(4):1643-8.

[74] Waalkes MP, Ward JM, Liu J, Diwan BA. Transplacental carcinogenicity of inorganic arsenic in the drinking water: Induction of hepatic, ovarian, pulmonary, and adrenal tumors in mice. *Toxicol Appl Pharmacol*. 2003;186(1):7-17.

[75] Chowdhury TR, Basu GK, Mandal BK, Biswas BK, Samanta G, Chowdhury UK, Chanda CR, Lodh D, Roy SL, Saha KC, Roy S, Kabir S, Quamruzzaman Q, Chakraborti D. Arsenic poisoning in the Ganges delta. *Nature*. 1999;401(6753):545-6; discussion 546-7.

[76] Smedley PL, Kinniburgh DG. A review of the source, behaviour and distribution of arsenic in natural waters. *Appl Geochemistry*. 2002;17(5):517-68.

[77] Sengupta MK, Mukherjee A, Hossain MA, Ahamed S, Rahman MM, Lodh D. Groundwater arsenic contamination in the Ganga-Padma-Meghna-Brahmaputra plain of India and Bangladesh. *Arch Environ Health*. 2003;58(11):701-2.

[78] Mishra S, Mattusch J, Wennrich R. Accumulation and transformation of inorganic and organic arsenic in rice and role of thiol-complexation to restrict their translocation to shoot. *Sci Rep*. 2017;7:405-22.

[79] Chatterjee A, Mukherjee A. Hydrogeological investigation of ground water arsenic contamination in south Calcutta. *Sci Total Environ*. 1999;225(3):249-62.

[80] Thornton I, Farago M. The geochemistry of arsenic. *Arsenic*. 1997;1–16.

[81] Mirza N, Mahmood Q, Maroof Shah M, Pervez A, Sultan S. Plants as useful vectors to reduce environmental toxic arsenic content. *ScientificWorldJournal*. 2014;2014:921581.

[82] Straskraba V, Moran RE. Environmental occurrence and impacts of arsenic at gold mining sites in the western United States. *Int J mine water association*. 1990;9(1):181-91.

[83] Zubair M, Ahmad M, Qureshi ZI. Review on arsenic-induced toxicity in male reproductive system and its amelioration. *Andrologia*. 2017;49(9):e12791.

[84] Shaji E, Santosh M, Sarath K V., Prakash P, Deepchand V, Divya B V. Arsenic contamination of groundwater: A global synopsis with focus on the Indian Peninsula. *Geosci Front.* 2021;12(3):101079.

[85] Brammer H, Ravenscroft P. Arsenic in groundwater: a threat to sustainable agriculture in South and South-east Asia. *Env Int.* 2009;35(3):647-54.

[86] Sorg TJ, Chen ASC, Wang L. Arsenic species in drinking water wells in the USA with high arsenic concentrations. *Water Res.* 2014;48(1):156-69.

[87] Vahter ME. Interactions between arsenic-induced toxicity and nutrition in early life. *J Nutr.* 2007;137(12):2798-804.

[88] Faita F, Cori L, Bianchi F, Andreassi MG. Arsenic-Induced Genotoxicity and Genetic Susceptibility to Arsenic-Related Pathologies. *Int J Environ Res Public Health.* 2013;10(4):1527.

[89] Yamanaka K, Hoshino M, Okamoto M, Sawamura R, Hasegawa A, Okada S. Induction of DNA damage by dimethylarsine, a metabolite of inorganic arsenics, is for the major part likely due to its peroxyl radical. *Biochem Biophys Res Commun.* 1990;168(1):58-64.

[90] Acharyya N, Deb B, Chattopadhyay S, Maiti S. Arsenic-Induced Antioxidant Depletion, Oxidative DNA Breakage, and Tissue Damages are Prevented by the Combined Action of Folate and Vitamin B12. *Biol Trace Elem Res.* 2015;168(1):122-32.

[91] Chattopadhyay S, Ghosh D. Role of dietary GSH in the amelioration of sodium arsenite-induced ovarian and uterine disorders. *Reprod Toxicol.* 2010;30(3):481-8.

[92] Gamble MV, Liu X, Ahsan H, Pilsner JR, Ilievski V, Slavkovich V, Parvez F, Chen Y, Levy D, Factor-Litvak P, Graziano JH. Folate and arsenic metabolism: a double-blind, placebo-controlled folic acid-supplementation trial in Bangladesh. *Am J Clin Nutr.* 2006;84(5):1093-101.

[93] Jomova K, Jenisova Z, Feszterova M, Baros S, Liska J, Hudecova D, Rhodes CJ, Valko M. Arsenic: toxicity, oxidative stress and human disease. *J Appl Toxicol.* 2011;31(2):95-107.

[94] Hall AH. Chronic arsenic poisoning. *Toxicol Lett.* 2002;128(1–3):69-72.

[95] Mittal M, Flora SJ. Vitamin E supplementation protects oxidative stress during arsenic and fluoride antagonism in male mice. *Drug Chem Toxicol.* 2007;30(3):263-81.

[96] Rahman MM, Chowdhury UK, Mukherjee SC, Mondal BK, Paul K, Lodh D, Biswas BK, Chanda CR, Basu GK, Saha KC, Roy S, Das R, Palit SK, Quamruzzaman Q, Chakraborti D. Chronic arsenic toxicity in Bangladesh and West Bengal, India--a review and commentary. *J Toxicol Clin Toxicol.* 2001;39(7):683-700.

[97] Hughes MF, Beck BD, Chen Y, Lewis AS, Thomas DJ. Arsenic exposure and toxicology: a historical perspective. *Toxicol Sci.* 2011;123(2):305-32.

[98] Fatoki JO, Badmus JA. Arsenic as an environmental and human health antagonist: A review of its toxicity and disease initiation. *J Hazard Mater Adv.* 2022;5:100052.

[99] Bustaffa E, Stoccoro A, Bianchi F, Migliore L. Genotoxic and epigenetic mechanisms in arsenic carcinogenicity. *Arch Toxicol.* 2014;88(5):1043-67.

[100] Biswas S, Kumar Mukhopadhyay P. Casein- and pea-enriched high-protein diet can take care of the reprotoxic effects of arsenic in male rats. *Andrologia.* 2020;52(5):e13560.

[101] Thakur M, Rachamalla M, Niyogi S, Datusalia AK, Flora SJS. Molecular Mechanism of Arsenic-Induced Neurotoxicity including Neuronal Dysfunctions. *Int J Mol Sci.* 2021;22(18):10077.

[102] Soni M, Prakash C, Sehwag S, Kumar V. Protective effect of hydroxytyrosol in arsenic-induced mitochondrial dysfunction in rat brain. *J Biochem Mol Toxicol.* 2017;31(7).

[103] Renu K, Madhyastha H, Madhyastha R, Maruyama M, Vinayagam S, Valsala Gopalakrishnan A. Review on molecular and biochemical insights of arsenic-mediated male reproductive toxicity. *Life Sci.* 2018;212:37-58.

[104] Wares MA, Awal MA, Das SK, Hannan MA, Anas MA, Latif MA, Masud N. Chronic natural arsenic exposure affecting histoarchitecture of gonads in Black Bengal goats (Capra aegagrushircus). *J Adv Vet Anim Res.* 2015;2(2):128-33.

[105] Khan AG.; Malik JK. ST. Arsenic-induced oxidative stress, apoptosis and alterations in testicular steroidogenesis and spermatogenesis in wistar rats: ameliorative effect of cucurmin. *Wudpecker J Pharm Pharmacol.* 2013;2(3):33-48.

[106] Chiou TJ, Chu ST, Tzeng WF, Huang YC, Liao CJ. Arsenic trioxide impairs spermatogenesis via reducing gene expression levels in testosterone synthesis pathway. *Chem Res Toxicol.* 2008;21(8):1562-9.

[107] Pant N, Murthy RC, Srivastava SP. Male reproductive toxicity of sodium arsenite in mice. *Hum Exp Toxicol.* 2004;23(8):399-403.

[108] Maharjan M, Watanabe C, Ahmad SA, Umezaki M, Ohtsuka R. Mutual interaction between nutritional status and chronic arsenic toxicity due to groundwater contamination in an area of Terai, lowland Nepal. *J Epidemiol Community Health.* 2007;61(5):389.

[109] Sarkar S, Hazra J, Upadhyay S, Singh R, Amal R. Arsenic induced toxicity on testicular tissue of mice. *Indian J Physiol Pharmacol.* 2008;52(1):84-90.

[110] Jana K, Jana S, Samanta PK. Effects of chronic exposure to sodium arsenite on hypothalamo-pituitary-testicular activities in adult rats: possible an estrogenic mode of action. *Reprod Biol Endocrinol.* 2006;4:9.

[111] Sarkar M, Chaudhuri GR, Chattopadhyay A, Biswas NM. Effect of sodium arsenite on spermatogenesis, plasma gonadotrophins and testosterone in rats. *Asian J Androl.* 2003;5(1):27-31.

[112] Kim YJ, Kim JM. Arsenic Toxicity in Male Reproduction and Development. *Dev Reprod.* 2015;19(4):167.

[113] Momeni HR, Oryan S, Eskandari N. Effect of vitamin E on sperm number and testis histopathology of sodium arsenite-treated rats. *Reprod Biol.* 2012;12(2):171-81.

[114] Reddy PS, Rani GP, Sainath SB, Meena R, Supriya C. Protective effects of N-acetylcysteine against arsenic-induced oxidative stress and reprotoxicity in male mice. *J Trace Elem Med Biol.* 2011;25(4):247-53.

[115] Ghosh D, Ghosh S, Sarkar S, Ghosh A, Das N, Das Saha K, Mandal AK. Quercetin in vesicular delivery systems: evaluation in combating arsenic-induced acute liver toxicity associated gene expression in rat model. *Chem Biol Interact.* 2010;186(1):61-71.

[116] Kitchin KT. Recent advances in arsenic carcinogenesis: modes of action, animal model systems, and methylated arsenic metabolites. *Toxicol Appl Pharmacol.* 2001;172(3):249-61.

[117] Ramanathan K, Balakumar BS, Panneerselvam C. Effects of ascorbic acid and α-tocopherol on arsenic-induced oxidative stress. *Human and Experimental Toxicology.* 2002;21(12):675-80.

[118] Wang TS, Kuo CF, Jan KY, Huang H. Arsenite induces apoptosis in chinese hamster ovary cells by generation of reactive oxygen species. *J Cell Physiol.* 1996;169(2):256-68.

[119] Chattopadhyay S, Ghosh S, Chaki S, Debnath J, Ghosh D. Effect of sodium arsenite on plasma levels of gonadotrophins and ovarian steroidogenesis in mature albino rats: Duration-dependent response. *J Toxicol Sci.* 1999;24(5):425-31.

[120] Mondal S, Mukherjee S, Chaudhuri K, Kabir SN, Mukhopadhyay PK. Prevention of arsenic-mediated reproductive toxicity in adult female rats by high protein diet. *Pharm Biol.* 2013;51(11):1363-71.

[121] Akram Z, Jalali S, Shami SA, Ahmad L, Batool S, Kalsoom O. Adverse effects of arsenic exposure on uterine function and structure in female rat. *Exp Toxicol Pathol.* 2010;62(4):451-9.

[122] Biswas P, Mukhopadhyay A, Kabir SN, Mukhopadhyay PK. High-Protein Diet Ameliorates Arsenic-Induced Oxidative Stress and Antagonizes Uterine Apoptosis in Rats. *Biol Trace Elem Res.* 2019;192(2):222-33.

[123] Banu SA, Kile ML, Christiani DC, Qumruzzaman Q. Study of prenatal arsenic exposure and reproductive health outcome in Bangladesh. *Bangladesh J Obstet Gynecol.* 2013;28(2):76-81.

[124] Yu H, Kuang M, Wang Y, Rodeni S, Wei Q, Wang W, Mao D. Sodium Arsenite Injection Induces Ovarian Oxidative Stress and Affects Steroidogenesis in Rats. *Biol Trace Elem Res.* 2019;189(1):186-193.

[125] Souza ACF, Ervilha LOG, Coimbra JLP, Bastos DSS, Guimarães SEF, Machado-Neves M. Reproductive disorders in female rats after prenatal exposure to sodium arsenite. *J Appl Toxicol.* 2020;40(2):214-23.

[126] Jana K, Jana S, Samanta PK. Effects of chronic exposure to sodium arsenite on hypothalamo-pituitary-testicular activities in adult rats: Possible an estrogenic mode of action. *Reprod Biol Endocrinol.* 2006;4:9.

[127] Chattopadhyay S, Ghosh S, Debnath J, Ghosh D. Protection of sodium arsenite-induced ovarian toxicity by coadministration of L-ascorbate (vitamin C) in mature wistar strain rat. *Arch Environ Contam Toxicol.* 2001;41(1):83-9.

[128] Ghosh D, Chattopadhyay S, Debnath J. Effect of sodium arsenite on adrenocortical activity in immature female rats: Evidence of dose dependent response. *J Environ Sci.* 1999;11(4):419-22.

[129] Bodwell JE, Kingsley LA, Hamilton JW. Arsenic at very low concentrations alters glucocorticoid receptor (GR)-mediated gene activation but not GR-mediated gene repression: complex dose-response effects are closely correlated with levels of activated GR and require a functional GR DNA binding domain. *Chem Res Toxicol.* 2004;17(8):1064-76.

[130] Davey JC, Bodwell JE, Gosse JA, Hamilton JW. Arsenic as an endocrine disruptor: Effects of arsenic on estrogen receptor-mediated gene expression *in vivo* and in cell culture. *Toxicol Sci.* 2007;98(1): 75-86.

[131] Tandon N, Roy M, Roy S, Gupta N. Protective effect of Psidium guajava in arsenic-induced oxidative stress and cytological damage in rats. *Toxicol Int.* 2012;19(3)245-9.

[132] Wang XN, Zhang CJ, Diao HL, Zhang Y. Protective effects of curcumin against sodium arsenite-induced ovarian oxidative injury in a mouse model. *Chin Med J (Engl).* 2017;130(9):1026-32.

[133] Dash M, Maity M, Dey A, Perveen H, Khatun S, Jana L, Chattopadhyay S. The consequence of NAC on sodium arsenite-induced uterine oxidative stress. *Toxicol Rep.* 2018;5:278-287.

[134] Mondal R, Mukhopadhyay A, Chattopadhyay A, Bandyopadhyay A. Ovarian follicular atresia and uterine toxicity after subchronic oral exposure of postpubertal rats to sodium arseniteo Title. *Comp Clin Path.* 2022;31(4).

[135] Salmon TB, Evert BA, Song B, Doetsch PW. Biological consequences of oxidative stress-induced DNA damage in Saccharomyces cerevisiae. *Nucleic Acids Res.* 2004;32(12):3712-23.

[136] Akram Z, Jalali S, Shami SA, Ahmad L, Batool S, Kalsoom O. Genotoxicity of sodium arsenite and DNA fragmentation in ovarian cells of rat. *Toxicol Lett.* 2009;190(1):81-5.

[137] Dash M, Dey A, Chattopadhyay S. Mitigation of arsenic driven utero-ovarian malfunction and changes of apoptotic gene expression by dietary NAC. *Ecotoxicol Environ Saf.* 2020;199:110675.

[138] Ommati MM, Shi X, Li H, Zamiri MJ, Farshad O, Jamshidzadeh A, Heidari R, Ghaffari H, Zaker L, Sabouri S, Chen Y. The mechanisms of arsenic-induced ovotoxicity, ultrastructural alterations, and autophagic related paths: An enduring developmental study in folliculogenesis of mice. *Ecotoxicol Environ Saf.* 2020;204:110973.

[139] Shukla JP, Pandey K. Impaired ovarian functions in arsenic-treated freshwater fish, Colisa fasciatus (BL. and SCH.). *Toxicol Lett.* 1984;20(1):1-3.

[140] Chattopadhyay S, Pal Ghosh S, Ghosh D, Debnath J. Effect of dietary co-administration of sodium selenite on sodium arsenite-induced ovarian and uterine disorders in mature albino rats. *Toxicological Sciences.* 2003:75(2):412-22.

[141] Chatterjee A, Chatterji U. Arsenic abrogates the estrogen-signaling pathway in the rat uterus. *Reprod Biol Endocrinol.* 2010;8:80.

[142] Davila-Esqueda ME, Jimenez-Capdeville ME, Delgado JM, De la Cruz E, Aradillas-Garcia C, Jimenez-Suarez V. Effects of arsenic exposure during the pre- and postnatal development on the puberty of female offspring. *Exp Toxicol Pathol.* 2012;64(1–2):25-30.

[143] Elshawarby AM, Saleh HA, Attia AAEM, Negm EA. Arsenic-induced toxicity in the endometrium of adult albino rat and the possible role of human chorionic gonadotropin hormone: A histological study. *Egypt J Histol.* 2014;37(2):327-38.

[144] Wares M, Awal M, Das S, Alam J. Environmentally persistant toxicant arsenic affects uterus grossly and histologically. *Bangladesh J Vet Med.* 2014;11(1):61-8.

[145] Guo Z, Yang W, Guo H, Han X, Xia Y. Effects of chronic arsenic exposure on mRNA expression of estrogen receptor-binding fragment-associated gene 9 and estrogen-responsive finger protein in female rat's uterus and ovary tissue. *Chinese J Endem.* 2014;33(1).

[146] Li X, Sun Z, Manthari RK, Li M, Guo Q, Wang J. Effect of gestational exposure to arsenic on puberty in offspring female mice. *Chemosphere.* 2018;202:119-26.

[147] Pan W, Ye X, Zhu Z, Li C, Zhou J, Liu J. A case-control study of arsenic exposure with the risk of primary ovarian insufficiency in women. *Environ Sci Pollut Res.* 2020;27(20):25220-9.

[148] Nava-Rivera LE, Betancourt-Martínez ND, Lozoya-Martínez R, Carranza-Rosales P, Guzmán-Delgado NE, Carranza-Torres IE, Delgado-Aguirre H, Zambrano-Ortíz JO, Morán-Martínez J. Transgenerational effects in DNA methylation, genotoxicity and reproductive phenotype by chronic arsenic exposure. *Sci Rep.* 2021;11(1):8276.

[149] Ali K, Iqbal A, Bukhari SM, Safdar S, Raiz A, Ali W, Hussain A, Javid A, Hussain M, Ali MM, Mahmud A, Iqbal MJ, Nasir MF, Mubeen I, Kanwal S, Sughra F, Khattak A, Saleem M. Amelioration potential of Moringa oleifera extracts against sodium arsenate induced embryotoxicity and genotoxicity in mouse (Mus musculus). *Braz J Biol.* 2021;83:e248022.

[150] Bloom MS, Fitzgerald EF, Kim K, Neamtiu I, Gurzau ES. Spontaneous pregnancy loss in humans and exposure to arsenic in drinking water. Vol. 213, *International Journal of Hygiene and Environmental Health.* 2010;213(6):401-13.

[151] Wang H, Li J, Zhang X, Zhu P, Hao JH, Tao FB, Xu DX. Maternal serum arsenic level during pregnancy is positively associated with adverse pregnant outcomes in a Chinese population. *Toxicol Appl Pharmacol.* 2018;356:114-119.

[152] Hill DS, Wlodarczyk BJ, Finnell RH. Reproductive consequences of oral arsenate exposure during pregnancy in a mouse model. *Birth Defects Res Part B - Dev Reprod Toxicol.* 2008;83(1):40-7.

[153] He W, Greenwell RJ, Brooks DM, Calderón-Garcidueñas L, Beall HD, Coffin JD. Arsenic exposure in pregnant mice disrupts placental vasculogenesis and causes spontaneous abortion. *Toxicol Sci.* 2007;99(1):244-53.

[154] Sohel N, Vahter M, Ali M, Rahman M, Rahman A, Streatfield PK, Kanaroglou PS, Persson LA. Spatial patterns of fetal loss and infant death in an arsenic-affected area in Bangladesh. *Int J Health* Geogr. 2010;9:53.

[155] Yang CY, Chang CC, Tsai SS, Chuang HY, Ho CK, Wu TN. Arsenic in drinking water and adverse pregnancy outcome in an arseniasis-endemic area in northeastern Taiwan. *Environ Res.* 2003;91(1):29-34.

[156] Rahman A, Vahter M, Ekström EC, Persson LÅ. Arsenic exposure in pregnancy increases the risk of lower respiratory tract infection and diarrhea during infancy in Bangladesh. *Environ Health Perspect.* 2011;119(5):719-24.

[157] Hopenhayn-Rich C, Browning SR, Hertz-Picciotto I, Ferreccio C, Peralta C, Gibb H. Chronic arsenic exposure and risk of infant mortality in two areas of Chile. *Environ Health Perspect.* 2000;108(7):667-73.

[158] Laine JE, Bailey KA, Rubio-Andrade M, Olshan AF, Smeester L, Drobná Z, Herring AH, Stýblo M, García-Vargas GG, Fry RC. Maternal arsenic exposure, arsenic methylation efficiency, and birth outcomes in the Biomarkers of Exposure to ARsenic (BEAR) pregnancy cohort in Mexico. *Environ Health Perspect.* 2015;123(2):186-92.

[159] Dhar P, Mohari N, Mehra RD. Preliminary morphological and morphometric study of rat cerebellum following sodium arsenite exposure during rapid brain growth (RBG) period. *Toxicology.* 2007;234(1–2):10-20.

[160] Aschencrau A, Zierler S, Cohen A. Quality of community drinking water and the occurrence of spontaneous abortion. *Arch Environ Health.* 1989;44(5):283-90.

[161] Ahmad SA, Sayed MH, Barua S, Khan MH, Faruquee MH, Jalil A, Hadi SA, Talukder HK. Arsenic in drinking water and pregnancy outcomes. *Environ Health Perspect.* 2001;109(6):629-31.

[162] Cherry N, Shaikh K, McDonald C, Chowdhury Z. Stillbirth in rural Bangladesh: Arsenic exposure and other etiological factors: A report from Gonoshasthaya Kendra. *Bull World Health Organ.* 2008;86(3):172-7.

[163] Fry RC, Navasumrit P, Valiathan C, Svensson JP, Hogan BJ, Luo M, Bhattacharya S, Kandjanapa K, Soontararuks S, Nookabkaew S, Mahidol C, Ruchirawat M, Samson LD. Activation of inflammation/NF-kappaB signaling in infants born to arsenic-exposed mothers. *PLoS Genet.* 2007;3(11):e207.

[164] Kwok RK, Kaufmann RB, Jakariya M. Arsenic in drinking-water and reproductive health outcomes: a study of participants in the Bangladesh Integrated Nutrition Programme. *J Heal Popul Nutr.* 2006;24(2):190-205.

[165] Rahman A, Vahter M, Ekström EC, Rahman M, Golam Mustafa AH, Wahed MA, Yunus M, Persson LA. Association of arsenic exposure during pregnancy with fetal loss and infant death: a cohort study in Bangladesh. *Am J Epidemiol.* 2007;165(12):1389-96.
[166] von Ehrenstein OS, Guha Mazumder DN, Hira-Smith M, Ghosh N, Yuan Y, Windham G, Ghosh A, Haque R, Lahiri S, Kalman D, Das S, Smith AH. Pregnancy outcomes, infant mortality, and arsenic in drinking water in West Bengal, India. *Am J Epidemiol.* 2006;163(7):662-9.

Chapter 19

# Rethinking Endometriosis Pathophysiology: Judging the Oxidative Balance

**Pratip Chakraborty[1,*], PhD**
**Imon Mitra[2]**
**and Ratna Chattopadhyay[1], PhD**
[1]Department of Assisted Reproduction, Institute of Reproductive Medicine, Kolkata, India
[2]School of Medical Science and Technology, Indian Institute of Technology Kharagpur, India

## Abstract

Endometriosis is classically defined as a chronic, gynaecological disease characterized by endometrial-like tissue present outside of the uterus. Currently, the possible hypothesized etiologies of endometriosis are immunological, inflammatory, genetic and endocrinological, however, the precise mechanisms underlying origin and development remain mainly unknown. The disease affects 5–10% of women of reproductive age globally. However, despite its prevalence, diagnosis is typically delayed by years, and misdiagnosis is common, with delivery of effective therapy is prolonged. Reactive oxygen species (ROS), dysregulation of the immune system and cellular architectural disruption constitute the crucial mechanisms that detrimentally affect oocyte and embryo developmental potential. Endogenous ROS is correlated with increased cellular proliferation and is the upstream driver for progression of inflammation in endometriosis. Before a definitive diagnosis is made, women often endure symptoms for years while negative effects on well-being and quality of life. In summary, an oxidative environment leads to stimulation of few signaling pathways including ERK, Nrf2 and PI3K/AKT/mTOR that facilitate endometriotic lesion progression through adhesion, angiogenesis, and proliferation; the focal regulatory cellular functions in endometriosis. Here, we discuss the epidemiology, pathophysiology, and pathogenesis of endometriosis, as well as diagnosis and treatment, with an emphasis on how understanding the oxidative balance of the disease can inform development of better diagnostic and treatment options.

**Keywords:** endometriosis, folliculogenesis, ROS, antioxidants, inflammation, infertility

---

* Corresponding Author's Email: pratip_2011@yahoo.com.

In: The Role of Reactive Oxygen Species in Health and Disease
Editor: Kuladip Jana
ISBN: 979-8-89113-283-2

## 1. Introduction

Endometriosis is a chronic, debilitating disease associated with pelvic pain and infertility having a prevalence of about 10% [1]. The most common types of pelvic endometriosis are superficial peritoneal lesions, deep-infiltrating endometriosis, and ovarian endometrioticcysts (endometriomas) [2]. Although a definitive diagnosis of endometriosis is often madeon the basis of histological criteria, visual diagnosis is commonly used in practice. However, visualisation can be challenging as a result of the heterogeneity in lesion location and highly variable appearance [3]. The conundrum on how the reproductive potential of a patient with endometriosis is compromised has been thoroughly discussed in the literature. Since the onset, endometriosis-associated pain symptoms are most often reported during adolescence and young adulthood, with the pathophysiological steps of initiation and developmentusually occurring earlier in the life span. Evidence of genetic loci has also been associated with a risk of endometriosis [4]. Several studies have suggested that fertilization rate, the number of competent embryos, blastocyst formation rate, implantation rate and pregnancy rate are impaired in cases of diagnosed endometriosis [4]. Consequently, adhesion, chronic inflammation, oxidative stress, perturbed folliculogenesis, luteal phase defect, anti-endometrial antibodies, spermatozoa defects, progesterone resistance and disrupted uterotubal motility constitute some of the proposed mechanisms [5]. Of these, oxidative stress (OS), characterized by an imbalance between pro-oxidant molecules including reactiveoxygen and nitrogen species, and antioxidant defenses, resulting damage to cellular lipids, proteins, or DNA, has been identified to play negative effects on female reproductive abilities. The link between pro-inflammation; a condition often observed in endometriosisand OS is widely documented [6, 7]. Nevertheless, the effect of reactive oxygen species on developmental dynamics of the reproductive tissue has been little showcased. Hence, a shift in our perspective towards acknowledging endometriosis as a systemic disease is required. Importance of oxygen: the rise of reactive oxygen species Oxygen is a necessary element ofaerobic life, and oxidative metabolism represents a principal source of energy. Under normalcircumstances, two types of antioxidants in the body are present: non-enzymatic antioxidants (superoxide dismutase, glutathione peroxidase etc.) and enzymatic antioxidants(synthetic or dietary supplements, including vitamin C, vitamin E, etc.). OS occurs with the generation of excessive ROS or when the antioxidants' defense mechanisms are weakened [8]. The most important biologically ROS are superoxide anion (O 2•), hydroxyl radical (•OH), peroxyl (ROO•), alkoxyl (RO•) and hydroperoxyl (HO 2•). OS is considered to be responsible for the initiation or development of pathological processes affecting female reproductive processes [9], such as embryonic resorption, recurrent pregnancy loss, preeclampsia, intrauterine growth restriction (IUGR) and fetal death [10] including endometriosis [8, 9]. Although, normal level of ROS plays a chief regulatory role through various signalling transduction pathways in folliculogenesis, corpus luteum oocytematuration and feto-placental development; however, ROS can exert damaging effects when overabundant. Nonetheless, the relationship between ROS-induced OS and diseases isunclear and cannot be adequately investigated in human pregnancies because of self-evident ethical reasons. ROS drives the way to a vicious circle of inflammation. Several mechanisms such as metaplasia of the coelomic epithelium, environmental factors, genetic predisposition, immunological abnormalities, and chronic inflammation have been proposed to explain the pelvic development of endometriosis [11]. Indeed, pro-inflammatory cytokinesand growth factors were found to be elevated in the peritoneal fluid and serum of women with endometriosis and could contribute to the proliferation of endometriotic implants and neo-angiogenesis [12, 13, 14]. Chronic inflammation has also been documented with increased OS [15]. Incidentally, erythrocytes, apoptotic endometrial tissue and cell debris in the peritoneal cavity by menstrual reflux and macrophages are potential inducers of OS. Pro-inflammatory cytokines may impact the recruitment of macrophages, which are one of the main producers of ROS [16]. The peritoneal fluid is rich in lipoproteins, which generates oxidized lipid components in a macrophage-rich inflammatory environment. The oxidants exacerbate the growth of endometriosis by inducing chemo-attractants such as MCP-1 and endometrial cell growth-promoting activity. Presence of oxidative stress in the peritoneal cavity of women with endometriosis, stimulates non-scavenging properties of macrophages that are non-adherent, "switching on" the synergistic interaction between macrophages, oxidative stress, and the endometrial cells [17]. A critical component of inflammation is the infiltration of inflammatory cells, like neutrophils, monocytes, and lymphocytes, to the site ofstimulus. The infiltration of leukocytes to the site of inflammation is a highly coordinated process involving margination, rolling, and adhesion of leukocytes to the vascular endothelium, transmigration across the endothelium, and migration toward a chemotactic stimulus. The participation of a number of adhesion molecules, including selectins, intercellular adhesion molecule-1 (ICAM-1), and vascular cell adhesion molecule-1 (VCAM-1),and their respective leukocyte receptors and chemokines like monocyte chemo attractant protein 1 (MCP-1) or interleukin-8 (IL-8) is crucial for the inflammatory cellular infiltration [18]. At the site of inflammation the activated inflammatory cells release many enzymes (neutral proteases, elastase, collagenase, acid hydrolases, phosphatases, lipases, etc.), reactive species (superoxide, hydrogen peroxide, hydroxyl radical, hypochlorous acid, etc.), and chemical mediators (eicosanoids, complement components, cytokines, chemokines,nitric oxide, etc.) ultimately inducing tissue damage and oxidative stress [19]. Increased inflammatory mediators in the peritoneum, endometrium, and serum of endometriosis cases have been recently observed in several studies [20, 21]. Interleukin-1β (IL-1β) enhances theproliferation of endometriotic cells. This proliferative effect does not occur in healthy cells of endometrium. IL-1β also triggers the production of IL-6 and IL-8 in endometriotic tissues, leading to more proliferation [21, 22] and decreased apoptotic rate [21]. In endometriotic cells, TNF-α is associated with pluripotency mediation and the production of inflammatory cytokines, especially IL-8 [23]. NF-κB, an important regulator of various inflammatory and pathologic responses induces cell proliferation and inflammation and inhibits apoptoticprocess in endometriosis [24]. Recently, it has been revealed that NF-κB–IκB complex creation increased the inflammatory mediator, such as IL-6 and IL-8 in endometriosis [23]. Current investigations on the NF-κB inhibitors in endometriotic cells have shown that the inhibition of NF-κB could reduce endometriosis maintenance and development [25]. High ROS concentrations in endometriosis increase DNA fragmentation, modulate NF-κB expression, and stimulate numerous genes encoding angiogenic growth, and pro- inflammatory factors, cyclooxygenases, and adhesion molecule (Figure 1). Furthermore, elevated expression and production of metalloproteinase and IL-8 leads to increased endometrial stroma adhesion to extracellular matrix proteins [21]. The elevated levels of IL- 6 and IL-8 are related to the resistance to progesterone found in endometriosis [26].Furthermore, the inflammatory cytokine IL-6 has been found to produce ROS through increased expression of NADPH oxidase 4 (NOX4) in non-small cell lung cancer [27]. The NOX4 overexpression has also been found to enhance IL-6 production, and a positive reciprocal feedback loop has been found between IL-6 and NOX4, the two mediators of inflammation and oxidative stress, respectively [27] (Figure 1).

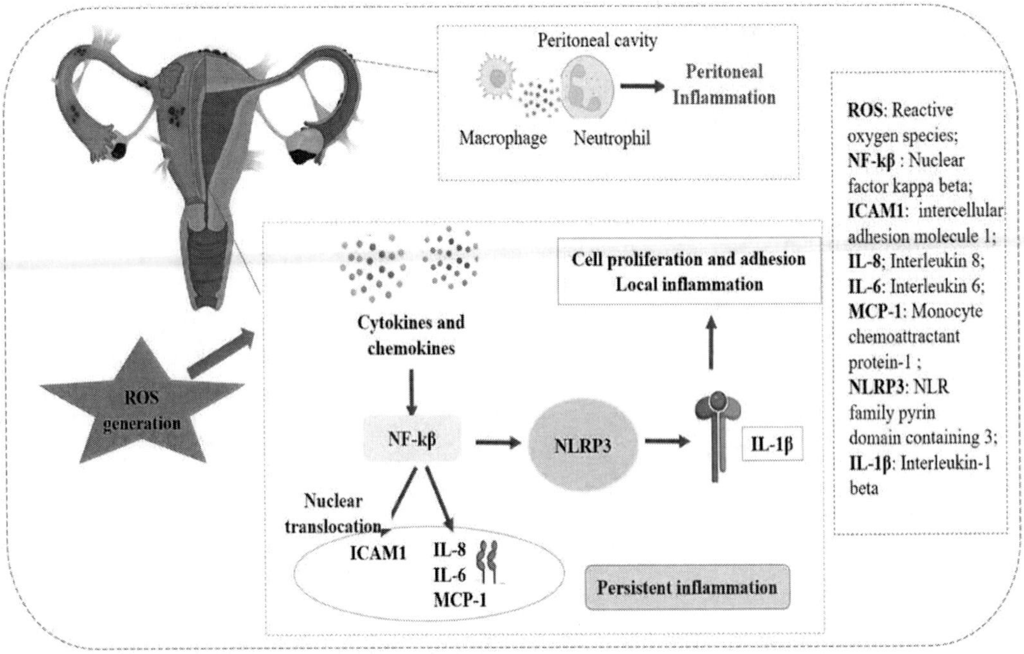


**Figure 1.** The reason behind persistent inflammation in endometriosis: the generation of reactive oxygen species at the hindsight.

## 2. Oxidative Stress and Inflammation : Relation and Dependence

As the inflammatory process can induce oxidative stress, the oxidative stress can also induce inflammation through activation of multiple pathways. The molecular mechanisms involved in the pathogenesis of endometriosis are not completely understood; however, inflammation plays a significant role in the pathophysiology of the disease, mainly by altering the function of immune cells (macrophages, natural killer, and T cells). These immune alterations inhibit apoptotic pathways and promote adhesion and proliferation of endometriotic cells, as well as angiogenesis and neurogenesis in endometriotic lesions. In addition, IL-33 has emerged as a possible biomarker in a syngeneic mouse model of endometriosis [28]. NLRP3 inflammasome has emerged as a key player in innate immunity as well as inflammation. Abnormal inflammasome activation, in absence of detectable infectious causes, might stand as one of the molecular mechanisms involved in establishing an abnormal endometrium that triggers innate immune defenses through the maturation of pro-inflammatory cytokines like IL-1β and IL-18 potentially leading to endometriosis [29]. The ROS released from damaged mitochondria has been shown to activate NLRP3 inflammasomes leading to IL-1β secretion and localized inflammation [29]. Oxidized mitochondrial DNA has also been found to activateNLRP3 inflammasomes during apoptosis [30]. Furthermore, in conditions of oxidative stress the ROS causes the thioredoxin-interacting protein, an inhibitor of endogenous antioxidant thioredoxin, to dissociate from thioredoxin and to bind with NLRP3 leading to activation of NLRP3 inflammasome [31] (Figure 1). Overproduction of chemokines, prostaglandins, with recruitment of immune cells like activated macrophages, natural killer cells, and neutrophils to endometriotic sites have been rigorously reviewed in literature with regard to local inflammatory milieu; however, the role of neutrophils in the biology of endometriosis has not been clearly defined. Future research may throw optimum light on this issue in conjunction to ligand stimulated externalisation of DNA and neutrophil-extracellular-traps (NET) production to look into the inflammatory status [32, 33]. Body, and its signaling system in respect to oxidative stress in endometriosis the body's signalling pathway will manifest a series of changes following exposure to the dual impact of OS and inflammation. Excessive ROS cause an imbalance between oxidation and anti-oxidation, leading to OS involving p38 MAPK pathway, the Kelch-like ECH-associated protein 1 (Keap1)-Nuclear factor erythroid 2-related factor 2 (Nrf2) pathway, the Jun N-terminal kinase (JNK) pathway, the fork-head transcription factors of the O class (FOXO) family, and apoptosis [34]. Nrf2 is a key molecule activated in response to OS, which regulates antioxidant response to protect cell function. Primarily, Nrf2 binds to Keap1; then sequestered in cytoplasm, and subsequently degraded by a proteasome pathway. Post activation, it is transferred to the nucleus to activate a large number of antioxidant defense genes and restoration of vascular redox homeostasis as and when OS erupt/s. It is implied that Nrf2 deficiency cause fetal DNA damage and neurological deficits, to underlie inflammation-induced trophoblastic apoptosis. Studies have highlighted the role of Nrf2 in antioxidant defense in vascular dysfunction and oxidative damage in endometriosis [35]. Moreover, suitable OS increase Nrf2 and the expression of downstream targets, such as heme oxygenase 1 (HO-1), NAD(P)H: quinone-oxidoreductase (NQO1), and glutamate-cysteine ligase subunit catalysis (GCLC), etc., to resist OS [36]. Related studies have shown that FOXO3 participates in the interaction between Keap1 and Nrf2 [37] and loss of FOXO3 cues to inactivation of Keap1; hence, failing to activate Nrf2 due to presence of OS. Henceforth, if OS induces inflammation the changes in FOXO3 affect the interaction betweenKeap1 and Nrf2, which could be a marker of damaging OS [34] (Figure 2).

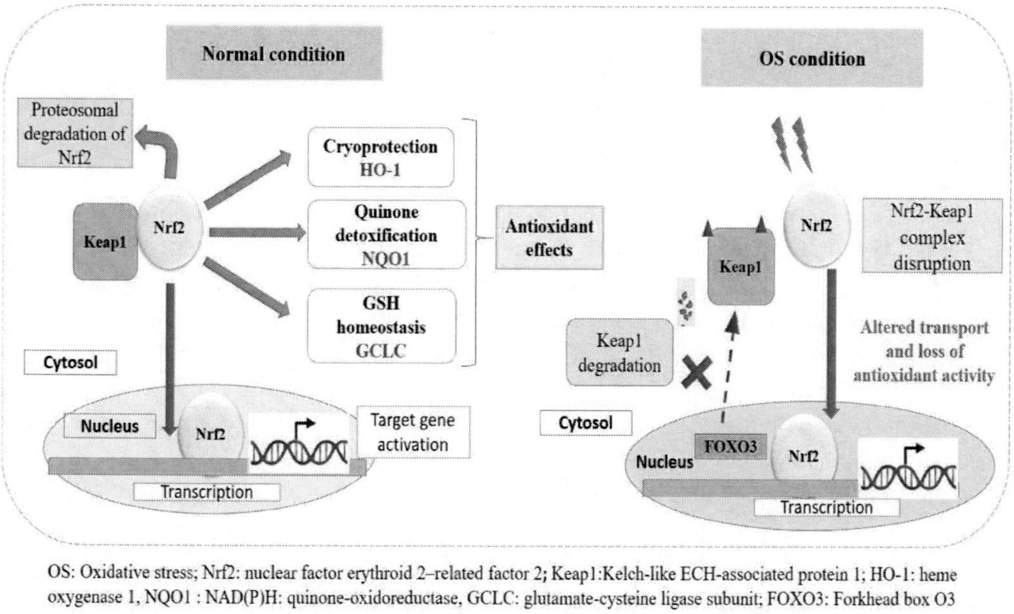


**Figure 2.** Body, and its signalling system in respect to oxidative stress in endometriosis.

## 3. Treatment Horizon

Given the *in vitro* studies showing that antioxidant supplementation of peritoneal and follicular fluids mitigates damage to human sperm, oocytes and embryos, it is logical to predict that *in vivo* treatments with antioxidants will improve parameters of fertility in endometriosis. Of these, Vitamin C and E [38], N-acetyl cysteine (NAC) [39], melatonin [40] and resveratrol [41] have variously been shown to reduce the numbers and sizes ofendometrotic implants in animal studies. Although according to a publication search by Pubmed, a single study has been cited with inadequate statistics, however, in a SCID mousemodel, melatonin induced regression of transplanted human endometriotic lesions [42]. To date studies with N-Acetyl Cysteine (NAC) have progressed the most in terms of therapeutic potential. A dual combination of antioxidant/s (NAC and α-lipoic acid) with bromelainprevents upregulation of an inflammatory marker (VCAM1), promotes apoptosis of endometriotic but not normal uterine cells, and reduces the size and number of endometriotic lesions in mice [43]. Another human study demonstrated decreased diameter and volume, decreased number, and even disappearance of endometriotic lesions after treatment for 3 months with NAC in women with endometriosis [44]. These studies demonstrate ROS implement the growth and/or maintenance of endometriotic lesions. In this way, ROS have at least an indirect impact on the quality of gametes and embryos. Furthermore, NAC and other antioxidants prevent chromosome and spindle misalignments and aneuploidy in mouse oocytes and embryos [45]. Future studies will further elucidate therole/s of ROS in gametic and embryonic dysfunction and aneuploidy in infertility caused by endometriosis.

## Conclusion

A double-edged sword of ROS in health and pathology is being increasingly appreciated, although the underlying mechanisms are still incompletely understood. While ROS production is up-regulated in endometriotic cells, the origin of this dysregulation is still unclear. Impaired detoxification process led to excess ROS and OS, and may be involved inincreased cellular proliferation and inhibition of apoptosis in endometriotic cells. Investigating the mechanisms underlying oxidative stress associated with endometriosis may well prove useful for determining its specific pathways may be essential in future. In the future, a strategy to reinforce the antioxidant defense system and target mitochondria will be a huge step. To increase the antioxidant capacity of the endometrium, we must decrease the production of ROS from the mitochondrial electron transport chain that occurs in response to inflammation, endometriotic cells and fatty acid levels and decrease ROS production without significantly affecting ATP production. At the same time, we should increase the degradation of intracellular ROS and increase the bioavailability of antioxidants, with their passage through the barriers must be considered. Interestingly, published data suggest that their detrimental effect does not always correlate with severity of endometriosis. Considering the aforementioned, and in the era of precision and personalizedmedicine, it appears that future studies should be focused on discovering sensitive-novel tools reporting on both oxidative stress status as well as on the inflammation system status. Additionally, these biomarkers may prove an essential tool for identifying patients who may benefit from receiving adjuvant treatments prior to addressing endometriosis related infertility, including antioxidant treatment. To conclude, effectively profiling endometriosis patients may potentially be the missing link in the equation of managing these perplexing cases. The future in the era of precision medicine should encompass individualized treatment plans and avoidance of futile IVF attempts.

## References


[1] Shafrir A. L., Farland L. V., Shah D. K., Harris H. R., Kvaskoff M., Zondervan K., Missmer S. A. Risk for and consequences of endometriosis: A critical epidemiologic review. *Best Pract Res Clin Obstet Gynaecol.* 2018; 51:1-15. doi: https://doi.org/10.1016/j.bpobgyn.2018.06.001.

[2] Bulun S. E. Endometriosis. *N Engl J Med.* 2009; 360(3):268-79. doi: https://doi.org/10.1056/NEJMra0804690.

[3] Fauconnier A., Chapron C. Endometriosis and pelvic pain: epidemiological evidence of the relationship and implications. *Hum Reprod Update.* 2005; 11(6):595-606. doi: https://doi.org/10.1093/humupd/dmi029.

[4] Máté G., Bernstein L. R., Török A. L. Endometriosis is a cause of infertility. Does reactiveoxygen damage to gametes and embryos play a key role in the pathogenesis of infertility caused by endometriosis? *Front Endocrinol (Lausanne).* 2018; 9:725. doi: https://doi.org/10.3389/fendo.2018.00725.

[5] Tanbo T., Fedorcsak P. Endometriosis-associated infertility: aspects of pathophysiological mechanisms and treatment options. *Acta Obstet Gynecol Scand.* 2017; 96(6):659-667. doi: https://doi.org/10.1111/aogs.13082.

[6] Samimi M., Pourhanifeh M. H., Mehdizadehkashi A., Eftekhar T., Asemi Z. The role of inflammation, oxidative stress, angiogenesis, and apoptosis in the pathophysiology ofendometriosis: Basic science and new insights based on gene expression. *J Cell Physiol.* 2019; 234(11):19384-19392. doi: https://doi.org/10.1002/jcp.28666.

[7] Simopoulou M., Rapani A., Grigoriadis S., Pantou A., Tsioulou P., Maziotis E., TzanakakiD., Triantafyllidou O., Kalampokas T., Siristatidis C., Bakas P., Vlahos N. Getting to KnowEndometriosis-Related Infertility Better: A Review on How Endometriosis Affects Oocyte Quality and Embryo Development. *Biomedicines.* 2021; 9(3):273. doi: https://doi.org/10.3390/biomedicines 9030273.

[8] Lu J., Wang Z., Cao J., Chen Y., Dong Y. A novel and compact review on the role of oxidative stress in female reproduction. *Reprod Biol Endocrinol.* 2018; 16(1):80. doi: https://doi.org/10.1186/s12958-018-0391-5.

[9] Gupta S., Agarwal A., Krajcir N., Alvarez J. G. Role of oxidative stress in endometriosis. *Reprod Biomed Online.* 2006; 13(1):126-34. doi: https://doi.org/10.1016/s1472-6483(10)62026-3.

[10] Gupta S., Agarwal A., Banerjee J., Alvarez J. G. The role of oxidative stress in spontaneous abortion and recurrent pregnancy loss: a systematic review. *Obstet Gynecol Surv.* 2007 May;62(5):335-47; quiz 353-4.

[11] Giudice L. C., Kao L. C. Endometriosis. *Lancet.* 2004; 364(9447):1789-99. doi: https://doi.org/10.1016/S0140-6736(04)17403-5.

[12] Healy D. L., Rogers P. A., Hii L., Wingfield M. Angiogenesis: a new theory for endometriosis. *Hum Reprod Update.* 1998; 4(5):736-40. doi: https://doi.org/10.1093/humupd/4.5.736.

[13] Oner-Iyidoğan Y., Koçak H., Gürdöl F., Korkmaz D., Buyru F. Indices of oxidative stressin eutopic and ectopic endometria of women with endometriosis. *Gynecol Obstet Invest.* 2004; 57(4):214-7. doi: https://doi.org/10.1159/000076691.

[14] Szczepańska M., Koźlik J., Skrzypczak J., Mikołajczyk M. Oxidative stress may be a piece in the endometriosis puzzle. *Fertil Steril.* 2003; 79(6):1288-93. doi: https://doi.org/10.1016/s0015-0282(03)00266.

[15] Cacciottola L., Donnez J., Dolmans M. M. Can endometriosis-related oxidative stress pave the way for new treatment targets? *Int J Mol Sci.* 2021; 22(13):7138. doi: https://doi.org/10.3390/ijms22137138.

[16] Ross E. A., Devitt A., Johnson J. R. Macrophages: The good, the bad, and the gluttony. *Front Immunol.* 2021; 12:708186. doi: https://doi.org/10.3389/fimmu.2021.708186.

[17] Santanam N., Murphy A. A., Parthasarathy S. Macrophages, oxidation, and endometriosis. *Ann N Y Acad Sci.* 2002; 955:183-98; discussion 19-200, 396-406. doi: https://doi.org/10.1111/j.1749-6632.2002.tb02779.x.

[18] Collins T. Acute and chronic inflammation. *Robbins pathologic basis of disease.* 1999:83-4.

[19] Sandhir R., Halder A., Sunkaria A. Mitochondria as a centrally positioned hub in the innate immune response. *Biochim Biophys Acta Mol Basis Dis.* 2017; 1863(5):1090-1097. doi: https://doi.org/10.1016/j.bbadis.2016.10.020.

[20] Fan Y. Y., Chen H. Y., Chen W., Liu Y. N., Fu Y., Wang L. N. Expression of inflammatory cytokines in serum and peritoneal fluid from patients with different stages of endometriosis. *Gynecol Endocrinol.* 2018; 34(6):507-512. doi: https://doi.org/10.1080/09513590.2017.1409717.

[21] Sikora J., Smycz-Kubańska M., Mielczarek-Palacz A., Kondera-Anasz Z. Abnormal peritoneal regulation of chemokine activation-The role of IL-8 in pathogenesis of endometriosis. *Am J Reprod Immunol.* 2017; 77(4). doi: https://doi.org/10.1111/aji.12622.

[22] Bersinger N. A., Günthert A. R., McKinnon B., Johann S., Mueller M. D. Dose-responseeffect of interleukin (IL)-1β, tumour necrosis factor (TNF)-α, and interferon-γ on the *in vitro* production of epithelial neutrophil activating peptide-78 (ENA-78), IL-8, and IL-6 by human endometrial stromal cells. *Arch Gynecol Obstet.* 2011; 283(6):1291-6.doi: https://doi.org/10.1007/s00404-010-1520-3.

[23] Bina F., Soleymani S., Toliat T., Hajimahmoodi M., Tabarrai M., Abdollahi M., Rahimi R. Plant-derived medicines for treatment of endometriosis: A comprehensive review of molecular mechanisms. *Pharmacol Res.* 2019; 139:76-90. doi: https://doi.org/10.1016/j.phrs.2018.11.008.

[24] Taniguchi F., Uegaki T., Nakamura K., Mon K. Y., Harada T., Ohbayashi T., Harada T. Inhibition of IAP (inhibitor of apoptosis) proteins represses inflammatory status via nuclear factor-kappa B pathway in murine endometriosis lesions. *Am J Reprod Immunol.* 2018; 79(1). doi: https://doi.org/10.1111/aji.12780.

[25] Zhu F., Liu M., Pan Y., Wang X., Chen Y. [Small hairpin RNA targeting inhibition of NF- κB gene in endometriosis therapy of Macaca fascicularis]. *Zhonghua Fu Chan Ke Za Zhi.* 2015;50(1):48-53.

[26] Barragan F., Irwin J. C., Balayan S., Erikson D. W., Chen J. C., Houshdaran S., Piltonen T. T., Spitzer T. L., George A., Rabban J. T., Nezhat C., Giudice L. C. Human endometrial fibroblasts derived from mesenchymal progenitors inherit progesterone resistance and acquire an inflammatory phenotype in the endometrial niche in endometriosis. *Biol Reprod.* 2016; 94(5):118. doi: https://doi.org/10.1095/biolreprod.115.136010.

[27] Li J., Lan T., Zhang C., Zeng C., Hou J., Yang Z., Zhang M., Liu J., Liu B. Reciprocal activation between IL-6/STAT3 and NOX4/Akt signalings promotes proliferation and survival of non-small cell lung cancer cells. *Oncotarget.* 2015; 6(2):1031-48. doi: https://doi.org/10.18632/oncotarget.2671.

[28] Ghosh S., Chakraborty P., Chatterjee S., Chattopadhyay R., Goswami S. K., Ghosh P., Chowdhury K. D., Chakravarty M., Chakravarty B. Interleukin-33 modulates inflammation and may be treated as a biomarker in endometriosis. *Hum Reprod.* 2019;34:291-292.

[29] D'Ippolito S., Tersigni C., Marana R., Di Nicuolo F., Gaglione R., Rossi E. D., Castellani R., Scambia G., Di Simone N. Inflammosome in the human endometrium: further step in the evaluation of the "maternal side." *Fertil Steril.* 2016;105(1):111-8.

[30] Kelley N., Jeltema D., Duan Y., He Y. The NLRP3 Inflammasome: An overview of mechanisms of activation and regulation. *Int J Mol Sci.* 2019; 20(13):3328. doi: https://doi.org/10.3390/ijms20133328.

[31] Zhang W., Li G., Luo R., Lei J., Song Y., Wang B., Ma L., Liao Z., Ke W., Liu H., Hua W. Cytosolic escape of mitochondrial DNA triggers cGAS-STING-NLRP3 axis-dependent nucleus pulposus cell pyroptosis. *Exp Mol Med.* 2022;54(2):129-42.

[32] Zhou R., Tardivel A., Thorens B., Choi I., Tschopp J. Thioredoxin-interacting protein links oxidative stress to inflammasome activation. *Nat Immunol.* 2010; 11(2):136-40.doi: https://doi.org/10.1038/ni.1831.

[33] Klopf J., Brostjan C., Eilenberg W., Neumayer C. Neutrophil extracellular traps andtheir implications in cardiovascular and inflammatory disease. *Int J Mol Sci.* 2021; 22(2):559. doi: https://doi.org/10.3390/ijms22020559.

[34] Lu J., Wang Z., Cao J., Chen Y., Dong Y. A novel and compact review on the role of oxidative stress in female reproduction. *Reprod Biol Endocrinol.* 2018; 16(1):80. doi:https://doi.org/10.1186/s12958-018-0391-5.

[35] Liu Y., Zhang Z., Lu X., Meng J., Qin X., Jiang J. Anti-nociceptive and anti-inflammatory effects of sulforaphane on sciatic endometriosis in a rat model. *Neurosci Lett.* 2020; 723:134858. doi: https://doi.org/10.1016/j.neulet.2020.134858.

[36] Guan L., Zhang L., Gong Z., Hou X., Xu Y., Feng X., Wang H., You H. FoxO3 inactivation promotes human cholangiocarcinoma tumorigenesis and chemoresistance through Keap1-Nrf2 signaling. *Hepatology.* 2016; 63(6):1914-27. doi: https://doi.org/10.1002/hep.28496.

[37] Lim R., Barker G., Lappas M. The transcription factor Nrf2 is decreased after spontaneous term labour in human fetal membranes where it exerts anti- inflammatory properties. *Placenta.* 2015; 36(1):7-17. doi: https://doi.org/10.1016/j.placenta.2014.11.004.

[38] Durak Y., Kokcu A., Kefeli M., Bildircin D., Çelik H., Alper T. Effect of vitamin C on the growth of experimentally induced endometriotic cysts. *J Obstet Gynaecol Res.* 2013; 39(7):1253-8. doi: https://doi.org/10.1111/jog.12050.

[39] Pittaluga E., Costa G., Krasnowska E., Brunelli R., Lundeberg T., Porpora M. G., Santucci D., Parasassi T. More than antioxidant: N-acetyl-L-cysteine in a murine model of endometriosis. *Fertil Steril.* 2010; 94(7):2905-8. doi: https://doi.org/10.1016/j.fertnstert.2010.06.038.

[40] Cetinkaya N., Attar R., Yildirim G., Ficicioglu C., Ozkan F., Yilmaz B., Yesildaglar N. The effects of different doses of melatonin treatment on endometrial implants in an oophorectomized rat endometriosis model. *Arch Gynecol Obstet.* 2015; 291(3):591-8.doi: https://doi.org/10.1007/s00404-014-3466-3.

[41] Ozcan Cenksoy P., Oktem M., Erdem O., Karakaya C., Cenksoy C., Erdem A., Guner H., Karabacak O. A potential novel treatment strategy: inhibition of angiogenesis and inflammation by resveratrol for regression of endometriosis in an experimental rat model. *Gynecol Endocrinol.* 2015; 31(3):219-24. doi: https://doi.org/10.3109/09513590.2014.976197.

[42] Yesildaglar N., Yildirim G., Yildirim O. K., Attar R., Ozkan F., Akkaya H., Yilmaz B. The effects of melatonin on endometriotic lesions induced by implanting human endometriotic cells in the first SCID-mouse endometriosis-model developed in Turkey. *Clin Exp Obstet Gynecol.* 2016;43(1):25-30.

[43] Agostinis C., Zorzet S., De Leo R., Zauli G., De Seta F., Bulla R. The combination of N- acetyl cysteine, alpha-lipoic acid, and bromelain shows high anti-inflammatory properties in novel *in vivo* and *in vitro* models of endometriosis. *Mediators Inflamm.* 2015; 918089.

[44] Porpora M. G., Brunelli R., Costa G., Imperiale L., Krasnowska E. K., Lundeberg T., Italo Nofroni, Maria Grazia Piccioni, Eugenia Pittaluga, Adele Ticino, Tiziana Parasassi. A promise in the treatment of endometriosis: an observational cohort study on ovarian endometrioma reduction by N-acetylcysteine. *Evid Based Complement Alternat Med.* 2013; 240702.

[45] Harlev A., Gupta S., Agarwal A. Targeting oxidative stress to treat endometriosis. *Expert Opin Ther Targets.* 2015; 19:1447–64.

## Chapter 20

# The Role of ROS-Responsive Biomaterials in Bone-Related Diseases

**Yu Wu**[1,2]
**Wenhui Hu**[2]
**and Shiwu Dong**[2,*]
[1]Department of Pharmacy, School of Pharmacy and Bioengineering, Chongqing University of Technology, Chongqing, China
[2]Department of Biomedical Materials Science, School of Biomedical Engineering, Army Medical University, Chongqing, China

## Abstract

Reactive oxygen species (ROS) consisting of reactive ions or free radicals are produced in many physiological processes in humans and have been also recognized as one of conventional biomarkers for disease progression. Excessive levels of ROS in tissue microenvironment may cause oxidative stress, which disrupts cellular homeostasis (attacking membranes, proteins, DNAs and lipids of cells) and leads to a series of pathological disorders. In addition, a series of chronic diseases (such as infection, inflammation, bone-related diseases, metabolic disorders, cancer) show robust evidence for ROS involvement. More recently, ROS-responsive biomaterials have been identified as a type of promising therapeutic substances. According to their working mechanisms, ROS-responsive biomaterials are divided into two types: a vehicle triggered by ROS to realize drug release and an antioxidant to scavenge ROS. Studies have revealed that ROS-responsive biomaterials (such as nanoparticles, microspheres, scaffolds, and hydrogels) have been widely applied in the tissue engineering and treatment of cancer, skin wounds and other bone-related diseases. In the study of metabolic bone diseases, in vitro and vivo findings revealed that ROS-responsive biomaterials have the ability to scavenge ROS, inhibit bone resorption and carry drugs. In this chapter, we summarize ROS-responsive biomaterials for therapeutic strategies in bone-related diseases, such as osteoporosis (OP), rheumatoid arthritis (RA) and osteoarthritis (OA), highlight their mechanisms and advantages and discuss current limitations and challenges for the clinic application.

**Keywords:** ROS, oxidative stress, ROS-responsive biomaterials, osteoarthritis, rheumatoid arthritis, bone-related diseases

---

* Corresponding Author's E-mail: dongshiwu@163.com.

In: The Role of Reactive Oxygen Species in Health and Disease
Editor: Kuladip Jana
ISBN: 979-8-89113-283-2

## 1. Introduction

Reactive oxygen species (ROS) are a collective term which includes oxygen radicals (e.g., superoxide, hydroxyl radicals) and non-radicals (e.g., hydrogen peroxide, singlet oxygen), produced in metabolic and physiological processes [1, 2]. Nicotinamide adenine dinucleotide phosphate (NADPH) oxidases (NOX) can produce ROS, which play a pivotal role in mediating tissue microenvironments and tissue regeneration, as well as initiating oxidative stress response, which can directly or indirectly impact on various reactions in the body [2, 3]. It is believed that ROS is closely tied to bone-related diseases. Recent studies have demonstrated that ROS has a significant correlation with OA progression, due to their ability to induce DNA damage and apoptosis of chondrocytes and activate matrix metalloproteinases (MMPs) to degrade extracellular matrix (ECM) [4-6]. Relative researches have confirmed that RA patients showed a marked increase in ROS production, lipid peroxidation, protein oxidation and DNA damage [7]. Some evidence has demonstrated that overproduced ROS has detrimental effects on the functions of bone cells [6-9]. ROS are recognized as an emerging triggered event similar to specific stimuli found in pathological microenvironments, such as acidic pH, overproduced enzymes and temperature. Over the past decades, tremendous efforts have been devoted to developing ROS-responsive materials, which are a type of promising therapeutic agents, and used as a vehicle triggered by inflammatory diseases to release drugs under pathological oxidative microenvironments or with multiple stimuli [10-13]. In addition, ROS-responsive nanocarriers also can overcome the restriction of multidrug resistance and off-target toxicity of some therapeutic agents. Thus, it is important to understand the mechanisms of these responsive materials based on abnormal ROS levels and design effective ROS-responsive biomaterials for the treatment of bone-related diseases.

## 2. ROS-Responsive Biomaterials

ROS are inevitably produced by cellular oxidative metabolism and they play an important part in regulating physiological processes. However, the overproduction of ROS usually leads to oxidation stress which has a close relationship with a series of diseases such as inflammation, cancer, cardiovascular and neurological disorders. Therefore, how to inhibit ROS formation or scavenge ROS is the key point of current research. Over the past decades, due to the sensitivity to ROS, ROS-responsive biomaterials have drawn considerable attention and been widely applied in biomedical fields. ROS-responsive biomaterials have been successfully used as carriers to control the release of payloads in cells or tissues existing with the excessive ROS. In this section, we will systematically summarize the working mechanisms of ROS-responsive functional groups, and the applications of ROS-responsive materials, in the form of nanoparticles, scaffolds and hydrogel, for drug delivery system (DDS) and tissue engineering.

### 2.1 The Working Mechanisms of ROS-Responsive Functional Groups

According to the literature, ROS have been regarded as a target or an indicator for various treatments. ROS-responsive biomaterials, especially ROS-responsive polymers, are considered advanced functional materials, which can be used as carriers in DDS or effective ROS-scavenging agents. Based on the reaction to oxidation, ROS-responsive biomaterials can release drug through degradation, structural change or physicochemical property switch when ROS are over expressed, as well as eliminate ROS. The achievements of ROS-responsive biomaterials are attributed to the ROS-sensitive functional groups which works as the following three roles: i) the linkage of the repeating monomers of polymeric backbones; ii) the terminal cross linkers between hydrophobic parts and hydrophilic parts of copolymer; iii) the linker between polymer monomer and drug (polymer prodrug). As is shown in Figure 1, the representative ROS-responsive biomaterials can be roughly divided into three categories, including sulfur-containing polymers, Se, Te, B-containing polymers and other polymers [12, 14-21].

### 2.1.1. Thioether-Containing Polymers

Thioether-based polymers are one type of the most investigated ROS-responsive biomaterials, which can undergo a phase change from hydrophobic to hydrophilic states in oxidative environments. This is mainly because thioether groups are easily converted into sulfoxide under mild conditions by $H_2O_2$ and turned into sulfones under strong oxidative conditions. In addition, suspensions of thioether-containing biomaterials are stable in water for several weeks at room temperature; however, after inducing by ROS, these materials undergo degradation and disruption eventually. Therefore, various thioether-containing polymers composed the main backbone, and the side chains or the tail chains have been synthesized and developed for the treatment of cancer, inflammation and other oxidation related disorders.

Poly (propylene sulfide) (PPS) is a common thioether-containing polymer which is frequently used as a precursor for the preparation of thioether-based oxidation-sensitive materials. By taking advantage of their sensitivity to ROS, multiple kinds of PPS-based ROS-responsive materials have been developed for drug delivery, prodrug or mediate the microenvironments of overproduced ROS [22-24]. Block co-polymers synthesized from PPS and other polymers can form nanoparticles through self-assembly. These nanoparticles can be used as vehicles to load drugs with low water solubility and poor bioavailability, and then accurately release payloads through the response to the local excessive ROS. For one research, the hydrophobic andrographolide was encapsulated by micelles which are assembled from the block copolymer of PEG-PPS [25]. In response to ROS, the drug-loaded micelles can not only achieve accurate release of payloads, but also scavenge ROS by itself. This is attributed to the hydrophobic block PPS can react with ROS and synchronically be converted into hydrophilicity after oxidation, and the loss of hydrophobic interactions may lead to micellar decomposition and drug release. *In vitro* and *in vivo* results showed that andrographolide-loaded ROS-responsive polymeric micelles can simultaneously alleviate inflammation and oxidative stress, which is a novel strategy against atherosclerosis [25].

It is well known that tumor cells or inflammatory tissues may overproduce ROS, which may be more than one order of magnitude higher than healthy. Based on this condition, researchers have fabricated thioether-containing polymeric prodrug/nanocarriers for the treatment/diagnosis of cancer or inflammation through local stimuli-responsive systems. For example, researchers designed a prodrug for 7-ethyl-10-hydroxyl-camptothecin (OEG-2S-SN38), which could be not only tumor-specific but also sensitive to the tumor GSH/ROS heterogeneity. The prodrug is able to fast release drug in various tumor cells via enhanced hydrolysis due to thioether groups [26]. For another example, the core-shell structured micelles (TPP@PMM) with particle size of 57.5 nm, have been formed by amphipathic copolymer PMPC-PMEMA through self-assembling [27]. In the presence of overexpressed ROS, TPP@PMM can realize the accurate release of carried drug and the inflammation therapy, because of the degradation of ROS-sensitive linker. The results demonstrated that TPP-PMM exhibited high-resolution inflammation diagnosis and high therapeutic efficiency to lung injury, arthritis and atherosclerosis. Therefore, thioether-containing ROS-responsive biomaterials are the promising candidates for acute and chronic inflammation diagnosis and therapy.

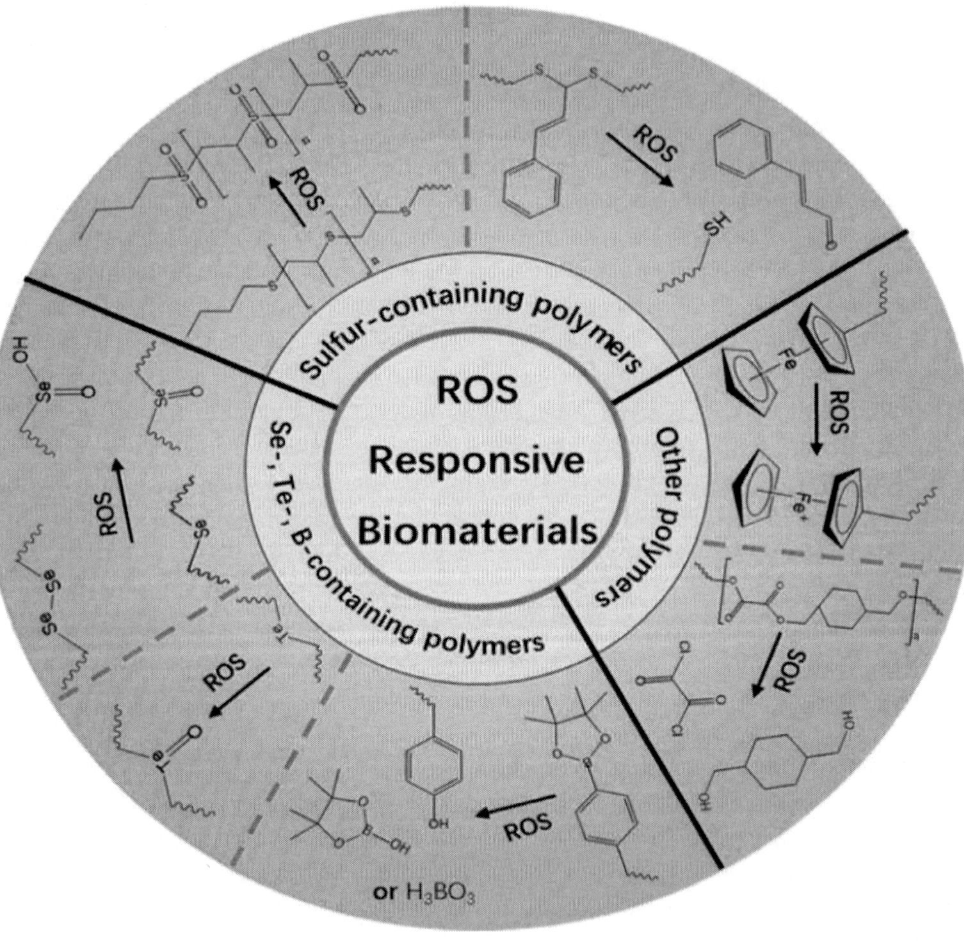


**Figure 1.** The representative ROS-responsive biomaterials and their oxidation products.

### 2.1.2. Poly(thioketal)-Containing Polymers

Due to the stability of thioketals to acid and base, thioketals have been used as the carbonyl protecting groups in organic synthesis. However, the thioketal bonds can be destroyed following ROS induction to produce ketones and organic thiols, and then lead to the scission of poly(thioketal)-containing polymers chain [28]. Meanwhile, oxidation of thioketal bonds endow polymers with higher polarity and water solubility. Due to this property, endogenous and exogenous ROS can serve as biological triggers to program drug release from poly(thioketal)-containing polymeric carriers. For example, model drugs can be copolymerized with ROS-responsive thioketal-containing linkers to prepare polymer prodrugs. After incubating with ROS, polymer prodrugs undergo a ROS-responsive elimination reaction, which induces a chain breakage of prodrugs and the release of model drugs [29]. In another example, photosensitizer and chemotherapeutic drug were simultaneously loaded in a ROS-responsive polymeric micelles, which were formed by self-assembly of amphiphilic copolymers with degradable thioketal linkages. The polymeric micelles can be rapidly cleaved by ROS generated from NIR-activated photosensitizer, and then release the chemotherapeutic drug [30]. The by-products of thioketal cleavage are easily metabolized. Therefore, thioketal is a significant component in the design of ROS-responsive biomaterials.

### 2.1.3. Selenium- or Tellurium-Containing Polymers

Selenium and Tellurium belong to the chalcogen group, which are easily oxidized by $H_2O_2$. According to literature, the bond energies of C-Se, Se-Se, C-Te are below 300 kJ $mol^{-1}$, which make these elements-based biomaterials sensitive to $H_2O_2$ [31]. Selenium is an important mineral nutrient for humans, which plays an essential role in improving the activity of seleno-enzymes by scavenging free-radicals [32]. Similar to sulfur, selenide undergoes hydrophobic to hydrophilic transition under oxidant environments. Therefore, selenium-based ROS-responsive biomaterials have been wildly used in biomedicine for treating inflammation, cancer or other disorders. For instance, RNA interference therapy has drawn much more attention in the treatment of disease. However, the specificity and efficiency of controlled siRNA release still have challenges, because of the complexity of disease sites. To overcome these challenges, multilayer nanoparticles formed from diselenide-containing polymers have been designed to carry siRNA [33]. In this multilayer nanoparticle, photosensitizer (red bengal, RB) conjugated polymer (PEI 600) was used as the inner coating layer; for middle layer, ROS-sensitive diselenide linked PEI 600 and bound to therapeutic siRNA and cell-penetrating peptide R8 via electrostatic adsorption; negatively charged agent (hyaluronic acid, HA) was used as the outer coating layer. The outer coating layer can prevent siRNA leakage during the delivery process and target diseased sites through receptor-ligand interaction. Under NIR irradiation, the encapsulated RB was activated to promote the production of ROS, thereby accelerating the decomposition of PEI-SeSe, and finally inducing a fast and specific siRNA release [33]. Due to the lower electronegativity of tellurium, tellurium-based polymers are more sensitive to ROS. Moreover, the cytotoxic of tellurium-based polymers is lower than that of selenium-based polymers. In summary, tellurium-based polymers have more advantages in the field of biomedicine [34]. According to the research, a variety of ultra-sensitive ROS-responsive tellurium-based biomaterials have been synthesized, and these ROS-sensitive biomaterials can respond to 100 μM $H_2O_2$, which is a typically biologically relevant concentration [35, 36]. To expand the application of selenium- and tellurium-based biomaterials in the field of biomedicine, multi-hierarchical responsive polymers containing selenium- and tellurium have been developed to investigate the sensitivity to both chemical and electrochemical stimuli [37]. The results showed that a selenium- and tellurium-containing amphiphilic block copolymer could spontaneously self-assemble into micelles in aqueous solution, and they can be stepwise oxidized by chemical methods and electrochemical methods. These advanced selenium- and tellurium-based biomaterials provide a novel and promising way for biomedicine.

### 2.1.4. Phenylboronic Acid and Ester-Containing Polymers

Due to the lack of polymer systems which can response sensitively to biologically relevant concentration of $H_2O_2$. Phenylboronic acid/ester-containing polymers also have been developed to address this problem. Phenylboronic acid/ester-containing polymers are frequently used to prepare ROS-responsive biomaterials, as phenylboronic acid and ester bonds can be selectively degraded by $H_2O_2$ to produce phenol and boronic acid [38]. This feature makes these ROS-responsive biomaterials more widely used, such as cancer diagnosis, targeted drug delivery [18, 39]. Generally, phenylboronic acid/ester functional groups exist in the side chain of ROS-responsive polymers. As drug carriers, these ROS-responsive biomaterials are able to degrade and release the guest molecules in the presence of $H_2O_2$. What makes it superior to the conventional drug delivery systems is that phenylboronic acid/ester-containing polymeric nanocarriers can crumble into small molecules upon stimuli and enable the payloads to release without any carrier adsorption [40]. In addition, due to the specific binding affinities between phenylboronic acid/ester functional groups and diols, amino alcohols, cyanide and fluoride, the phenylboronic acid/ester functional polymeric nanocarriers can release drug by the solubility switch strategy. For example, water-soluble polysaccharide can convert into fat-soluble after modifying with phenylboronic ester, which can facilitate drug encapsulation. Polysaccharides modified by phenylboronic ester could be returned into water-soluble state through reversing the original modification under over produced ROS conditions, and release the payload concurrently [41]. Based on the specific binding affinities to diols, tannic acid was mixed with polymeric phenylboronate to form a polymeric phenylboronic acid-tannic acid hydrogel (PTNG) [42]. After studying the reaction of PTNG with $H_2O_2$, the results demonstrated that the as-prepared PTNG could be applied as an antioxidant to scavenge $H_2O_2$. Under an inflammation model *in vitro*, PTNG showed high biocompatibility and strong anti-inflammatory effects. Moreover, in the enzyme-induced peritonitis mouse model, PTNG can significantly decrease neutrophil recruitment and pro-inflammatory cytokines, which indicated the successful alleviation of inflammation [42]. Therefore, as the emerging oxidation-responsive functional groups, phenylboronic acid/ester groups can be exploited in clinical applications for further research.

### 2.1.5. Other-Linkers Polymers

Polyoxalate contains -$O(CO)_2O$- functional groups, which can be degraded into oxalic acid and corresponding diols under aqueous and ROS conditions [43]. It also possesses good biocompatibility which makes it an emerging candidate for biomedicine. For example, a self-immolative biodegradable polyoxalate prodrug has been synthesized through one-pot step-growth polymerization [21]. This prodrug can release chemotherapeutic drug by eliminating the link between polymer chains and chemotherapeutic drugs in ROS-rich cancerous tissue. It is more specific for drug release and minimizes toxic side effects to healthy cells. Besides polyoxalate-containing ROS-responsive biomaterials, ferrocene-containing polymers attract rapid interests in biomedical area, because of the reversible redox activity and stability. Ferrocene groups are hydrophobic, which can be quickly converted into hydrophilic state in the presence of ROS. Based on this property, an amphiphilic block ferrocene-containing copolymer micelle has been synthesized, which held a relative higher encapsulation efficiency of model drug and swelled to release model drug in response to ROS [44]. The research demonstrated that this polymeric micelle could reversibly self-assemble under the action of different oxidants and reductants, and the toxicity of this polymeric micelle is low even at high concentrations.

In summary, the ROS-responsive functional groups endow ROS-responsive biomaterials with much more properties, which make these biomaterials have more advantages in treating oxidative stress-related diseases, mediating tissue microenvironments and tissue regeneration.

## 2.2. ROS-Responsive Polymeric Nanoparticles, Hydrogels and Scaffolds

Recently, stimuli-responsive biomaterials have been proposed to undergo changes of structure or property in external or internal environmental factors, such as pH, ROS, enzyme and temperature. Particularly, ROS-responsive polymeric materials have drawn extensive attention in biomedical application, and they can assemble into nanoparticles, hydrogels, and scaffolds. These ROS-responsive biomaterials are capable to efficiently target pathological sites to release payloads or mediate the microenvironments of diseased sites.

Nanoparticles are the main form of ROS-responsive biomaterials which have been widely used in biomedicine. Sulfur-, selenium-, tellurium-, phenylboronic acid/ester-, ferrocene-, aryloxalate- and other ROS-responsive functional groups-containing polymers have been designed to self-assemble into nanoparticles to carry model drugs or synthesize polymer prodrugs for the treatment of oxidative stress-related diseases [45]. The responsive mechanisms of ROS-responsive polymeric nanoparticles are mainly divided into two types: ROS-induced non-cleavable hydrophobic-hydrophilic transition, namely solubility switch, and ROS-induced structural cleavage. For example, amino acid-based poly(ester amide) (AA-PEA) is a new family of biodegradable materials and the synthetic routes of AA-PEA are environmentally-friendly without very toxic reagents and byproducts [46]. L-methionine possesses high anti-oxidative property *in vivo*, and it can be oxidized into methionine sulfoxide by ROS. Therefore, Met-based PEA has been conjugated to PEG to form an amphiphilic polymer which can self-assemble into nanoparticles. The size and morphology of these nanoparticles changed rapidly, when they were induced by $H_2O_2$. In addition, after loading with therapeutics, these nanoparticles showed a $H_2O_2$ concentration induced time-dependent release behavior [46].

Hydrogels are 3D polymeric networks with high water retention rate and tunable physiochemical properties, and the driving forces for hydrogels formation include ionic, H-bonding, hydrophobic interactions or chemical crosslinking [47]. Because of their hydrophilicity and ability to mimic the natural extracellular matrix, hydrogels have been widely applied in biomedical fields, such as drug delivery and tissue engineering [48, 49]. For drug delivery, ROS-responsive hydrogels contain two categories: hydrogels with solubility switch units and hydrogels with cleavable units. For example, a series of thermoresponsive, ABC triblock copolymer-based hydrogels have been engineered to degrade and release drug [50]. This ABC triblock copolymers are comprised of "A" block (oxidation-sensitive polymer), "B" block (hydrophilic polymer) and "C" block (theroresponsive polymer). At ambient temperature, these copolymers form micelles in aqueous solutions which can preload therapeutic agents. These solutions can quickly transform into stable hydrogels when they are further heated to 37 °C. These as-prepared hydrogels can be selectively degraded by $H_2O_2$, due to the oxidative solubility switch of "A" block. *In vitro* study, the ROS-responsive hydrogels are noncytotoxic. After injecting into mice, the ROS-responsive hydrogels loaded with Nile Red, showed slow but potentially environmentally responsive and more consistent release over time. In addition, the results *in vivo* administration also demonstrated these implants promote robust cellular infiltration and didn't cause any excessive inflammatory response. For another example, ROS-responsive functional groups-modified therapeutic agent has been covalently cross-linked with poly(vinyl alcohol) (PVA) to fabricate ROS-responsive hydrogel for local delivery of anti-programmed cell death protein ligand-1 antibodies. In the presence of endogenous ROS, the as-prepared ROS-responsive hydrogel prodrug underwent scission and released free SN 38 and antibodies. The released SN 38 can elicit antitumor immune responses by triggering immunogenic cell death of cancer cells, and the simultaneously released antibodies can bind to programmed cell death protein ligand-1 expressed on cancer cells to facilitate T-cell-mediated immune responses. This ROS-responsive hydrogel exhibited high efficiency in stimulating immune system and inhibiting tumor growth [51]. For tissue engineering, ROS-responsive polymeric hydrogels have been synthesized, and these hydrogels act as ROS-scavenging agents to heal wounds by decreasing the level of ROS around wounds. As we know, ROS generated by wounds or bacterial infection could further hinder wound healing. To overcome this difficulty, a ROS-scavenging hydrogel has been synthesized through ROS-responsive links, which has loaded with an antibiotic and a growth factor to promote tissue regeneration. The ROS-scavenging hydrogel can release therapeutic agents to kill bacteria and accelerate the wound closure in response to ROS [52].

Macroscopic scaffolds have been widely used in tissue engineering, because they can provide appropriate sites for cell adhesion and adequate space for cell proliferation and migration. Generally, biodegradable scaffolds can be degraded only by hydrolysis without capacity to interact with cells and tissues. The degradation rate can't match the rate of tissue in-growth, thus limiting their application. In recent years, ROS-responsive and biodegradable scaffolds have attracted much more interest, which could be selectively degraded by ROS for specific therapeutic purposes [53]. Poly(thioketal)-, poly(thioether)-, poly(boronate)-containing scaffolds have been engineered and used for tissue engineering and drug delivery to spread their applications [54-56]. In addition, the therapeutic efficacy of direct stem cell injections into the damaged tissues is limited, which is another challenge for tissue engineering. Therefore, a series of ROS-responsive polymeric porous scaffolds have been fabricated for carrying cells. For example, ROS-degradable polymeric scaffolds have been developed for carrying mesenchymal stem cells (MSCs). According to the increased number of micropores in ROS-degradable scaffolds, it could be confirmed that these cross-linked scaffolds respond to physiological levels of ROS and underwent degradation over 3 weeks *in vivo*. The expression of Flk1 significantly increased in ROS-responsive scaffolds compared to the non-cross-linked ones. These results demonstrated that the as-prepared cross-linked scaffolds played an essential role in MSCs differentiation to endothelial cells by a ROS-mediated mechanism and/or involvement of host macrophages [57]. For drug delivery, ROS-degradable scaffolds have been engineered to carry therapeutic agents for the treatment of tumors. For example, an in situ formed ROS-responsive scaffold has been developed through crosslinking PVA with ROS-responsive links to carry gemcitabine and checkpoint inhibitor. This ROS-degradable scaffold can sustainably release payloads in the overproduced ROS microenvironment of tumor. *In vivo* studies showed that the drug-loaded scaffold has an obvious tumor suppression effect on mice and half of the mice survived over 60 days after injection. The results of immunofluorescence and flow cytometry demonstrated that the as-prepared scaffold can trigger anti-tumor immune response induced by T cells. Therefore, this ROS-responsive scaffold can not only work as a carrier for drugs delivery, but also as a ROS scavenger to enhance the immunogenic phenotypes. It is a promising method for patients with low-immunogenic tumors responds poorly to immune checkpoint blockade [58].

## 3. Application of ROS-Responsive Biomaterials for Bone-Related Diseases

Mounting preclinical and clinical research evidence demonstrates that numerous pathological conditions, including cancer, inflammatory diseases, especially bone-related diseases, such as osteoarthritis (OA), rheumatoid arthritis (RA), are accompanied by excessive ROS [59, 60]. Moreover, anti-inflammatory or anti-tumor drugs have some limitations, such as poor aqueous solubility, low bioavailability, and off-target effects. To address these challenges, ROS has been used as the trigger for the development of ROS-responsive biomaterials loaded with these therapeutic agents. In recent years, due to the sensitivity and biocompatibility of ROS-responsive biomaterials, they have become the promising stimuli-responsive materials in the treatment of bone-related diseases. The classification of the representative ROS-responsive biomaterials used for the treatment of bone-related diseases is shown in Figure 2.

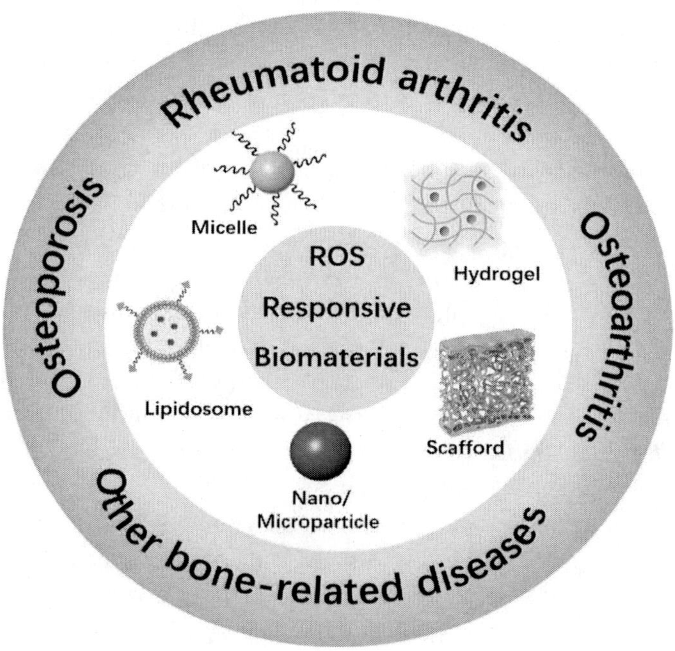


**Figure 2.** The classification of the representative ROS-responsive biomaterials for the therapy for bone-related diseases.

### 3.1. Rheumatoid Arthritis

RA is a non-curable disease, which has been known as an autoimmune and chronic inflammatory disease in the joints which causes cartilage destruction, bone erosion, synovitis and loss of joint function [61]. Traditional therapies require frequent and long-term administration of anti-rheumatic drugs, causing obvious side effects. To address the difficulties of traditional therapies, some advanced biomaterials have been developed to carry anti-rheumatic drugs, especially ROS-responsive polymeric nanocarriers. The overproduced ROS in RA joints can alter the physical/chemical properties or disrupt the structure of ROS-responsive polymeric nanocarriers to trigger the release of payloads. The site-specific and targeted drug delivery is attributed to the detection and response of the ROS-responsive nanocarrier to micro- or macroenvironment. Taking advantage of their response to ROS, smart nanomedicines based on ROS-responsive biomaterials have been fabricated to treat RA. Celastrol (Cel) is an anti-rheumatic drug with poor water solubility and frequently administered by parenteral delivery. To boost its curative effect for RA, thioether-containing polymeric micelles have been developed to carry Cel [62]. The average hydrodynamic diameter and the distribution index of the blank ROS-responsive micelles (B-PEPS) increased in the presence of $H_2O_2$, because of the oxidation of thiother. The results *in vitro* showed that Cel-loaded ROS-responsive micelles (C-PEPS) were safe at the concentrations from 28 to 226 ng/mL, C-PEPS could remarkably decrease pro-inflammatory cytokines in a dose-dependent manner, due to their capability of effective down-regulating the transcription of iNOS, TNF-α and IL-6 genes. According to the therapeutic effect of C-PEPS and B-PEPS to collagen induced arthritis (CIA) animal model, the C-PEPS-treated group presented a better improvement in bone erosion and bone structure than B-PEPS-treated group. The C-PEPS-treated group also showed more body weight loss than the untreated CIA group, because of the anti-obesity role of Cel. As we know, RA patients are always accompanied with obesity, and obesity is related to a more rapid progression of disability in RA. The above results indicated that C-PEPS may be a promising drug candidate for the treatment of obese patients with RA. For another example, ROS-responsive berberine polymeric micelles (BPseP) have been assembled by selenocystamine-containing polymer, which can effectively suppress the inflammation of RA by targeting mitochondria [63]. *In vitro* study, the cumulative release of berberine of BPseP micelles can quickly achieve more than 80% in mimic RA microenvironment, due to the sensitivity of diselenium bond to ROS. In addition, the research evidence demonstrated that BPseP micelles can remarkably increase the uptake of berberine in RA fibroblast and the efficacy of BPseP micelles was ten times higher than that of free berberine *in vitro* and *in vivo* studies. After intraperitoneal injection into adjuvant-induced arthritis rats, the BPseP micelles can inhibit cytokines generation and bone destruction. Compared to free berberine, BPseP showed better efficacy even at one-tenth concentration. Consequently, ROS-responsive polymeric nanomedicines represent an effective approach to preferentially release drugs in inflammatory tissues and improve therapeutic efficacy of RA. Moreover, to enhance their specificity and cellular uptake, surface functionalization or surface coating has been adopted. For example, to specifically deliver dexamethasone (Dex) to inflammatory sites and reduce the administration frequency, folic acid modified Dex-loaded ROS-responsive nanoparticles (Dex/FA-Oxi-αCD) have been developed. Due to the overexpressed folate receptors in the activated macrophages, these nanoparticles can be taken by the activated macrophages through receptor-ligand interaction. Compared to free Dex, the therapeutic efficiency of Dex/FA-Oxi-αCD has been significantly increased [64]. Furthermore, folate acid has been conjugated to polyethylene glycol and further prepared methotrexate (MTX) and catalase (CAT) co-capsulated liposomes (FOL-MTX&CAT-L) which can target to activated macrophages [65]. When intracellular $H_2O_2$ infiltrate into liposomes, $H_2O_2$ can be decomposed into water and oxygen under the catalysis of CAT, and then liposomes were destroyed by oxygen to release the encapsulated MTX. *In vitro* results demonstrated that FOL-MTX&CAT-L was sensitive to ROS and exhibited enhanced cellular uptake via folate-mediated endocytosis. *In vivo* results indicated that FOL-MTX&CAT-L could prolong blood circulation time, enhance the accumulation of MTX in diseased tissue of CIA model, improve therapeutic effect, and reduce the toxicity to major organs. Therefore, surface functionalization of ROS-responsive biomaterials can not only enhance their specificity but also improve their therapeutic effect.

### 3.2. Osteoarthritis

OA is caused by the imbalance between synthesis and decomposition of chondrocytes, ECM and subchondral bone, under the influence of mechanical and biological factors [66]. The features of OA are pain, cartilage degradation, and inflammation. Similar to the treatment of RA, OA is also treated with anti-inflammatory drugs that can be released in a short time, but the side effects are obvious. To address the problems of traditional therapies, stimuli-responsive, targeted, controlled release drug delivery system may be the promising approach to prolong relief and minimize side effects. As we know, the overproduced ROS in joints has a positive correlation with OA. Numerous antioxidants have been developed to relieve symptoms of OA, such as drug-loaded and drug-free ROS-responsive polymers.

For instance, thioether-containing polymer, poly(propylene sulfide) (PPS), has been used to prepare microspheres (PPS-MS) for treating post-traumatic osteoarthritis (PTOA). *In vitro* studies indicated that PPS-MS possessed the capability of cleaning multiple ROS which are implicated in disease progression. *In vivo* studies demonstrated that PPS-MS could hinder the destruction of articular cartilage in PTOA models [67]. In addition, the polarized phenotype of macrophage has been demonstrated to be associated with OA [68]. Numerous studies have focused on the direct inhibition of M1 or promotion of M2 polarization. However, macrophages are present in all tissues and organs, and the systemic drug exposure may induce immunosuppression to increase the risk of serious infections. Therefore, a boronate-stabilized polyphenol-poloxamer 188 assembled dexamethasone nanomedicine has been prepared to modulate M2 polarization, and it also had the capacity of drug release induced by ROS and ROS scavenging. In lipopolysaccharide-activated RAW264.7 macrophages, this nanomedicine can efficiently inhibit the production of ROS and modulate macrophages M2 polarization with a much lower concentration than free dexamethasone. Furthermore, this nanomedicine also can restore the monosodium iodoacetate-induced OA mice to normal state, and it also could inhibit the cartilage degradation and bone erosion in the joints [69]. This nanomedicine with excellent immunomodulation properties can supply a new strategy for other inflammatory diseases which are associated with ROS and macrophage polarization. As a matter of fact, few polymeric carriers are sufficiently sensitive to selectively deliver drug at physiological concentration of $H_2O_2$ (50-100 μM). To enhance the sensitivity of ROS-responsive biomaterials, an ultrasensitive ROS-responsive polymeric hollow microsphere (HM) has been prepared, which encapsulated dexamethasone, ethanol, $FeCl_2$ and $NaHCO_3$ [70]. In inflammatory joint, endogenic $H_2O_2$ diffused into HM, and then the encapsulated ethanol was oxidized into acetic acid in the presence of $Fe^{2+}$ by the Fenton reaction [71]. Subsequently, $NaHCO_3$in HM reacted with acetic acid to produce $CO_2$ bubbles that can destroy the shell of HM and finally, dexamethasone was released in high concentration at the inflammatory joint. Due to the Fenton reaction, the as-prepared HM showed a good response to 50 μM of $H_2O_2$ and exhibited highly effective local anti-inflammatory activity in the treatment of OA. The ingeniously designed ROS-responsive polymeric carrier opens a new way for the treatment of OA.

### 3.3. Other Bone-Related Diseases

ROS-responsive biomaterials have also been used for treating other bone-related diseases, such as osteoporosis (OP), intervertebral disc degeneration (IDD), tumor-induced bone diseases. OP is one of the epidemics, which is caused by the dysregulation of bone remodeling. OP is characterized by systemic damage of bones mass and micro-architecture, which causes bones to be brittle and liable to fracture [72]. Interestingly, oxidative stress is closely linked to bone remodeling, and ROS have opposite effects on osteoblasts and monocyte/pre-osteoclast cells. Therefore, a "tandem" nanomedicine with dual functions of ROS clearance and drug release has been proposed to treat OP by inhibiting the differentiation of inflammatory cells into osteoclasts [73]. In this work, copolymer synthesized from PEGylated and polysulfide was used to form the ROS-responsive polymeric micelles to carry rapamycin. These micelles can not only eliminate ROS by the oxidation of thioether bond, but also inhibit mTOR through the release of rapamycin. Synergic therapy is considerably more powerful than monotherapy, and it is the new approach for treating OP.

For IDD, a natural progression of aging, and its manifestation is pain of back. Currently, surgical treatment is a relatively effective approach for treating disc degeneration. However, this strategy not only fails to keep the function of disc, but also increases the stress on adjacent discs. From a long-term perspective, it is urgent to develop a more effective therapy to treat disc degeneration. It is generally believed that oxidative stress and inflammation are the common pathological mechanisms for the initiation and progression of IDD [74]. Therefore, regulating local inflammatory microenvironment is the key point for disc tissue regeneration. A rapamycin-loaded ROS-scavenging hydrogel scaffold has been constructed, which can not only eliminate ROS, but also achieve the controlled release of rapamycin, after local injection into the intervertebral disc of the model rats [75]. *In vitro* and *in vivo* studies showed that this drug-loaded hydrogel scaffold possessed the capability of promoting the polarization of microphages to M2 type, which can further promote tissue repair. To conclude, this method may inspire a new way for tissue regeneration through mediating inflammatory microenvironment and macrophage activation. For another example, a thermosensitive, ROS-responsive hydrogel with MR409 has been developed to improve disc degeneration [76]. *In vivo* studies showed that MR409-loaded ROS-responsive hydrogel achieved local controlled release. Due to the ability of MR409 to suppress secretory autophagy, this dual responsive drug-loaded hydrogel can alleviate needle puncture-induced disc degeneration in rats via the inhibition of secretory autophagy.

Moreover, pH and ROS dual-responsive nanoparticles (NPs) have been used to load curcumin, and the therapeutic efficacy of these NPs has been evaluated through a lipopolysaccharide-induced acute ankle inflammation model [10]. *In vitro* and *in vivo* studies results showed that the release of curcumin from the formulated NPs relied on excessive ROS and reduced pH. In addition, *in vivo* anti-inflammatory activity studies indicated that the curcumin-loaded NPs can significantly decrease the level of ROS in inflammatory joints and cure the lipopolysaccharide-induced inflamed ankle.

ROS-responsive biomaterials have been used as nanocarriers for systemic delivery of therapeutic agents to inhibit tumor-induced bone disease as well. GANT58 is the antagonist of Gli, which is able to inhibit Gli2 nuclear translocation and PTHrP expression in tumor cells. But the low water solubility and poor bioavailability of GANT58 limit its therapeutic efficacy *in vivo*. Therefore, GANT58-loaded micellar nanoparticles (GANT58-NPs) have been developed to improve the therapeutic efficacy of GANT58 [77]. The ROS-responsive poly (propylene sulfide) (PPS) was used as the hydrophobic core to sequester GANT58. After inducing by ROS, PPS could be converted into hydrophilic state and degrade to release payloads. In breast and lung cancer bone metastasis models, GANT58-NPs possessed the capability of decreasing bone lesion area and number and increasing trabecular bone volume. Importantly, the non-cytotoxic GANT58-NPs can reduce tumor cell proliferation without altering mesenchymal stem cell proliferation or osteoblast mineralization. Furthermore, GANT58-NPs can prolong the circulation time and specifically target tumorous tissues, thereby, GANT58-NPs is a promising strategy for reducing tumor induced-bone diseases.

## Conclusion

ROS are recognized as a double-edged sword, because normal level of ROS is the key modulator of cell function, while excessive ROS may be responsible for various diseases. ROS-responsive biomaterials have increasingly attracted attention, due to their capability of scavenging ROS in diseased sites [78]. With the rapid development of nanotechnology, the types and preparation strategies of ROS-responsive biomaterials have expanded, and these biomaterials have been widely used in biomedical fields [79]. Mounting evidence has demonstrated that ROS-responsive biomaterials play an important role in the treatment of cancers by precisely controlling the rate of materials degradation and drug release [80, 81]. Therefore, this type of biomaterials can also be used as an effective strategy towards bone-related diseases caused by excessive ROS. RA, OA, OP and other bone-related diseases are the common chronic diseases which are closely related to overproduced ROS. However, the traditional therapies of these diseases are costly, and require a long-term administration accompanied with adverse side effects. Therefore, there is an urgent requirement for more effective treatments tailored for these diseases in clinical practice.

Unfortunately, the transformation of preclinical research into clinic still faces huge challenges. The possible reasons have been described as below: i) the selectivity and bioavailability of ROS-responsive biomaterials for bone tissues are insufficient; ii) the structures and formulations of ROS-responsive biomaterials are sophisticated and difficult to scale up; iii) due to the variations and complexities of patients, the endogenous triggers are difficult to control; iv) the batch-to-batch reproducibility is poor under the current Good Manufacturing Practices; v) the cost of transferring preclinical experimental models to clinic practice is high [82, 83]. To enhance the therapeutic efficacy of ROS-responsive biomaterials, the following aspects should be considered: i) full understanding of the unique biological barrier posed by the skeletal architecture; ii) simplifying the structures and formulations of ROS-responsive biomaterials; iii) designing dual/multi-responsive synergetic systems (pH+ROS, thermal+ROS, light+ROS, ultrasound+ROS, etc); iv) considering the specificity of patients and personalized medicine. To conclude, ROS-responsive biomaterials have made a huge contribution to preclinical studies. In order to achieve more potential for commercialization and realistic clinical application, further studies are needed in materials, medicine and biology to develop promising ROS-responsive biomaterials.

## Acknowledgements

This work was supported by the National Nature Science Foundation of China (81930067).

## References

[1] Altindag O, Erel O, Aksoy N, Selek S, Celik H, Karaoglanoglu M. Increased oxidative stress and its relation with collagen metabolism in knee osteoarthritis. *Rheumatol Int.* 2007;27(4):339-344.

[2] Phull AR, Nasir B, Haq IU, Kim SJ. Oxidative stress, consequences and ROS mediated cellular signaling in rheumatoid arthritis. *Chem Biol Interact.* 2018;281:121-136.

[3] Panagiotis L, Athanasios GP. ROS/oxidative stress signaling in osteoarthritis. *Biochim Biophys Acta Mol Basis Dis.* 2016;1862:576-591.

[4] Drevet S, Gavazzi G, Grange L, Dupuy C, Lardy B. Reactive oxygen species and NADPH oxidase 4 involvement in osteoarthritis. *Exp Gerontol.* 2018;111:107-117.

[5] Bolduc JA, Collins JA, Loeser RF. Reactive oxygen species, aging and articular cartilage homeostasis. *Free Radic Biol Med.* 2019;132:73-82.

[6] Arra M, Swarnkar G, Ke K, Otero JE, Ying J, Duan X, Maruyama T, Rai MF, O'Keefe RJ, Mbalaviele G, Shen J, Abu-Amer Y. LDHA-mediated ROS generation in chondrocytes is a potential therapeutic target for osteoarthritis. *Nat Commun.* 2020;11(1):3427.

[7] Somaiya M, Shagufta M, Abdul QK, Atif Z, Naureen F. Increased reactive oxygen speciesformation and oxidativestress in rheumatoid arthritis. *PloS ONE.* 2016;11(4):e0152925- e0152939.

[8] Agidigbi TS, Kim C. Reactive oxygen species in osteoclast differentiation and possible pharmaceutical targets of ROS-mediated osteoclast diseases. *Int J MolSci.* 2019;20(14):3576-3591.

[9] Callaway DA, Jiang JX. Reactive oxygen species and oxidative stress in osteoclastogenesis, skeletal aging and bone diseases. *J Bone Miner Metab.* 2015;33(4):359-370.

[10] Pu HL, Chiang WL, Maiti B, Liao ZX, Ho YC, Shim MS, Chuang EY, Xia Y, Sung HW. Nanoparticles with dual responses to oxidative stress and reduced ph for drug release and anti-inflammatory applications. *ACS Nano.* 2014;8(2):1213-21.

[11] Kim AY, Ha JH, Park SN. Selective release system for antioxidative and anti-inflammatory activities using $H_2O_2$-responsive therapeutic nanoparticles. *Biomacromolecules.* 2017;18(10):3197-3206.

[12] Tian J, Chen J, Ge C, Liu X, He J, Ni P, Pan Y. Synthesis of PEGylated Ferrocene Nanoconjugates as the Radiosensitizer of Cancer Cells. *Bioconjug Chem.* 2016;27(6):1518-24.

[13] Zhang ZQ, Lu ZN, Yuan Q, Zhang C, Tang YL. ROS-Responsive and active targeted drug delivery based on conjugated polymer nanoparticles for synergistic chemo-/photodynamic therapy. *J Mater Chem B.* 2021;9(9):2240-2248.

[14] Dai L, Yu Y, Luo Z, Li M, Chen W, Shen X, Chen F, Sun Q, Zhang Q, Gu H, Cai K. Photosensitizer enhanced disassembly of amphiphilic micelle for ROS-response targeted tumor therapy in vivo. *Biomaterials.* 2016;104:1-17.

[15] Xu L, Zhao M, Zhang H, Gao W, Guo Z, Zhang X, Zhang J, Cao J, Pu Y, He B. Cinnamaldehyde-Based Poly(ester-thioacetal) To Generate Reactive Oxygen Species for Fabricating Reactive Oxygen Species-Responsive Nanoparticles. *Biomacromolecules.* 2018;19(12):4658-4667.

[16] WangL, Zhu KX, Cao W, Sun CX, Lu CJ, Xu H. ROS-triggered degradation of selenide-containing polymers based on selenoxide elimination. *PolymChem.* 2019;10(16):2039-2046.

[17] Deepagan VG, Kwon S, You DG, Nguyen VQ, Um W, Ko H, Lee H, Jo DG, Kang YM, Park JH. *In situ* diselenide-crosslinked polymeric micelles for ROS-mediated anticancer drug delivery. *Biomaterials.* 2016;103:56-66.

[18] Li F, Li T, Cao W, Wang L, Xu H. Near-infrared light stimuli-responsive synergistic therapy nanoplatforms based on the coordination of tellurium-containing block polymer and cisplatin for cancer treatment. *Biomaterials.* 2017;133:208-218.

[19] Stubelius A, Lee S, Almutairi A. The chemistry of boronic acids in nanomaterials for drug delivery. *Acc Chem Res.* 2019;52(11):3108-3119.

[20] Cui Y, Zhang M, Du FS, Li ZC. Facile synthesis of $H_2O_2$-cleavable polyester-amides by Passerini multicomponent polymerization. *ACS Macro Lett.* 2017;6(1):11-15.

[21] Höcherl A, Jäger E, Jäger A, Hrubý M, Konefał R, Janoušková O, Spěváček J, Jiang Y, Schmidt PW, Lodge TP, Štěpánek P. One-pot synthesis of reactive oxygen species (ROS)-self-immolative polyoxalate prodrug nanoparticles for hormone dependent cancer therapy with minimized side effects. *Polym Chem.* 2017;8(13):1999-2004.

[22] NapoliA, ValentiniM, TirelliN, Müller M, Hubbell JA. Oxidation-responsive polymeric vesicles. *Nat Mater.* 2004;3(3):183-189.

[23] Yang X, Shi X, Zhang Y, Xu J, Ji J, Ye L, Yi F, Zhai G. Photo-triggered self-destructive ROS-responsive nanoparticles of high paclitaxel/chlorin e6 co-loading capacity for synergetic chemo-photodynamic therapy. *J Control Release.* 2020;323:333-349.

[24] O'Grady KP, Kavanaugh TE, Cho H, Ye H, Gupta MK, Madonna MC, Lee J, O'Brien CM, Skala MC, Hasty KA, Duvall CL. Drug-Free ROS Sponge Polymeric Microspheres Reduce Tissue Damage from Ischemic and Mechanical Injury. *ACS Biomater Sci Eng.* 2018;4(4):1251-1264.

[25] Wu T, Chen X, Wang Y, Xiao H, Peng Y, Lin L, Xia W, Long M, Tao J, Shuai X. Aortic plaque-targeted andrographolide delivery with oxidation-sensitive micelle effectively treats atherosclerosis via simultaneous ROS capture and anti-inflammation. *Nanomedicine.* 2018;14(7):2215-2226.

[26] Wang J, Sun X, Mao W, Sun W, Tang J, Sui M, Shen Y, Gu Z. Tumor redox heterogeneity-responsive prodrug nanocapsules for cancer chemotherapy. *Adv Mater.* 2013;25(27):3670-6.

[27] Ma BX, Xu H, Zhuang WH, Wang YN, Li G, Wang Y. Reactive oxygen species responsive theranostic Nano platform for two-photon aggregation-induced emission imaging and therapy of acute and chronic inflammation. *ACS Nano.* 2020;14(5):5862-5873.

[28] Kim JS, Jo SD, Seah GL, Kim I, Nam YS. ROS-induced biodegradable polythioketal nanoparticles for intracellular delivery of anti-cancer therapeutics. *J Ind Eng Chem.* 2015;21:1137-1142.

[29] Xu X, Saw PE, Tao W, Li Y, Ji X, Bhasin S, Liu Y, Ayyash D, Rasmussen J, Huo M, Shi J, Farokhzad OC. ROS-Responsive Polyprodrug Nanoparticles for Triggered Drug Delivery and Effective Cancer Therapy. *Adv Mater.* 2017;29(33):10.

[30] Wang C, Huang B, Yang G, Ouyang Y, Tian J, Zhang W. NIR-triggered multifunctional and degradable nanoplatform based on an ROS-sensitive block copolymer for imaging-guided chemo-phototherapy. *Biomacromolecules.* 2019;20(11):4218-4229.

[31] Yang N, Xiao WY, Song XJ, Wang WJ, Dong XC. Recent advances in tumor microenvironment hydrogen peroxide-responsive materials for cancer photodynamic therapy. *Nanomicro Lett.* 2020;12(1):15.

[32] Qamar N, John P, Bhatti A. Emerging role of selenium in treatment of rheumatoid arthritis: an insight on its antioxidant properties. *J Trace Elem Med Biol.* 2021; 66:126737.

[33] He Y, Guo S, Wu L, Chen P, Wang L, Liu Y, Ju H. Near-infrared boosted ROS responsive siRNA delivery and cancer therapy with sequentially peeled up conversion nano-onions. *Biomaterials.* 2019;225:119501.

[34] Cunha RLO, Gouvea IE, Juliano L. A glimpse on biological activities of tellurium compounds. *An Acad Bras Cienc.* 2009;81(3):393-407.

[35] Cao W, Gu YW, Li TY, Xu HP. Ultra-sensitive ROS-responsive tellurium-containing polymers. *Chem Commun* (Camb). 2015;51(32):7069-7071.

[36] Wang L, Fan FQ, Cao W, Xu H. Ultrasensitive ROS-responsive coassemblies of tellurium-containing molecules and phospholipids. *ACS Appl Mater Interfaces.* 2015;7(29):16054-16060.

[37] Wang L, Wang WC, Cao W, Xu HP. Multi-hierarchical responsive polymers: stepwise oxidation of a selenium- and tellurium-containing block copolymer with sensitivity to both chemical and electrochemical stimuli. *Polym Chem.* 2017;8(31):4520-4527.

[38] Deng Z, Qian Y, Yu Y, Liu G, Hu J, Zhang G, Liu S. Engineering intracellular delivery nanocarriers and nanoreactors from oxidation-responsive polymersomes via synchronized bilayer cross-linking and permeabilizing inside live cells. *J AmChemSoc.* 2016;138(33):10452-10466.

[39] Li C, Pan R, Li P, Guan Q, Ao J, Wang K, Xu L, Liang X, Jin X, Zhang C, Zhu X. Hydrogen Peroxide-Responsive Nanoprobe Assists Circulating Tumor Cell Identification and Colorectal Cancer Diagnosis. *Anal Chem.* 2017;89(11):5966-5975.

[40] de Gracia Lux C, Joshi-Barr S, Nguyen T, Mahmoud E, Schopf E, Fomina N, Almutairi A. Biocompatible polymeric nanoparticles degrade and release cargo in response to biologically relevant levels of hydrogen peroxide. *J Am Chem Soc.* 2012;134(38):15758-64.

[41] Broaders KE, Grandhe S, Fréchet JMJ. A biocompatible oxidation-triggered carrier polymer with potential in therapeutics. *J Am Chem Soc.* 2011;133(4):756-758.

[42] Yeo J, Lee J, Yoon S, Kim WJ. Tannic acid-based nanogel as an efficient anti-inflammatory agent. *Biomater Sci.* 2020;8(4):1148-1159.

[43] Swami Vetha BSS, Adam AG, Aileru A. Redox responsive Copolyoxalate smart polymers for inflammation and other aging-associatedd iseases. *Int J Mol Sci.* 2021;22(11):5607-5620.

[44] Xu F, Li H, Luo YL, Tang W. Redox-responsive self-assembly micelles from poly(n-acryloylmorpholine-block-2-acryloyloxyethylferrocene-carboxylate) amphiphilic block copolymers as drug release carriers. *ACS Appl Mater Interfaces.* 2017;9(6):5181-5192.

[45] SaravanakumarG, KimJ, KimWJ. Reactive-oxygen-species-responsive drug delivery systems: promises and challenges. *Adv Sci (Weinh).* 2017;4(1):1600124.

[46] Xu Q, Chu CC. Development of ROS-responsive amino acid-based poly (ester amide) nanoparticle for anticancer drug delivery. *J Biomed Mater Res* A. 2021;109(4):524-537.

[47] Xu Q, Venet M, Wang W, Creagh-Flynn J, Wang X, Li X, Gao Y, Zhou D, Zeng M, Lara-Sáez I, A S, Tai H, Wang W. Versatile Hyperbranched Poly(β-hydrazide ester) Macromers as Injectable Antioxidative Hydrogels. *ACS Appl Mater Interfaces.* 2018;10(46):39494-39504.

[48] Zhang M, Song CC, Du FS, Li ZC. Supersensitive oxidation-responsive biodegradable peg hydrogels for glucose-triggered insulin delivery. *ACS Appl Mater Interfaces.* 2017;9(31):25905-25914.

[49] Zhu Y, Matsumura Y, Velayutham M, Foley LM, Hitchens TK, Wagner WR. Reactive oxygen species scavenging with a biodegradable, thermally responsive hydrogel compatible with soft tissue injection. *Biomaterials.* 2018;177:98-112.

[50] Gupta MK, Martin JR, Dollinger BR, Hattaway ME, Duvall CL. Thermogelling, ABC triblock copolymer platform for resorbable hydrogels with tunable, degradation-mediated drug release. *AdvFunctMater.* 2017;27(47):1704107.

[51] Gong Y, Chen M, Tan Y, Shen J, Jin Q, Deng W, Sun J, Wang C, Liu Z, Chen Q. Injectable Reactive Oxygen Species-Responsive SN38 Prodrug Scaffold with Checkpoint Inhibitors for Combined Chemoimmunotherapy. *ACS Appl Mater Interfaces.* 2020;12(45):50248-50259.

[52] Zhao H, Huang J, Li Y, Lv X, Zhou H, Wang H, Xu Y, Wang C, Wang J, Liu Z. ROS-scavenging hydrogel to promote healing of bacteria infected diabetic wounds. *Biomaterials.* 2020;258:120286.

[53] Xu QH, He CL, Xiao CS, Chen XS. Reactive oxygen species (ROS) responsive polymers for biomedical applications. *Macromol Biosci.* 2016;16(5):635-646.

[54] Martin JR, Gupta MK, Page JM, Yu F, Davidson JM, Guelcher SA, Duvall CL. A porous tissue engineering scaffold selectively degraded by cell-generated reactive oxygen species. *Biomaterials.* 2014;35(12):3766-76.

[55] Kramer JR, Deming TJ. Glycopolypeptides with a redox-triggered helix-to-coil transition. *J Am Chem Soc.* 2012;134(9):4112-4115.

[56] Lou J, Best MD. Reactive oxygen species-responsive liposomes via boronate-caged phosphatidyl ethanolamine. *Bioconjug Chem.* 2020;31(9):2220-2230.

[57] Lee SH, Zachman AL, Deskins DL, Young PP, Sung HJ. ROS-responsive scaffold for angiogenic differentiation of mesenchymal stem cells. *Summer Bioengineering Conference,* Oregon, USA, June 26 to 29; vol 14553; 2013.

[58] Wang C, Wang J, Zhang X, Yu S, Wen D, Hu Q, Ye Y, Bomba H, Hu X, Liu Z, Dotti G, Gu Z. In situ formed reactive oxygen species-responsive scaffold with gemcitabine and checkpoint inhibitor for combination therapy. *Sci Transl Med.* 2018;10(429):eaan3682.

[59] Wang C, Wang J, Zhang X, Yu S, Wen D, Hu Q, Ye Y, Bomba H, Hu X, Liu Z, Dotti G, Gu Z. In situ formed reactive oxygen species-responsive scaffold with gemcitabine and checkpoint inhibitor for combination therapy. *Sci Transl Med.* 2018;10(429):eaan3682.

[60] Deng ZY, Liu SY. Inflammation-responsive delivery systems for the treatment of chronic inflammatory diseases. *Drug Deliv Transl Res.* 2021;11(4):1475-1497.

[61] Dou Y, Li CW, Li LL, Guo JW, Zhang JX. Bioresponsive drug delivery systems for the treatment of inflammatory diseases. *J Control Release.* 2020;327:641-666.

[62] An L, Li Z, Shi L, Wang L, Wang Y, Jin L, Shuai X, Li J. Inflammation-Targeted Celastrol Nanodrug Attenuates Collagen-Induced Arthritis through NF-κB and Notch1 Pathways. *Nano Lett.* 2020;20(10):7728-7736.

[63] Fan XX, Xu MZ, Leung ELH, Jun C, Yuan Z, Liu L. ROS-responsive berberine polymeric micelles effectively suppressed the inflammation of rheumatoid arthritis by targeting mitochondria. *Nano Micro Lett.* 2020;12(1):76.

[64] Ni R, Song G, Fu X, Song R, Li L, Pu W, Gao J, Hu J, Liu Q, He F, Zhang D, Huang G. Reactive oxygen species-responsive dexamethasone-loaded nanoparticles for targeted treatment of rheumatoid arthritis via suppressing the iRhom2/TNF-α/BAFF signaling pathway. *Biomaterials.* 2020;232:119730.

[65] Chen M, Amerigos J C KD, Su Z, Guissi NEI, Xiao Y, Zong L, Ping Q. Folate Receptor-Targeting and Reactive Oxygen Species-Responsive Liposomal Formulation of Methotrexate for Treatment of Rheumatoid Arthritis. *Pharmaceutics.* 2019;11(11):582.

[66] Yao Y, Zhang H, Wang Z, Ding J, Wang S, Huang B, Ke S, Gao C. Reactive oxygen species (ROS)-responsive biomaterials mediate tissue microenvironments and tissue regeneration. *J Mater Chem B.* 2019;7(33):5019-5037.

[67] O'Grady KP, Kavanaugh TE, Cho H, Ye H, Gupta MK, Madonna MC, Lee J, O'Brien CM, Skala MC, Hasty KA, Duvall CL. Drug-Free ROS Sponge Polymeric Microspheres Reduce Tissue Damage from Ischemic and Mechanical Injury. *ACS Biomater Sci Eng.* 2018;4(4):1251-1264.

[68] Chen Y, Jiang W, Yong H, He M, Yang Y, Deng Z, Li Y. Macrophages in osteoarthritis: pathophysiology and therapeutics. *Am J Transl Res.* 2020;12(1):261-268.

[69] Li X, Wang X, Liu Q, Yan J, Pan D, Wang L, Xu Y, Wang F, Liu Y, Li X, Yang M. ROS-Responsive Boronate-Stabilized Polyphenol-Poloxamer 188 Assembled Dexamethasone Nanodrug for Macrophage Repolarization in Osteoarthritis Treatment. *Adv Healthc Mater.* 2021;10(20):e2100883.

[70] Chung MF, Chia WT, Wan WL, Lin YJ, Sung HW. Controlled release of an anti-inflammatory drug using an ultrasensitive ROS-responsive gas-generating carrier for localized inflammation inhibition. *J Am Chem Soc.* 2015;137(39):12462-12465.

[71] BhangeVP, WilliamSP, SharmaA, GabhaneJ, VaidyaAN, Wate SR. Pretreatment of garden biomass using Fenton's reagent: influence of $Fe^{2+}$ and $H_2O_2$ concentrations on lignocellulose degradation. *J Environ Health Sci Eng.* 2015;13:12.

[72] QinD, ZhangHJ, ZhangHF, SunTY, ZhaoHB, Lee WH. Anti-osteoporosis effects of Osteoking via reducing reactive oxygen species. *J Ethnopharmacol.* 2019;244:112045.

[73] El Mohtadi F, d'Arcy R, Burke J, Rios De La Rosa JM, Gennari A, Marotta R, Francini N, Donno R, Tirelli N. "Tandem" Nanomedicine Approach against Osteoclastogenesis: Polysulfide Micelles Synergically Scavenge ROS and Release Rapamycin. *Biomacromolecules.* 2020;21(2):305-318.

[74] SrivastavaA, IsalL, RooneyP, PanditA. Bioengineered three-dimensional diseased intervertebral disc model revealed inflammatory crosstalk. *Biomaterials.* 2017;123:127-141.

[75] Bai J, Zhang Y, Fan Q, Xu J, Shan H, Gao X, Ma Q, Sheng L, Zheng X, Cheng W, Li D, Zhang M, Hao Y, Feng L, Chen Q, Zhou X, Wang C. Reactive Oxygen Species-Scavenging Scaffold with

Rapamycin for Treatment of Intervertebral Disk Degeneration. *Adv Healthc Mater.* 2020; 9(3):e1901186.

[76] Zheng Q, Shen H, Tong Z, Cheng L, Xu Y, Feng Z, Liao S, Hu X, Pan Z, Mao Z, Wang Y. A thermosensitive, reactive oxygen species-responsive, MR409-encapsulated hydrogel ameliorates disc degeneration in rats by inhibiting the secretary autophagy pathway. *Theranostics.* 2021;11(1):147-163.

[77] Vanderburgh JP, Kwakwa KA, Werfel TA, Merkel AR, Gupta MK, Johnson RW, Guelcher SA, Duvall CL, Rhoades JA. Systemic delivery of a Gli inhibitor via polymeric nanocarriers inhibits tumor-induced bone disease. *J Control Release.* 2019;311-312:257-272.

[78] Dou Y, Tao H, Zhang D, Wang R, Li X, Zhang J. Structure-property correlations of reactive oxygen species-responsive and hydrogen peroxide-eliminating materials with anti-oxidant and anti-inflammatory activities. *Chem Mater.* 2017;29(19):8221-8238.

[79] Tao WH, He Z. ROS-responsive drug delivery systems for biomedical applications. *Asian J Pharm Sci.* 2018;13(2):101-112.

[80] Yuan Y, Zhao L, Shen C, He Y, Yang F, Zhang G, Jia M, Zeng R, Li C, Qiao R. Reactive oxygen species-responsive amino acid-based polymeric nanovehicles for tumor-selective anticancer drug delivery. *Mater Sci Eng C Mater Biol Appl.* 2020;106:110159.

[81] Yu LY, Su GM, Chen CK, Chiang YT, Lo CL. Specific cancer cytosolic drug delivery triggered by reactive oxygen species-responsive micelles. *Biomacromolecules.* 2016;17(9):3040-3047.

[82] Liu D, Yang F, Xiong F, Gu N. The smart drug delivery system and its clinical potential. *Theranostics.* 2016;6(9):1306-1323.

[83] Lavrador P, Gaspar VM, Mano JF. Stimuli-responsive nanocarriers for delivery of bone therapeutics-barriers and progresses. *J Control Release.* 2018;273:51-67.

# Chapter 21

# Stem Cell: Ageing and Senescence Mechanisms

**Ankita Das**
**Chandrayee Ghatak**
**and Ananya Barui***
Centre for Healthcare Science and Technology, IIEST Shibpur, Indian Institute of Engineering Science and Technology, Shibpur, Howrah, West Bengal, India

## Abstract

Stem cells (MSCs) are exploited for their properties of totipotency, multipotency and pluripotency depending upon the type of stem cells. For example, mesenchymal stem cells, one of the widely used cell types with immense therapeutic potential and clinical significance is multipotent in nature. Interest in different stem cells arises due to their potential for self-renewal and differentiation. Though, it was deemed initially that several tissue-specific stem cells are immortal in nature, phenomenon of natural aging similar to that observed in somatic cell types have also been discovered in stem cells. The aging phenomenon results in gradual loss of regenerative efficacy accompanied by enhancement of dysfunction of cells. In many reports, such cell dysfunction and exhaustion of replicative potential has been attributed to functional attrition of organisms and is a marker for aging process. So, the current paradigm has emerged that stem cells do not undergo infinite replication but exhibit limited number of self-renewal cycles prior to senescing. Therefore, characterization of aging and senescence has important implication in clinical practice as expansion to sufficient number of cells is a pre-requisite. Such characterization will also help in determining the age of donor for stem cell transplantation and effect of age on viability, functionality, and integrity of stem cells. Further, alterations in micro-environment can have different effects of stem cell behaviour of different ages. It has also been seen that stem cell regenerative potential can be influenced by systemic alterations like immune system performance. This chapter reviews the molecular-level mechanisms affecting senescence of stem cells, phenotypic and functional properties of stem cells, and the techniques that are currently used to identify signs of stem cells aging.

**Keywords:** mesenchymal stem cells, replicative senescence, signalling pathway, reprogramming, epigenetic changes, donor efficiency, transplantation, tissue engineering

---

* ananya.pariksha@gmail.com; ananyabarui@chest.iiests.ac.in.

In: The Role of Reactive Oxygen Species in Health and Disease
Editor: Kuladip Jana
ISBN: 979-8-89113-283-2

## 1. Introduction

The age-old concept of stem cells has an immense role to play in the clinical translation based on cell therapy. Stem cells have the capability to self-renew and give rise to differentiated cells of different types. The three different types of stem cells are embryonic stem cells (ESCs), adult stem cells (ASCs) and induced pluripotent stem cells (iPSCs). After fertilisation, the blastocyst is formed after approximately 5 days. The cells derived from the blastocyst are referred to as the embryonic stem cells. The blastocyst is a hollow mass of cells in which the peripheral cells or the trophoblast produces the embryonic membrane and placenta whereas the fetus arises from the inner cell mass. They basically act as a stem cell source which can form any cell type of an adult organism due to their pluripotent nature. There are several ethical constraints associated with the procurement of the ESCs as it requires the destruction of the blastocyst. Another alternate approach is the accumulation of only a single cell from the inner cell mass and permitting the rest for subsequent implantation, but again, this raises the question of development of a normal human being. ESCs in their undifferentiated state can be maintained up to several hundred passages. The common markers associated with ESCs are: CD9, CD24, alkaline phosphatase, Rex-1, Oct-4, Sox-2, LIN-28, Thy-1, Nanog and SSE-3 and 4. Though, the ESCs have the advantage of limitless expansion *ex-vivo*, there are a lot of safety issues involved in their isolation and cultivation. Therefore, compared to the ESCs, the ASCs are a much better choice for application in the field of medicine. ASCs have a finite self-renewal and differentiation potential. The concept of ASCs arises from the fact that almost all adult tissue comprises a compartment of stem cells which can replenish the old and damaged cells. Moreover, there are different stem cell sources based on which they can be further classified. Hematopoietic stem cells (HSCs) procured from sources such as bone marrow and peripheral blood are often used in bone marrow transplantation. These precursor cells have the capability to form different types of blood cells, including red blood cells, white blood cells and platelets. The most interesting one is the mesenchymal stem cells (MSCs), which can be obtained from numerous sources such as bone-marrow, umbilical cord tissue, amniotic fluid, placenta, dental pulp, and adipose tissue. They are present in almost every tissue in an organism. These stem cells are attractive sources for clinical translation due to their ready availability along with self-renewal and differentiation capability. MSCs are stromal in origin and due to their self-renewal and differentiation ability, their cell division are classified into symmetric and asymmetric types. Fate of the stem cells is highly dependent on their surrounding microenvironment or niche. Basically, the interaction of cells with their niche drives stem cell fate determination. Since, niches are usually composed of extracellular matrix along with signalling molecules, they often determine the stemness of the cells. Apart from the extrinsic factors, several intrinsic factors of the stem cells also help in regulating their fate. Therefore, by understanding the relationship of cells with their niche in *in vitro* culture, it is possible to augment the adult stem cells. As, in the case of any other somatic cell, ASCs also have a finite life span. After a few number of cell divisions they enter a state of senescence [1].

## 2. Replicative Senescence in Stem Cells

MSCs have a great capacity in the area of cellular therapy. Generally, before their application, these cells are initially isolated and cultured. The MSCs have been reported to maintain functional capabilities up to 7-12 passages only. The cells demonstrated distortions in morphology, decreased expression of cellular markers and finally cell cycle arrest. Leonard Hayflick first described this phenomenon in the 1960s terming it replicative senescence or *in vitro* aging. Replicative senescence is governed by both intrinsic and extrinsic factors, and the process is deemed to be progressive and irreversible. The understanding of the mechanism of senescence among ASCs is of utmost importance for quality control in cell-based therapy. The aging status of the cell directly reflects their stemness potential. Usually, early passage stem cells are preferred for clinical expansion as the therapeutic potential of the late passage stem cells becomes attenuated [2]. For effective cellular therapy, it is important to monitor the stemness status. These changes occur subsequently in later stages observed by the downregulation of genes linked with DNA replication and DNA repair, accumulation of the cyclin-dependent kinase inhibitor p16INK4a and oxidative stress. The MSCs obtained from aged donors' demonstrated parameters observed in senescent cells marked by flat, enlarged morphology and a high amount of senescence associated β-galactosidase cells along with lower proliferation rate. The molecular mechanism associated with cellular senescence can either be gene driven or it could be a result of chance events or a combination of both. In ESCs, high levels of telomerase are considered a crucial factor in their limitless expansion in culture whereas in ASCs, the gradual shortening of the telomeres has been associated with one of the causes of replicative senescence [3].

## 3. Characteristics of Senescent Stem Cells

Down-regulation of transcription factors of E2F family, cyclin A, [4] heterochromatin-related histone markers and increased level of H3K9me3 and H3K27me3 were observed in senescent stem cells. Increased level of H3K9me3 and H3K27me3 in the gene promoter results in reduced gene expression, which is one of the crucial marker for replicative senescence [5]. Transcriptomic studies by mRNA expression profiling revealed changes at different passages in the gene expression signature of MSCs [6]. Decrease in age-related genes during senescence, specifically, lysine-specific demethylases (KDM6a-b, KDM3a-b and KDM5d), DNMT1, and DNMT3B were observed [7, 8]. Down regulation of Oct4, Nanog, and Tert genes also occur during replicative senescence. Using techniques such as chromatin immune-precipitation along with whole genome sequencing, epigenetic mapping can also allow detection of cellular senescence.

Senescence-associated secretory phenotype (SASP) namely, interleukin-1 (IL-1), IL-6, IL8, TNF, matrix metalloproteinase 1 (MMP1) and vascular endothelial growth factor (VEGF) may also regulate the aging characteristics of MSCs. These are known to cause chronic inflammation that leads to several age-related phenotypes [9]. Exosomes derived from the MSCs tend to contain growth factors, cytokines, mRNA and regulatory miRNA. The content and composition of the exosomes can differ with the age of the MSCs. The miRNA expression profile *in vitro* showed upregulation of hsa-mir-369-5P, hsa-mir-371,hsa-mir-499, hsa-mir-29c, and hsa-let-7f. Moreover, exosomes taken from aged bone marrow exhibited higher levels of miR-183 clusters and miR-31a-5p, resulting in reduced MSC proliferation along with osteogenic differentiation, positively affecting MSC aging [10]. Furthermore, in senescent MSCs, upregulation of miR-146a, miR-17, miR-34, miR-183 p-5p and miR-335 p-5p were also observed [11]. These miRNAs inhibit pro-apoptotic genes along with regulation of processes such as cell senescence, telomere erosion and stem cell depletion by manoeuvring tissue-specific adult stem cell function depletion during aging.

## 4. Mechanisms of Replicative Senescence

Replicative senescence is broadly classified into stress-induced senescence, oncogene-induced senescence, and developmental senescence. Telomere abrasion is a senescence hallmark because it limits MSC division, disrupting tissue formation.

### 4.1. Telomere Shortening

Telomerase in mammalian cells is a type of enzyme composed of telomerase RNA component (TERC) along with telomerase reverse transcriptase component (TERT). New telomeric repeats are synthesized by the catalytic action of TERT. Telomerase introduces repeated TTAGGG to the ends of chromosome, arresting telomere abrasion and inducing telomere elongation. Despite this phenomenon, telomerase activity is not sufficient to sustain cell replication indefinitely, as adult stem cells show a persistent decrease in telomere length with age, as shown by certain studies [12].

It was von Patel et al. who observed that the mean lifespan of mice can increase with the overexpression of TERT [13]. The limited proliferation capacity of cells cultured *in vitro* (replicative senescence or Hayflick limit) can be explained by exhaustion of the telomeres [14]. Telomere length detection could therefore act as a biomarker in assessing MSC senescence. Techniques involving quantification of telomere length has gained quite a focus in the study of senescence. Telomere-based tests have their use in the diagnosis and management of senescence of cells [15]. Delaying physiological aging without increasing cancer incidence in adult wild-type mice using methods like pharmacological activation and systemic viral transduction to alter telomerase activity is also possible [16]. In extracts derived from tissue culture cells and normal and cancerous tissues, the correlation between human TERT mRNA expression and telomerase activity further indicates that the rate-limiting agent of telomerase activity tends to be the human TERT expression factor.

### 4.2. Dysfunctioning of Mitochondria

Mitochondrial dysregulation and stem cell aging may also be related directly due to reactive oxygen species (ROS) generation through leakage of intermediates from electron transport chain, disruption of $Ca^{2+}$ homeostasis, along with higher level of cellular apoptosis. These processes positively affect age-related diseases. The free radical theory of aging given by

Harman in the 1950s, hypothesized that accumulation of endogenous ROS lead to deterioration of mitochondria and cellular damage resulting in aging and cell death [17]. Back in the year 1972, it was hypothesized that life span was dictated by the mitochondrial damage due to free radicals. Oxidative stress in aerobic organisms is due to the endogenous metabolites or exogenous sources such as UV- radiation, and ROS accumulation. This may occur due to differences arising between free radical production and antioxidant defences, namely, superoxide dismutase, responsible for superoxide anion scavenging [18]. Young stem cells show functional antioxidant mechanisms, whereas aged stem cells show less antioxidant defence linked with functional inability which includes decreased feedback to physical environmental cues and lower resistance to oxidative stress [19]. Oxidative stress being one of the crucial factors of aging process, leading to dysregulation of certain cellular pathways, toxic aggregate accumulation, cellular debris, and activation of cell death or survival pathways resulting in necrosis, apoptosis or autophagy [19]. Backing the hypothesis that ROS generation aids aging of stem cell, it has been observed that elevated ROS level in aged human mesenchymal stem cells, and rate of hematopoietic stem cells (HSCs) with low ROS level decreases in mice with age [20]. Moreover, in mice, increased cellular ROS concentrations in HSCs and neural stem cells (NSCs) lead to malignancy, irregular proliferation and compromised stem cell self-renewal. It was observed that knocking down transcription factors such as FoxO1, FoxO3 and FoxO4, downstream effectors of insulin and insulin-like growth factor 1 (IGF-1) signaling pathways in mouse increases ROS cumulation in HSCs. Similarly, regeneration capacity of HSCs with conditional ablation of phosphatase and tensin homolog (PTEN) and protein kinase B (AKT) 1 and 2 deficiency indicated that the signaling by PTEN–AKT–mammalian target of rapamycin (mTOR) pathway, upstream of FoxO tends to sense and control ROS and ultimately regulates HSC self-renewal and survival in the organism.

### 4.3. Effect of Free Radicals

Impaired cellular antioxidant mechanisms are associated with stem cell dysfunction and ROS. Mice in the absence of gene encoding superoxide dismutase 2 (Sod2), FoxO3 regulated mitochondrial antioxidant enzyme, resulted in oxidative stress–mediated hematopoietic dysfunctionality [21]. The protective role of SOD2 and its overexpression, upregulated the life expectancy of mouse NSCs in both *in vitro* and *in vivo* environment. HSC function declination with aging is related to age-dependent changes of SOD2 due to FoxO3 and DNA damage sensing serine/threonine protein kinase ATM (ataxia-telangiectasia mutated) [22-24]. Inhibition of oxidative stress regulated by Atm−/− can self-renew HSC in mice, suggesting ROS accumulation driven stem cell aging [25]. Oxidative metabolism regulation associating sirtuin family of NAD-dependent protein deacetylases also lead to stem cell aging. Growth and differentiation of mesenchymal stem cell is maintained by SIRT1 which tends to decline with age [26, 27]. During HSC aging, ROS production is controlled by SIRT3 and its external induction increases antioxidant activity of SOD2 which favours HSC function [28, 29]. Changing the metabolic and redox state can reverse aging of stem cells, which might also affect the intracellular accumulation of ROS [30].

### 4.4. DNA Damage and Mutation

Earliest senescence theory known as mutation accumulation theory states that DNA damage and subsequent mutation cumulation occur during stem cell senescence. HSC DNA damage is evident through histone H2AX phosphorylation and comet tail, which increases with age [31, 32]. Aging HSCs exhibit replication stress along with diminished DNA helicase expression as well as chromosomal aneuploidies, copy-number variations for large chromosomal anomalies [33, 34]. HSCs are indulging pre-cancerous mutation increases in number during senescence [35, 36, 37]. DNA damage due to internal factors like errors in DNA replication, ROS or external factors such as ionization radiation, UV-radiation can cause DNA lesions leading to modification of base or sugar residue, DNA strand crosslinking, formation of DNA adducts or single and double-strand breaks. Moreover, DNA damage response (DDR) favours stem cell senescence when it fails to repair DNA damage, resulting in cell senescence and apoptosis which facilitates high genome instability. DDR-related genes, namely, ATM, ATR, Chk1, and Chk2 in the intestinal cells minimized DDR and enhanced cell death and senescence when depleted, influencing Drosophila lifespan [38]. Moreover, DNA damage aggregation can lead to MSC senescence through various mechanisms, such as inducing endogenous IFNβ, amplifying DDR through related signal pathways, shortening telomere length, activating p53 signalling pathway [39]. In mice, accelerated aging due to DNA repair malfunctioning is observed, along with various human progeroid syndromes viz. Bloom syndrome, Werner syndrome, and Xerodermapigmentosum [40-42]. To counter this, repair mechanisms like artificial reinforcement of nuclear DNA can be employed to delay aging [43]. Nuclear lamina defects have shown to change the nuclear architecture which leads to genomic instability [44].

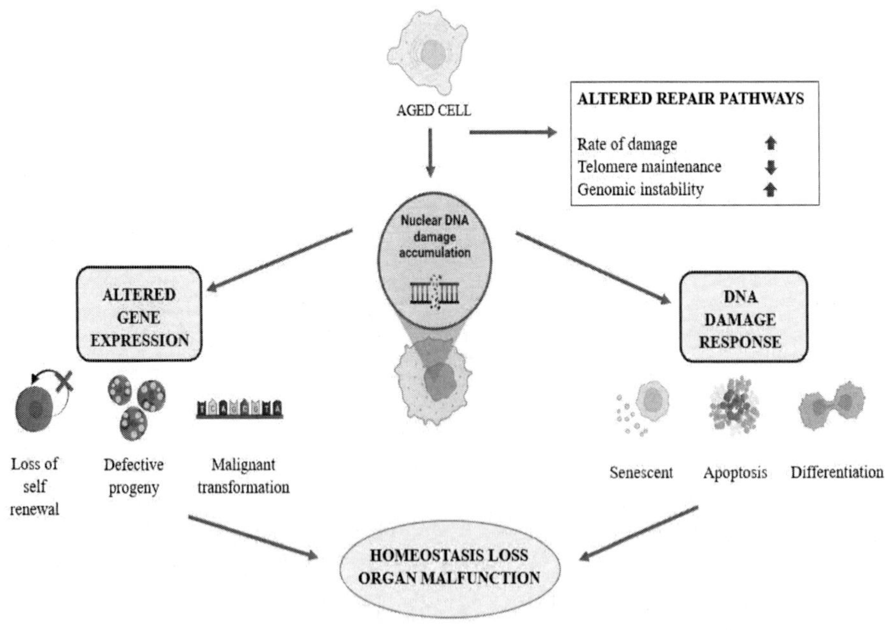


**Figure 1.** DNA damage in the aging cell inflicting various internal changes in the stem cell.

Normal aging shows alterations in the nuclear lamina and prelamin A isoform (progerin) levels in humans. Telomere dysfunction enhances progerin production in human fibroblasts, confirming a link between them [45, 46]. Additionally, reduced lamin B1 level can be a useful biomarker of cell senescence [47]. Figure 3 represents the overview of the different endogenous changes which occur during stem cell aging (Figure 1).

## 4.5. Epigenetic Changes

Stem cell functioning requires adequate chromatin state regulation. Moreover, Waddington's theory of epigenetic landscape mentions that stem cells remain undifferentiated at epigenetic summit above multiple cell fate [48]. Differentiation and aging of cells cause chromatin and gene expression alteration on account of cell stress or DNA damage signals. Epigenetic modification alteration of chromatin and histone such as histone acetylation and DNA methylation causes dysregulation of gene expression. Histone expression level is likely to decrease with age. Aging causes H3K4me3 levels to rise at loci regulating HSC self-renewal and H4K16Ac decreases with age [49]. Restoration of H4K16Ac levels up to young HSCs by CDC42 inhibition also tends to reverse HSC aging phenotypes in transplantation assays [50]. Stem cell functioning declines with age due to variation in chromatin modifiers (SWI-SNF and PRC complexes), HDACs comprising sirtuins, along with DNA methyltransferases expression levels [51- 53]. Furthermore, EZH2 overexpression helps in repopulating HSCs as observed in a study 54]. Based on chromosomal location of aged HSCs, gene cluster expression level increases, suggesting that epigenetic dysregulation could result in a regional loss of transcriptional silencing [51]. Modulation of stem cell function without excessive and deregulated proliferation will moreover increase tissue lifespan and regenerative capacity. The epigenetic modulation taking place in an aging stem cell is represented in Figure 2.

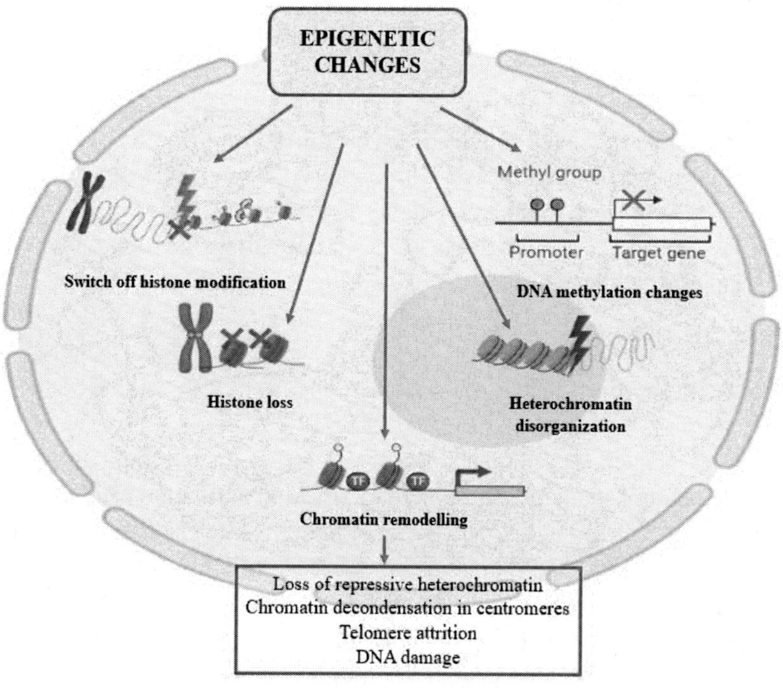


**Figure 2.** Epigenetic changes during aging, including loss of core histones, changes related to histone modification and chromatin remodelling, resulting in negative after effects.

### 4.6. Niche Regulation

Aging can greatly affect the niche of stem cell as was observed in Drosophila where the number of stem cell forming the niche decreased with age [55, 56] and adhesion of glioblastoma stem cells (GSCs) to the microenvironment was also observed to be affected in the process [55, 57]. Moreover, morphogenesis and growth factor signalling of niche cell declines with age influencing the cell fate direction. Expression of MDM2 through P13K/AKT signalling is reduced by fibroblast growth factor (FGF) which negatively regulates p53, inhibiting cell senescence by regulating Cyclin G expression [58]. In mice muscle, FGF2 expression increases in satellite cell niche after aging impairing the self-renewal capacity [59]. Additionally, high Leptin and NAP-2 levels in systemic lupus erythematosus patient's microenvironment contributes to aging and can be reduced by treatment with P13K/AKT inhibitor LY294002 [60]. Furthermore, inflammatory markers in aging niche negatively affects stem cell functioning [61]. Homeostasis in adult stem cell microenvironment is governed by transcription or translation system regulation together with circadian rhythm [62]. Glucose concentration in both internal and external microenvironment influences differentiation, cell proliferation, senescence, and apoptosis. Increased glucose concentration in adipose MSCs increases miR-486, represses SIRT1 expression and leads to cell senescence [63]. Moreover, resveratrol induces high-sugar environment by inhibiting SIRT1 which can protect MSCs from senescence [64]. O-nitroacetylglucosamine (O-GlcNAc) glycosylation modification induced by high glucose level can impede cellular proliferation, arrest cell cycle, and enhance cell apoptosis and senescence [65]. Various exogenous elements encompassing circulating factors like insulin growth factor (IGF) can contribute to senescence if altered. Rejuvenation of senescent cells through blood or serum of either young or calorie-limited animals has also been observed. In one of the studies, mice with growth hormone receptor knockout showed extremely low levels of circulating IGF-1 and additionally, calorie restrictions didn't promote further lifespan [66, 67]. Other studies suggest that intestinal environment in Drosophila can be improved by blocking insulin-IGF signal leading to increased lifespan, and decrease in IGF-1 level positively affects HSCs self-renewal as well [68, 69]. NSC and GSC numbers in aging animal can be up regulated by insulin, IGF-1, and growth hormones [70-72] but self-renewal ability might be affected. Cyclic factor TGF-β tends to be higher in the serum of both aging mice and humans affecting satellite cell and NSCs functions [73, 74]. Growth differentiation factor 11 improves satellite cells and NSCs, however they decrease with age [75, 76].

### 4.7. Senescence Associated Secretory Phenotypes

The senescence-associated secretory phenotypes (SASPs) alter the functional capabilities of the neighbouring healthy stem cells through autocrine and paracrine mechanism. This depends on various specific transcription factor activation, namely, NF-κB, C/EBP β, mTOR, GATA4, Notch1 signaling molecules, p38MAPK, MLL1 (KMT2A), HMGB2, H2A.J, and MacroH2A; several other cytokines, growth factor, chemokines and matrix metalloproteinases [77, 78]. p16INK4a, a marker of stem cell senescence ameliorates with aging among the cells. The $p16^{INK4a}$ repression ameliorates the phenomenon of replicative senescence [79-84].

### 4.8. Cell Cycle Changes

The asymmetric isolation of damaged proteins, DNA and organelles can be utilised by stem cells to maintain protein homeostasis. Since, the daughter stem cells usually tend to survive longer than differentiated cells, therefore, undamaged components will likely be in higher amount in the progeny stem cells [85-87]. Stem cell senescence can also be altered by loss of polarity among the GSCs [88]. In particular, the Wnt signaling pathway is related with aging in HSCs ensuing loss of polarity [89]. During active cell cycle, the proteasome and autophagic activity can reduce the quantity of cellular waste generation [90]. Previous studies have shown the connection between autophagy damage, stem cell depletion and aging on muscle and hematopoietic stem cells [91]. The various signalling pathways affected by stem cell aging are depicted in Figure 3.

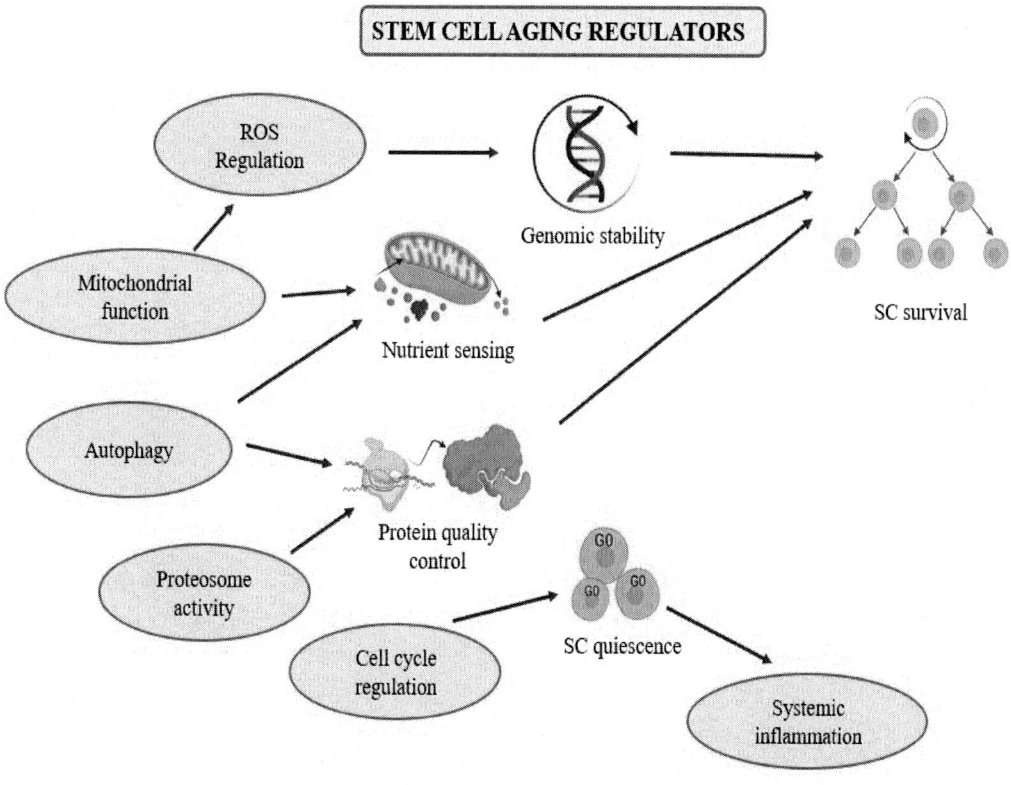


**Figure 3.** Stem cell aging related signalling pathways according to the functions that they regulate.

## 5. Age of Donor and Transplantation Efficiency

Tissue derived factors released by transplanted stem cells such as extracellular vesicles (EVs) encompassing proteins, RNAs, and lipid components promote cell plasticity and tissue regeneration [92-96]. Interestingly, *in vitro* conditions have shown that younger mesenchymal stroma cells (MSCs) derived EVs rejuvenate aged endothelial progenitor cells (EPCs) favourably when compared to old MSCs. Rejuvenation of senescent EPCs in vitro by miRNA-

126-loaded EVs has been also noted [97]. After hematopoietic stem cell transplantation in recipient, multiple tissues such as muscle, liver, myocardium and endothelium go through transdifferentiating into dysfunctional recipient tissue and support organ function [98-100]. Studies have shown that in patients undergoing matched unrelated or haploidentical stem cell transplantation, higher donor age gives adverse outcomes [101, 102]. In another study relating hematopoietic stem cell transplantation (HSCT) with donor age, increased neutrophil recovery and less non-relapse mortality was observed, along with decreased patient overall survival (OS) with older donors [103]. Conventional-intensity allogeneic hematopoietic stem cell transplantation (CI-HSCT) outcomes have poorer results when combined with treatment-related mortality (TRM) and relapse in older donors [104-106]. Older donors (>45 years) have shown greater relapse risk, OS and disease-free survival negatively affected by increased LDH, refractory disease and poor performance status. In hematologic malignancy related sub-myeloablative allogeneic hematopoietic stem cell transplantation (HSCT), young donors showed better outcome because of lower treatment-related mortality and relapse [107].

In a study conducted on human mesenchymal stem cell transplantation, more blood vessels in hearts were found after a myocardial infarction due to improved myocardial perfusion that received young hMSCs rather than old hMSCs. Vascular density increment due to VEGF upregulation, higher TIMP-3 levels, and lower MMP-9 levels was also noted in animals with young hMSCs, however, old hMSCs had less plasticity, weaker funtionality when compared with young hMSCs [108].

Erythropoietic stem cells transplantation is detrimental to stem cell-repopulating abilities, as has been seen in single serial transplantation of marrow cell lines that led to a decline in repopulating ability. Donors identified by hemoglobin type showed greater transplantation effect implying that erythroid precursor cells are affected by transplantation. Single transplantation declined repopulating abilities of phytohemagglutinin-responsive cell precursors, along with erythroid cell precursors. Transplantation tends to accelerate normal aging process and serial transplantation of stem cells can be done five times before they start to detriment functionally, so this implies that stem cells can normally function up to 15-50 life spans [109].

In a study conducted, cells taken from child donors (0 to 12 years) showed fibroblast-like morphology up to a high number of passages along with higher proliferation rate than from adult donors (25 to 50 years) and old donors (over 60 years). Moreover, osteogenic, adipogenic, along with neurogenic differentiation potential declined with age but chondrogenic potential remained unaffected. In adult hMSCs, with passage number topo IIβ expression levels and neural marker elevated but topo IIα level reduced in both adult hMSCs and child hMSCs. This hints that age of donor and long-term passage both modulate characteristics of hMSCs even in standardized culture conditions [110].

iPSC-derived cell functionality in humans is not influenced by age of donor as observed in a study where both iPSCs and iPSC-derived erythroid cells taken from fetal liver mesenchymal stem cells, embryonic cells, and adult fibroblasts cells resulted in no relevant difference between the various age groups [111]. In a similar study, iPSC-derived fibroblasts taken from young and old women expressed nearly identical pluripotency markers, namely, Sox-2, Nanog, Oct-4, SSEA-4, Tra-1-60,hTERT, and Tra-1-81, senescent markers p53 and p21, along with the apoptosis marker Bax. The donor age had no negative impact on iPSC-derived fibroblasts and β-galactosidase activity suggesting senescence is not affected [112].

## 6. Reprogramming

Reprogramming for regenerative medicine has been established using somatic cells taken from both mouse and human in which they generated induced pluripotent stem cells (iPSCs) through overexpression of transcription factors Oct4, Sox2, Klf4, and c-Myc (OSKM) [113-114]. The observation of reprogramming was reported *in vivo* in mouse blastocysts implanted with HSC-derived iPSCs which ultimately differentiated back to HSCs [115-116]. The transplantation of bone marrow from the iPSCs generated in the chimeric mice was successful and it showed the same functional capacity of the iPSC-derived HSCs from young mice. Similar observation was made in a recent study for the human MSCs isolated from old individuals (60 to 70years) [117]. It was observed that the reprogrammed MSCs possessed functional potential comparable to that of young MSCs and they showed rejuvenation property. Gene profile showed OSKM expression in a progeria transgenic model performed in LAKI-4F mice which alleviated hallmarks of aging such as decrease in IL-6, MMP13, p16INK4a, along with senescence-associated β-galactosidase activity [118].

## 7. Rejuvenation of Aged Stem Cells

Caloric restriction study conducted on mesenchymal stem cells performed for 12 weeks on both young and old C57BL/6 mice showed increased number of stem cells along with improved muscle healing capability [119]. Calorie restricted young donor muscle stem cells (MuSCs) showed higher myofiber generation compared to ad libitum (AL) fed mice's stem cells as observed in transplant assays. Low level of glucose-cultured MSCs enhanced osteogenic differentiation along with decreased lactate production, increased oxygen consumption, increased electron transport chain complex expression, as well as better antioxidant defense functionality [120]. Another study reported that elevate dosteogenic differentiation potential and reduced β-galactosidase activity with low glucose concentration can be found in bone-marrow derived MSCs [121]. In another study, fasting for 48 hours in C57BL/6J mice before chemotherapy resulted in decreased apoptosis in primitive hematopoietic cells and higher number of lineage-balanced HSCs [69]. Young plasma injections, heterochronic blood exchange or circulating factors such as α-Klotho and GDF11 from young to old mice, resulted in an improved aged wound healing in aged MuSCs [122]. In a study it was also observed that secretome of young MSCs promoted cardiac repair [76], and MuSC differentiation into functional muscle tissue. Additionally, secretome obtained from umbilical cord-derived MSCs has the potential to enhance differentiation along with self-renewal of aged MSCs [123]. Favourable microenvironment to the aged stem cells can also rejuvenate their functional capabilities such as seeding of fetal MSCs onto a decellularized extracellular matrix (ECM) increased proliferation and differentiation capacity, pluripotency marker expression, and chondrogenic potential, compared to later passage MSCs [124]. Direct injection of ECM molecules within older stem cell niche can also be applied for modification of the ECM. Injection of various ECM components such as Y-27632, collagen-I or apocynin in young C57BL/6 mice showed better wound healing capacity in full-thickness skin wounds further improving stem cell function [125]. Other molecules such as the NAD+, flavonoids, rapamycin, metformin, cucurmin and senolytics like ABT263, 17-DMAG have also reduced senescence among aging stem cells [126-130]. The different target pathways which can be utilised for the restoration of aged stem cell are depicted in Figure 4.

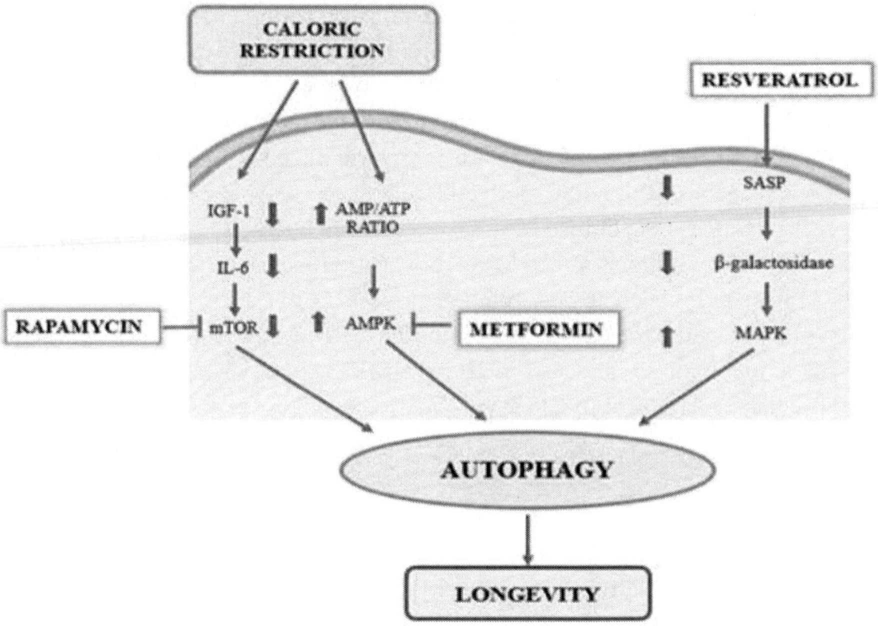


**Figure 4**. Targeted pathways by treatment type affecting the restoration of stem cell.

## 8. Tissue Engineering in Vitiating Senescence

It was observed in a study concerning soft polyacrylamide (PAA) gel cultured human mesenchymal stem cells (hMSCs) that passaging for a long time tend to preserve the stemness and cellular morphology of hMSCs. Moreover, delaying of senescence and proliferation capacity enhancement is observed as indicated by level of β-gal and BrdU expressions, along with loss of Lamin B marker and procurement of Lamin A marker. Additionally, PPAR-gamma expression favoured differentiation of cells into adipogenic lineage when cultured on gel compared to cells growing on tissue culture plastic (TCP) [131]. Similarly, another study including hMSCs on two-dimensional soft hydrogels with imprinted microscale aligned grooves showed that dimensions of the topographical features allows less cellular spreading and cytoskeletal tension, while enhancing osteogenic and adipogenic differentiation potential, multipotency, and repressing cellular senescence [132]. Xu et al. used immunocompatible poly (ether imide) (PEI) substrate with low roughness (Low-R) and high roughness (High-R) on the PEI surface bottom by injection molding. It was observed that Low-R PEI promoted MSC proliferation, repressing apoptosis and senescence, while favouring the proliferation of neighboring cells via paracrine effect [133]. One of the studies on umbilical cord-derived mesenchymal stem cells (UCDMSC) using poly viny alcohol (PVA)-honey matrices showed enhanced cell proliferation and reduced β-gal senescent markers compared to cells on PVA substrates. Moreover, honey can decrease the ROS-related senescence as it has antioxidant and anti-inflammatory components [134]. Furthermore, study related to polydopamine (PDA) coated substrate have been seen to reduce the oxidative stress and mitochondrial damage in replicative senescent MSCs. Suppression of β-galactosidase expression along with lower level of interleukin 6 (IL-6) was observed on PDA coated substrate. p53 and p21 cellular senescent genes were down regulated and stemness marker OCT4 was found to be up- regulated on the PDA-coated substrate [135]. Ogle et al. used poly (ethylene glycol) hydrogel substrates to address the effect of culture surfaces on pro-regenerative secretome. MSCs cultured on 30 kPa hydrogels secreted immunomodulatory and regenerative factors when compared to stiffer 100 kPa or tissue culture plastic surfaces. Moreover, serial culture on 100 kPa RGD hydrogel surfaces reduced senescence in MSCs than in tissue culture plastic, while enhancing cell secretome in response to 30 kPa surfaces [136]. Similarly, another group of researchers showed that electrospun nano fibrous scaffolds have physical and mechanical properties similar to that of native tissue. In this particular study, culturing MSCs on electrospun scaffolds reduced proliferation, decreased cellular senescence, and enhanced stemness due to decreased contractility and YAP nuclear localization compared to cells on tissue culture plastic (TCP). Additionally, cells pre-cultured on electrospun scaffolds showed greater response to mechanical stimuli and higher chondrogenesis in methacrylated hyaluronic acid hydrogels [137].

## Conclusion

Aging stem cells experience a gradual decline in their homeostatic and regenerative capacities. This has been attributed to the degenerative changes in the tissue-specific stem cells, stem cell niches and biochemico-physical cues that regulate stem cell activity. In this chapter, we have outlined the plethora of reasons of stem cell senescence and the several molecular mechanisms involved in this age-dependent deterioration of stem cell function. The essential factors enlisted are critical and demands constant monitoring for developing new stem-cell based therapies. We have included the possible remedies along with several transplantation case studies. Further research on the cellular mechanisms leading to the aging of stem cells will pave the way to designing future anti-aging procedures necessary to cope up with the ever-increasing demand of young and healthy stem cells in the field of translational research globally.

## References


[1] Chagastelles PC, Nardi NB. Biology of stem cells: an overview. *Kidney Int Suppl* (2011). 2011;1(3):63-7.

[2] Kološa K, Leskovšek A, Rajar T, Lah T. Fast assay to predict multipotent mesenchymal stromal cell replicative senescence dynamics. *BioTechniques*. 2022;72(3):90-9.

[3] Wagner W, Horn P, Castoldi M, Diehlmann A, Bork S, Saffrich R. Replicative senescence of mesenchymal stem cells: a continuous and organized process. *PLOS ONE*. 2008;3(5):e2213.

[4] Narita M, Nũnez S, Heard E, Narita M, Lin AW, Hearn SA, Spector DL, Hannon GJ, Lowe SW. Rb-mediated heterochromatin formation and silencing of E2F target genes during cellular senescence. *Cell*. 2003;113(6):703-16.

[5] Paluvai H, Di Giorgio E, Brancolini C. The histone code of senescence. *Cells*. 2020;9(2).

[6] Xie W, Baylin SB, Easwaran H. DNA methylation in senescence, aging and cancer. *Oncoscience*. 2019;6(1-2):291-3.

[7] Cakouros D, Gronthos S. Epigenetic Regulation of Bone Marrow Stem Cell Aging: revealing Epigenetic Signatures associated with hematopoietic and Mesenchymal Stem Cell Aging. *Aging Dis.* 2019;10(1):174-89.

[8] Childs BG, Durik M, Baker DJ, van Deursen JM. Cellular senescence in aging and age-related disease: from mechanisms to therapy. *Nat Med.* 2015;21(12):1424-35.

[9] Reitinger S, Schimke M, Klepsch S, de Sneeuw S, Yani SL, Gaßner R. Systemic impact molds mesenchymal stromal/stem cell aging. *Transfus Apher Sci.* 2015;52(3):285-9.

[10] Davis C, Dukes A, Drewry M, Helwa I, Johnson MH, Isales CM, Hill WD, Liu Y, Shi X, Fulzele S, Hamrick MW. MicroRNA-183-5p Increases with Age in Bone-Derived Extracellular Vesicles, Suppresses Bone Marrow Stromal (Stem) Cell Proliferation, and Induces Stem Cell Senescence. *Tissue Eng Part A.* 2017;23(21-22):1231-1240.

[11] Ahmadi M, Rezaie J. Ageing and mesenchymal stem cells derived exosomes: molecular insight and challenges. *Cell Biochem Funct.* 2021;39(1):60-6.

[12] Vaiserman A, Krasnienkov D. Telomere length as a marker of biological age: state-of-the-art, open issues, and future perspectives. *Front Genet.* 2020;11:630186.

[13] Patel PL, Suram A, Mirani N, Bischof O, Herbig U. Derepression of hTERT gene expression promotes escape from oncogene-induced cellular senescence. *Proc Natl AcadSci U S A.* 2016 Aug 23;113(34):E5024-33.

[14] Victorelli S, Passos JF. Telomeres and cell senescence - size matters not. *EBiomedicine.* 2017;21:14-20.

[15] Baxter MA, Wynn RF, Jowitt SN, Wraith JE, Fairbairn LJ, Bellantuono I. Study of telomere length reveals rapid aging of human marrow stromal cells following in vitro expansion. *Stem Cells.* 2004;22(5):675-82.

[16] Hong J, Yun CO. Telomere gene therapy: polarizing therapeutic goals for treatment of various diseases. *Cells.* 2019;8(5).

[17] Gladyshev VN. The free radical theory of aging is dead. Long live the damage theory*! Antioxid Redox Signal.* 2014;20(4):727-31.

[18] Sharifi-Rad M, Anil Kumar NV, Zucca P, Varoni EM, Dini L, Panzarini E, Rajkovic J, TsouhFokou PV, Azzini E, Peluso I, Prakash Mishra A, Nigam M, El Rayess Y, Beyrouthy ME, Polito L, Iriti M, Martins N, Martorell M, Docea AO, Setzer WN, Calina D, Cho WC, Sharifi-Rad J. Lifestyle, Oxidative Stress, and Antioxidants: Back and Forth in the Pathophysiology of Chronic Diseases. *Front Physiol.* 2020;11:694.

[19] Chen F, Liu Y, Wong NK, Xiao J, So KF. Oxidative stress in stem cell aging. *Cell Transplant.* 2017;26(9):1483-95.

[20] Mejia-Ramirez E, Florian MC. Understanding intrinsic hematopoietic stem cell aging. *Haematologica.* 2020;105(1):22-37.

[21] Oh J, Lee YD, Wagers AJ. Stem cell aging: mechanisms, regulators and therapeutic opportunities. *Nat Med.* 2014;20(8):870-80.

[22] Miyamoto K, Araki KY, Naka K, Arai F, Takubo K, Yamazaki S, Matsuoka S, Miyamoto T, Ito K, Ohmura M, Chen C, Hosokawa K, Nakauchi H, Nakayama K, Nakayama KI, Harada M, Motoyama N, Suda T, Hirao A. Foxo3a is essential for maintenance of the hematopoietic stem cell pool. *Cell Stem Cell.* 2007;1(1):101-112.

[23] Ito K, Hirao A, Arai F, Matsuoka S, Takubo K, Hamaguchi I, Nomiyama K, Hosokawa K, Sakurada K, Nakagata N, Ikeda Y, Mak TW, Suda T. Regulation of oxidative stress by ATM is required for self-renewal of haematopoietic stem cells. *Nature.* 2004;431(7011):997-1002.

[24] Ito K, Hirao A, Arai F, Takubo K, Matsuoka S, Miyamoto K, Ohmura M, Naka K, Hosokawa K, Ikeda Y, Suda T. Reactive oxygen species act through p38 MAPK to limit the lifespan of hematopoietic stem cells. *Nat Med.* 2006 Apr;12(4):446-51. doi: 10.1038/nm1388. Epub 2006 Mar 26. Erratum in: *Nat Med.* 2010;16(1):129.

[25] Liang R, Ghaffari S. Stem cells, redox signaling, and stem cell aging. *Antioxid Redox Signal.* 2014;20(12):1902-16.

[26] Yuan HF, Zhai C, Yan XL, Zhao DD, Wang JX, Zeng Q, Chen L, Nan X, He LJ, Li ST, Yue W, Pei XT. SIRT1 is required for long-term growth of human mesenchymal stem cells. *J Mol Med* (Berl). 2012;90(4):389-400.

[27] Chen H, Liu X, Chen H, Cao J, Zhang L, Hu X, Wang J. Role of SIRT1 and AMPK in mesenchymal stem cells differentiation. *Ageing Res Rev.* 2014;13:55-64.

[28] Brown K, Xie S, Qiu X, Mohrin M, Shin J, Liu Y, Zhang D, Scadden DT, Chen D. SIRT3 reverses aging-associated degeneration. *Cell Rep.* 2013;3(2):319-27.

[29] Kim HS, Patel K, Muldoon-Jacobs K, Bisht KS, Aykin-Burns N, Pennington JD, van der Meer R, Nguyen P, Savage J, Owens KM, Vassilopoulos A, Ozden O, Park SH, Singh KK, Abdulkadir SA, Spitz DR, Deng CX, Gius D. SIRT3 is a mitochondria-localized tumor suppressor required for maintenance of mitochondrial integrity and metabolism during stress. *Cancer Cell.* 2010;17(1):41-52.

[30] Le Belle JE, Orozco NM, Paucar AA, Saxe JP, Mottahedeh J, Pyle AD, Wu H, Kornblum HI. Proliferative neural stem cells have high endogenous ROS levels that regulate self-renewal and neurogenesis in a PI3K/Akt-dependant manner. *Cell Stem Cell.* 2011;8(1):59-71.

[31] Sousa-Victor P, Gutarra S, García-Prat L, Rodriguez-Ubreva J, Ortet L, Ruiz-Bonilla V, Jardí M, Ballestar E, González S, Serrano AL, Perdiguero E, Muñoz-Cánoves P. Geriatric muscle stem cells switch reversible quiescence into senescence. *Nature.* 2014;506(7488):316-21.

[32] Beerman I, Seita J, Inlay MA, Weissman IL, Rossi DJ. Quiescent hematopoietic stem cells accumulate DNA damage during aging that is repaired upon entry into cell cycle. *Cell Stem Cell.* 2014;15(1):37-50.

[33] Forsberg LA, Rasi C, Razzaghian HR, Pakalapati G, Waite L, Thilbeault KS, Ronowicz A, Wineinger NE, Tiwari HK, Boomsma D, Westerman MP, Harris JR, Lyle R, Essand M, Eriksson F, Assimes TL, Iribarren C, Strachan E, O'Hanlon TP, Rider LG, Miller FW, Giedraitis V, Lannfelt L, Ingelsson M, Piotrowski A, Pedersen NL, Absher D, Dumanski JP. Age-related somatic structural changes in the nuclear genome of human blood cells. *Am J Hum Genet.* 2012;90(2):217-28.

[34] Laurie CC, Laurie CA, Rice K, Doheny KF, Zelnick LR, McHugh CP, Ling H, Hetrick KN, Pugh EW, Amos C, Wei Q, Wang LE, Lee JE, Barnes KC, Hansel NN, Mathias R, Daley D, Beaty TH, Scott AF, Ruczinski I, Scharpf RB, Bierut LJ, Hartz SM, Landi MT, Freedman ND, Goldin LR, Ginsburg D, Li J, Desch KC, Strom SS, Blot WJ, Signorello LB, Ingles SA, Chanock SJ, Berndt SI, Le Marchand L, Henderson BE, Monroe KR, Heit JA, de Andrade M, Armasu SM, Regnier C, Lowe WL, Hayes MG, Marazita ML, Feingold E, Murray JC, Melbye M, Feenstra B, Kang JH, Wiggs JL, Jarvik GP, McDavid AN, Seshan VE, Mirel DB, Crenshaw A, Sharopova N, Wise A, Shen J, Crosslin DR, Levine DM, Zheng X, Udren JI, Bennett S, Nelson SC, Gogarten SM, Conomos MP, Heagerty P, Manolio T, Pasquale LR, Haiman CA, Caporaso N, Weir BS. Detectable clonal mosaicism from birth to old age and its relationship to cancer. *Nat Genet.* 2012;44(6):642-50.

[35] Corces-Zimmerman MR, Hong WJ, Weissman IL, Medeiros BC, Majeti R. Preleukemic mutations in human acute myeloid leukemia affect epigenetic regulators and persist in remission. *Proc Natl Acad Sci U S A.* 2014;111(7):2548-53.

[36] Genovese G, Kähler AK, Handsaker RE, Lindberg J, Rose SA, Bakhoum SF, Chambert K, Mick E, Neale BM, Fromer M, Purcell SM, Svantesson O, Landén M, Höglund M, Lehmann S, Gabriel SB, Moran JL, Lander ES, Sullivan PF, Sklar P, Grönberg H, Hultman CM, McCarroll SA. Clonal hematopoiesis and blood-cancer risk inferred from blood DNA sequence. *N Engl J Med.* 2014;371(26):2477-87.

[37] Jaiswal S, Fontanillas P, Flannick J, Manning A, Grauman PV, Mar BG, Lindsley RC, Mermel CH, Burtt N, Chavez A, Higgins JM, Moltchanov V, Kuo FC, Kluk MJ, Henderson B, Kinnunen L, Koistinen HA, Ladenvall C, Getz G, Correa A, Banahan BF, Gabriel S, Kathiresan S, Stringham HM, McCarthy MI, Boehnke M, Tuomilehto J, Haiman C, Groop L, Atzmon G, Wilson JG, Neuberg D, Altshuler D, Ebert BL. Age-related clonal hematopoiesis associated with adverse outcomes. *N Engl J Med.* 2014;371(26):2488-98.

[38] Park JS, Jeon HJ, Pyo JH, Kim YS, Yoo MA. Deficiency in DNA damage response of enterocytes accelerates intestinal stem cell aging in *Drosophila*. *Aging* (Albany NY). 2018 Mar 7;10(3):322-338.

[39] Yu Q, Katlinskaya YV, Carbone CJ, Zhao B, Katlinski KV, Zheng H, Guha M, Li N, Chen Q, Yang T, Lengner CJ, Greenberg RA, Johnson FB, Fuchs SY. DNA-damage-induced type I interferon promotes senescence and inhibits stem cell function. *Cell Rep.* 2015;11(5):785-797.
[40] Murga M, Bunting S, Montaña MF, Soria R, Mulero F, Cañamero M, Lee Y, McKinnon PJ, Nussenzweig A, Fernandez-Capetillo O. A mouse model of ATR-Seckel shows embryonic replicative stress and accelerated aging. *Nat Genet.* 2009;41(8):891-8.
[41] Hoeijmakers JH. DNA damage, aging, and cancer. *N Engl J Med.* 2009;361(15):1475-85.
[42] Gregg SQ, Gutiérrez V, Robinson AR, Woodell T, Nakao A, Ross MA, Michalopoulos GK, Rigatti L, Rothermel CE, Kamileri I, Garinis GA, Stolz DB, Niedernhofer LJ. A mouse model of accelerated liver aging caused by a defect in DNA repair. *Hepatology.* 2012;55(2):609-21.
[43] López-Otín C, Blasco MA, Partridge L, Serrano M, Kroemer G. The hallmarks of aging. *Cell.* 2013 Jun 6;153(6):1194-217.
[44] Dechat T, Pfleghaar K, Sengupta K, Shimi T, Shumaker DK, Solimando L, Goldman RD. Nuclear lamins: major factors in the structural organization and function of the nucleus and chromatin. *Genes Dev.* 2008;22(7):832-53.
[45] Ragnauth CD, Warren DT, Liu Y, McNair R, Tajsic T, Figg N, Shroff R, Skepper J, Shanahan CM. Prelamin A acts to accelerate smooth muscle cell senescence and is a novel biomarker of human vascular aging. *Circulation.* 2010; 121(20):2200-10.
[46] Cao K, Blair CD, Faddah DA, Kieckhaefer JE, Olive M, Erdos MR, Nabel EG, Collins FS. Progerin and telomere dysfunction collaborate to trigger cellular senescence in normal human fibroblasts. *J Clin Invest.* 2011; 121(7):2833-44.
[47] Freund A, Laberge RM, Demaria M, Campisi J. Lamin B1 loss is a senescence-associated biomarker. *Mol Biol Cell.* 2012;23(11):2066-75.
[48] Gilbert SF. Commentary: "The epigenotype" by C.H. Waddington. *Int J Epidemiol.* 2012;41(1):20-3.
[49] Sun D, Luo M, Jeong M, Rodriguez B, Xia Z, Hannah R, Wang H, Le T, Faull KF, Chen R, Gu H, Bock C, Meissner A, Göttgens B, Darlington GJ, Li W, Goodell MA. Epigenomic profiling of young and aged HSCs reveals concerted changes during aging that reinforce self-renewal. *Cell Stem Cell.* 2014;14(5):673-88.
[50] Florian MC, Dörr K, Niebel A, Daria D, Schrezenmeier H, Rojewski M, Filippi MD, Hasenberg A, Gunzer M, Scharffetter-Kochanek K, Zheng Y, Geiger H. Cdc42 activity regulates hematopoietic stem cell aging and rejuvenation. *Cell Stem Cell.* 2012;10(5):520-30.
[51] Chambers SM, Shaw CA, Gatza C, Fisk CJ, Donehower LA, Goodell MA. Aging hematopoietic stem cells decline in function and exhibit epigenetic dysregulation. *PLOS Biol.* 2007;5(8):e201.
[52] Kofman AE, Huszar JM, Payne CJ. Transcriptional analysis of histone deacetylase family members reveal similarities between differentiating and aging spermatogonial stem cells. *Stem Cell Rev Rep.* 2013;9(1):59-64.
[53] Rossi DJ, Bryder D, Zahn JM, Ahlenius H, Sonu R, Wagers AJ, Weissman IL. Cell intrinsic alterations underlie hematopoietic stem cell aging. *Proc Natl AcadSci U S A.* 2005;102(26):9194-9.
[54] Kamminga LM, Bystrykh LV, de Boer A, Houwer S, Douma J, Weersing E, Dontje B, de Haan G. The Polycomb group gene Ezh2 prevents hematopoietic stem cell exhaustion. *Blood.* 2006;107(5):2170-9.
[55] Pan L, Chen S, Weng C, Call G, Zhu D, Tang H, Zhang N, Xie T. Stem cell aging is controlled both intrinsically and extrinsically in the Drosophila ovary. *Cell Stem Cell.* 2007;1(4):458-69.
[56] Toledano H, D'Alterio C, Czech B, Levine E, Jones DL. The let-7-Imp axis regulates ageing of the Drosophila testis stem-cell niche. *Nature.* 2012;485(7400):605-10.
[57] Boyle M, Wong C, Rocha M, Jones DL. Decline in self-renewal factors contributes to aging of the stem cell niche in the Drosophila testis. *Cell Stem Cell.* 2007;1(4):470-8.
[58] Coutu DL, François M, Galipeau J. Inhibition of cellular senescence by developmentally regulated FGF receptors in mesenchymal stem cells. *Blood.* 2011;117(25):6801-12.
[59] Chakkalakal JV, Jones KM, Basson MA, Brack AS. The aged niche disrupts muscle stem cell quiescence. *Nature.* 2012;490(7420):355-60.
[60] Chen H, Shi B, Feng X, Kong W, Chen W, Geng L. Leptin and neutrophil-activating Peptide 2 promote mesenchymal stem cell senescence through activation of the phosphatidylinositol 3-kinase/Akt pathway in patients with systemic lupus erythematosus. *Arthritis Rheumatol.* 2015;67(9):2383-93.

[61] Doles J, Storer M, Cozzuto L, Roma G, Keyes WM. Age-associated inflammation inhibits epidermal stem cell function. *Genes Dev.* 2012;26(19):2144-53.

[62] Benitah SA, Welz PS. Circadian regulation of adult stem cell homeostasis and aging. *Cell Stem Cell.* 2020;26(6):817-31.

[63] Kim YJ, Hwang SH, Lee SY, Shin KK, Cho HH, Bae YC. miR-486-5p induces replicative senescence of human adipose tissue-derived mesenchymal stem cells and its expression is controlled by high glucose. *Stem Cells Dev.* 2012;21(10):1749-60.

[64] Lei LT, Chen JB, Zhao YL, Yang SP, He L. Resveratrol attenuates senescence of adipose-derived mesenchymal stem cells and restores their paracrine effects on promoting insulin secretion of INS-1 cells through Pim-1. *Eur Rev Med Pharmacol Sci.* 2016;20(6):1203-13.

[65] Maury JJ, Chan KK, Zheng L, Bardor M, Choo AB. Excess of O-linked N-acetylglucosamine modifies human pluripotent stem cell differentiation. *Stem Cell Res.* 2013;11(2):926-37.

[66] Arum O, Bonkowski MS, Rocha JS, Bartke A. The growth hormone receptor gene-disrupted mouse fails to respond to an intermittent fasting diet. *Aging Cell.* 2009;8(6):756-60.

[67] Bonkowski MS, Rocha JS, Masternak MM, Al Regaiey KA, Bartke A. Targeted disruption of growth hormone receptor interferes with the beneficial actions of calorie restriction. *Proc Natl Acad Sci U S A.* 2006;103(20):7901-5.

[68] Biteau B, Karpac J, Supoyo S, Degennaro M, Lehmann R, Jasper H. Lifespan extension by preserving proliferative homeostasis in Drosophila. *PLOS Genet.* 2010;6(10):e1001159.

[69] Cheng CW, Adams GB, Perin L, Wei M, Zhou X, Lam BS, Da Sacco S, Mirisola M, Quinn DI, Dorff TB, Kopchick JJ, Longo VD. Prolonged fasting reduces IGF-1/PKA to promote hematopoietic-stem-cell-based regeneration and reverse immunosuppression. *Cell Stem Cell.* 2014; 14(6):810-23.

[70] Blackmore DG, Golmohammadi MG, Large B, Waters MJ, Rietze RL. Exercise increases neural stem cell number in a growth hormone-dependent manner, augmenting the regenerative response in aged mice. *Stem Cells.* 2009;27(8):2044-52.

[71] Hsu HJ, Drummond-Barbosa D. Insulin levels control female germline stem cell maintenance via the niche in Drosophila. *Proc Natl Acad Sci U S A.* 2009;106(4):1117-21.

[72] How CM, Yen PL, Wei CC, Li SW, Liao VH. Early life exposure to di(2-ethylhexyl)phthalate causes age-related declines associated with insulin/IGF-1-like signaling pathway and SKN-1 in Caenorhabditis elegans. *Environ Pollut.* 2019;251:871-8.

[73] Carlson ME, Conboy MJ, Hsu M, Barchas L, Jeong J, Agrawal A, Mikels AJ, Agrawal S, Schaffer DV, Conboy IM. Relative roles of TGF-beta1 and Wnt in the systemic regulation and aging of satellite cell responses. *Aging Cell.* 2009;8(6):676-89.

[74] Pineda JR, Daynac M, Chicheportiche A, Cebrian-Silla A, Sii Felice K, Garcia-Verdugo JM, Boussin FD, Mouthon MA. Vascular-derived TGF-β increases in the stem cell niche and perturbs neurogenesis during aging and following irradiation in the adult mouse brain. *EMBO Mol Med.* 2013;5(4):548-62.

[75] Katsimpardi L, Litterman NK, Schein PA, Miller CM, Loffredo FS, Wojtkiewicz GR, Chen JW, Lee RT, Wagers AJ, Rubin LL. Vascular and neurogenic rejuvenation of the aging mouse brain by young systemic factors. *Science.* 2014;344(6184):630-4.

[76] Loffredo FS, Steinhauser ML, Jay SM, Gannon J, Pancoast JR, Yalamanchi P, Sinha M, Dall'Osso C, Khong D, Shadrach JL, Miller CM, Singer BS, Stewart A, Psychogios N, Gerszten RE, Hartigan AJ, Kim MJ, Serwold T, Wagers AJ, Lee RT. Growth differentiation factor 11 is a circulating factor that reverses age-related cardiac hypertrophy. *Cell.* 2013;153(4):828-39.

[77] Coppé JP, Kauser K, Campisi J, Beauséjour CM. Secretion of vascular endothelial growth factor by primary human fibroblasts at senescence. *J Biol Chem.* 2006;281(40):29568-74.

[78] Kuilman T, Peeper DS. Senescence-messaging secretome: SMS-ing cellular stress. *Nat Rev Cancer.* 2009;9(2):81-94.

[79] Cosgrove BD, Gilbert PM, Porpiglia E, Mourkioti F, Lee SP, Corbel SY, Llewellyn ME, Delp SL, Blau HM. Rejuvenation of the muscle stem cell population restores strength to injured aged muscles. *Nat Med.* 2014;20(3):255-64.

[80] Janzen V, Forkert R, Fleming HE, Saito Y, Waring MT, Dombkowski DM, Cheng T, DePinho RA, Sharpless NE, Scadden DT. Stem-cell ageing modified by the cyclin-dependent kinase inhibitor p16INK4a. *Nature.* 2006;443(7110):421-6.

[81] Molofsky AV, Slutsky SG, Joseph NM, He S, Pardal R, Krishnamurthy J, Sharpless NE, Morrison SJ. Increasing p16INK4a expression decreases forebrain progenitors and neurogenesis during ageing. *Nature.* 2006;443(7110):448-52.

[82] Nishino J, Kim I, Chada K, Morrison SJ. Hmga2 promotes neural stem cell self-renewal in young but not old mice by reducing p16Ink4a and p19Arf Expression. *Cell.* 2008;135(2):227-39.

[83] Signer RA, Montecino-Rodriguez E, Witte ON, Dorshkind K. Aging and cancer resistance in lymphoid progenitors are linked processes conferred by p16Ink4a and Arf. *Genes Dev.* 2008; 22(22):3115-20

[84] García-Prat L, Martínez-Vicente M, Perdiguero E, Ortet L, Rodríguez-Ubreva J, Rebollo E, Ruiz-Bonilla V, Gutarra S, Ballestar E, Serrano AL, Sandri M, Muñoz-Cánoves P. Autophagy maintains stemness by preventing senescence. *Nature.* 2016 Jan 7;529(7584):37-42.

[85] Bufalino MR, DeVeale B, van der Kooy D. The asymmetric segregation of damaged proteins is stem cell-type dependent. *J Cell Biol.* 2013;201(4):523-30.

[86] Katajisto P, Döhla J, Chaffer CL, Pentinmikko N, Marjanovic N, Iqbal S, Zoncu R, Chen W, Weinberg RA, Sabatini DM. Stem cells. Asymmetric apportioning of aged mitochondria between daughter cells is required for stemness. *Science.* 2015;348(6232):340-3.

[87] Rujano MA, Bosveld F, Salomons FA, Dijk F, van Waarde MA, van der Want JJ. Polarised asymmetric inheritance of accumulated protein damage in higher eukaryotes. *PLOS Biol.* 2006;4(12):e417.

[88] Köhler A, Schmithorst V, Filippi MD, Ryan MA, Daria D, Gunzer M, Geiger H. Altered cellular dynamics and endosteal location of aged early hematopoietic progenitor cells revealed by time-lapse intravital imaging in long bones. *Blood.* 2009;114(2):290-8.

[89] Florian MC, Nattamai KJ, Dörr K, Marka G, Uberle B, Vas V, Eckl C, Andrä I, Schiemann M, Oostendorp RA, Scharffetter-Kochanek K, Kestler HA, Zheng Y, Geiger H. A canonical to non-canonical Wnt signalling switch in haematopoietic stem-cell ageing. *Nature.* 2013;503(7476):392-6.

[90] Salemi S, Yousefi S, Constantinescu MA, Fey MF, Simon HU. Autophagy is required for self-renewal and differentiation of adult human stem cells. *Cell Res.* 2012;22(2):432-5.

[91] Revuelta M, Matheu A. Autophagy in stem cell aging. *Aging Cell.* 2017;16(5):912-5.

[92] Takasugi M. Emerging roles of extracellular vesicles in cellular senescence and aging. *Aging Cell.* 2018;17(2).

[93] Vader P, Breakefield XO, Wood MJA. Extracellular vesicles: emerging targets for cancer therapy. *Trends Mol Med.* 2014;20(7):385-93.

[94] Lonati C, Bassani GA, Brambilla D, Leonardi P, Carlin A, Maggioni M, Zanella A, Dondossola D, Fonsato V, Grange C, Camussi G, Gatti S. Mesenchymal stem cell-derived extracellular vesicles improve the molecular phenotype of isolated rat lungs during ischemia/reperfusion injury. *J Heart Lung Transplant.* 2019;38(12):1306-1316.

[95] Harting MT, Srivastava AK, Zhaorigetu S, Bair H, Prabhakara KS, Toledano Furman NE. Inflammation-stimulated mesenchymal stromal cell-derived extracellular vesicles attenuate inflammation. *Stem Cells.* 2018;36(1):79-90.

[96] McBride JD, Rodriguez-Menocal L, Guzman W, Candanedo A, Garcia-Contreras M, Badiavas EV. Bone marrow mesenchymal stem cell-derived CD63+ exosomes transport Wnt3a exteriorly and enhance dermal fibroblast proliferation, migration, and angiogenesis in vitro. *Stem Cells Dev.* 2017;26(19):1384-98.

[97] Wang L, Wei J, Da Fonseca Ferreira A, Wang H, Zhang L, Zhang Q. Rejuvenation of senescent endothelial progenitor cells by extracellular vesicles derived from mesenchymal stromal cells. *JACC Basic Transl Sci.* 2020;5(11):1127-41.

[98] Lagasse E, Connors H, Al-Dhalimy M, Reitsma M, Dohse M, Osborne L, Wang X, Finegold M, Weissman IL, Grompe M. Purified hematopoietic stem cells can differentiate into hepatocytes in vivo. *Nat Med.* 2000;6(11):1229-34.

[99] Orlic D, Kajstura J, Chimenti S, Jakoniuk I, Anderson SM, Li B, Pickel J, McKay R, Nadal-Ginard B, Bodine DM, Leri A, Anversa P. Bone marrow cells regenerate infarcted myocardium. *Nature.* 2001; 410(6829):701-5.

[100] Jackson KA, Majka SM, Wang H, Pocius J, Hartley CJ, Majesky MW, Entman ML, Michael LH, Hirschi KK, Goodell MA. Regeneration of ischemic cardiac muscle and vascular endothelium by adult stem cells. *J Clin Invest.* 2001;107(11):1395-402.

[101] Kollman C, Spellman SR, Zhang MJ, Hassebroek A, Anasetti C, Antin JH, Champlin RE, Confer DL, DiPersio JF, Fernandez-Viña M, Hartzman RJ, Horowitz MM, Hurley CK, Karanes C, Maiers M, Mueller CR, Perales MA, Setterholm M, Woolfrey AE, Yu N, Eapen M. The effect of donor characteristics on survival after unrelated donor transplantation for hematologic malignancy. *Blood.* 2016;127(2):260-7.
[102] Canaani J, Savani BN, Labopin M, Huang X, Ciceri F, Arcese W. Donor age determines outcome in acute leukemia patients over 40 undergoing haploidentical hematopoietic cell transplantation. *Am J Hematol.* 2018;93(2):246-53.
[103] Galamidi-Cohen E, Joyce A, Simantov R. Impact of donor age on outcomes in hematopoietic stem cell transplantation. *Stem Cells Transl Med.* 2020;9(S1):S7-.
[104] Atkinson K, Farrell C, Chapman G, Downs K, Penny R, Biggs J. Female marrow donors increase the risk of acute graft-versus-host disease: effect of donor age and parity and analysis of cell subpopulations in the donor marrow inoculum. *Br J Haematol.* 1986;63(2):231-9.
[105] Kollman C, Howe CW, Anasetti C, Antin JH, Davies SM, Filipovich AH, Hegland J, Kamani N, Kernan NA, King R, Ratanatharathorn V, Weisdorf D, Confer DL. Donor characteristics as risk factors in recipients after transplantation of bone marrow from unrelated donors: the effect of donor age. *Blood.* 2001;98(7):2043-51.
[106] Mehta J, Singhal S, Gee AP, Chiang KY, Godder K, Rhee FvFv, DeRienzo S, O'Neal W, Lamb L, Henslee-Downey PJ. Bone marrow transplantation from partially HLA-mismatched family donors for acute leukemia: single-center experience of 201 patients. *Bone Marrow Transplant.* 2004;33(4):389-96.
[107] Mehta J, Gordon LI, Tallman MS, Winter JN, Evens AM, Frankfurt O, Williams SF, Grinblatt D, Kaminer L, Meagher R, Singhal S. Does younger donor age affect the outcome of reduced-intensity allogeneic hematopoietic stem cell transplantation for hematologic malignancies beneficially? *Bone Marrow Transplant.* 2006;38(2):95-100.
[108] Fan M, Chen W, Liu W, Du GQ, Jiang SL, Tian WC. The effect of age on the efficacy of human mesenchymal stem cell transplantation after a myocardial infarction. *Rejuvenation Res.* 2010;13(4):429-38.
[109] Harrison DE, Astle CM. Loss of stem cell repopulating ability upon transplantation. Effects of donor age, cell number, and transplantation procedure. *J Exp Med.* 1982;156(6):1767-79.
[110] Zaim M, Karaman S, Cetin G, Isik S. Donor age and long-term culture affect differentiation and proliferation of human bone marrow mesenchymal stem cells. *Ann Hematol.* 2012;91(8):1175-86.
[111] Chang CJ, Mitra K, Koya M, Velho M, Desprat R, Lenz J. Production of embryonic and fetal-like red blood cells from human induced pluripotent stem cells. *PLOS ONE.* 2011;6(10):e25761.
[112] Wen Y, Wani P, Zhou L, Baer T, Phadnis SM, Reijo Pera RA. Reprogramming of fibroblasts from older women with pelvic floor disorders alters cellular behavior associated with donor age. *Stem Cells Transl Med.* 2013;2(2):118-28.
[113] Takahashi K, Yamanaka S. Induction of pluripotent stem cells from mouse embryonic and adult fibroblast cultures by defined factors. *Cell.* 2006;126(4):663-76.
[114] Takahashi K, Tanabe K, Ohnuki M, Narita M, Ichisaka T, Tomoda K, Yamanaka S. Induction of pluripotent stem cells from adult human fibroblasts by defined factors. *Cell.* 2007;131(5):861-72.
[115] Wahlestedt M, Erlandsson E, Kristiansen T, Lu R, Brakebusch C, Weissman IL. Clonal reversal of ageing-associated stem cell lineage bias via a pluripotent intermediate. *Nat Commun.* 2017;8(1):14533.
[116] Wahlestedt M, Norddahl GL, Sten G, Ugale A, Frisk MA, Mattsson R, Deierborg T, Sigvardsson M, Bryder D. An epigenetic component of hematopoietic stem cell aging amenable to reprogramming into a young state. *Blood.* 2013;121(21):4257-64.
[117] Spitzhorn LS, Megges M, Wruck W, Rahman MS, Otte J, Degistirici Ö, Meisel R, Sorg RV, Oreffo ROC, Adjaye J. Human iPSC-derived MSCs (iMSCs) from aged individuals acquire a rejuvenation signature. *Stem Cell Res Ther.* 2019;10(1):100.
[118] Ocampo A, Reddy P, Martinez-Redondo P, Platero-Luengo A, Hatanaka F, Hishida T, Li M, Lam D, Kurita M, Beyret E, Araoka T, Vazquez-Ferrer E, Donoso D, Roman JL, Xu J, Rodriguez Esteban C, Nuñez G, NuñezDelicado E, Campistol JM, Guillen I, Guillen P, Izpisua Belmonte JC. In Vivo

Amelioration of Age-Associated Hallmarks by Partial Reprogramming. *Cell.* 2016;167(7):1719-1733.e12.

[119] Cerletti M, Jang YC, Finley LW, Haigis MC, Wagers AJ. Short-term calorie restriction enhances skeletal muscle stem cell function. *Cell Stem Cell.* 2012;10(5):515-9.

[120] Lo T, Ho JH, Yang MH, Lee OK. Glucose reduction prevents replicative senescence and increases mitochondrial respiration in human mesenchymal stem cells. *Cell Transplant.* 2011;20(6):813-25.

[121] Kim HJ, Ji BR, Kim JS, Lee HN, Ha DH, Kim CW. Proteomic analysis of proteins associated with cellular senescence by calorie restriction in mesenchymal stem cells. *In Vitro Cell Dev Biol Anim.* 2012;48(3):186-95.

[122] Sahu A, Mamiya H, Shinde SN, Cheikhi A, Winter LL, Vo NV. Age-related declines in α-klotho drive progenitor cell mitochondrial dysfunction and impaired muscle regeneration. *Nat Commun.* 2018;9(1):4859.

[123] Liang M, Liu W, Peng Z, Lv S, Guan Y, An G. The therapeutic effect of secretome from human umbilical cord-derived mesenchymal stem cells in age-related osteoporosis. *Artif Cells Nanomed Biotechnol.* 2019;47(1):1357-66.

[124] Li J, Hansen KC, Zhang Y, Dong C, Dinu CZ, Dzieciatkowska M, Pei M. Rejuvenation of chondrogenic potential in a young stem cell microenvironment. *Biomaterials.* 2014;35(2):642-53.

[125] Liu N, Matsumura H, Kato T, Ichinose S, Takada A, Namiki T, Asakawa K, Morinaga H, Mohri Y, De Arcangelis A, Geroges-Labouesse E, Nanba D, Nishimura EK. Stem cell competition orchestrates skin homeostasis and ageing. *Nature.* 2019;568(7752):344-350.

[126] Braidy N, Liu Y. NAD+ therapy in age-related degenerative disorders: A benefit/risk analysis. *Exp Gerontol.* 2020;132:110831.

[127] Sousa-Victor P, Perdiguero E, Muñoz-Cánoves P. Geroconversion of aged muscle stem cells under regenerative pressure. *Cell Cycle.* 2014;13(20):3183-90.

[128] Neumann B, Baror R, Zhao C, Segel M, Dietmann S, Rawji KS, Foerster S, McClain CR, Chalut K, van Wijngaarden P, Franklin RJM. Metformin Restores CNS Remyelination Capacity by Rejuvenating Aged Stem *Cells. Cell Stem Cell.* 2019;25(4):473-485.e8.

[129] Pirmoradi S, Fathi E, Farahzadi R, Pilehvar-Soltanahmadi Y, Zarghami N. Curcumin affects adipose tissue-derived mesenchymal stem cell aging through tert gene expression. *Drug Res* (Stuttg). 2018;68(4):213-21.

[130] Xu M, Pirtskhalava T, Farr JN, Weigand BM, Palmer AK, Weivoda MM, Inman CL, Ogrodnik MB, Hachfeld CM, Fraser DG, Onken JL, Johnson KO, Verzosa GC, Langhi LGP, Weigl M, Giorgadze N, LeBrasseur NK, Miller JD, Jurk D, Singh RJ, Allison DB, Ejima K, Hubbard GB, Ikeno Y, Cubro H, Garovic VD, Hou X, WerohaSJ, Robbins PD, Niedernhofer LJ, Khosla S, Tchkonia T, Kirkland JL. Senolytics improve physical function and increase lifespan in old age. *Nat Med.* 2018;24(8):1246-1256.

[131] Kureel SK, Mogha P, Khadpekar A, Kumar V, Joshi R, Das S. Soft substrate maintains proliferative and adipogenic differentiation potential of human mesenchymal stem cells on long-term expansion by delaying senescence. *Biol Open.* 2019;8(4).

[132] Abbasi A, Imaichi S, Ling V, Shukla A. Mesenchymal stem cell behavior on soft hydrogels with aligned surface topographies. *ACS Appl Bio Mater.* 2022;5(5):1890-900.

[133] Xu X. Fine-tuning of rat mesenchymal stem cell senescence via microtopography of polymeric substrates. *MRS Adv.* 2020;5(12):643-53.

[134] Das A, Datta P, Chowdhury AR, Barui A. Honey-incorporated nanofibre reduces replicative senescence of umbilical cord-derived mesenchymal stem cells. *IET Nanobiotechnology.* 2020;14(9):870-80.

[135] Deng Z, Wang W, Xu X, Nie Y, Liu Y, Gould OEC. Biofunction of polydopamine coating in stem cell culture. *ACS Appl Mater Interfaces.* 2021;13(9):10748-59.

[136] Ogle ME, Doron G, Levy MJ, Temenoff JS. Hydrogel culture surface stiffness modulates mesenchymal stromal cell secretome and alters senescence. *Tissue Eng Part A.* 2020;26(23-24):1259-71.

[137] Heo SJ, Szczesny SE, Kim DH, Saleh KS, Mauck RL. Expansion of mesenchymal stem cells on electrospun scaffolds maintains stemness, mechano-responsivity, and differentiation potential. *J Orthop Res.* 2018;36(2):808-15.

Chapter 22

# Superoxide ($O_2^-$): A Biological Prodigy for Developing and Curing Disease

**Asis Bala[1,*]**
**Motlalepula Gilbert Matsabisa[2]**
**Pallab Kanti Haldar[3]**
**Subir Biswas[4]**
**and Ashis Kumar Mukherjee[1]**

[1]Division of Life Sciences, Institute of Advanced Study in Science and Technology (IASST), An Autonomous Institute under the Department of Science & Technology (Govt. of India) Vigyan Path, Guwahati, Assam, India

[2]Department of Pharmacology, Faculty of Health Sciences, School of Clinical Medicines, University of Free State, Bloemfontein, South Africa

[3]Division of Pharmacology, Department of Pharmaceutical Technology, Jadavpur University, Kolkata, West Bengal, India

[4]Division of Physical Sciences, Institute of Advanced Study in Science and Technology (IASST), An Autonomous Institute under the Department of Science & Technology (Govt. of India) Vigyan Path, Guwahati, Assam, India

## Abstract

The highly reactive superoxide (HRS; $O_2\bullet^-$) radical known as dioxidanidyl hyperoxid dioxide (1-) is of great significance as an immediate reactive oxygen radical species generated in cells that implicated in diverse biochemical reactions. However, the twisting knowledge of $O_2\bullet^-$ for developing and curing disease rather not well understood. This chapter aimed to enumerate the role of $O_2\bullet^-$ as a biological prodigy based on current development in knowledge insights into mechanisms and products of $O_2\bullet^-$ reactions. Excessive production and/or insufficient scavenging of HRS have been implicated in the pathogenesis of many diseases. In contrary, reactive oxygen species (ROS) including HRS exert its multi modality effects on acute and chronic inflammation and play a pivotal role in controlling immune responses and curing diseases. Hence, comprehending the significance of this prodigy in human health and disease can aid biomedical scientists in their pursuit of devising and directing drugs in the foreseeable future.

[*] Corresponding Author's Email: asisbala_ju@yahoo.co.in/ asisbala@iasst.gov.in.

In: The Role of Reactive Oxygen Species in Health and Disease
Editor: Kuladip Jana
ISBN: 979-8-89113-283-2

**Keywords:** highly reactive superoxide radical; reactive oxygen species a biological prodigy, development of disease, cure of disease, mechanism of action, inflammation, and immunity

## 1. Introduction

A Superoxide ($O_2^-$) like free radical is defined as a molecular species capable of its independent existence that always has an unpaired electron in an atomic orbital [1]. Free radicals generally include reactive oxygen and nitrogen species including OH•, superoxide ($O_2$•–), alkoxy radical (RO•), peroxyl (ROO•), and lipid peroxyl (LOO•), nitric oxide (NO•) and nitrogen dioxide ($NO_2$•) [2,3]. The by-product free radical like singlet oxygen ($^1O_2$), ozone ($O_3$), $H_2O_2$, hypochlorous acid (HOCl), hypobromous acid (HOBr), lipid peroxide (LOOH), nitrous acid ($HNO_2$), dinitrogen trioxide ($N_2O_3$), and peroxynitrite (ONOOH) are considered as toxic non-radicals. These are also collectively referred to as oxidants and can readily comprise free radical reactions in biological systems and show several health implications [2-4]. In contrary, both free radical and free radical mediated non radicals are also considered the main culprit for oxidative damage to cells specially protein, lipid and also DNA [5]. The highly reactive superoxide (HRS; $O_2$•$^-$) radical is considered as the first generated free radical also known as dioxidanidyl hyperoxid dioxide (1-) that implicated in diverse biochemical reactions in cells [5-6].

These oxygen and nitrogen species are produced throughout normal cellular metabolism and or due to endogenous and exogenous stimulus or stress sources [2-3]. When cells utilize oxygen to produce energy, these species are released by mitochondrial adenosine triphosphate (ATP) synthesis. As by-products of the cellular redox process, reactive nitrogen species (RNS) as well as reactive oxygen species (ROS) will be released. The vital aspect of life is maintaining a subtle equilibrium between these two species' antagonistic effects. An adequate quantity of RNS as well as ROS could exert favourable actions on different cellular responses, signalling pathways, mitogenic responses, redox regulation, and immune function [7-8]. Both ROS and RNS exert oxidative stress or insult as well as nitrosative stress and producing possible deleterious effects to the biomolecules, including lipids, proteins, and nucleic acids [9-10]. Overload of ROS/RNS, as well as a deficit of antioxidants (enzymatic as well as non-enzymatic), can produce oxidative and nitrosative stress.

The role of ROS or RNS in cellular homeostasis is well evidenced by the fact that [9-11]-

I. Aberrantly low levels of ROS or RNS induce the cell cycle arrest.
II. Normal or cellular basal ROS or RNS levels maintain the normal cell functioning and homeostasis.
III. Elevated amount of ROS/RNS induces number of transcription factors to drive a cell-mediated defence response.
IV. Excessive ROS induction activates pro-apoptotic proteins for subsequent induction of cell death via apoptosis and in extreme cases lead cellular necrosis [9-11].

These $O_2^-$ and $O_2^-$ mediated ROS or RNS play a pivotal role in normal physiological processes, such as controlling the redox regulation of protein phosphorylation, ion channels, and various transcription factors. These are also required for biosynthetic processes, including thyroid hormone production and crosslinking of extracellular matrix [12]. Interestingly, recent study revealed that the beneficial or detrimental role of $O_2^-$ and $O_2^-$ mediated ROS or RNS is determined by its concentration [13-14]. The production and/or insufficient scavenging of HRS have been implicated in the pathogenesis of many diseases like cancer autoimmune disease and lifestyle diseases [15].

In contrary, reactive oxygen species (ROS) including HRS exert its multimodality effects on acute and chronic inflammation and play a pivotal role in controlling immune responses and curing diseases [16]. Orientation of disproportionality in the basal homeostasis by enlarged accumulation of free radicals i.e., ROS and RNS can support the treatment of disease like cancer [10, 17]. Hence, fine-tuning the ROS level in the cells by novel therapeutics are considered as modern approaches for treatments. This chapter aimed to enumerate the role of $O_2^{\bullet-}$ as a biological prodigy based on current development in knowledge insights into mechanisms and products of $O_2^{\bullet-}$ reactions. Number of schematic illustration and tables on this aspect to understand the role of this prodigy in human health and disease hopefully will facilitate the biomedical scientists for developing and targeting drugs in future.

**Table 1.** Classification of reactive oxygen and nitrogen species and endogenous antioxidant present inside cells to neutralize the reactive oxygen and nitrogen species i.e., both free radical and non-radical species

| **Reactive Oxygen Species** | | |
|---|---|---|
| **Free Radicals** | **Symbol (Formula)** | **Antioxidant Defence Mechanism** |
| Hydroxyl | $OH^{\cdot}$ | Reduce Glutathione (GSH) |
| Superoxide | $O_2^{\cdot}-$ | Reduce Glutathione (GSH) |
| Alkoxyl radical | $RO^{\cdot}$ | Reduce Glutathione (GSH) |
| Peroxyl radical | $ROO^{\cdot}$ | Reduce Glutathione (GSH) |
| Lipid peroxyl | $LOO^{\cdot}$ | Reduce Glutathione (GSH) |
| **Non-radicals** | **Symbol (Formula)** | **Antioxidant Defence Mechanism** |
| Ozone | $O_3$ | Reduce Glutathione (GSH) |
| Singlet oxygen | $^1O_2$ | Reduce Glutathione (GSH) |
| Hydrogen peroxide | $H_2O_2$ | Catalase, Reduce Glutathione (GSH) |
| Organic / Lipid peroxide | ROOH / LOOH | Reduce Glutathione (GSH) |
| Hypochlorous acid | HOCl | Reduce Glutathione (GSH) |
| Hypobromous acid | HOBr | Reduce Glutathione (GSH) |
| Aldehydes | HCOR | Reduce Glutathione (GSH) |

| Reactive Nitrogen Species | | |
|---|---|---|
| **Free Radicals** | **Symbol (Formula)** | **Antioxidant Defence Mechanism** |
| Nitric oxide | $NO^{\cdot}$ | Reduce Glutathione (GSH) |
| Nitrogen dioxide | $NO_2^{\cdot}$ | Reduce Glutathione (GSH) |
| **Non-radicals** | **Symbol (Formula)** | **Antioxidant Defence Mechanism** |
| Nitrosyl cation | $NO^+$ | Reduce Glutathione (GSH) |
| Nitroxyl anion | NO- | Reduce Glutathione (GSH) |
| Dinitrogen trioxide | $N_2O_3$ | Reduce Glutathione (GSH) |
| Dinitrogen tetraoxide | $N_2O_4$ | Reduce Glutathione (GSH) |
| Nitrous acid | $HNO_2$ | Reduce Glutathione (GSH) |
| Peroxynitrite | ONOO- | Reduce Glutathione (GSH) |
| Peroxynitrous acid | ONOOH | Reduce Glutathione (GSH) |
| Nitronium ion | $NO_2^+$ | Reduce Glutathione (GSH) |
| Nitryl chloride | $NO_2Cl$ | Reduce Glutathione (GSH) |
| Nitrosyl cation | $NO^+$ | Reduce Glutathione (GSH) |

## 2. Type of Free Radicals

Generally, the free radicals are derived from oxygen, known as ROS and free radicals are generated when a chemical bond is broken, and each fragment retains an electron, when one radical is cleaved to yield a new radical, or when a redox reaction occurs. ROS and RNS are commonly classified into two groups of compounds namely, radicals and non-radicals as tabulated in table 1 [18]. Radical species contain one unpaired electron in the shells around the atomic nucleus and are always capable of autonomous existence. Free radicals include ROS such as OH•, superoxide ($O_2$•–), alkoxy radical (RO•), peroxyl (ROO•), and lipid peroxyl (LOO•) [19-20]. Other hand, RNS are as nitric oxide (NO•) and nitrogen dioxide ($NO_2$•). Whereas ROS such as singlet oxygen ($^1O_2$), ozone ($O_3$), $H_2O_2$, hypochlorous acid (HOCl), hypobromous acid (HOBr), lipid peroxide (LOOH), and RNS such as nitrous acid ($HNO_2$), dinitrogen trioxide ($N_2O_3$), and peroxynitrite (ONOOH) are non- radicals [19-21] (Table 1).

### 2.1. Sources of Superoxide

#### *2.1.1. NADPH Oxidases (NOXs) Multiprotein Complex*

Cellular NADPH oxidases (known as NOXs) are membrane bound multiprotein complexes that catalyze the transfer of electrons from NADPH to oxygen to produce superoxide [22]. The NOXs complex are consisting of membrane- bound heterodimer consisting of flavin adenine dinucleotide (FAD; flavocytochrome b558 and p22) and during any stimulation by stress condition or inflammatory condition the cytoplasmic components in turn recruited to the membrane [22-23]. Further the membrane bound NADPH becomes oxidized and transfers individual electrons (NAD+ $H^+$) to produce superoxide upon acceptance by oxygen on the external surface (Figure 1).

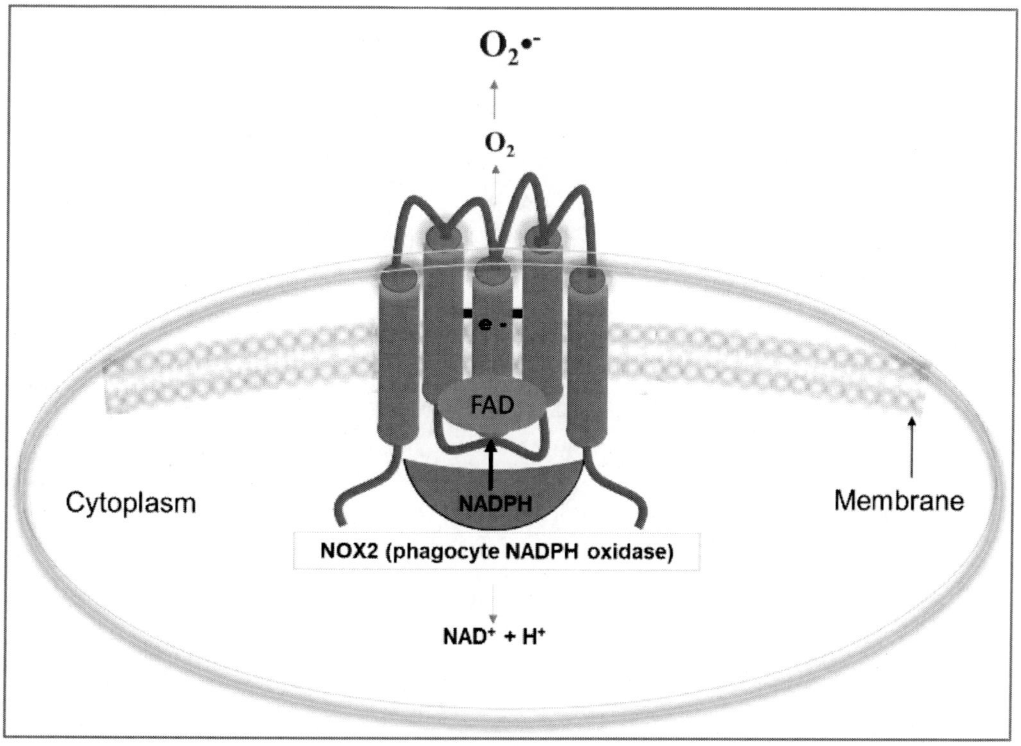


**Figure 1.** NADPH oxidases (NOXs) Multiprotein Complex: NOXs complex are consisting of membrane-bound heterodimer consisting of flavin adenine dinucleotide (FAD). Any stimulation by exogenous or endogenous stress or inflammatory condition the cytoplasmic components engage the membrane NADPH for oxidation and transfer the electrons to oxygen for producing the superoxide.

### *2.1.2. Mitochondrial Electron Transport Chain*

The mitochondrial electron transport chain (METC)-derived superoxide anion radical has recently received much attention for progression of disease and understanding the disease mechanism. [24]. Generation of superoxide mainly takes place at the electron transport chain systems that are located on the inner mitochondrial membrane during the oxidative phosphorylation [25] (Figure 2).

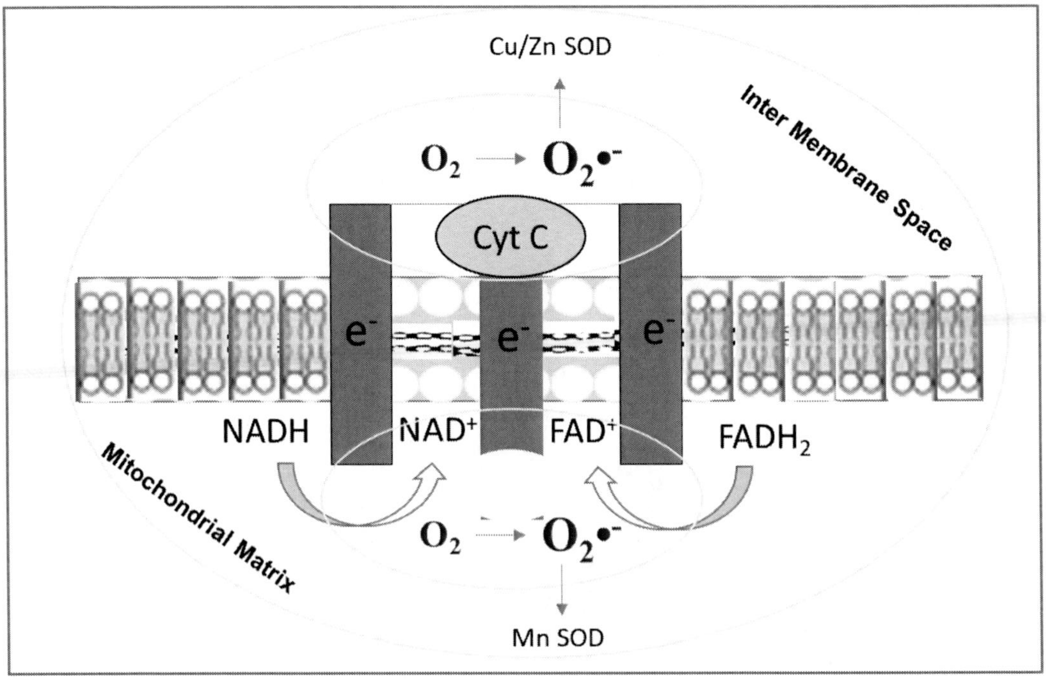


**Figure 2**. The mitochondrial electron transport chain (METC)-derived superoxide anion radical formation: Superoxide generation occur at the electron transport chain systems located on the inner mitochondrial membrane during the oxidative phosphorylation. Superoxide radicals are generated within the cellular matrix through electron-transferring flavoprotein-ubiquinone oxidoreductase and flavin adenine dinucleotide (FADH2) and Nicotinamide adenine dinucleotide (NADH). Superoxide radicals within the matrix immediately reacts with Mn-SOD (Superoxide Dismutase), and the superoxide radicals that released into the intermembrane space reacts with Cu/Zn-SOD (Superoxide Dismutase) for the formation of hydrogen peroxide ($H_2O_2$).

The superoxide radicals are generated within the cellular matrix through electron-transferring flavoprotein-ubiquinone oxidoreductase and flavin adenine dinucleotide (FADH2) and Nicotinamide adenine dinucleotide (NADH) [26]. Further the superoxide radicals that are generated within the matrix immediately reacts with Mn-SOD (Superoxide Dismutase), and the superoxide radicals that released into the intermembrane space reacts with Cu/Zn-SOD (Superoxide Dismutase) for the formation of hydrogen peroxide ($H_2O_2$). Although the overall rates of mitochondrial $H_2O_2$ production differed when mitochondria were oxidizing different substrates as fuel molecules (Table 2) [27].

**Table 2.** Major Products of $O_2\bullet^-$ reaction. The major reaction is tabularized with the rate limiting or functioning enzymes and its implications

| Reactions | Enzyme(s)/ Rate Limiting factors | Implications | Reference(s) |
|---|---|---|---|
| $O^2 + e^- \rightarrow O_2\bullet^-$ | NADPHOxidase (NADPHO), Xanthine Oxidase (XOD) | Formation of toxic reactive superoxide radicals | 38-39 |
| $O_2\bullet^- \rightarrow H_2O_2$ | Superoxide Dismutase (SOD) | Comparatively less toxic hydrogen peroxide ($H_2O_2$) formed | 19-20 |
| $H_2O_2 \rightarrow H_2O$ | Catalase (CAT); Glutathione peroxidases (GPXs); peroxiredoxins [Prxs] | Catalase convert hydrogen peroxide ($H_2O_2$) to neural water ($H_2O$) molecules | 19-21 |
| $H_2O_2 \rightarrow OH^- +$ OH• <br> $H_2O_2 \rightarrow$ HOCl | Myeloperoxidase (MPO), $Cl^-$, $Fe^{++}$ | The Fenton reaction-the formation of hydroxide (OH−) and hydroxyl radical by a reaction between Iron (II) (Fe2+) and hydrogen peroxide ($H_2O_2$) in presence of chlorine ion ($Cl^-$ )Myeloperoxidase uses hydrogen peroxide as substrate for the formation of highly toxic halogenated ROS such as hypochlorite (OCl−). | 40-41 |
| $O_2\bullet^-$ + NO (Nitric Oxide) → ONOO− | L-Arginine, Nitric Oxide Synthase (NOS) | Peroxynitrite (ONOO−): Superoxide ($O_2$-.) with reaction to nitric oxide (NO) form the highly toxic and reactive product peroxynitrite (ONOO-) | 42-43 |

### *2.1.3. Xanthine Oxidase*

Xanthine oxidase (XO)is a form of xanthine oxidoreductase was the first enzyme discovered that generate superoxide anion [28]. Xanthine oxidase is from the molybdenum family that contains flavin adenine dinucleotide, one molybdenum, and four iron-sulfur attachments [29]. XO enzyme originated in numerous species, including humans, and the main functions is the catabolism of purine nucleic acids [30]. As Xanthine oxidase catalyses the oxidation of xanthine and hypoxanthine to uric acid by molecular oxygen, which is further converted to superoxide radicals and hydrogen peroxide ($H_2O_2$) [28-31].

## 2.2. Major Products of $O_2\bullet^-$ Reaction

An unavoidable consequence of aerobic metabolism is production of reactive oxygen species (ROS) as discussed in table 1. ROS include free radicals such as superoxide anion (O2•−), hydroxyl radical (•OH), as well as non-radical molecules like hydrogen peroxide ($H_2O_2$), singlet oxygen ($^1O_2$), [32] and so forth as discussed in table 1. Major Products of O2•- reaction is tabularized with the rate limiting or functioning enzymes and its implications in Table 2.

Preliminary free radical reaction started with formation of toxic reactive superoxide radicals by the action of rate limiting enzymes like NADPH Oxidase (NADPHO), Xanthine Oxidase (XOD) [33]. Secondly, comparatively less toxic hydrogen peroxide ($H_2O_2$) is formed due to the action of Superoxide Dismutase (SOD) [34]. Importantly, Superoxide radicals within the matrix immediately reacts with Mn-SOD (Superoxide Dismutase), and the superoxide radicals that released into the intermembrane space reacts with Cu/Zn-SOD (Superoxide Dismutase) for the formation of hydrogen peroxide ($H_2O_2$) [35]. The multiple SOD proteins present in all aerobic organisms show slow diffusion capability and strong affinity to multiple sources of their substrate superoxide [34-35].

The antioxidant Catalase and Glutathione peroxidases (GPXs) enzyme convert hydrogen peroxide ($H_2O_2$) to neural water (H2O) molecules [36]. Although the neutralization of comparatively toxic $H_2O_2$ is highly delimited by abundant enzymatic systems (e.g., catalase, glutathione peroxidases, and peroxiredoxins [Prxs] but this might be for the purpose of controlling the cellular signalling capabilities. Catalase is a ubiquitous tetrameric heme-containing enzyme that show maximum dismutation of two molecules of $H_2O_2$ into water and oxygen [37].

The Fenton reaction is the formation of hydroxide (OH−) and hydroxyl radical by a reaction between Iron (II) (Fe2+) and hydrogen peroxide ($H_2O_2$) in presence of chlorine ion (Cl-) [44]. Hydroxyl radical (•OH) are considered as potent reactive oxygen species. It is able to degrade hazardous organic compounds, kill bacteria, and inactivate viruses [45]. Myeloperoxidase uses hydrogen peroxide as substrate for the formation of highly toxic halogenated ROS such as hypochlorite (HOCl) [46]. Reaction of hypochlorite with hydrogen peroxide will result in the formation of singlet oxygen (1O2) [45-47]. Importantly Superoxide (O2-.) with reaction to nitric oxide (NO) form the highly toxic and reactive product peroxynitrite (ONOO-) whereas the NO supplies by the enzymatic action of Nitric Oxide Synthase utilizing L-Arginine.

Therefore, under so many physiological states, ROS are produced as intermediates, and their cellular concentration are strongly triggered by various detoxifying enzymes, such as superoxide dismutase (SODs), glutathione peroxidase (GPX), and catalase (CAT), or by different antioxidants, including flavonoids, ascorbic acids, vitamin E, and glutathione (GSH) [38-43].

## 2.3. Role of $O_2^{\bullet -}$ on Developing Disease

The superoxide mediated ROS / RNS are formed as products under normal physiological conditions due to the partial or complete reduction of molecular oxygen [48]. ROS/ RNS arises in many ways inside or outside the cells, as a product of many photochemical and enzymatic reactions due to endogenous stress like hypoxia, biochemical imbalance or due to exogenous stress or injury like exposure to UV light, ionizing radiation, exposure of heavy metal ions and or inflammation or injury [49].

Superoxide is generated directly from the reduction of oxygen and then dismutase to hydrogen peroxide. Hydrogen peroxide is a molecule with low reactivity, but it can readily penetrate cell's membranes and generate the most reactive form of oxygen, the hydroxyl radical, via Fenton's reaction ($H_2O_2 + Fe^{2}+ \rightarrow Fe^{3}+ + OH^- + OH\bullet$) [50].

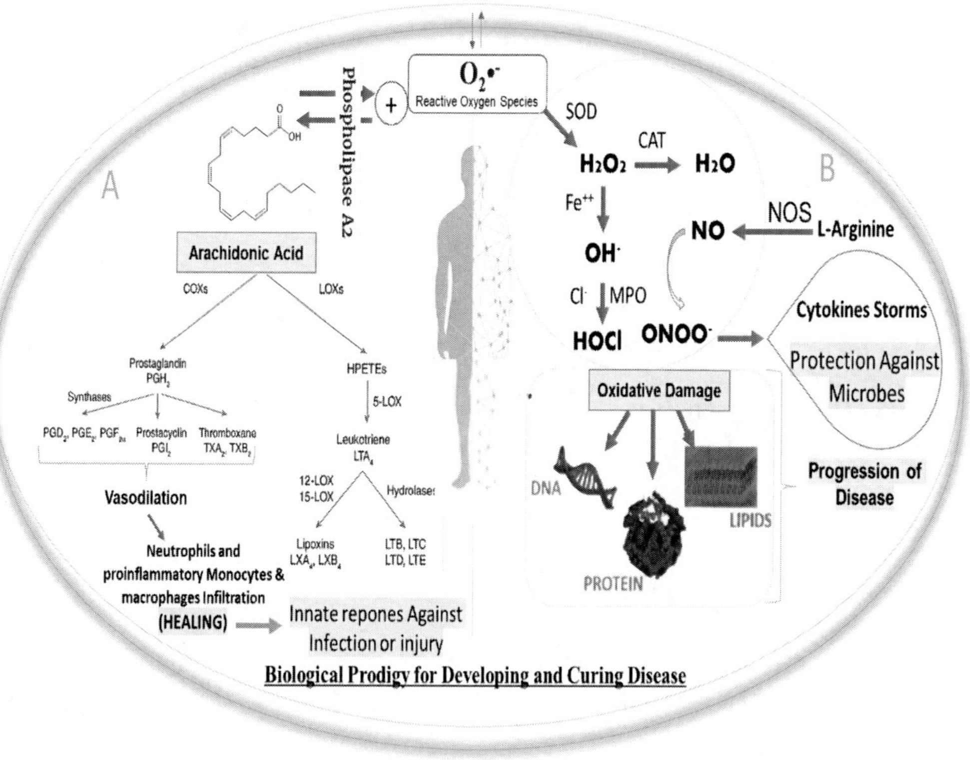


**Figure 3**. A. Role of $O_2^{\bullet-}$on Curing Disease:NADPH oxidase-derived ROS have emerged as key regulators of host immune responses and neutrophilic inflammation. Activation of phospholipase A2 in human neutrophils and other inflammatory cells by polyunsaturated fatty acids and its role in stimulation of superoxide production triggered the innate immune reactions. Increased production of superoxide may also activate the Arachidonic acid pathways. Prostaglandin E2 (PGE2), a lipid signalling molecule involved in immune modulator for regulating the pain and inflammation that might potentiate tissue regeneration and repair following injury in diverse organ systems B. Role of $o_2^{\bullet-}$ on developing disease: Oxygen undergoes a series of univalent reductions, leading sequentially to the production of O2•−, hydrogen peroxide (H2O2), and H2O. Nitric oxide (NO) always shows its anti-inflammatory and anticoagulant properties as well as vasodilator effect, can further be inactivated by reaction with O2•− and leading to the production of the strong oxidant peroxynitrite (ONOO−).It can participate directly in one- and two-electron oxidation reactions with biological macromolecules including protein, lipid and DNA.

Superoxide anion ($O_2^{\bullet-}$), play a pivotal role in the pathogenesis of many diseases, including cancer, cardiovascular complication, auto immune disease, atherosclerosis etc. [38-39]. In the process of normal cellular metabolism, oxygen undergoes a series of univalent reductions, leading sequentially to the production of $O_2^{\bullet-}$, hydrogen peroxide ($H_2O_2$), and $H_2O$ [40-41] as shown in figure 3. B. Nitric oxide (NO) always shows its anti-inflammatory and anticoagulant properties as well as vasodilator effect, can further be inactivated by reaction with $O_2^{\bullet-}$ and leading to the production of the strong oxidant peroxynitrite ($ONOO^-$)[40-42, 51]. The same reaction is important in common conditions leading to endothelial and mitochondrial dysfunction, including hypercholesterolemia, hypertension, diabetes, and aging, in which vascular production of $O_2^{\bullet-}$ is increased [51-52]. Peroxynitrite that are coming from the reaction of nitric oxide and superoxide anion are strong oxidants capable of damaging lipids, proteins and DNA [53]. The specific reactivity of peroxynitrite is extremely high and is highly pH-dependent and both peroxynitrite anion ($ONOO^-$) and peroxynitrous acid (ONOOH) can participate directly in one- and two-electron oxidation reactions with biological macromolecules including protein, lipid and DNA [54]. The proton-catalysed decomposition of ONOO− to form •OH and •NO2 radicals becomes relevant in hydrophobic phases and results in the initiation of protein damage and lipid peroxidation processes [55, 56].

Furthermore, reactive oxygen species like $ONOO^-$ have been reported to play a pivotal role in the activation of nuclear factor kappa beta (NF-κB), a transcription factor that regulates cytokine production, inflammation and innate immunity [57]. NF-κB signalling modification marks in a number of downstream cellular effects controlling immune and inflammatory responses [58]. NF-κB also plays a protagonist in cell existence by antagonizing apoptosis associated with tumor necrosis factor-α receptors mediated stimulation of Jun N-terminal kinase (JNK) [59]. The JNK pathway is one of the signaling cascades of the mitogen activated protein kinases which controls cell proliferation, cytokine production and apoptosis [58-59] (Figure 3).

### 2.4. Role of $O_2{\bullet}^-$ on Curing Disease

The majority of the ROS in neutrophils is produced by initiation of the superoxide-generating enzyme nicotinamide adenine dinucleotide phosphate (NADPH) oxidase [60]. Independent of their anti-microbial function, NADPH oxidase-derived ROS have emerged as key regulators of host immune responses and neutrophilic inflammation. Low levels of either intracellular or extracellular ROS (e.g., superoxide and H2O2) play major roles in many biochemical processes, including intracellular signaling, defense against microorganisms, and cell function [61]. Activation of phospholipase A2 in human neutrophils and other inflammatory cells by polyunsaturated fatty acids and its role in stimulation of superoxide production triggered the innate immune reactions [62]. In contrast, increased production of superoxide may also activate the Arachidonic acid pathways [62].

Tissue repair and regeneration after injury is essential in all living organisms. In recent years, some evidence has noted that prostaglandin E2 (PGE2), a lipid signalling molecule involved in immune modulator for regulating the pain and inflammation that might potentiate tissue regeneration and repair following injury in diverse organ systems [63]. PGE2 production as influenced by ROS/RNS being an inflammatory mediator and a fibroblast modulator is always considered as promising therapeutic candidate for improving tissue repair and regeneration [64-65].

## Conclusion

ROS play a crucial role either to develop or to cure many different diseases. It also helps in endothelial and mitochondrial function by inhibiting oxidative inactivation of bioavailable NO, and thus preventing $ONOO^-$ formation. This can be stated that, optimal levels of ROS are required but excess or lower amounts of ROS are inhibitory for biological responses. ROS or RNS always play a vital role in the biological process however, based on the concentration of the ROS or RNS level; they can be beneficial or detrimental to tumour cell growth. In recent development it is well established that ROS act as a double-edged sword not only for promoting tumorigenesis and proliferation but also triggering the cell death by enhancing intracellular oxidative stress.

The chapter was aimed to enumerate the role of $O_2$•⁻ as a biological prodigy based on current development in knowledge insights into mechanisms and products of $O_2$•⁻ reactions. This review provides a comprehensive analysis, supported by numerous figures and tables, of the crucial role that superoxide plays in human health and disease. It is hoped that this review will serve as a valuable resource for biomedical scientists in their efforts to develop and target drugs that leverage ROS machinery in the future.

## Acknowledgment

This work is supported by the Institute of Advanced Study in Science and Technology funded by the In-House Project (Ref. IASST/R&D/ICP/IHP-23/2023-24/1338-1347) to Dr. Asis Bala.

## References

[1] Lobo V, Patil A, Phatak A, Chandra N. Free radicals, antioxidants and functional foods: Impact on human health. *Pharmacogn. Rev.* 2010 Jul;4(8):118-26. doi: 10.4103/0973-7847.70902. PMID: 22228951; PMCID: PMC3249911.

[2] Sur D, Haldar P K, Maji H S, Balaraman A K, Roy L, Bala A. Role of GSK-3 in the Regulation of Insulin Release and Glucose Metabolism, *Free Radical Biology of Endocrine, Metabolic & Immune Disorders* (2022) 1: 42.

[3] Khandelwal B, Gupta C, Thomas L, Miraj S S, Rao M, Dey R, Dolma K G, Bhutia R D, Singh V P. Free Radical Biology of Diabetes Mellitus, *Free Radical Biology of Endocrine, Metabolic & Immune Disorders* (2022) 1: 1.

[4] Genestra M. Oxyl radicals, redox-sensitive signalling cascades and antioxidants. Review. *Cell Signal.* 2007;19(9):1807-19.

[5] Lobo V, Patil A, Phatak A, Chandra N. Free radicals, antioxidants and functional foods: Impact on human health. *Pharmacogn. Rev.* 2010 Jul;4(8):118-26.

[6] Winterbourn, C. C. Biological chemistry of superoxide radicals. *ChemTexts*.2020; 6: 7.

[7] Indo H P, Yen H C, Nakanishi I, Matsumoto K, Tamura M, Nagano Y, Matsui H, Gusev O, Cornette R, Okuda T, Minamiyama Y, Ichikawa H, Suenaga S, Oki M, Sato T, Ozawa T, Clair D K, Majima H J. A mitochondrial superoxide theory for oxidative stress diseases and aging. *J. Clin. Biochem. Nutr.* 2015 Jan;56(1):1-7.

[8] Keyer K, Imlay J A. Superoxide accelerates DNA damage by elevating free-iron levels. *Proc. Natl. Acad. Sci. U. S. A.* 1996 Nov 26;93(24):13635-40.

[9] Fukai T, Ushio-Fukai M. Superoxide dismutases: role in redox signaling, vascular function, and diseases. *Antioxid. Redox. Signal.* 2011 Sep 15;15(6):1583-606.

[10] Di Meo S, Reed T T, Venditti P, Victor V M. Role of ROS and RNS Sources in Physiological and Pathological Conditions. *Oxid. Med. Cell Longev.* 2016;2016:1245049.

[11] Dunnill C, Patton T, Brennan J, Barrett J, Dryden M, Cooke J, Leaper D, Georgopoulos N T. Reactive oxygen species (ROS) and wound healing: the functional role of ROS and emerging ROS-modulating technologies for augmentation of the healing process. *Int. Wound J.* 2017 Feb;14(1):89-96.

[12] Brieger K, Schiavone S, Miller F J Jr, Krause K H. Reactive oxygen species: from health to disease. *Swiss Med. Wkly.* 2012 Aug 17; 142:w13659.

[13] Li Y, Yang J, Sun X. Reactive Oxygen Species-Based Nanomaterials for Cancer Therapy. *Front. Chem.* 2021 Apr 22; 9:650587.

[14] Kalyane D, Choudhary D, Polaka S, Goykar H, Karanwad T, Rajpoot K, Tekade R K. Reactive oxygen nano-generators for cancer therapy, *Progress in Materials Science*, 2022; 130: 100974.

[15] Bhattacharyya A, Chattopadhyay R, Mitra S, Crowe S E. Oxidative stress: an essential factor in the pathogenesis of gastrointestinal mucosal diseases. *Physiol. Rev.* 2014 Apr;94(2):329-54.

[16] Mittal M, Siddiqui M R, Tran K, Reddy S P, Malik A B. Reactive oxygen species in inflammation and tissue injury. *Antioxid. Redox. Signal*. 2014 Mar 1;20(7):1126-67.

[17] Liou G Y, Storz P. Reactive oxygen species in cancer. *Free Radic. Res*. 2010 May;44(5):479-96.

[18] Phaniendra A, Jestadi D B, Periyasamy L. Free radicals: properties, sources, targets, and their implication in various diseases. *Indian J. Clin. Biochem*. 2015 Jan;30(1):11-26.

[19] Sur D, Dutta A, Mondal C, Banerjee A, Haldar P K, Maji H S, Bala A. Repurposing monoamine oxidase inhibitors (MAOI) for the treatment of rheumatoid arthritis possibly through modulating reactive oxidative stress mediated inflammatory cytokines. *Inflammopharmacology*. 2022 Apr;30(2):453-463.

[20] Bala A, Mondal C, Haldar P K, Khandelwal B. Oxidative stress in inflammatory cells of patient with rheumatoid arthritis: clinical efficacy of dietary antioxidants. *Inflammopharmacology*. 2017 Dec;25(6): 595-607.

[21] Bala A, Haldar P K, Kar B, Naskar S, Mazumder U K. Carbon tetrachloride: a hepatotoxin causes oxidative stress in murine peritoneal macrophage and peripheral blood lymphocyte cells. *Immunopharmacol. Immunotoxicol*. 2012 Feb;34(1):157-62.

[22] Vermot A, Petit-Härtlein I, Smith S M E, Fieschi F. NADPH Oxidases (NOX): An Overview from Discovery, Molecular Mechanisms to Physiology and Pathology. *Antioxidants* (Basel). 2021 Jun 1;10(6):890.

[23] Giardino G, Cicalese M P, Delmonte O, Migliavacca M, Palterer B, Loffredo L, Cirillo E, Gallo V, Violi F, Pignata C. NADPH Oxidase Deficiency: A Multisystem Approach. *Oxid. Med. Cell Longev*. 2017;2017:4590127.

[24] Li Y, Zhu H, Kuppusamy P, Zweier J L, Trush M A. Mitochondrial Electron Transport Chain-Derived Superoxide Exits Macrophages: Implications for Mononuclear Cell-Mediated Pathophysiological Processes. *React. Oxyg. Species* (Apex). 2016;1(1):81-98.

[25] Murphy M P. How mitochondria produce reactive oxygen species. *Biochem. J*. 2009 Jan 1;417(1):1-13.

[26] Zorov D B, Juhaszova M, Sollott S J. Mitochondrial reactive oxygen species (ROS) and ROS-induced ROS release. *Physiol. Rev*. 2014 Jul;94(3):909-50.

[27] Wong H S, Dighe P A, Mezera V, Monternier P A, Brand M D. Production of superoxide and hydrogen peroxide from specific mitochondrial sites under different bioenergetic conditions. *J. Biol. Chem*. 2017 Oct 13; 292(41):16804-16809.

[28] Masaichi-Chang-il Lee, Murugesan Velayutham, Tomoko Komatsu, Russ Hille, and Jay L. Zweier. Measurement and Characterization of Superoxide Generation from Xanthine Dehydrogenase: A Redox-Regulated Pathway of Radical Generation in Ischemic Tissues. *Biochemistry* 2014 53 (41), 6615-6623.

[29] Brondino C D, Romão M J, Moura I, Moura J J. Molybdenum and tungsten enzymes: the xanthine oxidase family. *Curr. Opin. Chem. Biol*. 2006 Apr;10(2):109-14.

[30] Meganathan V, Hamilton C E, Natarajan K, Keshava S, Boggaram V. NADPH and xanthine oxidases control induction of inflammatory mediator expression by organic dust in the lung. *FASEB J*. 2022 Jul;36(7):e22381.

[31] Aziz N, Jamil R T. Biochemistry, Xanthine Oxidase. [Updated 2021 Aug 1]. In: *Stat Pearls. Treasure Island* (FL): StatPearls Publishing; Jan, 2023.

[32] Hasanuzzaman M, Bhuyan M H M B, Zulfiqar F, Raza A, Mohsin S M, Mahmud J A, Fujita M, Fotopoulos V. Reactive Oxygen Species and Antioxidant Defense in Plants under Abiotic Stress: Revisiting the Crucial Role of a Universal Defense Regulator. *Antioxidants* (Basel). 2020 Jul 29;9(8):681.

[33] Chamulitrat W. Activation of the superoxide-generating NADPH oxidase of intestinal lymphocytes produces highly reactive free radicals from sulfite. *Free Radic. Biol. Med*. 1999 Aug;27(3-4):411-21.

[34] Fukai T, Ushio-Fukai M. Superoxide dismutases: role in redox signaling, vascular function, and diseases. *Antioxid. Redox. Signal*. 2011 Sep 15;15(6):1583-606.

[35] Wang Y, Branicky R, Noë A, Hekimi S. Superoxide dismutases: Dual roles in controlling ROS damage and regulating ROS signaling. *J. Cell Biol*. 2018 Jun 4;217(6):1915-1928.

[36] Lubos E, Loscalzo J, Handy D E. Glutathione peroxidase-1 in health and disease: from molecular mechanisms to therapeutic opportunities. *Antioxid. Redox. Signal.* 2011 Oct 1;15(7):1957-97.

[37] Day B J. Catalase and glutathione peroxidase mimics. *Biochem. Pharmacol.* 2009 Feb 1;77(3):285-96.

[38] Bus J S, Gibson J E. Mechanisms of superoxide radical-mediated toxicity. *J. Toxicol. Clin. Toxicol.* 1982 Aug;19(6-7):689-97.

[39] Wang Y, Branicky R, Noë A, Hekimi S. Superoxide dismutases: Dual roles in controlling ROS damage and regulating ROS signaling. *J. Cell Biol.* 2018 Jun 4;217(6):1915-1928.

[40] Winterbourn C C. Toxicity of iron and hydrogen peroxide: the Fenton reaction. *Toxicol. Lett.* 1995 Dec;82-83:969-74.

[41] Lloyd R V, Hanna P M, Mason R P. The origin of the hydroxyl radical oxygen in the Fenton reaction. *Free Radic. Biol. Med.* 1997;22(5):885-8.

[42] Pacher P, Beckman J S, Liaudet L. Nitric oxide and peroxynitrite in health and disease. *Physiol. Rev.* 2007 Jan;87(1):315-424.

[43] Varela-Chinchilla C D, Farhana A. Biochemistry, Superoxides. 2022 Feb 10. In: *StatPearls. Treasure Island* (FL): StatPearls Publishing; 2022 Jan–.

[44] Fischbacher A, von Sonntag C, Schmidt T C. Hydroxyl radical yields in the Fenton process under various pH, ligand concentrations and hydrogen peroxide/Fe(II) ratios. *Chemosphere.* 2017 Sep;182:738-744.

[45] Forooshani P K, Pinnaratip R, Polega E, Tyo A G, Pearson E, Liu B, Folayan T O, Pan L, Rajachar R M, Heldt C L, Lee B P. Hydroxyl Radical Generation Through the Fenton-Like Reaction of Hematin- and Catechol-Functionalized Microgels. *Chem. Mater.* 2020 Oct 13;32(19):8182-8194.

[46] Kato Y. Neutrophil myeloperoxidase and its substrates: formation of specific markers and reactive compounds during inflammation. *J. Clin. Biochem. Nutr.* 2016 Mar;58(2):99-104.

[47] Panasenko O M, Arnhold J, Schiller J. Hypochlorite reacts with an organic hydroperoxide forming free radicals, but not singlet oxygen, and thus initiates lipid peroxidation. *Biochemistry* (Mosc). 1997 Sep; 62(9):951-9.

[48] Di Meo S, Reed T T, Venditti P, Victor V M. Role of ROS and RNS Sources in Physiological and Pathological Conditions. *Oxid. Med. Cell Longev.* 2016; 2016:1245049.

[49] Nita M, Grzybowski A. The Role of the Reactive Oxygen Species and Oxidative Stress in the Pathomechanism of the Age-Related Ocular Diseases and Other Pathologies of the Anterior and Posterior Eye Segments in Adults. *Oxid. Med. Cell Longev.* 2016; 2016:3164734.

[50] Zorov D B, Juhaszova M, Sollott S J. Mitochondrial reactive oxygen species (ROS) and ROS-induced ROS release. *Physiol. Rev.* 2014 Jul; 94(3):909-50.

[51] Pacher P, Beckman J S, Liaudet L. Nitric oxide and peroxynitrite in health and disease. *Physiol. Rev.* 2007 Jan; 87(1):315-424.

[52] Petrie J R, Guzik T J, Touyz R M. Diabetes, Hypertension, and Cardiovascular Disease: Clinical Insights and Vascular Mechanisms. *Can. J. Cardiol.* 2018 May;34(5):575-584.

[53] Oxidative stress and damages to biomolecules (lipids, proteins, DNA)]. *Ann. Pharm. Fr.* 2006 Nov;64(6):383-9.

[54] Islam B U, Habib S, Ahmad P, Allarakha S, Moinuddin, Ali A. Pathophysiological Role of Peroxynitrite Induced DNA Damage in Human Diseases: A Special Focus on Poly(ADP-ribose) Polymerase (PARP). *Indian J. Clin. Biochem.* 2015 Oct;30(4):368-85.

[55] DeFronzo R A., Cooke C R, Andres R, Faloona G R, Davis P J. The effect of insulin on renal handling of sodium, potassium, calcium, and phosphate in man. *J. Clin. Invest.* 1975;55:845–855.

[56] Ferrannini E., Mari A. β-cell function in type 2 diabetes. *Metabolism.* 2014;63:1217–1227.

[57] Forcados G E, Muhammad A, Oladipo O O, Makama S and Meseko C A (2021) Metabolic Implications of Oxidative Stress and Inflammatory Process in SARSCoV-2 Pathogenesis: Therapeutic Potential of Natural Antioxidants. *Front. Cell. Infect. Microbiol.* 11:654813.

[58] Liu T, Zhang L, Joo D, Sun S C. NF-κB signaling in inflammation. *Signal Transduct Target Ther.* 2017; 2:17023.

[59] Tang F, Tang G, Xiang J, Dai Q, Rosner M R, Lin A. The absence of NF-kappaB-mediated inhibition of c-Jun N-terminal kinase activation contributes to tumor necrosis factor alpha-induced apoptosis. Mol. Cell Biol. 2002 Dec;22(24):8571-9.

[60] Zeng M Y, Miralda I, Armstrong C L, Uriarte S M, Bagaitkar J. The roles of NADPH oxidase in modulating neutrophil effector responses. *Mol. Oral Microbiol.* 2019 Apr; 34(2):27-38.

[61] Fukai T, Ushio-Fukai M. Superoxide dismutases: role in redox signaling, vascular function, and diseases. *Antioxid. Redox. Signal.* 2011 Sep 15;15(6):1583-606.

[62] Robinson B S, Hii C S, Ferrante A. Activation of phospholipase A2 in human neutrophils by polyunsaturated fatty acids and its role in stimulation of superoxide production. *Biochem. J.* 1998 Dec 15;336 (Pt 3)(Pt 3):611-7.

[63] Cheng H, Huang H, Guo Z, Chang Y, Li Z. Role of prostaglandin E2 in tissue repair and regeneration. *Theranostics*. 2021 Aug 13;11(18):8836-8854.

[64] Zhang S, Liu Y, Zhang X, Zhu D, Qi X, Cao X, Fang Y, Che Y, Han ZC, He ZX, Han Z, Li Z. Prostaglandin E2 hydrogel improves cutaneous wound healing via M2 macrophages polarization. *Theranostics*. 2018 Oct 22;8(19):5348-5361.

[65] Cheng H, Huang H, Guo Z, Chang Y, Li Z. Role of prostaglandin E2 in tissue repair and regeneration. *Theranostics*. 2021 Aug 13;11(18):8836-8854.

# Index

## A



## B



## C

## S



## T



## U



## V



## W



## X